# Interdisciplinary Applied Mathematics

## Volume 41

Problems in engineering, computational science, and the physical and biological sciences are using increasingly sophisticated mathematical techniques. Thus, the bridge between the mathematical sciences and other disciplines is heavily traveled. The correspondingly increased dialog between the disciplines has led to the establishment of the series: Interdisciplinary Applied Mathematics.

The purpose of this series is to meet the current and future needs for the interaction between various science and technology areas on the one hand and mathematics on the other. This is done, firstly, by encouraging the ways that mathematics may be applied in traditional areas, as well as point towards new and innovative areas of applications; and secondly, by encouraging other scientific disciplines to engage in a dialog with mathematicians outlining their problems to both access new methods as well as to suggest innovative developments within mathematics itself.

The series will consist of monographs and high-level texts from researchers working on the interplay between mathematics and other fields of science and technology.

More information about this series at http://link.springer.com/bookseries/1390

Paul C. Bressloff

# Stochastic Processes in Cell Biology

Volume II

**Second Edition**

 Springer

Paul C. Bressloff
Department of Mathematics
University of Utah
Salt Lake City, UT, USA

ISSN 0939-6047          ISSN 2196-9973   (electronic)
Interdisciplinary Applied Mathematics
ISBN 978-3-030-72521-1          ISBN 978-3-030-72519-8   (eBook)
https://doi.org/10.1007/978-3-030-72519-8

Mathematics Subject Classification: 35K57, 35Q84, 35Q92, 35B25, 60G07, 60J10, 60J60, 60J65, 60K20, 60K37, 82B05, 82B31, 82C03, 82C31, 82C26, 82D60, 92C05, 92C10, 92C37, 92C40, 92C45, 92C15

This Springer imprint is published by the registered company Springer Nature Switzerland AG
The registered company address is: Gewerbestrasse 11, 6330 Cham, Switzerland

To Alessandra and Luca

# Preface to 2nd edition

This is an extensively updated and expanded version of the first edition. I have continued with the joint pedagogical goals of (i) using cell biology as an illustrative framework for developing the theory of stochastic and nonequilibrium processes, and (ii) providing an introduction to theoretical cell biology. However, given the amount of additional material, the book has been divided into two volumes, with

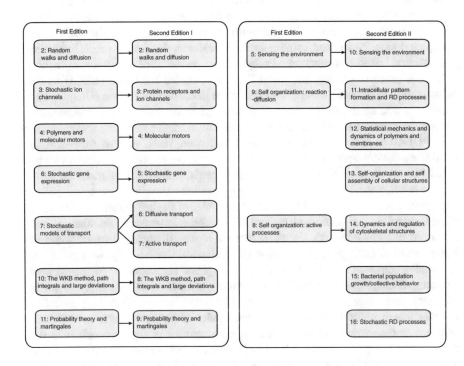

Mapping from the 1st to the 2nd edition

Volume I mainly covering molecular processes and Volume II focusing on cellular processes. The latter also includes significantly expanded material on nonequilibrium systems: intracellular pattern formation and reaction–diffusion processes, statistical physics, and the dynamics/self-organization of cellular structures. The mapping from the first to the second edition is shown in the diagram. In Volume I, the chapter on intracellular transport processes has been split into two chapters, covering diffusive and active processes, respectively. There are four completely new chapters in Volume II: statistical mechanics of polymers and membranes; self-organization and assembly of cellular structures; bacterial population growth and collective behavior; stochastic reaction–diffusion processes. The other three chapters have been significantly expanded.

Major new topics include the following: theory of continuous-time Markov chains (Chapter 3); first passage time problems with (nucleating) sticky boundaries (Chapter 4); genetic oscillators, the repressilator, the degrade-and-fire model, delay differential equations, theory of chemical reaction networks, promoter dynamics, transcriptional bursting and queuing theory, epigenetics, gene expression, and morphogen gradients (Chapter 5); molecular crowding and homogenization theory, percolation theory, narrow capture problems, extreme statistics, diffusion in randomly switching environments, and stochastically gated gap junctions (Chapter 6); reversible vesicular transport in axons, distribution of resources across multiple targets and queuing theory, and stochastic resetting (Chapter 7); metastability in gene networks, Brownian functionals, large deviation theory, generalized central limit theorems, and Levy stable distributions (Chapter 8); phosphorylation–dephosphorylation cycles and ultrasensitivity, Goldbeter–Koshland model, photoreceptors and phototransduction, Poisson shot noise, linear response theory, eukaryotic gradient sensing, the Local Excitation/Global Inhibition (LEGI) model of adaptation in gradient sensing, and maximum likelihood estimation (Chapter 10); robustness and accumulation times of protein gradients, non-classical mechanisms for protein gradient formation, pattern formation in mass-conserving systems, coupled PDE-ODE systems, cell polarization in fission yeast, pattern formation in hybrid reaction-transport systems, pattern formation on growing domains, synaptogenesis in *C. elegans*, protein clustering in bacteria, multi-spike solutions far from pattern onset, reaction-diffusion models of intracellular traveling waves, and pulled and pushed fronts (Chapter 11); elastic rod model of flexible polymers, worm-like chains, curvature and torsion, stress and strain tensors, membrane fluctuations and curvature, polymer networks, viscoelasticity and reptation, nuclear organization, and Rouse model of DNA dynamics (Chapter 12); classical theories of phase separation, spinodal decomposition and Ostwald ripening, phase separation of biological condensates, Becker–Döring model of molecular aggregation, self-assembly of phospholipids, and active membranes (Chapter 13); doubly stochastic Poisson model of flagellar length control, diffusion–secretion model of filament length control, cell adhesion, motor-clutch model of crawling cells, growth of focal adhesions, variational method for free energy minimization, and cytoneme-based morphogen gradients (Chapter 14); age-structured models of population growth and cell size regulation, bacterial persistence and phenotypic switching, stochastic models of population extinction, bacterial quorum sensing, synchronization of genetic oscillators, and biofilms (Chapter 15); stochastic reaction–diffusion processes, stochastic Turing patterns, non-normality and noise-induced pattern amplification, statistical field theory, diagrammatic expansions and the renormalization group, and stochastic traveling waves (Chapter 16).

Meaning no disrespect to vegetarians, I do not explicitly cover plant cells. However, many of the mechanisms and concepts developed in this book would still apply. Chapter 15 on bacterial population growth suggests another natural extension of the current book, namely, stochastic and nonequilibrium processes at the multicellular and tissue levels, including biological neural networks, immunology, collective cell migration, cell development, wound healing, and cancer. This would involve additional topics such as cell-to-cell signaling, the propagation of intercellular signals, nonlocal differential and integral equations, physical properties of the extracellular matrix, and network theory. Clearly ripe themes for a possible third volume!

## *Acknowledgements*

There are many applied mathematicians, physical scientists, and life scientists upon whose sturdy shoulders I have stood during the writing of this book, and whose work is featured extensively in the following pages. I apologize in advance if I have excluded anyone or didn't do proper justice to their contributions. It should also be noted that the relatively large number of self-citations is not a reflection of the significance of my own work in the field, but a consequence of the fact that I am most familiar with my own work! Finally, I would like to thank my wife Alessandra and son Luca (the Shmu) for their continuing love and support.

Salt Lake City, USA                                            Paul C. Bressloff

# Preface to 1st edition

In recent years there has been an explosion of interest in the effects of noise in cell biology. This has partly been driven by rapid advances in experimental techniques, including high-resolution imaging and molecular-level probes. However, it is also driven by fundamental questions raised by the ubiquity of noise. For example, how does noise at the molecular and cellular levels translate into reliable or robust behavior at the macroscopic level? How do microscopic organisms detect weak environmental signals in the presence of noise? Have single-cell and more complex organisms evolved to exploit noise to enhance performance? In light of the above, there is a growing need for mathematical biologists and other applied mathematicians interested in biological problems to have some background in applied probability theory and stochastic processes. Traditional mathematical courses and textbooks in cell biology and cell physiology tend to focus on deterministic models based on differential equations such as the Hodgkin-Huxley and FitzHugh-Nagumo equations, chemical kinetic equations, and reaction-diffusion equations. Although there are a number of well-known textbooks on applied stochastic processes, they are written primarily for physicists and chemists or for population biologists. There are also several excellent books on cell biology written from a biophysics perspective. However, these assume some background in statistical physics and a certain level of physical intuition. Therefore, I felt that it was timely to write a textbook for applied mathematicians interested in learning stochastic processes within the context of cell biology, which could also serve as an introduction to mathematical cell biology for statistical physicists and applied probabilists.

I started my interest in stochastic cell biology, as distinct from my work in mathematical neuroscience, around eight years ago when I volunteered to teach a course in biophysics for the mathematical biology graduate program at Utah. I was immediately fascinated by the molecular processes underlying the operation of a cell, particularly the mechanisms for transporting proteins and other macromolecules to the correct subcellular targets at the correct times. Such an issue is particularly acute for neurons, which are amongst the largest and most complex cells in biology. In healthy cells, the regulation of protein trafficking within a neuron provides an important mechanism for modifying the strength of synaptic connections between neurons, and synaptic plasticity is generally believed to be the cellular substrate of learning and memory. On the other hand, various types of dysfunction in protein trafficking appear to be a major contributory factor to a number of neurodegenerative diseases associated with memory loss including Alzheimer's disease.

In writing this book, I have gone back to my roots in theoretical physics, but refracted through the lens formed by many years working in applied mathematics. Hence, the book provides extensive coverage of analytical methods such as initial boundary value problems for partial differential equations, singular perturbation theory, slow/fast analysis and quasi-steady-state approximations, Green's functions, WKB methods and Hamilton-Jacobi equations, homogenization theory and multi-scale analysis, the method of characteristics and shocks, and reaction-diffusion equations. I have also endeavored to minimize the use of statistical mechanics, which is not usually part of a mathematician's tool-kit and requires a certain level of physical intuition. It is not possible to avoid this topic completely, since many experimental and theoretical papers in cell biology assume some familiarity with terms such as entropy, free energy and chemical potential. The reason is that microscopic systems often operate close to thermodynamic equilibrium or asymptotically approach thermodynamic equilibrium in the long-time limit. This then imposes constraints on any model of the underlying stochastic process. In most cases, one can understand these constraints by considering the Boltzmann-Gibbs distribution of a macromolecule in thermodynamic equilibrium, which is the approach I take in this book.

There are two complementary approaches to modeling biological systems. One involves a high level of biological detail and computational complexity, which means that it is usually less amenable to mathematical analysis than simpler reduced models. The focus tends to be on issues such as parameter searches and data fitting, sensitivity analysis, model reductions, numerical convergence, and computational efficiency. This is exemplified by the rapidly growing field of systems biology. The other approach is based on relatively simple conceptual or "toy" models, which are analytically tractable and, hopefully, capture essential features of the phenomena of interest. In this book I focus on the latter for pedagogical reasons and because of my own personal tastes. In the introductory chapter, I summarize some of the basic concepts in stochastic processes and non-equilibrium systems that are used throughout the book, describe various experimental methods for probing noise at the molecular and cellular levels, give a brief review of basic probability theory and statistical mechanics, and then highlight the structure of the book. In brief, the book is divided into two parts: Part I (Foundations) and Part II (Advanced Topics). Part I provides the basic foundations of both discrete and continuous stochastic processes in cell biology. It's five chapters deal with diffusion, random walks and the Fokker-Planck equation (chapter 2), discrete Markov processes and chemical reaction networks (chapter 3), polymers and molecular motors (chapter 4), gene expression and regulatory networks (chapter 5), and biochemical signaling and adaptation (chapter 6). Part II covers more advanced topics that build upon the ideas and techniques from part I. Topics include transport processes in cells (chapter 7), self-organization in reaction-diffusion models (chapter 8), self-organization of the cytoskeleton (chapter 9), WKB methods for escape problems (chapter 10), and some more advanced topics in probability theory (chapter 11). The chapters are supplemented by additional background material highlighted in gray boxes, and numerous exercises that reinforce the analytical methods and models introduced in the main body of the text. I have attempted to make the book as self-contained as possible. However, some introductory background in partial differential equations, integral transforms, and applied probability theory would be advantageous.

Finally, this book should come with a "government health warning." That is, throughout most of the book, I review the simplest mechanistic models that have been constructed in order to investigate a particular biological phenomenon or illustrate a particular mathematical method. Although I try to make clear the assumptions underlying each model, I do not carry out a comparative study of different models in terms of the degree of quantitative agreement with experimental data. Therefore, the reader should be cautioned that the models are far from the last word on a given phenomenon, and the real biological system is usually way more complicated than stated. However, it is hoped that the range of modeling and analytical techniques presented in this book, when combined with efficient numerical methods, provide the foundations for developing more realistic, quantitative models in stochastic cell biology.

Salt Lake City, USA                                                                              Paul C. Bressloff

# Organization of volumes I and II

## Volume I: Molecular processes

The first volume begins with a short introduction to probability theory and statistical mechanics (Chapter 1). Chapter 2 presents two microscopic theories of diffusion in cells, one based on random walks and the other on overdamped Brownian motion. The latter leads to the theory of continuous Markov processes. Two complementary approaches to formulating continuous Markov process are developed, one in terms of the sample paths generated by a Stochastic Differential Equation (SDE) or Langevin equation, and the other in terms of the Fokker–Planck (FP) equation describing the evolution of the probability density of possible paths. In the former case, a basic introduction to stochastic calculus is given, focusing on the rules for integrating an SDE in order to obtain an expression that can be used to generate moments of the stochastic process. The distinction between Ito and Stratonovich interpretations of multiplicative noise is explained in some detail. It is also shown how, in the case of linear SDEs, Fourier methods can be used to determine the power spectrum, which is important in quantifying the linear response properties of a noisy system. The FP equation, which is a deterministic Partial Differential Equation (PDE) that generalizes the diffusion equation, is then analyzed using standard methods in the theory of linear PDEs: separation of variables, transform methods, Green's functions, and eigenfunction expansions. Many quantities measured by experimentalists can be interpreted mathematically in terms of the solution to a First Passage Time (FPT) problem. Using the fact that the distribution of first passage times satisfies a backward FP equation, the mean FPT is shown to satisfy a boundary value problem. This is then used to derive the classical Kramer's rate formula for escape across a potential barrier. Noise-induced changes in the effective potential (quasi-potential) in the presence of multiplicative noise are also discussed. Finally, some numerical methods for solving SDEs are reviewed.

Chapter 3 covers some of the main molecular players in cell signaling and transduction, namely, receptors and ion channels. After briefly summarizing the most common types of receptors, some simple kinetic models of cooperative binding are introduced, including the Monod-Wyman-Changeaux model and the Ising model. These provide one mechanism for a cell to amplify signals from the extracellular environment. Following a description of various single ion channel models, the stochastic dynamics of an ensemble of independent ion channels is formulated in terms of a birth–death process. The latter is an example of a discrete Markov process or Markov chain. It is shown how the probability distribution for the number of open ion channels evolves according to a corresponding birth–death master equation. Two models of stochastic ion channels are then explored, a conductance-based model of spontaneous action potential generation in a neuron, which is driven by the random opening and closing of voltage-gated ion channels, and the spontaneous release of calcium puffs and sparks by ligand-gated ion channels. In both cases, the occurrence of spontaneous events can be analyzed in terms of a FPT problem. Finally, the general theory of continuous-time Markov chains is reviewed, including a discussion of the Perron–Frobenius theorem and an introduction to Poisson processes. There are a number of systems considered in subsequent chapters where the signal received by a biochemical sensor involves a sequence of discrete events that can be modeled as a Poisson process. Examples include the arrival of photons at photoreceptors of the retina, and the arrival of action potentials (spikes) at the synapse of a neuron. Another type of event is the random arrival of customers at some service station, resulting in the formation of a queue. However, these processes are typically non-Markovian.

Chapter 4 describes how random walks and SDEs are used to model polymerization and molecular motor dynamics. Polymerization plays a major role in the self-organization of cytoskeletal structures, whereas molecular motors "walking" along polymer filaments constitute a major active component of intracellular transport. The analysis of polymerization focuses on the Dogterom–Leibler model of microtubule catastrophes, which takes the form of a two-state velocity-jump process for the length of a microtubule. The effects of nucleation and constrained growth are taken into account, and FPT problems with "sticky" boundaries are analyzed using the theory of conditional expectations, stopping times, and strong Markov processes. The FP equation for a Brownian particle moving in a periodic ratchet (asymmetric) potential is then analyzed. It is shown that the mean velocity of the Brownian particle is zero, which implies that the periodicity of the potential must be broken for a molecular motor to perform useful work against an applied load. One such mechanism is to rectify the motion, as exemplified by polymerization and translocation ratchets. A qualitative model of processive molecular motors is then introduced, based on a flashing Brownian ratchet. It is shown how useful work can be generated if the motor switches between different conformational states (and corresponding potentials) at rates that do not satisfy detailed balance; this is achieved via the hydrolysis of Adenosine Triphosphate (ATP). The theory of molecular motors is further developed by considering two examples of the collective motion of an ensemble of molecular motors: (i) the tug-of-war model of bidirectional vesicular transport by opposing groups of processive motors; (ii) a model of interacting motors attached to a rigid cytoskeletal backbone.

Chapter 5 covers the basics of stochastic gene expression and chemical reaction networks. First, various deterministic rate models of gene regulatory networks are described, including autoregulatory networks, the toggle switch, the *lac* operon, the repressilator, NK-$\beta$B oscillators, and the circadian clock. Brief reviews of linear stability analysis, Hopf bifurcation theory, and oscillations in delay differential equations are also given. The analysis of molecular noise associated with low copy numbers is then developed, based on the chemical master equation. Since chemical master equations are difficult to analyze directly, a system-size expansion is used to approximate the chemical master equation by an FP equation and its associated chemical Langevin equation. Gillespie's stochastic simulation algorithm for generating exact sample paths of a continuous-time Markov chain is also summarized. Various effects of molecular noise on gene expression are then explored, including translational bursting, noise-induced switching, and noise-induced oscillations. One of the assumptions of many stochastic models of gene networks is that the binding/unbinding of transcription factors at promoter sites is faster than the rates of synthesis and degradation. If this assumption is relaxed, then there exists another source of intrinsic noise known as promoter noise. The latter is modeled in terms of a stochastic hybrid system, also known as a piecewise deterministic Markov process. This involves the coupling between a continuous-time Markov chain and a continuous process that may be deterministic or stochastic. The evolution of the system is now described by a differential Chapman–Kolmogorov (CK) equation, which is a mixture of a master equation and an FP equation. In the limit of fast switching, a quasi-steady-state approximation is used to reduce the CK equation to an effective FP equation. This is analogous to the system-size expansion of chemical master equations. Various examples of networks with promoter noise are presented, including a stochastic version of the toggle switch. It is shown how one of the major effects of promoter noise, namely transcriptional bursting, can be analyzed using queuing theory. Some time-limiting steps in gene regulation are then described, including kinetic proofreading based on enzymatic reactions, and DNA transcription times. The penultimate section consists of a brief introduction to epigenetics. This concerns phenotypic states that are not encoded as genes, but as inherited patterns of gene expression originating from environmental factors, and maintained over multiple cell generations when the original environmental stimuli have been removed. A number of epigenetic mechanisms are discussed, including the infection of *E. coli* by the $\lambda$ phage DNA virus, and local mechanisms such as DNA methylation and gene silencing by nucleosome modifications. Finally, the role of gene expression in interpreting morphogen gradients during early development is discussed.

Chapter 6 and 7 consider various aspects of intracellular transport, focusing on diffusive and active transport, respectively. Chapter 6 begins by describing the anomalous effects of molecular crowding and trapping, where the differences in diffusive behavior at multiple time scales are highlighted. The classical Smoluchowski theory of diffusion-limited reactions is then developed, with applications to chemoreception and to facilitated diffusion, which occurs when a protein searches for specific DNA binding sites. Extensions of the classical theory to stochastically gated diffusion-limited reactions and ligand rebinding in enzymatic reactions are also considered. Next it is shown how Green's functions and singular perturbation theory can be used to analyze narrow escape and narrow capture problems. The former concerns the escape of a particle from a bounded domain through small openings in the boundary of the domain, whereas the latter refers to a diffusion–trapping problem in which the interior traps are much smaller than the size of the domain. An alternative measure of the time scale for diffusive search processes is then introduced, based on the FPT of the fastest particle to find a target among a large population of independent Brownian particles, which is an example of an extreme statistic. This leads to the so-called "redundancy principle," which provides a possible explanation for the apparent redundancy in the number of molecules involved in various cellular processes, namely, that it accelerates search processes. In certain examples of diffusive search, regions of a boundary may randomly switch between open and closed states, which requires the analysis of PDEs in randomly switching environments. In particular, it is shown how a common switching environment can induce statistical correlations between noninteracting particles. The analysis of randomly switching environments is then extended to the case of molecular diffusion between cells that are coupled by stochastically gated gap junctions. Finally, diffusive transport through narrow membrane pores and channels is analyzed using the Fick–Jacobs equation and models of single-file diffusion. Applications to transport through the nuclear pore complex are considered.

Chapter 7 begins by considering population models of axonal transport in neurons. The stochastic dynamics of a single motor complex is then modeled in terms of a velocity-jump process, which focuses on the transitions between different types of motions (e.g., anterograde vs. retrograde active transport and diffusion vs. active transport) rather than the microscopic details of how a motor performs a single step. Transport on a 1D track and on higher-dimensional cytoskeletal networks is considered, including a model of virus trafficking. Next, the efficiency of transport processes in delivering vesicular cargo to a particular subcellular domain is analyzed in terms of the theory of random search-and-capture processes. The latter describe a particle that randomly switches between a slow search phase (e.g., diffusion) and a faster non-search phase (e.g., ballistic transport). In certain cases it can be shown that there exists an optimal search strategy, in the sense that the mean time to find a target can be minimized by varying the rates of switching between the different phases. The case of multiple search-and-capture events, whereby targets accumulate resources, is then analyzed using queuing theory. Another example of a random search process is then introduced, in which the position of a particle (searcher) is reset randomly in time at a constant rate. One finds that the MFPT to find a target is finite and has an optimal value as a function of the resetting rate. Stochastic resetting also arises in models of cell adhesion and morphogen gradient formation. Finally, it is shown how the effects of molecular crowding of motors on a filament track can be modeled in terms of asymmetric exclusion processes. In the mean-field limit, molecular crowding can be treated in terms of quasi-linear PDEs that support shock waves.

Chapters 8 and 9 cover more advanced topics. Chapter 8 focuses on methods for analyzing noise-induced transitions in multistable systems, such as Wentzel–Kramers–Brillouin (WKB) methods, path integrals, and large deviation theory. First, WKB theory and asymptotic methods are used to analyze noise-induced escape in an SDE with weak noise. It is shown how the most likely paths of escape can be interpreted in terms of least action paths of a path integral representation of the SDE. An analogous set of analyses are also carried out for birth–death processes and stochastic hybrid systems, which are illustrated using the examples of an autoregulatory gene network and a conductance-based neuron model. The path integral representation of an SDE is then used to derive the Feynman–Kac formula for Brownian functionals. The latter are random variables defined by some integral measure of a Brownian path. Chapter 8 ends with a brief introduction to large deviation theory, as well as a discussion of generalized central limit theorems and Lévy stable distributions. Finally, Chapter 9 briefly reviews the theory of martingales and applications to branching processes and counting processes.

## Volume II: Cellular processes

Chapter 10 explores the general problem of detecting weak signals in noisy environments. Illustrative examples include photoreceptors and shot noise, inner hair cells and active mechanotransduction, and cellular chemotaxis. Various mechanisms for signal amplification and adaption are described, such as phosphorylation–dephosphorylation cycles, ultrasensitivity, and receptor clustering. The basic principles of linear response theory are also introduced. The fundamental physical limits of cell signaling are developed in some detail, covering the classical Berg–Purcell analysis of temporal signal integration, and more recent developments based on linear response theory, and maximum likelihood estimation. One of the useful features of the latter approach is that it can be extended to take into account temporal concentration changes, such as those that arise during bacterial chemotaxis. Bacteria are too small to detect differences in concentrations across their cell bodies, so they proceed by measuring and comparing concentrations over time along their swimming trajectories. Some simple PDE models of bacterial chemotaxis, based on velocity-jump processes, are also considered. In contrast to bacterial cells, eukaryotic cells such as the social amoeba *Dictyostelium discoideum* are sufficiently large so that they can measure the concentration differences across their cell bodies without temporal integration. Various models of spatial gradient sensing in eukaryotes are investigated, including the Local Excitation, Global Inhibition (LEGI) model, which takes into account the fact that cells adapt to background concentrations.

Chapter 11 explores intracellular pattern formation based on reaction–diffusion processes. First, various mechanisms for the formation of intracellular protein concentration gradients are considered, and the issue of robustness is discussed. Next, after reviewing the general theory of Turing pattern formation, two particular aspects are highlighted that are specific to intracellular pattern formation: (i) mass conservation and (ii) the dynamical exchange of proteins between the cytoplasm and plasma membrane. Various examples of mass-conserving reaction–diffusion models of cell polarization and division are then described, including Min protein oscillations in *E. coli*, cell polarization in budding and fission yeast, and cell polarization in motile eukaryotic cells. An alternative mechanism for intracellular pattern formation is then introduced, based on a hybrid transport model where one chemical species diffuses and the other undergoes active transport. Evolving the model on a slowly growing domain leads to a spatial pattern that is consistent with the distribution of synaptic puncta during the development of *C. elegans*. Next, asymptotic methods are used to study the existence and stability of multi-spike solutions far from pattern onset; the latter consist of strongly localized regions of high concentration of a slowly diffusing activator. The theory is also applied to a model of the self-positioning of Structural Maintenance of Chromosomes (SMC) protein complexes in *E. coli*, which are required for correct chromosome condensation, organization, and segregation. Finally, various examples of intracellular traveling waves are analyzed, including polarization fronts in motile eukaryotic cells, mitotic waves, and CaMKII translocation waves in dendrites. An introduction to the theory of bistable and unstable waves is also given.

Chapter 12 presents an introduction to the statistical mechanics and dynamics of polymers, membranes, and polymer networks such as the cytoskeleton. First, the statistical mechanics of single polymers is considered, covering random walk models such as the freely jointed chain, and elastic rod models (worm-like chains). The latter type of model treats a polymer as a continuous curve, whose free energy contributions arise from the stretching, bending, and twisting of the polymer. The continuum mechanics of elastic rods is briefly reviewed in terms of curvature and torsion in the Frenet—Serret frame. A generalized worm-like chain model is used to account for experimentally obtained force–displacement curves for DNA. The statistical mechanics of membranes is then developed along analogous lines to flexible polymers, by treating membranes as thin elastic sheets. In order to construct the bending energy of the membrane, some basic results from membrane elasticity are reviewed, including stress and strain tensors, bending/compression moduli, and the theory of curved surfaces. The corresponding partition function is used to estimate the size of thermally driven membrane fluctuations. Since the membrane is modeled as an infinite-dimensional continuum, the partition function takes the form of a path integral whose associated free energy is a functional. The analysis of statistical properties thus requires the use of functional calculus. The next topic is the statistical dynamics of systems at or close to equilibrium. This is developed by generalizing the theory of Brownian motion to more complex structures with many internal degrees of freedom. Various results and concepts from classical nonequilibrium statistical physics are introduced, including Onsager's reciprocal relations, nonequilibrium forces, time correlations and susceptibilities, and a general version of the fluctuation–dissipation theorem. The theory is illustrated by deriving Langevin equations for fluctuating polymers and membranes. The chapter then turns to polymer network models, which are used extensively by biophysicists to understand the rheological properties of the cytoskeleton. Only the simplest classical models are considered: the rubber elasticity of a cross-linked polymer network, swelling of a polymer gel, and the macroscopic theory of viscoelasticity in uncross-linked polymer fluids. Reptation theory, which is used to model the dynamics of entangled polymers, is also briefly discussed. Finally, the dynamics of DNA within the nucleus is considered. After describing some of the key features of nuclear organization, a classical stochastic model of a Gaussian polymer chain (the Rouse model) is introduced. The latter is used to model the subdiffusive motion of chromosomal loci, and to explore mechanisms for spontaneous DNA loop formation. The mean time to form a loop requires solving an FP equation with nontrivial absorbing boundary condition. The Wilemski–Fixman theory of diffusion-controlled reactions is used to solve the problem by replacing the boundary condition with a sink term in the FP equation.

Chapter 13 considers the self-organization and assembly of a number of distinct cellular structures. First, there is a detailed discussion of the theory of liquid–liquid phase separation and the formation of biological condensates. This introduces various classical concepts in nonequilibrium systems, such as coexistence curves, spinodal decomposition, nucleation and coarsening, Ostwald ripening, and Onsager's principle. Recent developments that are specific to biological condensates are also described, including the effects of nonequilibrium chemical reactions and protein concentration gradients. The chapter then turns to the Becker and Döring model of molecular aggregation and fragmentation, which provides a framework for investigating the processes of nucleation and coarsening. An application of the model to the self-assembly of phospholipids in the plasma membrane is also included. Finally, a model for the cooperative transport of proteins between cellular organelles is introduced, which represents a self-organizing mechanism for organelles to maintain their distinct identities while constantly exchanging material.

Chapter 14 considers various models for the dynamics and regulation of the cytoskeleton. First, several mechanisms for filament length regulation are presented, including molecular motor-based control, protein concentration gradients, and diffusion-based secretion in bacterial flagella. The role of Intraflagellar Transport (IFT) in the length control of eukaryotic flagella is analyzed in terms of a doubly stochastic Poisson process. The dynamics of the mitotic spindle during various stages of cell mitosis is then described, including the search-and-capture model of microtubule–chromosome interactions and force-balance equations underlying chromosomal oscillations. Next, various models of biophysical mechanisms underlying cell motility are considered. These includes the tethered ratchet model of cell protrusion and the motor-clutch mechanism for crawling cells. The latter describes the dynamical interplay between retrograde flow of the actin cytoskeleton and the assembly and disassembly of focal adhesions. The resulting dynamics exhibits a number of behaviors that are characteristic of physical systems involving friction at moving interfaces, including biphasic force–velocity curves and stick-slip motion. A mean-field analysis is used to show how these features can be captured by a relatively simple stochastic model of focal adhesions. In addition, a detailed model of the force-induced growth of focal adhesions is analyzed using a variational method for free energy minimization. Finally, a detailed account of cytoneme-based morphogensis is given. Cytonemes are thin, actin-rich filaments that can dynamically extend up to several hundred microns to form direct cell-to-cell contacts. There is increasing experimental evidence that these direct contacts allow the active transport of morphogen to embryonic cells during development. Two distinct models of active transport are considered. The first involves active motor-driven transport of morphogen along static cytonemes with fixed contacts between a source cell and a target cell. The second is based on nucleating cytonemes from a source cell that dynamically grow and shrink until making temporary contact with a target cell and delivering a burst of morphogen. The delivery of a single burst is modeled in terms of a FPT problem for a search process with stochastic resetting, while the accumulation of morphogen following multiple rounds of cytoneme search-and-capture and degradation is analyzed using queuing theory.

Chapter 15 presents various topics related to bacterial population growth and collective behavior. First, a continuum model of bacterial population growth is developed using an age-structured evolution equation. Such an equation supplements the continuously varying observational time by a second time variable that specifies the age of an individual cell since the last division. Whenever a cell divides, the age of the daughter cells is reset to zero. Although the total number of cells grows exponentially with time, the normalized age distribution approaches a steady state. The latter determines the effective population growth rate via a self-consistency condition. The age-structured model is then extended in order to keep track of both the age and volume distribution of cells. This is used to explore various forms of cell length regulation, including timer, sizer, and adder mechanisms. Further aspects of cell size regulation are analyzed in terms of a discrete-time stochastic map that tracks changes across cell generations. The chapter then turns to another important issue, namely, to what extent single-cell molecular variation plays a role in population-level function. This is explored within the context of phenotypic switching in switching environments, which is thought to be an important factor in the phenomenon of persistent bacterial infections following treatment with antibiotics. At the population level, phenotypic switching is modeled in terms of a stochastic hybrid system. The chapter then turns to a discussion of bacterial Quorum Sensing (QS). This is a form of collective cell behavior that is triggered by the population density reaching a critical threshold, which requires that individual cells sense their local environment. The next topic is an analysis of synchronization in a population of synthetic gene oscillators that are dynamically coupled to an external medium via a QS mechanism. In particular a continuity equation for the distribution of oscillator phases is constructed in the thermodynamic limit, and various methods of analysis are presented, including the Ott–Antonsen dimensional reduction ansatz. The chapter ends with a review of some mathematical models of bacterial biofilms.

Chapter 16 discusses various analytical methods for studying stochastic reaction–diffusion processes. First, the effects of intrinsic noise on intracellular pattern formation are investigated using the notion of a reaction–diffusion master equation.

The latter is obtained by discretizing space and treating spatially discrete diffusion as a hopping reaction. Carrying out a linear noise approximation of the master equation leads to an effective Langevin equation, whose power spectrum provides a means of extending the definition of a Turing instability to stochastic systems, namely, in terms of the existence of a peak in the power spectrum at a nonzero spatial frequency. It is also shown how the interplay between intrinsic noise and transient growth of perturbations can amplify the weakly fluctuating patterns. The source of transient growth is the presence of a non-normal matrix in the linear evolution operator. Next, using the canonical example of pair annihilation with diffusion, various well-known techniques from statistical field theory are used to capture the dimension-dependent asymptotic decay of the system. These include moment generating functionals, diagrammatic perturbation expansions (Feynman diagrams), and the renormalization group. Finally, a formal perturbation method is used to analyze bistable front solutions of a stochastic reaction–diffusion equation, which exploits a separation of time scales between fast fluctuations of the front profile and a slowly diffusing phase shift in the mean location of the front.

---

At the end of each chapter, there is a set of exercises that further develop the mathematical models and analysis introduced within the body of the text. Additional comments and background material are scattered throughout the text in the form of framed boxes.

Chapters 1–5 can be used to teach a one-semester advanced undergraduate or graduate course on "Stochastic processes in cell biology." Subsequent chapters develop more advanced material on intracellular transport (Chapters 6 and 7), noise-induced transitions (Chapter 8), chemical sensing (Chapter 10), cellular self-organization and pattern formation (Chapters 11–14 and 16), and bacterial population dynamics (Chapter 15).

---

## Volume II: Cellular

**15. Bacterial population growth and collective behavior**
- age-structured models • cell size control
- bacterial persistence/phenotypic switching
- extinction in bacterial populations
- bacterial quorum sensing • biofilms
- synchronization of genetic oscillators

**16. Stochastic RD processes**
- RD master equation
- stochastic Turing patterns
- path integral of RD master equation
- statistical field theory
- diagramatic expansions
- renormalization theory and scaling
- stochastic traveling waves

**14. Regulation of the cytoskeleton**
- filament length control
- intraflagellar transport
- doubly stochastic poisson processes
- cell mitosis • cell motility • cell adhesion
- cytoneme-based morphogenesis

**12. Statistical mechanics/dynamics of polymers and membranes**
- random walk models of polymers
- elastic rod models of polymers
- stress/strain tensors, membrane curvature
- elastic plate model of flluid membranes
- fluctuation-dissipation theorem for Brownian particles, Onsager relations, susceptibilities
- polymer networks, viscoelsticity, reptation
- DNA dynamics in the nucleus

**13. Self-organization and assembly of cellular structures**
- phase separation/biological condensates
- active membranes
- nucleation and growth of molecular clusters
- self-assembly of micelles

**11. Intracellular pattern formation and reaction-diffusion processes**
- intracellular protein gradients
- Turing pattern formation
- mass-conserving systems
- coupled PDE/ODE systems
- cell polarization and division
- hybrid reaction-transport models
- intracellular traveling waves

**10. Sensing the environment**
- phosphorylation and ultrasensitivity
- photoreceptors and shot noise
- linear response theory
- hair cell mechanotransduction
- bacterial chemotaxis
- physical limits of chemical sensing
- spatial gradient sensing

## Volume I: Molecular

**9. Probability theory and martingales**
- filtrations, martingales and stopping times
- branching processes
- counting process and biochemical networks

**8. WKB method, path integrals,...**
- WKB method for noise-induced escape
- path integral for SDEs
- Doi-Peliti path integral of a birth death process
- path integral for hybrid systems
- local and occupation times
- large deviation theory

**5. Stochastic gene expression**
- genetic switches and oscillators
- molecular noise and master equations
- system-size expansion • translational bursting
- noise-induced switching and oscillations
- promoter noise/stochastic hybrid systems
- transcriptional bursting and queuing theory
- time limiting steps in DNA transcription
- epigenetics
- morphogen gradients/gene expression

**7. Active motor transport**
- axonal transport
- PDE models of active transport
- transport on microtubular networks
- virus trafficking
- random intermittent search
- stochastic resetting
- exclusion processes

**3. Protein receptors and ion channels**
- receptor-ligand binding and cooperativity
- stochastic ion channels
- Markov chains and single channel kinetics
- birth-death process for channel ensembles
- voltage-gated ion channels
- calcium sparks in myocytes
- Poisson processes

**6. Diffusive transport**
- anomalous diffusion
- diffusion-limited reactions
- narrow capture and escape problems
- protein search for DNA target sites
- extreme statistics
- diffusion in switching environments
- gap junctions
- diffusion in channels and pores
- nuclear transport • diffusion on trees

**2. Random walks and Brownian motion**
- random walks and diffusion
- Wiener process and Ito stochastic calculus
- Langevin and Fokker-Planck equations
- first passage times
- Kramers escape rate
- multiplicative noise

**4. Molecular motors**
- polymerization
- microtubular catastrophes
- Brownian ratchets
- tethered ratchet and cell motility
- processive molecular motors
- collective motor transport

# Contents

# Boxes

# Chapter 10
# Sensing the environment

One important requirement of sensory eukaryotic cells and single-cell organisms such as bacteria is detecting weak signals in noisy extracellular environments. As we briefly discussed in Chap. 6 within the context of bacterial chemoreception, there are fundamental limits to the strength of the signal that can be detected. However, even if a weak signal is detected, it is necessary for some form of amplification to occur in order that the signal is not lost in subsequent stages of processing within the cell. Moreover, it is advantageous for a cell to be able to shift its response so that it always operates in a regime of maximal gain, that is, it is able to respond to small changes in signal irrespective of the mean strength of the signal—a process known as adaptation. In this chapter, we explore these issues in more detail. (For an excellent exploration of these themes from the perspective of biophysics, see the book by Bialek [78].)

We begin in Sect. 10.1 by considering an alternative to cooperativity (Chap. 3) as a mechanism for signal amplification, based on zero-order ultrasensitive phosphorylation–dephosphorylation cycles. We then consider two different examples of sensory cells in multicellular organisms. First, in Sect. 10.2 we consider photoreceptors and phototransduction. This introduces the basic issue of detecting weak signals (single photons) in the presence of noise and minimizing detection errors. The response of a photoreceptor to multiple photons distributed in time according to a Poisson process provides an example of Poisson shot noise, which involves the convolution of the input photon "spike train" with a linear response function. We also review some basic linear response theory (Box 10A). Second, in Sect. 10.3 we consider how amplification and adaptation occur in hair cells of the inner ear via active mechanotransduction. In the latter case, the interactions between the mechanical properties of transduction elements, the action of myosin motors, and $Ca^{2+}$ signaling allow a hair cell to operate close to a Hopf bifurcation point for the onset of spontaneous oscillations. This, in turn, provides the basis for active signal processing such as amplification and frequency tuning. In Sect. 10.4 we turn to the problem of bacterial chemotaxis, which is a canonical system used to explore the sensitivity of biochemical sensors to environmental signals. We describe some of

© Springer Nature Switzerland AG 2021

P. Bressloff, *Stochastic Processes in Cell Biology*, Interdisciplinary Applied Mathematics 41, https://doi.org/10.1007/978-3-030-72519-8_10

the biochemical signaling networks responsible for amplification via receptor clustering (cooperativity) and for adaptation. We also analyze some simple PDE models of bacterial run-and-tumble motion in response to a chemotactic gradient, which are based on velocity- jump processes analogous to those previously encountered in models of microtubule catastrophes (Chap. 4) and bidirectional motor transport (Chap. 7). We also briefly discuss the Navier–Stokes equation and swimming at a low Reynolds number (Box 10B).

In Sect. 10.5 we derive the classical Berg–Purcell limit for a chemical sensor to estimate the concentration of some signaling molecule via temporal integration, and then consider more recent extensions of the theory based on linear response theory, and maximum likelihood estimation (Box 10C). One of the useful features of the latter approach is that it can be extended to take into account temporal concentration changes, such as those that arise during bacterial chemotaxis. Bacteria are too small to detect differences in concentrations across their cell bodies, so they proceed by measuring and comparing concentrations over time along their swimming trajectories. In contrast to bacterial cells, eukaryotic cells such as the social amoeba *Dictyostelium discoideum* are sufficiently large so that they can measure the concentration differences across their cell bodies without temporal integration. Finally, in Sect. 10.6 we consider various models of spatial gradient sensing in eukaryotes including the local excitation, global inhibition (LEGI) model, which takes into account the fact that cells adapt to background concentrations.

## 10.1 Phosphorylation–dephosphorylation cycles and ultrasensitivity

In Chap. 3 we described the classical Monod-Wyman-Changeux (MWC) model of cooperative receptor–ligand binding, and showed how cooperativity can sharpen the stimulus–response properties of receptors to ligand concentration. (For a general review of cooperativity in cell signaling, see Ref. [58].) In particular, the sharpness of the sigmoidal response is related to the number of interacting sites. A characteristic feature of amplification via cooperative binding is that it occurs at thermodynamic equilibrium. In this section, we consider an alternative, nonequilibrium mechanism for the generation of sigmoidal response curves, based on a driven phosphorylation–dephosphorylation cycle (PdPC) first introduced by Goldbeter and Koshland in 1981 [68, 351]. The system consists of a protein that can exist in an unmodified (unphosphorylated) form known as the substrate and a modified (phosphorylated) form known as the product. Interconversion of the inactivated and activated protein states is catalyzed by two enzymes, kinases that phosphorylate the inactivated protein and phosphatases that dephosphorylate the activated protein. PdPC is a very common signaling mechanism within cells, and also occurs within the context of intracellular protein gradients (see Sect. 11.1). The analysis of PdPCs proceeds along analogous lines to the classical derivation of Michaelis–Menten kinetics in a single enzymatic pathway (Chap. 1). A schematic illustration of the different signal transduction schemes is provided in Fig. 10.1.

Goldbeter and Koshland [351] also coined the phrase zero-order ultrasensitivity. "Ultrasensitivity" refers to the fact that the PdPC can exhibit an output response that is more sensitive to changes in stimulus than the Michaelis–Menten form, while "zero-order" indicates that the enzymes operate close to saturation. Following their original model, a variety of biological systems have been found to exhibit zero-order ultrasensitivity in signal transduction, including mitogen-activated protein kinase (MAPK) cascades, glucose mobilization, cell division/apoptosis, and membrane transport. Moreover, ultrasensitive components can play an important role in the generation of bistability via positive feedback, biochemical oscillations via negative feedback loops, and signaling cascades. For a detailed review of zero-order sensitivity, see Refs. [304–306].

The cooperative nature of the underlying nonequilibrium steady (or stationary) state (NESS) has been explored in considerable detail by Qian and collaborators [51, 339, 340, 441, 740–742]. In particular, they highlight how the PdPC can be considered as a form of temporal cooperativity. Substrate proteins compete for kinase such that, as the reaction proceeds, the substrate number decreases and the competition lessens. Thus, earlier turnovers (phosphorylation) help the later turnovers so that the substrates are temporally cooperative. However, one also has to take account of the competition from the product (phosphorylated proteins) for the same enzyme. This is precisely the role of the driven system: If the products were equally likely to compete for the enzymes, i.e., the enzymatic reactions were fully reversible, then the temporal cooperativity would disappear. The system is maintained out of equilibrium by ATP. (In the original analysis of PdPCs, the final step in the enzymatic reaction schemes was taken to be irreversible [351].)

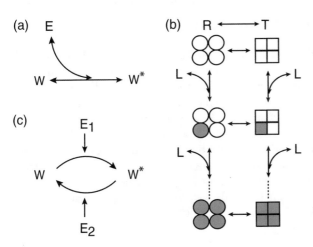

Fig. 10.1: Three examples of signal transduction schemes. (a) Michaelis–Menten type enzymatic reaction. (b) MWC cooperative binding reaction. Here each subunit has two conformational states with different binding affinities for ligand $L$. (c) Goldbeter–Koshland enzymatic reaction scheme.

## 10.1.1 Goldbeter–Koshland model

Ultrasensitivity in a PdPC was originally analyzed by Golbeter and Koshland [351], who assumed that the action of kinases and phosphatases are irreversible. Suppose that a protein exists in the unmodified form $W$ and the modified form $W^*$. Interconversion of the two forms is catalyzed by two enzymes $E_1$ and $E_2$ according to the reaction schemes

$$W + E_1 \underset{d_1}{\overset{a_1}{\rightleftharpoons}} WE_1 \overset{k_1}{\rightarrow} W^* + E_1,$$

$$W^* + E_2 \underset{d_2}{\overset{a_2}{\rightleftharpoons}} W^* E_2 \overset{k_2}{\rightarrow} W + E_2.$$

Introducing the concentrations $w = [W]$, $w^* = [W^*]$, $z_j = [E_j]$, $w_1 = [WE_1]$, and $w_2^* = [W^*E_2]$, the corresponding kinetic equations are

$$\frac{dw}{dt} = -a_1 w z_1 + d_1 w_1 + k_2 w_2^*, \tag{10.1.1a}$$

$$\frac{dw_1}{dt} = a_1 w z_1 - (d_1 + k_1) w_1, \tag{10.1.1b}$$

$$\frac{dw^*}{dt} = -a_2 w^* z_2 + d_2 w_2^* + k_1 w_1, \tag{10.1.1c}$$

$$\frac{dw_2^*}{dt} = a_2 w^* z_2 - (d_2 + k_2) w_2^*. \tag{10.1.1d}$$

These equations are supplemented by the conservation equations

$$W_T = w + w^* + w_1 + w_2^*, \quad E_{1T} = z_1 + w_1, \quad E_{2T} = z_2 + w_2^*. \tag{10.1.2}$$

Proceeding along analogous lines to the derivation of Michaelis–Menten kinetics (Chap. 1), we assume that the concentration of $W$ and $W^*$ is much larger than that of the kinase and phoshphotase, that is, $W_T \gg E_{1T} + E_{2T}$ or equivalently $W_T = w + w^*$. This implies that the time scale for the dynamics of the enzymes $E_1$ and $E_2$ is much faster than that for the dynamics of $W$ and $W^*$. Performing a separation of time scales, we can treat the concentrations $w$ and $w^*$ as constants when analyzing equations (10.1.1b,d), while we can take the steady-state values of the concentrations $z_1, z_2$ when solving equations (10.1.1a,c). Hence, setting $dw_1/dt = 0$ and $z_1 = E_{1T} - w_1$ in (10.1.1b), we can solve for $w_1$ in terms of $w$. Similarly, setting $dw_2^*/dt = 0$ and $z_2 = E_{2T} - w_2^*$ in (10.1.1d), we can solve for $w_2^*$ in terms of $w^*$. We thus obtain the reduced kinetic scheme (see Ex. 10.1)

$$W \underset{v_2(w^*)}{\overset{v_1(w)}{\rightleftharpoons}} W^*, \tag{10.1.3}$$

with

$$v_1(w) = \frac{V_1 w / K_1}{1 + w / K_1}, \quad v_2(w^*) = \frac{V_2 w^* / K_2}{1 + w^* / K_2}, \tag{10.1.4}$$

and

$$V_1 = k_1 E_{1T}, \ V_2 = k_2 E_{2T}, \ K_1 = \frac{d_1 + k_1}{a_1}, \ K_2 = \frac{d_2 + k_2}{a_2}.$$

Imposing the conservation condition $W_T = w + w^*$ thus yields the single independent kinetic equation

$$\frac{dw^*}{dt} = v_1(W_T - w^*) - v_2(w^*).  \tag{10.1.5}$$

The steady-state fraction of the product protein $W^*$, $\phi = w^*_{eq}/W_T$ is thus obtained by solving the equation $v_1(W_T - w^*_{eq}) = v_2(w^*_{eq})$, which yields a quadratic equation for $w^*_{eq}$ with a unique positive solution (monostability)

$$\phi = \frac{-B + \sqrt{B^2 + 4AC}}{2A},  \tag{10.1.6}$$

for $V_1 \neq V_2$, where

$$A = \frac{V_1}{V_2} - 1, \quad B = \frac{1}{W_T}\left(K_1 + K_2 \frac{V_1}{V_2}\right) - \left(\frac{V_1}{V_2} - 1\right), \quad C = \frac{K_2}{W_T}\left(\frac{V_1}{V_2}\right).$$

A plot of the steady-state molar fraction of $W^*$, $\phi$, as a function of the ratio $V_1/V_2$ is shown in Fig. 10.2 for $K_1 = K_2$. Note that $V_1/V_2$ represents the ratio of the kinase activity to phosphatase activity, which characterizes the stimulus into the PdPC. At low values of $K_1$ and $K_2$, there is a sharp change from low to high levels of modified protein over a very small change in the $V_1/V_2$ ratio (zero-order ultrasensitivity); this corresponds to a regime in which the two enzymes are saturated. On the other hand, for large values of $K_1$ and $K_2$, the curve is relatively shallow, and one obtains a response similar to first-order kinetics. In order to understand the origins of zero-order ultrasensitivity, suppose that $W_T \gg K_1, K_2$. If $w$ and $w^*$ were both $O(W_T)$, then $v_1(w) \approx V_1$ and $v_2(w^*) \approx V_2$ so that

$$\frac{dw^*}{dt} \approx V_1 - V_2.  \tag{10.1.7}$$

The steady-state solution only exists in this regime provided that $V_1 = V_2$, that is, $k_1 E_{1T} = k_2 E_{2T}$. If $V_1 \neq V_2$ then either $w \approx 0$ or $w^* \approx 0$.

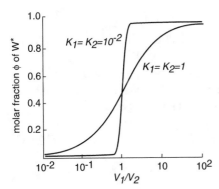

Fig. 10.2: Molar fraction of modified protein $W^*$ at steady state as a function of the modification rates.

**Reversible kinetic model.** One of the major simplifications of the Goldbeter–Koshland model is the irreversibility of the distinct phosphorylation and dephosphorylation processes. In order to develop a deeper understanding of kinetic models of PdPCs and the nature of the nonequilibrium steady state, it is necessary to treat these processes as reversible and to include the effects of ATP [740, 741]. Explicitly incorporating the reversible hydrolysis reaction $ATP \rightleftharpoons ADP + P$, we have

$$W + E_1 + ATP \underset{d_1}{\overset{a_1^0}{\rightleftharpoons}} W \cdot E_1 \cdot ATP \underset{q_1^0}{\overset{k_1}{\rightleftharpoons}} W^* + E_1 + ADP,$$

$$W^* + E_2 \underset{d_2}{\overset{a_2}{\rightleftharpoons}} W^* E_2 \underset{q_2^0}{\overset{k_2}{\rightleftharpoons}} W + E_2 + P.$$

At constant concentrations for ADT, ADP, and P, which are assumed to be sufficiently below the saturation levels of their respective enzymes, we can eliminate the dynamics of hydrolysis, and obtain the reduced reversible reaction scheme

$$W + E_1 \underset{d_1}{\overset{a_1}{\rightleftharpoons}} WE_1 \underset{q_1}{\overset{k_1}{\rightleftharpoons}} W^* + E_1, \tag{10.1.8}$$

$$W^* + E_2 \underset{d_2}{\overset{a_2}{\rightleftharpoons}} W^* E_2 \underset{q_2}{\overset{k_2}{\rightleftharpoons}} W + E_2, \tag{10.1.9}$$

with effective first-order rate constants

$$a_1 = a_1^0[\text{ATP}], \quad q_1 = q_1^0[\text{ADP}], \quad q_2 = q_2^0[\text{P}].$$

If the concentrations of ATP, ADP, and P were allowed to reach thermodynamic equilibrium, then the law of mass action would yield the following equilibrium constant for ATP hydrolysis:

$$\frac{[\text{ATP}]_{\text{eq}}}{[\text{ADP}]_{\text{eq}}[\text{P}]_{\text{eq}}} = \frac{d_1 q_1^0 d_2 q_2^0}{a_1^0 k_1 a_2 k_2} = e^{-\Delta G^0 / k_B T},$$

where $\Delta G^0$ is the standard free energy change for the ATP hydrolysis reaction (see also Sect. 4.6). However, since these concentrations are maintained out of equilibrium, we find that summing up the changes in free energy associated with the four reversible reactions gives

$$\Delta G = k_B T \ln \gamma \equiv k_B T \ln \frac{a_1 k_1 a_2 k_2}{d_1 q_1 d_2 q_2} = \Delta G_0 + k_B T \ln \left( \frac{[\text{ATP}]}{[\text{ADP}][\text{P}]} \right). \tag{10.1.10}$$

Thermodynamic equilibrium is recovered when $\Delta G = 0$, that is, $\gamma = 1$

The analysis of the reversible model proceeds along similar lines to the Goldbeter–Koshland model. We now obtain a reduced reaction scheme of the form

$$W \underset{f_2(w^*, w)}{\overset{f_1(w, w^*)}{\rightleftharpoons}} W^*, \tag{10.1.11}$$

with

$$f_1(w, w^*) = \frac{V_1 w/K_1}{1 + w/K_1 + w^*/K_1^*} + \frac{V_2^* w/K_2^*}{1 + w/K_2^* + w^*/K_2}, \tag{10.1.12}$$

$$f_2(w^*, w) = \frac{V_2 w^*/K_2}{1 + w/K_2^* + w^*/K_2} + \frac{V_1^* w/K_1^*}{1 + w/K_1 + w^*/K_1^*}, \tag{10.1.13}$$

and

$$V_1^* = d_1 E_{1T}, \; V_2^* = d_2 E_{2T}, \; K_1^* = \frac{d_1 + k_1}{q_1}, \; K_2^* = \frac{d_2 + k_2}{q_2}.$$

The steady-state molar fraction of protein $W^*$ is then determined by solving the quadratic equation arising from [740]

$$f_1(W_T - w_{eq}^*, w_{eq}^*) = f_2(w_{eq}^*, w_T - w_{eq}^*).$$

Note the results for the Golbeter–Koshland model are recovered in the limits $q_1, q_2 \to 0$, for which $K_1^*, K_2^* \to \infty$, $\gamma \to \infty$, and

$$f_1(w, w^*) \to v_1(w), \quad f_2(w^*, w) \to v_2(w^*).$$

For finite $q_1, q_2$, one still observes ultrasensitivity in the saturated regime where $K_1, K_2$ are sufficiently small.

Recall that cooperative receptor–ligand binding is based on a high-order reaction scheme (see Sect. 3.2),

$$R + nL \overset{K^n}{\rightleftharpoons} R_n,$$

where $K$ is the dissociation constant for a receptor binding a single ligand, and $n$ is the number of binding sites. The equilibrium fraction of receptors with all their sites bound is then

$$Y = \frac{[R_n]}{[R_n] + [R]} = \frac{[L]^n}{K^n + [L]^n},$$

and the steepness of the response curve increases with $n$. In particular,

$$n = 2 \frac{d \ln Y}{d \ln[L]} \bigg|_{Y=1/2}.$$

On the other hand, ultrasensitivity in a PdPC occurs in a regime where the enzymes are saturated, which means that the steady state is only weakly dependent on the concentrations $w_{eq}, w_{eq}^*$. Defining an effective Hill coefficient by

$$n_H = 2 \frac{d \ln \phi}{d \ln \sigma} \bigg|_{\phi=1/2}, \quad \sigma = \frac{V_1}{V_2},$$

we find that $n_H \gg 1$. A Hill function would be obtained from the corresponding steady-state equation $k_p w_{eq}^v = k_d w_{eq}^{*\,v}$, where $k_p$ and $k_d$ are the rates of phosphorylation and dephosphorylation, and $v = 1/n_H \approx 0$. In a standard single-step reaction, the order $v$ of the reaction represents the number of molecules converted (stoichiometry) in a single step, so that for $v \approx 0$ we have a "zeroth-order" reaction. Another way to view this is that converting one protein $W$ to $W^*$ will require on average $n_H$ phosphorylation cycles, that is, $n$ binding sites in equilibrium cooperative binding has been replaced by $n_H$ temporal cycles in a nonequilibrium process, leading to the notion of temporal cooperativity.

## 10.1.2 Multisite phosphorylation

Several theoretical studies have explored how multisite phosphorylation combined with feedback loops can generate more complex dynamics such as bistability and oscillations [187, 305, 306, 590]. Consider, for example, the dual phosphorylation scheme shown in Fig. 10.3(a). The substrate $W$ has both a single ($W^*$) and a double ($W^{**}$) phosphorylation state mediated by the action of kinases and phosphatases. The kinetics can either be processive or distributive. In the case of processive catalysis, after binding to the substrate, the kinase or phosphatase induces two phosphorylations or dephosphorylations before releasing the final product. On the other hand, distributive catalysis releases the intermediate monophosphorylated product, and requires a second reaction step in order to convert the intermediate product to its final form. Experimentally it is found that in MAPK cascade, dual PdPCs occur via a distributive mechanism, as indicated in Fig. 10.3(a). The basic phosphorylation and dephosphorylation reaction schemes are as follows [590]:

$$W + EK \underset{k_{-1}}{\overset{k_1}{\rightleftharpoons}} W \cdot EK \overset{k_2}{\rightarrow} W^* + EK,$$

$$W^* + EK \underset{k_{-3}}{\overset{k_3}{\rightleftharpoons}} W^* \cdot EK \overset{k_4}{\rightarrow} W^{**} + EK, \qquad (10.1.14)$$

$$W^{**} + EP \underset{h_{-1}}{\overset{h_1}{\rightleftharpoons}} W^{**} \cdot EP \overset{h_2}{\rightarrow} W^* \cdot EP \underset{h_{-3}}{\overset{h_3}{\rightleftharpoons}} W^* + EP,$$

$$W^* + EP \underset{h_{-4}}{\overset{h_4}{\rightleftharpoons}} W^* \cdot EP \overset{k_5}{\rightarrow} W \cdot EP \underset{h_{-6}}{\overset{h_6}{\rightleftharpoons}} W + EP.$$

The one major difference between the phosphorylation and dephosphorylation pathways is that in the former case the catalysis and product dissociation steps are lumped together, whereas in the latter case the release of the product from the phosphatase is reversible. The product is said to sequester the phosphatase.

Another assumption of the model is that both the kinase and phosphatase are saturated by their respective substrates $W$ and $W^{**}$, and this effectively adds a pair of inhibitory feedback loops into the reaction network, see Fig. 10.3(b). First, the sequestration of kinase by $W$ in equation (10.1.14d) means that less kinase is available to phosphorylate $W^*$, that is, the phosphorylation of $W^*$ is suppressed. Second, the sequestration of phosphatase by $W^{**}$ implies that less phosphatase is around to dephosphorylate $W^*$, resulting in the dephosphorylation of $W^*$ being suppressed.

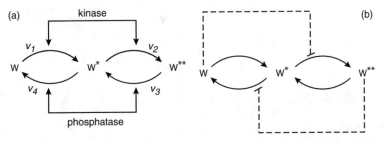

Fig. 10.3: Dual phosphorylation scheme with three states of increasing phosphorylation $W, W^*, W^{**}$. (b) The substrate is acted on by a kinase $(EK)$ and a phosphatase $(EP)$ with the latter reversibly sequestered. (b) Implicit negative feedback loops arising from sequestration effects.

The combination of two negative feedback loops is equivalent to the positive feedback of $W^{**}$, and this is the source of bistability.

Carrying out a reduction of the corresponding mass-action kinetic equations along similar lines to the standard Golbeter–Koshland model yields Michaelis–Menten kinetics for the concentrations $w = [W]$ and $w^{**} = [W^{**}]$. These take the form [590]

$$\frac{dw}{dt} = v_4 - v_1, \quad \frac{dw}{dt} = v_2 - v_3, \quad w^* \equiv [W^*] = W_T - w - w^{**}, \qquad (10.1.15)$$

with

$$v_1 = \frac{\kappa_1 [EK] w / K_1}{1 + w/K_1 + w^*/K^2}, \quad v_2 = \frac{\kappa_2 [EK] w^* / K_2}{1 + w/K_1 + w^*/K^2}, \qquad (10.1.16a)$$

$$v_3 = \frac{\kappa_3 [EP] w^{**} / K_3}{1 + w/K_5 + w^*/K_4 + w^{**}/K_3}, \quad v_4 = \frac{\kappa_4 [EP] w^* / K_4}{1 + w/K_5 + w^*/K_4 + w^{**}/K_3}. \qquad (10.1.16b)$$

Here $[EK]$ and $[EP]$ denote the total concentrations of kinase and phosphatase. The various effective reaction rates are

$$\kappa_1 = k_2, \ \kappa_2 = k_4, \ \kappa_3 = \frac{h_2}{1 + h_2/h_3}, \ \kappa_4 = \frac{h_5}{1 + h_5/h_6 + h_{-3}(h_{-4} + h_5)/(h_3 h_4)},$$

$$K_1 = \frac{k_{-1} + k_2}{k_1}, \quad K_2 = \frac{k_{-3} + k_4}{k_3}, \quad K_3 = \frac{h_{-1} + h_2}{h_1 + h_1 h_2 / h_3},$$

and

$$K_4 = \frac{h_{-4} + h_5}{h_4} \frac{1}{1 + h_5/h_6 + h_{-3}(h_{-4} + h_5)/(h_3 h_4)}, \quad K_5 = \frac{h_6}{h_{-6}}.$$

The equilibria of the deterministic dynamics are obtained from the two algebraic conditions $v_4 = v_1$ and $v_2 = v_3$ together with the conservation condition. One finds

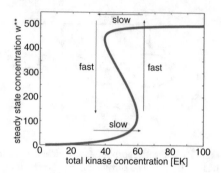

Fig. 10.4: Bistability in the dual phosphorylation reaction scheme. Steady-state fixed points for $w^{**}$ are plotted as a function of the total kinase concentration $[EK]$. Parameter values can be found in [187].

that there exists a parameter regime for which the steady-state concentrations exhibit bistability as the total concentration of kinase is varied, see Fig. 10.4.

Finally, note that the bistable dual PdPC can be converted to a relaxation oscillator by adding an equation for the production of kinase:

$$\frac{d[EK]}{dt} = \frac{\kappa}{1 + w^{**}/K} - \gamma[EK]. \tag{10.1.17}$$

Suppose that the networks start on the lower branch of Fig. 10.4. Since $w^{**}$ is relatively small, the production of kinase slowly increases and the quasi-equilibrium slowly moves along the lower branch to the right. When it reaches the saddle node, it rapidly jumps to the upper branch, where $w^{**}$ is relatively large and kinase production is small. The system thus moves leftward on the upper branch until it reaches the other saddle node, where it rapidly drops to the lower branch. This results in a relaxation oscillator analogous to ones previously explored for conductance-based neuron models (Sect. 3.4) and genetic oscillators (Sect. 5.1).

### 10.1.3 Noise amplification.

A number of studies have investigated noise signal amplification in ultrasensitive signal transduction pathways based on stochastic versions of equation (10.1.5), in which the concentration of kinases or phosphatases fluctuates (intrinsic noise) or there is a fluctuating source term (extrinsic noise) [68, 339, 555, 818]. A full analysis would need to start from a stochastic version of the full system of equations (10.1.1a–d). Here, however, we will follow along similar lines to previous authors and consider the effects of fluctuations by applying linear response theory to equation (10.1.5); see also Sect. 2.3. For concreteness, suppose that the concentration of kinases fluctuates by writing $E_{1T} \rightarrow E_{1T} + R(t)$ with $R(t)$ evolving according to the Langevin equation

$$\frac{dR}{dt} = -\gamma_r R(t) + \sigma\xi(t), \tag{10.1.18}$$

with $\xi(t)$ a white noise process

$$\langle \xi(t) \rangle = 0, \quad \langle \xi(t)\xi(t') \rangle = \delta(t - t'),$$

and $\sigma$ the noise intensity. Here $\gamma_r$ is the rate of relaxation to the equilibrium kinase concentration $E_{1T}$. It is convenient to rewrite equation (10.1.5) as

$$\frac{dw^*}{dt} = \mathscr{F}(w^*, R) \equiv \frac{k_1[E_{1T} + R](W_T - w^*)}{K_1 + [W_T - w^*]} - \frac{k_2 E_{12} w^*}{K_2 + w^*}. \tag{10.1.19}$$

Linearizing equation (10.1.19) about the fixed point solution by setting $w^* = w^*_{eq} + X$, $w^*_{eq} = \phi(V_1/V_2)W_T$, gives

$$\frac{dX}{dt} = -\beta_1 X + \beta_2 R, \tag{10.1.20}$$

with

$$\beta_1 = -\left. \frac{\partial \mathscr{F}}{\partial w^*} \right|_{eq} = \frac{k_1 K_1 E_{1T}}{[K_1 + [W_T - w^*_{eq}]]^2} + \frac{k_2 K_2 E_{2T}}{[K_2 + w^*_{eq}]^2}, \tag{10.1.21}$$

and

$$\beta_2 = \left. \frac{\partial \mathscr{F}}{\partial R} \right|_{eq} = \frac{k_1(W_T - w^*_{eq})}{K_1 + [W_T - w^*_{eq}]}. \tag{10.1.22}$$

Note that the gain of the equilibrium system, which is proportional to the slope of the input–output curves in Fig. 10.2 is

$$g = \frac{\Delta w^*/w^*_{eq}}{\Delta R/E_{1T}} = \frac{\beta_2}{\beta_1} \frac{E_{1T}}{w^*_{eq}}. \tag{10.1.23}$$

In order to determine the variance of the concentration $w^*$, we use Fourier transforms. Taking

$$\widehat{R}(\omega) = \int_{-\infty}^{\infty} e^{i\omega t} R(t)dt,$$

etc., we Fourier transform the linear equations (10.1.18) and (10.1.20) to obtain

$$\widehat{X}(\omega) = \frac{\beta_2}{\beta_1 - i\omega} \widehat{R}(\omega), \quad \widehat{R}(\omega) = \frac{\sigma}{\gamma_r - i\omega} \widehat{\xi}(\omega),$$

where $\widehat{\xi}(\omega)$ is the Fourier transform of a white noise process with

$$\langle \widehat{\xi}(\omega) \rangle = 0, \quad \langle \widehat{\xi}(\omega)\overline{\widehat{\xi}(\omega')} \rangle = 2\pi\delta(\omega - \omega').$$

Using the Wiener–Khinchin theorem (Sect. 2.3), the variance of the kinase concentration is given by the integral of the power spectrum defined by

$$2\pi S_R(\omega)\delta(\omega - \omega') = \langle \widehat{R}(\omega)\overline{\widehat{R}(\omega')} \rangle.$$

That is,

$$\sigma_R^2 = \int_{-\infty}^{\infty} S_R(\omega) \frac{d\omega}{2\pi} = \int_{-\infty}^{\infty} \frac{\sigma^2}{\gamma_r^2 + \omega^2} \frac{d\omega}{2\pi} = \frac{\sigma^2}{2\gamma_r}. \tag{10.1.24}$$

Similarly, the variance of the product concentration is

$$\begin{aligned}
\sigma_X^2 &= \int_{-\infty}^{\infty} S_X(\omega) \frac{d\omega}{2\pi} = \int_{-\infty}^{\infty} \frac{\beta_2^2}{\beta_1^2 + \omega^2} \frac{\sigma_0^2}{\gamma_r^2 + \omega^2} \frac{d\omega}{2\pi} \\
&= \frac{\beta_2^2}{\beta_1^2 - \gamma_r^2} \frac{\sigma^2}{2\gamma_r} + \frac{\beta_2^2}{2\beta_1} \frac{\sigma^2}{\gamma_r^2 - \beta_1^2} = \frac{\beta_2^2 \sigma^2}{2\beta_1 \gamma_r} \frac{1}{\gamma_r + \beta_1}.
\end{aligned} \tag{10.1.25}$$

If we interpret $\sigma_X / w_{eq}^*$ as the relative noise intensity of the output and $\sigma_R / E_{1T}$ as the relative noise intensity of the input, then the noise amplification of the PdPC in response to receptor fluctuations is defined by [818]

$$G = \frac{\sigma_X / w_{eq}^*}{\sigma_R / E_{1T}} = \frac{E_{1T}}{w_{eq}^*} \sqrt{\frac{\beta_2^2}{\beta_1} \frac{1}{\gamma_r + \beta_1}} = g \sqrt{\frac{\beta_1}{\gamma_r + \beta_1}}. \tag{10.1.26}$$

It follows that if the relaxation rate $\gamma_r$ of the receptor noise fluctuations is much slower than the relaxation rate of the protein $W^*$, then the noise amplification $G$ approaches the deterministic gain $g$. This implies that ultrasensitivity produces high noise amplification in the domain around $V_1 = V_2$. On the other hand, if the relaxation rate of the fluctuations is relatively fast, then $G \approx g \sqrt{\beta_1 / \gamma_r} \ll g$.

**Noise-induced bimodality.** The amplification of external noise-driven fluctuations can qualitatively change the PdPC dynamics, as demonstrated by Samoilov et al. [789]. The basic idea is that multiplicative noise can generate a bimodal stationary probability density, such that single stochastic trajectories of the substrate concentrations exhibit oscillations. Following [789], consider a stochastic version of equation (10.1.5) of the Ito form

$$\begin{aligned}
dw = -dw^* &= -A(w)dt + B(w)dW(t) \\
&= -\left[ \frac{k_1 E_{1T} w}{K_1 + w} - \frac{k_2 E_{2T}(W_T - w)}{K_2 + W_T - w} \right] + \frac{\sigma k_1 w}{K_1 + w} dW(t),
\end{aligned} \tag{10.1.27}$$

with $\sigma = \sigma(E_{1T})$. (In order not to confuse the stochastic variable $w(t)$ with the Wiener process $W(t)$, we write the former in lower case.) The corresponding FP equation for $p(w,t)$ is

$$\frac{\partial p}{\partial t} = \frac{\partial}{\partial w} \left[ A(w)p + \frac{1}{2} \frac{\partial}{\partial w} D(w)p \right], \tag{10.1.28}$$

with $D(w) = B(w)^2$. The stationary density is then given by

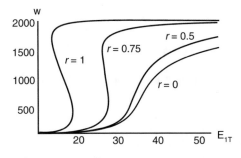

Fig. 10.5: Sketch of maxima and minima of the steady-state probability density $p_s(w)$ as a function of the total kinase concentration $E_{1T}$ and various coefficients $r$ with $\sigma(E_{1T}) = \sigma_0 E_{1T}^r$. Here $\sigma_0 = 0.2, a_1 = 40, d_1 = 10^4, k_1 = 10^4, a_2 = 200, d_2 = 100, k_2 = 5 \times 10^3$, and $E_{1T} = 20, E_{2t} = 50, W_T = 2000$ [Redrawn from Samoilov et. al [789].]

$$p_s(w) = \frac{C}{D(w)} e^{-\Phi(w)}, \quad \Phi(w) = -2 \int^w \frac{A(x)}{D(x)} dx, \qquad (10.1.29)$$

which has the explicit form

$$p_s(w) = \tilde{C} \left( 1 + \frac{K_1}{w} \right)^2 \exp \left[ -\frac{2}{(\sigma k_1)^2} [k_1 E_{1T} F_+(w) - k_2 E_{2T} F_-(w)] \right],$$

where $\tilde{C}$ is a normalization constant,

$$F_+(w) = w + K_1 \ln(w/K_1),$$

and

$$F_-(w) = w - \frac{K_2(K_1 + K_2 + W_T)^2}{(K_2 + W_T)^2} \ln \left[ \frac{w}{K_2 + W_T - w} \right] + (2K_1 + K_2) \ln \left[ \frac{w}{K_1} \right]$$
$$- \frac{K_1^2 W_T}{K_2 + W_T} \frac{1}{w}.$$

In order to determine whether or not the stationary solution is bimodal, one has to find maxima of the probability density, which first requires finding roots of the equation

$$0 = \frac{\partial p_s(w)}{\partial w} = -\frac{2p_s(w)}{(\sigma k_1)^2} \left( k_1 E_{1T} \frac{\partial F_+(w)}{\partial w} - k_2 E_{2T} \frac{\partial F_-(w)}{\partial w} + \frac{(\sigma k_1)^2 K_1}{w(K_1 + w)} \right).$$

Taking $\sigma(E) = \sigma_0 E^r$, one finds that bimodality occurs for a range of values of $r$ and $E_{1T}$, as illustrated in Fig. 10.5.

## 10.1.4 Noise attenuation in a signaling cascade

Consider a generic model of a stochastic signaling cascade consisting of molecular species labeled $i = 0, \ldots, n$ with corresponding concentrations $y_i$ [870]; see

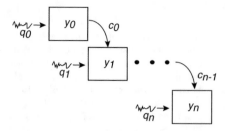

Fig. 10.6: Schematic diagram of a linearized stochastic cascade.

Fig. 10.6. Suppose that the rate of production of species $i$ only depends on the concentration $y_{i-1}$ of the species at the previous level of the cascade, and that it degrades at a fixed rate $\gamma_i$. Take these concentrations to evolve according to a linear system of SDEs of the form

$$\dot{Y}_i + \gamma_i Y_i = c_{i-1} Y_{i-1} + \eta_i, \tag{10.1.30}$$

where

$$\langle \eta_i \rangle = 0, \quad \langle \eta_i(t) \eta_j(t') \rangle = q_i \delta_{i,j} \delta(t - t'),$$

and $c_i$ is rate of production of species $i$. Both $\gamma_i$ and $c_i$ have units of $t^{-1}$. For convenience, set $\gamma_i = 1$ for all $i$. Following Ref. [870], we will analyze how noise propagates through the cascade using linear response theory and power spectra (see Sect. 2.2).

Fourier transforming the equation for $\dot{Y}_i$ we have

$$-i\omega \widehat{Y}_i(\omega) + \widehat{Y}_i(\omega) = c_{i-1} \widehat{Y}_{i-1}(\omega) + \widehat{\eta}_i,$$

which on rearranging gives

$$\widehat{Y}_i(\omega) = \frac{1}{1 - i\omega} \left[ c_{i-1} \widehat{Y}_{i-1}(\omega) + \widehat{\eta}_i \right].$$

Hence

$$\left\langle \widehat{Y}_i(\omega) \widehat{Y}_i(\omega') \right\rangle = \frac{1}{1 - i\omega} \frac{1}{1 - i\omega'} \left[ c_{i-1}^2 \left\langle \widehat{Y}_{i-1}(\omega) \widehat{Y}_{i-1}(\omega') \right\rangle + \left\langle \widehat{\eta}_i(\omega) \widehat{\eta}_i(\omega') \right\rangle \right].$$

Setting

$$\left\langle \widehat{Y}_i(\omega) \widehat{Y}_i(\omega') \right\rangle = 2\pi \delta(\omega + \omega') S_i(\omega)$$

gives the recurrence relation

$$S_i(\omega) = \frac{c_{i-1}^2 S_{i-1}(\omega) + q_i}{1 + \omega^2}. \tag{10.1.31}$$

Iterating this equation shows that

$$S_n(\omega) = \frac{q_n}{1 + \omega^2} + \frac{c_{n-1}^2}{1 + \omega^2} \frac{c_{n-2}^2 S_{n-2}(\omega) + q_{n-1}}{1 + \omega^2},$$

that is,

$$S_n(\omega) = \alpha_n + \beta_{n-1}\alpha_{n-1} + \beta_{n-1}\beta_{n-2}\alpha_{n-2} + \ldots + \beta_{n-1}\ldots\beta_0 S_0(\omega), \quad (10.1.32)$$

where

$$\alpha_j = \frac{q_j}{1+\omega^2}, \quad \beta_j = \frac{c_j^2}{1+\omega^2}.$$

Now define

$$\alpha = \frac{q}{1+\omega^2}, \quad \beta = \frac{c^2}{1+\omega^2}, \quad q = \max_i\{q_i\}, \quad c = \max_i\{c_i\}.$$

It follows from equation (10.1.32) that

$$S_n(\omega) \leq \alpha + \beta\alpha + \beta^2\alpha + \ldots + \beta^n S_0(\omega).$$

Since $c_{-1} = 0$, we have $S_0(\omega) \leq \alpha$,

$$S_n(\omega) \leq \alpha(1+\beta+\beta^2+\ldots+\beta^n)(\omega),$$

and thus

$$\lim_{n\to\infty} S_n(\omega) \leq \frac{\alpha}{1-\beta}. \quad (10.1.33)$$

Substituting for $\alpha, \beta$, taking the inverse Fourier transform and using the Wiener–Khinchin theorem (Sect. 2.2), we obtain the result (for $t > 0$)

$$\lim_{n\to\infty}\langle Y_n^2(t)\rangle \leq \int_{-\infty}^{\infty} \frac{q e^{-i\omega t}}{1+\omega^2 - c^2}\frac{d\omega}{2\pi}$$

$$= \int_C \frac{q e^{-i\omega t}}{(\omega - i\sqrt{1-c^2})(\omega + i\sqrt{1-c^2})}\frac{d\omega}{2\pi}$$

$$= \frac{q}{2\sqrt{1-c^2}}e^{-\sqrt{1-c^2}\,t} \leq \frac{q}{2\sqrt{1-c^2}}. \quad (10.1.34)$$

We have used the calculus of residues, after closing the contour $C$ in the lower-half complex $\omega$-plane. This establishes that fluctuations in the output of the signaling cascaded will be bounded provided that $|c_i| \leq |c| < 1$.

*Finite cascade.* Now consider a finite cascade of length $n$ with $c_i = c < 1$ for all $i = 0,\ldots,n$, $q_i = q$ for all $i = 1,\ldots,n$, and $q_0 > q$. Thus the noise at the input level is higher than in successive levels of the cascade. Equation (10.1.31) reduces to

$$S_n(\omega) = \frac{c^2 S_{n-1}(\omega) + q}{1+\omega^2}, \quad n > 0, \quad S_1(\omega) = \frac{c^2 S_0(\omega) + q_0}{1+\omega^2}.$$

Iterating these equations with $S_0(\omega) = q_0/(1+\omega^2)$ yields

$$\langle S_n(\omega)\rangle = q\sum_{j=0}^{n-1}\frac{c^{2j}}{(1+\omega^2)^{j+1}} + q_0\frac{c^{2n}}{(1+\omega^2)^{n+1}}. \quad (10.1.35)$$

Taking inverse Fourier transforms and using the Wiener-Khinchin theorem,

$$\langle Y_n^2(t)\rangle \le \left\| \int_{-\infty}^{\infty} \left[ q \sum_{j=0}^{n-1} \frac{c^{2j}}{(1+\omega^2)^{j+1}} + q_0 \frac{c^{2n}}{(1+\omega^2)^{n+1}} \right] \frac{d\omega}{2\pi} \right\| .$$

From contour integration, we have

$$\int_{-\infty}^{\infty} \frac{1}{(1+\omega^2)^{n+1}} \frac{d\omega}{2\pi} = \int_{-\infty}^{\infty} \frac{1}{(\omega-i)^{n+1}(\omega+i)^{n+1}} \frac{d\omega}{2\pi}$$

$$= \frac{1}{n!} \frac{d^n}{da^n} \int_{-\infty}^{\infty} \frac{1}{(\omega-a)(\omega+i)^{n+1}} \frac{d\omega}{2\pi} \bigg|_{a=i} = \frac{1}{n!} \frac{d^n}{da^n} \frac{1}{(a+i)^{n+1}} \bigg|_{a=i}$$

$$= (-1)^n \frac{(n+1)(n+2)\dots 2n}{n!} \frac{i}{(2i)^{2n+1}} = \frac{2n!}{n!n!} \frac{i}{2^{2n+1}} .$$

It follows that

$$\langle Y_n^2(t)\rangle \le q \sum_{j=0}^{n-1} \frac{(2j)!}{j!j!} \frac{c^{2j}}{2^{2j+1}} + q_0 \frac{(2n)!}{n!n!} \frac{c^{2n}}{2^{2n+1}} . \qquad (10.1.36)$$

Applying Stirling's approximation

$$j! \approx \left(\frac{j}{e}\right)^j \sqrt{2\pi j},$$

we see that

$$\langle Y_n^2(t)\rangle \le \frac{q}{2} \left( 1 + \sum_{j=1}^{n-1} \frac{c^{2j}}{\sqrt{\pi j}} \right) + \frac{q_0}{2} \frac{c^{2n}}{\sqrt{\pi n}} .$$

The first term represents an increase in intrinsic fluctuations with cascade length $n$, whereas the second represents a faster than exponential decrease in the input noise with cascade length. There exists an optimal cascade length for minimizing the total noise, which is given by [870]

$$n_{\text{opt}} = \left[ \frac{1}{1-(q_0-q)^2/(q_0^2 c^4)} \right]_- ,$$

with $[x]_-$ denoting the greatest integer less than $x$.

## 10.2 Photoreceptor cells and phototransduction

Phototransduction is the process by which a photon of light generates an electrical response in a photoreceptor cell, which is then transmitted to downstream cells in the retina and beyond to the thalamus and cerebral cortex. There are two main classes of photoreceptor cells in mammalian eyes, rods and cones, each contributing information used by the visual system to form a representation of the visual world. The rods are narrower than the cones and are distributed differently across the retina. However, the chemical processes that underlie phototransduction are similar. (A third class of mammalian photoreceptor cells, intrinsically photosensitive retinal ganglion cells, do not contribute to vision directly, but are thought to support circadian rhythms and the pupil light reflex.) Rods are much more light sensitive than cones and can be triggered by a single photon. Hence, at very low light levels, visual information is based solely on the signal from rod cells. Cones require a larger number of photons to produce a signal and, in contrast to rods, are color sensitive. For example, in humans, there are three different types of cone cells, referred to as S-cones, M-cones, and L-cones, which respond more strongly to the light of short, medium, and long wavelengths, respectively. Since the firing of a cone cell depends on the number of photons absorbed, it follows that the frequency-response profiles of the three types of cone cells are determined by the likelihoods that their respective photoreceptor proteins will absorb photons of different wavelengths. For example, an L-cone cell more readily absorbs long wavelengths of light, which means that higher light intensities are required to elicit the same response at short wavelengths compared to long wavelengths.

In this section, we focus on phototransduction in vertebrate rod cells [31, 152, 291, 762, 968]. For a review of phototransduction in invertebrates, specifically, *drosophila*, see [398, 968]. Chap. 2 of the book by Bialek [78] has an extensive review of the biophysics of phototransduction and the historical development of the subject, while the physiology of photoreceptors and light adaptation is covered in Chap. 19 of Keener and Sneyd [497].

### 10.2.1 The rod photoreceptor

A schematic diagram of the structure of a rod photoreceptor cell is shown in Fig. 10.7(a). (Cones have the same basic structure.) Rod photoreceptors in vertebrates are found on the outermost layer of the retina, that is, the layer most distant from the eye. We will describe the basic components of the cell, starting with those closest to the visual field (and farthest from the brain), and working backwards. The most proximal component is the synapse at the axon terminal, which releases the excitatory neurotransmitter glutamate to bipolar cells. Farther back is the cell body, which contains the nucleus, followed by the inner segment, a specialized part of the cell that is rich in mitochondria. The main role of the inner segment is to provide ATP (energy) for a sodium–potassium ion pump. Finally, the most distal component is the outer segment, which is the part of the photoreceptor responsible for absorbing light. The outer segment is essentially a modified cilia containing rows of unconnected disk-like vesicles whose plasma membranes are filled with opsin, the molecule that absorbs photons, as well as voltage-gated sodium channels. The membrane-bound photoreceptor protein opsin contains a pigment molecule called retinal, which together form the complex known as rhodopsin. (In cone cells, there are different types of opsins that combine with retinal to form color-sensitive pigments called photopsins.)

When a rhodopsin protein in the disk membrane of the outer segment absorbs a photon, the configuration of the retinal molecule inside the protein changes shape from a cis-form to a trans-form, see Fig. 10.7(b), initiating a signaling cascade (see

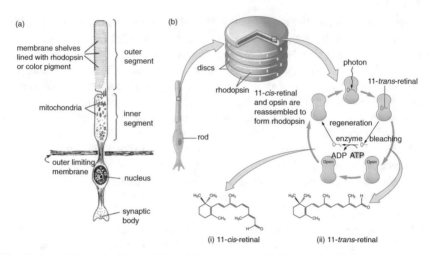

(i) 11-*cis*-retinal                (ii) 11-*trans*-retinal

Fig. 10.7: (a) Schematic diagram of a rod photoreceptor cell. See text for details. (b) Schematic diagram showing how absorption of a photon causes the retinal molecule inside an opsin protein to change from its 11-cis-retinal isomer into its 11-trans-retinal isomer. This change in shape of retinal pushes against the outer opsin protein to begin a signaling cascade. Retinal is then reloaded into its 11-cis configuration by ATP phosphorylation and the cycle can begin again [Public domain figures downloaded from Wikipedia Commons.]

below) that ultimately leads to hyperpolarization of the outer segment membrane due to the closing of light-sensitive ion channels that are permeable to $Na^+$ and to a lesser extent $Ca^{2+}$. This change in the cell's membrane potential then causes voltage-gated calcium channels within the inner segment to close, which leads to a decrease in the influx of calcium ions into the cell and a reduction in the intracellular calcium ion concentration. The calcium-induced exocytosis of vesicles at the axon terminal of the photoreceptor is subsequently reduced, which means that less glutamate is released to the downstream bipolar cell. Less neurotransmitter in the synaptic cleft between a photoreceptor and bipolar cell either excites (depolarizes) or inhibits (hyperpolarizes) the bipolar cell, depending on the type of postsynaptic receptors. (This provides the basis for distinct ON and OFF pathways in the early visual system.) Finally, ATP provided by the inner segment fuels a $Na^+$-$K^+$ pump and a $Na^+$-$Ca^{2+}$, $K^+$ ion exchanger, which is necessary in order to maintain ionic balance.

The signaling cascade transducing the absorption of a photon to hyperpolarization of the photoreceptor is illustrated in Fig. 10.8. In the absence of light, rod cells are depolarized at about -40 mV due to the constant influx of sodium (and calcium) ions in the outer segment, which is balanced by potassium ions flowing out through channels in the inner segment. This so-called dark current of around $20 pA$ is maintained by the opening of ligand-gated ion channels via the binding of cyclic guanosine 3'-5' monophosphate (cGMP), which is present at relatively high concentrations in the absence of light stimulation. When absorption of a photon induces a

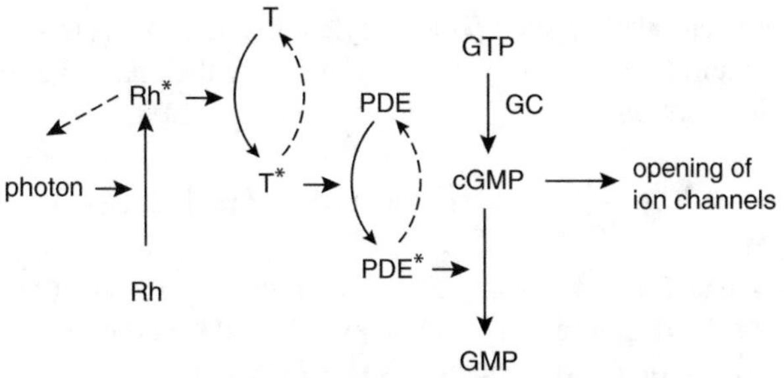

Fig. 10.8: Signaling cascade underlying the transduction of photon absorption to the closing of light-sensitive ion channels in the outer segment. Here asterisks denote activated states of the relevant molecules. The symbols represent rhodopsin (Rh), the G-protein transducin (T), the enzyme phosphodiesterase (PDE), cyclic guanosine monophosphate (cGMP), and guanylate cyclase (GC), which synthesizes cGMP from guanosine triphosphate (GTP).

structural change in the retinal, it leads to the activation of a regulatory G protein called transducin. Each transducin then activates the enzyme cGMP-specific phosphodiesterase (PDE), which catalyzes the hydrolysis of cGMP to 5' GMP. (The inactivated form of PDE also catalyzes hydrolysis, but at a slower rate.) The net concentration of intracellular cGMP is reduced, resulting in the closure of the cyclic nucleotide-gated ion channels and hyperpolarization of the photoreceptor outer segment. There are at least three amplification stages. (i) Photoactivated rhodopsin triggers activation of about 100 transducins; (ii) A single PDE hydrolyses about 1000 cGMP molecules; (iii) cGMP binds cooperatively to sodium channels with an effective Hill coefficient of $n \sim 3$ (see Sect. 3.2).

The fact that a sequence of enzymatic reactions produces signal amplification is not surprising. A more striking feature of phototransduction is the reproducibility of the rod response to single photons [50]. Recall from Sect. 10.1 that a single biochemical amplifier also tends to amplify background noise, although this can be mitigated by dividing the amplification into multiple steps with smaller gains per step. Even prior to the amplification stages, given that the lifetime of the active state of rhodopsin is stochastic, it is not clear how the molecule produces a reliable single-photon response. More specifically, suppose that a rhodopsin molecule is activated at time $t = 0$ and that the rate of decay to the inactive state is $k$. The probability $P_0(t)$ that it is still active at time $t$ satisfies the equation $\dot{P}_0 = -kP_0$, with $P_0(0) = 1$. The probability density that the molecule is active for exactly a time $t$ is then $kP_0(t)$. Denoting the mean and variance of the activation time by $\mu$ and $\sigma$, respectively, one finds that $\sigma/\mu = 1$, consistent with a Poisson process. One possible mechanism for enhanced reliability is based on the observation that rhodopsin is inactivated in a series of phosphorylation steps according to the reaction scheme [394, 496]

$$R_0 \xrightarrow{k_0} R_1 \xrightarrow{k_1} R_2 \cdots \xrightarrow{k_{n-2}} R_{n-1} \xrightarrow{k_{n-1}} \text{inactive rhodopsin},$$

where $R_j$ is the state with $j$ sites phosphorylated. The rhodopsin is inactivated as soon as a maximum of $n$ sites are phosphorylated. We then have the following hierarchy of kinetic equations:

$$\frac{dP_0}{dt} = -k_0 P_0, \quad \frac{dP_j}{dt} = k_{j-1}P_{j-1} - k_j P_j, \quad j = 1,\ldots,n-1, \qquad (10.2.1)$$

with $P_0(0) = 1$ and $P_j(0) = 0$ for $j > 0$. This multistep inactivation can increase the reliability of the response. This can be established by Fourier transforming the system of equations for $j > 0$, which gives (for $k_j = k$, $j = 1,\ldots,n-1$)

$$P_{n-1}(t) = \int_{-\infty}^{\infty} \frac{k^{n-1}}{(k-i\omega)^n} e^{-i\omega t} \frac{d\omega}{2\pi} = t^{n-1} e^{-kt} H(t),$$

where $H(t)$ is the Heaviside function (see Ex. 10.2). We now note that $kP_{n-1}(t)$ is the probability density that the rhodopsin molecule remains active for $t$ seconds, and define the mean and variance of the activation time according to

$$\mu = k \int_0^{\infty} P_{n-1}(t) t \, dt, \quad \sigma^2 = k \int_0^{\infty} P_{n-1}(t) t^2 dt - \mu^2.$$

Substituting for $P_{n-1}(t)$ then establishes that $\sigma/\mu = 1/\sqrt{n}$. In other words, increasing the number of phosphorylation steps $n$ increases the reliability of the single-photon response.

## 10.2.2 Light adaptation

Another important property of phototransduction is light adaptation [291, 497], which is the ability of a photoreceptor to adapt to varying levels of background light intensity. In response to a small flash of light, the membrane potential of a photoreceptor first decreases (hyperpolarizes) and then returns to rest. Let $V(I,I_0)$ be the maximum decrease in the voltage in response to a light flash of magnitude $I$ superimposed on a background light level $I_0$. The peak sensitivity of the photoreceptor is defined as

$$S(I_0) = \left. \frac{\partial V}{\partial I} \right|_{I=I_0}. \qquad (10.2.2)$$

Experimentally, one finds that the sensitivity $S(I_0)$ is approximately inversely proportional to the intensity $I_0$. This physiological property $S(I_0) \sim 1/I_0$ is related to the Weber–Fechner law of psychophysics, which describes the fact that the stimulus threshold necessary to elicit an observable response increases as the background light level increases. It immediately follows that $dV/d\ln I|_{I=I_0}$ is independent of the background intensity $I_0$.

Light adaptation plays two fundamentally important roles. First, it allows the retina to handle light levels spanning 10 orders of magnitude, ranging from a starlit night to bright sunlight. Without any adaptation, this large operating regime would conflict with the high sensitivity of photoreceptors in the dark, since all of the photoreceptors would be saturated in bright light and all features of an image would be "whited out." Although the range of light sensitivity is increased by the fact that rods and cones operate under dim and bright light conditions, respectively, this is not sufficient to account for the observed range of light sensitivity. The second role of light adaptation is to filter out background light levels so that only the contrast of a scene is transmitted to the visual cortex. The contrast only depends on the reflectances of the objects being observed so that, for example, we can perceive black text on a white page under various conditions of illumination.

A useful way to illustrate light adaptation is to plot the peak response $V(I, I_0)$ as a function of flash intensity for different background levels $I_0$. It turns out that experimentally measured intensity response curves can be fitted to the so-called Naka–Rushton equation

$$\frac{V(I, I_0)}{V_{max}} = \frac{I}{I + \sigma(I_0)}, \tag{10.2.3}$$

with $\sigma$ an increasing function of $I_0$, see Fig. 10.9(a). As the background light level is increased, the response curve maintains its sigmoidal shape but is shifted to higher light levels and moved slightly downward. (The latter is not modeled explicitly in the Naka–Rushton equation.) Substituting the Naka–Rushton equation into equation (10.2.2) yields

$$S(I_0) = V_{max} \frac{\sigma(I_0)}{(I_0 + \sigma(I_0))^2}.$$

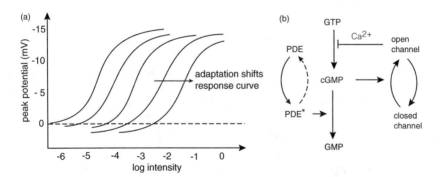

Fig. 10.9: (a) Sketch of typical intensity response curves in vertebrate photoreceptors based on the Naka–Rushton model. The peak of the response to a light flash is plotted against the log of the intensity of the flash for different background light levels. The sigmoid response curves are shifted across and slightly downward at higher background light levels. (b) Positive feedback loop that contributes to light adaptation. $Ca^{2+}$ inhibits the enzyme guanylate cyclase and thus reduces the production of cGMP. When the light intensity is increased, the light-sensitive ion channels close resulting in a reduction of $[Ca^{2+}]$ and disinhibition of the production of cGMP, thus reopening the channels.

Taking $S_D$ to be the sensitivity in the dark, with $S_D = V_{max}/\sigma(0)$, and taking $\sigma(I_0) = \sigma_D + kI_0$, we see that

$$\frac{S(I_0)}{S_D} = \frac{(\sigma_D + kI_0)\sigma_D}{[I_0(1+k) + \sigma_D]^2} \sim \frac{1}{I_0}$$

when $kI_0 \gg \sigma_D$. We thus recover the Weber–Fechner law.

It turns out that many features of light adaptation in rods and cones can be explained in terms of feedback of $Ca^{2+}$ on the activity of guanylate cyclase. When the light-sensitive channels close, the entry of $Ca^{2+}$ into the outer segment is restricted, resulting in a reduction of the intracellular $Ca^{2+}$ concentration. This then increases the activity of the enzyme guanylate cyclase responsible for making cGMP from GTP. Hence, a decrease in $Ca^{2+}$ results in an increase in the rate of production of cGMP, which reopens the light-sensitive channels and completes the feedback loop, see Fig. 10.9(b). For a recent detailed model of various stages of phototransduction, including $Ca^{2+}$-based adaptation, see [395]. In Ex. 10.3, we consider a simple linear cascade model of adaptation [78, 497].

### 10.2.3 Photon counting and shot noise

We now turn to some general aspects of detecting photons in the presence of noise, including the issue of Bayesian statistics, following along similar lines to Chap. 2 of [78]. Suppose that the number $n$ of photons arriving at the retina in a time interval of length $T \leq 100ms$ is given by a Poisson process with rate $r$ (Sect. 3.2). Given that the mean number of counts is $M = rT$, we have

$$P(n|M) = e^{-M}\frac{M^n}{n!}.$$

The rate $r$ will be proportional to the intensity $I_0$ of the light flash so that $M = \alpha I_0 T$, where the constant $\alpha$ lumps together all of the details concerning what happens to the light during its journey to the retina. Suppose that a minimum number $K$ of photons needs to be counted in order to detect the light flash. The probability of detection as a function of the product $I = I_0 T$ is then given by

$$P_{detect}(I) = \sum_{n=K}^{\infty} P(n|M = \alpha I) = e^{-\alpha I}\sum_{n=K}^{\infty}\frac{(\alpha I)^n}{n!}.$$

Fitting psychophysical data from different human subjects detecting dim light flashes to the above probability distribution establishes that most people have a similar count threshold of $K$ ranging from 5 to 7, whereas $\alpha$ varies according to age [404]. Given that only around six photons distributed across the whole retina are needed to detect a dim light flash, it follows that individual photoreceptors can reliably respond to single photons.

**Minimizing errors in photon detection.** Recall that in the dark, a photoreceptor emits a current of around 20 pA. In response to a photon of light, the current is reduced by around 1 pA and this reduction must be detected in the presence of background noise of around 0.1pA. In the case of a steady stream of photons, both false positives (high responses in the absence of a photon) and false negatives (low responses when a photon is absorbed) can occur. We would like to estimate the probabilities for such events and understand how to minimize any errors. First consider the case of no light. Let's assume that the background noise is Gaussian with some variance $\sigma_0^2$ so that in the absence of photons ($n = 0$), the peak current $I$ is drawn from the probability distribution

$$p(I|n = 0) = (2\pi\sigma_0^2)^{-1/2}e^{-I^2/2\sigma_0^2}.$$

Similarly, we assume that the current probability density in response to $n$ photons is

$$p(I|n) = (2\pi(\sigma_0^2 + n\sigma_1^2))^{-1/2}e^{-(I-nI_1)^2/2(\sigma_0^2+n\sigma_1^2)}.$$

That is, each photon adds a contribution $I_1$ to the mean peak current and a contribution $\sigma_1^2$ to the variance. Given that the photon count is generated by a Poisson distribution, we have

$$P(I) = \sum_{n=0}^{\infty} P(I|n)P(n) = \sum_{n=0}^{\infty} P(I|n)e^{-\bar{n}}\frac{\bar{n}^n}{n!}, \qquad (10.2.4)$$

where $\bar{n}$ is the mean photon count. An illustration of the first three components is given in Fig. 10.10. Now suppose that we want to set a threshold $\theta$ for distinguishing between the cases $n = 0$ (no photon) and $n = 1$ (single photon), say, based on a measurement of the current $I$ and the underlying probability distributions $P(I|n), P(n)$. An initial guess would be to take $\theta = 0.5pA$, which is the location of the trough in $P(I)$ between the first two peaks. Is there a more principled way to set the threshold that minimizes any errors? First, we define the total error as the sum of false positives and false negatives,

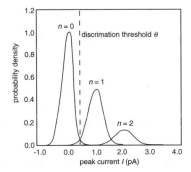

Fig. 10.10: First three terms in the sum on the right-hand side of equation (10.2.4) for $I_1 = 1, \sigma_0 = 0.1, \sigma_1 = 0.2$ and $\bar{n} = 1$.

$$P_{\text{error}}(\theta) = P(\text{decide } n = 1|n = 0)P(n = 0) + P(\text{decide } n = 0|n = 1)P(n = 1)$$

$$= P(0) \int_\theta^\infty P(I|0)dI + P(1) \int_{-\infty}^\theta P(I|1)dI.$$

The optimal discrimination threshold $\theta^*$ is then determined by minimizing $P_{\text{error}}(\theta)$. Since,

$$\frac{dP_{\text{error}}(\theta)}{d\theta} = P(0)\frac{d}{d\theta} \int_\theta^\infty P(I|0)dI + P(1)\frac{d}{d\theta} \int_{-\infty}^\theta P(I|1)dI$$

$$= P(1)P(\theta|1) - P(0)P(\theta|0).$$

It follows that $\theta^*$ is determined by the maximum likelihood equation

$$P(\theta^*|0)P(0) = P(\theta^*|1)P(1). \qquad (10.2.5)$$

We can also interpret this result within the context of Bayes' theorem (Sect. 1.2), which in the current case can be expressed as

$$P(n|I) = \frac{P(I|n)P(n)}{P(I)}.$$

Hence, $P(I|n)P(n)$ is proportional to the *a posteriori* probability that the observed current was generated by the counting of $n$ photons. The condition for the optimal threshold $\theta^*$ is thus equivalent to equating the *a posteriori* probabilities for $n = 0$ and $n = 1$. In other words, for each observable current $I$, one should compute the probability $p(n|I)$ according to Bayes' theorem and take the best guess for the photon count to be the one that maximizes this probability.

**Shot noise.** Now consider a train of photons being detected by a single photorecep-tor. Let $T_n$ denote the arrival time of the $n$th photon, and suppose that each photon elicits a voltage response of the form $\phi(t - T_n)$, where $\phi$ is a positive function and $\phi(t) = 0$ for $t < 0$. Assuming that there are no refractory effects due to the need for rhodopsin molecules to recharge following photon absorption, the accumulative

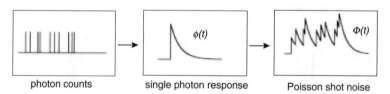

Fig. 10.11: Illustration of shot noise in the case of a train of photons being detected and filtered by a photoreceptor. The interarrival times of the photons are exponentially distributed, since the total photon count is a Poisson process. The $n$th photon generates a voltage response $\phi(t - T_n)$, where $T_n$ is the corresponding photon arrival time, and the total response $\Phi(t)$ of the photoreceptor to the train of photons is obtained by summing the individual responses to produce Poisson shot noise.

response of the photoreceptor to a train of photons can be written as

$$\Phi(t) = \sum_n \phi(t - T_n). \tag{10.2.6}$$

Given that the photon number is a Poisson process, so that the interarrival times $T_{n+1} - T_n$ are exponentially distributed, it follows that $\Phi(t)$ is an example of Poisson shot noise [760], see Fig. 10.11. Although shot noise was originally analyzed within the context of electronic devices and photon detection [788, 863], it has subsequently arisen in applications to neural spike trains [864] and cell length regulation [121].

Before proceeding with the analysis of shot noise, note that we can rewrite equation (10.2.6) as the convolution

$$\Phi(t) = \int_{-\infty}^{\infty} \phi(t - \tau)h(\tau)d\tau, \tag{10.2.7}$$

where $h(\tau)$ represents the photon input to the photoreceptor in the form of a train of Dirac delta functions,

$$h(t) = \sum_n \delta(t - T_n). \tag{10.2.8}$$

Hence, within the context of linear response theory (see Box 10A), we may interpret $\phi$ as the linear response function or causal Green's function of the photoreceptor, which acts as a linear filter.

The statistics of a shot noise process can be analyzed using generating functions; see [214, 510] for a more rigorous treatment. For the moment, consider a homogeneous Poisson process with constant rate $\lambda_0$. We are interested in the large-time limit $t \to \infty$ for which the process is stationary. Suppose that exactly $N$ events occur in the time interval $[0, T]$ with $T$ large but finite. The event times are uniformly distributed in $[0, T]$ so that if $T_n$ is the time of the $n$th event, then the probability that $T_n$ lies in the interval $[t, t + dt]$ is $dt/T$ irrespective of the arrival of the other pulses. (Note that the labeling of the $N$ pulses is arbitrary.) It is assumed that $T$ is much larger than the range of values of $t$ for which $\phi(t)$ is significantly away from zero. For example, if $\phi(t)$ decays exponentially with time constant $\tau_p$ then $T \gg \tau_p$. Introduce the generating function

$$G_N(s, T) = \mathbb{E}_N[e^{s\Phi(T)}], \tag{10.2.9}$$

where expectation is taken with respect to the random event times $T_n$, conditioned on the total number of events $N$. Substituting for $\Phi$ and using the independence of the $T_n$, we have

$$G_N(s, T) = \mathbb{E}[e^{s \sum_{n=1}^{N} \phi(T-T_n)}] = \int_0^T \frac{dT_1}{T} \cdots \int_0^T \frac{dT_N}{T} \prod_{n=1}^{N} e^{s\phi(T-T_n)}$$

$$= \left( \frac{1}{T} \int_0^T e^{s\phi(T-\tau)}d\tau \right)^N.$$

We now have to average over all possible values of $N$, assuming that $P_N(T)$ is given by a homogeneous Poisson distribution with constant rate $\lambda_0$:

$$G(s,T) = \sum_{N=0}^{[T]} G_N(s,T)P_N(T) = \sum_{N=0}^{[T]} \frac{(\lambda_0 T)^N}{N!}e^{-\lambda_0 T}\left(\frac{1}{T}\int_0^T e^{s\phi(T-\tau)}d\tau\right)^N.$$

Finally, taking the limit $T \to \infty$ and using $\phi(t) = 0$ for $t < 0$ yields the generating function for the stationary shot noise process:

$$G(s) \equiv \int_0^\infty e^{s\Phi}P(\Phi)d\Phi = \exp\left(\lambda_0\int_{-\infty}^\infty\left(e^{s\phi(t)}-1\right)dt\right). \qquad (10.2.10)$$

Taylor expanding with respect to $s$ with $s \to 0$ gives

$$1 + s\langle\Phi\rangle + \frac{s^2}{2}\langle\Phi^2\rangle + O(s^3) = 1 + \lambda_0 s\int_{-\infty}^\infty\phi(t)dt + \frac{1}{2}\left(\lambda_0 s\int_{-\infty}^\infty\phi(t)dt\right)^2$$

$$+ \frac{1}{2}\lambda_0 s\int_{-\infty}^\infty\phi(t)^2 dt + O(s^3).$$

Comparing both sides, we deduce that

$$\langle\Phi\rangle = \lambda_0\int_{-\infty}^\infty\phi(t)dt, \quad \text{Var}[\Phi] = \lambda_0\int_{-\infty}^\infty\phi(t)^2 dt. \qquad (10.2.11)$$

These classical results are a particular version of Campbell's theorem [510]; see also Ex. 10.4.

One way to extend the above analysis to calculate various statistical correlations is to consider so-called characteristic functionals [214]; see Box 14A. Here we will use a more direct method to calculate two-point correlations. As before, suppose that the number of events in the interval $[0,T]$ is $N$. Then

$$R_N(\tau) \equiv \mathbb{E}_N[\Phi(t)\Phi(t+\tau)] = \sum_{k=1}^N\sum_{n=1}^N\int_0^T\frac{dT_1}{T}\cdots\int_0^T\frac{dT_N}{T}\phi(t-T_k)\phi(t+\tau-T_n).$$

In this expression there are $N$ diagonal terms for which $k = n$ and $N^2 - N$ off-diagonal terms for which $k \neq n$. In the latter case, $T_k$ and $T_n$ are independent so that the average of the product reduces to the product of averages. Hence, in the large $T$ limit

$$R_N(\tau) = \frac{N(T)}{T}\int_{-\infty}^\infty\phi(t)\phi(t+\tau)dt + \frac{N(T)(N(T)-1)}{T^2}\left(\int_{-\infty}^\infty\phi(t)dt\right)^2.$$

Now summing with respect to the number of events $N$ using the Poisson distribution, we have $\langle N(T)\rangle = \lambda_0 T$ and $\langle N(T)^2\rangle = \lambda_0 T + (\lambda_0 T)^2$, so that the stationary two-point correlation function is

$$R(\tau) = \mathbb{E}[\Phi(t)\Phi(t+\tau)] = \lambda_0 \int_{-\infty}^{\infty} \phi(t)\phi(t+\tau)dt + \left(\lambda_0 \int_{-\infty}^{\infty} \phi(t)dt\right)^2. \quad (10.2.12)$$

Given the two-point correlation function, we can determine the power spectrum of the shot noise $\Phi(t)$ using the Wiener–Khinchin theorem (see Sect. 2.3). In particular,

$$S_\Phi(\omega) = \int_{-\infty}^{\infty} e^{i\omega\tau} R(\tau)d\tau = \lambda_0|\widehat{\phi}(\omega)|^2 + \lambda_0^2|\widehat{\phi}(0)|^2\delta(\omega). \quad (10.2.13)$$

**Signal-to-noise ratio.** Now suppose that the Poisson rate is modulated by some time-dependent signal $I(t)$. That is, the events are distributed according to an inhomogeneous Poisson process with

$$\lambda(t) = \lambda_0(1 + I(t)).$$

Repeating the previous analysis, one finds that

$$\langle \Phi(t) \rangle = \lambda_0 \left(1 + \int_{-\infty}^{\infty} I(\tau)\phi(t-\tau)d\tau\right). \quad (10.2.14)$$

Setting $\langle \Delta\Phi(t) \rangle = \langle \Phi(t) \rangle - \lambda_0$, and using Fourier transforms, we obtain the following response to the input $I(t)$ in the frequency domain:

$$\langle \Delta\widehat{\Phi}(\omega) \rangle = \lambda_0\widehat{I}(\omega)\widehat{\phi}(\omega). \quad (10.2.15)$$

The so-called transfer function $\widehat{T}(\omega)$ is then given by

$$\widehat{T}(\omega) := \frac{\langle \Delta\widehat{\Phi}(\omega) \rangle}{\widehat{I}(\omega)} = \lambda_0\widehat{\phi}(\omega). \quad (10.2.16)$$

The signal-to-noise ratio is now defined to be the ratio of the amplitude square of the (complex-valued) transfer function and the spectrum in the absence of a signal:

$$\text{SNR}(\omega) := \frac{|\widehat{T}(\omega)|^2}{S_\Phi(\omega)} = \frac{\lambda_0^2|\widehat{\phi}(\omega)|^2}{\lambda_0|\widehat{\phi}(\omega)|^2} = \lambda_0. \quad (10.2.17)$$

Note that both the transfer function $\widehat{T}$ and noise spectrum $\widehat{S}$ are dependent on the form of the linear response function $\phi(t)$. Since the latter has compact support, it follows that the transfer function becomes smaller at higher frequencies. On the other hand, the signal-to-noise ratio is independent of the frequency $\omega$, as it simply recovers the photon counting rate $\lambda_0$. It is important to point out that an underlying assumption of the above analysis is that the photoreceptor acts as a perfect photon counter, since every photon arrival event is counted in equation (10.2.6). Amazingly, experimental studies of fly photoreceptors show that they act as ideal photon counters over a wide dynamic range [236, 237].

Finally, note that once a photoreceptor has detected a photon, it has to transmit this information downstream via synaptic connections with bipolar cells. Let $X(t)$

represent the light intensity and $Y(t)$ the current produced by a rod cell. Suppose that we write the input–output relationship of the rod cell as a noisy linear filter

$$Y(t) = \int_{-\infty}^{\infty} g(t')X(t-t')dt' + \xi(t),  \tag{10.2.18}$$

where $g(t)$ is the linear response function and $\xi(t)$ is a noise term. Suppose that the synaptic connection to a bipolar cell can also be treated as a linear filter, whose role is to separate signal from noise. That is, its output is an estimate of $X(t)$:

$$x_{\text{est}}(t) = \int_{-\infty}^{\infty} f(t')Y(t-t')dt'.  \tag{10.2.19}$$

One finds that at low light intensities, the latter acts like an optimal filter in the sense that it minimizes the mean-square error

$$E = \left\langle \int_{-\infty}^{\infty} dt \left( X(t) - \int_{-\infty}^{\infty} f(t')Y(t-t')dt' \right)^2 \right\rangle.  \tag{10.2.20}$$

Using Fourier analysis, it can be shown that (see Ex. 10.5 and [78])

$$\widehat{f}(\omega) = \frac{\widehat{g}^*(\omega)S_X(\omega)}{|\widehat{g}^*(\omega)|^2 S_X(\omega) + S_\xi(\Omega)},$$

where $S_X$ and $S_\xi$ are the power spectra of the signal and noise, respectively.

---

### Box 10A. Linear response theory.

A common way of characterizing cells in a variety of sensory systems is in terms of their linear response to small changes in input. In the case of autonomous systems, the output of the cell can typically be written as the convolution of the input with a function known as the causal Green's function or linear response function $G$. The latter is determined by the intrinsic properties of the cell, which effectively acts as an input–output device or linear filter. Using Fourier transforms and the convolution theorem (Box 2A), one can express the output as the product of the input and a frequency-dependent transfer function $\chi$, which is the Fourier transform of Green's function. Moreover, methods from complex analysis can be used to extract various features of the linear response. It turns out that there is a fundamental connection between the linear response properties of thermodynamic (and quantum) physical systems and statistical fluctuations, which will be explored further in Sect. 10.5. Here we review some basic results of linear response theory; see [530] for a more extensive discussion.

Suppose that $x(t) \in \mathbb{R}$ represents the linear response of some system to a small external input $I(t)$, such that

$$x(t) = \int_{-\infty}^{\infty} G(t - \tau)I(\tau)d\tau, \quad G(t) = 0 \text{ for } t < 0, \qquad (10.2.21)$$

where $G(t)$ is known as the linear response function or causal Green's function. In a mechanical system, $x(t)$ would represent a physical displacement and $I$ an applied force, whereas in a magnetic system $x(t)$ would represent magnetization and $I$ an applied magnetic field. In the latter case, $G(t)$ is known as the magnetic susceptibility. Introducing the Fourier transform of a function $f(t)$,

$$\widehat{f}(\omega) = \int_{-\infty}^{\infty} f(t)e^{i\omega t}dt, \quad f(t) = \int_{-\infty}^{\infty} \widehat{f}(\omega)e^{-i\omega t}\frac{d\omega}{2\pi},$$

and applying the convolution theorem, see Box 2A, we have

$$\widehat{x}(\omega) = \chi(\omega)\widehat{I}(\omega), \qquad (10.2.22)$$

where $\chi(\omega) = \widehat{G}(\omega)$ is known as the transfer function.

**Analyticity and causality.** Suppose that we decompose the transfer function as

$$\chi(\omega) = \text{Re } \chi(\omega) + i\text{Im } \chi(\omega) := \chi'(\omega) + i\chi''(\omega).$$

(Here $'$ does not denote differentiation.) The imaginary part can be decomposed as

$$\chi''(\omega) = -\frac{i}{2}[\chi(\omega) - \overline{\chi}(\omega)] = -\frac{i}{2}\int_{-\infty}^{\infty} G(t)[e^{i\omega t} - e^{-i\omega t}]dt$$

$$= -\frac{i}{2}\int_{-\infty}^{\infty} e^{i\omega t}[G(t) - G(-t)].$$

It follows that $\chi''(\omega)$ contains information about the part of the response function that is not invariant under time reversal, $t \to -t$. In other words, it arises from dissipative processes, and is often called the dissipative or absorptive component of the response function. On the other hand, the real part is not sensitive to the "arrow of time,"

$$\chi'(\omega) = \frac{1}{2}\int_{-\infty}^{\infty} e^{i\omega t}[G(t) + G(-t)],$$

and is often called the reactive part. Also note that $\chi'(\omega) = \chi'(-\omega)$ and $\chi''(\omega) = -\chi''(-\omega)$.

Recall that $G(t) = 0$ for $t < 0$, which is an expression of causality. This has implications for the transfer function $\chi(\omega)$, since

$$G(t) = \int_{-\infty}^{\infty} e^{-i\omega t} \chi(\omega) \frac{d\omega}{2\pi}.$$

When $t < 0$, we can evaluate the integral by closing in the upper-half plane, since $(i\omega(i|t|)) \to -\infty$ as $\omega$ goes to infinity. The result has to be zero, which means that $\chi(\omega)$ is analytic for Im $\omega > 0$. Analyticity in the upper-half plane implies that there is a relationship between $\chi'$ and $\chi''$ known as the Kramers–Kronig relation. In order to obtain such a relation, it is first necessary to recall a few results from contour integration.

**Discontinuous functions and principal values.** Consider a general meromorphic function $\rho(\omega)$, which means that it is analytic except at isolated poles. Define a new function according to

$$f(\omega) = \frac{1}{\pi i} \int_a^b \frac{\rho(\omega')}{\omega' - \omega} d\omega', \tag{10.2.23}$$

with $a, b \in \mathbb{R}$. Note that when $\omega \in [a, b]$, the integral diverges at $\omega = \omega'$. In order to avoid this singularity, deform the contour of the integral into the complex plane, either by forming a small semicircle above or below the singularity, see Fig. 10.12. This is equivalent to shifting the singularity according to $\omega \to \omega \mp i\varepsilon$. However, the two choices yield different answers, which are determined by Cauchy's residue theorem.

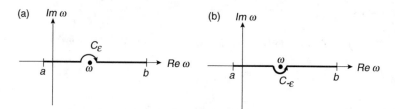

Fig. 10.12: Contours used in the construction of a principal value. (a) Contour defining $f(\omega - i\varepsilon)$. (b) Contour defining $f(\omega + i\varepsilon)$.

That is,

$$\frac{1}{2}[f(\omega + i\varepsilon) - f(\omega - i\varepsilon)] = \frac{1}{2\pi i} \left[ \int_{C_{-\varepsilon}} - \int_{C_\varepsilon} \right] \frac{\rho(\omega')}{\omega' - \omega} d\omega' = \rho(\omega). \tag{10.2.24}$$

This means that the function $f(\omega)$ is discontinuous across the real axis for $\omega \in [a, b]$, and the discontinuity is a branch cut if $\rho(\omega)$ is everywhere analytic. The principal value of the function $f(\omega)$, which is denoted by $\mathscr{P}$, is defined as the average of the two functions on either side of the singularity:

$$\frac{1}{2}[f(\omega + i\varepsilon) + f(\omega - i\varepsilon)] = \frac{1}{\pi i}\mathscr{P}\int_a^b \frac{\rho(\omega')}{\omega' - \omega}d\omega'. \tag{10.2.25}$$

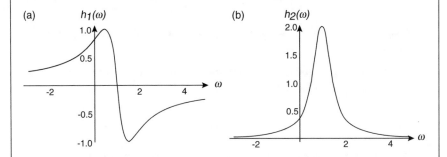

Fig. 10.13: Plots of (a) real part $h_1(\omega)$ and (b) imaginary part $h_2(\omega)$ of the function $(1 - \omega - i\varepsilon)^{-1}$ for $\varepsilon = 0.5$.

Further insight into the meaning of the principal part can be obtained by noting that

$$\frac{1}{\omega' - (\omega \pm i\varepsilon)} := h_1(\omega) + ih_2(\omega) = \frac{\omega' - \omega}{(\omega' - \omega)^2 + \varepsilon^2} \pm \frac{i\varepsilon}{(\omega' - \omega)^2 + \varepsilon^2}.$$

Thus the principal part corresponds to the real part $h_1(\omega)$. It replaces a small segment of the function around the singularity at $\omega$, which has divergent positive and negative parts that cancel, by a smooth function through zero. On the other hand, the discontinuity is determined by the imaginary part $h_2(\omega)$, which approaches a Dirac delta function as $\varepsilon \to 0$, see Fig. 10.13.

**Kramers–Kronig relations.** Let us now apply the notion of a principal value to the transfer function $\chi(\omega)$ by setting $\rho(\omega) = \chi(\omega)$ in equation (10.2.23) with $a = -\infty$, $b = +\infty$, and $\chi$ analytic in the upper-half plane:

$$f(\omega) = \frac{1}{\pi i}\int_{-\infty}^{\infty} \frac{\chi(\omega')}{\omega' - \omega}d\omega'.$$

We also assume that $\chi(z)$ falls off faster than $1/|z|$ at infinity. It follows that $f(\omega - i\varepsilon) = 0$, since there are no poles in the upper-half plane and, hence, subtracting equations (10.2.24) and (10.2.25) yields

$$\chi(\omega) = \frac{1}{\pi i}\mathscr{P}\int_{-\infty}^{\infty} \frac{\chi(\omega')}{\omega' - \omega}d\omega'.$$

Taking real and imaginary parts then yields the Kramers–Kronig relations

$$\mathrm{Re}\,\chi(\omega) = \mathscr{P}\int_{-\infty}^{\infty} \frac{\mathrm{Im}\,\chi(\omega')}{\omega'-\omega}\frac{d\omega'}{\pi}, \qquad (10.2.26a)$$

$$\mathrm{Im}\,\chi(\omega) = \mathscr{P}\int_{-\infty}^{\infty} \frac{\mathrm{Re}\,\chi(\omega')}{\omega'-\omega}\frac{d\omega'}{\pi}. \qquad (10.2.26b)$$

Therefore, if one knows the dissipative part of the transfer function, then one knows everything.

**Example: Damped harmonic oscillator.** Consider the damped harmonic oscillator driven by a force $F(t)$:

$$\frac{d^2x}{dt^2} + \gamma\frac{dx}{dt} + \omega_0^2 x = F(t).$$

Substituting the inverse Fourier transform into the left-hand side of this equation gives

$$\int_{-\infty}^{\infty}[-\omega^2 - i\gamma\omega + \omega_0^2]e^{-i\omega t}\widehat{x}(\omega)\frac{d\omega}{2\pi} = F(t).$$

In Fourier space, we have

$$\widehat{x}(\omega) = \chi(\omega)\widehat{F}(\omega) = \chi(\omega)\int_{-\infty}^{\infty} e^{i\omega t'}F(t')dt'.$$

Using the identity

$$\int_{-\infty}^{\infty} e^{-i\omega(t-t')}\frac{d\omega}{2\pi} = \delta(t-t'),$$

we deduce that

$$\chi(\omega) = \frac{1}{-\omega^2 - i\gamma\omega + \omega_0^2}.$$

The poles of $\chi(\omega)$ are located in the lower-half plane at

$$\omega_\pm = -\frac{i\gamma}{2} \pm \sqrt{\omega_0^2 - \gamma^2/4}.$$

There are two distinct parameter regimes. (i) Underdamped ($\omega_0^2 > \gamma^2/4$) for which the poles have a real and imaginary part, and (ii) overdamped ($\omega_0^2 < \gamma^2/4$) with poles on the positive imaginary axis.

The real and imaginary parts of the response function are plotted in Fig. 10.14 for the underdamped case, and have the explicit form

$$\mathrm{Re}\,\chi(\omega) = \frac{\omega_0^2 - \omega^2}{(\omega_0^2 - \omega^2)^2 + \gamma^2\omega^2}, \qquad \mathrm{Im}\,\chi(\omega) = \frac{\omega\gamma}{(\omega_0^2 - \omega^2)^2 + \gamma^2\omega^2}.$$

$$(10.2.27)$$

It is instructive to see how the imaginary part of the transfer function is related to dissipation. The rate of change of work done on the system is

$$\frac{dW}{dt} = F(t)\dot{x}(t) = F(t)\frac{d}{dt}\int_{-\infty}^{\infty} G(t-t')F(t')dt'$$

$$= F(t)\int_{-\infty}^{\infty}\left[\int_{-\infty}^{\infty}(-i\omega)e^{-i\omega(t-t')}\chi(\omega)\frac{d\omega}{2\pi}\right]F(t')dt'$$

$$= \int_{-\infty}^{\infty}\int_{-\infty}^{\infty}(-i\omega)\chi(\omega)\widehat{F}(\omega)\widehat{F}(\omega')e^{-i(\omega+\omega')t}\frac{d\omega}{2\pi}\frac{d\omega'}{2\pi}.$$

Suppose that the system is driven at a particular frequency $\Omega$, so that

$$F(t) = F_0\cos(\Omega t).$$

It follows that

$$\widehat{F}(\omega) = 2\pi F_0[\delta(\omega-\Omega)+\delta(\omega+\Omega)].$$

Substituting this into the expression for $dW/dt$,

$$\frac{dW}{dt} = -iF_0^2\Omega[\chi(\Omega)e^{-i\Omega t}-\chi(-\Omega)e^{i\Omega t}][e^{-i\Omega t}+e^{i\Omega t}].$$

Finally, integrating over one cycle of the external drive gives

$$\frac{dW_{\mathrm{av}}}{dt} = \frac{\Omega}{2\pi}\int_0^{2\pi/\Omega}\frac{dW}{dt}dt = -iF_0^2\Omega[\chi(\Omega)-\chi(-\Omega)] = 2F_0^2\Omega\,\mathrm{Im}\,\chi(\Omega).$$

$$(10.2.28)$$

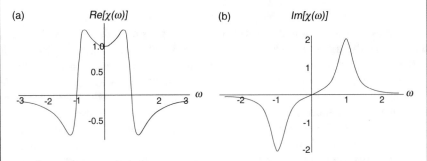

Fig. 10.14: Plots of (a) real and (b) imaginary parts of the transfer function for the damped harmonic oscillator in the underdamped regime, with natural frequency $\omega_0 = 1$ and damping coefficient $\gamma = 0.5$.

## 10.3 Hair cells and active mechanotransduction

In contrast to photoreceptors, hair cells of the vertebrate inner ear use a mechanical active process to amplify their inputs [443, 584, 596]. When sound reaches the cochlea—a spiraled, hollow, conical chamber of bone in the inner ear along which waves propagate—it elicits mechanical vibrations that stimulate hair cell receptors. These receptors transduce the vibrations into an electrical signal via mechanotransduction, simultaneously performing work that amplifies the mechanical signal resulting in positive feedback. The hair cells of all vertebrates share a similar structure and transduce mechanical stimuli according to the same basic mechanism [444]. On the top of each hair cell is a cluster of 20–300 actin-based cylindrical structures called stereocilia, which is known as the hair bundle, see Fig. 10.15. The stereocilia develop in such a way that there is a specific variation of their lengths across the hair bundle giving the latter a beveled shape, see Fig. 10.16. The mechanical stimulus induced by sound reaching the ear deflects the hair bundles, with their component stereocilia bending at their base. This deflection causes a shearing motion between neighboring stereocilia, which is detected by mechanosensitive ion channels located near the stereociliary tips. This transduction is mediated by cadherin-based adhesive tip links that couple adjacent stereocilia and can open the ion channels under tension. This allows $K^+$ to enter the hair cell. The resulting depolarization opens voltage-gated $Ca^{2+}$ channels and the intracellular $Ca^{2+}$ concentration rises. This in turn opens $Ca^{2+}$-sensitive $K^+$ channels through which $K^+$ can flow out of the cell and the cell returns to rest. The given sequence of events can result in an action potential being produced by the hair cell.

One major feature of the hair bundle is that the tension of the tip link can be adjusted by myosin motors that walk up and down the stereocilia [842, 845], which then allows the hair cell to adapt to a sustained deflection of the hair bundle. It is thought that, at least in the case of nonmammalian tetrapods (four-legged vertebrates), the interaction of the molecular motors with the mechanical properties of the hair bundle form the basis of active processes in the inner ear, which include signal amplification, enhanced frequency selectivity, and spontaneous oscillatory acoustic emissions. Moreover, from a dynamical systems perspective, these characteristics emerge naturally if the transduction process operates near a Hopf bifurcation [157, 191, 597, 660, 908]. In this section, we review some of the models that have been developed to explore active processes in hair cells.

Fig. 10.15: Electron micrograph showing stereocilia of an inner hair cell of the bullfrog [Public domain figure downloaded from Wikipedia Commons.]

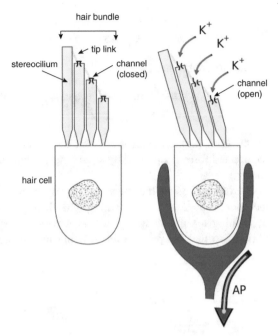

Fig. 10.16: Schematic diagram of a hair cell, illustrating how a mechanical stimulus deflects the bundle of stereocilia surmounting the cell resulting in the opening of mechanosensitive ion channels and the influx of $K^+$ and subsequent influx of $Ca^{2+}$. This can lead to the firing of an action potential (AP) [Public domain figure downloaded from Wikipedia Commons.]

### 10.3.1 Gating-spring model

Our starting point is a gating-spring model of mechanotransduction in hair cells, as reviewed in [591]. Suppose that the hair bundle is modeled as a collection of $N$ elastic units in parallel. Each unit consists of a mobile element of width $\Delta A$ that is attached to two fixed walls by a pair of springs in series, see Fig. 10.17(b), which is a simplified model of the mechanical properties of a pair of linked stereocilia, see Fig. 10.17(a). The left spring represents the tip link or gating spring (gs) and is attached to the element via a hinge or trapdoor, whereas the right spring (sp) represents the stereociliary pivots. Assume, for the moment, that the trapdoor is closed. Let $x$ be the distance of the mobile element from the left-hand wall, and denote the lengths of the gating spring and pivots by $a_{gs}$ and $a_{sp}$, respectively. If the distance between the walls is $A$, then $a_{gs} = x$, $a_{sp} = A - \Delta A - x$. Suppose that each spring has an equilibrium length (no tension) denoted by $\bar{a}_{gs}$ and $\bar{a}_{sp}$. It follows that we can express the displacements from equilibrium of the two springs as

$$\Delta a_{gs} = a_{gs} - \bar{a}_{gs} = x - \bar{x}_{gs}, \quad \Delta a_{sp} = a_{sp} - \bar{a}_{sp} = -(x - \bar{x}_{sp}),$$

with $\bar{x}_{gs} = \bar{a}_{gs}$ and $\bar{x}_{sp} = A - \Delta A - \bar{a}_{sp}$. When the trapdoor is open, it is assumed that the left spring's length is reduced by an amount $\delta$, so that $\Delta a_{gs} \to \Delta a_{gs} - \delta$. Now suppose that an external force $f$ in the positive $x$-direction is applied to a single unit, and that we have Hookean springs with spring constants $k_{gs}$ and $k_{sp}$, respectively. The displacement $x$ is then determined by the force-balance equations

$$f = f_c(x) \equiv k_{gs}\Delta a_{gs} - k_{sp}\Delta a_{sp} = k_{gs}(x - \bar{x}_{gs}) + k_{sp}(x - \bar{x}_{sp})$$
$$= K(x - \bar{x}) \tag{10.3.1}$$

for the closed trapdoor, and

$$f = f_o(x) \equiv k_{gs}(\Delta a_{gs} - \delta) - k_{sp}\Delta a_{sp} = k_{gs}(x - \bar{x}_{gs} - \delta) + k_{sp}(x - \bar{x}_{sp})$$
$$= K(x - \bar{x}) - k_{gs}\delta \tag{10.3.2}$$

for the open trapdoor, where $K = k_{gs} + k_{sp}$ and $\bar{x} = [k_{gs}\bar{x}_{gs} + k_{sp}\bar{x}_{sp}]/K$. (Note that the applied force is opposed by stretching the left spring and compressing the right spring.)

Having looked at the mechanical properties of the springs, we now have to incorporate the stochastic opening and closing of the trapdoor. This is achieved by treating the gating-spring unit as a two-state system in thermodynamic equilibrium, so that the probability of being in an open or closed state is given by a

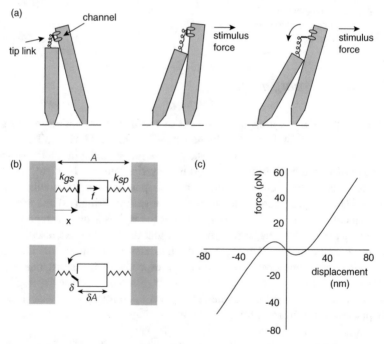

Fig. 10.17: Gating-spring model of mechanotransduction. (a) Schematic diagram of two stereo-cilia connected by a tip link that is attached to a transduction ion channel. Deflection of the bundle by a positively directed stimulus force bends the stereocilia, stretches the tip link, and consequently opens the ion channel, allowing K and $Ca^{2+}$ to enter the cytoplasm and depolarize the hair cell. (b) Simplified mechanical model of coupled stereocilia. (c) Sketch of force–displacement curve similar to one found experimentally in hair cells of the bullfrog [597]. A region of negative spring stiffness can be seen.

Boltzmann–Gibbs distribution (Sect. 1.3). That is $P_c(x) = 1 - P_o(x)$, with $P_o(x) = (1 + e^{\Delta E(x)/k_B T})^{-1}$ and $\Delta E(x)$ is the energy difference between the open and closed states. There are two contributions to this energy difference. First, there is an increase $\Delta E_0$ in the configuration energy of the trapdoor when it jumps to the open state. Second, there is a change in the potential energy stored by the left spring when it shifts by an amount $\delta$, which is given by $-\delta k_{gs}(x - \bar{x}_{gs})$. Hence,

$$P_o(x) = \frac{1}{1 + e^{-[\delta k_{gs}(x - \bar{x}_{gs}) - \Delta E_0]/k_B T}} = \frac{1}{1 + e^{-z(x - x_0)/k_B T}}, \tag{10.3.3}$$

where $z = \delta k_{gs}$ and $x_0 = \bar{x}_{gs} + \Delta E_0/z$. Now let us consider $N$ identical gating-spring units in parallel. Since they are in parallel, each unit experiences the same applied force $f/N$. If $N$ is sufficiently large, then fluctuations in the fraction of open and closed gates can be ignored. It follows that we have the force-balance equation

$$\begin{aligned} f = f(x) &\equiv N\left[f_o(x)P_o(x) + f_c(x)P_c(x)\right] \\ &= N\left[(K(x - \bar{x}) - k_{gs}\delta)P_o(x) + K(x - \bar{x})P_c(x)\right] \\ &= N\left[K(x - \bar{x}) - k_{gs}\delta P_o(x)\right] = K_{tot}x + f_0 - \frac{Nz}{1 + e^{-z(x - x_0)/k_B T}}, \end{aligned} \tag{10.3.4}$$

where $K_{tot} = NK$, $f_0 = -NK\bar{x}$. Fitting the model to experimental data from bull-frog hair cells yields the following example set of parameter values [596]: $N = 65$, $K_{tot} = 10^3 \, \mu N/m$, $z = 0.72$ pN, $f_0 = 25$ pN, and $x_0 = -2.2$ nm. If one plots the force–displacement function $f(x)$ for these values at room temperature, one obtains the curve shown in Fig. 10.17(c). It can be seen that for sufficiently large displacements in the positive or negative $x$ directions, the system acts like an ordinary spring, that is, it has approximately constant stiffness. However, within $\pm 20$ nm of the resting position, the stiffness $df/dx$ varies significantly with displacement. Even more striking is that, within $\pm 10$ nm of the resting position, the stiffness slope is negative and displacement of the bundle in a particular direction requires a force in the opposite direction—the hair bundle is said to have channel compliance. It is important to note that no active processes have been included in the model, since there is no net consumption of energy. However, when the force–displacement characteristics of the form shown in Fig. 10.17(c) are combined with the action of myosin motors, the active features of hair cells can be reproduced [596, 598].

Before considering active processes, however, we briefly consider what happens when $N$ is small so that fluctuations in the number of open ion channels cannot be ignored. From the analysis of Sect. 3.3, we know that the probability $P_N(n|x)$ of there being $n$ open channels for a given displacement $x$ is given by a Binomial distribution of the form

$$P_N(n|x) = P_o(x)^n (1 - P_o(x))^{N-n} \frac{N!}{(N - n)!n!}.$$

It follows that the mean and variance of the number of open channels are

$$\langle n \rangle = N P_o(x), \quad \sigma^2 = N P_o(x)(1 - P_o(x)).$$

For a given displacement $x$, the necessary force on the hair bundle can be written as, see equation (10.3.4), $f(x) = f_0(x) - zn$, where $f_0(x) = NK(x - \bar{x})$ and $n$ is the stochastic number of open ion channels. Thus $f_0(x)$ is the force needed to hold the bundle at $x$ when all the channels are closed. We can now determine the mean and variance of the force:

$$\langle f - f_0 \rangle = -z \langle n \rangle = -N z P_o(x),$$

and

$$\mathrm{Var}[f - f_0] = z^2 \sigma^2 = N z^2 P_o(x)(1 - P_o(x)).$$

## 10.3.2 Channel compliance, myosin motors, and spontaneous oscillations

The channel compliance (negative stiffness) of the hair bundle implies that the bundle can operate in a bistable regime for sufficiently small applied forces. However, the force–displacement curve can be shifted by the $Ca^{2+}$-regulated action of myosin molecular motors that move up and down a stereocilium, altering the stiffness of the tip link [596, 598]. The myosin motors adapt the response of the hair bundle so that if a negative (positive) displacement has been maintained for some time, the force–displacement curve of Fig. 10.17(c) is shifted to the left and downward (to the right and upward). This has the effect of moving the negative stiffness region toward the offset point. One possible mechanism for $Ca^{2+}$ regulation of myosin is that $Ca^{2+}$ simply reduces the probability that myosin motors bind to actin filaments, thus allowing the transduction element to lose tension [444]. Alternatively, $Ca^{2+}$ could alter the equilibrium between different bound conformational states of the motor. It has also been suggested that $Ca^{2+}$ might regulate hair bundle dynamics in a myosin-independent fashion [191, 444]. For example, the energy associated with binding of $Ca^{2+}$ directly to the channel or to an associated protein could reduce the open probability. However, the myosin-dependent mechanism is currently thought to be more likely.

As shown by Martin et al. [596], the above adaptation mechanism can also result in spontaneous oscillations, which is illustrated in Fig. 10.18. Suppose that the bundle occupies a negative displacement equilibrium when $f = 0$ (black curve in Fig. 10.18(a)). The ion channels are then in a low open probability state and the $Ca^{2+}$ concentration is kept at a low level by $Ca^{2+}$ pumps. This upregulates the myosin motors, resulting in an increased stiffness of the tip link and a downward/leftward shift of the force–displacement curve (dashed curve in Fig. 10.18 (a)). A sufficient shift leads to the disappearance of the negative fixed point (i) and the system jumps to the corresponding positive fixed point (ii). The ion channels are now in a high open probability state, $Ca^{2+}$ flows into the cell and downregulates the myosin

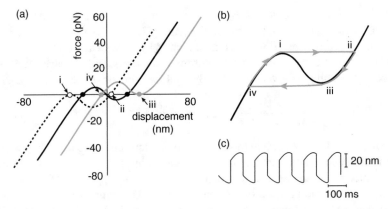

Fig. 10.18: Possible mechanism for spontaneous oscillations of a hair bundle. (a) Shifts in the force–displacement curve due to the action of myosin motors. See text for details. (b) Hair bundle acts like a relaxation oscillator with fast jumps ($i \rightarrow ii$ and $iii \rightarrow iv$) alternating with slow shifts in the fixed point due to adaptation ($ii \rightarrow iii$ and $iv \rightarrow i$). (c) Sketch of variation of displacement $x$ with time [Redrawn from Martin et al. [596].]

motors, and the force–displacement curve shifts in an upward/rightward direction. Eventually, the positive fixed point (iii) disappears (gray curve in Fig. 10.18(a)), and the system jumps back to a negative fixed point (iv). The cycle then repeats itself resulting in a periodic solution, see Fig. 10.18(b,c). Since the jumps are much faster than adaptation, the hair bundle acts like a relaxation oscillator, analogous to conductance-based models of a neuron; see Sect. 3.4. (Indeed, it is possible to linearly transform the resulting model equations, see below, to obtain equations similar in structure to those of the FitzHugh-Nagumo model [199].) It is thought that these oscillations are responsible for the spontaneous acoustic emissions observed in nonmammalian vertebrates. For example, Martin et al. [596] measured the power spectrum of spontaneous oscillations emitted by the hair bundle of a bullfrog hair cell, and found a sharp peak at around 8 Hz and a half-width of around 3 Hz; the spectral broadening is a result of thermal noise. The typical range of spontaneous oscillation frequencies is 5–50 Hz. The existence of spontaneous oscillations provides a possible mechanism for signal amplification and frequency tuning, namely, a nonlinear resonance effect when the stimulus frequency is sufficiently close to the natural frequency of spontaneous oscillations and the system operates close to a Hopf bifurcation point [157, 191, 266, 597, 660, 908].

For the sake of illustration, we will consider a dynamical model of spontaneous oscillations in hair bundles based on the combined action of $Ca^{2+}$ and myosin motors on the gating-spring model [598, 660]; see also Ref. [908]. Neglecting inertial effects, the position of a single gating-spring element evolves according to the equation

$$\xi \frac{dx}{dt} = -N\left[k_{gs}(x - x_M - P_o(x)\delta) + k_{sp}x\right] + F_{ext}, \qquad (10.3.5)$$

where $\xi$ is a friction coefficient, $F_{ext}$ is an external force, and $\delta$ is scaled by a geometrical factor that takes into account the gain of stereociliary shear motion. The effective equilibrium position of the gating spring is now denoted by $x_M$, in order to indicate that it depends on the action of myosin motors, and we have set $\bar{x}_{sp} = 0$. The open probability is given by a slightly modified version of equation (10.3.3):

$$P_o(x) = \frac{1}{1 + Ae^{-[k_{gs}\delta(x-x_M)]/k_BT}}, \quad A = e^{[\Delta E_0 + k_{gs}\delta^2/(2N)]/k_BT}. \quad (10.3.6)$$

Following the idea that myosin motors modify the tension in the tip link of a channel, the effective equilibrium position $x_M$ of the gating spring is maintained by $N_M$ myosin motors pulling against the force $f(x) = k_{gs}(x - x_M) - P_o(x)\delta$. That is, we identify $x_M$ with the position of the motor cluster. The dynamics of the cluster is assumed to satisfy a linear force–velocity relation with slope $\xi_M$ (see also Sect. 4.6):

$$\xi_M \frac{dx_M}{dt} = Nk_{gs}[x - x_M - P_o(x)\delta] - N_M f p(C), \quad (10.3.7)$$

where $C$ is the intracellular $Ca^{2+}$ concentration. Here the force exerted by the motors is taken to be proportional to the force $f$ generated by a single motor and the probability $p$ that a motor is bound to an active filament. The active force produced by the molecular motors corresponds to the motors climbing up the stereocilia, $dx_M/dt < 0$, which increases the tension of the gating springs and thus increases the open probability $P_o$ of the ion channels. The binding probability $p = p(C)$ is assumed to be a monotonically decreasing function of $C$. Ignoring nonlinearities in $p(C)$, the binding probability can be written as $p(C) \approx p_0 - p_1 C$ with $p_{0,1} > 0$, provided that $C < p_0/p_1$. Finally, the intracellular $Ca^{2+}$ dynamics is modeled as

$$\tau \frac{dC}{dt} = -(C - C_0) + [C_M - C_0]P_o(x), \quad (10.3.8)$$

where the decay term represents the effects of $Ca^{2+}$ pumps, and the other term on the right-hand side is the total flux through the open ion channels. When all the channels are closed, $C$ returns to the background concentration $C_0$. A crucial aspect of the model is a separation of time scales—the channel kinetics are assumed to be much faster than the $Ca^{2+}$ dynamics, which are themselves assumed to be faster than the bundle and motor dynamics.

Since $\tau \ll \xi, \xi_M$ (fast $Ca^{2+}$ dynamics), one can use a slow–fast analysis and set $\tau = 0$. This yields an effective planar dynamical system for $(x, x_M)$ given by equations (10.3.5) and (10.3.7) with $p(C) = p_0 - p_1 C_M P_o(x)$ for $C_0 = 0$. (Vilfan and Duke [908] also use a slow–fast decomposition, but treat the motors as the slow system by fixing $x_M$ and consider the planar dynamics of $(x, C)$.) One can then determine the existence and stability of fixed points for $F_{ext} = 0$, and derive conditions for the occurrence of a Hopf bifurcation along the lines of Box 5A. Nadrowski et al. [660] constructed a bifurcation diagram in terms of two parameters: the maximal force $f_{max} = \gamma N_m f p_0$ exerted by the molecular motors, and the dimensionless strength $S = C_M p_1/p_0$ of the negative $Ca^{2+}$ feedback. The final term on the

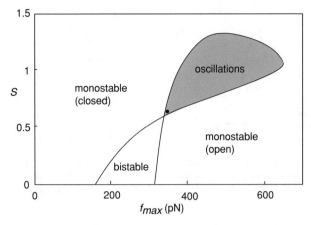

Fig. 10.19: Bifurcation diagram of hair bundle model given by equations (10.3.5)–(10.3.8) for $\tau = 0$. Different dynamical regimes are shown as a function of the maximal motor force $f_{\max}$ and the strength $S$ of $Ca^{2+}$ feedback on motor activity. There are two monostable domains where either most of the ion channels are closed or most are open. The shaded region indicates where stable oscillations exist, and Hopf bifurcations occur along the boundary of this region. For smaller $f_{\max}$ (toward the bistable region) the Hopf bifurcation is subcritical, whereas for larger $f_{\max}$ (away from the bistable regime) the Hopf bifurcation is supercritical. (The black dot indicates the point $f_{\max} = 50.3$ pN, $S = 0.65$ where stochastic simulations produced the best quantitative agreement with experiments; see text for details.) Other parameter values are as follows: $\xi = 2.8\,\mu\text{N s m}^{-1}$, $\xi_M = 10\,\mu\text{N s m}^{-1}$, $N = 50$, $N_M = 3000$, $\delta = 61$ nm, $k_{gs} = 15\,\mu\text{N m}^{-1}$, $k_{sp} = 12\,\mu\text{N m}^{-1}$, $C_0 = 0$, $\Delta G = 10 k_B T$, and $T = 300$ K [Redrawn from Nadrowski et al. [660].]

right-hand side of (10.3.7) becomes $-f_{\max}(1 - SP_o(x))$. The bifurcation diagram is sketched in Fig. 10.19, and consists of different dynamical regimes as indicated. In particular spontaneous oscillations occur at intermediate values of the maximal force and the strength of $Ca^{2+}$ feedback; in other regions the system is either in a monostable or bistable regime; see Ex. 10.6.

One simplification of the above model is that it ignores the effects of noise. Nadrowski et al. [660] also considered a stochastic version of the model by introducing white noise terms $\eta, \eta_M$ and $\eta_C$ on the right-hand side of equations (10.3.5), (10.3.7), and (10.3.8), respectively. The Gaussian random variables are taken to have zero mean and autocorrelations

$$\langle \eta(t)\eta(0) \rangle = 2k_B T \xi \delta(t), \tag{10.3.9a}$$

$$\langle \eta_M(t)\eta_M(0) \rangle = 2k_B T_M \xi_M \delta(t), \tag{10.3.9b}$$

$$\langle \eta_C(t)\eta_C(0) \rangle = 2N^{-1} C_M^2 P_o(1 - P_o)\tau_C \delta(t). \tag{10.3.9c}$$

The major source of noise for the hair bundle is Brownian motion in the surrounding fluid, and one can use the Einstein relation to determine the noise strength in terms of the friction coefficient $\xi$. Although the motors also undergo Brownian motion, there are additional sources of noise due to the random binding and binding

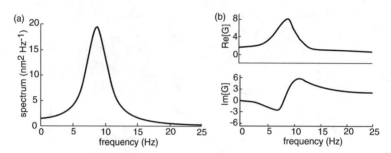

Fig. 10.20: Stochastic version of hair bundle model. (a) Power spectrum for spontaneous oscillations. (b) Real and imaginary parts of linear response function in response to sinusoidal forcing at
an amplitude of 1 pN. Parameter values given by black dot in Fig. 10.19 and $\tau_C = 1$ ms [Redrawn
from Nadrwoski et al. [660].]

to filament tracks. This leads to an effective temperature $T_M \approx 1.5\,T$. Finally, the
main source of noise for $Ca^{2+}$ dynamics is the random opening and closing of ion
channels, which can be described by a Binomial distribution. Assuming the channel
kinetics relaxation time $\tau_C$ is very fast, one can approximate the channel noise by
white noise. Simulations of the stochastic model for $\tau_C = 1$ ms, $f_{max} = 50.3$ pN, and
$S = 0.65$ (indicated by the black dot in the bifurcation diagram of Fig. 10.19) generate a spectrum of spontaneous oscillations that agrees quantitatively with experiments. An example of a spectrum obtained by Nadrowski et al. [660] is sketched
in Fig. 10.20, together with the corresponding real and imaginary parts of the linear
response function in the frequency domain.

### 10.3.3 Active amplification close to a Hopf bifurcation

A number of theoretical studies have suggested that many of the active properties of
a hair bundle can be reproduced by assuming that it operates close to a Hopf bifurcation [157, 191, 266, 596]. This is also consistent with the dynamical model of spontaneous oscillations considered by Nadrowski et al. [660], see Fig. 10.19, although
certain care needs to be taken since the Hopf bifurcation may be subcritical, so that
there is a rapid transition to a large amplitude relaxation oscillator. Following Refs.
[157, 266], we now consider the generic behavior of a forced oscillator close to a
supercritical Hopf bifurcation. Let $\mu$ denote some bifurcation parameter of the system, which could be related to the activity of myosin motors or the concentration of
intracellular $Ca^{2+}$ in the case of a hair bundle. Suppose that the system is in a stable
stationary state for $\mu < 0$, whereas it exhibits spontaneous oscillations for $\mu > 0$ due
to a supercritical Hopf bifurcation at the critical value $\mu_c = 0$. Recall from Box 5A
that close to a Hopf bifurcation, the dynamics of an unforced oscillator with natural
frequency $\omega_0$ can be represented (after an appropriate change of variables) by the

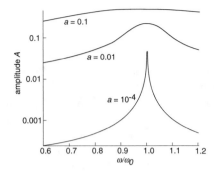

Fig. 10.21: Amplitude in response to a stimulus of strength $a$ and frequency $\omega$ for $\mu = 0.2$.

normal form

$$\frac{dx}{dt} = \mu x - \omega_0 y - x(x^2 + y^2), \quad \frac{dy}{dt} = \omega_0 x + \mu y - y(x^2 + y^2),$$

which can be recast in complex form by setting $z = x + iy$:

$$\frac{dz}{dt} = (\mu + i\omega_0)z - |z|^2 z. \qquad (10.3.10)$$

Now suppose that we drive the oscillator with a forcing term $ae^{i\omega t}$, and look for solutions of the form $z = Ae^{i(\omega t + \theta)}$. Substituting into the complex version of the normal form gives

$$\left(iA(\omega - \omega_0) - \mu A + A^3\right)e^{i\theta} = a.$$

The relevant quantity in terms of amplification is the amplitude of the response, so taking the modulus of both sides we have

$$A^6 - 2\mu A^4 + [\mu^2 + (\omega - \omega_0)^2]A^2 = a^2. \qquad (10.3.11)$$

Solving this equation for fixed stimulus strength $a$, we can plot the amplitude $A$ as a function of the stimulus frequency. The results are shown in Fig. 10.21 for various input amplitudes $a$. it can be seen that when $a$ is small, there is significant amplification and sharp frequency tuning around the resonant frequency $\omega_0$. The amplitude response curves are qualitatively similar to those seen experimentally [444, 597]. At the resonant frequency ($\omega = \omega_0$), the amplitude equation reduces to the simpler form $A^3 - \mu A = a$, which establishes that at the Hopf bifurcation point where $\mu = 0$, we have $A \sim a^{1/3}$. This is a highly compressive nonlinearity that boosts weak signals much more strongly than strong signals. On the other hand, if the stimulus frequency differs significantly from $\omega_0$ then the cubic term in the amplitude equation can be neglected, and the system operates in a linear regime for which $A \sim a/(\omega - \omega_0)$, and there is a $90^o$ phase lag. Finally, note that in order for the above nonlinear resonance to be realized by hair bundles, there has to be some feedback mechanism that keeps the system close to Hopf bifurcation point for a range of different natural frequencies—a process known as self-tuning [157, 266].

## 10.4 Bacterial chemotaxis

We now turn to chemical sensing in a single-cell organism, *E. coli*, which is an example of a motile cell that undergoes chemotaxis. That is, its motion is biased in a direction corresponding to a gradient of increasing or decreasing concentration of a signaling molecule. Bacterial chemotaxis was briefly mentioned in Sect. 6.3, within the context of diffusion-limited reactions and sensitivity to chemical gradients. *E. coli* is one of the most studied organisms in systems biology, exhibiting a number of important signaling mechanisms, including signal amplification, adaptation, and robustness to noise. For a comprehensive discussion of some of the issues covered here, see Chap. 4 of Bialek [78] and the reviews [877, 888]. Many bacteria, including *E. coli*, possess flagella, which are helical polymer filaments that are turned by molecular motors embedded in the cell's membrane. (The axial-asymmetric helical structure of flagella provides a mechanism for swimming at low Reynolds number; see Box 10B.) When all of the flagellar motors are rotating counterclockwise, the helical filaments bundle together and efficiently drive the bacterium in a straight line comprising a single run. On the other hand, if the motors reverse direction, the flagellar bundle flies apart and the bacterium rotates in a random fashion called a tumble. This is illustrated in Fig. 10.22(a). Over longer time scales, the motion of the bacterium looks like a sequence of straight line trajectories arranged at random angles

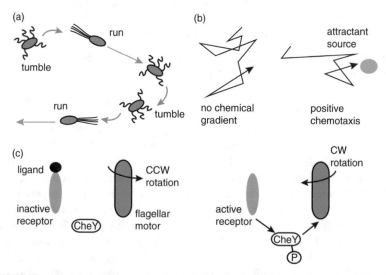

Fig. 10.22: Bacterial chemotaxis. (a) A schematic showing the motion of a bacterium that consists of a series of runs and tumbles. (b) The sequence of runs and tumbles can be altered by an external chemical gradient so that the motion is biased toward (away from) an attractant (a repellant). (c) The switching of a flagellar motor from counterclockwise to clockwise rotation, resulting in a switch from running to tumbling, is controlled by a signaling pathway in which unbinding of a ligand (attractant molecule) from a chemoreceptor in the cell membrane leads to the phosphorylation of CheY, which subsequently binds to the motor and induces the switch.

to each other, see Fig. 10.22(b). Tuning of the swimming behavior by environmental signaling molecules allows the bacterium to swim either toward a food source (chemoattractant) or away from a noxious toxin (chemorepellant). These signaling molecules bind to chemoreceptors in the cell membrane that induce dephosphorylation of a downstream signaling molecule CheY, which tends to switch the flagellar motors from clockwise to counterclockwise rotation, see Fig. 10.22(c).

---

**Box 10B. Swimming at low Reynolds number.**

---

**Reynolds number.** The flagellar-based swimming mechanism of *E coli* is one strategy for moving in a fluid at a low Reynolds number. In order to understand what this means, it is necessary to consider a little fluid mechanics [66, 202, 738]. Consider a flat body such as a spoon moving in a fluid such as air or water. Roughly speaking, the force necessary to keep the object moving at a constant speed $v_0$ is $F \sim \eta A dv/dy$, where $A$ is the surface area of the spoon and $v(y)$ is the velocity of different cross sections of fluid at a perpendicular distance $y$ from the object. The constant $\eta$ is known as the viscosity of the fluid. (The linear relationship between $F$ and the velocity gradient is characteristic of a Newtonian fluid such as air or water.) Suppose that $l_0$ is a characteristic size of the object. Using dimensional analysis, the viscous force $\eta A dv/dy$ will scale as $\eta l_0 v_0$, whereas the inertial force $mdv/dt$ due to the fluid's momentum will scale as $\rho l_0^2 v_0^2$ where $\rho$ is the density of the fluid. The ratio of these two forces is characterized by a single dimensionless parameter known as the Reynolds number (Re):

$$\mathrm{Re} = \frac{\rho l_0^2 v_0^2}{\eta l_0 v_0} = \frac{\rho l_0 v_0}{\eta}. \tag{10.4.1}$$

When $\mathrm{Re} \gg 1$ inertial forces dominate, whereas viscous forces dominate when $\mathrm{Re} \ll 1$. Using typical length and velocity scales for humans and bacteria swimming in water, one finds that $\mathrm{Re} \sim 10^4$ for humans and $\mathrm{Re} \sim 10^{-3} - 10^{-5}$ for bacteria. Note that the Reynolds number can also be obtained from the Navier–Stokes equation, which is the governing equation in fluid dynamics and is associated with the conservation of momentum, see below.

**Incompressibility.** Let $\rho(\mathbf{x},t)$ denote the density of the fluid and $\mathbf{v}(\mathbf{x},t)$ the velocity field. The mass conservation equation takes the form

$$\frac{\partial \rho}{\partial t} + \nabla \cdot (\rho \mathbf{v}) = 0, \tag{10.4.2}$$

since $\rho \mathbf{v}$ is the fluid flux. If the fluid is incompressible, then the total time derivative of the density is zero:

$$\frac{d\rho}{dt} = \frac{\partial\rho}{\partial t} + \mathbf{v} \cdot \nabla\rho. \tag{10.4.3}$$

Comparison with the mass conservation equation implies that the velocity field of an incompressible fluid is divergence free:

$$\nabla \cdot \mathbf{v} = 0. \tag{10.4.4}$$

**The stress tensor.** Consider a force in the fluid $d\mathbf{F}$ acting across a plane of area $dA$ normal to the unit vector $\mathbf{n}$. The force can be expressed in terms of the so-called stress tensor $\sigma_{ij}$ according to (see also Box 12B)

$$dF_i = \sum_{j=1,2,3} \sigma_{ij}n_j dS. \tag{10.4.5}$$

That is, $\sigma_{ij}$ is the $i$th component of the force acting across a plane of unit area normal to the $j$-axis. In the case of a Newtonian fluid, the stress tensor can be expressed in terms of the pressure $p$ and the velocity gradient $\kappa_{ij} = \partial v_i / \partial x_j$ according to the constitutive equation

$$\sigma_{ij} = \eta(\kappa_{ij} + \kappa_{ji}) - p\delta_{i,j}. \tag{10.4.6}$$

The total force acting on the surface $\partial\Omega$ of a region $\Omega$ is written as

$$F_i = \int_{\partial\Omega} \sum_j \sigma_{ij}n_j dA. \tag{10.4.7}$$

From the divergence theorem, this can be rewritten as

$$F_i = \int_\Omega \sum_j \frac{\partial\sigma_{ij}}{\partial x_j} d^3\mathbf{x}. \tag{10.4.8}$$

**Navier–Stokes equation.** Applying Newton's law of motion to the fluid in the domain $\Omega$ gives

$$\frac{d}{dt}\int_\Omega \rho v_i d^3\mathbf{x} = F_i - \int_{\partial\Omega}(\rho v_i)\mathbf{v} \cdot \mathbf{n}dA, \tag{10.4.9}$$

where the left-hand side is the rate of change of the momentum within $\Omega$ and the second term on the right-hand side is the loss of momentum due to flux through the surface $\partial\Omega$. Applying the divergence theorem to the latter and using the fact that $\Omega$ is arbitrary, we obtain the equation of motion

$$\frac{\partial\rho v_i}{\partial t} + \nabla \cdot (\rho v_i\mathbf{v}) = \sum_j \frac{\partial\sigma_{ij}}{\partial x_j}.$$

Finally, imposing mass conservation, the condition $\nabla \cdot \mathbf{v} = 0$, and using the constitutive relation for the stress tensor yield the Navier–Stokes equation

$$-\frac{\partial p}{\partial x_i} + \eta \sum_j \frac{\partial}{\partial x_j} \left( \frac{\partial v_i}{\partial x_j} + \frac{\partial v_j}{\partial x_i} \right) = \rho \frac{\partial v_i}{\partial t} + \rho \sum_j v_j \frac{\partial v_i}{\partial x_j}. \qquad (10.4.10)$$

The terms on the left-hand side represent pressure and viscous terms, and the terms on the right-hand side correspond to inertial terms. After non-dimensionalizing the Navier–Stokes equation, the right-hand side is multiplied by Re, so that for Re$\ll$ 1, the Navier–Stokes equation reduces to the time-independent Stokes equation

$$\eta \nabla^2 \mathbf{v} = \nabla p. \qquad (10.4.11)$$

The latter can be interpreted as a force-balance equation between viscous forces and pressure. In the derivation of these equations, gravity and other external forces have been ignored.

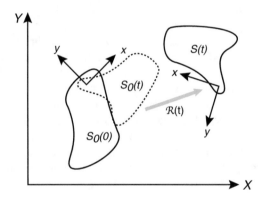

Fig. 10.23: Representing a shape $S(t)$ in physical space $(X,Y)$ by a shape $S_0(t)$ at a fixed location and orientation, which is then shifted and reoriented by a rigid body transformation $\mathcal{R}(t)$.

**Scallop theorem.** One of the immediate consequences of swimming at low Reynolds number is that the net forces acting on a body must at all times be zero, since they cannot be counterbalanced by an inertial force (mass times acceleration). Consider, for example, a microscopic swimmer that moves by changing its shape. Clearly the sum of all internal forces must be zero, i.e., the organism cannot "bootstrap" its own motion. However, changing its shape elicits reactive resistive forces from the fluid which themselves have to sum to zero. It turns out that the sequence of shape changes is uniquely determined by the requirement that the resistive forces cancel, and can result in net motion of the swimmer.

The requirement that there is net motion then constrains the allowed sequence of motions, as illustrated by the so-called scallop theorem formulated by Purcell [738]. Consider a scallop that moves in water at high Re by slowly opening and rapidly closing its shell. The latter action expels a jet of water that propels the scallop in the opposite direction, whereas the drag associated with reopening can be reduced by opening slowly. In contrast, at low Reynolds number, the flow of water into and out of the scallop over one cycle would be the same, regardless of the speed, implying that a scallop would make no net progress at low Re. This theorem reflects the fact that the Navier–Stokes equation in the limit Re→ 0 is time reversal symmetric, that is, it doesn't change under the transformations $t \rightarrow -t$ and $\mathbf{v} \rightarrow -\mathbf{v}$. The motion is independent of the speed and is determined by the sequence of body configurations. Since the opening and closing of a scallop's shell is time reversible, there is no net progress at low Re. The helical structure of flagella motors clearly breaks time reversal symmetry and thus allows *E coli* to generate net motion at low Re.

**How to swim at low Re.** We now give a more abstract mathematical description of how to achieve net motion at low Re due to Shapere and Wilczek [813]; see also [202]. Consider a cyclic sequence of shape changes of a microscopic swimmer at low Re in a Newtonian fluid. Let $S(t)$ denote the shape of the swimmer at time $t$ at the actual orientation and location in space. $S(t)$ can be decomposed in terms of a displacement and orientation operator $\mathscr{R}(t)$ acting on a shape function $S_0(t)$, where $\{S_0\}$ denotes the set of all possible shapes at a fixed location and orientation. This is illustrated in Fig. 10.23, which shows two shapes $S_0(0)$ and $S_0(t)$ determined by a fixed local coordinate system $(x,y)$. The actual physically located shape $S(t)$ is obtained by displacing and rotating the local coordinate system by the rigid body transformation $\mathscr{R}(t)$. For example, suppose that $S(\sigma)$ is a Simply connected shape in $\mathbb{R}^3$, which is treated as a map from the two-sphere $S^2$ to $\mathbb{R}^3$ with $\sigma \in S^2$. Then $\mathscr{R}$ can be represented as a $4 \times 4$ matrix

$$\mathscr{R} = \begin{pmatrix} \mathbf{R} \ \mathbf{d} \\ 0 \ 1 \end{pmatrix},$$

where $\mathbf{R}$ is a standard $3 \times 3$ rotation matrix and $\mathbf{d} = (d_1, d_2, d_3)^T$ represents displacements of the global Cartesian coordinates $(X,Y,Z)$. The operator $\mathscr{R}$ operates on the 4- vector $(\mathbf{S}_0(\sigma), 1)^T$ with $S_0(\sigma) \in \mathbb{R}^3$. It follows that under the rigid body transformation, $\mathbf{S}_0(\sigma) \rightarrow \mathbf{R}\mathbf{S}_0(\sigma) + \mathbf{d}$.

It should be noted that the physical shape $S(t)$ is independent of the choice of local coordinates used to define the shapes $S_0$—changing the local coordinate system changes $\mathscr{R}(t)$ but the sequence of motions is invariant under these transformations. Introduce a matrix $\mathbf{A}(t)$ for infinitesimal motions according to

$$\frac{d\mathscr{R}}{dt} = \mathscr{R}(t)\mathbf{A}(t).$$

This can be formally integrated to give

$$\mathscr{R}(t) = \mathbb{T} \exp \left( \int_0^t \mathbf{A}(t')dt' \right).$$

Here $\mathbb{T}$ denotes the time-ordering operator. That is, on Taylor expanding the exponential, we obtain a series of multiple integrals involving products of the operators $\mathbf{A}(t)$ at different times. The time-ordered product means that operators at later times appear to the right of operators at earlier times—in general the operators don't commute:

$$\mathbb{T} \exp \left( \int_0^t \mathbf{A}(t')dt' \right) = 1 + \int_0^t \mathbf{A}(t')dt' + \int_0^t \int_0^{t''} \mathbf{A}(t')\mathbf{A}(t'')dt'dt'' + \dots$$

Finally, one can express $\mathbf{A}(t)$ in a time-independent manner by setting $\mathbf{A}(t)dt \equiv \mathbf{A}[S_0(t)]dS_0$ such that

$$\mathscr{R}(t) = \mathbb{T} \exp \left( \int_{S_0(0)}^{S_0(t)} \mathbf{A}[S_0]dS_0 \right).$$

In the case of a cyclic sequence of shape changes, the net rotation and displacement per cycle period $\Delta$ is

$$\mathscr{R}(\Delta) = \mathbb{T} \exp \left( \oint \mathbf{A}[S_0]dS_0 \right).$$

Of course, in order to calculate this explicitly one still has to solve the fluid dynamics equations at low Re to determine $\mathbf{A}[S_0]$. Various examples can be found in [813].

### 10.4.1 Receptor clustering and signal amplification

The main components of the signaling transduction pathways involved in *E. coli* chemotaxis are shown in Fig. 10.24 [877, 888]. Each chemoreceptor forms a complex with kinase CheA via an adaptor protein CheW. (Recall from Sect. 10.1 that a protein kinase is an enzyme that modifies other proteins by chemically adding phosphate groups to them.) The autophosphorylation (self-activation) of CheA is suppressed (enhanced) when a chemoattractant (chemorepellant) binds to the associated receptor. In the activated state, CheA transfers a phosphate group to the motor regulator CheY thus counteracting dephosphorylation by CheZ. The phosphorylated form of CheY then diffuses away and binds with a flagellar motor, which then increases the motor's clockwise bias and hence the cell's probability of

tumbling. As with other biological sensory systems, the bacterial chemotaxis pathway allows the cell to adapt to persistent chemical stimuli. Adaptation is mediated
by the methylation and demethylation of the chemoreceptors by the enzymes CheR
and CheB*, where CheB* is the phosphorylated form of CheB that is also targeted
by the activated form of CheA.

There are more than 10,000 chemoreceptors in a single *E. coli* cell, and they
tend to form clusters around the cell pole. As first hypothesized by Bray *et al.* [103]
and later confirmed experimentally [614, 836], one important function of receptor
clustering is signal amplification due to cooperativity, analogous to cooperativity
between multiple binding sites on a ligand-gated ion channel (see Sect. 3.2). Let us
first consider a single chemoreceptor, which can bind to a single-ligand molecule.
Suppose that the kinase activity of the chemoreceptor (via CheA) has two discrete
states, active (a) and inactive (i), and that the equilibrium constants for ligand bind-
ing/unbinding are different in the two states. We thus have a version of the Monod-
Wyman-Changeux (MWC) model with a single binding site ($n = 1$). It immediately
follows (see Sect. 3.2) that the steady-state probability of being in the active state is

$$p_a = \frac{Y_0(1 + K_a[L])}{Y_0(1 + K_a[L]) + (1 + K_i[L])},$$
(10.4.12)

where $Y_0$ is the equilibrium constant for $i \rightleftharpoons a$ and $K_i, K_a$ are the equilibrium
constants for ligand binding in the inactive and active states, respectively, with
$K_a < K_i$. From equilibrium thermodynamics we know that $Y_0 = e^{-\Delta E/k_B T}$, where
$\Delta E = E_a - E_i$ is the free energy difference between the active and inactive states in
the absence of a ligand. Given equation (10.4.12), we can also define an effective
free energy difference $\Delta \overline{E}$ between the two states that is "averaged" with respect to
the binding state by setting

$$p_a = \frac{1}{1 + e^{\Delta \overline{E}/k_B T}}.$$

Comparison with (10.4.12) shows that

$$\Delta \overline{E}([L]) = \Delta E + k_B T \ln \frac{1 + K_i[L]}{1 + K_a[L]}.$$
(10.4.13)

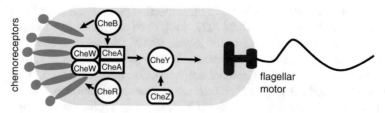

Fig. 10.24: Schematic diagram of major signaling pathways in *E. coli* chemotaxis. See text for
details.

Since $K_i > K_a$, increasing the ligand concentration $[L]$ increases the effective free energy and thus decreases the probability of being in the active state.

There are two basic models of receptor clustering. One involves dividing the receptors into a set of independent subclusters. Within each subcluster, all the receptors are tightly coupled and always in the same state (active or inactive), which is uncorrelated with the collective state of any other subcluster. However, the binding state of each receptor within a subcluster varies independently as in the single-receptor case. This all-or-none activation state of a subcluster of $N$ receptors is thus equivalent to the MWC model with $N$ binding sites, and the effective free energy is $E_N = N\Delta\overline{E}$ and $p_a = (1 + e^{E_N/k_BT})^{-1}$. On using (10.4.13), we have

$$p_a = \left[1 + \exp\left(N\left(\Delta\overline{E} + k_BT\ln\frac{1+K_i[L]}{1+K_a[L]}\right)\right)\right]^{-1}$$

$$= \frac{Y_0(1+K_a[L])^N}{Y_0(1+K_a[L])^N + (1+K_i[L])^N}. \tag{10.4.14}$$

If $K_a \ll [L]^{-1} \ll K_i$ then we obtain a Hill function of order $n$:

$$p_a = \frac{Y_0}{Y_0 + K^n[L]^n}, \tag{10.4.15}$$

where $K = K_i$. It follows that

$$\frac{dp_a}{d[L]} = -n[L]^{n-1}\frac{Y_0}{(Y_0 + K^n[L]^n)^2} = -\frac{n}{[L]}p_a(1 - p_a).$$

This suggests that maximal sensitivity will occur if the system is kept in a regime where $p_a \approx 0.5$.

An alternative model of receptor clustering is to take the receptors to be distributed on some form of lattice with nearest neighbor interactions [260]. Let $m = 1,\ldots,N$ be a receptor label and denote the state of the $m$th receptor by $a_m$ with $a_m = 1$ (active) or $a_m = 0$ (inactive). Let $\mathbf{a} = (a_1,\ldots,a_N)$ denote a given cluster state and take the corresponding free energy to be

$$H(\mathbf{a}) = -J\sum_{\langle m,n\rangle}(2a_m - 1)(2a_n - 1) + \Delta\overline{E}([L])\sum_m a_m. \tag{10.4.16}$$

The first term on the right-hand side represents interactions between nearest neighbor pairs on the lattice, denoted by $\langle n,m\rangle$, with $J$, $J > 0$, a coupling strength. Such coupling favors neighboring receptors to be in the same activation state (1 or 0). The second term takes into account the internal energetics of individual receptors. In steady state, the probability $P(\mathbf{a})$ of the cluster state $\mathbf{a}$ is given by

$$P(\mathbf{a}) = \frac{1}{Z}e^{-H(\mathbf{a})/k_BT}, \quad Z = \sum_{\mathbf{a}}e^{-H(\mathbf{a})/k_BT}.$$

The second model of receptor clustering is identical in structure to the Ising model of cooperative binding (Sect. 3.2) and of a polymer (Sect. 12.1.1). For example, the latter is obtained under the transformations $2a_m - 1 \rightarrow \sigma_m = \pm 1$ and $\Delta \overline{E} \rightarrow -\alpha$, where $\sigma_m$ is the orientation of the $m$th link in the chain and $\alpha$ represents an applied force at one end of the chain. An exact solution for the mean number of active receptors can then be obtained along identical lines to the derivation of the mean extension of the polymer; see Ex. 10.7:

$$a \equiv \frac{1}{N} \left\langle \sum_m a_m \right\rangle = \frac{1}{2} \left[ 1 - \frac{\sinh \Delta \overline{E}([L])}{\sqrt{\sinh^2 \Delta \overline{E}([L]) + e^{-4J}}} \right]. \tag{10.4.17}$$

It turns out that the flagellar motors are very sensitive to changes in the concentration of the phosphorylated form of CheY. Since ligand binding to chemoreceptors reduces the level of phosphorylation, this provides another mechanism for signal amplification. In Fig. 10.25, we show a model for the modulation of motor rotation bias due to binding of phosphorylated CheY (denoted by CheY*) [78]. Each flagellar motor is a rotary engine with a ring-like structure. The CheY molecules bind independently to multiple sites distributed around the ring with the binding affinity greater when the motor is rotating clockwise (CW) rather than counterclockwise (CCW), that is, the associated equilibrium constants satisfy $K_{CW} > K_{CCW}$. When all sites are empty, the equilibrium is biased toward CCW rotation. However, as more sites become occupied the equilibrium shifts to CW rotation. It follows that increasing the concentration of CheY* will favor the latter state. From the perspective of binding reactions, this model is also identical to the Monod-Wyman-Changeux (MWC) model of a ligand-gated ion channel with $n$ binding sites. That is, we can map the open ($R$) and closed ($T$) states of the ion channel to the CW and CCW states of the molecular motor. Taking $Y_0$ to be the equilibrium constant for the switching between the CW and CCW states with all sites empty, we see that the probability $P_{CW}$ that the motor is in the CW state is

$$P_{CW} = \frac{Y_0(1 + K_{CW}c)^n}{Y_0(1 + K_{CW}c)^n + (1 + K_{CCW}c)^n}, \tag{10.4.18}$$

where $c$ denotes the concentration of CheY*.

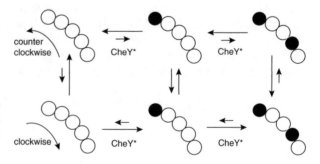

Fig. 10.25: Model of motor as a sensor of CheY. See text for details. Redrawn from [78].

## 10.4.2 Adaptation in signal transduction pathways

A sudden increase in the concentration of a chemoattractant results in a decrease in the cell's tumbling frequency, but over a longer time scale the frequency recovers to its prestimulus level [63]. This frequency adaptation occurs over a wide range of stimulus strengths. At the molecular level, adaptation is mediated by the enzymes CheR and CheB*, which are responsible for the methylation and demethylation of chemoreceptors. The level of methylation $v$, say, can be incorporated into the receptor clustering models by taking the difference in free energies between the active and inactive receptor states to depend on $v$, that is, $\Delta E = \Delta E(v)$. As highlighted by Barkai and Leibler [44], the level of parameter fine-tuning that would be needed to account for the observed adaptation is unrealistic given the presence of noise. Therefore, they proposed a robust adaptation mechanism that doesn't need any fine-tuning. The basic idea is to assume that the relatively slow process of demethylation (CheB*) counteracts the shift in tumbling frequency induced by changes in the level of kinase activity (CheA*). More specifically, the Barkai–Leibler model assumes the following: (i) The rate of catalysis of the methylation enzyme CheR is at its maximum and is thus independent of any concentrations; (ii) The action of the demethylation enzyme CheB* on active receptors is given by a Hill function of index $n = 1$:

$$\frac{dv}{dt} = F(a) \equiv \Gamma_R - \frac{\Gamma_B a}{K_B + a}, \qquad (10.4.19)$$

where $\Gamma_{R,B}$ are maximum catalytic rates and $a$ is the average receptor activity. In general, this is a nonlinear equation for $v$ since $a$ is a function of $v$ via its dependence on $\Delta E(v)$.

In order to illustrate how adaptation occurs, let us return to the MWC model. Equation (10.4.14) shows that the average receptor activity $a$ is a function of the level of methylation $v$ according to

$$a = a([L], v) \equiv \left[1 + \exp\left(N\left(\Delta E(v) + k_B T \ln \frac{1 + K_i[L]}{1 + K_a[L]}\right)\right)\right]^{-1}. \qquad (10.4.20)$$

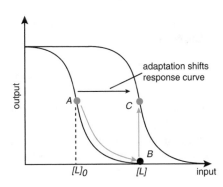

Fig. 10.26: Illustration of the Barkai–Leibler adaptation mechanisms. A sudden change in input (ligand concentration) $[L]_0 \to [L]$ induces a fast response from receptor activity state A to state B. Over longer time scales, the response curve shifts to a higher attractant concentration as the system adapts its methylation level until it reaches the adapted state C with the same activity as the prestimulus state A. As a result of the response curve shift, the high response sensitivity of adapted state C in the new environment is identical to that of state A. Redrawn from [888].

Suppose that there is a fixed background ligand concentration $[L] = L_0$ and we denote the equilibrium receptor activity by $a_0$. This must also correspond to the steady-state solution of (10.4.19) with $F(a_0) = 0$, that is, $[a_0 = K_B \Gamma_R / (\Gamma_B - \Gamma_R)$, which is independent of ligand concentration. (The fixed point is globally stable as $F'(a) < 0$ and $da/dv > 0$.) Given this solution for $a_0$, equation (10.4.20) implies that $a_0 = a(v, L_0)$, which can be inverted to yield the steady-state methylation level $v_0 = v(a_0, L_0)$. Now suppose that there is a sudden change in ligand concentration, $[L] = L_1$. The average receptor activity rapidly changes to give the new equilibrium solution for $v_0$ fixed, $a_0 \to a_1 = a(L_1, v_0)$. Suppose that $a_0 \approx 1/2$ so the system initially operates in the sensitive region of the response curve, whereas $a_1$ is outside this domain, see Fig. 10.26. However, over a longer time scale, the methylation level adapts according to equation (10.4.19) so that $a_1 \to a_0$ with $v_0 \to v_1 = v(a_0, L_1)$, see Fig. 10.26. The linear response of the Barkai–Leibler model to a small oscillatory input is considered in Ex. 10.8.

### 10.4.3 Run-and-tumble model of bacterial motility

One of the challenges in modeling bacterial chemotaxis is understanding how extracellular biochemical signals are transduced into behavioral changes at the macroscopic level illustrated in Fig. 10.22(b). Early models tended to be phenomenological in nature, representing the dynamics of cells in terms of an advection–diffusion equation for the cell density $n(\mathbf{x}, t)$, in which the velocity is taken to depend on the concentration gradient of some chemotactic substance [499, 500, 655]. An alternative, stochastic formulation of bacterial motion has been developed in terms of a velocity-jump process, in which the velocity of the cell can randomly jump according to a discrete or continuous Markov process [8, 241, 412, 414, 696, 697]. It is also usually assumed that jumps are instantaneous, that is, the cell spends negligible time in a tumbling state—experimentally it is an order of magnitude faster than a typical run length. In addition, if the different velocity states have the same speed and the direction of motion following a tumble is randomly selected from a uniform distribution, then one obtains the so-called run-and-tumble particle (RTP) model.

Let us begin by considering a 1D RTP model of bacterial motion in the absence of a chemotactic signal. There are then just two velocity states $v_{\pm} = \pm v$. Let $p_{\pm}(x, t)$ denote the probability densities of a cell being at $(x, t)$ and moving to the right $(+)$ and left $(-)$, respectively. The corresponding CK equation is

$$\frac{\partial p_+}{\partial t} = -v\frac{\partial p_+}{\partial x} - kp_+ + kp_-, \tag{10.4.21a}$$

$$\frac{\partial p_-}{\partial t} = v\frac{\partial p_-}{\partial x} + kp_+ - kp_-, \tag{10.4.21b}$$

where $k$ is a constant turning rate, with $1/k$ measuring the mean run length between velocity jumps. These equations were originally introduced by Goldstein [352] and

Kac [484] to describe a correlated random walk. They are a symmetric version of the Dogterom–Leibler model of microtubule catastrophe [249], see Sect. 4.2, where $x(t)$ is the position of the tip of a microtubule and $\pm v$ represent the rates of growth and shrinkage, respectively. Equations of the form (10.4.21) were also used to describe unbiased bidirectional motor-driven transport in Chap. 7, where they were shown to be equivalent to the telegrapher equation. Thus many of the analytical results obtained from these applications can be adapted to the case of bacterial run-and-tumble, including FPT problems and sticky boundary conditions [21] (see Ex. 10.9 and Ex. 10.10).

In order to model 1D chemotaxis, it is necessary to introduce some bias into the stochastic switching (tumbling) between the velocity states $\pm v$ that depends on the extracellular concentration gradient $c$. For example, Erban and Othmer [277] have developed a detailed 1D model of chemotaxis that incorporates aspects of the biochemical signal transduction pathways described in earlier parts of this section. Here we consider a simpler phenomenological model [78], which assumes that the rate of tumbling depends on the time derivative of the concentration $c(t) = c(x(t))$ along the bacterial trajectory. Since $\dot{c} = \pm v\,dc/dx$ in the case of two velocity states $\pm v$, equation (10.4.21) becomes

$$\frac{\partial p_+}{\partial t} = -v\frac{\partial p_+}{\partial x} - k_+(x)p_+ + k_-(x)p_-, \tag{10.4.22a}$$

$$\frac{\partial p_-}{\partial t} = v\frac{\partial p_-}{\partial x} + k_+(x)p_+ - k_-(x)p_-, \tag{10.4.22b}$$

$$k_\pm(x) = k_0 \pm k_1 v c'(x). \tag{10.4.22c}$$

We are assuming that when the bacterium tumbles there is an equal probability of moving in either direction. Another simplification is to take the tumble rate to depend on instantaneous values of the concentration gradient rather than a time-averaged change in concentration. The steady-state probability densities satisfy the pair of equations

$$v\frac{dp_+}{dx} = k_-(x)p_-(x) - k_+(x)p_+(x),$$

$$-v\frac{dp_-}{dx} = k_+(x)p_+(x) - k_-(x)p_-(x).$$

Adding these two equations gives

$$v\frac{dp_+}{dx} - v\frac{dp_-}{dx} = 0,$$

which implies that the difference $p_+(x) - p_-(x) = \text{constant}$. Assuming that $-\infty < x < \infty$, normalizability of the probability densities requires this constant to be zero. Hence, $p_\pm(x) = p(x)/2$ with $p(x)$ satisfying the single equation

$$v\frac{dp}{dx} = [k_-(x) - k_+(x)]\,p(x) = -2k_1 v c'(x)p(x).$$

This has the straightforward solution

$$p(x) = \mathcal{N}e^{-2k_1 c(x)}, \tag{10.4.23}$$

where $\mathcal{N}$ is a normalization factor. If the signaling molecules correspond to a chemoattractant, then the rate of tumbling decreases in the direction for which $\dot{c} > 0$, that is, $k_1 < 0$, and maxima of the steady-state solution (10.4.23) coincide with maxima of the concentration $c(x)$. Conversely, $k_1 > 0$ for a chemorepellant and maxima of $P(x)$ coincide with minima of the concentration.

Let us return to the unbiased RTP model (10.4.21) and suppose that both the speed and tumbling rate are space-dependent:

$$\frac{\partial p_+}{\partial t} = -\frac{\partial v(x)p_+}{\partial x} - k(x)p_+ + k(x)p_-, \tag{10.4.24a}$$

$$\frac{\partial p_-}{\partial t} = \frac{\partial v(x)p_-}{\partial x} + k(x)p_+ - k(x)p_-. \tag{10.4.24b}$$

Following Refs. [172, 801], we will derive a diffusion approximation of equations (10.4.24) by rewriting them in terms of the total probability $p = p_+ + p_-$ and the flux $J = v[p_+ - p_-]$. We also perform the rescalings $t \to \varepsilon^2 t$ and $x \to \varepsilon x$. Adding and subtracting the rescaled pair of equations then yields

$$\varepsilon^2 \frac{\partial p}{\partial t} = -\varepsilon \frac{\partial J}{\partial x},$$

$$\varepsilon^2 \frac{\partial J}{\partial t} = -\varepsilon v(x) \frac{\partial v(x)p}{\partial x} - 2k(x)J.$$

The diffusion approximation entails dropping the $\partial_t J$ term in the limit $\varepsilon \to 0$. This gives the FP equation[1]

$$\frac{\partial p}{\partial t} = -\frac{\partial J}{\partial x}, \quad J(x) = -D(x)\frac{\partial p(x)}{\partial x} + V(x)p(x), \tag{10.4.25}$$

with

$$D(x) = \frac{v(x)^2}{2k(x)}, \quad V(x) = -\frac{v(x)v'(x)}{2k(x)}. \tag{10.4.26}$$

Given the minus sign in the definition of the effective drift $V(x)$, we see that the RTP tends to drift toward regions of slower speed. The equilibrium solution of the reduced equation is given by $J(x) = 0$, that is,

$$p(x) = \frac{p_0 v_0}{v(x)}, \tag{10.4.27}$$

---

[1] The steady-state equation $\partial_x J(x) = 0$ is exact, since it follows from setting all time derivatives to zero in equation (10.4.25).

where $p_0$ is the density at a reference point $x_0$ for which $v(x_0) = v_0$. Note that the steady-state density is independent of the tumbling rate $k(x)$, which means that the density is spatially uniform when $v(x) = v_0$ for all $x$. (Recall that we are dealing with a symmetric run-and-tumble model.) On the other hand, if the particle speeds up in regions of high chemoattractant concentration, then cells will tend to aggregate in low concentration domains (a form of reverse chemotaxis).

Finally, suppose that there is an external potential $U(x)$ such that the velocities of the left and right moving states are different: $v_R(x) = v_0 + f(x)$ and $v_L(x) = v_0 - f(x)$, where $f(x) = -\xi^{-1}U'(x)$. It can be shown that the effective flux becomes [801]

$$J(x) = -\frac{v_R(x) + v_L(x)}{2k(x)} \frac{\partial}{\partial x}\left[\frac{v_R(x)v_L(x)p}{v_R(x) + v_L(x)}\right] + \frac{v_R(x) - v_L(x)}{2}p,$$

and the equilibrium density is

$$p(x) = p_0 \frac{v_R(x_0)v_L(x_0)}{v_R(x_0) + v_L(x_0)} \frac{v_R(x) + v_L(x)}{v_R(x)v_L(x)} \exp\left(\int_{x_0}^x k(y)\frac{v_R(y) - v_L(y)}{v_R(y)v_L(y)}dy\right).$$

Substituting for $v_{L,R}$ in terms of the external force, we thus find

$$p(x) = p_0 \frac{D(x)}{D(x_0)} \exp\left(\int_{x_0}^x \frac{f(y)}{D(y)}dy\right), \quad D(x) = \frac{v_0^2 - f^2(x)}{k(x)}. \tag{10.4.28}$$

As an explicit example, suppose that there is a harmonic potential well with stiffness $\kappa$, i.e., $U(x) = \kappa x^2/2$ and $k(x) = k_0$ for all $x$. In the case of passive diffusion with $D(x) \to D_0$, the equilibrium density would be a Gaussian as predicted from the Boltzmann–Gibbs distribution of equilibrium thermodynamics. On the other hand, the RTP is confined to the region $|x| < v_0\xi/\kappa$ with a probability density [857]

$$p(x) = p_0\left[1 - \left(\frac{\kappa x}{v_0\xi}\right)^2\right]^a, \quad a = \frac{\xi k_0}{\kappa} - 1. \tag{10.4.29}$$

The latter can exhibit large deviations away from a Gaussian and, in particular, becomes bimodal for $k_0 < \kappa/\xi$, such that the RTP tends to be localized at the edges of the allowed region. (Analogous behavior was found in the stochastic hybrid model of a two-state gene network with promoter noise [490]; see Chap. 5.)

*Remark 10.1.* There has been considerable recent interest in studying RTPs within the statistical physics community, since they exhibit novel statistical properties at both the individual and population levels [172, 239, 640, 856, 857], see also Sect. 15.7. An individual RTP can be viewed as a self-propelled element that consumes energy in order to move. There are a wide class of models of active particles that can be written in the general form [313, 751]

$$d\mathbf{X}(t) = \mathbf{v}dt - \frac{1}{\xi}\nabla U + \sqrt{2D_0}d\mathbf{W}(t), \tag{10.4.30}$$

where $\mathbf{v}$ is a random velocity state that represents the self-propulsion, $\xi$ is a friction coefficient, $U$ is an external potential, and $\mathbf{W}(t)$ is a vector-valued Wiener process with isotropic

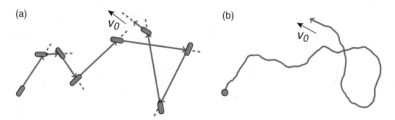

Fig. 10.27: Two examples of a self-propelled particle. (a) Run-and-tumble particle (RTP). (b) Active Brownian particle (ABP).

diffusivity $D_0$. (For simplicity we set $D_0 = 0$ and $U = 0$ in the above analysis.) The directionality of self-propulsion, that is, the tendency to move in a constant direction for some mean persistence time $\tau$ is incorporated by assuming that correlations decay exponentially in time,

$$\langle v_i(t)v_j(0)\rangle = \delta_{i,j}\frac{v_0^2}{d}e^{-|t|/\tau}, \tag{10.4.31}$$

where $d$ is the spatial dimension and $v_0$ is the speed. Models of active particles differ in terms of the higher-order statistics.

In the particular case of an RTP, changes in direction are instantaneous and completely isotropic. Moreover, $\tau^{-1}$ is equal to the tumbling rate $k$, and trajectories consist of random lengths of mean size $v_0/k$. Another well-known example of self-propulsion is an active Brownian particle (ABP). Again the particle moves at a constant speed $v_0$, whereas the angular direction is determined by a diffusive process. In 2D, the self-propulsion is written as

$$\mathbf{u} = v_0(\cos\theta, \sin\theta), \quad d\theta(t) = \sqrt{D_{\text{ang}}}dW_{\text{ang}}. \tag{10.4.32}$$

Now $\tau^{-1} = (d-1)D_{\text{ang}}$.

## 10.4.4 Higher-dimensional transport equation

The analysis of run-and-tumble models in higher dimensions is considerably more involved, since there is now a continuum of velocity orientations rather than just two. Here we describe a theory of transport that was developed in Refs. [412, 414, 696, 697]. Let $p(\mathbf{x},\mathbf{v},t)$ denote the probability density of cells at position $\mathbf{x} \in \mathbb{R}^d$ and velocity $\mathbf{v} \in \mathcal{V} \subset \mathbb{R}^d$ at time $t$. Then $p$ evolves according to an equation of the form

$$\frac{\partial}{\partial t}p(\mathbf{x},\mathbf{v},t) + \mathbf{v}\cdot\nabla p(\mathbf{x},\mathbf{v},t) = -kp(\mathbf{x},\mathbf{v},t) + k\int_{\mathcal{V}}T(\mathbf{v},\mathbf{v}')p(\mathbf{x},\mathbf{v}',t)d\mathbf{v}'. \tag{10.4.33}$$

Here $k$ is a constant turning rate, with $1/k$ measuring the mean run length between velocity jumps. For simplicity the time spent in the tumbling state is neglected. The so-called turning kernel $T(\mathbf{v},\mathbf{v}')$ is the conditional probability of a velocity jump from $\mathbf{v}'$ to $\mathbf{v}$ given that a jump occurs. The kernel $T$ is assumed to have the following properties:

$$T(\mathbf{v},\mathbf{v}') \geq 0; \quad \int_V T(\mathbf{v},\mathbf{v}')d\mathbf{v} = 1; \quad \int_V \int_V T^2(\mathbf{v},\mathbf{v}')d\mathbf{v}'d\mathbf{v} < \infty. \tag{10.4.34}$$

These insure that $T(\mathbf{v},\mathbf{v}')$ for fixed $\mathbf{v}'$ is a nonnegative probability density on $V$. The marginal particle density and flux are given by

$$\bar{p}(\mathbf{x},t) = \int_V p(\mathbf{x},\mathbf{v},t)d\mathbf{v}, \quad J(\mathbf{x},t) = \int_V p(\mathbf{x},\mathbf{v},t)\mathbf{v}d\mathbf{v}. \tag{10.4.35}$$

*Remark 10.2.* From a mathematical perspective one can express the right-hand side of equation (10.4.33) in terms of a linear integral operator $T$ acting on the Hilbert space $L^2(V)$. That is, $T : L^2(V) \rightarrow L^2(V)$ with

$$T\phi(\mathbf{v}) = \int_V T(\mathbf{v},\mathbf{v}')\phi(\mathbf{v}')d\mathbf{v}'. \tag{10.4.36}$$

It follows that (10.4.33) becomes

$$\frac{\partial}{\partial t}p + \mathbf{v}\cdot\nabla p = -kp + kTp. \tag{10.4.37}$$

Hence, one mathematical tool for analyzing the transport equation is to investigate the spectral properties of the linear operator $T$. This, in turn, requires imposing additional mathematical constraints on the integral operator, see the cited references for details. One example that satisfies these additional conditions is the Pearson random walk.

The simplest example of a turning kernel is to take $V$ to be the sphere $\mathbb{S}^{d-1}$ with radius $v_0$, and to assume that the particle's direction of movement is uniformly distributed. In this case $T(\mathbf{v},\mathbf{v}') = 1/|V|$ and equation (10.4.33) reduces to

$$\frac{\partial}{\partial t}p(\mathbf{x},\mathbf{v},t) + \mathbf{v}\cdot\nabla p(\mathbf{x},\mathbf{v},t) = -kp(\mathbf{x},\mathbf{v},t) + \frac{k}{|V|}\int_V p(\mathbf{x},\mathbf{v}',t)d\mathbf{v}'. \tag{10.4.38}$$

The associated stochastic process is known as a Pearson random walk, and is the natural higher-dimensional extension of the symmetric 1D run-and-tumble model [801]. The equilibrium steady-state solution is given by $p(\mathbf{x},\mathbf{v}) = p_0$, since $\int_V \mathbf{v}d\mathbf{v} = 0$ for $|\mathbf{v}| = v_0$. Another steady-state solution is one for which the particle gradient $\nabla p$ is a constant. In this case

$$p(\mathbf{x},\mathbf{v}) = \frac{\bar{p}(\mathbf{x})}{|V|} - \frac{\nabla\bar{p}\cdot\mathbf{v}}{k|V|}.$$

The corresponding particle flux has the Fickian form

$$\mathbf{J} = -\frac{v_0^2}{kd}\nabla\bar{p}.$$

Another common example in both cell biology and ecology describes movement of a cell or an animal along a fiber or road network. Now turning decisions are based on the local environment. Again $V = \mathbb{S}^{d-1}$ but the turning kernel may now depend on spatial position. Under the simplifying assumption that the newly selected direction of movement is independent of the previous direction, which essentially ignores the effects of inertia, we have $T(\mathbf{v}, \mathbf{v}', \mathbf{x}) = q(\mathbf{v}, \mathbf{x})$ for $\mathbf{v} \in \mathbb{S}^{d-1}$ and

$$\frac{\partial}{\partial t} p(\mathbf{x}, \mathbf{v}, t) + \mathbf{v} \cdot \nabla p(\mathbf{x}, \mathbf{v}, t) = -kp(\mathbf{x}, \mathbf{v}, t) + kq(\mathbf{v}, \mathbf{x}) \int_{\mathcal{V}} p(\mathbf{x}, \mathbf{v}', t) d\mathbf{v}'. \quad (10.4.39)$$

This equation is almost identical to the one used to model intracellular motor-driven transport on a cytoskeletal network; see Sect. 7.2.3.

**Diffusion limit.** As in the case of a 1D RTP, it is possible to carry out a formal diffusion limit of the transport equation by rescaling time and space [412, 414, 697]. (As highlighted in Sect. 7.2, the choice of scaling differs from the one used to perform the quasi-steady-state reduction of intracellular transport models.) Characteristic parameter values for *E. coli* are speeds $v_0 = 10\,\mu\text{m/s}$, colony sizes $L = 10^{-3} - 10^{-2}\text{m}$, observation times $\tau = 1 - 10\text{h}$, and turning rates $k = 1 - 10/\text{s}$. Non-dimensionalizing by taking $L = \tau = 1$ means that $v = O(1/\varepsilon)$ and $k = O(1/\varepsilon^2)$ with $\varepsilon \sim 10^{-2}$. Hence, the non-dimensionalized transport equation can be written as

$$\varepsilon^2 \frac{\partial p}{\partial t} + \varepsilon \mathbf{v} \cdot \nabla p = -kp + \int_{\mathcal{V}} T p d\mathbf{v}' \equiv \mathbb{L}p. \quad (10.4.40)$$

In order to determine the diffusion limit, introduce the perturbation expansion

$$p = p_0 + \varepsilon p_1 + \varepsilon^2 p_2 + O(\varepsilon^3).$$

Substituting into equation (10.4.40) and matching powers of $\varepsilon$ yields a hierarchy of equation, which to $O(\varepsilon^2)$ take the form

$$\mathbb{L}p_0 = 0, \quad (10.4.41\text{a})$$
$$\mathbb{L}p_1 = \mathbf{v} \cdot \nabla p_0, \quad (10.4.41\text{b})$$
$$\mathbb{L}p_2 = \frac{\partial p_0}{\partial t} + \mathbf{v} \cdot \nabla p_1. \quad (10.4.41\text{c})$$

We will assume that the null space of the linear operator $\mathbb{L}$ is spanned by the unique eigenfunction $\phi_0(\mathbf{v}) = 1$ so that we can set

$$p_0 = \bar{p}(\mathbf{x}, t) = \int_{\mathcal{V}} p(\mathbf{x}, \mathbf{v}, t) d\mathbf{v}. \quad (10.4.42)$$

(This can be proven in the case of the Pearson random walk, for example, [414].) Applying the Fredholm alternative theorem to the $O(\varepsilon)$ equation, we require that the inner product of $\mathbf{v} \cdot \nabla p_0$ with $\phi_0 = 1$ vanishes. This follows from the radial symmetry of $\mathcal{V}$:

$$\int_{\mathcal{V}} \mathbf{v} \cdot \nabla p_0 d\mathbf{v} = \nabla \bar{p} \cdot \int_{\mathcal{V}} \mathbf{v} d\mathbf{v} = \nabla \bar{p} \cdot \int_{\mathcal{V}} \mathbf{v} d\mathbf{v} = 0.$$

Hence, under the additional constraint $\int p_1 d\mathbf{v} = 0$, we can express the solution $p_1$ in terms of the pseudoinverse of $\mathbb{L}$:

$$p_1(\mathbf{x}, \mathbf{v}, t) = \mathbb{L}^\dagger \mathbf{v} \cdot \nabla \overline{p}. \tag{10.4.43}$$

The analysis of the $O(\varepsilon^2)$ equation is a little more involved. Integrating both sides gives

$$\int_{\mathcal{V}} \left( \frac{\partial p_0}{\partial t} + \mathbf{v} \cdot \nabla p_1 \right) d\mathbf{v} = 0.$$

Substituting the solutions for $p_0$ and $p_1$ gives

$$\int_{\mathcal{V}} \left( \frac{\partial \overline{p}}{\partial t} + \mathbf{v} \cdot \nabla \mathbb{L}^\dagger (\mathbf{v} \cdot \nabla \overline{p}) \right) d\mathbf{v} = |\mathcal{V}| \frac{\partial \overline{p}}{\partial t} + \nabla \cdot \int_{\mathcal{V}} \mathbf{v} \mathbb{L}^\dagger (\mathbf{v} \cdot \nabla \overline{p}) d\mathbf{v} = 0.$$

We thus obtain the diffusion equation

$$\frac{\partial \overline{p}}{\partial t} = \sum_{i,j} \frac{\partial}{\partial x_i} D_{ij} \frac{\partial \overline{p}}{\partial x_j}, \quad D_{ij} = -\frac{1}{|\mathcal{V}|} \int_{\mathcal{V}} v_i \mathbb{L}^\dagger v_j d\mathbf{v}. \tag{10.4.44}$$

In the case of a Pearson random walk, the diffusion tensor can be calculated explicitly. First note that the inhomogeneous equation $\mathbb{L}\phi = f$ with $\int_{\mathcal{V}} \phi d\mathbf{v} = 0$ becomes

$$f(\mathbf{v}) = \mathbb{L}\phi(\mathbf{v}) = -k\phi(\mathbf{v}) + \frac{k}{|\mathcal{V}|} \int_{\mathcal{V}} \phi(\mathbf{v}') d\mathbf{v}' = -k\phi(\mathbf{v}),$$

so that writing $\phi(\mathbf{v}) = \mathbb{L}^\dagger f(\mathbf{v})$ implies $\mathbb{L}^\dagger = -1/k$. It follows that

$$D_{ij} = \frac{1}{k|\mathcal{V}|} \int_{\mathcal{V}} v_i v_j d\mathbf{v}. \tag{10.4.45}$$

In two dimensions ($d = 2$), we recover the diffusion tensor derived in Sect. 7.2.3:

$$\mathbf{D} = \frac{v_0^2}{2\pi k} \int_0^{2\pi} \begin{pmatrix} \cos^2 \phi & \cos \phi \sin \phi \\ \cos \phi \sin \phi & \sin^2 \phi \end{pmatrix} d\phi = \frac{v_0^2}{2k} \mathbf{I}. \tag{10.4.46}$$

**Chemotaxis.** Now suppose there exists a weak chemotactic signal $S(x,t)$ that modifies the turning kernel

$$T(\mathbf{v}, \mathbf{v}', S) = T(\mathbf{v}, \mathbf{v}') + \varepsilon \alpha(S) \mathbf{v} \cdot \nabla S, \tag{10.4.47}$$

which implements the rule that the cell is more likely to choose a new direction that is toward $\nabla S$ (climbing up the chemotactic gradient). It follows that

$$\mathbb{L}\phi(\mathbf{v}) = -k\phi(\mathbf{v}) + k \int_{\mathcal{V}} T(\mathbf{v}, \mathbf{v}')\phi(\mathbf{v}') d\mathbf{v}' + \varepsilon k\alpha(S) \int_{\mathcal{V}} (\mathbf{v} \cdot \nabla S)\phi(\mathbf{v}') d\mathbf{v}'$$
$$= \mathbb{L}_0 \phi(\mathbf{v}) + \varepsilon \mathbb{L}_1 \phi(\mathbf{v}),$$

where $\mathbb{L}_0$ is the unperturbed operator and

$$\mathbb{L}_1 = k\alpha(S)(\mathbf{v} \cdot \nabla S)\overline{\phi}.$$

Substituting into the scaled transport equation (10.4.40) gives

$$\varepsilon^2 \frac{\partial p}{\partial t} + \varepsilon \mathbf{v} \cdot \nabla p = \mathbb{L}_0 p + \varepsilon \mathbb{L}_1 p, \tag{10.4.48}$$

which can again be analyzed using regular perturbation theory [414]. The $O(1)$ equation is as before so that $p_0 = \overline{p}(\mathbf{x}, t)$. The $O(\varepsilon)$ equation becomes

$$\mathbf{v} \cdot \nabla p_0 = \mathbb{L}_0 p_1 + \mathbb{L}_1 p_0,$$

which has the explicit form

$$\mathbb{L}_0 p_1 = \mathbf{v} \cdot \nabla \overline{p}(\mathbf{x}, t) - k\alpha(S)(\mathbf{v} \cdot \nabla S)|\mathcal{V}|\overline{p}. \tag{10.4.49}$$

It can be checked that the right-hand side is orthogonal to the constant function, so that by the Fredholm alternative theorem

$$p_1 = \mathbb{L}_0^\dagger \left( \mathbf{v} \cdot \nabla \overline{p} - k\alpha(S)(\mathbf{v} \cdot \nabla S)|\mathcal{V}|\overline{p} \right). \tag{10.4.50}$$

The $O(\varepsilon^2)$ equation takes the form

$$\mathbb{L}_0 p_2 = \frac{\partial \overline{p}}{\partial t} + \mathbf{v} \cdot \nabla p_1, \tag{10.4.51}$$

since we take $\overline{p}_1 = 0$. Substituting for $p_1$ and integrating with respect to $\mathbf{v}$ then gives

$$|\mathcal{V}| \frac{\partial \overline{p}}{\partial t} + \nabla \cdot \int_{\mathcal{V}} \mathbf{v} \mathbb{L}_0^\dagger (\mathbf{v} \cdot \nabla \overline{p}) d\mathbf{v} - k|\mathcal{V}|\nabla \cdot \left( \int_{\mathcal{V}} \mathbf{v} \mathbb{L}_0^\dagger (\mathbf{v} \cdot \nabla S) d\mathbf{v} \right) \alpha(S)\overline{p} = 0.$$

We thus obtain the chemotaxis equation

$$\frac{\partial \overline{p}}{\partial t} = \sum_{i,j} \frac{\partial}{\partial x_i} D_{ij} \left[ \frac{\partial \overline{p}}{\partial x_j} - k|\mathcal{V}|\alpha(S) \frac{\partial S}{\partial x_j} \overline{p} \right], \tag{10.4.52}$$

with the diffusion tensor defined in equation (10.4.44) for $\mathbb{L} \to \mathbb{L}_0$. In the particular case of a Pearson random walk, with $T(\mathbf{v}, \mathbf{v}') = 1/|\mathcal{V}|$ and $D = v_0^2 \mathbf{I}/kd$, we recover the classical isotropic chemotaxis model

$$\frac{\partial \overline{p}}{\partial t} = \nabla \cdot [D_0 \nabla \overline{p} - \chi(S)\overline{p}\nabla S], \quad D_0 = \frac{v_0^2}{kd}, \quad \chi(S) = \frac{|\mathcal{V}|\alpha(S)v_0^2}{d}. \tag{10.4.53}$$

*Remark 10.3.* In cases where bacteria (or eukaryotic cells) secrete their own chemoattractant, the above equation is coupled to a diffusion equation for the spatiotemporal evolution of the secreted chemical. The best known example of such a system is the Keller-Segel model [499, 500], which is widely used to study collective effects such as waves and patterns in cell populations [655]:

$$\frac{\partial n}{\partial t} = \nabla \cdot (D\nabla n - n\chi(c)\nabla c), \quad \frac{\partial c}{\partial t} = \nabla^2 c - \gamma c + \sigma n. \tag{10.4.54}$$

Here $n(\mathbf{x},t)$ is the cell density and $c(\mathbf{x},t)$ is the concentration of chemoattractant released by the cells themselves. The factor $\chi(c)$ is known as the chemotactic sensitivity function. This type of model has also attracted considerable interest with mathematicians due to the fact that it can exhibit interesting behaviors such as blow-up in finite times [723], see Ex. 10.11.

## 10.5 Physical limits of chemical sensing

One of the most remarkable features of living cells is that they can sense changes in their environment with high precision [78, 868]. For example, receptors in the human visual system can detect single photons [762] (see Sect. 10.2), the olfactory system of insects can detect single molecules [88], bacteria swimming in a chemo-tactic concentration gradient can respond to the binding and unbinding of only a limited number of molecules [64, 879], and eukaryotic cells can respond to $O(10)$ differences in the number of molecules between the front and the back of the cell [894]. Moreover, experimental studies suggest that the precision of the embryonic development of the fruit fly *Drosophila* almost matches the limit set by the available number of regulatory proteins [257, 281, 375, 879]. This has motivated many theoretical studies concerned with investigating what sets the fundamental limit to the precision of chemical sensing both at the single-cell level [25, 26, 64, 76, 77, 274, 485, 637], and at the multicellular level [296, 644, 779, 831].

Cells measure chemical concentrations via the binding of ligand molecules to receptor proteins (Chap. 3), which can be located either at the cell surface or inside the cell. These measurements are inherently noisy, particularly at low concentrations. There are two major sources of noise: (i) the stochastic transport of ligand molecules to the receptor via diffusion, and (ii) the stochastic binding of ligand molecules to the receptor after they have arrived at its surface. The presence of fluctuations sets fundamental physical limits on the precision of cell sensing, an issue that was first addressed in a seminal paper by Berg and Purcell [64].

Berg and Purcell assumed that, irrespective of the specific details of the biological sensory mechanism, one can essentially treat a cell as a device that counts molecules entering its local neighborhood. They then used a simple scaling argument to estimate the precision of sensing. Suppose that there is a uniform concentration $c_0$ and that the cell has a characteristic length scale $a$. It follows that the average number of molecules within a cell volume scales as $\bar{n} \sim c_0 a^3$. However, since the molecules are subject to diffusion, there will be Poisson-like fluctuations about the mean with $\sigma_n^2 = \bar{n}$. Equating the precision or relative error with the coefficient of variation $\sigma_n/\bar{n}$, it follows that $\sigma_n/\bar{n} \sim 1/\sqrt{\bar{n}} = 1/\sqrt{c_0 a^3}$. Berg and Purcell then reasoned that the variance can be reduced if the cell makes multiple measurements of the molecule number, provided that there is sufficient time between measurements to ensure statistical independence. The associated characteristic time scale is $a^2/D$, which is the approximate time for a molecule to diffuse out of the cell volume. Therefore, given a total sampling time $T$, the cell can make $DT/a^2$ independent measurements, thus

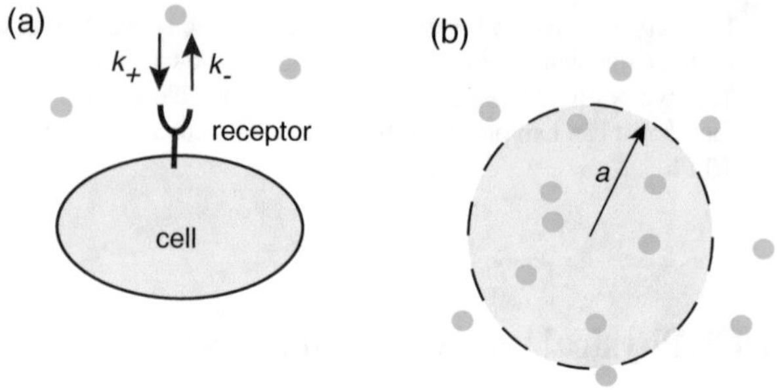

(a)

$k_+ \downarrow \uparrow k_-$

receptor

cell

(b)

a

Fig. 10.28: Two models of chemical sensing. (a) Binding/unbinding of a molecule to a receptor on the surface of a cell. (b) "The perfect instrument."

reducing the variance to $\sigma_n^2 \sim c_0 a^3/(DT/a^2) = c_0 a^5/DT$. The precision will now be

$$\frac{\sigma_n}{\bar{n}} \sim \frac{1}{\sqrt{aT c_0 D}}. \tag{10.5.1}$$

We see that sensory precision increases (the CV decreases) with cell size $a$, measurement time $T$, diffusivity $D$, and background concentration $c_0$.

In this section, we describe some specific models and mathematical methods for investigating the fundamental limits of chemical sensing. These yield a variety of prefactors and additional terms in the formula for sensory precision, but the basic scaling of equation (10.5.1) is preserved. We begin with two alternative formulations considered by Berg and Purcell [64], see Fig. 10.28. The first involves calculating the autocorrelation function for the occupancy of a single receptor, while the second treats a cell as a "perfect instrument" that counts the number of molecules inside its volume. In the latter case, the precision is calculated using Langevin dynamics and linear response theory [296]. We end the section by reviewing an alternative approach to analyzing the limits of chemical sensing based on so-called maximum likelihood estimation [26, 274, 637]. This method yields a more precise estimate of the concentration than the classical Berg–Purcell limit, although it remains to be established how a cell could perform such an estimate. Maximum likelihood is then used to extend the theory to the sensing of temporal concentration gradients [143, 273, 405, 637]. The latter is particularly relevant within the context of bacterial chemotaxis, since bacteria are too small to detect spatial gradients across their bodies. Instead, they detect spatial gradients indirectly by measuring temporal changes in concentration along its swimming trajectory. Direct spatial gradient sensing in Eukaryotic cells will be considered in Sect. 10.6.

### 10.5.1 Berg–Purcell limit for a single receptor

We begin by reviewing the original Berg–Purcell limit for the precision with which the external concentration $c$ of some ligand such as a chemoattractant can be inferred from the time-averaged occupancy of a single receptor embedded in the cell membrane [64, 65]. In order to develop the basic theory, we initially ignore the effects of diffusion by focusing on noise due to binding reactions; see also [386, 918]. Denote the time-dependent state of a receptor by $N(t)$ with $N(t) = 1$ ($N(t) = 0$) if the receptor's binding site is occupied (unoccupied) by a single-ligand molecule. In thermodynamic equilibrium at some given concentration $c_0$, the average receptor occupancy $\bar{n}$ is

$$\bar{n} = \frac{c_0}{c_0 + K_d}, \tag{10.5.2}$$

where $K_d = k_-/k_+$ is the dissociation constant for ligand binding/unbinding (see Sect. 3.2). If $\bar{n}$ were known then one could determine the uniform concentration $c_0$ according to

$$c_0 = \frac{K_d \bar{n}}{1 - \bar{n}}. \tag{10.5.3}$$

Clearly the instantaneous state $N(t)$ of the receptor is a poor estimator for $c$, since the instantaneous guess $C(t) = K_d N(t)/(1 - N(t))$ fluctuates between zero and $\infty$, and has infinite variance. Following Berg and Purcell [64, 65], we will assume that a cell estimates the concentration by averaging the occupancy state of a receptor over some period $T$,

$$n_T = \frac{1}{T} \int_0^T N(t)dt, \tag{10.5.4}$$

and taking $c_T = K_d n_T/(1 - n_T)$.

The error in the estimate of the concentration $c$ can be expressed in terms of the variance $\sigma_T^2 = \mathrm{Var}[n_T]$ according to

$$\frac{\delta c}{c_0} = \frac{1}{c_0}\left(\frac{dc_T}{dn_T}\right)_{n_T = \bar{n}} \delta n_T = \frac{\sigma_T}{\bar{n}(1 - \bar{n})}. \tag{10.5.5}$$

By definition, $\sigma_T^2$ takes the form

$$\sigma_T^2 := \frac{1}{T^2}\int_0^T\int_0^T \langle N(t)N(s)\rangle\,ds\,dt - \bar{n}^2 = \frac{2}{T^2}\int_0^T\int_0^t \langle N(t)N(s)\rangle\,ds\,dt - \bar{n}^2$$
$$= \frac{2}{T^2}\int_0^T\int_0^t \langle N(t)N(t - \tau)\rangle\,d\tau\,dt - \bar{n}^2 = \frac{2}{T^2}\int_0^T\int_0^t C(\tau)\,d\tau\,dt, \tag{10.5.6}$$

where $C(\tau)$ is the correlation function

$$C(\tau) := \langle N(t + \tau)N(t)\rangle - \bar{n}^2. \tag{10.5.7}$$

Under the assumption that $T \gg \tau_c$, where $\tau_c$ is the relaxation or correlation time of the stochastic process, we can take the upper limit of the $\tau$- integral to be $\infty$ so that

$$\sigma_T^2 = \frac{2}{T}\int_0^\infty C(\tau)d\tau = \frac{S_n(0)}{T}. \tag{10.5.8}$$

The final expression follows from the relationship between the correlation function and the power spectrum, see Sect. 2.2,

$$S_n(\omega) = \int_{-\infty}^{\infty} C(\tau) e^{i\omega\tau} d\tau = 2\mathrm{Re}\left[\int_0^{\infty} C(\tau) e^{i\omega\tau} d\tau\right].$$

The correlation time is then defined according to $2\tau_c \sigma_n^2 = S(0)$. Finally, combining equations (10.5.5) and (10.5.8) yields the general result

$$\frac{\delta c}{c_0} = \frac{1}{\bar{n}(1-\bar{n})} \sqrt{\frac{2\tau_c \sigma_n^2}{T}}. \tag{10.5.9}$$

The root mean-square (RMS) error for the estimated concentration will depend on the particular model for the occupancy $N(t)$. In the case of simple binding/unbinding reactions for a single receptor, $N(t)$ is a two-state Markov process, also known as a dichotomous noise process. Hence, its correlation function $C(\tau)$ is given by (see Sect. 5.3)

$$C(\tau) := \langle N(t+\tau)N(t)\rangle - \bar{n}^2 = \bar{n}(1-\bar{n})e^{-|\tau|/\tau_c}, \tag{10.5.10}$$

where

$$\tau_c := \frac{1}{k_+c_0 + k_-} = \frac{1-\bar{n}}{k_-} = \frac{\bar{n}}{k_+c_0} \tag{10.5.11}$$

is the relaxation or correlation time of the binding/unbinding process. Substituting for $C(\tau)$ into equation (10.5.6) gives

$$\begin{aligned}
\sigma_T^2 &= \frac{2\bar{n}(1-\bar{n})}{T^2} \int_0^T \int_0^t e^{-\tau/\tau_c} d\tau dt \\
&= \frac{2\bar{n}(1-\bar{n})\tau_c}{T}\left[1 - \frac{\tau_c}{T}(1-e^{-T/\tau_c})\right].
\end{aligned} \tag{10.5.12}$$

It follows that (for $T \gg \tau_c$)

$$\sigma_T^2 = \frac{2\bar{n}(1-\bar{n})\tau_c}{T} = \frac{2\tau_c}{T}\frac{c_0 K_d}{(c_0+K_d)^2}. \tag{10.5.13}$$

Finally, we have

$$\frac{\delta c}{c_0} = \sqrt{\frac{2\tau_c}{\bar{n}(1-\bar{n})T}} = \sqrt{\frac{2}{k_+c_0(1-\bar{n})T}}. \tag{10.5.14}$$

This is the Berg–Purcell limit when diffusion effects are ignored ($D \to \infty$). Note that a major feature of the model in the fast diffusion limit is that once a ligand unbinds it immediately returns to the bulk so that rebinding effects are negligible, and each binding event can be treated as independent. Since $1 - \bar{n}$ is the probability that the receptor is in an unbound state, we can interpret $[k_+c_0(1-\bar{n})]^{-1}$ as the mean time $\tau_u$ between binding events (average duration of unbound intervals) so that $T/\tau_u$ is the mean number $\mathcal{N}$ of independent measurements of the concentration $c$ in the time interval $T$. Therefore, we can write

$$\left(\frac{\delta c}{c_0}\right)^2 = \frac{2}{\mathcal{N}}. \tag{10.5.15}$$

The dependence on $\mathcal{N}^{-1}$ is precisely what one would expect from Poisson counting noise.

The original analysis of Berg and Purcell proceeded along similar lines, but focused on the diffusion-limited regime for which $k_{\pm} \rightarrow \infty$ for fixed $K_d$, and $D$ is finite [64, 65]. Hence, when a molecule encounters the receptor it is immediately absorbed, that is, noise due to binding reactions can be ignored. From Sect. 6.3, we know that the effective flux into a spherical target of radius $a$ is $4\pi Dac_0$, where $c_0$ is the equilibrium concentration of ligand. The effective binding rate is thus $k_{\mathrm{on}} = 4\pi Da$. The effective dissociation or escape rate $k_{\mathrm{off}}$ is then obtained by assuming that the detailed balance condition (10.5.3) is preserved, which implies that $k_{\mathrm{off}} = 4\pi DaK_d$. One can then proceed as before using the two-state receptor model, with $k_+ c$ replaced by $4\pi Dac_0$, to obtain the Berg–Purcell limit

$$\frac{\delta c}{c_0} = \sqrt{\frac{2}{4\pi Dac_0(1-\bar{n})T}}. \tag{10.5.16}$$

This is also consistent with the general form (10.5.15), since the number of counts that can be made in time $T$ is given by the number of molecules in the vicinity of the receptor, which is of order $c_0 a^3$, multiplied by the number of times diffusion renews these molecules, $T/\tau$, where $\tau \sim a^2/D$. That is, $\mathcal{N} \sim aDTc_0(1-\bar{n})$.

Berg and Purcell argued that their result also holds for reactions that are not diffusion-limited (finite $k_+$) since, if a ligand molecule fails to bind then it will rapidly keep re-colliding with the receptor until it does eventually bind. Such a process could be captured by rescaling the radius $a$. However, this ignores the possibility that after unsuccessfully binding, the specific ligand molecule diffuses back into the bulk, and a different ligand molecule subsequently binds. Similarly, a ligand molecule that has just dissociated from the receptor could either rapidly rebind or diffuse away into the bulk. Recently, a detailed study of the Berg–Purcell problem outside the diffusion-limited regime has been carried out by Kaizu et al. [485] using the theory of reversible diffusion-dependent reactions [2], who show that there are now two contributions to the RMS error:

$$\frac{\delta c}{c_0} = \sqrt{\frac{1}{2\pi Dac_0(1-\bar{n})T} + \frac{2}{k_+ c_0(1-\bar{n})T}}. \tag{10.5.17}$$

The first term recovers the fundamental limit of Berg and Purcell, whereas the second term takes into account the variability that results from the receptor–ligand binding kinetics; the latter vanishes in the limit $k_+ \rightarrow \infty$. A similar result has been obtained by Bialek and Setayeshgar [76] using a very different approach, which is based on a version of the fluctuation–dissipation (FD) theorem for chemical reactions to determine the spectrum $S_n(0)$ in equation (10.5.8):

$$\frac{\delta c}{c_0} = \sqrt{\frac{1}{\pi D a c_0 T} + \frac{2}{k_+ c_0 (1 - \bar{n}) T}}. \tag{10.5.18}$$

The contribution to uncertainty from binding kinetics agrees with the Kaizu et al. result [485], but the contribution from diffusion differs from both the latter and Berg–Purcell. Numerical simulations suggest that equation (10.5.17) is more accurate [485]. On the other hand, the FD theorem provides a relatively simple, intuitive method for addressing the physical limits of biochemical signaling, without requiring details concerning the underlying stochastic processes. If the latter are known, then linear response theory (Box 10A) can be used to calculate the spectrum; see below.

## 10.5.2 The perfect instrument

The second model considered by Berg and Purcell [64] consists of a spherical cell that is permeable to molecules and counts the number within its volume. Rather than following their original derivation of the precision or relative error, we use an alternative approach based on Langevin dynamics and linear response theory. The latter is more easily extended to cell sensing in multicellular systems [296, 644]. The basic idea is to include a spatiotemporal white noise term in the diffusion equation:

$$\frac{\partial c(\mathbf{x}, t)}{\partial t} = D \nabla^2 c(\mathbf{x}, t) + \eta(\mathbf{x}, t), \tag{10.5.19}$$

with $\langle \eta(\mathbf{x}, t) \rangle = 0$ and

$$\langle \eta(\mathbf{x}, t) \eta(\mathbf{x}', t') \rangle = 2 D c_0 \delta(t - t') \nabla_x \cdot \nabla_{x'} \delta^3(\mathbf{x} - \mathbf{x}'). \tag{10.5.20}$$

Here $c_0$ is the mean concentration inside the cell. The structure of the noise $\eta(\mathbf{x}, t)$ can be derived by considering the chemical master equation of a spatially discretized diffusion process, in which hopping between neighboring points on the spatial lattice are treated as chemical reactions. Carrying out a system-size expansion of the resulting diffusion master equation then yields an FP equation and an associated Langevin equation. Formally taking the continuum limit of the latter then yields equation (10.5.20); see [333]. (Treating spatially discrete diffusion as a set of hopping reactions plays a major role in the analysis of stochastic reaction–diffusion processes; see Chap. 16.)

We wish to determine how fluctuations in the concentration $c$ map to fluctuations in the total number of molecules in the cell, after averaging over a time $T$:

$$n_T = \frac{1}{T} \int_0^T n(t) \, dt, \quad n(t) = \int_V c(\mathbf{x}, t) \, d\mathbf{x}. \tag{10.5.21}$$

Setting $c(\mathbf{x}, t) = c_0 + \delta c(\mathbf{x}, t)$ and $n(t) = c_0 V + \delta n(t)$, we have

$$n_T = Vc_0 + \frac{1}{T}\int_0^T \delta n(t)dt. \tag{10.5.22}$$

Clearly $\bar{n}_T = Vc_0$ so that the variance of $n_T$ is given by

$$\sigma_T^2 = \frac{1}{T^2}\int_0^T\int_0^T \langle \delta n(t)\delta n(t')\rangle dt'dt = \frac{1}{T}\lim_{\omega\to 0} S_n(\omega), \tag{10.5.23}$$

where $S_n(\omega)$ is the power spectrum of $n$, that is,

$$S_n(\omega) = \int_{-\infty}^{\infty} \langle \delta\hat{n}(\omega)\delta\hat{n}^*(\omega')\rangle \frac{d\omega'}{2\pi}, \quad \delta\hat{n}(\omega) = \int_{-\infty}^{\infty} e^{i\omega t}\delta n(t)dt. \tag{10.5.24}$$

The last equality in equation (10.5.23) is obtained along identical lines to the derivation of equation (10.5.8). In order to find the spectrum $S_n(\omega)$, we substitute $c = c_0 + \delta c$ into equation (10.5.19) and Fourier transform with respect to $t$ and $\mathbf{x}$:

$$-i\omega\delta\hat{c}(\mathbf{k},\omega) = -Dk^2\delta\hat{c}(\mathbf{k},\omega) + \hat{\eta}(\mathbf{k},\omega), \tag{10.5.25}$$

where

$$\delta\hat{c}(\mathbf{k},\omega) = \int_{-\infty}^{\infty}\int_V e^{i\omega t}e^{i\mathbf{k}\cdot\mathbf{x}}\delta c(\mathbf{x},t)d\mathbf{x}dt.$$

Rearranging (10.5.25) gives

$$\delta\hat{c}(\mathbf{k},\omega) = \frac{\hat{\eta}(\mathbf{k},\omega)}{Dk^2 - i\omega}.$$

Hence,

$$\delta\hat{n}(\omega) = \int_V \left[\int \delta\hat{c}(\mathbf{k},\omega)e^{-i\mathbf{k}\cdot\mathbf{x}}\frac{d\mathbf{k}}{(2\pi)^3}\right]d\mathbf{x} = \int_V \eta(\mathbf{x},\omega)d\mathbf{x}, \tag{10.5.26}$$

where

$$\eta(\mathbf{x},\omega) = \int \frac{e^{-i\mathbf{k}\cdot\mathbf{x}}\eta_c(\mathbf{k},\omega)}{Dk^2 - i\omega}\frac{d\mathbf{k}}{(2\pi)^3}. \tag{10.5.27}$$

It follows that the power spectrum for $n(t)$ can be written as

$$S_n(\omega) = \int_{-\infty}^{\infty}\left[\int_V\int_V \langle\eta(\mathbf{x},\omega)\eta^*(\mathbf{x}',\omega')\rangle d\mathbf{x}'d\mathbf{x}\right]\frac{d\omega'}{2\pi}. \tag{10.5.28}$$

In Fourier space, equation (10.5.20) becomes

$$\langle\eta(\mathbf{k},\omega)\eta^*(\mathbf{k}',\omega')\rangle = 2Dc_0k^2[2\pi\delta(\omega-\omega')][(2\pi)^3\delta^3(\mathbf{k}-\mathbf{k}')].$$

Therefore,

$$\langle \eta(\mathbf{x}, \omega) \eta^*(\mathbf{x}', \omega') \rangle = \int \frac{\langle \eta(\mathbf{k}, \omega) \eta^*(\mathbf{k}', \omega') \rangle}{(Dk^2 - i\omega)(Dk'^2 + i\omega')} e^{i(\mathbf{k}' \cdot \mathbf{x}' - \mathbf{k} \cdot \mathbf{x})} \frac{d\mathbf{k} \, d\mathbf{k}'}{(2\pi)^6}$$

$$= 2c_0[2\pi\delta(\omega - \omega')] \int \frac{Dk^2}{(Dk^2)^2 + \omega^2} e^{i\mathbf{k} \cdot (\mathbf{x}' - \mathbf{x})} \frac{d\mathbf{k}}{(2\pi)^3}.$$

Substituting into equation (10.5.28) then yields the result

$$S_n(\omega) = \int_V \int_V \left[ \int \frac{2c_0 Dk^2}{(Dk^2)^2 + \omega^2} e^{i\mathbf{k} \cdot (\mathbf{x}' - \mathbf{x})} \frac{d\mathbf{k}}{(2\pi)^3} \right] d\mathbf{x}' \, d\mathbf{x}. \qquad (10.5.29)$$

The integral in the limit $\omega \to 0$ can be evaluated explicitly using spherical polar coordinates. Assuming that we can change the order of integration,

$$S_n(0) = \frac{2c_0}{D} \int \frac{1}{k^2} \left[ \int_0^a \int_\Omega r^2 e^{-i\mathbf{k} \cdot \mathbf{x}} dr \, d\Omega \right] \left[ \int_0^a \int_{\Omega'} r'^2 e^{i\mathbf{k}' \cdot \mathbf{x}'} dr' \, d\Omega' \right] \frac{d\mathbf{k}}{(2\pi)^3},$$

where $d\Omega = \sin\theta \, d\theta \, d\phi$ is the infinitesimal solid angle on the sphere and $a$ is the cell radius. Using properties of spherical Bessel functions, it can be shown that

$$\int_\Omega e^{\pm i\mathbf{k} \cdot \mathbf{x}} d\Omega = 4\pi j_0(kr).$$

Therefore,

$$S_n(0) = \frac{2c_0}{D} \int_0^\infty \frac{1}{k^2} \left[ \int_0^a 4\pi r^2 j_0(kr) dr \right]^2 \frac{4\pi k^2}{(2\pi)^3} dk$$

$$= \frac{16c_0}{D} \int \frac{1}{k^6} \left[ \int_0^{ka} z^2 j_0(z) dz \right]^2 dk = \frac{16\pi c_0 a^5}{15D}. \qquad (10.5.30)$$

Finally, from equation (10.5.23) we find that the precision is given by

$$\frac{\sigma_T}{\bar{n}_T} = \frac{3}{5\pi} \frac{1}{aT c_0 D}. \qquad (10.5.31)$$

This recovers the result originally obtained by Berg and Purcell [64].

### 10.5.3  Maximum likelihood estimation and the chemical sensing limit

Maximum likelihood estimation is a statistical method for fitting a mathematical model to data [493] (see Box 10C). That is, given a particular data set and an underlying parameterized model, maximum likelihood determines the values of the parameters that are most likely given the data. In the case of a single receptor sensing a fixed local concentration gradient $c_0$, the data is the time series $\Gamma = \{t_i^+, t_i^-\}$ of duration $\mathcal{T}$, where $t_i^+$ is the $i$th ligand binding event and $t_i^-$ is the $i$th ligand unbinding event, while $c_0$ is the model parameter [274], see Fig. 10.29. Following the classical Berg–Purcell theory, it is assumed that diffusion is sufficiently fast to remove recently unbound ligand molecules from the neighborhood of the receptor, so that the effects of ligand rebinding can be ignored. We will calculate the maximum likelihood estimate for $c_0$, following the analysis of Endres and Wingreen [274].

Given a particle concentration $c_0$, the probability density for a particular time series to occur is

$$P(\Gamma|c_0) = \prod_i p_b(t_i^+, t_i^-) p_-(t_i^-) p_u(t_i^-, t_{i+1}^+) p_+(t_{i+1}^+). \tag{10.5.32}$$

The first factor on the right-hand side is the probability that the receptor remains bound by a ligand from $t_i^+$ to $t_i^-$,

$$p_b(t_i^+, t_i^-) = e^{-k_-(t_i^- - t_i^+)},$$

and the third factor is the probability that the receptor remains unbound from $t_-^i$ to $t_{i+1}^+$,

$$p_u(t_i^+, t_i^-) = e^{-k_+ c_0(t_{i+1}^+ - t_i^-)}.$$

The other two factors are the probability density of binding in the infinitesimal time interval $[t_{i+1}^+, t_{i+1}^+ + d\tau]$, $p_+(t_{i+1}^+) = k_+ c_0$, and the probability density of unbinding in the infinitesimal time interval $[t_i^-, t_i^- + d\tau]$, $p_-(t_i^-) = k_-$. It follows that

$$P(\Gamma|c) \approx e^{-k_- T_b} e^{-k_+ c_0 T_u} k_-^n (k_+ c)^n, \tag{10.5.33}$$

where $n$ is the number of binding or unbinding events over the integration time $T$. (The number of these two types of events may differ by at most 1, which can be ignored when $n \gg 1$.) The total bound and unbound time intervals satisfy

$$T_b = \sum_{i=1}^n (t_i^- - t_i^+), \quad T_u = \sum_{i=1}^n (t_{i+1}^+ - t_i^-).$$

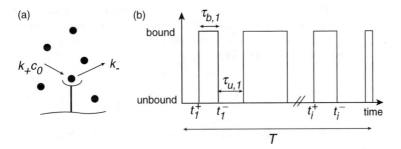

Fig. 10.29: Schematic diagram of ligand–receptor binding. See text for details.

The maximum likelihood estimate (MLE) $\hat{c}$ for $c$ is obtained by the maximizing the log likelihood $\ln P(\Gamma|c)$ with respect to $c$:

$$0 = \frac{\partial \ln P(\Gamma|c_0)}{\partial c_0} = -k_+ T_u + \frac{n}{c_0},$$

which implies that

$$\hat{c} = \frac{n}{k_+ T_u}. \tag{10.5.34}$$

Hence, the best estimate of the concentration is determined by the average duration of unbound intervals $T_u/n$.

The uncertainty in the MLE is given by the variance $(\delta\hat{c})^2$. To simplify the analysis, we assume that $n$ is fixed by allowing the total integration $\mathcal{T}$ time to vary. An upper limit of the variance is given by the Cramer–Rao bound (see Box 10C), which in the case of a single parameter is given by the inverse of the second derivative of $\log P$ with respect to $c$, averaged over the probability density of the time series. Since the MLE is unbiased in the case of long time series, the Cramers–Rao bound is realized so that

$$\frac{(\delta\hat{c})^2}{c_0^2} = -\frac{1}{c_0^2}\left\langle \frac{\partial^2 \ln P}{\partial c^2} \right\rangle_{c_0}^{-1} = \frac{1}{n}. \tag{10.5.35}$$

Since we can identify $n$ as the number of independent measurements $\mathcal{N}$, comparison with the Berg–Purcell limit (10.5.15) shows that the MLE is more accurate by a factor of 2. It immediately follows that in the case of diffusion-limited binding for which $k_+ = 4\pi Da$, where $a$ is the receptor radius, we have

$$\frac{\delta c}{\bar{c}} = \sqrt{\frac{1}{4\pi Da\bar{c}(1-\bar{n})\mathcal{T}}}. \tag{10.5.36}$$

One of the useful features of the maximum likelihood approach is that it can be extended to take into account temporal concentration changes [143, 273, 637], such as those that arise during bacterial chemotaxis. Bacteria are too small to detect differences in concentrations across their cell bodies, so they proceed by measuring and comparing concentrations over time along their swimming trajectories. For the sake of illustration, suppose that the concentration has the time-dependent form $c(t) = c_0 + c_1 t$, which could be due to the motion of a cell up a concentration gradient. The likelihood function (10.5.33) becomes

$$P(\Gamma|c_0, c_1) = -e^{k_- T_b} \exp\left(-k_+ \sum_i \int_{t_i^-}^{t_{i+1}^+} c(t)dt\right) k_-^n \prod_{i=1}^n k_+ c(t_i^+). \tag{10.5.37}$$

The concentration and gradient parameters $c_0, c_1$ can then be determined using MLE, as outlined in [637].

---

**Box 10C. Maximum likelihood estimation.**

---

We briefly review the theory of maximum likelihood estimation. See the book by Kay [493] for more details. Let $X_1, \ldots, X_n$ be i.i.d. random variables with probability density function $f(x_i; \theta)$ where $\theta$ denotes a set of $k$ parameters $\theta_1, \ldots, \theta_k$. (One can view the $X_i$ as sample data points.) From independence, the joint density is $f(\mathbf{x}; \theta) = \prod_{i=1}^{n} f(x_i; \theta)$, where $\mathbf{x} = (x_1, \ldots x_n)$. The likelihood function is defined as the joint density treated as a function of the parameters of $\theta$,

$$L(\theta|\mathbf{x}) = f(\mathbf{x}; \theta) = \prod_{i=1}^{n} f(x_i; \theta). \tag{10.5.38}$$

Note that as a function of $\theta$, the likelihood function is not a probability density function, since

$$\int \cdots \int L(\theta|\mathbf{x}) d\theta_1 \cdots d\theta_k \neq 1.$$

Suppose that we have a random sample of $n$ data points from the density $f(x_i; \theta)$ and we would like to estimate $\theta$. The maximum likelihood estimator (MLE), denoted by $\widehat{\theta}$, is defined to be the value of $\theta$ that maximizes the likelihood function. It turns out that it is usually much easier to maximize the log-likelihood function $\ln L(\theta|\mathbf{x})$, which is equivalent to maximizing $L(\theta|\mathbf{x})$ since $\ln(\cdot)$ is monotonic. Hence, we define $\widehat{\theta}$ as the value of $\theta$ that solves

$$\max_{\theta} \ln L(\theta|\mathbf{x}), \quad \ln L(\theta|\mathbf{x}) = \sum_{i=1}^{n} \ln f(x_i; \theta).$$

In order to ensure that the maximization procedure is well defined, we impose certain sufficiency conditions on the density $f(x_i; \theta)$:

1. The support of the random variable $X$, $S_X = \{x : f(x; \theta) > 0\}$, is independent of $\theta$.
2. $f(x; \theta)$ is at least three times differentiable with respect to $\theta$.
3. The true value of $\theta$ lies in a compact set.

Under these conditions, the density $f$ is said to be regular, and we can determine the MLE by differentiation:

$$0 = \left. \frac{\partial \ln L(\theta|\mathbf{x})}{\partial \theta_a} \right|_{\theta = \widehat{\theta}}, \quad a = 1, \ldots, k. \tag{10.5.39}$$

**Example: Gaussian sampling.** In the case of Gaussian sampling, $\theta = (\mu, \sigma^2)$ and

$$f(x_i; \theta) = \frac{1}{\sqrt{2\pi\sigma^2}} \exp\left(-(x_i - \mu)^2 / 2\sigma^2\right).$$

The corresponding likelihood function is

$$L(\theta|\mathbf{x}) = \frac{1}{(2\pi\sigma^2)^{n/2}} \exp\left(-\frac{1}{2\sigma^2}\sum_{i=1}^{n}(x_i - \mu)^2\right).$$

It follows that

$$\ln L(\theta|\mathbf{x}) = -\frac{n}{2}\ln(2\pi) - \frac{n}{2}\ln(\sigma^2) - \frac{1}{2\sigma^2}\sum_{i=1}^{n}(x_i - \mu)^2. \qquad (10.5.40)$$

Therefore,

$$\frac{\partial \ln L(\theta|\mathbf{x})}{\partial \mu} = \frac{1}{\sigma^2}\sum_{i=1}^{n}(x_i - \mu),$$

$$\frac{\partial \ln L(\theta|\mathbf{x})}{\partial \sigma^2} = -\frac{n}{2\sigma^2} + \frac{1}{2\sigma^4}\sum_{i=1}^{n}(x_i - \mu)^2.$$

We thus obtain the following equations for the MLE:

$$0 = \frac{1}{\widehat{\sigma}^2}\sum_{i=1}^{n}(x_i - \widehat{\mu}),$$

$$0 = -\frac{n}{2\widehat{\sigma}^2} + \frac{1}{2\widehat{\sigma}^4}\sum_{i=1}^{n}(x_i - \widehat{\mu})^2.$$

Solving the first equation for $\widehat{\mu}$ gives

$$\widehat{\mu} = \frac{1}{n}\sum_{i=1}^{n}x_i = \bar{x}. \qquad (10.5.41)$$

Hence, the sample average is the MLE for $\mu$. Setting $\widehat{\mu} = \bar{x}$ in the second equation for $\widehat{\sigma}^2$ then shows that

$$\widehat{\sigma}^2 = \frac{1}{n}\sum_{i=1}^{n}(x_i - \bar{x})^2. \qquad (10.5.42)$$

(Note that $\widehat{\sigma}^2$ is not the sample variance, since the latter involves dividing by $n-1$ rather than $n$.)

**Information matrix.** Let us now introduce the score vector $S(\theta|\mathbf{x})$ with components

$$S_a(\theta|\mathbf{x}) = \frac{\partial \ln L(\theta|\mathbf{x})}{\partial \theta_a}, \quad a = 1,\dots,k. \qquad (10.5.43)$$

The MLE is thus the solution to $S(\widehat{\theta}|\mathbf{x}) = 0$. Under the assumption of independent sampling, we have

$$S_a(\theta|\mathbf{x}) = \sum_{i=1}^{n} S_a(\theta|x_i), \quad S_a(\theta|x_i) = \frac{\partial \ln f(x_i; \theta)}{\partial \theta_a}.$$

Consider the Hessian matrix $H(\theta|\mathbf{x})$ with entries

$$H_{ab}(\theta|\mathbf{x}) = \frac{\partial^2 \ln L(\theta|\mathbf{x})}{\partial \theta_a \partial \theta_b}. \tag{10.5.44}$$

The information matrix for $n$ observations is then defined as

$$I^{(n)}(\theta) = -\mathbb{E}[H(\theta|\mathbf{x})],$$

where expectation is taken with respect to $\mathbf{x}$. Again under the assumption of independent sampling,

$$H_{ab}(\theta|\mathbf{x}) = \sum_{i=1}^{n} \frac{\partial^2 \ln f(x_i; \theta)}{\partial \theta_a \partial \theta_b} = \sum_{i=1}^{n} H_{ab}(\theta|x_i),$$

and thus

$$I^{(n)}(\theta) = -\sum_{i=1}^{n} \mathbb{E}[H(\theta|x_i)] = -n\mathbb{E}[H(\theta|x_i)] = -nI(\theta), \tag{10.5.45}$$

where $I(\theta)$ is the information matrix for a single observation. It turns out that

$$\mathbb{E}[S_a(\theta|x_i)] = 0, \quad \mathbb{E}[S_a(\theta|x_i)S_b(\theta|x_i)] = I_{ab}(\theta|x_i). \tag{10.5.46}$$

The first result follows from

$$\mathbb{E}[S_a(\theta|x_i)] = \int S_a(\theta|x_i)f(x_i; \theta)dx_i = \int \frac{\partial \ln f(x_i; \theta)}{\partial \theta_a} f(x_i; \theta)dx_i$$

$$= \int \frac{\partial f(x_i; \theta)}{\partial \theta_a} dx_i = \frac{\partial}{\partial \theta_a} \int f(x_i; \theta)dx_i = 0.$$

For simplicity, we establish the second result in the case of a single parameter ($k = 1$). First, we have

$$\mathbb{E}[S(\theta|x_i)^2] = \int S(\theta|x_i)^2 f(x_i; \theta)dx_i = \int \left(\frac{\partial \ln f(x_i; \theta)}{\partial \theta}\right)^2 f(x_i; \theta)dx_i$$

$$= \int \frac{1}{f(x_i; \theta)} \left(\frac{\partial f(x_i; \theta)}{\partial \theta}\right)^2 dx_i.$$

From the chain rule,

$$\frac{\partial^2}{\partial \theta^2} \ln f(x_i; \theta) = \frac{\partial}{\partial \theta} \left( \frac{1}{f(x_i; \theta)} \frac{\partial}{\partial \theta} f(x_i; \theta) \right)$$

$$= -\frac{1}{f(x_i; \theta)^2} \left( \frac{\partial}{\partial \theta} f(x_i; \theta) \right)^2 + \frac{1}{f(x_i; \theta)} \frac{\partial^2}{\partial^2 \theta} f(x_i; \theta).$$

Hence,

$$I(\theta) = -\mathbb{E}[H(\theta|x_i)] = -\int \frac{\partial^2}{\partial \theta^2} \ln f(x_i; \theta) dx_i$$

$$= -\int \left[ -\frac{1}{f(x_i; \theta)} \left( \frac{\partial}{\partial \theta} f(x_i; \theta) \right)^2 + \frac{\partial^2}{\partial \theta^2} f(x_i; \theta) \right] dx_i$$

$$= \mathbb{E}[S(\theta|x_i)^2] - \frac{\partial^2}{\partial \theta^2} \int f(x_i; \theta) dx_i = \mathbb{E}[S(\theta|x_i)^2].$$

**Precision of MLE and the Cramer–Rao bound.** The information matrix is directly related to the precision of the MLE, as expressed by the Cramers–Rao inequality, which we now state. Let $X_1, \ldots, X_n$ be iid samples from the regular probability density function $f(x; \theta)$. Let $\hat{\theta}$ be an unbiased estimator of $\theta$, that is, $\mathbb{E}[\hat{\theta}] = \theta$. Then

$$\mathrm{Var}[\hat{\theta}] \geq I^{(n)}(\theta|\mathbf{x})^{-1}, \qquad (10.5.47)$$

where $I^{(n)}(\theta|\mathbf{x}) = -\mathbb{E}[H(\theta|\mathbf{x})$ is the sample information matrix.

In the case of Gaussian sampling, the Hessian matrix with $\theta_1 = \mu$ and $\theta_2 = \sigma^2$ is

$$H(\mu, \sigma^2|x_i) = \begin{pmatrix} -\sigma^{-2} & -\sigma^{-4}(x_i - \mu) \\ -\sigma^{-4}(x_i - \mu) & \sigma^{-4}/2 - \sigma^{-6}(x_i - \mu)^2 \end{pmatrix}.$$

Since

$$\mathbb{E}[(x_i - \mu)] = 0, \quad \mathbb{E}[(x_i - \mu)^2] = \sigma^2,$$

it follows that

$$I(\mu, \sigma^2) = -\mathbb{E}[H(\mu, \sigma^2|x_i)] = \begin{pmatrix} \sigma^{-2} & 0 \\ 0 & \sigma^{-4}/2 \end{pmatrix},$$

and $I^{(n)}(\mu, \sigma^2) = nI(\mu, \sigma^2)$. Hence, the Cramers–Rao lower bound $(B)$ is

$$B = \begin{pmatrix} \sigma^2/n & 0 \\ 0 & 2\sigma^4/n \end{pmatrix}.$$

The MLE $\hat{\mu} = \bar{x}$ is unbiased, since

$$\mathbb{E}[\hat{\mu}] = \frac{1}{n} \sum_{i=1}^{n} \mathbb{E}[x_i] = \frac{1}{n} \sum_{i=1}^{n} \mu = \mu.$$

Moreover,

$$\text{Var}[\widehat{\mu}] = \mathbb{E}[(\widehat{\mu} - \mu)^2] = \frac{1}{n^2}\mathbb{E}[\sum_{i=1}^{n}(x_i - \mu)^2] = \frac{1}{n}\mathbb{E}[(x_i - \mu)^2] = \frac{\sigma^2}{n},$$

that is, the MLE of the mean realizes the Cramer–Rao lower bound. On the other hand,

$$\mathbb{E}[\widehat{\sigma}^2] = \frac{n-1}{n}\sigma^2,$$

which implies that the MLE for the variance is biased, and the Cramer–Rao inequality does not apply.

## 10.6 Spatial gradient sensing in Eukaryotes

In contrast to bacterial cells, eukaryotic cells are sufficiently large so that they can measure the concentration differences across their cell bodies without temporal integration [385, 711]. One of the best characterized model systems is the social amoeba *Dictyostelium discoideum* [753]. Chemoattractants in the surrounding medium are detected by binding to specific G-protein coupled receptors on the cell membrane of *Dictyostelium*. Spatial variations in the chemoattractant concentration result in an asymmetric distribution of ligand-occupied receptors over the cell surface. This then activates multiple second messenger pathways inside the cell that ultimately drives the extension of pseudopods preferentially in the direction of the chemoattractant gradient. However, the receptor signal is inherently noisy due to fluctuations in the binding of ligand to the chemoreceptors (see also Chap. 3), which has been demonstrated using single-molecule imaging [893]. Nevertheless, *Dictyostelium* cells exhibit a very high sensitivity to chemical gradients, being able to detect a 1%−2% difference in chemical concentration across the cell length [326]. In such shallow gradients, the difference in receptor occupancy between the front and back halves of a cell can be as low as 10−30, which naturally raises the issue of how weak signal detection in the presence of a noisy receptor signal is achieved.

### 10.6.1 Physical limits of spatial gradient sensing

We begin by reviewing one particular approach to analyzing the physical limits of spatial gradient sensing in eukaryotes due to Hu *et al.* [435–437, 439]. These authors focus on a single cell in the presence of a non-fluctuating concentration gradient, and explore the effects of intrinsic noise due to the activation/inactivation of surface receptors. It is assumed that the response of all the receptors can

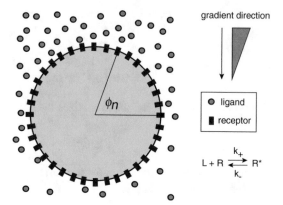

Fig. 10.30: Schematic illustration of the model due to Hu *et al.* [435]. $N$ receptors $R$ are uniformly distributed around the membrane of a circular cell of diameter $L$. The $n$th receptor has angular coordinate $\phi_n = 2\pi n/N$.

be instantaneously pooled. In particular, consider a circular cell of diameter $L$ with $N$ receptors uniformly distributed around the membrane of the cell, see Fig. 10.30. Let $\phi_n = 2\pi n/N$, $n = 1,\ldots,N$, denote the angular coordinate of the $n$th receptor. Based on studies of *Dictyostelium*, the concentration gradient is taken to have an exponential profile such that the local ligand concentration at the $n$th receptor is

$$C_n = C_0 \exp\left(\frac{A}{2}\cos(\phi_n - \theta)\right),$$

where $C_0$ is the background concentration, $A$ is the steepness of the gradient, and $\theta$ is the gradient direction. Note that $A$ determines the percentage concentration change across the cell according to $A = L|\nabla C|/C_0$. Following the analysis of the Ising model of cooperative binding (Sect. 3.2), we assume that each receptor independently switches between two states: an active or bound state ($s_n = +1$), and an inactive or unbound state ($s_n = -1$). (Hu *et al.* use Ising spin notation rather than the occupancy notation $\sigma_n = (s_n + 1)/2$.) Let $\mp\varepsilon_n$ denote the free energy (in units of $k_BT$) associated with the state $s_n = \pm$. It follows that the probability $P_n$ of the $n$th receptor being active is given by the Boltzmann distribution

$$P_{\text{on}} = \frac{e^{\varepsilon_n}}{e^{\varepsilon_n} + e^{-\varepsilon_n}} = \frac{C_n}{C_n + K_d},$$

where $K_d = k_-/k_+$ is the dissociation constant. The final equality follows from the equilibrium law of mass action The free energy $\varepsilon_n$ can thus be expressed in terms of the local concentration:

$$\varepsilon_n = \frac{1}{2}\ln\frac{C_n}{K_d} = \frac{1}{2}\ln\frac{C_0}{K_d} + \frac{A}{4}\cos(\phi_n - \theta). \qquad (10.6.1)$$

The total free energy of a given configuration $\mathbf{s} = (s_1,\ldots,s_N)$ is given by

$$E[\mathbf{s}] = -\sum_{n=1}^{N} s_n \varepsilon_n = -\alpha_0 \sum_{n=1}^{N} s_n - \frac{\alpha_C}{2} \sum_{n=1}^{N} s_n \cos \phi_n - \frac{\alpha_S}{2} \sum_{n=1}^{N} s_n \sin \phi_n, \quad (10.6.2)$$

where

$$\alpha_0 = \frac{1}{2} \ln \frac{C_0}{K_d}, \quad \alpha_C = A \cos \theta, \quad \alpha_S = A \sin \theta. \quad (10.6.3)$$

We have used the identity $\cos(\phi_n - \theta) = \cos\theta \cos\phi_n + \sin\theta \sin\phi_n$. The corresponding Boltzmann–Gibbs distribution for the full system is $P(\mathbf{s}) = Z^{-1} e^{-E[\mathbf{s}]}$, where $Z$ is the partition function,

$$Z = \sum_{s_1 = \pm 1} \cdots \sum_{s_N = \pm 1} e^{-E[\mathbf{s}]} = \prod_{n=1}^{N} \left( e^{\varepsilon_n} + e^{-\varepsilon_n} \right) = \prod_{n=1}^{N} 2 \cosh \left( \alpha_0 + \frac{A}{4} \cos(\phi_n - \theta) \right).$$

*Remark 10.4. Perturbation expansion of Z.* Suppose that $A$ is sufficiently small (shallow concentration gradient) so that we can Taylor expand the following exponentials

$$e^{\alpha_0 + A \cos(\phi_n - \theta)/4} = e^{\alpha_0} \left( 1 + \frac{A}{4} \cos(\phi_n - \theta) + \frac{A^2}{32} \cos^2(\phi_n - \theta) + \ldots \right),$$

$$e^{-\alpha_0 - A \cos(\phi_n - \theta)/4} = e^{-\alpha_0} \left( 1 - \frac{A}{4} \cos(\phi_n - \theta) + \frac{A^2}{32} \cos^2(\phi_n - \theta) + \ldots \right).$$

It follows that

$$\ln Z = \sum_{n=1}^{N} \ln 2 \cosh \left( \alpha_0 + \frac{A}{4} \cos(\phi_n - \theta) \right) = \sum_{n=1}^{N} \ln(2 \cosh \alpha_0)$$

$$+ \sum_{n=1}^{N} \ln \left[ 1 + \frac{A}{4} \frac{\sinh \alpha_0}{\cosh \alpha_0} \cos(\phi_n - \theta) + \frac{A^2}{32} \cos^2(\phi_n - \theta) + \ldots \right].$$

Using $\ln(1 + x) = x - x^2/2 + \ldots$, we have

$$\ln Z = N \ln(2 \cosh \alpha_0) + \sum_{n=1}^{N} \left[ \frac{A}{4} \frac{\sinh \alpha_0}{\cosh \alpha_0} \cos(\phi_n - \theta) + \frac{A^2}{32} \cos^2(\phi_n - \theta) \right.$$

$$\left. - \frac{A^2}{32} \frac{\sinh^2 \alpha_0}{\cosh^2 \alpha_0} \cos^2(\phi_n - \theta) + \ldots \right] = N \ln(2 \cosh \alpha_0)$$

$$+ \sum_{n=1}^{N} \left[ \frac{A}{4} \frac{\sinh \alpha_0}{\cosh \alpha_0} \cos(\phi_n - \theta) + \frac{A^2}{32} \frac{1}{\cosh^2 \alpha_0} \cos^2(\phi_n - \theta) + \ldots \right].$$

The final step is to note that for large $N$, we can replace each sum over $n$ by an integral with respect to a continuous angular coordinate. That is,

$$\sum_{n=1}^{N} f(\phi_n) = \sum_{n=1}^{N} f(2\pi n/N) = \frac{N}{2\pi} \sum_{n=1}^{N} f(\phi_n) \Delta \phi_n \approx \frac{N}{2\pi} \int_0^{2\pi} \cos(\phi) d\phi.$$

Since,

$$\int_0^{2\pi} \cos(\phi - \theta) d\phi = 0, \quad \int_0^{2\pi} \cos^2(\phi - \theta) d\phi = \pi,$$

we have

$$\ln Z = N \ln(2 \cosh \alpha_0) + \frac{NA^2}{64 \cosh^2 \alpha_0} + O(A^4). \quad (10.6.4)$$

The partition function contains all of the statistical information of the system given the gradient parameters $A, \theta$. Of particular interest are the statistical quantities

$$z_0 = \sum_{n=1}^{N} s_n, \quad z_C = \frac{1}{2} \sum_{n=1}^{N} s_n \cos\phi_n, \quad z_S = \frac{1}{2} \sum_{n=1}^{N} s_n \sin\phi_n. \tag{10.6.5}$$

Here $z_0$ is a measure of overall receptor activity, whereas $z_C, z_S$ quantify the asymmetry in receptors states. Equation (10.6.2) shows that the total free energy can be expressed as

$$E[s] = -\alpha_0 z_0 - \frac{1}{2}(\alpha_C z_C + \alpha_S z_S).$$

Hence, from equation (10.6.4),

$$\mathbb{E}[z_{C,S}] = 2\frac{\partial \ln Z}{\partial \alpha_{C,S}} = \frac{\alpha_{C,S} N C_0 K_d}{4(C_0 + K_d)^2} + O(A^3),$$

and

$$\mathrm{Var}[z_{C,S}] = 4\frac{\partial^2 \ln Z}{\partial \alpha_{C,S}^2} = \frac{N C_0 K_d}{2(C_0 + K_d)^2} + O(A^2).$$

We have used the relations $A^2 = \alpha_C^2 + \alpha_S^2$ and

$$\cosh\alpha_0 = \frac{1}{2}\left[e^{\ln\sqrt{C_0/K_d}} + e^{-\ln\sqrt{C_0/K_d}}\right] = \frac{1}{2}\left[\sqrt{C_0/K_d} + \sqrt{K_d/C_0}\right].$$

One also finds that $\mathrm{Cov}[z_C, z_S] = 0$ so that for small $A$, the joint probability density for $z_C$ and $z_S$ is

$$p(z_c, z_S | \alpha_C, \alpha_S) \approx \frac{1}{2\pi\sigma^2}\exp\left[-\frac{(z_C - \mu\alpha_C)^2 + (z_S - \mu\alpha_S)^2}{2\sigma^2}\right], \tag{10.6.6}$$

with

$$\mu = \frac{N}{16\cosh^2\alpha_0} = \frac{N C_0 K_d}{4(C_0 + K_d)^2}, \quad \sigma^2 = 2\mu. \tag{10.6.7}$$

**Estimation of gradient parameters.** We would like to estimate the parameters of the concentration gradient based on an instantaneous sampling of the receptor states. First, rewrite the Gaussian density as

$$p(z'_c, z'_S | \alpha_C, \alpha_S) \approx \frac{\mu^2}{2\pi\sigma^2}\exp\left[-\frac{(z'_C - \alpha_C)^2 + (z'_S - \alpha_S)^2}{2\sigma^2/\mu^2}\right], \tag{10.6.8}$$

where $z'_{C,S} = z_{C,S}/\mu$. The maximum likelihood estimator (MLE) for the means $\alpha_C$ and $\alpha_S$ are then (see Box 10C)

$$\widehat{\alpha}_{C,S} = \frac{z_{C,S}}{\mu}. \tag{10.6.9}$$

These are unbiased estimators, which realize their Cramers–Rao lower bounds,

$$\text{Var}[\hat{\alpha}_{C,S}] = \frac{\sigma^2}{\mu^2} = \frac{2}{\mu} = \frac{8(C_0 + K_d)^2}{NK_dC_0}. \tag{10.6.10}$$

Since $A$ and $\phi$ are smooth functions of $\alpha_{C,S}$, we have the corresponding unbiased MLEs

$$\hat{A} = \sqrt{\hat{\alpha}_C^2 + \hat{\alpha}_S^2} = \mu^{-1}\sqrt{z_C^2 + z_S^2}, \quad \hat{\theta} = \arctan(\hat{\alpha}_S/\hat{\alpha}_C) = \arctan(z_S/z_C). \tag{10.6.11}$$

Moreover, the error in the estimates are [435],

$$\sigma_a^2 := \text{Var}[\hat{A}] = \frac{8(C_0 + K_d)^2}{NK_dC_0}, \quad \sigma_\theta^2 := \text{Var}[\hat{\theta}] = \frac{1}{A^2}\text{Var}[\hat{A}]. \tag{10.6.12}$$

Note that the error in the direction estimate depends on the gradient steepness $A$ according to $\sigma_\theta^2 \sim A^{-2}$. Since $A \sim L$, it follows that larger cells are able to sense the gradient with greater accuracy. Finally, both errors are minimized when $C_0 = K_d$.

There have been various extensions of the above analysis, including short-range cooperative interactions between receptors [435], see Ex. 10.12, and other geometric shapes such as elliptical cells [437]. Here we briefly discuss another extension, namely, the inclusion of temporal integration [435]. In the above analysis, we considered gradient sensing based on a single snapshot of $N$ receptors. However, as we established in Sect. 10.5, greater sensitivity can be achieved if the cell integrates the receptor signals over some time interval $\mathcal{T}$. Following the arguments used in the derivation of equation (10.5.13), the cell can make roughly $\mathcal{T}/2\tau_c$ independent measurements, where $\tau_c = 1/(k_- + k_+C_0)$. The corresponding variance is reduced according to

$$\sigma_{a,\mathcal{T}}^2 \approx \frac{2\tau_c}{\mathcal{T}}\sigma_a^2 = \frac{16}{N\mathcal{T}k_-}(1 + K_d/C_0).$$

## 10.6.2 Local excitation, global inhibition (LEGI) model of adaptation in spatial gradient sensing

Chemotactic cells such as *Dictyostelium* respond to changes in environmental stimuli by activating specific intracellular signaling pathways. However, in the case of spatially uniform changes, the response is only transient, that is, the system relaxes back to prestimulus conditions. This form of adaptation allows cells to ignore the background concentration of a chemoattractant, and simply respond to the gradient. A simple model that can account for both the adaptation and gradient sensing properties exhibited by eukaryotic cells is the so-called local excitation, global inhibition (LEGI) model [553, 706, 711].

Suppose that there exists a signaling molecule that determines the response of a chemotactic cell to changes in the external stimulus, and which can exist in either an active state $R^*$ or an inactive state $R$. Analogous to the Golbeter and Koshland model of phosphorylation–dephosphorylation [351], see Sect. 10.1, it is assumed that conversion between these two states is mediated by two active enzymes $E^*$ and $I^*$, corresponding to excitation and inhibition, respectively. In *Dictyostelium*, the intermediate signaling molecule $R$ can be identified with the lipid phosphatidylinositol bisphosphate (PIP$_2$), whose active form is phosphatidylinositol triphosphate (PIP$_3$). Similarly, the excitatory and inhibitory enzymes correspond to phosphoinositide 3-kinase (PI3K) and PTEN, respectively [468].

The basic reaction scheme for excitation and inhibition is

$$R + E^* \underset{d_1}{\overset{a_1}{\rightleftharpoons}} RE^* \overset{k_1}{\rightarrow} R^* + E^*,$$

$$R^* + I^* \underset{d_2}{\overset{a_2}{\rightleftharpoons}} R^* I^* \overset{k_2}{\rightarrow} R + I^*.$$

Introducing the concentrations $r = [R]$, $r^* = [R^*]$, $x = [E^*]$, $y = [I^*]$, $u = [RE^*]$, and $v = [R^* I^*]$, the corresponding kinetic equations are

$$\frac{dr}{dt} = -a_1 rx + d_1 u + k_2 v, \qquad \frac{du}{dt} = a_1 rx - (d_1 + k_1)u, \qquad (10.6.13a)$$

$$\frac{dr^*}{dt} = -a_2 r^* y + d_2 v + k_1 u, \qquad \frac{dv}{dt} = a_2 r^* y - (d_2 + k_2)v. \qquad (10.6.13b)$$

These equations are supplemented by the conservation equation $R_T = r + r^* + u + v \approx r + r^*$. In contrast to the Goldbeter–Koshland model, it is assumed that both enzymes operate far from saturation so that we can take $[E^*]$ and $[I^*]$ to be the total active enzyme concentrations. Performing a quasi-steady-state approximation, we set $du/dt = 0$ and $dv/dt = 0$ so that equation (10.6.13b) for $dr^*/dt$ reduces to

$$\frac{dr^*}{dt} = -a_2 r^* y + \frac{a_2 d_2}{d_2 + k_2} r^* y + \frac{a_1 k_1}{d_1 + k_1} rx \equiv -k_{-r} r^* y + k_r rx.$$

Activation of the two enzymes is controlled by the local level of receptor occupancy, which is proportional to the local external stimulus $S$. For the moment, we will assume that the latter is uniform so that we can ignore spatial effects. We thus have the additional reactions

$$E + S \underset{k_{-e}}{\overset{k'_e}{\rightleftharpoons}} E^*, \quad I + S \underset{k_{-i}}{\overset{k'_i}{\rightleftharpoons}} I^*,$$

with corresponding kinetic equations

$$\frac{dx}{dt} = -k_{-e} x + k'_e (E_{\text{tot}} - x)S, \qquad \frac{dy}{dt} = -k_{-i} y + k'_i (I_{\text{tot}} - y)S.$$

Finally, it is assumed that $E_{\text{tot}} \gg [E^*]$, $I_{\text{tot}} \gg [I^*]$ so that we have the simplified system of equations

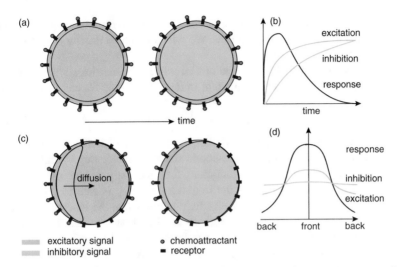

Fig. 10.31: Schematic illustration of the LEGI model. (a,b) Transient response to a uniform increase in the external stimulus. Excitation rises faster than inhibition, but is eventually balanced by the latter so that the response returns to baseline. (c,d) Nonuniform response to a chemical gradient.

$$\frac{dr^*}{dt} = -k_{-r}r^*y + k_r x(R_T - r^*), \tag{10.6.14a}$$

$$\frac{dx}{dt} = -k_{-e}x + k_e S, \quad \frac{dy}{dt} = -k_{-i}y + k_i S, \tag{10.6.14b}$$

where $k_e = k'_e E_{tot}$ and $k_i = k'_i I_{tot}$. Equations (10.6.14a-b) describe the basic LEGI model in the case of spatially uniform external stimuli. It is clear that the steady-state response is independent of the stimulus concentration $S$. In particular, if $r^* \ll R_T$, then

$$\frac{r^*}{R_T} = \frac{k_r x}{k_{-r} y} = \frac{k_r}{k_{-r}} \frac{k_e}{k_{-e}} \frac{k_{-i}}{k_i}. \tag{10.6.15}$$

However, there is a transient response to sudden uniform changes in $S$, under the assumption that excitation has a faster response than inhibition, $k_{-e} > k_{-i}$, which is illustrated in Fig. 10.31(a,b).

So far, we have not considered the local–global distinction between excitation and inhibition in the LEGI model. This refers to the fact that the inhibitory enzyme is assumed to diffuse freely throughout the cell and thus averages over the total receptor occupancy (global response), whereas the excitatory enzyme cannot diffuse and thus samples the local receptor occupancy (local response). This difference plays a major role in allowing the cell to detect spatial gradients, as explained in Fig. 10.31(c,d). The basic idea is that a stronger signal at the front of a cell, say, will lead to an increase in both excitation and inhibition in this region. However, diffusion of the inhibitory enzyme will tend to reduce the asymmetry in the inhibitory response, resulting in excitation being dominant at the front and inhibition being dominant at the rear, thus amplifying the response. We will now make this more quantitative, following along the lines of [706].

Consider a two-dimensional model of a circularly symmetric cell of radius $L$ and suppose that there is a chemical gradient in the $y$-direction. We can then represent the stimulus as

$$S(\theta) = \bar{S}(1 + \gamma \sin \theta), \tag{10.6.16}$$

where $\bar{S}$ is the mean concentration of chemoattractant outside the cell, and $\gamma$ is the spatial gradient. Since the membrane-bound signaling molecules $R^*$ and excitatory enzymes $E^*$ are localized in the membrane, they will only depend on the angular coordinate $\theta$. On the other hand, the inhibitory enzyme is free to diffuse in the cytoplasm and membrane. (More precisely, it is likely that the inactive form of the enzyme diffuses in the cytoplasm, binds to the membrane, and is then converted to the active form that diffuses in the membrane. Since the inactive form will be spatially uniform in the steady state, we ignore its dynamics in the following.) The LEGI model can thus be written in polar coordinates $(\rho, \theta)$ as

$$\frac{\partial r^*(\theta,t)}{\partial t} = -k_{-r} r^*(\theta,t) y(\theta,t) + k_r R_T x(\theta,t), \tag{10.6.17a}$$

$$\frac{\partial x(\theta,t)}{\partial t} = -k_{-e} x(\theta,t) + k_e S(\theta), \tag{10.6.17b}$$

$$\frac{\partial y(\theta,t)}{\partial t} = -k_{-i} y(\theta,t) + k_i S(\theta) + \frac{D}{L^2} \frac{\partial^2 y(\theta,t)}{\partial \theta^2}. \tag{10.6.17c}$$

We will determine the steady-state solution. Substituting for $S(\theta)$ into the steady-state equation for $y(\theta)$ gives

$$\frac{D}{L^2} \frac{d^2 y(\theta)}{d\theta^2} - k_{-i} y(\theta,t) + k_i \bar{S}(1 + \gamma \sin \theta) = 0.$$

Using a Fourier series expansion, we obtain the solution

$$y(\theta) = \frac{\bar{S}}{K_i}\left(1 + \frac{\gamma}{1 + \lambda^2} \sin \theta\right), \quad \lambda = \sqrt{D/L^2 k_{-i}}. \tag{10.6.18}$$

The steady-state response is then

$$\frac{r^*(\theta)}{R_T} = \frac{k_r x(\theta)}{k_{-r} y(\theta)} = \frac{k_e k_{-i} k_r}{k_{-e} k_i k_{-r}} \frac{1 + \gamma \sin \theta}{1 + (\gamma/(1+\lambda^2)) \sin \theta}. \tag{10.6.19}$$

In particular, if diffusion in the membrane is relatively fast, $D \gg k_{-i} L^2$, then the concentration of the active signaling molecule $R^*$ is proportional to the signal $S(\theta)$, that is, the gradient is detected. On the other hand, if there is no diffusion, then the response is $\theta$-independent.

One potential mechanism for improving the precision of spatial gradient sensing is cell–cell communication [159, 271, 296, 644, 855]. (Analogously, receptor coupling can enhance gradient sensing at the single-cell level; see Ex. 10.12.) That is, cell–cell coupling could reduce fluctuations in the response by averaging individual responses over multiple cells. It could also mitigate sensory noise by extending the spatial range of the sensing. For example, consider a perfect detector of length $A$ aligned along the direction of a spatial concentration gradient of slope $g$. Suppose that there is a cell of size $a$ at either end of the detector, which estimates the local concentration. The concentrations at the left ($L$) and right ($R$) ends are $c_L = c_{1/2} - gA/2$ and $c_R = c_{1/2} + gA/2$, where $c_{1/2}$ is the concentration at the center of the detector (which may fluctuate about a mean $\bar{c}_{1/2}$). From classical Burg–Purcell theory (Sect. 10.5), the relative error in the concentration measurement (due to counting errors) is of the form

$$\left( \frac{\delta c_{L,R}}{\bar{c}_{L,R}} \right)^2 \sim \frac{1}{\bar{c}_{L,R}}.$$

Suppose one cell knows instantly and perfectly the measurement of the other cell. The gradient could then be estimated by the difference in the concentration measurements made by the two cells. That is, setting $\Delta = (c_L - c_R)$, we have

$$\bar{\Delta} = (\bar{c}_R - \bar{c}_L) = gA, \quad \langle \delta \Delta^2 \rangle = A^2 \langle \delta g^2 \rangle = \langle \delta c_L^2 \rangle + \langle \delta c_R^2 \rangle \sim \bar{c}_L + \bar{c}_R = 2\bar{c}_{1/2},$$

assuming the measurements are independent. Hence

$$\frac{\langle \delta \Delta^2 \rangle}{\bar{\Delta}^2} = \frac{\langle \delta g^2 \rangle}{g^2} \sim \frac{2\bar{c}_{1/2}}{(gA)^2}. \tag{10.6.20}$$

This implies that the perfect detector could arbitrarily reduce measurement errors by increasing its size $A$. (For a more rigorous analysis, see [275]). However, this cannot be a fundamental limit because it assumes an instantaneous, error-free communication between cells. Therefore, one needs to develop an explicit model of collective gradient sensing that takes proper account of cell–cell communication. One approach to modeling the opposing effects of cell–cell coupling on gradient sensing has been developed by Levchenko and collaborators by considering a multicellular version of the LEGI model. [271, 644]. These authors consider the branching morphogenesis of epithelial tissue in mammary gland cultures, in which the major chemoattractant is epidermal growth factor (EGF).

## 10.7 Exercises

**Problem 10.1. Phosphorylation–dephosphorylation cycles.**
(a) Fill in the details of the analysis of the Golbeter–Koshland model by deriving equations (10.1.3) and (10.1.6).

(b) Generalize the analysis of the Golbeter–Koshland model, in order to derive the effective Michaelis–Menten equation (10.1.15) for the dual phosphorylation model.

(c) Evaluate the integral in equation (10.1.29) and thus establish the existence of bimodality as shown in Fig. 10.5.

**Problem 10.2. Biochemical amplification in photoreceptors.** Consider the series of phosphorylation steps for rhodopsin whose mass-action kinetics are given by equation (10.2.1) with $k_j = k$, $k = 1,\ldots,n-1$. Assume that each state $R_j$ elicits the same response. The relevant quantity is then the probability density that the molecule exits the state $R_{n-1}$ exactly at time $t$, which is given by $kP_{n-1}(t)$. By Fourier transforming the kinetic equations with respect to time $t$, show that

$$P_{n-1}(t) = \int_{-\infty}^{\infty} \frac{k^{n-1}}{(k-i\omega)^n} e^{-i\omega t} \frac{d\omega}{2\pi}.$$

Using the identity

$$\int_0^{\infty} t^{n-1} e^{-kt} e^{i\omega t} dt = \frac{(n-1)!}{(k-i\omega)^n},$$

and reversing the order of the resulting double integral, show that $P_{n-1}(t) = t^{n-1} e^{-kt} \Theta(t)$, where $\Theta(t)$ is the Heaviside function. Determine the mean and variance defined according to

$$\mu = k \int_0^{\infty} P_{n-1}(t) t \, dt, \quad \sigma^2 = k \int_0^{\infty} P_{n-1}(t) t^2 dt - \mu^2.$$

Hence, deduce that

$$\frac{\sigma}{\mu} = \frac{1}{\sqrt{n}}.$$

This establishes that increasing the number of phosphorylation steps $n$ increases the reliability of the single-photon response.

**Problem 10.3. Cascade model of adaptation.** Suppose that are is a set of molecular species such that, when one component is activated, it acts as a catalyst for the production of the next component of the cascade. In the case of three species with concentrations in the active form denoted by $A, B, C$, we have

$$\frac{dA}{dt} = -\frac{A}{\tau} + \kappa R^*(t),$$

$$\frac{dB}{dt} = -\frac{B}{\tau} + \kappa A,$$

$$\frac{dC}{dt} = -\frac{C}{\tau} + \kappa B,$$

where $\kappa$ is the catalysis rate and $\tau$ is the decay time, which for simplicity are taken to be the same at all stages of the cascade. The input to the cascade is the transient concentration of light-activated rhodopsin, $R^*(t)$, in response to a light flash at $t = 0$.

We can interpret the three chemical components as different intermediate states of transducin. As a further simplification, take $R^*(t) = R^*(0)e^{-t/\tau}$, which is the response to a light impulse $\delta(t)$ of size $R^*(0)$.

(a) Fourier transform the above equations to obtain the transfer function of the system, defined according to

$$\widehat{C}(\omega) := \widehat{H}(\omega)R^*(0).$$

(b) Using contour integration to evaluate the inverse Fourier transform of $\widehat{H}(\omega)$, show that the casual linear response function is

$$H(t) = \frac{\kappa^3 \tau^3}{3!}\left(\frac{t}{\tau}\right)^3 e^{-t/\tau}\Theta(t),$$

where $\Theta(t)$ is the Heaviside function. Alternatively, check by evaluating the Fourier transform of $H(t)$.

(c) Define the gain $g$ of the cascade to be the integrated output response divided by the impulse amplitude $R^*(0)$, that is, $g = \int_0^\infty H(t)dt$. Assuming that the decay time $\tau$ of the photoreceptor decreases as the background light level increases, show that the gain also decreases.

**Problem 10.4. Shot noise process.** Consider a slight generalization of the shot noise process (10.2.6) given by

$$\Phi(t) = \sum_n a_n \phi(t - T_n),$$

where the random amplitudes $a_n$ are generated from some probability density $q(a)$. Introduce the generating function $G_N(s, T) = \mathbb{E}_N[e^{s\Phi(T)}]$. Using similar arguments to the derivation of equation (10.2.10), obtain the following expression for the generating function:

$$G(s) \equiv \int_0^\infty e^{s\Phi}P(\Phi)d\Phi = \exp\left(v\int_{-\infty}^\infty\int_{-\infty}^\infty q(a)\left(e^{s\phi(t)} - 1\right)dadt\right).$$

Hence, show that

$$\mathbb{E}[\Phi] = va_1\int_{-\infty}^\infty \phi(t)dt, \quad \mathrm{Var}[\Phi] = va_2\int_{-\infty}^\infty \phi(t)^2 dt,$$

where $a_1 = \langle a \rangle$ and $a_2 = \langle a^2 \rangle$.

**Problem 10.5. Optimal filter.** Consider the linear process

$$Y(t) = \int_{-\infty}^\infty g(t')X(t - t')dt' + \xi(t),$$

where $X(t)$ is the input signal, $Y(t)$ is the output signal, $g(t)$ is the linear response function, and $\xi(t)$ is a zero mean noise term. (We also assume $\langle X(t)\xi(t)\rangle = 0$.) In order to separate the signal $X(t)$ from the noise $\xi(t)$ based on observations of $Y(t)$, we introduce a second filter $f$ and set

$$x_{\text{est}}(t) = \int_{-\infty}^{\infty} f(t')Y(t-t')dt'.$$

The optimal filter $f$ is obtained by minimizing the mean-square error

$$E = \left\langle \int_{-\infty}^{\infty} dt \left( X(t) - \int_{-\infty}^{\infty} f(t')Y(t-t')dt' \right)^2 \right\rangle.$$

(a) Using Parseval's theorem

$$\int_{-\infty}^{\infty} f^2(t)dt = \int_{-\infty}^{\infty} |\widehat{f}(\omega)|^2 \frac{d\omega}{2\pi},$$

and the convolution theorem for Fourier transforms, show that

$$E = \left\langle \int_{-\infty}^{\infty} \frac{d\omega}{2\pi} |\widehat{X}(\omega) - \widehat{f}(\omega)\widehat{Y}(\omega)|^2 \right\rangle.$$

Assuming that we can reverse the order of taking expectations and integration, minimize $E$ by setting

$$\frac{\partial}{\partial \widehat{f}(\omega)} \langle |\widehat{X}(\omega) - \widehat{f}(\omega)\widehat{Y}(\omega)|^2 \rangle = 0.$$

(Alternatively, solve the functional derivative equation $\delta E[\widehat{f}]/\delta \widehat{f}(\omega) = 0$; see Box 12C.) Hence show that the optimal filter in the Fourier domain satisfies

$$\widehat{f}(\omega) = \frac{\langle \widehat{X}(\omega)\widehat{Y}^*(\omega) \rangle}{\langle |\widehat{Y}(\omega)|^2 \rangle}.$$

Why doesn't one also need to separately have to differentiate with respect to $\widehat{f}^*(\omega)$?
(b) By Fourier transforming the original integral equation show that

$$\langle \widehat{Y}^*(\omega)\widehat{X}(\omega) \rangle = \widehat{g}^*(\omega)\langle |\widehat{X}(\omega)|^2 \rangle$$

$$\langle |\widehat{Y}(\omega)|^2 \rangle = |\widehat{g}^*(\omega)|^2 \langle |\widehat{X}(\omega)|^2 \rangle + \langle |\widehat{\xi}(\omega)|^2 \rangle.$$

Using the Wiener–Khinchin theorem of Sect. 2.2, show that

$$\widehat{f}(\omega) = \frac{\widehat{g}^*(\omega)S_X(\omega)}{|\widehat{g}^*(\omega)|^2 S_X(\omega) + S_\xi(\Omega)},$$

where $S_X$ and $S_\xi$ are the power spectra of the signal and noise, respectively. What happens in the small noise limit?

**Problem 10.6. Dynamical model of hair cell oscillations.** Consider the stochastic version of the planar model of hair cell oscillations given by equations (10.3.5) and (10.3.7) in the quasi-steady-state limit of fast $Ca^{2+}$ dynamics:

$$\xi \frac{dx}{dt} = -N \left[ k_{sp}(x - x_M - \delta P_o(x)/\gamma) + k_{sp}x \right] + \eta(t),$$

and

$$\xi_M \frac{dx_M}{dt} = N k_{gs} \left[ x - x_M - \delta P_o(x)/\gamma \right] - f_{max}(1 - SP_o(x)) + \eta_M(t),$$

with the open channel probability $P_o(x)$ given by equation (10.3.6) and Gaussian white noise terms satisfying

$$\langle \eta(t)\eta(0) \rangle = 2k_B T \xi \delta(t) \quad \langle \eta_M(t)\eta_M(0) \rangle = 2k_B T_M \xi_M \delta(t)$$

for $T_M = 1.5T$.

(a) First consider the zero-noise case. Investigate the existence and stability of any fixed points as a function of $f_{max}$ for the parameter values given in Fig. 10.19 in the two cases (a) $S = 0.5$ and (b) $S = 1.0$. Show that the results are consistent with the bifurcation diagram shown in Fig. 10.19.

(b) Using the numerical methods outlined in Sect. 2.6, simulate the stochastic version of the model for $f_{max} = 50.3pn$ and $S = 0.65$, and plot the power spectrum.

**Problem 10.7. Ising model of receptor clustering.** Consider the Ising model of receptor clustering with partition function

$$Z = \sum_{\mathbf{a}} e^{-H(\mathbf{a})/k_B T}, \quad H(\mathbf{a}) = -J \sum_{\langle m,n \rangle} (2a_m - 1)(2a_n - 1) + F([L]) \sum_m a_m.$$

Using the identity

$$\left\langle \sum_m a_m \right\rangle = -\frac{\partial \ln Z}{\partial F},$$

and the analysis of the Ising model in Chap. 3, show that the mean level of kinase activity per receptor is

$$\langle a \rangle = \frac{1}{2} \left[ 1 - \frac{\sinh F([L])}{\sqrt{\sinh^2 F([L]) + e^{-4J}}} \right],$$

with

$$F([L]) = \Delta E + k_B T \ln \frac{1 + K_i[L]}{1 + K_a[L]}.$$

Given that $K_i > K_a$, describe how $\langle a \rangle$ changes as $[L]$ increases from 0 to $\infty$.

**Problem 10.8. Linear response in the Barkai–Leibler model.** Consider the Barkai–Leibler model of adaptation in bacterial chemotaxis. The methylation level evolves as

$$\frac{dv}{dt} = F(a) \equiv \Gamma_R - \frac{\Gamma_B a}{K_B + a},$$

where $\Gamma_{R,B}$ are maximum catalytic rates and the average receptor activity $a$ is given by the MWC model (with $k_B T = 1$):

$$a = a([L], v) \equiv \left[1 + \exp\left(N\left(\Delta E(v) + \ln\frac{1 + K_i[L]}{1 + K_a[L]}\right)\right)\right]^{-1}.$$

Let $a_0, v_0$ be the steady state at a background ligand concentration $L_0$. Suppose that there is a small oscillatory modulation of the ligand concentration

$$[L](t) = L_0 e^{A\cos(\omega t)},$$

where $A$ is the amplitude, $A \ll 1$, and $\omega$ is the modulation frequency. Finally, assume that $K_a[L] \ll 1 \ll K_i[L]$ so that

$$\ln\frac{1 + K_i[L]}{1 + K_a[L]} \approx \ln([L]/K_i).$$

(a) Linearizing about the steady state $(a_0, v_0)$, show that

$$\frac{d\Delta v}{dt} = F'(a_0)\Delta a,$$

with $\Delta v = v - v_0$,

$$\Delta a = a - a_0 = N a_0(1 - a_0)[\alpha \Delta v - A\cos(\omega t)],$$

and $\alpha = \Delta E'(v_0)$.

(b) Setting

$$\Delta v = \mathrm{Re}[A_m e^{i\omega t}], \quad \Delta a = \mathrm{Re}[A_a e^{i\omega t}],$$

use part (a) to solve for the complex amplitudes $A_m, A_a$:

$$A_a = \frac{i\omega c_a}{i\omega + \omega_m}A, \quad A_m = \frac{\omega_m c_m}{i\omega + \omega_m}A,$$

where

$$c_a = N a_0(1 - a_0), \quad c_m = \alpha^{-1}, \quad \omega_m = -\alpha F'(a_0)N a_0(1 - a_0).$$

(c) The linear response of receptor activity can be characterized by the amplitude $|A_a|$ and phase $\phi_a = \pi/2 + \tan^{-1}(v/v_m)$. Plot $|A_a|/|A_a|_{\max}$ and $\phi/\pi$ as a function of $v/v_m$. (A typical value of $v_m$ is around $5 \times 10^{-3}$ Hz).

**Problem 10.9. First passage time problem for bacterial run-and-tumble.** Consider a bacterium confined to a domain $0 \le x \le L$ and whose probability densities $p_\pm(x,t)$ evolve according to equation (10.4.21), supplemented by a reflecting boundary at $x = 0$ and an absorbing boundary at $x = L$:

$$p_+(0,t) = p_-(0,t), \quad p_-(L,t) = 0.$$

The goal is to calculate the MFPT to hit the wall at $x = L$ having started at position $y$ and in velocity state $m$. It can be shown that the backward CK equation takes the form

$$\frac{\partial q_+}{\partial t} = v\frac{\partial q_+}{\partial y} - k[q_+ - q_-],$$

$$\frac{\partial q_-}{\partial t} = -v\frac{\partial q_-}{\partial y} + k[q_+ - q_-],$$

where $q_m(y,t) = p(x,t|y,m,0)$ with $p(x,0|y,m,0) = \delta(x,y)\sigma_m$ and $\sigma_m$ the probability that the initial velocity sate is $m$. For the sake of illustration, we take $\sigma_m = \delta_{m.,+}$. The boundary conditions are

$$q_+(0,t) = q_-(0,t), \quad q_+(L,t) = 0.$$

Let $S_m(y,t)$ be the survival probability

$$S_m(y,t) = \int_0^L p(x,t|y,m,0)dx.$$

(a) Show that the MFPT is

$$\tau_m(y) = \int_0^\infty S_m(y,t)dt.$$

(b) Integrating the backward equations with respect to $x$ and $t$, obtain the following pair of equations

$$-1 = v\frac{d\tau_+}{dy} - k(\tau_+ - \tau_-), \quad 0 = -v\frac{d\tau_-}{dy} + k[\tau_+ - \tau_-],$$

which are supplemented by the boundary conditions

$$\tau_+(0) = \tau_-(0), \quad \tau_+(L) = 0.$$

(c) Solve the pair of equations in part (b) to obtain the results

$$\tau_-(y) = \frac{L}{v} + \frac{k(L^2 - y^2)}{2v^2}, \quad \tau_+(y) = \frac{L-y}{v} + \frac{k(L^2 - y^2)}{2v^2}.$$

[Hint: first solve for $\tau(y) = \tau_-(y) - \tau_+(y)$ by adding the pair of equations. You will need to impose the various boundary conditions.]

(d) Recall that for sufficiently large $k$ (fast switching), the unbiased velocity-jump process reduces to pure diffusion with diffusion coefficient $D = v^2/2k$. Verify that, to leading order, the formula for the MFPT in part (c) is consistent with the diffusion approximation.

**Problem 10.10. Bacterial run-and-tumble with sticky boundary conditions.**
Consider a bacterium confined to a domain $-L/2 \leq x \leq L/2$ and whose probability

densities $p_\pm(x,t)$ evolve according to equation (10.4.21), but now supplemented by sticky boundary conditions at both ends:

$$vp_+(-L/2,t) = k'Q_-(t), \quad vp_-(L/2,t) = k'Q_+(t),$$

with

$$\frac{dQ_-}{dt} = -J(-L/2,t), \quad \frac{dQ_+}{dt} = J(L/2,t), \quad J(x,t) = vp_+(x,t) - vp_-(x,t),$$

and the conservation condition

$$\int_{-L/2}^{L/2} P(x,t)dx + Q_+(t) + Q_-(t) = 1 \quad P(x,t) = p_+(x,t) + p_-(x,t).$$

Finally, take the initial conditions to be

$$P(x,0) = \delta(x), \quad J(x,0) = 0, \quad Q_-(0) = Q_+(0) = 0.$$

(a) Reexpress equation (10.4.21) and the boundary conditions in terms of the variables

$$P = p_+ + p_-, \quad J = v(p_+ - p_-).$$

(b) First consider the case $k = k'$. Using Laplace transforms, obtain the general solution

$$\widehat{P}(x,s) = A_1 e^{c(s)|x|} + A_2 e^{-c(s)|x|}, \quad c(s) = \sqrt{\frac{s(s+2k)}{v^2}},$$

and determine the coefficients $A_{1,2}$ by imposing the boundary conditions and the conservation equation. Hence show that the probability that bacteria are stuck to a given boundary is (in Laplace space)

$$\widehat{Q}(s) = \frac{1}{2} \frac{1}{s\cosh(cL/2) + vc\sinh(cL/2)}.$$

(c) Show that $Q(t) = H(t - L/2v)/2$ when $k = 0$, where $H$ is the Heaviside function. Interpret the result. For $k > 0$ determine the steady-state probability $Q_\infty$ ising

$$\lim_{t \to \infty} Q(t) = \lim_{s \to 0} s\widetilde{Q}(s),$$

(d) Show that when $k \neq k'$, the final result of (b) becomes

$$\widehat{Q}(s) = \frac{1}{2s} \frac{k(s)}{k(s)\cosh(cL/2) + vc\sinh(cL/2)}, \quad k(s) = \frac{s(s+2k)}{s+2k'}.$$

**Problem 10.11. Blow-up of the Keller–Segel model.** Consider the following non-dimensionalized version of the Keller–Segel (KS) model of chemotaxis in 2D:

$$\frac{\partial n}{\partial t} = \nabla \cdot (D\nabla n - n\chi\nabla c),$$

$$\varepsilon\frac{\partial c}{\partial t} = \nabla^2 c + n.$$

Here $n(\mathbf{x},t)$ is the cell density and $c(\mathbf{x},t)$ is the concentration of chemoattractant released by the cells themselves. The chemotactic sensitivity $\chi$ is taken to be a constant and the dynamics of the chemoattractant is assumed to be much faster than that of the cells so we can set $\varepsilon = 0$. The total number of cells is conserved so that with $\int_{\mathbb{R}^2} n(\mathbf{x},t)d\mathbf{x} = N$.

(a) Using the 2D Green's function on $\mathbb{R}^2$,

$$G(\mathbf{x},\mathbf{y}) = \frac{1}{2\pi}\ln|\mathbf{x}-\mathbf{y}|, \quad \nabla^2 G(\mathbf{x},\mathbf{y}) = \delta(\mathbf{x}-\mathbf{y}),$$

solve the quasi-steady-state equation $\nabla^2 c + n = 0$ and hence show that

$$\nabla c(\mathbf{x},t) = -\frac{1}{2\pi}\int_{\mathbb{R}^2}\frac{\mathbf{x}-\mathbf{y}}{|\mathbf{x}-\mathbf{y}|^2}n(\mathbf{y},t)d\mathbf{y}.$$

(b) Define the second moment

$$m_2(t) = \int_{\mathbb{R}^2}|\mathbf{x}|^2 n(\mathbf{x},t)d\mathbf{x}.$$

Differentiate both sides with respect to $t$ and use the KS equations to derive the equation

$$\frac{dm_2}{dt} = 4N\left(1 - \frac{\chi}{8\pi}N\right).$$

Hence show that if $N > 8\pi/\chi$ then $m_2$ becomes negative in finite time, which is impossible since $n$ is positive. This suggests that the solution cannot remain smooth until that time.

**Problem 10.12. Spatial gradient sensing and cooperativity.** In this problem, we investigate the possible effects of receptor cooperativity on spatial gradient sensing using the model of Hu et al [436]. Intuitively, short-range interactions allow receptors to collectively sharpen the asymmetry of receptor signals by dampening fluctuations. Receptor cooperativity is incorporated into the model of Fig. 10.30 by including a nearest neighbor interaction $J$ (in units of $k_B T$). This leads to an Ising model whose free energy takes the form

$$E[\mathbf{s}] = -\sum_{n=1}^{N}[Js_n s_{n+1} + \varepsilon_n(s_n + s_{n+1})/2],$$

with $\varepsilon_n$ given by equation (10.6.1). The free energy has been written in a symmetric form by exploiting the fact that $\varepsilon_n \approx \varepsilon_{n\pm 1}$. The corresponding partition function can be written as

$$Z = \sum_{s_1=\pm 1} \cdots \sum_{s_N=\pm 1} e^{-E_0[s]+E_1[s},$$

where

$$E_0 = -\sum_{n=1}^{N} [J s_n s_{n+1} + \alpha_0 (s_n + s_{n+1})/2]$$

is the isotropic part of the free energy, and

$$E_1 = -\frac{A}{4} \sum_n s_n \cos(\phi_n - \theta).$$

We will assume that $N$ is large and $A$ is small.

(a) Using transfer matrices to analyze the Ising model on a ring (Chap. 3), show that for $A = 0$ (uniform concentration) $Z_N = Z_N^{(0)}$ with

$$Z_N^{(0)} = \lambda_+^N + \lambda_-^N, \approx \lambda_+^N, \quad \lambda_\pm = e^J \cosh \alpha_0 \pm \sqrt{e^{-2J} + e^{2J} \sinh^2 \alpha_0}.$$

(b) Show that the full partition function can be expressed as

$$Z_N = Z_N^{(0)} \langle e^{-E_1} \rangle,$$

where expectation is taken with respect to the Boltzmann distribution $e^{-E_0}/Z_N^{(0)}$.

(c) Carry out a perturbation expansion to second order in $A$ to show that

$$Z_N \approx \lambda_+^N \left[ 1 + \frac{A}{4} \sum_n \langle s_n \rangle + \frac{A^2}{32} \sum_{n,m} \langle s_m s_n \rangle \cos \theta_n \cos \theta_m \right],$$

where we have set $\theta_n = \phi_n - \theta$. From properties of the classical Ising model,

$$\langle s_n \rangle = \text{constant}, \quad \langle s_m s_n \rangle = \cos^2 2\omega + \gamma^{|n-m|} \sin^2 \omega, \quad \gamma = \lambda_-/\lambda_+,$$

and the parameter $\omega$ is the unique solution to

$$\cot(2\omega) = e^{2J} \sinh \alpha_0, \quad 0 < \omega < \pi/2.$$

Using the fact that $\sum_n \cos \theta_n = 0$ for large $N$, establish that

$$Z_N \approx \lambda_+^N \left[ 1 + \frac{A^2}{32} \sin^2 \omega \sum_{n,m} \gamma^{|n-m|} \cos \theta_n \cos \theta_m \right].$$

For large $N$, the double sum can be approximated as [436]

$$\sum_{n,m} \gamma^{|n-m|} \cos \theta_n \cos \theta_m \approx \frac{N(1+2\xi)}{2}, \quad \xi = [\ln(\lambda_+/\lambda_-)]^{-1},$$

where $\xi$ is the correlation length of the classical Ising chain.

(d) Combining the various approximations of part (c), obtain the result

$$\ln Z_N \approx N \ln \lambda_+ + \frac{NA^2(1+2\xi)}{64(1+e^{4J}\sinh^2\alpha_0)} + O(A^3).$$

Show that as $J \to 0$, we recover equation (10.6.4).

(e) Parameter estimation proceeds as in Sect. 10.6.1 except that we have to replace $\mu = N/16\cosh^2\alpha_0$ by

$$\tilde{\mu} = \frac{N(1+2\xi)}{16(1+e^{4J}\sinh^2\alpha_0)}.$$

In particular, the error estimates are now

$$\tilde{\sigma}_a^2 = \frac{2}{\tilde{\mu}}, \quad \tilde{\sigma}_\theta^2 = \frac{2}{A^2\tilde{\mu}}.$$

Plot $\tilde{\sigma}_\theta^2$ as a function of $\ln(C_0/K_d)$ for $J = 0, 0, 2, 0.4, 0.6$ with $A = 0.08$ and $N = 80,000$. Hence establish that cooperativity improves the precision of gradient sensing around the minima at $C_0 = K_d$, but enlarges the error when $C_0$ is far away from $K_d$.

# Chapter 11
# Intracellular pattern formation and reaction–diffusion processes

A fundamental question in modern cell biology is how cellular and subcellular structures are formed and maintained given their particular molecular components. How are the different shapes, sizes, and functions of cellular organelles and molecular complexes determined, and why are specific structures formed at particular locations and stages of the life cycle of a cell? In order to address these questions it is necessary to consider the theory of self-organizing nonequilibrium systems [623, 809, 935]. Within the context of cell biology, self-organization can be defined as [623] "the capacity of a macromolecular complex or organelle to determine its own structure based on the functional interactions of its components." In particular, there is no underlying architectural blueprint for a self-organized structure. Instead, a stable stationary state is maintained via the continual dynamical exchange of energy and materials between different regions of the system and/or the environment. Consequently, self-organized systems can display a range of spatiotemporal dynamics, including spatially periodic patterns, oscillations, and waves.

Self-organization is typically contrasted with self-assembly, which involves the assembly of a stable, static structure in thermodynamic equilibrium. In the latter case, ordered structures can form via phase transitions. However, the distinction isn't so clear once the kinetics of phase transitions is taken into account. A classical example is phase separation, where an unstable or metastable homogeneous state evolves into an inhomogeneous state of coexisting phases via dissipative processes such as diffusion. Moreover, the dynamics can be regulated by various active processes. Phase separation is thought to play a major role in the formation of various membrane-less organelles [42, 70, 293, 451] and lipid membranes [403], see Sect. 13.1.

One major mechanism for self-organization within cells (and between cells) is the interplay between diffusion and nonlinear chemical reactions. Historically speaking, the idea that a reaction–diffusion (RD) system can spontaneously generate spatiotemporal patterns was first introduced by Turing in his seminal 1952 paper [889]. Turing considered the general problem of how organisms develop their structures during the growth from embryos to adults. He established the principle that two

© Springer Nature Switzerland AG 2021
P. Bressloff, *Stochastic Processes in Cell Biology*, Interdisciplinary Applied Mathematics 41, https://doi.org/10.1007/978-3-030-72519-8_11

nonlinearly interacting chemical species differing significantly in their rates of diffusion can amplify spatially periodic fluctuations in their concentrations, resulting in the formation of a stable periodic pattern. The Turing mechanism for morphogenesis was subsequently refined by Gierer and Meinhardt [346], who showed that one way to generate a Turing instability is to have an antagonistic pair of molecular species known as an activator–inhibitor system, which consists of a slowly diffusing chemical activator and a quickly diffusing chemical inhibitor. Over the years, the range of models and applications of the Turing mechanism expanded dramatically [655][1], in spite of the fact that most experimental findings suggested that morphogenesis was often guided by explicit spatial cues, based on the localization of specific proteins or RNA [950]. Indeed, for many years the only direct experimental evidence for spatiotemporal patterning of molecular concentrations came from the inorganic Belousov–Zhabotinsky reaction [973], until Kondo and Asai demonstrated the occurrence of the Turing mechanism in studies of animal coat patterning [520]. Recent advances in live cell imaging and gene knockout protocols are now allowing for a closer connection between theories of pattern formation and developmental cell biology. There are also a growing number of examples of spatiotemporal patterns of signaling molecules at the intracellular level [431, 505], which will be the focus of this chapter. Such patterns typically regulate downstream structures such as the cytoskeleton and cell membrane, which then drive various mechanical processes including cell mitosis and cell motility. The regulation of cytoskeletal structures will be considered in Chap. 14.

We begin by considering the formation of intracellular protein gradients, which are characterized by spatial variations in the activity state (e.g., phosphorylation state) of a protein due to some localized source of activation or deactivation (Sect. 11.1). We address the robustness of intracellular gradients in the presence of intrinsic or extrinsic noise fluctuations and characterize the approach to steady state in terms of the accumulation time. In Sect. 11.2 we introduce the theory of Turing pattern formation. We first review the predominant mechanism for Turing pattern formation, which is based on deterministic activator–inhibitor RD equations. In particular, conditions for a Turing instability are derived using linear stability analysis. Although activator–inhibitor systems are prevalent in developmental biology, they do not apply to many of the known examples of intracellular patterning [391, 392]. We discuss the first of two major differences, namely, that intracellular processes tend to be mass conserving, at least on the time scales for patterns to initially form and stabilize. In Sect. 11.3 we turn to the second characteristic of intracellular pattern formation. Instead of two or more chemical species diffusing in the same medium and mutually affecting their rates of production and degradation, intracellular patterns typically involve the dynamical exchange of proteins between the cytoplasm and plasma membrane, resulting in associated changes of conformational state and a spatial redistribution of mass. We illustrate this by presenting various examples of

---

[1] Biological pattern formation has also been studied within the context of ecology [554, 764] and systems neuroscience [108, 283]; in the latter case nonlocal synaptic interactions drive pattern forming instabilities rather than diffusion.

mass-conserving RD models of cell polarization and division, including Min protein oscillations in *E. coli* prior to cell division, and cell polarization in budding and fission yeast.

In Sect. 11.4 we consider another mechanism for Turing pattern formation, which involves two interacting chemical species, where one is passively diffusing and the other is actively trafficked by molecular motors [136, 137]. Such a system provides a possible explanation for the regular spacing of synaptic puncta along the ventral cord of *C. elegans* during development [769]. In Sect. 11.5, we show how asymptotic analysis can be used to study the existence and stability of multi-spike solutions far from pattern onset. The latter consists of strongly localized regions of high concentration of a slowly diffusing activator and can be analyzed using singular perturbation theory. We develop the theory using the classical Gieier–Meinhardt model and then describe a recent application to an RD model of the self-positioning of structural maintenance of chromosomes (SMC) protein complexes in *E. coli*, which are required for correct chromosome condensation, organization, and segregation in prokaryotes. Finally, in Sect. 11.6, we consider another major application of RD systems, namely, the modeling of intracellular traveling waves. We first explore the role of bistable traveling fronts in the cell polarization of motile eukaryotic cells and in the early development of cell division in eggs that are laid externally. We then consider the propagation of CaMKII waves along the dendrites of neurons as an example of an unstable wave or pulled front. Note that in this chapter we primarily focus on deterministic RD equations; the theory of stochastic RD processes will be considered in Chap. 16.

## 11.1 Intracellular protein concentration gradients

It has been known for some time that concentration gradients play a crucial role in the spatial regulation of patterning during development [32, 541, 543, 823, 865, 925, 949, 950]. That is, a spatially varying concentration of a morphogen protein drives a corresponding spatial variation in gene expression through some form of concentration thresholding mechanism. For example, in regions where the morphogen concentration exceeds a particular threshold, a specific gene is activated (see also Sect. 5.7). Hence, a continuously varying morphogen concentration can be converted into a discrete spatial pattern of differentiated gene expression across a cell population. The classical mechanism for morphogen gradient formation involves a localized source of protein production within the embryo, combined with diffusion away from the source and subsequent degradation. Recently, however, an alternative mechanism for delivering morphogens to embryonic cells has been proposed, which involves actin-rich dynamic cellular extensions known as cytonemes that make direct contacts with target cells [522, 523], see Sect. 14.5.

There is emerging experimental evidence that concentration gradients not only arise within the context of embryonic development, but are also found within individual cells, typically taking the form of a concentration gradient in some active protein [431, 505]. An important difference between intracellular gradients and mul-

ticellular morphogen gradients is that degradation does not play a significant role in
the formation of intracellular gradients. This is a consequence of the fact that the
lifetime of a typical protein exceeds the duration of the cellular process regulated
by the presence of a gradient. Instead, some modification in the protein, such as its
phosphorylation state, changes as it moves away from the catalytic source of the
modification.

Irrespective of the particular details regarding a protein concentration gradient,
there are several important issues that are common to all types of gradient. (i) The
concentration profile should be robust to fluctuations in the rate of protein production or changes in conformational state, and to fluctuations in the diffusive or
active transport of the proteins. This is particularly important since the spatial profile encodes information such as cell fate in embryos or the size of individual cells.
(ii) The accumulation time associated with forming a steady-state concentration gradient should be consistent with the relevant biological time scales. (iii) The downstream process that decodes gradient information should itself be robust to fluctuations. In this section we consider these issues primarily within the context of
intracellular protein concentration gradients.

The existence of an intracellular gradient was first predicted theoretically by
Brown and Kholodenko [140], and has subsequently been found to play a role in a
wide range of cellular processes, including cell division, polarity, and mitotic spindle dynamics. We begin by highlighting some of these processes.

1. An important component of many signal transduction pathways is the reversible
   cycling between an inactive and an active protein state, which is catalyzed by
   opposing activator and deactivator enzymes. For example, a kinase and phosphatase acting on phosphoproteins or a GEF (guanine nucleotide-exchange fac-

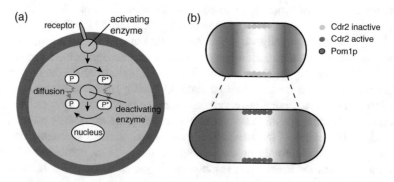

Fig. 11.1: (a) Cartoon of a protein modification cycle, in which an inactive form $P$ is converted
to an active form $P^*$ at the plasma membrane. Both forms diffuse in the cytoplasm, resulting in
deactivation of $P^*$ by cytoplasmic enzymes. (b) Pom1p concentration gradient in fission yeast is
highest at the poles and lowest in the midcell region where Cdr2 concentrates in cortical nodes. In
early interphase (short cells), Pom1p at the midcell is present at a sufficient concentration to inhibit
Cdr2 and the transition to mitosis. As cells grow, the midcell Pom1p concentration decreases until
it crosses a threshold that relieves Cdr2 inhibition thereby promoting mitosis.

tor) and GAP (GTPase-activating protein) acting on small G-proteins such as proteins of the Ras and Rho families. A concentration gradient in these signaling cycles can then be generated by the spatial segregation of the opposing enzymes [505, 653]. One such mechanism is the phosphorylation of proteins by a membrane-bound kinase, which are then dephosphorylated by a cytosolic phosphatase, see Fig. 11.1(a). This results in a gradient of the phosphorylated protein, with a high concentration close to the cell membrane and a low concentration within the interior of the cell. As the cell grows in size, the surface-to-volume ratio decreases and membrane-activated proteins have to diffuse over longer distances in order to reach their target such as the nucleus. Hence, the proteins become progressively deactivated toward the cell interior, thus providing a mechanism for coupling cell growth with the cell cycle. Indeed, activated spatial gradients have been observed during cell mitosis [173, 486, 684] involving for example the small GTPase Ran, which assists in the formation of the mitotic spindle by biasing microtubule growth toward the chromosomes, see also Sect. 14.3. Activity gradients may also play a role in the localization of the Rho GTPase cell division control protein 42 (Cdc42), which is a regulator of actin polymerization, resulting in a coupling between cell shape and protein activation [618].

2. One of the best studied intracellular gradients involves the Dual-specificity tyrosine phosphorylation-regulated kinase (DYRK) Pom1p in fission yeast, see also Sect. 11.1.2. Pom1p forms a concentration gradient within the rod-shaped yeast cell, with the highest concentrations at the cell tips and the lowest concentrations at the cell center. This is achieved by a combination of localized binding to the membrane at the cell tips, diffusive spreading within the membrane, followed by membrane unbinding, and diffusion in the cytoplasm until rebinding at the tips. It is thought that Pom1p inhibits cell division, acting to localize the cell division factor Mid1 toward the cell mid-plane [174, 702]. Pom1p also phosphorylates and suppresses the activity of Cdr2p, which is a promoter of cell mitosis that localizes to nodes within the membrane at the center [599, 642]. In a short cell, the concentration of Pom1p at the cell center is relatively high so that Cdr2 activity is inhibited. On the other hand, as the cell grows, the midcell concentration of Pom1p decreases resulting in activation of Cdr2 and initiation of mitosis, see Fig. 11.1(b). Thus, the formation of a spatial concentration gradient again provides a mechanism for coupling cell growth with the cell cycle. (Note, however, that recent experimental observations have case doubt on Pom1 concentration gradients as the main cell size control mechanism [290, 707].)

3. Spatial concentration gradients also appear to play a role in cell division in small bacterial cells with spatial extents of only a few microns [766, 869]. One well-studied example is the oscillatory dynamics of Min proteins in *E. coli*, which creates a time-averaged concentration gradient that directs localization of the cell division machinery [430, 440, 528, 573, 612]. Such an oscillation is the result of a dynamic instability of the underlying RD system, see Sect. 11.3.3.

4. In the early developing one-cell *C. elegans* embryo (zygote), interactions between cell polarity proteins and the cytoskeleton set up an initial polarity along the anterior–posterior cell axis [353, 748]. In particular, transient flows of a thin layer of mechanically active actomyosin in the cell cortex instruct the patterning of a conserved cell polarity pathway, which consists of two groups of partitioning-defective (PAR) proteins that mutually exclude one another from the cell membrane [350, 654]. Initially, anterior PARs (aPARs: PAR-3, PAR-6, and atypical protein kinase C) are distributed across the entire cell membrane. Asymmetry is then generated by the entry of a sperm, which donates microtubule-organizing center-derived centrosomes in the posterior region. This then causes the flows generated by cortical actomyosin to be oriented away from the posterior-localized centrosome, which induces the segregation of aPARs. Microtubules at the centrosome are also thought to contribute to the formation of a domain of posterior PARs (pPARs: PAR-1, PAR-2, and LGL) [643]. Once the asymmetric localization of PARs has been established, downstream signaling molecules form concentration gradients via the heterogeneous switching between different diffusive states [127, 957]. The latter will be explored further in Sect. 11.1.3.

5. An intracellular concentration gradient of the soluble protein stathmin plays a role in the regulation of microtubule (MT) growth at the lamellipodium of protruding cells, through interaction with the Rho GTPase Rac1 [945, 974]. Stathmin is known to inhibit MT growth, either by sequestering free tubulin or by direct interaction with the polymer chain, resulting in an increase in the catastrophe rate [171, 384]. Activation of membrane-bound Rac1 in the leading edge of a protruding cell by polymerizing MTs leads to local deactivation of stathmin via phosphorylation. The net result is a positive feedback loop, consisting of one positive interaction (activation of Rac1) and a doubly negative interaction (inhibition of stathmin by Rac1 and inhibition of MT growth by stathmin). Mathematical modeling suggests that this could lead to a bimodal distribution of MT lengths [974], see Sect. 14.1.2.

### 11.1.1 Diffusion equation for gradient formation

We begin by considering the simplest system capable of generating a stationary concentration gradient, consisting of an activating enzyme located in the cell membrane and a deactivating enzyme freely diffusible in the cytoplasm [140, 505], see Fig. 11.1(a). For simplicity, consider an effective 1D geometry representing, for example, a cylindrical bacterial cell of length $L$. Suppose that a kinase is localized to a pole at $x = 0$, which generates a flux of phosphorylated protein at a rate $J$, whereas the phosphatase is distributed throughout the cytoplasm and deactivates each protein at a rate $v_-$. Assume that the phosphatase is far from saturation so that $v_- = kc$, where $c$ is the active protein concentration and $k$ is the deactivation rate. It follows that $c(x,t)$ evolves according to the diffusion equation[2]

$$\frac{\partial c}{\partial t} = D\frac{\partial^2 c}{\partial x^2} - kc(x,t), \quad -D\frac{\partial c}{\partial x}\bigg|_{x=0} = J, \quad \frac{\partial c}{\partial x}\bigg|_{x=L} = 0. \quad (11.1.1)$$

This has the steady-state solution

$$c(x) = c(0)\left(\frac{e^{x/\lambda} + e^{2L/\lambda}e^{-x/\lambda}}{1 + e^{2L/\lambda}}\right), \quad \lambda = \sqrt{\frac{D}{k}}. \quad (11.1.2)$$

The concentration $c(0)$ can be determined from the boundary condition at $x = 0$. Note that when $L \ll \lambda$, the gradient is approximately linear, whereas when $L \gg \lambda$ it decays exponentially with length constant $\lambda$. In the latter case,

$$c(x) = \frac{J\lambda}{D}e^{-x/\lambda}. \quad (11.1.3)$$

A straightforward generalization of the above model is to allow the diffusing active proteins to reversibly bind to some substrate, which acts as an immobile trap [59, 210]. Let $c_1(x,t)$ and $c_0(x,t)$ denote the concentrations of the mobile and immobile active protein, respectively. Equation (11.1.1) becomes, see also Ex. 11.1,

$$\frac{\partial c_1(x,t)}{\partial t} = D\frac{\partial^2 c_1(x,t)}{\partial x^2} - k_1 c_1(x,t) + \alpha c_0(x,t) - \beta c_1(x,t), \quad (11.1.4a)$$

$$\frac{\partial c_0(x,t)}{\partial t} = -k_0 c_0(x,t) - \alpha c_0(x,t) + \beta c_1(x,t). \quad (11.1.4b)$$

Here $k_j$, $j = 0,1$ specify the dephosphorylation rates in the mobile and immobile states, and $\alpha, \beta$ are the transitions rates for switching between the two states. Equations (11.1.4a,b) are similar in form to the first moment equations for switching diffusion processes studied in Chap. 6. In the fast switching limit, $\alpha, \beta \to \infty$ with $\alpha/\beta$ fixed, we have $c_n = \rho_n C$ and the total protein concentration $C = c_0 + c_1$ evolves according to the scalar equation

$$\frac{\partial C(x,t)}{\partial t} = D_{\text{eff}}\frac{\partial^2 C(x,t)}{\partial x^2} - k_{\text{eff}}C(x,t), \quad (11.1.5)$$

where

$$D_{\text{eff}} = \rho_1 D, \quad k_{\text{eff}} = \rho_0 k_0 + \rho_1 k_1, \quad \rho_1 = \frac{\alpha}{\alpha + \beta} = 1 - \rho_0.$$

It follows that the effective characteristic length constant $\lambda_{\text{eff}}$ on the semi-infinite line with a source at $x = 0$ satisfies

---

[2] In the case of diffusion-based morphogenesis, $c(x,t)$ represents the extracellular morphogen concentration gradient along the body axis of a developing embryo, and $k$ is an effective degradation or removal rate due to binding of morphogen to cell surface receptors.

$$\frac{1}{\lambda_{\text{eff}}} = \sqrt{\frac{\beta}{\alpha}\frac{1}{\lambda_0} + \frac{1}{\lambda_1}}, \quad \lambda_j = \frac{D}{k_j}. \tag{11.1.6}$$

**Spatially distributed signaling cascades.** For biophysically reasonable parameter values, one typically finds that the protein gradient is very steep, which means that it is unlikely to reach the nucleus of large cells such as developing neurons. Several mechanisms have been suggested that could produce longer range signal transduction, including spatially distributed signaling cascades, active transport, and traveling waves of protein phosphorylation [653]. Here we will consider the particular example of signaling cascades as developed in Ref. [652]. That is, suppose there exists a cascade of protein modification cycles, in which each cycle involves transitions between inactive and active forms of a signaling protein. At each level of the cascade, the active form of the protein catalyzes the activation of the protein at the next downstream level, see Fig. 11.2(a). Let $c_n(x,t)$ and $\bar{c}_n(x,t)$ denote the concentration of activated and deactivated protein, respectively, at the $n$th level of the cascade. Assume that the total concentration of protein at each each cascade level is fixed at $c_n^{\text{tot}}$ so $\bar{c}_n(x,t) = c_n^{\text{tot}} - c_n(x,t)$. Then

$$\frac{\partial c_1}{\partial t} = D\frac{\partial^2 c_1}{\partial x^2} - v_1^-(x,t), \tag{11.1.7a}$$

$$\frac{\partial c_n}{\partial t} = D\frac{\partial^2 c_n}{\partial x^2} + v_n^+(x,t) - v_n^-(x,t), \quad n = 2,\ldots,N, \tag{11.1.7b}$$

where $v_n^+$ and $v_n^-$ are the phosphorylation and dephosphorylation rates, respectively. Equation (11.1.7) is supplemented by the boundary conditions

$$-D\frac{\partial c_1}{\partial x}\bigg|_{x=0} = v_1^+, \quad \frac{\partial c_1}{\partial x}\bigg|_{x=L} = 0, \tag{11.1.8a}$$

$$\frac{\partial c_n}{\partial x}\bigg|_{x=0} = 0 = \frac{\partial c_n}{\partial x}\bigg|_{x=L}, \quad n = 2,\ldots,N. \tag{11.1.8b}$$

Assuming the Michaelis–Menten kinetics for the enzymatic reactions (see Sect. 1.3),

$$v_n^- = V_n^- \frac{c_n(x,t)}{K_n^- + c_n(x,t)}, \quad n = 1,\ldots,N, \tag{11.1.9a}$$

$$v_1^+ = V_1^+ \frac{c_1^{\text{tot}} - c_1(0,t)}{K_1^+ + c_1\text{tot} - c_1(0,t)}, \tag{11.1.9b}$$

$$v_n^+ = \widehat{V}_n^+ c_{n-1}(x,t) \frac{c_n^{\text{tot}} - c_n(x,t)}{K_n^+ + c_n^{\text{tot}} - c_n(x,t)}, \quad n = 2,\ldots,N. \tag{11.1.9c}$$

Suppose, for simplicity, that all parameters are independent of the cascade level $n$: $K_n^- = K_-, V_n^- = V_-, c_n^{\text{tot}} = c_{\text{tot}}$ for $n = 1,\ldots,N$ and $K_n^+ = K_+, \widehat{V}_n^+ c_{\text{tot}} = V_+$ for $n = 2,\ldots N$. We also assume that the Michaelis–Menten kinetics operate in the linear regime (small concentrations). Non-dimensionalizing the equations by setting

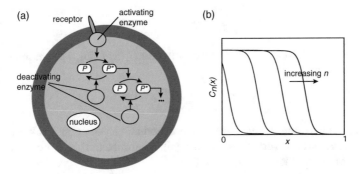

Fig. 11.2: Spatial propagation of activated forms for a cascade of protein modification cycles. (a) Phosphorylation cycles within a cell. (b) Schematic illustration of steady-state concentration profiles at successive levels $n$ of the cascade for $\gamma < 1$ in non-dimensionalized version of the model.

$$C_n = c_n/c_{\text{tot}}, \quad x' = \sqrt{\frac{k_-}{D}}x, \quad t' = k_+t, \quad \gamma = \frac{k_-}{k_+},$$

with $k_\pm = V_\pm/K_\pm$, and dropping primes give

$$\frac{\partial C_1}{\partial t} = \gamma\frac{\partial^2 C_1}{\partial x^2} - \gamma C_1, \tag{11.1.10a}$$

$$\frac{\partial C_n}{\partial t} = \gamma\frac{\partial^2 C_n}{\partial x^2} - \gamma C_n + (1 - C_n)C_{n-1}, \quad n = 2,\dots,N, \tag{11.1.10b}$$

$$-\frac{\partial C_1}{\partial x}\bigg|_{x=0} = v(1 - C_1)_{x=0}, \quad D\frac{\partial C_1}{\partial x}\bigg|_{x=L} = 0, \quad v = \frac{V_1^+}{K_1^+}\sqrt{\frac{1}{Dk_-}}, \tag{11.1.10c}$$

$$\frac{\partial C_n}{\partial x}\bigg|_{x=0} = 0 = \frac{\partial C_n}{\partial x}\bigg|_{x=L}, \quad n = 2,\dots,N. \tag{11.1.10d}$$

The crucial parameter that determines the degree of spread of activity is $\gamma$, which is the ratio of the deactivation and activation rates. Numerically solving these equations shows that for $\gamma < 1$, an initial activation signal at the boundary $x = 0$ propagates into the domain, converging to a steady-state solution consisting of stationary front-like profiles that are shifted further into the domain at higher levels of the cascade [652]. This is illustrated schematically in Fig. 11.2(b). On the other hand, if $\gamma > 1$ then the activated proteins fail to propagate into the domain, and the concentrations of activated proteins decay rapidly close to the plasma membrane. In this latter regime, it is possible to obtain analytical approximations for the concentration profiles. First note that the steady-state equation for $C_1$ can be solved as in the single-level model. In particular, for a large domain, it has the exponential form $C_1(x) = C_1(0)e^{-x}$. The steady-state solution at successive levels can then be determined from the approximate recurrence relation [652]

$$\gamma \frac{d^2 C_n}{dx^2} - \gamma C_n + C_{n-1} = 0, \qquad\qquad (11.1.11)$$

assuming $C_n \ll 1$. When $\gamma > 1$, one can solve for $C_n(x)$ using a polynomial expansion of the form

$$C_n(x) = \left( \sum_{m=0}^{n-1} Q_n^{(m)} x^m \right) e^{-x}.$$

In the more interesting regime $\gamma < 1$, the polynomial construction breaks down near the boundary $x = 0$. Nevertheless, one can still determine the leading order behavior in the tail of the profiles, $x \gg 1$, for which

$$C_n(x) \approx \frac{C_1(0)}{(n-1)!(2\gamma)^{n-1}} x^{n-1} e^{-x}.$$

This can then be used to determine the spread of activated protein at the $n$th level [652]. To a first approximation, the front profile shifts an amount (in physical units)

$$\Delta X = (1 - \gamma) \ln(1/\gamma) \sqrt{k_-/D} \qquad\qquad (11.1.12)$$

from one cascade level to the next. Qualitatively similar behavior is found when the full Michaelis–Menten kinetics is used, except now the condition for propagation failure depends on $\gamma$ and the degree of saturation.

## 11.1.2 Robustness of concentration gradients

Intracellular protein gradients provide a mechanism for determining spatial position within a cell so that, for example, cell division occurs at the appropriate time and location [429, 431, 792]. Similarly, developmental morphogen gradients control patterns of gene expression so that each stage of cell differentiation occurs at the correct spatial location within an embryo [45, 90, 268, 542, 978]. For gradient mechanisms to be biologically effective, however, position determination has to be robust to both intrinsic and extrinsic noise fluctuations. Recall from Chap. 5 that extrinsic noise is usually associated with cell-to-cell variations in environmental factors, whereas intrinsic noise refers to fluctuations within a cell due to biochemical reactions involving small numbers of molecules. Both forms of noise can affect positional accuracy as illustrated in Fig. 11.3.

**Effects of intrinsic noise.** We begin by exploring the effects of intrinsic noise on the simple gradient producing systems shown in Fig. 11.1, following the analysis of Tostevin et al. [881, 882]. For simplicity, consider a cylindrical geometry with the dimension $d$ of the system given by $d = 2$ if the gradient is restricted to the membrane (as in Pom1p) or $d = 3$ if it is in the cytoplasm. Take the $x$-axis to be the axial coordinate and assume that the concentration is uniform in the transverse coordinates(s). Take the length of the cell to be $L$ with $0 < x < L$ and assume that

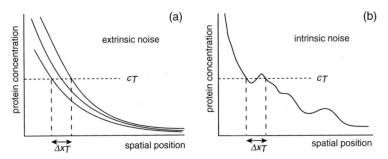

Fig. 11.3: Effect of noise on the positional information provided by concentration gradients. (a) Extrinsic noise in the protein gradient leads to a varying profile from one cell to another. Variation in the position at which the gradient concentration drops below a critical level $c_T$ leads to imprecision in the specification of position $x_T$. (b) Intrinsic noise within a single gradient also leads to imprecise positional information. Redrawn from Howard [431].

there is a source at $x = 0$ where proteins are produced at a rate $J$ per unit area or circumference. The concentration gradient is taken to be given by equation (11.1.3). Suppose that the concentration gradient has to identify a particular spatial location along its length. This location could be determined by the point $x_T$ at which the monotonically decreasing concentration profile crosses a threshold $c_T$. The system will then be divided into two domains: an active domain $0 \leq x < x_T$ for which $c(x) > c_T$ and an inactive domain $x_T \leq x \leq L$ where $c(x) \leq c_T$. Noise in the local protein concentration will cause fluctuations in the threshold position. Tostevin et al. [881] estimated the uncertainty in the position due to intrinsic noise. Recall from Sect. 5.2 that a simple reaction involving protein production and degradation exhibits Poisson statistics. In particular, if $n(\mathbf{x})$ is the random number of molecules in a volume $a^d$ centered at $\mathbf{x} \in \mathbb{R}^d$, then

$$\langle n(\mathbf{x})^2 \rangle - \langle n(\mathbf{x}) \rangle^2 = \langle n(\mathbf{x}) \rangle.$$

Dividing through by the volume, fluctuations in the concentration are given by

$$\langle c(\mathbf{x})^2 \rangle - \langle c(\mathbf{x}) \rangle^2 = \frac{\langle c(\mathbf{x}) \rangle}{a^d},$$

with $\langle c(\mathbf{x}) \rangle$ given by equation (11.1.3). We identify $a$ with the size of the region measuring the concentration, which could be the size of a receptor with which the gradient proteins interact.

The uncertainty $\Delta x$ in spatial location can now be estimated using

$$\Delta x |c'(x_T)| = \sqrt{\mathrm{Var}\, c(\mathbf{x})} = \sqrt{\frac{c(x_T)}{a^d}}, \qquad (11.1.13)$$

which implies that

$$\Delta x = \sqrt{\frac{\lambda D}{J a^d}} e^{x_T/2\lambda}.$$

(11.1.14)

The optimal decay rate $\lambda$ then depends on whether the flux $J$ is kept constant or the number of molecules $N$ is kept constant. In the former case, minimizing $\Delta x$ with respect to $\lambda$ shows that the optimal value is $\lambda = x_T/2$. In the latter case, the flux $J$ can be expressed in terms of the total number of proteins $N$ according to (for $d = 2$)

$$N = L_\perp \int_0^L \rho(x) dx = \frac{J L_\perp \lambda^2}{D} \left[ 1 - e^{-L/\lambda} \right] \approx \frac{J L_\perp \lambda^2}{D},$$

assuming $\lambda \ll L$ and $L_\perp$ is the circumference of the cell. This implies that $J = ND/(L_\perp \lambda^2)$ and thus

$$\frac{\Delta x}{L} = \frac{1}{L} \sqrt{\frac{\lambda^3 L_\perp}{N a^2}} e^{x_T/2\lambda}.$$

(11.1.15)

This formula implies that the uncertainty in position decreases with copy number according to $N^{-1/2}$ and is a nonlinear function of $\lambda$ with a global minimum given at $\lambda_{min} = x_T/3$, since

$$\frac{d\Delta x(\lambda)}{d\lambda} = \frac{3}{2\lambda} \Delta x - \frac{x_T}{2\lambda^2} \Delta x.$$

Typical parameter values for membrane gradients in bacteria such as fission yeast are cell length $L = 10 \mu m$, circumference $L_\perp = 6 \mu m$, diffusivity $D = 0.1 \mu m^2/s$, decay length $\lambda = 2 \mu m$, and detector size $a = 0.01 \mu m$. Taking $x_T = 4 \mu m$ and a reasonable copy number $N = 4000$ leads to the following estimate of the uncertainty in position: $\Delta x \approx 2L$. This implies that the cell must carry out some additional processing at the signal detection level in order to reduce the uncertainty in position. One mechanism for achieving high precision even for low copy numbers is time averaging [375, 881, 882]. (Alternative mechanisms of noise reduction are considered by [429, 431, 792] for intracellular gradients and [90, 268, 276] for developmental morphogen gradients.) Suppose that a receptor, for example, integrates the concentration of the gradient protein over a time interval of length $\tau$. The detector is then able to perform $N_\tau = \tau/\tau_D$ independent measurements of the concentration, where $\tau_D$ is the time for correlations in the concentration to decay. A rough estimate of $\tau_D$ (ignoring logarithmic corrections) is $\tau_D = a^2/D \approx 10^{-3}$ s. From the law of large numbers, we expect the uncertainty in position after time averaging becomes

$$\overline{\Delta x} = \frac{\Delta x}{\sqrt{N_\tau}} = \sqrt{\frac{\tau_D}{\tau}} \Delta x \approx \frac{0.1}{\sqrt{\tau}} L.$$

It follows that an integration time of 100 s would lead to uncertainty in position that is only 1% of the cell length.

In the above analysis we considered the accuracy of positional information based on a single concentration gradient. Another example of a non-uniform distribution of proteins, as proposed for Pom1p in fission yeast, involves two opposing gradients that are used to determine the position of the center of the cell in preparation for cell division [174, 599, 642, 702]. Cell division is initiated when the concentration

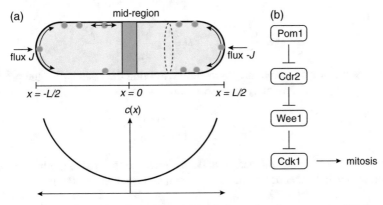

Fig. 11.4: (a) A schematic illustration of a model for the formation of two opposing Pom1p protein gradients within the membrane of a rod-like fission yeast cell [883]. Diffusing proteins within the cytoplasm bind to the membrane at the poles and then undergo lateral diffusion within the membrane. When this is combined with dissociation from the membrane, a symmetric protein concentration profile is set up with a minimum at the mid-plane. The cytoplasmic diffusion and membrane binding are represented as polar fluxes. (b) Cascade of kinases regulating mitosis in fission yeast.

drops below a critical threshold at the mid-plane, which occurs at a specific cell length. Pom1p regulates a mitotic trigger network based on a competing kinase–phosphatase pair, Wee1 and Cdc25, which in turn regulates the activity of the central mitotic regulator cyclin-dependent kinase Cdk1. More specifically, Pom1p inhibits the Wee1-inhibitory kinases, Cdr1 and Cdr2, which localize to specific nodes at the mid-plane[3].

The effects of intrinsic noise on the precision of specifying the mid-point of a fission yeast cell have been analyzed by Tostevin et al. [881, 883], along similar lines to the single gradient case. The basic model is illustrated in Fig. 11.4. It is assumed that Pom1p associates with the membrane at the cell poles, resulting in an effective polar flux $J/L_\perp$ at each end. Here $J$ is the number of molecules per second and $L_\perp$ is the length of the cell circumference. This naturally generates a symmetric concentration profile with a minimum at the center through a combination of lateral membrane diffusion and dissociation from the membrane. Assume that the membrane concentration around the circumference of the cell is uniform, and let $c(x)$ be the density of proteins (per unit area) as a function of the axial coordinate $x$, $x \in [-L/2, L/2]$, where $L$ is cell length. Then $c(x)$ evolves according to the RD

---

[3] Note, however, that recent observations [707] indicate that Pom1p levels at the mid-plane are approximately constant irrespective of cell length, casting doubt on the proposed sizer mechanism. Instead, an "activator accumulation model" has been suggested [290, 707], in which the mitosis activator Cdr2 concentrates within nodes at the mid-plane as a part of a sizer mechanism. That is, the total amount of Cdr2 increases proportionally with cell size (as is the case for many other proteins). Since most cellular Cdc2 is located within the nodes, and the nodes occupy a region of the cellular cortex whose area remains approximately constant, it follows that the midcell concentration of Cdr2 is proportional to the cell area and can thus act as a size indicator.

equation

$$\frac{\partial c}{\partial t} = D\frac{\partial^2 c}{\partial x^2} - \mu c, \quad D\frac{\partial c}{\partial x}\bigg|_{x=\pm L/2} = \pm\frac{J}{L_\perp}, \tag{11.1.16}$$

where $\mu$ is the rate of membrane dissociation. Solving the boundary value problem (see Ex. 11.2) yields the steady-state solution

$$c(x) = \frac{J\lambda}{DL_\perp}\frac{\cosh(x/\lambda)}{\sinh(L/2\lambda)}, \tag{11.1.17}$$

where $\lambda = \sqrt{D/\mu}$ is the characteristic decay length of the spatial gradient. Clearly the deterministic concentration has a minimum at the center $x = 0$ with

$$c(0) = \frac{J}{DL_\perp}\frac{\lambda}{\sinh(L/2\lambda)}. \tag{11.1.18}$$

Now suppose that intrinsic noise generates an uncertainty in the concentration at the center given by $\Delta c = c(0)/a^2$ with $a$ the size of the protein detector. Since the deterministic concentration has a minimum at $x = 0$, it is necessary to Taylor expand to second order, in order to determine the uncertainty $\Delta x$ in position, that is

$$\Delta c = \frac{1}{2}|c''(0)|\Delta x^2. \tag{11.1.19}$$

One thus finds that (see Ex. 11.2)

$$\Delta x = \left(\frac{4DL_\perp\lambda^3\sinh(L/2\lambda)}{Ja^2}\right)^{1/4}. \tag{11.1.20}$$

As in the case of a single concentration gradient, much greater precision can be obtained by time averaging. However, greater care has to be taken with regards to temporal correlations when estimating the minimum time $\tau_D$ required for independent measurements in two-dimensional domains such as the cell membrane, see [881] for details.

**Remark 11.1.** In actively proliferating cells, the size of a cell is determined via the coordination of cell growth and cell division (Chap. 15). In the case of symmetrically dividing cells, each cell approximately doubles in size before dividing. Studies of simple organisms such as bacteria and yeast have identified three general kinds of size control mechanisms that maintain a given size by modifying abnormal cell sizes within one or more cell cycles or generations [14, 290]: (i) "sizer" control, where cells monitor their size and divide at a fixed target size; (ii) "adder" control, involving the addition of a fixed size increment; (iii) "timer" control, in which cells grow for a fixed time duration. The first mechanism is the most efficient, in the sense that it only needs one generation to correct an abnormal cell size, but it also requires an energetically costly size-sensing system. Timer control does not require a size measurement provided that cell growth is linear, that is, cell volume $V$ evolves according to $\dot{V} = C$ (constant growth rate). In this case, which is also an adder mechanism, smaller cells grow proportionally more than larger cells, so that the population approaches a mean size over several generations without any size-sensing mechanism. On the other hand, for exponential growth, $\dot{V} = CV$, cells born larger grow proportionally more over a fixed time interval. Hence, some form of size-sensing mechanism is required to limit cell-cycle duration in larger cells. There are thought to be a number of different size-sensing mechanism utilized by cells, including external landmarks (e.g. axons reaching synaptic targets) and volume measurements based on titrating a constant-concentration sensor molecule against a fixed internal yardstick [14]. In the latter case, the total amount of the sensor molecule has to scale with the cell size. A third size-sensing mechanism exploits the geometry of a cell, under conditions where there is little cell-to-cell variation in the geometry. The most plausible candidates are rod-shaped bacteria and fission yeast, in which growth occurs along a single dimension, and size is determined by spatially varying protein concentrations along the lines illustrated in Fig. 11.4,

**Change in flux $J$.** Another way to characterize the sensitivity of a protein concentration gradient is to consider its response to a change in the flux $J$ at the end $x = 0$ [268]. For the sake of generality, we will allow the deactivation rate to be a nonlinear function $F$ of the local concentration. (In the case of morphogen gradients this would correspond to a nonlinear rate of binding to cell surface receptors.) Therefore, consider the steady-state diffusion equation for $c(x,t) = c(x)$:

$$0 = D\frac{d^2 c(x)}{dx^2} - kF(c(x)), \quad 0 < x < \infty, \tag{11.1.21}$$

supplemented by the boundary conditions

$$c(0) = c_0, \quad c(x) \to 0 \text{ as } x \to \infty. \tag{11.1.22}$$

Note that fixing the concentration at $x = 0$ is equivalent to having a constant flux $J$. Multiplying both sides of the steady-state equation by $c'(x)$ and integrating with respect to $x$ give

$$\lambda^2 c'(x)^2 - E(c) = 0, \quad E(c) = 2\int_0^c F(c')dc', \tag{11.1.23}$$

where $\lambda = \sqrt{D/k}$. Integrating again then yields the solution

$$x(c;c_0) = \lambda \int_c^{c_0} \frac{dc'}{\sqrt{E(c')}}. \tag{11.1.24}$$

Here $x(c;c_0)$ is the position at which $c(x) = c$, given that $c(0) = c_0$.

Now suppose that the flux is changed by modifying the protein concentration at $x = 0$, $c_0 \to c_0 + \Delta c_0$. It follows that there is a shift $\delta(c,c_0)$ in the position where $c(x) = c$, that is,

$$\delta(c,c_0) = x(c;c_0 + \Delta c_0) - x(c;c_0) = \lambda \int_{c_0}^{c_0 + \Delta c_0} \frac{dc}{\sqrt{E(c)}}. \tag{11.1.25}$$

It immediately follows that the shift is independent of the concentration $c$, that is, the protein gradient is shifted uniformly. (A similar result holds for a large but finite domain of length $L$, except for a region around the right-hand boundary [268].) To a first approximation,

$$\delta(c_0) \approx \frac{\lambda}{\sqrt{E(c_0)}} \Delta c_0 = \lambda \frac{c_0}{\sqrt{E(c_0)}} \Delta [\ln c_0].$$

In the linear case, $F(c_0) = c_0$, we see that $\delta(c_0) \approx \lambda \Delta [\ln(c_0)]$.

This simple analysis shows that the shift $\delta$ in the protein profile depends on the characteristic decay length of the gradient at the end $x = 0$. In the case of linear dependence on $c$, the decay length can be identified with $\lambda$ everywhere. This means that in order to make the gradient robust to changes of $c_0$ (small $\delta$), we should take $\lambda$ to be as small as possible. However, such a fast decay could prevent active proteins from extending over the length of the cell (or morphogen extending across the embryonic tissue). This dilemma can be resolved by taking a nonlinear deactivation rate such that the spatial decay rate is fast near to the protein source at $x = 0$ but slower elsewhere. Following [268], introduce the local decay length

$$\Lambda(x) := -\frac{dx}{d\ln c} = -C(x)\frac{dx}{dc} = -\frac{\lambda c(x)}{\sqrt{E(c(x))}}. \tag{11.1.26}$$

We have used equation (11.1.23). In the case of a linear degradation function, $F(c) = c$, we have $E(c) = c^2$ and thus $\Lambda(x) = \lambda$ for all $x$. Let $\Lambda_{av}$ denote the average decay length up to some threshold $c_h$ by

$$\Lambda_{av} = \frac{\displaystyle\int_{c_0}^{c_h} \Lambda(x) d\ln c}{\displaystyle\int_{c_0}^{c_h} d\ln c} = \frac{x_h}{\ln(c_0/c_h)}, \quad c(x_h) = c_h.$$

A possible measure of robustness is then $\mathcal{R} = \Lambda_{av}/\Lambda(0)$ with $\mathcal{R} > 1$ (short initial decay length but a longer decay length in the bulk of the domain).

For the sake of illustration, consider a nonlinear deactivation term of the form $kC^n$ with $n > 1$ and a constant flux $J$ at $x = 0$. At steady state the protein concentration takes the form of a power law rather than an exponential:

$$c(x) = \frac{A}{(x + \xi_0)^m}, \quad A = (m(m+1)D/k)^{1/(n-1)}, \quad \xi_0 = \left(\frac{mAD}{Q_0}\right)^{(n-1)/(n+1)}, \tag{11.1.27}$$

with $m = 2/(n-1)$. The derivation proceeds as follows:

$$\frac{d^2 c}{dx^2} - \frac{c^n}{\lambda^2} = 0,$$

which implies that

$$c'(x)^2 = \frac{2}{\lambda^2(n+1)} c^{n+1}(x).$$

Hence,

$$\int_{c_0}^{c} \frac{da}{a^{(n+1)/2}} = -x\sqrt{\frac{2}{\lambda^2(n+1)}},$$

which can be integrated to give

$$c^{-1/m} = \frac{x + \xi_0}{\sqrt{m(m+1)\lambda^2}}$$

for some integration constant $\xi_0$. We thus obtain the desired result, with $\xi_0$ determined by the boundary condition $-Dc'(0) = J$. Inverting the solution for $c(x)$, $x = (A/c)^{1/m} - \xi_0$, from which the following results hold:

$$\Lambda(x) = \frac{x + \xi_0}{m}, \quad \Lambda_{av} = \frac{x}{m\ln(1 + x/\xi_0)}, \quad \mathcal{R} = \frac{x/\xi_0}{\ln(1 + x/\xi_0)} > 1.$$

Note that as the flux $J$ is increased, $\xi_0$ decreases and robustness increases.

### 11.1.3 Accumulation time

It is important that the time for a concentration gradient to be established is consistent with the relevant biological time scales. One way to estimate the time to approach steady state is to consider an eigenfunction expansion of the solution. Suppose that equation (11.1.1) is written as

$$\frac{\partial c}{\partial t} = -\mathbb{L}c, \quad -D\partial_x c(0,t) = J, \tag{11.1.28}$$

where $\mathbb{L} = -D\partial_x^2 + k$ is a linear operator on the domain $[0, L]$, say. Suppose that there exists a complete set of orthogonal eigenfunctions $\phi_n(x)$ such that

$$\mathbb{L}\phi_n = \lambda_n \phi_n, \quad \int_0^L \phi_m(x)\phi_n(x)dx = 0, \ n \neq m,$$

with the eigenvalues ordered according to $0 = \lambda_0 < \lambda_1 < \lambda_2 \ldots$. The eigenfunction expansion of the concentration takes the form

$$c(x,t) = \sum_{n \geq 0} \phi_n(x)e^{-\lambda_n t}. \tag{11.1.29}$$

It is clear that $c(x,t) \to \phi_0(x)$ as $t \to \infty$ so we can identify $\phi_0(x)$ with the steady-state solution $c^*(x)$. Hence,

$$c(x,t) - c^*(x) = \sum_{n \geq 1} c_n \phi_n(x)e^{-\lambda_n t},$$

with the constant coefficients $c_n$ determined by initial conditions. If the positive eigenvalues are well separated, then the relaxation to steady state will be dominated by the term $\phi_1(x)e^{-\lambda_1 t}$, and we can identify $1/\lambda_1$ as an effective relaxation time. However, this does not account for differences in the relaxation rate at different spatial locations $x$ and relies on the assumption that the eigenvalues have large spectral gaps.

This has motivated an alternative way of characterizing the approach to steady state based on the accumulation time (see also Sect. 6.2). That is, consider the function

$$R(x,t) = 1 - \frac{c(x,t)}{c^*(x)}, \tag{11.1.30}$$

which represents the fractional deviation of the concentration from the steady state. Assuming that there is no overshooting, $1 - R(x,t)$ is the fraction of the steady-state concentration that has accumulated at $x$ by time $t$. It follows that $-\partial_t R(x,t)dt$ is the fraction accumulated in the interval $[t,t+dt]$. The accumulation time is then defined by analogy to MFPTs [60, 61, 359]

$$\tau(x) = \int_0^\infty t\left(-\frac{\partial R(x,t)}{\partial t}\right) dt = \int_0^\infty R(x,t)dt. \tag{11.1.31}$$

As a simple illustration of calculating $\tau(x)$, consider the time-dependent solution of equation (11.1.1) for $L \to \infty$, which is given by

$$c(x,t) = c^*(x)\left[1 - \frac{1}{2}\mathrm{erfc}\left(\frac{\sqrt{Dt}}{\lambda} - \frac{x}{2\sqrt{Dt}}\right) - \frac{e^{2x/\lambda}}{2}\mathrm{erfc}\left(\frac{\sqrt{Dt}}{\lambda} + \frac{x}{2\sqrt{Dt}}\right)\right],$$

where $\mathrm{erfc}(z)$ is the complementary error function. It follows that

$$R(x,t) = \frac{1}{2}\mathrm{erfc}\left(\frac{\sqrt{Dt}}{\lambda} - \frac{x}{2\sqrt{Dt}}\right) + \frac{e^{2x/\lambda}}{2}\mathrm{erfc}\left(\frac{\sqrt{Dt}}{\lambda} + \frac{x}{2\sqrt{Dt}}\right),$$

and [60]

$$\tau(x) = \frac{1}{2k}(1+x/\lambda). \tag{11.1.32}$$

**Nonlinear degradation.** In Sect. 11.1.2, we showed how having a deactivation rate that is a nonlinear function of the protein concentration can increase robustness of gradient formation to fluctuations in the flux $J$. However, this makes the calculation of the accumulation time more complicated. Here we address this issue following the approach of Gordon et al [359], which is based on the comparison principle for parabolic PDEs [308, 916], see Box 11A. Consider the RD equation

$$\frac{\partial c}{\partial t} = D\frac{\partial^2 c(x)}{\partial x^2} - k(c)c, \quad 0 < x < \infty, \tag{11.1.33}$$

with $c(x,0) = 0$ and the boundary conditions

$$-D\frac{\partial c}{\partial x}\bigg|_{x=0} = J, \quad c(x) \to 0 \text{ as } x \to \infty. \tag{11.1.34}$$

Recall that the accumulation time is given by

$$\tau(x) = \int_0^\infty R(x,t)dt = \frac{1}{c^*(x)} \int_0^\infty (c^*(x) - c(x,t))dt.$$

Hence, given the ordered sequence $\underline{c}(x,t) \le c(x,t) \le \overline{c}(x,t)$, we have

$$\frac{1}{c^*(x)} \int_0^\infty (c^*(x) - \overline{c}(x,t))dt \le \tau(x) \le \frac{1}{c^*(x)} \int_0^\infty (c^*(x) - \underline{c}(x,t))dt. \qquad (11.1.35)$$

For the sake of illustration let $D = 1$ and take $k(c) = c$. The steady-state solution is given by

$$c^*(x) = \frac{6}{(a+x)^2}, \quad a = (12/J)^{1/3}.$$

It is convenient to introduce the new variable $w = c^* - w$ such that

$$\frac{\partial w}{\partial t} = \frac{\partial^2 w}{\partial x^2} - (c + c^*)w, \quad 0 < x < \infty,$$

with

$$\partial_x w(0,t) = 0, \quad w(x,0) = c^*(x).$$

Super- and subsolutions are now constructed as follows:

$$\frac{\partial \overline{w}}{\partial t} = \frac{\partial^2 \overline{w}}{\partial x^2} - c^* \overline{w}, \quad 0 < x < \infty,$$

with

$$\partial_x \overline{w}(0,t) = 0, \quad \overline{w}(x,0) = c^*(x),$$

and

$$\frac{\partial \underline{w}}{\partial t} = \frac{\partial^2 \underline{w}}{\partial x^2} - 2c^* \underline{w}, \quad 0 < x < \infty,$$

with

$$\partial_x \underline{w}(0,t) = 0, \quad \underline{w}(x,0) = c^*(x).$$

Since $0 \le c(x,t) \le c^*(x)$ for all $x \ge$ and $t \ge 0$, it can be seen that $\overline{w}, \underline{w}$ satisfy the inequalities necessary for the comparison principle (see Box 11A). Hence,

$$\underline{w} \le w \le \overline{w},$$

for all $x \ge$ and $t \ge 0$, and

$$\frac{1}{c^*(x)} \int_0^\infty \underline{w}(x,t)dt \le \tau(x) \le \frac{1}{c^*(x)} \int_0^\infty \overline{w}(x,t)dt. \qquad (11.1.36)$$

We now exploit the fact that the equations for $\overline{w}, \underline{w}$ are linear so that we can use Laplace transforms. Introducing $W(x,s) = \int_0^\infty e^{-st} w(x,t)dt$, etc., we obtain the boundary value problems

$$\frac{\partial^2 \overline{W}(x,s)}{\partial x^2} - (c^*(x)+s)\overline{W}(x,s) = -c^*(x), \quad \partial_x \overline{W}(0,s) = 0,$$

$$\frac{\partial^2 \underline{W}(x,s)}{\partial x^2} - (2c^*(x)+s)\underline{W}(x,s) = -c^*(x), \quad \partial_x \underline{W}(0,s) = 0.$$

Moreover, in terms of the Laplace transform the bounds on $\tau(x)$ take the simple form

$$\frac{\underline{W}(x,0^+)}{c^*(x)} \leq \tau(x) \leq \frac{\overline{W}(x,0^+)}{c^*(x)}. \tag{11.1.37}$$

It turns out that one can obtain closed form analytical solutions for the Laplace transforms [359], which yield $\overline{W}(x,0^+) = 1$ and $\underline{W}(x,0^+) = 1/2$. Hence,

$$\frac{1}{12}(a+x)^2 \leq \tau(x) \leq \frac{1}{6}(a+x)^2. \tag{11.1.38}$$

---

**Box 11A. Comparison principle for parabolic PDEs.**

---

**Maximum principle.** Let $\Omega \subset \mathbb{R}^n$ be a smooth bounded domain and $f$ a function on $\Omega$. Consider the inhomogeneous diffusion equation

$$\frac{\partial u}{\partial t} = D\nabla^2 u + f, \quad x \in \Omega, \quad t \in (0,T),$$

with initial condition $u(x,0) = g(x)$, $x \in \Omega \cup \partial\Omega$. For the sake of illustration, we impose the Dirichlet boundary condition $u(x,t) = h(x,t)$ for $x \in \partial\Omega$ and $t \in (0,T)$. Take the functions $f,g,h$ to be smooth and define $Q_T = \Omega \times (0,T)$. Roughly speaking, the maximum principle reflects time irreversibility, that is, particles (or heat) flow from regions of high to low density (temperature). As we show below, this implies that depending on the sign of $f$, the solution $u(x,t)$ attains its maximum or minimum value on the boundary

$$\partial Q_T = (\overline{\Omega} \times \{t = 0\}) \cup (\partial\Omega \times (0,T)).$$

First, suppose $f < 0$ (particle sink within the interior). At any interior point the conditions for a local maximum are

$$\partial_t u = 0, \quad \nabla u = 0, \quad \nabla^2 u \leq 0.$$

This implies that $\partial_t u - \nabla^2 u \geq 0$, which contradicts the fact that $f < 0$. A similar argument holds if $t = T$. Hence, $u$ attains its maximum on $\partial Q_T$. In order to extend the argument to the case $f \leq 0$, define $v = u + \varepsilon|x|^2/2n$, where $n$ is the number of space dimensions. Then

$$\partial_t v - \nabla^2 v = f - \varepsilon < 0,$$

and $v$ attains its maximum on $\partial Q_T$. It follows that

$$u \leq v \leq \max_{\partial Q_T} v \leq \frac{\varepsilon a^2}{2n} + M,$$

where $M$ is the maximum value of $u$ on $\partial Q_T$ and $a$ is the largest value of $|x|^2$. Taking the limit $\varepsilon \to 0$ shows that $u \leq M$. Note, however, that $u$ can also take the maximum value in the interior $\Omega$ when $t > 0$ (weak maximum principle). Repeating the analysis for $f \geq 0$, we see that $u$ now attains its minimum value on $\partial Q_T$. Hence if $f = 0$ then $u$ attains both its maximum and minimum values on $\partial Q_T$,

$$\min_{\partial Q_T} u \leq u(x,t) \leq \max_{\partial Q_T} u \quad \text{for all } (x,t) \in Q_T.$$

**Comparison principle.** Let $u$ and $v$ be twice differentiable functions on $Q_T$

$$u_t = D\nabla^2 u + f_1, \quad v_t = D\nabla^2 v + f_2.$$

1. If $u \geq v$ on $\partial Q_T$ and $f_1 \geq f_2$, then $u \geq v$ in $Q_T$.
2. The following stability estimates hold:

$$\max_{\overline{Q}_T} |u - v| \leq \max_{\partial Q_T} |u - v| + T \max_{\overline{Q}_T} |f_1 - f_2|.$$

This ensures that the initial boundary value problem has at most one solution, which depends continuously on the data.

The first condition is a consequence of the maximum principle. That is, let $w = u - v$ so that

$$\partial_t w - D\nabla^2 w = f \equiv f_1 - f_2 \geq 0.$$

From the weak maximum principle, we see that $w$ attains its minimum on $\partial Q_T$. However, $w \geq 0$ on $\partial Q_T$ as $u \geq v$ on $\partial Q_T$. Therefore, $w(x,t) = u(x,t) - v(x,t) \geq 0$ for all $x \in \Omega$ and the result follows. In order to obtain the second condition, let $w = u - v - Mt$, where $M = \max_{\overline{Q}_T} |f_1 - f_2|$. Then

$$\partial_t w - \nabla^2 w = f_1 - f_2 - M \leq 0.$$

The weak maximum principle implies that

$$\max_{\overline{Q}_T} |w| \leq \max_{\partial Q_T} |w| \leq \max_{\partial Q_T} |u - v| + MT,$$

and the result follows. Finally, the comparison principle can be extended to reaction–diffusion equations for which $f$ depends on $u$. That is, consider

$$\partial_t u - \nabla^2 u = f(u,x,t).$$

Suppose that one can construct a supersolution $\bar{u}$ and a subsolution $\underline{u}$ such that

$$\bar{u} - DV^2\bar{u} \geq f(\bar{u},x,t), \quad \underline{u} - DV^2\underline{u} \leq f(\underline{u},x,t).$$

If $\bar{u} \geq u$ on $\partial Q_T$ then $\bar{u} \geq u$ in $Q_T$. Similarly, if $\underline{u} \leq u$ on $\partial Q_T$ then $\underline{u} \leq u$ in $Q_T$.

**Example.** Consider the reaction–diffusion equation

$$\partial u_t = \nabla^2 u + f(u), \quad f(u) = u(1-u)(u-a),$$

with $x \in \Omega, t > 0$, $0 < a < 1$, $u = 0$ on $\partial\Omega$, and $0 \leq u(x,0) \leq 1$ for all $x \in \Omega$. Introduce the supersolution $\bar{u} = 1$ and subsolution $\underline{u} = 0$. Since $f(0) = f(1) = 0$ it follows that both are trivial solutions of the reaction–diffusion equation, and $\bar{u} \geq u \geq \underline{u}$ on $\partial Q_T$. The comparison principle then implies that $0 \leq u(x,t) \leq 1$ for all $x \in \Omega$ and $0 < t < T$.

### 11.1.4 Protein concentration gradient formation in C. elegans via switching diffusions

A novel mechanism for intracellular gradient formation has been found in an experimental study of asymmetric division in the *C. elegans* zygote, which takes place after the initial phase of Par polarization [957]. A pair of RNA-binding proteins MEX-5 and PIE-1 form opposing subcellular concentration gradients in the absence of a local source due to a spatially heterogeneous switching process. That is, both proteins switch between fast-diffusing and slow-diffusing states on time scales that are much shorter (seconds) than the time scale of gradient formation (minutes). Moreover, the switching rates are strongly polarized along the anterior/posterior axis of the zygote such that fast-diffusing MEX-5 and PIE-1 proteins are approximately symmetrically distributed, whereas the corresponding slow-diffusing proteins are highly enriched in the anterior and posterior cytoplasm, respectively.

In this section we describe a theoretical analysis of protein gradient formation in *C. elegans* developed in Ref. [127], which extends previous theoretical work on spatially heterogeneous switching diffusions [122, 123], see also Sect. 6.5. Consider a population of diffusing particles that independently switch between two conformational states, labeled $n = 0, 1$, according to a two-state jump Markov process $N(t) \in \{0,1\}$, with $0 \underset{\beta}{\overset{\alpha}{\rightleftharpoons}} 1$, see Fig. 11.5. The diffusion coefficient is taken to depend on the conformational state, that is $D = D_n$ when $N(t) = n$. Let $C_n(x,t)$ denote the concentration of protein in state $N(t) = n$, where $t$ is the time from the onset of gradient formation and $x$, $0 < x < L$, is the distance from the anterior pole of the embryo whose size is $L$. Finally, the switching rates are taken to depend on $x$,

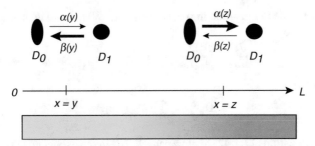

Fig. 11.5: Schematic illustration of heterogeneous diffusion due to temporal disorder. A Brownian particle randomly switches between two conformational states $n = 0, 1$ having different diffusivities such that $D_0 < D_1$. The switching rates $\alpha(x), \beta(x)$ are taken to be dependent on spatial position $x$, $0 \leq x \leq L$. In this particular example, the relative rate of switching $\alpha(x)/\beta(x)$ increases to the right as indicated by the color gradient. Hence, the particle spends more time in the slow-diffusing state at the end $x = 0$ and more time in the fast-diffusing state at the end $x = L$.

$\alpha = \alpha(x), \beta = \beta(x)$. We then have the system of equations

$$\frac{\partial C_0}{\partial t} = D_0 \frac{\partial^2 C_0}{\partial x^2} - \alpha(x)C_0 + \beta(x)C_1, \qquad (11.1.39a)$$

$$\frac{\partial C_1}{\partial t} = D_1 \frac{\partial^2 C_1}{\partial x^2} + \alpha(x)C_0 - \beta(x)C_1, \qquad (11.1.39b)$$

with boundary conditions

$$\partial_x C_0(0,t) = \partial_x C_1(0,t) = 0, \quad \partial_x C_0(L,t) = \partial_x C_1(L,t) = 0, \qquad (11.1.40)$$

and the initial conditions $C_n(x,0) = C_n^*$. In contrast to the standard mechanism of gradient formation, there are no localized sources of protein production.

We begin by considering the fast switching limit $\alpha(x), \beta(x) \to \infty$, see also Sect. 6.5. A typical length of C. elegans is around $L = 32\,\mu$m and the switching rates are of the order $0.1$ s$^{-1}$ [957]. Introducing the fundamental time scale $\tau = L^2/D$ with $D = 1\,\mu$m$^2$/s, we have $\tau \sim 1000$ s and thus the switching rates are at least two orders of magnitude faster than $\tau^{-1}$. Hence, we can rescale the transition rates according to $\alpha, \beta \to \alpha/\varepsilon, \beta/\varepsilon$, with $\alpha, \beta = O(1)$. For small but nonzero $\varepsilon$, one can use an adiabatic approximation to reduce the diffusion equations (11.1.39a,b) to a corresponding scalar diffusion equation for the total density $C(x,t) = \sum_{n=0,1} C_n(x,t)$ along similar lines to the steps outlined in Box 5D. First, decompose the density $C_n$ as

$$C_n(x,t) = C(x,t)\rho_n(x) + \varepsilon w_n(x,t), \qquad (11.1.41)$$

where $\sum_n w_n(x,t) = 0$ and

$$\rho_0(x) = \frac{\beta(x)}{\alpha(x) + \beta(x)}, \quad \rho_1(x) = 1 - \rho_0(x).$$

Substituting this decomposition into equations (11.1.39a,b) and then adding the pair of equations give

$$\frac{\partial C}{\partial t} = \frac{\partial^2 \overline{D}(x)C}{\partial x^2} + \varepsilon \sum_{n=0,1} D_n \frac{\partial^2 w_n}{\partial x^2}, \qquad (11.1.42)$$

where

$$\overline{D}(x) = \sum_{n=0,1} D_n \rho_n(x). \qquad (11.1.43)$$

Next we use equation (11.1.42) to eliminate $\partial C/\partial t$ in the expanded version of equations (11.1.39a,b). Introducing the asymptotic expansion $w_n \sim w_n^{(0)} + \varepsilon w_n^{(1)} + O(\varepsilon^2)$ and collecting the $O(1)$ terms then yield an equation for $w_n^{(0)}$, which has the following unique solution on imposing the condition $\sum_n w_n^{(0)}(x,t) = 0$,

$$w_n^{(0)} = \frac{1}{\alpha(x) + \beta(x)} \left[ D_n \frac{\partial^2 \rho_n(x)C}{\partial x^2} - \rho_n(x) \frac{\partial^2 \overline{D}(x)C}{\partial x^2} \right].$$

Finally, setting $w_n = w_n^{(0)} + O(\varepsilon)$ in equation (11.1.42) shows that to $O(\varepsilon)$

$$\frac{\partial C}{\partial t} = \frac{\partial^2}{\partial x^2}(\overline{D}(x)C) + \varepsilon(D_0 - D_1)\frac{\partial^2 w_0^{(0)}}{\partial x^2}. \qquad (11.1.44)$$

In the fast switching limit $\varepsilon \to 0$, we thus have a diffusion equation with effective space-dependent diffusivity $\overline{D}(x)$:

$$\frac{\partial C}{\partial t} = \frac{\partial^2}{\partial x^2}(\overline{D}(x)C), \qquad (11.1.45)$$

with no-flux boundary conditions $\partial_x \overline{D}(x)C(x,t) = 0$ at $x = 0, L$. It is now straightforward to establish that the bulk solution can take the form of a protein concentration gradient. The steady-state solution takes the form

$$C^*(x) = \frac{A}{\overline{D}(x)}, \quad A = L[C_0^* + C_1^*]\left[\int_0^L \frac{dx}{\overline{D}(x)}\right]^{-1}, \qquad (11.1.46)$$

with the constant $A$ determined by the normalization condition $\int_0^L C^*(x)dx = L[C_0^* + C_1^*]$. It is clear that regions of slow diffusion will have higher concentrations than regions of fast diffusion. Suppose for the sake of illustration that $D_0 < D_1$. This means that $\overline{D}(x)$ will be a monotonically increasing function of $x$ if $\beta(x)$ is a constant or a decreasing function of $x$ and $\alpha(x)$ is an increasing function of $x$; the resulting stationary concentration will be a monotonically decreasing function of $x$. Substituting the switching rates found in [957] for the RNA-binding proteins MEX-5 and PIE-1 yields the numerical and first-order asymptotic results shown in Fig. 11.6 [127]. Note that the first-order asymptotics yield a good approximation of the steady-state concentration.

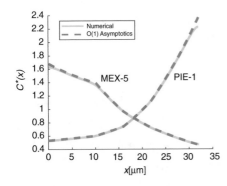

Fig. 11.6: Plot of numerical solution and first-order asymptotics for MEX-5 and PIE-1 spatial switching rates taken from [957].

*Boundary layer analysis.* There is one subtle point about the above analysis that needs to be highlighted. The original system given by equations (11.1.39a,b) involves two coupled diffusion equations so that at each boundary there are two boundary conditions, namely zero flux conditions for $C_0$ and $C_1$. On the other hand, the reduced diffusion equation (11.1.44) for the scalar $C = C_0 + C_1$ has a single boundary condition at each end. If we take this to be the linear combination $D_0 \partial_x C_0 + D_1 \partial_x C_1 = 0$ and substitute the decomposition (11.1.41), then we obtain the non-flux conditions

$$\partial_x [\overline{D}(x) C(x,t) + \varepsilon (D_0 - D_1) w_0^{(0)}]_{x=0} = 0, \tag{11.1.47}$$

$$\partial_x [\overline{D}(x) C(x,t) + \varepsilon (D_0 - D_1) w_0^{(0)}]_{x=L} = 0. \tag{11.1.48}$$

We thus have a singular perturbation problem, in which the solution to equation (11.1.44) represents an outer solution that is valid in the bulk of the domain, but has to be matched to an inner solution at each boundary. (An analogous situation holds in mathematical models of bidirectional motor transport [668, 989], see Sect. 7.1.)

In order to solve the steady-state singular perturbation problem, we introduce an $O(\sqrt{\varepsilon})$ boundary layer at $x = 0$ and similarly at $x = L$, which can capture rapid changes in spatial derivatives. We then construct an inner solution within each boundary layer that can then be matched to the outer solution of equation (11.1.39). For the sake of illustration, we focus on the boundary layer at $x = 0$; the analysis for the other boundary layer is very similar. Introduce the stretched coordinate $X = x/\sqrt{\varepsilon}$ and series expansions

$$\alpha(\sqrt{\varepsilon}X) \sim \alpha_0 + \alpha_1 \sqrt{\varepsilon}X + O(\varepsilon), \tag{11.1.49a}$$

$$\beta(\sqrt{\varepsilon}X) \sim \beta_0 + \beta_1 \sqrt{\varepsilon}X + O(\varepsilon). \tag{11.1.49b}$$

Denote the steady-state inner solution by $C_{in}^*(X)$, which is taken to have the series expansion

$$C_{in}^*(X) \sim c_{n,0}(X) + \sqrt{\varepsilon} c_{n,1}(X) + O(\varepsilon). \tag{11.1.50}$$

The steady-state version of equation (11.1.39) yields, to leading order, the following inner equations on the domain $X \in [0, \infty)$:

$$0 = D_0 \frac{d^2 c_{0,0}}{dX^2} - \alpha_0 c_{0,0} + \beta_0 c_{1,0}, \tag{11.1.51a}$$

$$0 = D_1 \frac{d^2 c_{1,0}}{dX^2} + \beta_0 c_{0,0} - \alpha_0 c_{1,0}, \tag{11.1.51b}$$

with boundary conditions $c_{0,0}'(0) = c_{1,0}'(0) = 0$. Adding equations (11.1.51a) and (11.1.51b) and imposing the boundary conditions shows that $\sum_{n=0,1} D_n c_{n,0}(X) = \Gamma_0$, where $\Gamma_0$ is a constant. Equation (11.1.51a) can thus be rewritten as

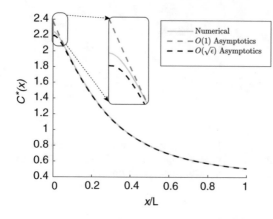

Fig. 11.7: Equilibrium density with switching rates given by $\alpha(x) = (x+0.65)^4$ and $\beta(x) = 2$. The diffusion coefficients were set to $D_0 = 0.5, D_1 = 5$, with $\varepsilon = 5 \times 10^{-4}$. The red and black dashed lines show the increased accuracy achieved by the $O(\sqrt{\varepsilon})$ terms.

$$0 = \frac{d^2c_{0,0}}{dX^2} - \left(\frac{\alpha_0}{D_0} + \frac{\beta_0}{D_1}\right)c_{0,0} + \frac{\beta_0\Gamma_0}{D_0D_1}. \tag{11.1.52}$$

This has the bounded solution

$$c_{0,0}(X) = A_0e^{-\gamma X} + \frac{\beta_0\Gamma_0}{D_0D_1\gamma}, \quad \gamma = \sqrt{\frac{\alpha_0}{D_0} + \frac{\beta_0}{D_1}} = \sqrt{\frac{\overline{D}(0)}{D_0D_1}}(\alpha_0 + \beta_0). \tag{11.1.53}$$

The boundary condition $c'_{0,0}(0)$ implies that $A_0 = 0$. Carrying out a similar analysis for $c_{1,0}$ we deduce that the lowest order terms are constants, that is, $c_{n,0}(X) = \overline{c}_n$ with $\overline{c}_n = \rho_n(0)\Gamma_0/\overline{D}(0)$.

Proceeding to the next order, we have

$$\alpha_1 X\overline{c}_0 - \beta_1 X\overline{c}_1 = D_0\frac{d^2c_{0,1}}{dX^2} - \alpha_0c_{0,1} + \beta_0c_{1,1}, \tag{11.1.54a}$$

$$-\alpha_1 X\overline{c}_0 + \beta_1 X\overline{c}_1 = D_1\frac{d^2c_{1,1}}{dX^2} + \alpha_0c_{0,1} - \beta_0c_{1,1}, \tag{11.1.54b}$$

with boundary conditions $c'_{0,1}(0) = c'_{1,1}(0) = 0$. Again adding equations (11.1.54a) and (11.1.54b) shows that $\sum_{n=0,1} D_nc_{n,1}(X) = \Gamma_1$ for some constant $\Gamma_1$. Hence, equation (11.1.54a) becomes

$$\frac{d^2c_{0,1}}{dX^2} - \left(\frac{\alpha_0}{D_0} + \frac{\beta_0}{D_1}\right)c_{0,1} + \frac{\beta_0\Gamma_1}{D_0D_1} = (\alpha_1\rho_0(0) - \beta_1\rho_1(0))\frac{\Gamma_0 X}{D_0\overline{D}(0)}.$$

The bounded solutions are

$$c_{0,1}(X) = B_0e^{-\gamma X} - \frac{D_1(\alpha_1\beta_0 - \beta_1\alpha_0)}{(\alpha_0 + \beta_0)^2}\frac{\Gamma_0 X}{\overline{D}(0)^2},$$

$$c_{1,1}(X) = B_1e^{-\gamma X} + \frac{D_0(\alpha_1\beta_0 - \beta_1\alpha_0)}{(\alpha_0 + \beta_0)^2}\frac{\Gamma_0 X}{\overline{D}(0)^2}.$$

Without loss of generality we have set $\Gamma_1 = 0$. The coefficients $B_0$ and $B_1$ can be determined in terms of $\Gamma_0$ by imposing the non-flux boundary conditions at $X = 0$.

Combining our various results and using the definition of $\overline{D}(x)$ leads to the following inner solution

$$C_{in}^*(X) \sim Be^{-\gamma X} + \Gamma_0 \left( \frac{1}{\overline{D}(0)} - \sqrt{\varepsilon} \frac{X\overline{D}'(0)}{\overline{D}(0)^2} \right) + O(\varepsilon) \sim Be^{-\gamma X} + \frac{\Gamma_0}{\overline{D}(\sqrt{\varepsilon}X)}.$$

$$(11.1.55)$$

The boundary condition $dC_{in}^*(X)/dX = 0$ at $X = 0$ shows that $B = -\Gamma_0 \overline{D}'(0)/(\gamma \overline{D}(0)^2)$. The composite solution then has the form

$$C^*(x) = \frac{A}{\overline{D}(x)} + \sqrt{\varepsilon} Be^{-\gamma x/\sqrt{\varepsilon}}, \qquad (11.1.56)$$

with the matching condition $\Gamma_0 = A$. Performing the same boundary analysis for the boundary layer at $x = L$, we find that the composite solution for the entire domain has the form

$$C^*(x) = \frac{A}{\overline{D}(x)} - \sqrt{\varepsilon} \frac{A\overline{D}'(0)}{\gamma \overline{D}(0)^2} e^{-\gamma x/\sqrt{\varepsilon}} + \sqrt{\varepsilon} \frac{A\overline{D}'(L)}{\mu \overline{D}(L)^2} e^{-\mu(L-x)/\sqrt{\varepsilon}}, \quad (11.1.57)$$

where we have defined

$$\mu = \sqrt{\frac{\alpha(L)}{D_0} + \frac{\beta(L)}{D_1}} = \sqrt{\frac{\overline{D}(L)}{D_0 D_1} (\alpha(L) + \beta(L))}. \qquad (11.1.58)$$

Note that the correction to the normalization constant $A$ given in (11.1.46) is of $O(\varepsilon)$, and hence does not need to be included here at $O(\sqrt{\varepsilon})$. For small $\varepsilon$, these terms result in an increased accuracy near the boundary by helping enforce the no-flux boundary conditions in the asymptotic solution. Fig. 11.7 highlights an example of this improvement.

## 11.2 Turing pattern formation

One of the necessary conditions for the formation of a concentration gradient is that there exists some local source of intracellular proteins or a preexisting spatial variation of upstream signaling molecules/mRNA in an embryo. In other words, an underlying spatial symmetry of the cell is broken explicitly, see Sect. 11.1. An alternative mechanism for the formation of spatial patterns, via spontaneous symmetry breaking, is the diffusion-driven instability first hypothesized by Turing [889]. Typically, such an instability is modeled in terms of a system of RD equations. In the absence of diffusion, the system converges to a homogeneous stable steady state, whereas the addition of diffusion can destabilize the homogeneous state resulting in a spatially varying pattern (provided that the physical domain is sufficiently large). One major biological application of the Turing mechanism has been in development, based on RD models of morphogens [655, 698]. More recently, however, the Turing mechanism has been used to account for various spatiotemporal patterns at the intracellular level (see Sect. 11.3), in particular, the spatial variation of membrane-bound proteins in polarizing and dividing cells [361, 615]. Before considering explicit applications to cell biology, we review the basic theory of pattern formation.

For concreteness, we will focus on a two-component system in two spatial dimensions, consisting of chemical concentrations $u(\mathbf{x}, t)$ and $v(\mathbf{x}, t)$ with $\mathbf{x} \in \mathbb{R}^2, t \in$

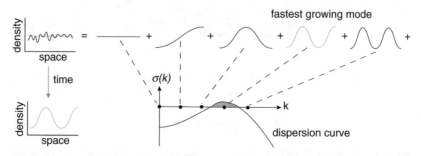

Fig. 11.8: Linear instability underlying the classical Turing mechanism. Any random perturbation of a spatially uniform state can be decomposed into the Fourier modes. Linear stability analysis yields the growth rate $\sigma(k)$ of the amplitude of all modes with wavenumber $k$. A plot of $\sigma(k)$ as a function of $k$ yields a dispersion curve. Unstable modes (marked blue) grow in amplitude and determine the initial pattern emerging out of the random perturbation. As the pattern continues to grow, nonlinearities of the system kick in and help to stabilize the resulting pattern. [Redrawn from Halatek, Brauns, and Frey [391].]

$\mathbb{R}^+$. We will assume that the system is restricted to a bounded square domain $\Omega$ of size $L$ so that $0 \leq x \leq L$ and $0 \leq y \leq L$. The standard RD model takes the form

$$\frac{\partial u}{\partial t} = D_u \nabla^2 u + f(u, v), \tag{11.2.1a}$$

$$\frac{\partial v}{\partial t} = D_v \nabla^2 v + g(u, v). \tag{11.2.1b}$$

Here $D_u$ and $D_v$ are the corresponding diffusion coefficients, and the nonlinear functions $f, g$ describe the chemical reactions. The RD system is typically supplemented by the no-flux boundary conditions

$$\mathbf{n} \cdot \nabla u = 0, \quad \mathbf{n} \cdot \nabla v = 0, \quad \text{for } \mathbf{x} \in \partial \Omega,$$

where $\mathbf{n}$ is the outward normal on the boundary $\partial \Omega$. Suppose that there exists a homogeneous stationary state $(u^*, v^*)$ for which $f(u^*, v^*) = g(u^*, v^*) = 0$. The basic idea of the Turing mechanism is that the stationary state is stable to perturbations in the absence of diffusion, but is unstable to spatially inhomogeneous perturbations when diffusion is present. This is illustrated in Fig. 11.8. Linear stability analysis can be used to identify the fastest growing perturbations, which can be expressed in terms of linear combinations of eigenmodes of the associated linear operator. However, in order to determine whether or not these growing patterns themselves stabilize, it is necessary to go beyond linear theory by using perturbation methods to derive a system of nonlinear ODEs for the amplitude of a given pattern.

For the moment, we will focus on classical mechanisms for Turing pattern formation such as activator–inhibitor systems [346, 610, 611], see Fig. 11.9, whose reactions involve a mixture of autocatalytic production and degradation and are typically non-mass conserving. However, it is well known that such systems tend to

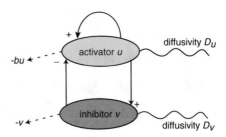

Fig. 11.9: Classical activator–inhibitor mechanism for Turing pattern formation, based on a combination of autocatalytic production and degradation.

exhibit spontaneous pattern formation over a relatively narrow range of parameters and often require a separation of time scales between the diffusion rates of the different chemical species. Moreover, as highlighted by Frey *et al.* [102, 391, 392], although these type of systems arise during cell development, they do not represent the typical intracellular processes underlying pattern formation during cell division and cell polarization, see Sect.11.3. The latter phenomena are more appropriately described in terms of the spatial redistribution of different chemical substrates in mass-conserving systems, involving the diffusive flux of cytosolic proteins, the exchange of proteins between the cytoplasm and the plasma membrane, and the autocatalytic activation of membrane-bound proteins. We will consider the general theory of pattern formation in mass-conserving systems in Sect. 11.3.1.

### 11.2.1 Linear stability analysis

Linearizing equation (11.2.1) about the homogeneous state $(u^*, v^*)$ by setting $U = u - u^*, V = v - v^*$, leads to the linear system

$$\frac{\partial}{\partial t}\begin{pmatrix} U \\ V \end{pmatrix} = \mathbb{L}\begin{pmatrix} U \\ V \end{pmatrix} \equiv \begin{pmatrix} f_u & f_v \\ g_u & g_v \end{pmatrix}_{u^*, v^*}\begin{pmatrix} U \\ V \end{pmatrix} + \begin{pmatrix} D_u & 0 \\ 0 & D_v \end{pmatrix}\begin{pmatrix} \nabla^2 U \\ \nabla^2 V \end{pmatrix}, \quad (11.2.2)$$

supplemented by no-flux boundary conditions. Here $f_u = \partial f / \partial u$, etc. Setting $\mathbf{U} = (U, V)^T$, the general solution of (11.2.2) can be written as

$$\mathbf{U}(\mathbf{x}, t) = \sum_{\mathbf{k}} \mathbf{c}_k e^{i\mathbf{k} \cdot \mathbf{x}} e^{\lambda(k)t},$$

where $\mathbf{c}_k e^{i\mathbf{k} \cdot \mathbf{x}}$ and $\lambda(k)$ form an eigenvalue pair of the linear operator $\mathbb{L}$, parameterized by the wavenumber $k = |\mathbf{k}|$. The $k$-dependence of the eigenvalue $\lambda(k)$ is known as a dispersion relation. Substitution of the general solution into (11.2.2) yields a characteristic equation for $\lambda(k)$:

$$|\mathbf{A} - \mathbf{D}k^2 - \lambda(k)\mathbf{I}| = 0, \quad (11.2.3)$$

with

$$\mathbf{A} = \begin{pmatrix} f_u & f_v \\ g_u & g_v \end{pmatrix}_{u^*,v^*}, \qquad \mathbf{D} = \begin{pmatrix} D_u & 0 \\ 0 & D_v \end{pmatrix}.$$

Evaluating the determinant, we obtain a quadratic equation for $\lambda$:

$$\lambda^2 + [(D_u + D_v)k^2 - f_u - g_v]\lambda + D_u D_v k^4 - k^2(D_v f_u + D_u g_v) + f_u g_v - f_v g_u = 0.$$

In the absence of diffusion ($D_u = D_v = 0$), this reduces to

$$\lambda^2 - [f_u + g_v]\lambda + f_u g_v - f_v g_u = 0,$$

and the requirement that the homogeneous state is stable in the absence of diffusion leads to the conditions

$$f_u + g_v < 0, \quad f_u g_v - f_v g_u > 0. \tag{11.2.4}$$

On the other hand, the requirement that the stationary state is unstable to perturbations in the presence of diffusion means that there exists a nonzero wavenumber $k$ for which $\lambda(k) = 0$. Setting $\lambda = 0$ in the quadratic equation yields

$$D_u D_v k^4 - k^2(D_v f_u + D_u g_v) + f_u g_v - f_v g_u = 0,$$

which will have a positive solution for some nonzero $k^2$ provided that

$$D_v f_u + D_u g_v > 4 D_u D_v (f_u g_v - f_v g_u) > 0. \tag{11.2.5}$$

Equations (11.2.4) and (11.2.5) give the conditions for a Turing instability. Clearly they cannot all be satisfied if $D_u = D_v$, which means that the chemical species have to diffuse at different rates. In particular, assuming $f_u > 0$ for the activator and $g_v < 0$, then we require $D_v > D_u$, that is the inhibitor diffuses faster than the activator.

One can use these to identify regions in parameter space where a Turing instability can occur. Suppose that one chooses a point in parameter space just outside this instability region such that plotting $\text{Re}[\lambda(k)]$ as a function of $k$ yields dispersion curves that lie below the horizontal axis. If one now varies an appropriate bifurcation parameter so that the system crosses a boundary of the instability region, then at least one of the dispersion curves crosses the axis at a critical wavenumber $k_c$, and spatially periodic patterns at the critical wavelength $2\pi/k_c$ start to grow; this is the onset of the Turing instability. Finally, note that in the given bounded domain, the linear operator $\mathbb{L}$ is compact and has a discrete spectrum (see Box 11G). In other words, the wave vectors are discrete with $k = (\pi/L)\sqrt{n_x^2 + n_y^2}$ for integers $n_x, n_y$. However, when discussing dispersion curves one often treats $k$ as a continuous variable.

**Example 11.1.** In order to illustrate the above ideas, we consider a non-dimensionalized activator-inhibitor RD system analyzed by Barrio et al. [47] (see also Ex. 11.3):

$$\frac{\partial u}{\partial t} = D\nabla^2 u + \kappa(u + av - uv^2 - Cuv), \qquad (11.2.6a)$$

$$\frac{\partial v}{\partial t} = \nabla^2 v + \kappa(-u + bv + uv^2 + Cuv). \qquad (11.2.6b)$$

This has a unique stationary state at $(u,v) = (0,0)$. Linearizing about this state leads to the characteristic equation

$$\left| \begin{pmatrix} \kappa - Dk^2 & \kappa a \\ -\kappa & b\kappa - k^2 \end{pmatrix} - \lambda(k)\mathbf{I} \right| = 0, \qquad (11.2.7)$$

which yields the quadratic equation

$$(\kappa - Dk^2 - \lambda)(b\kappa - k^2 - \lambda) + a\kappa^2 = 0.$$

When $k = 0$, we have

$$\lambda^2 - (1+b)\kappa\lambda + (b+a)\kappa^2 = 0,$$

with roots

$$\lambda = \frac{\kappa}{2}\left[(1+b) \pm \sqrt{(1+b)^2 - 4(b+a)}\right].$$

The fixed point $(0,0)$ is stable to uniform perturbations provided $b < -1$ and $b + a > 0$. The condition for a Turing instability is that there exists a positive solution for $k^2$ when $\lambda(k) = 0$:

$$Dk^4 - \kappa k^2(Db + 1) + \kappa^2(b+a) = 0.$$

At onset of the instability, the discriminant of the quadratic for $k^2$ vanishes, that is, $[\kappa(Db + 1)]^2 = 4D\kappa^2(b+a)$, which implies

$$a_c = \frac{(Db - 1)^2}{4D}, \ k_c^2 = \frac{\kappa(Db + 1)}{2D},$$

where $k_c$ is the critical wavenumber. Such a solution exists provided that $Db + 1 > 0$. In particular, $D < 1$, which indicates that the activator diffuses more slowly than the inhibitor. Note that the parameter $C$ plays no role in the linear analysis. The Turing instability region in the $(a,b)$-plane can now be determined, as illustrated in Fig. 11.10. One striking feature is that the Turing instability occupies a relatively small domain of parameter space - this is a common finding in the theory of Turing pattern formation, and is known as the fine-tuning problem. In Fig. 11.11 we sketch the dominant dispersion curves for two different sets of parameters, indicating a small band of unstable eigenmodes in both cases. One way to broaden the domain is to include the effects of noise, see Sect. 16.1.

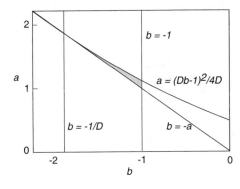

Fig. 11.10: Stability diagram for the example RD system given by equation (11.2.6). The shaded region indicates where in parameter space the homogeneous fixed point $(0,0)$ undergoes a Turing instability.

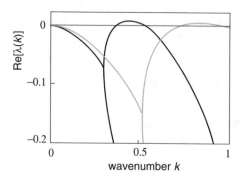

Fig. 11.11: The dispersion relation $\lambda(k)$ of the dominant eigenvalue for two different parameter sets. In both cases there is a small band of wavenumbers for which $\mathrm{Re}[\lambda(k)] > 0$. The critical wavenumber $k_c$ at the peak of the dispersion curve determines the wavelength of the emerging pattern.

**Fast diffusion limit.** Intuitively speaking, one would not expect a Turing instability to occur when the diffusivities of all chemical species are sufficiently large. This is indeed found to be the case for a general RD system of the form [655]

$$\frac{\partial \mathbf{u}}{\partial t} = \mathbf{D}\nabla^2 \mathbf{u} + \mathbf{f}(\mathbf{u}),$$

where $\mathbf{u} = (u_1, \ldots, u_n)$ and $\mathbf{D} = \mathrm{diag}(D_1, \ldots, D_n)$. We will consider a bounded domain $\Omega$ with no-flux boundary condition $(\mathbf{n} \cdot \nabla)\mathbf{u} = 0$ on $\partial\Omega$. Consider the so-called energy function

$$E(t) = \frac{1}{2} \int_{\Omega} \|\nabla \mathbf{u}\|^2 d\mathbf{x}, \quad \|\nabla \mathbf{u}\|^2 = \sum_{i=1}^{n} |\nabla u_i|^2. \tag{11.2.8}$$

Differentiating $E(t)$ and using the RD equation, we have

$$\frac{dE}{dt} = \sum_i \int_{\Omega} \nabla u_i \cdot \nabla \partial_t u_i d\mathbf{x} = \sum_i \int_{\Omega} \nabla u_i \cdot \nabla D_i \nabla^2 u_i d\mathbf{x} + \sum_i \int_{\Omega} \nabla u_i \cdot \nabla f_i(\mathbf{u}) d\mathbf{x}$$

$$= \sum_i \int_{\partial\Omega} (D_i \nabla^2 u_i) \nabla u_i \cdot d\sigma - \sum_i \int_{\Omega} (\nabla^2 u_i)(D_i \nabla^2 u_i) d\mathbf{x} + \sum_i \int_{\Omega} \nabla u_i \cdot \nabla f_i(\mathbf{u}) d\mathbf{x},$$

after integrating by parts. The first sum on the right-hand side vanishes due to the no-flux boundary conditions, whereas the second sum satisfies the inequality

$$-\sum_i \int_{\Omega} (\nabla^2 u_i)(D_i \nabla^2 u_i) d\mathbf{x} \leq -D_{\min} \int_{\Omega} (\nabla^2 u_i)^2 d\mathbf{x},$$

where $D_{\min} = \min_i D_i$. Finally, using the chain rule, we have

$$\sum_i \int_{\Omega} \nabla u_i \cdot \nabla f_i(\mathbf{u}) d\mathbf{x} = \sum_{i,\mu} \int_{\Omega} \partial_\mu u_i \partial_\mu f_i(\mathbf{u}) d\mathbf{x}$$

$$= \sum_{i,\mu} \int_{\Omega} \partial_\mu u_i \sum_j \frac{\partial f_i(\mathbf{u})}{\partial u_j} \partial_\mu u_j d\mathbf{x} = \sum_\mu \int_{\Omega} \mathbf{v}_\mu \cdot \mathbf{A} \mathbf{v}_\mu d\mathbf{x},$$

with

$$\mathbf{v}_\mu = \partial_\mu \mathbf{u}, \quad A_{ij} = \frac{\partial f_i(\mathbf{u})}{\partial u_j}.$$

Since

$$\mathbf{v}_\mu \cdot \mathbf{A} \mathbf{v}_\mu \le |\mathbf{v}_\mu \cdot \mathbf{A} \mathbf{v}_\mu| \le \sum_\mu |\mathbf{v}_\mu|^2 \|\mathbf{A}\|,$$

with

$$\|\mathbf{A}\| \equiv m = \max_{\mathbf{u}} \|\nabla_{\mathbf{u}} \mathbf{f}(\mathbf{u})\| = \max_{u_1, \dots u_n} \sqrt{\sum_{i,j=1}^n \left( \frac{\partial f_i}{\partial u_j} \right)^2},$$

it follows that

$$\sum_i \int_\Omega \nabla u_i \cdot \nabla f_i(\mathbf{u}) dx \le m \sum_\mu \int_\Omega |\mathbf{v}_\mu|^2 dx = m \sum_i \int_\Omega |\nabla u_i|^2 dx = mE.$$

Combining all of the results thus yields the inequality

$$\frac{dE}{dt} \le -D_{\mathrm{main}} \int_\Omega |\nabla^2 \mathbf{u}|^2 dx + mE. \tag{11.2.9}$$

The final step in establishing the non-existence of a Turing instability in the fast diffusion limit is to use the Poincare inequality (Box 11B). Taking $v = \partial_\mu u_i$ in (11.2.11) gives

$$\int_\Omega (\partial_\mu u_i)^2 dx \le \frac{1}{\mu_1} \int_\Omega \sum_v (\partial_v \partial_\mu u_i)^2 dx = \frac{1}{\mu_1} \int_\Omega \sum_v (\partial_v^2 u_i)(\partial_\mu^2 u_i) dx,$$

after integrating by parts. Summing both sides with respect to $i, \mu$ then gives

$$\int_\Omega \|\nabla \mathbf{u}\|^2 dx \le \frac{1}{\mu_1} \int_\Omega |\nabla^2 u(\mathbf{x})|^2 dx.$$

Substituting this result into equation (11.2.9), we obtain the inequality

$$\frac{dE}{dt} \le (m - 2\mu_1 D_{\mathrm{min}}) E. \tag{11.2.10}$$

It follows that in the fast diffusion limit with $D_{\mathrm{min}} > m/2\mu_1$, we have $E(t) \to 0$ as $t \to 0$, which implies that $\nabla \mathbf{u} \to 0$ as $t \to \infty$. In other words any spatial patterns disappear in the large-time limit.

---

**Box 11B. Poincare's inequality.**

---

Consider the Laplacian operator $-\nabla^2$ acting on a bounded domain $\Omega$ with the Neumann boundary condition $\nabla \phi \cdot \mathbf{n} = 0$, where $\mathbf{n}$ is the unit normal to the boundary $\partial\Omega$. The operator has a complete set of orthonormal eigenfunctions $\phi_k$ satisfying the equation

$$\nabla^2\phi_k + \mu_k\phi_k = 0, \quad \int_\Omega \phi_k(\mathbf{x})\phi_l(\mathbf{x})dx = \delta_{l,k}.$$

The eigenvalues $\mu_k$ are ordered such that $0 = \mu_0 < \mu_1 \le \mu_2 \le \mu_3 \ldots$. Suppose that $v(\mathbf{x})$ is a function in $\Omega$ that satisfies the Neumann boundary condition on $\partial\Omega$ and $\int_\Omega v(\mathbf{x})dx = 0$. We can then expand $v$ in terms of the generalized Fourier series

$$v(\mathbf{x}) = \sum_{k\ge 1} a_k\phi_k(\mathbf{x}), \quad a_k = \int_\Omega v(\mathbf{x})\phi_k(\mathbf{x})dx.$$

Note that $\phi_0(\mathbf{x}) = $ constant which means that $a_0 = 0$. The following result holds:

$$\int_\Omega |\nabla v|^2 dx = \int_\Omega \left[\sum_{k\ge 0} a_k\nabla\phi_k(\mathbf{x})\right] \cdot \left[\sum_{l\ge 0} a_l\nabla\phi_l(\mathbf{x})\right]$$

$$= \sum_{k,l\ge 0} a_k a_l \int_\Omega \nabla\phi_k(\mathbf{x})\cdot\nabla\phi_l(\mathbf{x})dx = \sum_{k,l\ge 0} a_k a_l \int_\Omega \left[\nabla\cdot(\phi_k\nabla\phi_l) - \phi_k\nabla^2\phi_l\right]dx$$

$$= \sum_{k,l\ge 0} a_k a_l \left[\int_{\partial\Omega}\phi_k\nabla\phi_l\cdot\mathbf{n}d\sigma + \int_\Omega \phi_k(\mathbf{x})\mu_l\phi_l(\mathbf{x})\right]dx$$

$$= \sum_{k\ge 1} a_k^2\mu_k \ge \mu_1\sum_{k\ge 1} a_k^2.$$

We have used the divergence theorem, the eigenvalue equation, and the Neumann boundary condition. Using a similar analysis, it is straightforward to show that

$$\int_\Omega v(\mathbf{x})^2 dx = \sum_{k\ge 1} a_k^2.$$

We thus obtain Poincare's inequality

$$\int_\Omega v(\mathbf{x})^2 dx \le \frac{1}{\mu_1}\int_\Omega |\nabla v(\mathbf{x})|^2 dx. \qquad (11.2.11)$$

### 11.2.2 Weakly nonlinear analysis and amplitude equations

It is important to note that although linear stability analysis can generate the conditions for the occurrence of a Turing instability, it cannot determine the selection and stability of the emerging patterns. That is, as the eigenmodes increase in amplitude,

Fig. 11.12: Illustration of a stripe pattern and a hexagonal spot pattern in the plane.

the linear approximation breaks down and nonlinear theory is necessary in order to investigate whether or not a stable pattern ultimately forms. Suppose that we take $\mu$ to denote a bifurcation parameter, such that the stationary state is stable for $\mu < \mu_c$ and undergoes a (supercritical) Turing instability at $\mu = \mu_c$. (In the model example given by equation (11.2.6), we could identify $\mu$ with the parameter $a$, say.) Sufficiently close to the bifurcation point, we can treat $\mu - \mu_c = \varepsilon$ as a small parameter and carry out a perturbation expansion in powers of $\varepsilon$. This generates a dynamical equation for the amplitude of the pattern that can be used to investigate pattern stability, at least in the weakly nonlinear regime. However, one immediate difficulty in carrying out this program is that a large number of eigenmodes could be excited beyond the bifurcation point. In the case of an unbounded domain, the wavenumber $k$ is continuous-valued, since the linear operator $\mathbb{L}$ has a continuous spectrum (see Box 11G), so that there will be a continuous band of growing modes in a neighborhood of $k = k_c$ when $\mu > \mu_c$. Moreover, the unbounded system is symmetric with respect to the action of the Euclidean group $\mathbb{E}(2)$—the group of rigid body translations, rotations, and reflections in the plane. This means that all eigenmodes $e^{i\mathbf{k}\cdot\mathbf{x}}$ lying on the critical circle $|\mathbf{k}| = k_c$ will be excited. Even though the number of excited eigenmodes becomes finite in a bounded domain, it can still be very large when the size of the domain satisfies $L \gg 2\pi/k_c$.

**Doubly periodic planforms.** One common property shared by many patterns observed in nature is that sufficiently close to the Turing bifurcation they tend to be relatively simple patterns such as stripes and hexagons. This is illustrated in Fig. 11.12 for the 2D activator–inhibitor RD system analyzed by Barrio et al. [47]. (Although the parameter $C$ of the model does not appear in the linear theory, it does contribute to pattern selection and stability.) This motivates the mathematical simplification of restricting the space of solutions of the RD system (11.2.1) to that of doubly periodic functions. (For simplicity, we neglect boundary effects by taking the domain to be $\mathbb{R}^2$.) That is, one imposes the conditions

$$u(\mathbf{x}+\boldsymbol{\ell},t) = u(\mathbf{x},t), \quad v(\mathbf{x}+\boldsymbol{\ell},t) = v(\mathbf{x},t)$$

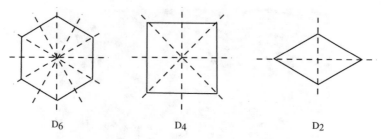

Fig. 11.13: Holohedries of the plane.

for every $\ell \in \mathcal{L}$ where $\mathcal{L}$ is some regular planar lattice. The lattice $\mathcal{L}$ is generated by two linearly independent vectors $\ell_1$ and $\ell_2$:

$$\mathcal{L} = \{(m_1\ell_1 + m_2\ell_2) : m_1, m_2 \in \mathbf{Z}\}, \tag{11.2.12}$$

with lattice spacing $d = |\ell_j|$. Let $\psi$ be the angle between the two basis vectors $\ell_1$ and $\ell_2$. We can then distinguish three types of lattice according to the value of $\psi$: square lattice ($\psi = \pi/2$), rhombic lattice ($0 < \psi < \pi/2$, $\psi \neq \pi/3$), and hexagonal ($\psi = \pi/3$, see Table 11.1. Restriction to double periodicity means that the original Euclidean symmetry group is now restricted to the symmetry group of the lattice, $\Gamma = D_n \dotplus \mathbf{T}^2$, where $D_n$ is the holohedry of the lattice, the subgroup of rotations and reflections $\mathbf{O}(2)$ that preserves the lattice, and $\mathbf{T}^2$ is the two torus of planar translations modulo the lattice. Thus, the holohedry of the rhombic lattice is $D_2$, the holohedry of the square lattice is $D_4$, and the holohedry of the hexagonal lattice is $D_6$, see Fig. 11.13. There are only a finite number of rotations and reflections to consider for each lattice (modulo an arbitrary rotation of the whole plane). Consequently, there is only a finite set of candidate excited eigenmodes.

Imposing double periodicity on the marginally stable eigenmodes restricts the lattice spacing such that the critical wavevector $\mathbf{k}_c$ lies on the *dual lattice* $\widehat{\mathcal{L}}$; the generators of the dual lattice satisfy $\hat{\ell}_i . \ell_j = \delta_{i,j}$ for $i, j = 1, 2$. In order to generate the simplest observed patterns, $d$ is chosen so that $k_c$ is the shortest length of a dual wave vector. Linear combinations of eigenmodes that generate doubly periodic solutions corresponding to dual wave vectors of shortest length are then given by $\mathbf{c}(\mathbf{x}) = \mathbf{c}_0 \phi(\mathbf{x})$ where

| Lattice | $\ell_1$ | $\ell_2$ | $\hat{\ell}_1$ | $\hat{\ell}_2$ |
|---------|----------|----------|----------------|----------------|
| Square | $(1,0)$ | $(0,1)$ | $(1,0)$ | $(0,1)$ |
| Hexagonal | $(1,0)$ | $\frac{1}{2}(1,\sqrt{3})$ | $(1,\frac{-1}{\sqrt{3}})$ | $(0,\frac{2}{\sqrt{3}})$ |
| Rhombic | $(1,0)$ | $(\cos\eta, \sin\eta)$ | $(1,-\cot\eta)$ | $(0,\csc\eta)$ |

Table 11.1: Generators for the planar lattices and their dual lattices in the case of unit lattice spacing ($d = 1$).

| $D_2$ | Action | $D_4$ | Action | $D_6$ | Action |
|---|---|---|---|---|---|
| $1$ | $(z_1,z_2)$ | $1$ | $(z_1,z_2)$ | $1$ | $(z_1,z_2,z_3)$ |
| $\xi$ | $(z_1^*,z_2^*)$ | $\xi$ | $(z_2^*,z_1)$ | $\xi$ | $(z_2^*,z_3^*,z_1^*)$ |
| $\kappa_\eta$ | $(z_2,z_1)$ | $\xi^2$ | $(z_1^*,z_2^*)$ | $\xi^2$ | $(z_3,z_1,z_2)$ |
| $\kappa_\eta\xi$ | $(z_2^*,z_1^*)$ | $\xi^3$ | $(z_2,z_1^*)$ | $\xi^3$ | $(z_1^*,z_2^*,z_3^*)$ |
| | | $\kappa$ | $(z_1,z_2^*)$ | $\xi^4$ | $(z_2,z_3,z_1)$ |
| | | $\kappa\xi$ | $(z_2^*,z_1^*)$ | $\xi^5$ | $(z_3^*,z_1^*,z_2^*)$ |
| | | $\kappa\xi^2$ | $(z_1^*,z_2)$ | $\kappa$ | $(z_1,z_3,z_2)$ |
| | | $\kappa\xi^3$ | $(z_2,z_1)$ | $\kappa\xi$ | $(z_2^*,z_1^*,z_3^*)$ |
| | | | | $\kappa\xi^2$ | $(z_3,z_2,z_1)$ |
| | | | | $\kappa\xi^3$ | $(z_1^*,z_3^*,z_2^*)$ |
| | | | | $\kappa\xi^4$ | $(z_2,z_1,z_3)$ |
| | | | | $\kappa\xi^5$ | $(z_3^*,z_2^*,z_1^*)$ |

Table 11.2: (Left) $D_2 \dot{+} \mathbf{T}^2$ action on rhombic lattice; (Center) $D_4 \dot{+} \mathbf{T}^2$ action on square lattice; (Right) $D_6 \dot{+} \mathbf{T}^2$ action on hexagonal lattice. In each case the generators of $D_n$ are a reflection and a rotation. For the square and hexagonal lattices, the generator $\kappa$ represents reflection across the $x$ axis, whereas for the rhombic lattice, the generator $\kappa_\eta$ represents reflections across the major diagonal. The counterclockwise rotation generator $\xi$ represents rotation through the angles $\pi$ (rhombic), $\frac{\pi}{2}$ (square), and $\frac{\pi}{3}$ (hexagonal).

$$\phi(\mathbf{x}) = \sum_{j=1}^{N} z_j e^{i\mathbf{k}_j \cdot \mathbf{x}} + \text{c.c.}, \qquad (11.2.13)$$

where the $z_j$ are complex amplitudes. Here $N = 2$ for the square lattice with $\mathbf{k}_1 = \mathbf{k}_c$ and $\mathbf{k}_2 = R_{\pi/2}\mathbf{k}_c$, where $R_\xi$ denotes rotation through an angle $\xi$. Similarly, $N = 3$ for the hexagonal lattice with $\mathbf{k}_1 = \mathbf{k}_c$, $\mathbf{k}_2 = R_{2\pi/3}\mathbf{k}_c$, and $\mathbf{k}_3 = R_{4\pi/3}\mathbf{k}_c = -\mathbf{k}_1 - \mathbf{k}_2$. It follows that the space of marginally stable eigenmodes can be identified with the $N$-dimensional complex vector space spanned by the vectors $(z_1, \ldots, z_N) \in \mathbb{C}^N$ with $N = 2$ for square or rhombic lattices and $N = 3$ for hexagonal lattices. It can be shown that these form irreducible representations of the group $\Gamma = D_n \dot{+} \mathbf{T}^2$ (see Box 11C for a definition of irreducibility), whose action on $\mathbb{C}^N$ is induced by the corresponding action of $\Gamma$ on $\phi(\mathbf{x})$. For example, on a hexagonal lattice, a translation $\phi(\mathbf{x}) \to \phi(\mathbf{r} - \mathbf{s})$ induces the action

$$\gamma \cdot (z_1,z_2,z_3) = (z_1 e^{-i\theta_1}, z_2 e^{-i\theta_2}, z_3 e^{i(\theta_1+\theta_2)}), \qquad (11.2.14)$$

with $\theta_j = \mathbf{k}_j \cdot \mathbf{s}$, a rotation $\phi(\mathbf{x}) \to \phi(R_{-2\pi/3}\mathbf{x})$ induces the action

$$\gamma \cdot (z_1,z_2,z_3) = (z_3,z_1,z_2), \qquad (11.2.15)$$

and a reflection across the $x$-axis (assuming $\mathbf{k}_c = k_c(1,0)$) induces the action

$$\gamma \cdot (z_1,z_2,z_3) = (z_1,z_3,z_2). \qquad (11.2.16)$$

The full action of $D_n \dot{+} \mathbf{T}^2$ on $\mathbb{C}^N$ for the various regular planar lattices is given in Table 11.2.

**Amplitude equation.** The selection and stability of patterns that emerge via a Turing instability can now be analyzed in terms of a system of nonlinear ODEs describing the slow dynamics of the amplitudes $\mathbf{z} = (z_1, \ldots, z_N)$ close to the bifurcation point, which can be derived using the method of multiple scales. The basic idea is that close to the Turing bifurcation point where $\mu - \mu_c = \varepsilon$, the critical eigenvalue $\lambda(k_c) = O(\varepsilon)$, which means that the excited eigenmodes grow slowly with respect to time. (We are assuming that $\lambda(k_c)$ is real; if it has a nonzero imaginary part, then a Turing–Hopf instability may lead to the formation of oscillatory patterns.) In order to pick out this slow exponential growth using perturbation theory, one introduces a slow time variable $\tau = \varepsilon t$ and substitutes the series expansion

$$\mathbf{u} = \mathbf{u}^* + \varepsilon^{1/2}\mathbf{u}_1(\mathbf{x}, \tau) + \varepsilon^{1/2}\mathbf{u}_2(\mathbf{x}, \tau) + \varepsilon^{3/2}\mathbf{u}_3(\mathbf{x}, \tau) + \ldots$$

into the full RD system (11.2.1). Taylor expanding the nonlinear functions $f(u, v)$ and $g(u, v)$ about the stationary solution $(u^*, v^*)$ and collecting terms having the same power of $\varepsilon$ lead to a hierarchy of equations of the general form

$$\mathbb{L}u_n(\mathbf{x}, \tau) = h_n(u_1, \ldots u_{n-1}),$$

where $\mathbb{L}$ is the linear operator defined in equation (11.2.2) and $h_n$ is a function of lower-order terms in the hierarchy. Since $h_1 \equiv 0$, it follows that the $O(\varepsilon^{1/2})$ solution $u_1$ is given by equation (11.2.13) with time-dependent amplitudes:

$$\mathbf{u}_1(\mathbf{x}, \tau) = \mathbf{c}_0 \sum_{j=1}^{N} z_j(\tau) e^{i\mathbf{k}_j \cdot \mathbf{x}} + \text{c.c.}$$

Applying the Fredholm alternative theorem (Box 2E) to the inhomogeneous higher-order equations then determines an amplitude equations for $\mathbf{z}(\tau)$ of the form

$$\frac{dz_j}{d\tau} = F_j(\mathbf{z}), \quad j = 1, \ldots, N, \tag{11.2.17}$$

where $F_j$ can be expanded as a polynomial in the $z_j$'s—close to the bifurcation point it is often sufficient to truncate the polynomials at cubic order in the amplitudes. In addition, the amplitude equations inherit the symmetries of the underlying planar lattice, which means symmetric bifurcation theory can be used to investigate the selection and stability of patterns (see Box 11C) [355, 433, 915].

*Pattern formation on a ring.* In order to illustrate the above theory, we develop the basic steps necessary to derive the amplitude equation in the case of an RD system (11.2.6) defined on a ring:

$$\frac{\partial u_1}{\partial t} = D\frac{\partial^2 u_1}{\partial \theta^2} + \kappa(u_1 + au_2 - u_1 u_2^2 - Cu_1 u_2), \tag{11.2.18a}$$

$$\frac{\partial u_2}{\partial t} = \frac{\partial^2 u_2}{\partial \theta^2} + \kappa(-u_1 + bu_2 + u_1 u_2^2 + Cu_1 u_2), \tag{11.2.18b}$$

with $\theta \in [0, 2\pi]$ and periodic concentrations $u_i(\theta + 2m\pi, t) = u_i(\theta, t)$ for all integers $m$. In this case the underlying symmetry group is $\mathbf{O}(2)$, consisting of rotations and reflections on the circle. Suppose that the onset of a Turing instability of the stationary state at $(u_1, u_2) = (0, 0)$ occurs at a critical parameter value $a = a_c$ due to a particular Fourier mode $v(\theta) = z e^{in\theta} + c.c.$ becoming marginally stable. This means that there exists a vector $\mathbf{c}$ such that

$$\mathbf{A}(n)\mathbf{c} \equiv \begin{pmatrix} \kappa - Dn^2 & \kappa a_c \\ -\kappa & b\kappa - n^2 \end{pmatrix} \mathbf{c} = 0. \tag{11.2.19}$$

Substitute into the RD equations the perturbation expansion

$$u_i = \varepsilon^{1/2} u_i^{(1)} + \varepsilon u_i^{(2)} + \varepsilon^{3/2} u_i^{(3)} + \cdots$$

where we have set $a - a_c = \varepsilon \Delta a$. The dominant temporal behavior just beyond bifurcation is the slow growth of the excited mode at a rate $e^{\varepsilon t}$. This motivates the introduction of a slow time-scale $\tau = \varepsilon t$ so that $\partial/\partial t \to \varepsilon \partial/\partial \tau$. Collecting terms with equal powers of $\varepsilon$ then leads to a hierarchy of equations of the form (see Ex. 11.4)

$$\mathbb{L}\mathbf{u}^{(1)} = 0, \quad \mathbb{L}\mathbf{u}^{(2)} = \mathbf{h}^{(2)}, \quad \mathbb{L}\mathbf{u}^{(3)} = \mathbf{h}^{(3)}, \ldots,$$

where

$$\mathbb{L}\mathbf{u} = \begin{pmatrix} \kappa & \kappa a_c \\ -\kappa & b\kappa \end{pmatrix} \begin{pmatrix} u_1 \\ u_2 \end{pmatrix} + \begin{pmatrix} D & 0 \\ 0 & 1 \end{pmatrix} \begin{pmatrix} \partial^2 u_1/\partial \theta^2 \\ \partial^2 u_2/\partial \theta^2 \end{pmatrix}, \tag{11.2.20}$$

and

$$h_1^{(2)} = \kappa C u_1^{(1)} u_2^{(1)} \quad h_2^{(2)} = -\kappa C u_1^{(1)} u_2^{(1)}, \tag{11.2.21a}$$

$$h_1^{(3)} = \frac{\partial u_1^{(1)}}{\partial \tau} - \kappa \left[ \Delta a u_2^{(1)} - u_1^{(1)} u_2^{(1)} u_2^{(1)} - C u_1^{(2)} u_2^{(1)} - C u_1^{(1)} u_2^{(2)} \right], \tag{11.2.21b}$$

$$h_2^{(3)} = \frac{\partial u_2^{(1)}}{\partial \tau} - \kappa \left[ u_1^{(1)} u_2^{(1)} u_2^{(1)} + C u_1^{(2)} u_2^{(1)} + C u_1^{(1)} u_2^{(2)} \right]. \tag{11.2.21c}$$

The $O(\varepsilon^{1/2})$ equation has a solution of the form

$$\mathbf{u}^{(1)} = \left[ z(\tau) e^{in\theta} + z^*(\tau) e^{-in\theta} \right] \mathbf{c}. \tag{11.2.22}$$

A dynamical equation for the complex amplitude $z(\tau)$ can be obtained by deriving solvability conditions for the higher-order equations, a method known as the Fredholm alternative. Defining the inner product of two periodic functions $u, v$ by

$$\langle u | v \rangle = \int_0^{2\pi} u^*(\theta) v(\theta) \frac{d\theta}{2\pi}, \tag{11.2.23}$$

we see that the adjoint operator $\mathbb{L}^\dagger$ has a one-dimensional complex null space spanned by $\bar{\mathbf{c}} e^{\pm in\theta}$, where $\bar{\mathbf{c}}$ is the left eigenvector of the matrix $\mathbf{A}(n)$, that is, $\bar{\mathbf{c}}^\top \mathbf{A}(n) = 0$ with $\bar{\mathbf{c}} \cdot \mathbf{c} = 1$. We thus have the solvability conditions

$$\sum_{i=1,2} \bar{c}_i \langle e^{in\theta} | h_i^{(m)} \rangle = 0, \quad m \geq 2. \tag{11.2.24}$$

Using the identity

$$\int_0^{2\pi} e^{im\theta} e^{im'\theta} \frac{d\theta}{2\pi} = \delta_{m+m',0},$$

one finds that the solvability condition is automatically satisfied for $\mathbf{h}^{(2)}$, since $\langle e^{in\theta}|u_1^{(1)}u_2^{(1)}\rangle = 0$. The $O(\varepsilon^{3/2})$ solvability condition then yields a cubic amplitude equation for $z(\tau)$:

$$\sum_{j=1,2}\bar{c}_j\left\langle e^{in\theta}\left|\frac{\partial u_j^{(1)}}{\partial\tau}\right.\right\rangle - \kappa\Delta a\,\bar{c}_1\langle e^{in\theta}|u_2^{(1)}\rangle$$

$$= \kappa(\bar{c}_2-\bar{c}_1)\left\langle e^{in\theta}\left|\left[u_1^{(1)}u_2^{(1)}u_2^{(1)}+Cu_1^{(2)}u_2^{(1)}+Cu_1^{(1)}u_2^{(2)}\right]\right.\right\rangle.$$

Substituting for $\mathbf{u}^{(1)}$ on the left-hand side gives

$$\text{l. h. s.} = \sum_{j=1,2}\bar{c}_jc_j\frac{dz}{d\tau} - \kappa\bar{c}_1c_2z(\tau).$$

In order to evaluate the first term on the right-hand side, we use

$$\langle e^{in\theta}|u_1^{(1)}u_2^{(1)}u_2^{(1)}\rangle = c_1c_2^2\int_0^{2\pi}\frac{d\theta}{2\pi}\left(ze^{in\theta}+z^*e^{-in\theta}\right)^3e^{-in\theta} = 3c_1c_2^2z|z|^2.$$

The next step is to determine $\mathbf{u}^{(2)}$. The explicit expression for $\mathbf{h}^{(2)}$ implies that

$$\mathbf{u}^{(2)}(\theta) = \mathbf{c}^+e^{2ni\theta}+\mathbf{c}^-e^{-2ni\theta}+\mathbf{c}^0+\zeta\mathbf{u}^{(1)}(\theta).$$

The constant vector $\zeta$ remains undetermined at this order of perturbation but does not appear in the amplitude equation for $z(\tau)$. Substituting for $\mathbf{u}^{(2)}$ into the equation $\mathbb{L}\mathbf{u}^{(2)} = \mathbf{h}^{(2)}$ gives the following equations for the vectors $\mathbf{c}^{\pm},\mathbf{c}^0$ (see Ex. 11.4):

$$\mathbf{A}(2n)\mathbf{c}^+ = \kappa Cc_1c_2z^2\begin{pmatrix}1\\-1\end{pmatrix}, \quad \mathbf{A}(2n)\mathbf{c}^- = \kappa Cc_1c_2z^{*2}\begin{pmatrix}1\\-1\end{pmatrix},$$

$$\mathbf{A}(0)\mathbf{c}^0 = 2\kappa Cc_1c_2|z|^2\begin{pmatrix}1\\-1\end{pmatrix}.$$

Hence

$$\langle e^{in\theta}|u_1^{(1)}u_2^{(2)}\rangle = c_1\int_0^{2\pi}\frac{d\theta}{2\pi}\left(ze^{in\theta}+z^*e^{-in\theta}\right)\left(c_2^+e^{2in\theta}+c_2^-e^{-2in\theta}+c_2^0\right)e^{-in\theta}$$

$$= c_1(c_2^+z^*+c_2^0z) = \kappa Cc_1^2c_2V_2z|z|^2,$$

and

$$\langle e^{in\theta}|u_2^{(1)}u_1^{(2)}\rangle = \kappa Cc_1c_2^2V_1z|z|^2,$$

where

$$\mathbf{V} = \mathbf{A}(2n)^{-1}\begin{pmatrix}1\\-1\end{pmatrix}+2\mathbf{A}(0)^{-1}\begin{pmatrix}1\\-1\end{pmatrix}.$$

Combining all of these results, we obtain the cubic amplitude equation

$$\frac{dz}{d\tau} = z(\tau)(\eta\Delta a - \Lambda|z(\tau)|^2), \tag{11.2.25}$$

after absorbing a factor of $\kappa$ into $\tau$, with

$$\eta = \bar{c}_1c_2, \quad \Lambda = (\bar{c}_1-\bar{c}_2)c_1c_2\left[3c_2+\kappa C^2(c_2V_1+c_1V_2)\right].$$

Since the uniform fixed point $(u_1, u_2) = (0,0)$ is unstable with respect to the modes $e^{\pm in\theta}$, it follows that $\eta > 0$. Therefore, a stable pattern will occur via a supercritical bifurcation provided that $\Lambda > 0$. In addition, just beyond the bifurcation point the amplitude is $\sqrt{\varepsilon}\sqrt{\eta(a-a_c)/\Lambda}$. Finally, note that the general form of the amplitude equation (11.2.25) can be determined using symmetry arguments along the lines of Box 11C. That is, consider the selected planform $u(\theta) = ze^{in\theta} + z^* e^{-in\theta}$. Under the action of the group $\mathbf{O}(2)$,

$$u(\theta + \xi) = ze^{in\xi} e^{2i\theta} + z^* e^{-in\xi} e^{-in\theta}, \quad u(-\theta) = ze^{-in\theta} + z^* e^{in\theta}.$$

It follows that the action of $\mathbf{O}(2)$ on $(z, z^*)$ is

$$\xi \cdot (z, z^*) = (ze^{in\xi}, z^* e^{-in\xi}), \quad \kappa \cdot (z, z^*) = (z^*, z).$$

Equivariance of the amplitude equation with respect to these transformations implies that quadratic and quartic terms are excluded, and the quintic term is of the form $z|z|^4$.

---

## Box 11C. Symmetric bifurcation theory and pattern formation [355]

---

**Group axioms.** A group $\Gamma$ is a set of elements $a \in \Gamma$ together with a group operation $\cdot$ that satisfies the following axioms: (i) If $a, b \in \Gamma$ then $a \cdot b \in \Gamma$ (closure); (ii) For all $a, b, c \in \Gamma$, we have $a \cdot (b \cdot c) = (a \cdot b) \cdot c$ (associativity); (iii) There exists an identity element 1 such that for all $a \in \Gamma$, we have $1 \cdot a = a \cdot 1 = a$ (identity element); (iv) For each $a \in \Gamma$, there exists an element $a^{-1}$ such that $a \cdot a^{-1} = a^{-1} \cdot a = 1$ (inverse element).

**Group representations.** A representation of a group $\Gamma$ acting on an $n$-dimensional vector space $V$ is a map $\rho : G \to GL(V)$, where $GL(V)$ is the general linear group on $V$, such that

$$\rho(a \cdot b) = \rho(a)\rho(b), \text{ for all } a, b \in \Gamma.$$

For a particular choice of basis set for $V$, $GL(V)$ can be identified with the group of invertible $n \times n$ matrices. An *irreducible representation* is one that has no proper closed sub-representations. In the language of matrices, this means that it is not possible to choose a basis set in which the matrix representation can be written in block diagonal form

$$D(a) = \begin{pmatrix} D^{(1)}(a) & 0 & \dots 0 \\ 0 & D^{(2)}(a) \dots 0 & \\ \vdots & \vdots & \ddots & \vdots \\ 0 & 0 & \dots & D^{(k)}(a) \end{pmatrix} = D^{(1)}(a) \oplus D^{(2)}(a) \dots \oplus D^{(k)}(a),$$

where $D^{(j)}(a)$ are submatrices.

**Group action on a function space.** Suppose that the map $\rho$ is a representation of a group $\Gamma$ acting on a finite-dimensional vector space $V$. Let $u : V \to \mathbb{R}$

be a function mapping elements of $V$ to the real line. (One could also consider complex-valued functions.) There is then a natural representation $\sigma$ of the group $\Gamma$ acting on the space of functions $C(V, \mathbb{R})$:

$$\sigma(\gamma) \cdot u(\mathbf{x}) = u(\rho(\gamma)^{-1}(\mathbf{x})), \text{ for all } \mathbf{x} \in V \text{ and } \gamma \in \Gamma.$$

It is necessary to use the inverse element $\rho(\gamma)^{-1}$ to ensure that $\sigma$ is a group representation. That is,

$$\begin{aligned}
\sigma(\gamma_1) \cdot \sigma(\gamma_2) \cdot u(\mathbf{x}) &= \sigma(\gamma_1) \cdot u(\rho(\gamma_2)^{-1}(\mathbf{x})) = \sigma(\gamma_1) \cdot u_2(\mathbf{x}) \\
&= u_2(\rho(\gamma_1)^{-1}(\mathbf{x})) = u(\rho(\gamma_2)^{-1}\rho(\gamma_1)^{-1}(\mathbf{x})) \\
&= u([\rho(\gamma_1)\rho(\gamma_2)]^{-1}(\mathbf{x})) = u([\rho(\gamma_1\gamma_2)]^{-1}(\mathbf{x})) \\
&= \sigma(\gamma_1\gamma_2) \cdot u(\mathbf{x}).
\end{aligned}$$

In the following we will use the same symbol $\gamma$ for an abstract group element and its corresponding group representation.

**Equivariance.** Suppose that $u \in C(\mathbb{R}^2, \mathbb{R})$ is the solution to a scalar reaction–diffusion equation of the form

$$\frac{\partial u(\mathbf{x}, t)}{\partial t} = \nabla_{\mathbf{x}}^2 u(\mathbf{x}, t) + f(u(\mathbf{x}, t)).$$

(One could equally well consider a system of RD equations.) This equation is equivariant with respect to the natural action of the Euclidean group $\mathbb{E}(2)$ on $C(\mathbb{R}^2, \mathbb{R})$. That is, if $u(\mathbf{x}, t)$ is a solution then so is $u(\gamma^{-1}\mathbf{x}, t)$ for all $\gamma \in \mathbb{E}(2)$. This is a consequence of the fact that the Laplacian operator is invariant with respect to the action of $\mathbb{E}(2)$: $\nabla_{\gamma^{-1}\mathbf{x}}^2 = \nabla_{\mathbf{x}}^2$. In other words, the operators $\gamma$ and $\nabla^2$ commute, $\gamma\nabla^2 = \nabla^2\gamma$. Equivariance then follows, since

$$\begin{aligned}
0 &= \gamma\left[\frac{\partial u(\mathbf{x}, t)}{\partial t} - \nabla_{\mathbf{x}}^2 u(\mathbf{x}, t) - f(u(\mathbf{x}, t))\right] \\
&= \frac{\partial u(\gamma^{-1}\mathbf{x}, t)}{\partial t} - \nabla_{\gamma^{-1}\mathbf{x}}^2 u(\gamma^{-1}\mathbf{x}, t) - f(u(\gamma^{-1}\mathbf{x}, t)) \\
&= \frac{\partial u(\gamma^{-1}\mathbf{x}, t)}{\partial t} - \nabla_{\mathbf{x}}^2 u(\gamma^{-1}\mathbf{x}, t) - f(u(\gamma^{-1}\mathbf{x}, t)).
\end{aligned}$$

Equivariance has major implications for the bifurcation structure of solutions to the PDE. In particular, suppose that $u_0$ is a homogeneous stationary solution. Such a solution preserves full Euclidean symmetry so that the PDE obtained by linearizing about the stationary solution is also equivariant with respect to $\mathbb{E}(2)$. Writing the linear PDE as $\partial_t u = \mathbb{L}u$, where $\mathbb{L}$ is a linear operator on $C(\mathbb{R}^2, \mathbb{R})$, we see that $\mathbb{L}$ commutes with the group elements $\gamma$. This implies that generically the eigenfunctions of $\mathbb{L}$ form irreducible representations of $\mathbb{E}(2)$. (Strictly speaking $\mathbb{L}$ has a continuous rather than a discrete spectrum. One could restrict the PDE to a bounded domain to obtain a discrete spectrum, but it would have to be sufficiently large so that Euclidean symmetry still approximately holds.) Applying the group action to the eigenvalue equation for $\mathbb{L}$ shows that

$$
\begin{aligned}
0 &= \gamma \cdot [\mathbb{L}\phi_\lambda(\mathbf{x}) - \lambda\phi_\lambda(\mathbf{x})] \\
&= \mathbb{L}\gamma\phi_\lambda(\mathbf{x}) - \lambda\gamma\phi_\lambda(\mathbf{x}) \\
&= \mathbb{L}\phi_\lambda(\gamma^{-1}\mathbf{x}) - \lambda\phi_\lambda(\gamma^{-1}\mathbf{x}).
\end{aligned}
$$

Thus, $\phi_\lambda(\gamma^{-1}\mathbf{x})$ for all $\gamma \in \Gamma$ have the same eigenvalue $\lambda$. This degeneracy is an immediate consequence of the underlying symmetry. In general, we don't expect any further degeneracy for the given eigenvalue, so the eigenfunctions $\phi_\lambda(\gamma^{-1}\mathbf{x})$ form an irreducible representation of the group.

The same analysis holds if we restrict solutions to doubly periodic functions, except that the resulting PDE is now equivariant with respect to the discrete group $\Gamma = D_n \dotplus T^2$. As we show below, bifurcations from a homogeneous stationary solution can now be analyzed in terms of a system of ODEs—amplitude equations—of the form

$$
\dot{\mathbf{z}} = F(\mathbf{z}), \tag{11.2.26}
$$

where $\mathbf{z}, F(\mathbf{z}) \in V$ with $V = \mathbb{R}^n$ or $\mathbb{C}^n$. These equations are also equivariant with respect to $\Gamma$, since

$$
\gamma \cdot F(z) = F(\gamma \cdot z)
$$

for all $\gamma \in \Gamma$. It immediately follows that if $z(t)$ is a solution to the system of ODEs, then so is $\gamma \cdot z(t)$. Moreover, since $F(0) = 0$, the origin is an equilibrium that is invariant under the action of the full symmetry group $\Gamma$. Thus linearizing about the fixed point $\mathbf{z} = 0$ generates a linear operator whose eigenvectors form irreducible representations of the group $\Gamma$.

**Isotropy subgroups.** The symmetries of any particular equilibrium solution $\mathbf{z}$ form a subgroup called the *isotropy* subgroup of $\mathbf{z}$ defined by

$$
\Sigma_{\mathbf{z}} = \{\sigma \in \Gamma : \sigma\mathbf{z} = \mathbf{z}\}. \tag{11.2.27}
$$

More generally, we say that $\Sigma$ is an isotropy subgroup of $\Gamma$ if $\Sigma = \Sigma_{\mathbf{z}}$ for some $\mathbf{z} \in V$. Isotropy subgroups are defined up to some conjugacy. A group $\Sigma$ is conjugate to a group $\widehat{\Sigma}$ if there exists $\sigma \in \Gamma$ such that $\widehat{\Sigma} = \sigma^{-1}\Sigma\sigma$. The *fixed-point subspace* of an isotropy subgroup $\Sigma$, denoted by $\mathrm{Fix}(\Sigma)$, is the set of points $\mathbf{z} \in V$ that are invariant under the action of $\Sigma$,

$$
\mathrm{Fix}(\Sigma) = \{\mathbf{z} \in V : \sigma\mathbf{z} = \mathbf{z} \,\forall\, \sigma \in \Sigma\}. \tag{11.2.28}
$$

Finally, the *group orbit* through a point $\mathbf{z}$ is

$$\Gamma \mathbf{z} = \{\sigma \mathbf{z} : \sigma \in \Gamma\}. \tag{11.2.29}$$

If $\mathbf{z}$ is an equilibrium solution of equation (11.2.26) then so are all other points of the group orbit (by equivariance). One can now adopt a strategy that restricts the search for solutions of equation (11.2.26) to those that are fixed points of a particular isotropy subgroup. In general, if a dynamical system is equivariant under some symmetry group $\Gamma$ and has a solution that is a fixed point of the full symmetry group then we expect a loss of stability to occur upon variation of one or more system parameters. Typically such a loss of stability will be associated with the occurrence of new solution branches with isotropy subgroups $\Sigma$ smaller than $\Gamma$. One says that the solution has spontaneously broken symmetry from $\Gamma$ to $\Sigma$. Instead of a unique solution with the full set of symmetries $\Gamma$ a set of symmetrically related solutions (orbits under $\Gamma$ modulo $\Sigma$) each with symmetry group (conjugate to) $\Sigma$ is observed.

**Equivariant branching lemma.** Suppose that equation (11.2.26) has a fixed point of the full symmetry group $\Gamma$. The *equivariant branching lemma* [355] states that generically there exists a (unique) equilibrium solution bifurcating from the fixed point for each of the axial subgroups of $\Gamma$ under the given group action—a subgroup $\Sigma \subset \Gamma$ is *axial* if $\dim \mathrm{Fix}(\Sigma) = 1$. The heuristic idea underlying this lemma is as follows. Let $\Sigma$ be an axial subgroup and $\mathbf{z} \in \mathrm{Fix}(\Sigma)$. Equivariance of $F$ then implies that

$$\sigma F(\mathbf{z}) = F(\sigma \mathbf{z}) = F(\mathbf{z}) \tag{11.2.30}$$

for all $\sigma \in \Sigma$. Thus $F(\mathbf{z}) \in \mathrm{Fix}(\Sigma)$ and the system of coupled ODEs (11.2.26) can be reduced to a single equation in the fixed point space of $\Sigma$. Such an equation is expected to support a codimension one bifurcation, in which new stationary solutions emerge whose amplitudes correspond to fixed points of axial isotropy subgroups. Since the codimension of a bifurcation corresponds generically to the number of parameters that need to be varied in order to induce the bifurcation, one expects the primary bifurcations to be codimension one. Thus one can systematically identify the various expected primary bifurcation branches by constructing the associated axial subgroups and finding their fixed points.

**Example.** For the sake of illustration, consider the full symmetry group $D_3$ of an equilateral triangle (see Fig. 11.14). The action is generated by the matrices (in an appropriately chosen orthonormal basis)

$$R = \begin{pmatrix} -1/2 & -\sqrt{3}/2 \\ \sqrt{3}/2 & -1/2 \end{pmatrix}, \quad S = \begin{pmatrix} -1 & 0 \\ 0 & 1 \end{pmatrix}.$$

Here $R$ is a rotation by $\pi/3$ and $S$ is a reflection about the $y$-axis. (The generators of a discrete group form a minimal set of group elements from which all other group elements can be obtained by combinations of group operations.) Clearly, $R$ fixes only the origin, while $S$ fixes any point $(x,0)$. We deduce that the isotropy subgroups are as follows: (i) the full symmetry group $D_3$ with single fixed point $(0,0)$; (ii) the two-element group $\mathbf{Z}_2(S)$ generated by $S$, which fixes the $x$-axis, and the groups that are conjugate to $\mathbf{Z}_2(S)$ by the rotations $R$ and $R^2$; and (iii) a trivial group formed by the identity matrix in which every point is a fixed point. The isotropy subgroups form the hierarchy $\{I\} \subset \mathbf{Z}_2(S) \subset D_3$. It follows that up to conjugacy the only axial subgroup is $\mathbf{Z}_2(S)$. Thus we expect the fixed point $(0,0)$ to undergo a symmetry breaking bifurcation to an equilibrium that has reflection symmetry. Such an equilibrium will be given by one of the three points $\{(x,0),R(x,0),R^2(x,0)\}$ on the group orbit generated by discrete rotations. Which of these states is selected will depend on initial conditions, that is, the broken rotation symmetry is hidden. Note that a similar analysis can be carried out for the symmetry group $D_4$ of the square (see Ex. 11.5). Now, however, there are two distinct types of reflection axes: those joining the middle of opposite edges and those joining opposite vertices. Since these two types of reflections are not conjugate to each other, there are now two distinct axial subgroups.

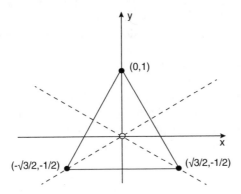

Fig. 11.14: Symmetry operations on an equilateral triangle.

## 11.3 Reaction–diffusion models of cell division and polarization

As highlighted by a number of authors [102, 390–392, 462, 699], the classical activator–inhibitor mechanism for Turing pattern formation in cell development does not apply to many of the known examples of intracellular patterning during cell polarization and division. First, these latter processes tend to be mass conserving, at least on the time scales for patterns to initially form and stabilize. Second, rather than two or more chemical species diffusing in the same medium and

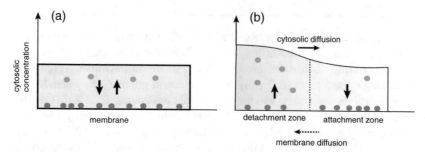

Fig. 11.15: Illustration of mass-conserving intracellular pattern formation based on the formation of spatially separated attachment and detachment zones within the membrane and the redistribution of proteins via cytosolic diffusion. (a) A random fluctuation in the distribution of membrane-bound proteins under a uniform background concentration of cytosolic proteins. (b) Formation of attachment and detachment zones that maintain a membrane-bound pattern. [Redrawn from Halatek, Brauns, and Frey [392].]

mutually affecting their rates of production and degradation, intracellular patterns typically involve the dynamical exchange of proteins between the cytoplasm and plasma membrane, combined with the membrane-associated switching between different protein conformational states. This naturally leads to a separation of time scales, since a typical diffusion coefficient in the membrane is $0.01\text{-}0.1\,\mu\text{m}^2\ \text{s}^{-1}$, whereas cytosolic proteins have diffusion coefficients of around $10\,\mu\text{m}^2\ \text{s}^{-1}$. The much slower rate of membrane diffusion means that redistribution is predominately mediated by a cytosolic concentration gradient, which is maintained by the formation of spatially separated attachment and detachment zones within reactive regions of the membrane. These emerge spontaneously due to the combination of cytosolic diffusion, protein interactions, and cell geometry. The basic scheme is illustrated in Fig. 11.15 for a single protein; in practice pattern formation involves the interactions and redistribution of several different proteins.

### 11.3.1 Mass-conserving systems

We begin by considering a general two-component RD process that conserves mass; we will turn to the second property of intracellular pattern formation in Sect. 11.3.2. Following [392, 462], consider a one-dimensional RD system of the form

$$\frac{\partial u}{\partial t} = D_u \frac{\partial^2 u}{\partial x^2} + f(u,v), \tag{11.3.1a}$$

$$\frac{\partial v}{\partial t} = D_v \frac{\partial^2 v}{\partial x^2} - f(u,v), \tag{11.3.1b}$$

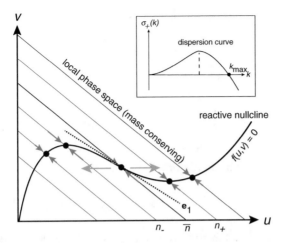

Fig. 11.16: Schematic diagram illustrating the linear stability analysis of a two-component mass-conserving system with $f_u < f_v$. In the absence of diffusion, the spatially uniform fixed point $(u^*(\bar{n}), v^*(\bar{n}))$ lies on the intercept between the reactive nullcline $f(u,v) = 0$ and the straight line $u + v = \bar{n}$, where $\bar{n}$ is the spatially uniform mass density. Since $f_u < f_v$, the fixed point is stable to uniform mass-preserving perturbations (red arrows) and marginally stable with respect to uniform perturbations along the eigenvector $\mathbf{e}_1$ tangential to the curve. Changes in $\bar{n}$ would move the fixed point along the curve $f(u,v) = 0$, whose associated generator is $\mathbf{e}_1$. When diffusion is included, the uniform state is unstable with respect to a spatially periodic perturbation of the form $\mathbf{e}_+(k)e^{ikx}e^{i\sigma_+(k)t}$ for $k \in [0, k_{\max}]$, with the dispersion curve $\sigma_+(k)$ shown in the inset; the dispersion curve $\sigma_-(k)$ lies beneath the $k$-axis. This means that certain spatial regions increase their local mass density ($n_+ > \bar{n}$), whereas other regions decrease their mass density ($n_- < \bar{n}$) such that the spatially averaged density is still equal to $\bar{n}$. The redistribution of local mass density as indicated by the blue arrows drives the emerging pattern toward a corresponding pattern of displaced local equilibria. [Redrawn from Halatek and Frey [392].]

with $x \in [0, L]$. Adding the pair of equations, integrating with respect to $x$, and imposing reflecting boundary conditions yield the mass conservation condition

$$N = \int_0^L n(x,t)dx, \quad n(x,t) = u(x,t) + v(x,t). \tag{11.3.2}$$

We first analyze the spatially uniform version of the model, for which

$$\frac{du}{dt} = f(u,v), \quad \frac{dv}{dt} = -f(u,v). \tag{11.3.3}$$

Let $(u^*, v^*)$ denote a fixed point of the system, which satisfies the pair of equations

$$f(u^*, v^*) = 0, \quad u^* + v^* = \bar{n} := \frac{N}{L}. \tag{11.3.4}$$

Linearizing about the fixed point and introducing the displacements

$$u(t) - u^* = Ue^{\sigma t}, \quad v(t) - v^* = Ve^{\sigma t},$$

yield the eigenvalue problem

$$\sigma \begin{pmatrix} U \\ V \end{pmatrix} = \begin{pmatrix} f_u & f_v \\ -f_u & -f_v \end{pmatrix} \begin{pmatrix} U \\ V \end{pmatrix}, \tag{11.3.5}$$

where

$$f_u = \left.\frac{\partial f}{\partial u}\right|_{u=u^*, v=v^*}, \quad f_v = \left.\frac{\partial f}{\partial v}\right|_{u=u^*, v=v^*}.$$

We thus find the following eigenpairs:

$$\sigma_1 = 0, \quad \mathbf{e}_1 = \begin{pmatrix} -f_v/f_u \\ 1 \end{pmatrix}; \quad \sigma_2 = f_u - f_v, \quad \mathbf{e}_2 = \begin{pmatrix} -1 \\ 1 \end{pmatrix}. \tag{11.3.6}$$

The first eigenpair shows that the system is marginally stable to perturbations that change the mass density, with the corresponding eigenvector tangential to the line of fixed points given by the reactive nullcline $f(u,v) = 0$, see Fig. 11.16. The second eigenpair determines the local stability of fixed points against mass-conserving perturbations.

Now let us include the effects of diffusion. Linearizing about the uniform steady state along analogous lines to Sect. 11.2.1, with

$$u(x,t) - u^* = U_k e^{ikx} e^{\sigma t}, \quad v(x,t) - v^* = V_k e^{ikx} e^{\sigma t},$$

leads to the matrix equation

$$\sigma(k) \begin{pmatrix} U_k \\ V_k \end{pmatrix} = \begin{pmatrix} f_u - D_u k^2 & f_v \\ -f_u & -f_v - D_v k^2 \end{pmatrix} \begin{pmatrix} U_k \\ V_k \end{pmatrix}. \tag{11.3.7}$$

For a given wavenumber $k$, we have a pair of eigenvalues

$$\sigma_\pm(k) = \frac{1}{2}\left(\tau(k) \pm \sqrt{\tau(k)^2 - 4\Delta(k)}\right), \tag{11.3.8}$$

where

$$\tau(k) = f_u - f_v - (D_u + D_v)k^2, \quad \Delta(k) = (D_u f_v - D_v f_u)k^2 + D_u D_v k^4.$$

If $k = 0$, then we recover the eigenvalues of the spatially uniform system with $(\sigma_+(0), \sigma_-(0)) = (0, f_u - f_v)$ for $f_u < f_v$ and $(\sigma_+(0), \sigma_-(0)) = (f_u - f_v, 0)$ for $f_u > f_v$. For the sake of illustration, we will focus on the case $f_u < f_v$, which means that the fixed point is stable with respect to spatially uniform, mass-preserving perturbations. In addition to the marginally stable mode at $k = 0$, there exists a second marginally stable mode at $k_{max}$ where $\Delta(k_{max}) = 0$ and thus $\sigma_-(k) = 0$, that is,

$$k_{max} = \sqrt{[D_v f_u - D_u f_v]/D_u D_v}, \tag{11.3.9}$$

provided that

$$D_v f_u > D_u f_v. \tag{11.3.10}$$

Assuming $f_u < f_v$, we require $D_v > D_u$. The corresponding dispersion curve displays a band of unstable modes in the interval $[0, k_{max}]$, see inset of Fig. 11.16. In the local $(u, v)$ phase plane, the eigenvector corresponds to a perturbation that either increases or decreases the local mass density. Hence, this will result in a periodically varying local increase or decrease of $n(x,t)$. In other words, there will be a spatially periodic redistribution of the mass density, resulting in corresponding displacements of the local equilibria, as illustrated in Fig. 11.16.

Numerically, one finds that the emerging pattern can be well approximated by the displaced local equilibria. That is, $(u(x,t), v(x,t)) \approx (u^*(x,t), v^*(x,t))$ with $f(u^*(x,t), v^*(x,t)) = 0$ and $n(x,t) = u^*(x,t) + v^*(x,t)$. This suggests that the total density $n(x,t)$ is the essential degree of freedom, since it controls the dynamics of the local equilibria that, in turn, form a scaffold for the emerging pattern [392]. Indeed, the condition (11.3.10) for a patterning forming instability can also be derived by replacing the local concentrations by the local equilibria [102]. That is, adding equations (11.3.1a,b) gives

$$\frac{\partial n}{\partial t} = D_u \frac{\partial^2 u}{\partial x^2} + D_v \frac{\partial^2 v}{\partial x^2} \approx D_u \frac{\partial^2 u^*(n)}{\partial x^2} + D_v \frac{\partial^2 v^*(n)}{\partial x^2}.$$

Using the chain rule of differentiation,

$$\frac{\partial n}{\partial t} = \frac{\partial}{\partial x} \left[ \left( D_u \frac{\partial u^*}{\partial n} + D_v \frac{\partial v^*}{\partial n} \right) \frac{\partial n}{\partial x} \right],$$

which is an inhomogeneous diffusion equation for the total density $n$. An instability of a uniform steady state will occur when the effective diffusion coefficient becomes negative:

$$D_u \frac{\partial u^*}{\partial n} + D_v \frac{\partial v^*}{\partial n} < 0.$$

Rearranging, this yields

$$\frac{\partial_n v^*}{\partial_n u^*} \equiv -\frac{\partial_u f}{\partial_v f} < -\frac{D_u}{D_v},$$

which recovers the inequality (11.3.10).

**Phase-space representation of stationary patterns.** The mass-redistribution Turing instability highlighted above is typically subcritical [102, 391, 392, 462, 641], in contrast to the supercritical Turing instability found in classical RD systems. This means that the linear stability analysis of a homogeneous state of a mass-conserving system is often a poor predictor of the final spatial pattern and its characteristic wavelength. In such cases, one finds that the initial pattern undergoes significant coarsening before reaching the full nonlinear pattern. Indeed, due to mass conservation, the solutions eventually approach a simple localized pattern or spike, after exhibiting long transient dynamics. However, it is possible to develop a geometric phase-space construction of a stationary pattern, which can be used to investigate various characteristics and bifurcations of patterns in the nonlinear regime [102]. Again, we will illustrate this using a simple two-component system. Any stationary pattern satisfies the steady-state RD equations

$$D_u \frac{d^2u}{dx^2} + f(u,v) = 0, \quad D_v \frac{d^2v}{dx^2} - f(u,v) = 0, \tag{11.3.11}$$

under the global constraint that total mass is conserved,

$$\bar{n} = \frac{1}{L} \int_0^L [u(x) + v(x)] dx. \tag{11.3.12}$$

Adding the pair of equations in (11.3.11), integrating once with respect to $x$, and using the no-flux boundary conditions show that

$$D_u \frac{du}{dx} = -D_v \frac{dv}{dx}$$

for all $x$. Performing a second integration then implies that any stationary pattern obeys the linear relation

$$\frac{D_u}{D_v} u(x) + v(x) = \eta_0 \tag{11.3.13}$$

for any point $x \in [0, L]$, with $\eta_0$ a constant that depends on the parameters $\bar{n}, L$. It immediately follows that any stationary pattern is confined to a linear subspace of the $(u, v)$-phase space, which is called the flux-balance subspace. This is illustrated in Fig. 11.17 for the particular example of a stationary front. (More generally, spatially periodic patterns can be constructed by sewing together a sequence of fronts of opposite polarity.) Since each point on the flux-balance subspace is displaced from its corresponding local equilibrium, except for the inflection point at $x_0$ and the endpoints, there is a reactive flux due to the fact that the local stationary values of $u$ and $v$ are not in chemical equilibrium, that is, $f(u, v) \neq 0$. The reactive flux is balanced by a corresponding diffusive flux, which maintains the pattern. Assuming that the fluxes are small, it follows that the displacements from local equilibria are also small. The latter then effectively scaffold the pattern, as shown in Fig. 11.17(a).

It remains to obtain an equation for the constant $\eta_0$. Multiplying both sides of equation (11.3.11a) by $u'(x)$ and then integrating with respect to $x \in [0, L]$ give

$$\begin{aligned}
0 &= \int_0^L \left[ D_u u'(x) \frac{d^2u}{dx^2} + u'(x) f(u(x), v(x)) \right] dx \\
&= \int_0^L \left[ \frac{D_u}{2} \frac{d[u'(x)^2]}{dx} + u'(x) f(u(x), \eta_0 - D_u u(x)/D_v) \right] dx \\
&= \int_{u(0)}^{u(L)} f(u, \eta_0 - D_u u/D_v) du, \tag{11.3.14}
\end{aligned}$$

where we have used the no-flux boundary conditions and equation (11.3.13). For large $L$, the flux-balance constant $\eta_0$ is approximately independent of the mean den-

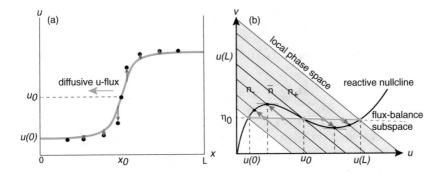

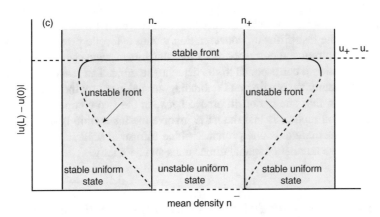

Fig. 11.17: Phase-space analysis of stationary patterns for large $L$. (a) Example of a stationary front linking $u(0)$ to $u(L)$ with an inflection point at $u_0$. In the absence of diffusion the local concentrations would tend to the local equilibria, as indicated by the red arrows. Diffusion maintains the displacements from the local equilibria, with the diffusive flux balancing the reactive flux arising from the fact that the system is out of chemical equilibrium. The system is in local equilibrium at the inflection point $x_0$ and the endpoints (due to the no-flux boundary conditions). (b) Embedding of spatial pattern in phase space. The concentrations lie on the flux-balance linear subspace, which is approximately horizontal when $D_u \ll D_v$. The slope of the reactive nullcline is more negative than that of the flux-balance subspace for all mean densities $n_- < \bar{n} < n_+$ (shaded green region). This means that the corresponding homogeneous state is unstable with respect to a Turing instability, resulting in the formation of a stationary front. On the other hand, if $\bar{n}$ lies in the shaded blue regions, then a stable front solution coexists with a stable homogeneous solution (and separated by an unstable front solution). Finally, the position of the front interface shifts as $\bar{n}$ is varied. (c) Bifurcation diagram of a stationary front for large $L$. [Redrawn from Brauns [102]].

sity $\bar{n}$ and $L$. This follows from the observation that $u(0) \approx u_-(\eta_0)$ and $u(L) \approx u_+(\eta_0)$, where $u_\pm(\eta_0)$ are points of intersection between the reactive nullcline and the flux-balance condition, see Fig. 11.17. That is, equation (11.3.14) reduces to a closed equation for $\eta_0$:

$$\int_{u_-(\eta_0)}^{u_+(\eta_0)} f(u, \eta_0 - D_u u/D_v)\,du = 0.$$

Finally, substituting equation (11.3.13) into (11.3.11a) yields

$$D_u \frac{d^2 u}{dx^2} + f(u, \eta_0 - D_u u/D_v) = 0, \tag{11.3.15}$$

which together with (11.3.14) completely determines the stationary pattern. The instability condition (11.3.10) is equivalent to requiring that the slope of the reactive nullcline be more negative than that of the flux-balance subspace at the homogeneous equilibrium.

Finally, note that the redistribution of membrane and cytosolic proteins shown in Fig. 11.15 can be related to the phase-space construction shown in Fig. 11.17. That is, we can identify the slowly diffusing concentration $u$ with membrane-bound proteins and the fast-diffusing concentration $v$ with cytosolic proteins. The region of high concentration $u$ corresponds to the attachment zone, while the region of low concentration $u$ corresponds to the detachment zone. The reactive fluxes associated with membrane binding and unbinding are balanced by the diffusive fluxes. Note, however, that one major difference between the two-component model and cell polarization models is that the latter involve the interaction between cytosolic proteins and membrane-bound proteins, which diffuse in separate spatial domains that differ in the number of spatial dimensions and geometry.

### 11.3.2 Coupling bulk diffusion with reactive membrane boundaries

From a mathematical perspective, if diffusion in the membrane is neglected then one has to deal with a coupled PDE-ODE system, where bulk diffusion in the cytoplasm is nonlinearly coupled to a reactive membrane via flux conservation conditions. This modeling paradigm is finding an increasing number of applications, both within the context of cell division and polarization [390–392, 963, 964], and more widely in terms of diffusively coupled active membranes or cells, where diffusion can induce collective synchronization [119, 366–368]. This type of problem also arises in the analysis of quorum sensing, see Sect. 15.4. One of the features of coupled PDE-ODE systems is that the steady state is typically non-uniform so linearization about the steady state generates a nontrivial eigenvalue problem. We will illustrate this by considering a simplified model and then turn to various models of cell division and polarization. Another example is developed in Ex. 11.6.

**Two diffusing protein species and a single reactive membrane.** The different types of coupled PDE-ODE systems are distinguished by the underlying dimension and shape of the bulk domain, the number of diffusing and membrane-bound protein species, the locations of the reactive and nonreactive boundaries (with the latter typically taken to be reflecting), the form of the nonlinear flux conditions at the

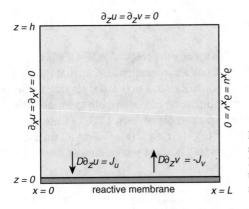

$\partial_z u = \partial_z v = 0$ (at $z = h$)

$\partial_x u = \partial_x v = 0$ (at $x = 0$)

$\partial_x u = \partial_x v = 0$ (at $x = L$)

$D\partial_z u = J_u$     $D\partial_z v = -J_v$

$z = h$     $z = 0$     $x = 0$     reactive membrane     $x = L$

Fig. 11.18: Coupled PDE-ODE model consisting of two protein species in a rectangular domain. The reactive membrane is taken to be at $z = 0$. All other boundaries are reflecting.

reactive boundaries, and the dynamics of membrane-bound proteins. As a simple example, consider the rectangular domain $[0, L] \times [0, h]$ shown in Fig. 11.18 with reflecting boundary conditions at $x = 0, L$, $z = h$, and a reactive membrane boundary at $z = 0$. Let $u$ and $v$ denote the concentrations of two chemical species A and B, evolving according to the RD system

$$\frac{\partial u}{\partial t} = D\nabla^2 u + \sigma v, \tag{11.3.16a}$$

$$\frac{\partial v}{\partial t} = D\nabla^2 v - \sigma v, \quad 0 < z < h. \tag{11.3.16b}$$

(Within the context of the self-organization of Min proteins in *E. coli*, see Sect. 11.3.3, $u$ and $v$ could represent the cytosolic concentrations of MinD-ATP and MinD-ADP, respectively [391] with $\sigma$ the rate of conversion of the ADP form back to the ATP form.) Equations (11.3.16a,b) are supplemented by the reflecting boundary conditions

$$\frac{\partial u}{\partial z} = \frac{\partial v}{\partial z} = 0, \quad z = h, \tag{11.3.17a}$$

$$\frac{\partial u}{\partial x} = \frac{\partial v}{\partial x} = 0, \quad x = 0, L, \tag{11.3.17b}$$

and the reactive boundary conditions

$$D\frac{\partial u}{\partial z}\bigg|_{z=0} = J_u \equiv f_{\text{on}}(c, u_0), \quad D\frac{\partial v}{\partial z}\bigg|_{z=0} = -J_v \equiv -k_{\text{off}}c. \tag{11.3.17c}$$

Here $u_0(x, t) = u(x, 0, t)$ is the local cytosolic concentration of protein A and $c(x, t)$ is the concentration of membrane-bound protein A. The function $f_{\text{on}}$ represents some nonlinear (autocatalytic) attachment process for protein A that depends on $u_0$ and $c$, whereas membrane proteins detach at a constant rate $k_{\text{off}}$ and immediately convert to protein B. From mass conservation it follows that the dynamical equation for $c$ is

$$\frac{\partial c}{\partial t} = f_{\text{on}}(c(x,t), u_0(x,t)) - k_{\text{off}}c(x,t).$$ (11.3.18)

Equations (11.3.16), (11.3.17), and (11.3.18) specify the coupled PDE-ODE system.

For simplicity suppose that all concentrations are $x$-independent, so that we have an effective 1D system with the boundary conditions at $x = 0, L$ not playing a role. (Note, however, that the additional spatial dimension can contribute to instabilities of any steady-state solutions, as shown in the example of Sect. 11.3.3.) Consider a steady-state solution of the form $u(z,t) = u^*(z), v(z,t) = v^*(z)$. Setting all time and $x$ derivatives to zero and imposing the reflecting boundary condition at $z = h$, we obtain the solution

$$u^*(z) = u_0^* + v_0^* \left( 1 - \frac{\cosh[(h-z)/\ell]}{\cosh(h/\ell)} \right),$$ (11.3.19a)

$$v^*(z) = v_0^* \frac{\cosh[(h-z)/\ell]}{\cosh(h/\ell)},$$ (11.3.19b)

where $\ell = \sqrt{D/\sigma}$, $u_0^* = u^*(0)$, and $v_0^* = v^*(0)$. The three unknown constants $u_0^*, v_0^*$, and $c^*$ are determined by the reactive boundary conditions (11.3.17c), which give

$$\frac{D}{\ell} \tanh(h/\ell)v_0^* = k_{\text{off}}c^*, \quad f_{\text{on}}(c^*, u_0^*) = k_{\text{off}}c^*,$$ (11.3.20)

together with mass conservation $\bar{n} = u_0^* + v_0^* + c^*/h$, where $\bar{n}$ is the total mass density. One can view $\ell$ as a penetration depth which sets a length scale for the nontrivial effects of bulk diffusion, namely, $\ell \ll h$. In this regime, the steady state has to be treated as spatially nonuniform. One consequence of this is that determining the stability of the steady state requires solving a nontrivial eigenvalue problem. In Sect. 11.3.3 we will consider a generalized version of the above model which has been used to explore the self-organization of Min proteins during cell division in *E. coli* [390, 391]. One finds from the associated eigenvalue equation that the model exhibits a Turing–Hopf bifurcation, resulting in the growth of a time-periodic $x$-dependent variation in the distribution of membrane proteins. Here we will focus on a simpler eigenvalue problem by considering linear stability with respect to $x$-independent perturbations.

*Linear stability analysis.* Consider the perturbations

$$u(z,t) = u^*(z) + U(z)e^{\lambda t}, \quad v(z,t) = v^*(z) + V(z)e^{\lambda t}, \quad c(t) = c^* + Ce^{\lambda t}.$$

Substituting into the time-dependent RD system (11.3.16) gives

$$D\frac{d^2U}{dz^2} - \lambda U + \sigma V = 0, \quad D\frac{d^2V}{dz^2} - (\lambda + \sigma)V = 0.$$ (11.3.21)

Imposing the no-flux boundary condition at $z = h$ shows that

$$U(z) = [U(0) + V(0)]\frac{\cosh([h-z]/\ell_\lambda(0))}{\cosh(h/\ell_\lambda(0))} - V(0)\frac{\cosh([h-z]/\ell_\lambda(\sigma))}{\cosh(h/\ell_\lambda(\sigma))}, \quad (11.3.22a)$$

$$V(z) = V(0)\frac{\cosh([h-z]/\ell_\lambda(\sigma))}{\cosh(h/\ell_\lambda(\sigma))}, \quad (11.3.22b)$$

where we now have a $\lambda$-dependent penetration depth:

$$\ell_\lambda(\sigma) = \sqrt{\frac{D}{\sigma+\lambda}}. \quad (11.3.23)$$

Substituting equations (11.3.22a-b) into the left-hand side of the linearized version of the reactive boundary conditions (11.3.17c) gives

$$-\alpha C - \beta h U(0) = \left[\frac{D}{\ell_\lambda(0)}\tanh(h/\ell_\lambda(0)) - \frac{D}{\ell_\lambda(\sigma)}\tanh(h/\ell_\lambda(\sigma))\right] V(0)$$
$$+ \frac{D}{\ell_\lambda(0)}\tanh(h/\ell_\lambda(0))U(0), \quad (11.3.24a)$$

$$k_{\mathrm{off}}C = \frac{D}{\ell_\lambda(\sigma)}\tanh(h/\ell_\lambda(\sigma))V(0), \quad (11.3.24b)$$

where

$$\alpha = \frac{\partial f_{\mathrm{on}}}{\partial c}\bigg|_{u=u^*,c=c^*}, \quad \beta = \frac{1}{h}\frac{\partial f_{\mathrm{on}}}{\partial u_0}\bigg|_{u=u^*,c=c^*}.$$

Note that $\beta$ and $\alpha$ have the same dimension of inverse time. Similarly, linearizing equation (11.3.18),

$$(\lambda + k_{\mathrm{off}})C = \alpha C + \beta h U(0). \quad (11.3.24c)$$

Combining equations (11.3.24a-c) yields the following transcendental equation for the eigenvalues $\lambda$:

$$\lambda + k_{\mathrm{off}} = \frac{k_{\mathrm{off}}\left[1 - \dfrac{\ell_\lambda(\sigma)}{\ell_\lambda(0)}\dfrac{\tanh(h/\ell_\lambda(0))}{\tanh(h/\ell_\lambda(\sigma))}\right] + \dfrac{\alpha}{\beta h}\dfrac{D}{\ell_\lambda(0)}\tanh(h/\ell_\lambda(0))}{1 + \dfrac{1}{\beta h}\dfrac{D}{\ell_\lambda(0)}\tanh(h/\ell_\lambda(0))} \equiv \Gamma(\lambda). \quad (11.3.25)$$

Eigenvalue equations analogous to equation (11.3.25) appear in various models of collective synchronization mediated by bulk diffusion [119, 366–368, 963, 964]. In order to allow for the possibility of collective oscillations, it is necessary to include additional molecular species within the reactive membrane (or to introduce discrete delays in the membrane dynamics, see Sect. 11.3.4). For example, suppose that equation (11.3.18) becomes (assuming $x$-independent concentrations)

$$\frac{dc_1}{dt} = f_{\mathrm{on}}(c_1, u_0) - k_{\mathrm{off}}c_1 + g_1(\mathbf{c}), \quad (11.3.26)$$

where $\mathbf{c} = (c_1, \ldots, c_M)$ with $c_1$ the membrane-bound version of species A, and $c_j$, $j > 1$, representing the concentrations of additional proteins evolving according to the system of ODEs

$$\frac{dc_j}{dt} = g_j(\mathbf{c}), \quad 1 < j \leq M. \quad (11.3.27)$$

Again one can look for a steady-state solution $(u(z,t), v(z,t), \mathbf{c}(t)) = (u^*(z), v^*(z), \mathbf{c}^*)$ and linearize about this solution to obtain an eigenvalue equation. In particular, equation (11.3.25) becomes the matrix equation

$$\sum_{j=1}^{M} (\lambda \delta_{i,j} - J_{ij}) C_j = [\Gamma(\lambda) - k_{\text{off}}] C_1 \delta_{i,1}, \quad J_{ij} = \left. \frac{\partial g_i(\mathbf{c})}{\partial c_j} \right|_{\mathbf{c}=\mathbf{c}^*}. \quad (11.3.28)$$

Suppose that the steady state becomes unstable due to a pair of complex conjugate eigenvalues crossing the imaginary axis as a model parameter is varied. The Hopf bifurcation theorem (Box 5A) can then be used to establish the existence of a limit cycle oscillation in the membrane concentration. Moreover, this can be related to the problem of collective synchronization by noting that if the domain in Fig. 11.18 is reflected about the line $z = h$, then one obtains a pair of reactive membranes (or cells) with identical dynamics coupled via bulk diffusion. Imposing the reflecting boundary condition at the midline $z = h$ for the linear stability problem then corresponds to an in-phase synchronization of the two membranes, whereas imposing the condition $u(h) = v(h) = 0$ corresponds to an anti-phase solution [366, 367]. It can thus be established that bulk diffusion induces synchronous oscillations in coupled membranes or cells in parameter regions that would not support oscillations in the absence of diffusive coupling. Such a mechanism is thought to play a role in oscillations of the signaling protein Cdc42 during the growth and division of fission yeast [223, 963, 964], see Sect. 11.3.4.

### 11.3.3 Self-organization of Min proteins in E. coli

The mechanism of cell division in bacteria differs significantly from eukaryotic cells; the latter is considered in Sect. 14.3. Bacterial cell division is initiated by the polymerization of the tubulin homolog FtsZ into the so-called Z-ring. In rod-shaped bacteria such as E. coli, formation of the Z-ring is usually restricted to the cell center, where it determines the site of cell division. The localization of the Z-ring is accurate to within 3% of the cell length, resulting in two daughter cells of almost equal size. A major process involved in the high precision of cell division is the regulatory system of Min proteins [526, 573]. The Min system consists of three proteins, MinC, MinD, and MinE. MinC inhibits Z-ring formation, whereas MinD and MinE act to confine MinC to the cell poles. A characteristic feature of Min protein dynamics is that the protein concentrations oscillate from pole to pole with a period of 1-2 minutes, which is much shorter than the cell cycle. Consequently, the time-averaged MinC concentration is maximized at the cell poles and minimized at the cell center, resulting in inhibition of Z-ring formation at the poles.

A more detailed picture of the mechanism underlying Min protein oscillations is shown in Fig. 11.19. MinC forms a complex with MinD and thus follows the spatiotemporal variation in MinD concentration, consistent with the finding that only MinD and MinE are essential for the occurrence of oscillations. The basic biochem-

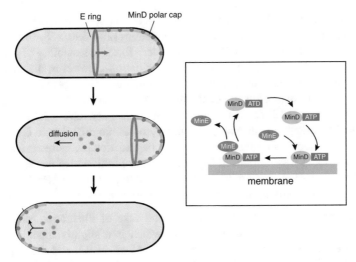

Fig. 11.19: Oscillatory patterns of Min protein system in *E. coli*. MinD·ATP (blue) binds to the membrane forming a polar cap. Min E (red) stimulates hydrolysis of MinD·ATP, which leads to protein release from the membrane. The polar cap shrinks and cytoplasmic MinD reverts to its ATP form and subsequently binds to the membrane at the opposite pole. Inset shows the cycle of biochemical reactions underlying the oscillations.

ical cycle is as follows: cytoplasmic MinD forms an ATPase MinD·ATP that binds cooperatively to the cell membrane, forming polymer filaments; MinE then binds to membrane MinD·ATP and stimulates ATP hydrolysis that causes MinD·ADP to be released from the membrane; and the cytoplasmic MinD·ADP is then converted back to MinD·ATP and rebinds to the membrane. If MinD·ATP initially binds to a polar region, then it forms a cap that extends toward the cell center and is flanked by a ring of MinE known as the E ring. The E ring stimulates the hydrolysis of Min·ATP in its neighborhood, leading to the release of MinD·ADP and MinE, with the latter rebinding to the shrinking MinD·ATP cap. This results in movement of the E ring toward the pole in the wake of the shrinking MinD·ATP cap. Meanwhile, the released MinD·ADP reconverts to MinD·ATP and rebinds to the membrane at the opposite pole, where the MinE concentration is lowest, forming a new MinD·ATP cap. Once the first cap has disappeared, the released MinE rebinds to form an E ring at the boundary of the new MinD·ATP cap. Iteration of this process underlies the observed Min protein oscillations.

The pole-to-pole oscillations observed in *E. coli* are intrinsically linked to its rod-like geometry. Various studies in other cells have further elucidated the dependence of Min protein dynamics on cell geometry. For example, in filamentous cells where cell division is inhibited, pole-to-pole oscillations develop additional nodes resulting in a standing wave pattern with a characteristic length scale [754]. Moreover, *in vitro* reconstitution of a Min system consisting of cytosol in a 3D box-like geometry on a supporting lipid bilayer has shown a wide variety of dynamic patterns, including target waves and spiral waves [572].

A number of models have been developed that describe the interactions between MinD and MinE in terms of a system of RD equations [297, 390, 428, 440, 528, 574, 609, 612]. All of the models can undergo a pattern forming instability of a homogeneous state, resulting in self-organized Min oscillations. However, the specifics of the molecular mechanism that generates the oscillations differ between the models and have not yet been resolved conclusively by experiments. For example, Meinhardt and de Boer [612] consider a phenomenological activator–inhibitor RD system with slow membrane diffusion and fast cytoplasmic diffusion; a crucial component of the model is protein synthesis and degradation. A more realistic mass-conserving model was introduced by Howard et al [428], which has subsequently been refined by various authors who emphasize the importance of MinD recruitment and/or aggregation within the membrane [390, 440, 528, 609].

**Mass-conserving recruitment model.** For the sake of illustration, consider the mass-conserving RD model of Ref. [390], which is a slightly modified version of Huang et al. [440]. Denote the interior of the cell by $\Omega$, and the cell boundary by $\partial\Omega$. We will assume that the boundary can be partitioned into a reactive cell membrane $\partial\Omega_m$, where exchange with the cytosol takes place, and a nonreactive part $\partial\Omega_0$. Let $u_D(\mathbf{r},t)$, $u_D^*(\mathbf{r},t)$, and $u_E(\mathbf{r},t)$, $\mathbf{r}\in\Omega$, denote the concentrations of MinD·ADP, MinD·ATP, and MinE in the cytoplasm, and let $u_d^*(\mathbf{r},t)$, $u_{de}(\mathbf{r},t)$, $\mathbf{r}\in\Omega_m$, denote the concentrations of MinD·ATP and MinD·ATP·MinE complexes in the reactive membrane. The various components of the model are as follows [390, 440]:

1. Conversion of cytoplasmic MinD·ADP to MinD·ATP at a rate $\sigma_D u_D$.

2. Hydrolysis of MinE-mediated membrane-bound MinD·ATP at a rate $\sigma_d u_{de}$; MinE and MinD·ADP are then immediately released from the membrane.

3. The recruitment of cytoplasmic MinD·ATP at a rate $[k_D + k_d u_d^*]U_D^*$, where $U_D^*$ is the concentration of cytoplasmic MinD·ATP close to the cell membrane.

4. The binding of cytoplasmic MinE to membrane-bound MinD·ATP at a rate $\sigma_E u_d^* U_E$, where $U_E$ is the concentration of cytoplasmic MinE close to the cell membrane.

5. All cytoplasmic proteins have the same diffusion coefficient $D$, whereas membrane diffusion is assumed to be negligible.

The resulting system of RD equations consists of the components

$$\frac{\partial u_D}{\partial t} = D\nabla^2 u_D - \sigma_D u_D, \tag{11.3.29a}$$

$$\frac{\partial u_D^*}{\partial t} = D\nabla^2 u_D^* + \sigma_D u_D, \quad \frac{\partial u_E}{\partial t} = D\nabla^2 u_E, \quad \mathbf{r}\in\Omega, \tag{11.3.29b}$$

$$\frac{\partial u_d^*}{\partial t} = -\sigma_E u_d^* U_E + [k_D + k_d u_d^*]U_D^*, \, \mathbf{r}\in\partial\Omega_m, \tag{11.3.29c}$$

$$\frac{\partial u_{de}}{\partial t} = -\sigma_d u_{de} + \sigma_E u_d^* U_E, \quad \mathbf{r}\in\partial\Omega_m. \tag{11.3.29d}$$

These equations are supplemented by nonlinear boundary conditions at the reactive cell membrane,

$$D\nabla_n u_D = \sigma_d u_{de}, \tag{11.3.30a}$$

$$D\nabla_n u_D^* = -[k_D + k_d u_d^*]U_D^*, \tag{11.3.30b}$$

$$D\nabla_n u_E = \sigma_d u_{de} - \sigma_E u_d^* U_E, \quad \mathbf{r} \in \partial\Omega_m, \tag{11.3.30c}$$

and no-flux boundary conditions at any nonreactive boundaries

$$D\nabla_n u_D = 0, \quad D\nabla_n u_D^* = 0, \quad D\nabla_n u_E = 0, \quad \mathbf{r} \in \partial\Omega_0. \tag{11.3.30d}$$

Here $\nabla_n$ denotes the derivative normal to the boundary. We also have the conservation equations

$$\text{MinD}_T = \int_\Omega (u_D + u_D^*)d\mathbf{r} + \int_{\partial\Omega_m} (u_d + u_{de})d\mathbf{s}, \tag{11.3.31a}$$

$$\text{MinE}_T = \int_\Omega u_E d\mathbf{r} + \int_{\partial\Omega_m} u_{de}d\mathbf{s}, \tag{11.3.31b}$$

where $\text{MinD}_T$ and $\text{MinE}_T$ are the total number of molecules of each protein and are specified by the initial conditions. Note that the above model is a generalization of the coupled PDE-ODE system considered in Sect. 11.3.2.

Numerical simulations of the system of equations (11.3.29) and (11.3.30) in various geometries reveal oscillations with similar characteristics to those found in experiments, including the growth and shrinkage of alternating polar caps and the formation of an E ring at the cell center [390, 392, 440]. For example, Halatek and Frey [390] considered a 2D elliptical cell of typical length 5 $\mu$m and width 1 $\mu$m. A detailed exploration of parameter space indicated that two necessary conditions for pole-to-pole oscillations to occur were $k_d < \sigma_E$ (recruitment of MinE faster than MinD) and $\text{MinE}_T < \text{MinD}_T$. The reason that MinD can then form a cap on the opposite pole is that it diffuses farther than MinE following release from the old polar cap, which is a consequence of the delay in converting MinD·ADP back to MinD·ATP. That is, MinE tends to locally rebind until recruitment by membrane-bound MinD ends. Once the original cap has disappeared, the newly released MinE proteins diffuse until they encounter the edge of the newly formed cap where they rapidly bind to form the E ring. As highlighted in [392] and Fig. 11.20, the Min oscillations can be interpreted in terms of the periodic exchange of membrane-bound attachment and detachment zones together with the cytosolic redistribution of proteins, exemplifying the general framework of mass-conserving intracellular pattern formation. Halatek and Frey [390] also established that increasing the cell length could result in the formation of oscillatory stripe patterns, consistent with the observations of [754]. However, this required at least a 2D model in which there is a separation of bulk and membrane dynamics.

*Linear stability analysis.* In order to analyze the mass-conserving RD model of Min oscillations, we will follow Halatek and Frey [392] by considering the simplified geometry of

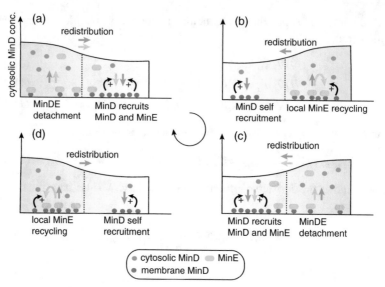

Fig. 11.20: Illustration of pole-to-pole Min oscillations in *E. coli* based on the formation of attachment and detachment zones and the redistribution of proteins via cytosolic diffusion. (a) MinD is
concentrated at the right pole and recruits cytosolic MinD and MinE via positive feedback. The
cytosolic fluxes are supplied by the detachment of MinDE at the left pole. (b) MinDE detaches
from the right pole but MinE locally rebinds, whereas dephosphorylated MinD is free to diffuse to
the left pole where it binds to the membrane and self-recruits. (c) MinE trapping ends at the right
pole when all minD has detached, and the new polar zone becomes an attachment zone for MinE.
(d) The process starts over at the left pole.

Fig. 11.18. (A 3D version of this geometry is used in experimentally reconstituted *in vitro*
Min systems [572].) The reactive membrane is taken to be $z = 0$ and the other boundaries
$(x = 0, L, z = h)$ are non-reactive. We first consider a solution that is spatially uniform across
the membrane, that is, the various densities are independent of $x$. The steady-state density
profiles in the bulk satisfy the equations

$$D\frac{d^2 u_D}{dz^2} - \sigma_D u_D = 0, \quad D\frac{d^2 u_D^*}{dz^2} + \sigma_D u_D = 0, \quad D\frac{d^2 u_E}{dz^2} = 0,$$

whereas $u_d^*(x) = \bar{u}_d^*, u_{de}(x) = \bar{u}_{de}$ with $\bar{u}_d^*, \bar{u}_{de}$ constants. Imposing the no-flux boundary
conditions at $z = h$ yields the solutions

$$u_D(z) = \bar{U}_D \frac{\cosh[(h-z)/l]}{\cosh(h/\ell)}, \tag{11.3.32a}$$

$$u_D^*(z) = \bar{U}_D^* + \bar{U}_D \left(1 - \frac{\cosh[(h-z)/\ell]}{\cosh(h/\ell)}\right), \quad u_E(z) = U_E, \tag{11.3.32b}$$

where $\ell = \sqrt{D/\sigma_D}$ determines the penetration depth into the cytosol, and $\bar{U}_D = u_D(x, 0)$
etc. Substituting these solutions into the reactive boundary conditions at the membrane with
$\nabla_n = -d/dz$ gives

$$\frac{D}{\ell}\bar{U}_D\tanh(h/\ell) = \sigma_d\bar{u}_{de},$$ (11.3.33a)

$$-\frac{D}{\ell}\bar{U}_D\tanh(h/\ell) = -[k_D + k_d\bar{u}_d^*)]\bar{U}_D^*, \quad 0 = \sigma_d\bar{u}_{de} - \sigma_E\bar{u}_d^*\bar{U}_E.$$ (11.3.33b)

Any solution to these equations automatically satisfies the steady-state version of equations (11.3.29c,d). Hence, equation (11.3.33) together with the mass conservation equations

$$M_D := \frac{\text{Min}D_T}{Lh} = \bar{U}_D + \bar{U}_D^* + \frac{\bar{u}_d + \bar{u}_{de}}{h}, \quad M_E := \frac{\text{Min}E_T}{Lh} = \bar{U}_E + \frac{\bar{u}_{de}}{h},$$ (11.3.34)

yield five equations in the five unknown constants $\bar{U}_D, \bar{U}_D^*, \bar{U}_E, \bar{u}_d^*, \bar{u}_{de}$. The corresponding equilibria represent $x$-independent steady-state solutions.

Suppose that such an equilibrium exists. We can then investigate the conditions for membrane patterning using linear stability analysis. That is, introduce the perturbed solutions

$$u_D(x,z,t) = u_D(z) + e^{\sigma(k)t}\cos(kx)w_D(z;k),$$ (11.3.35a)

$$u_D^*(x,z,t) = u_D^*(z) + e^{\sigma(k)t}\cos(kx)w_D^*(z;k),$$ (11.3.35b)

$$u_E(x,z,t) = u_E(z) + e^{\sigma(k)t}\cos(kx)w_E(z;k),$$ (11.3.35c)

$$u_d^*(x,t) = u_d^* + e^{\sigma(k)t}\cos(kx)W_d^*(k),$$ (11.3.35d)

$$u_{de}(x,t) = u_{de} + e^{\sigma(k)t}\cos(kx)W_{de}(k).$$ (11.3.35e)

The function $\cos(kx)$ satisfies the no-flux boundary conditions at $x = 0,L$ provided that $k = n\pi/L$ for integer $n$, and naturally arises when applying separation of variables to the diffusion equation. Substituting into equations (11.3.29a-c) yields

$$D\frac{d^2w_D}{dz^2} - (\sigma_D + \sigma(k) + Dk^2)w_D = 0, \quad D\frac{d^2w_D^*}{dz^2} - (\sigma(k) + Dk^2)w_D^* + \sigma_D w_D = 0,$$

and

$$D\frac{d^2w_E}{dz^2} - (\sigma(k) + Dk^2)w_E = 0.$$

Imposing the no-flux boundary condition at $z = h$ shows that

$$w_D(z;k) = W_D(k)\frac{\cosh([h-z]/\ell_k(\sigma_D))}{\cosh(h/\ell_k(\sigma_D))},$$ (11.3.36a)

$$w_D^*(z;k) = [W_D^*(k) + W_D(k)]\frac{\cosh([h-z]/\ell_k(0))}{\cosh(h/\ell_k(0))} - W_D(k)\frac{\cosh([h-z]/\ell_k(\sigma_D))}{\cosh(h/\ell_k(\sigma_D))},$$ (11.3.36b)

$$w_E(z;k) = W_E(k)\frac{\cosh([h-z]/\ell_k(0))}{\cosh(h/\ell_k(0))},$$ (11.3.36c)

where we now have a $k$-dependent penetration depth:

$$\ell_k(\sigma_D) = \sqrt{\frac{D}{\sigma_D + \sigma(k) + Dk^2}}.$$ (11.3.37)

Equations (11.3.35a-c) and (11.3.36) are now substituted into the left-hand side of the linearized version of the reactive boundary conditions (11.3.30):

$$\sigma_d W_{de} = \frac{D}{\ell_k(\sigma_D)} \tanh(h/\ell_k(\sigma_D)) W_D, \tag{11.3.38a}$$

$$-[k_D + k_d \bar{u}_d^*] W_D^* - k_d \bar{U}_D^* W_d^* = \left[ \frac{D}{\ell_k(0)} \tanh(h/\ell_k(0)) - \frac{D}{\ell_k(\sigma_D)} \tanh(h/\ell_k(\sigma_D)) \right] W_D$$
$$+ \frac{D}{\ell_k(0)} \tanh(h/\ell_k(0)) W_D^*, \tag{11.3.38b}$$

$$\sigma_d W_{de} - \sigma_E (\bar{u}_d^* W_E + \bar{U}_E W_d^*) = W_E \tanh(h/\ell_k(0)). \tag{11.3.38c}$$

Similarly, substituting equations (11.3.35d,e) into the linearized versions of equations (11.3.29c) and (11.3.29d) gives

$$(\sigma(k) + \sigma_E \bar{U}_E) W_d^* = -\sigma_E \bar{u}_d^* W_E + [k_D + k_d \bar{u}_d^*] W_D^* + k_d \bar{U}_D^* W_d^*, \tag{11.3.38d}$$

$$(\sigma(k) + \sigma_d) W_{de} = \sigma_E \bar{u}_d^* W_E + \sigma_E \bar{U}_E W_d^*. \tag{11.3.38e}$$

(If we included the effects of slow membrane diffusion, then $\sigma(k) \rightarrow \sigma(k) + D_m k^2$ in equation (11.3.38), with $D_m$ the membrane diffusion coefficient and $D_m \ll D$.) For a given wavenumber $k$, the system of linear equations (11.3.38a–e) can be rewritten as a fifth order matrix equation $\mathbf{M}(k)\mathbf{W} = 0$ with $\mathbf{W} = (W_D, W_D^*, W_E, W_d^*, W_{de})^\top$. Numerically solving $\det[\mathbf{M}(k)] = 0$ one can determine the dispersion relation $\max \mathrm{Re}[\sigma(k)]$ as a function of $k$ and thus identify parameter regions where there is a band of growing eigenmodes. An example dispersion curve is shown in Fig. 11.21. If $\mathrm{Im}[\sigma(k)] \neq 0$, then these modes represent standing wave oscillatory patterns. The details can be found in [392], where it is also shown how the emerging oscillatory patterns are slaved to moving local equilibria, whose displacements are controlled by the lateral redistribution of the total densities $M_D(x,t)$ and $M_E(x,t)$ with

$$M_D(x,t) = u_D(x,0,t) + u_D^*(x,0,t) + \frac{u_d^*(x,t) + u_{de}(x,t)}{h},$$

and

$$M_E(x,t) = u_E(x,0,t) + \frac{u_{de}(x,t)}{h}.$$

Recall from Sect. 11.2.1 that one of the features of classical Turing pattern formation is the emergence of a pattern with a characteristic length scale $2\pi/k_c$, where $k_c$ is the wavenumber of the fastest growing mode. More precisely, this is the length scale of the (possibly transient) pattern that initially forms from the uniform state according to linear stability analysis. It will only reflect the length scale of the final stable pattern when the growth of the pattern saturates at a small amplitude; in this case one can investigate pattern selection and stability using weakly nonlinear analysis. However, small amplitude patterns tend not to be robust so are unlikely to be prevalent in all biological systems. In the case of large amplitude patterns, the length

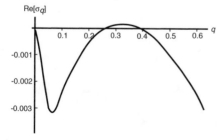

Fig. 11.21: Example dispersion relation for the Min system in a rectangular domain, showing the real part of the principal dispersion branch. Parameter values are $D = 60 \mu\mathrm{m}^2 \, \mathrm{s}^{-1}$, $k_D = 0.065 \, \mu\mathrm{m/s}$, $k_d = 0.098 \, \mu\mathrm{m}^3/\mathrm{s}$, $\sigma_E = 0.126 \, \mu\mathrm{m}^3/\mathrm{s}$, $\sigma_d = 0.34 \, \mathrm{s}^{-1}$, $\sigma_D = 6 \, \mathrm{s}^{-1}$, MinD mean total density $= 638 \mu\mathrm{m}^{-3}$, MinE mean total density $= 410 \mu\mathrm{m}^{-3}$, $L = 500 \mu\mathrm{m}$, and $h = 20 \mu\mathrm{m}$. [Redrawn from [392].]

scale of the final pattern is much harder to predict. In many cases, there is a form of coarsening or winner-takes-all dynamics, in which the length scale collapses to the system size or to some intermediate length scale that is independent of the initially selected length scale [70]. In the case of membrane-bound patterns during cell polarization, one often finds that there is a single region of enhanced protein concentration and a complementary region of reduced concentration, so that the pattern length scale is determined by the system size. On the other hand, it is possible to observe transient patterns with multiple peaks in various cell mutants and *in vitro*. In the above example of Min protein patterns in a rectangular geometry, one finds that a characteristic length scale independent of system size can emerge when the attachment and detachment zones are strongly coupled through cytosolic transport (so-called canalization) [390]. In this scenario, the magnitude of the fluxes onto and off the membrane are approximately balanced so that the fraction of cytosolic proteins is approximately constant. Canalized transport can lead to the emergence of a characteristic distance between the attachment and detachment zones.

## 11.3.4 Cell polarization in yeast

Many cellular processes depend critically on the establishment and maintenance of polarized distributions of signaling proteins on the plasma membrane [474]. These include cell motility, neurite growth and differentiation, epithelial morphogenesis, embryogenesis, and stem cell differentiation. Cell polarization typically occurs in response to some external spatial cue such as a chemical gradient. A number of features of the stimulus response are shared by many different cell types, including amplification of spatial asymmetries, persistence of polarity when the triggering stimulus is removed, and sensitivity to new stimuli whereby a cell can reorient when the stimulus gradient is changed. In many cases, cell polarity can also occur spontaneously, in the absence of preexisting spatial cues. Here we will focus on cell polarization in budding and fission yeast cells. Another important example, namely cell polarization in motile eukaryotic cells [422, 553, 589] will be considered in Sect. 11.6, within the context of intracellular traveling waves.

**Budding yeast.** One of the most studied model systems of cell polarization is the budding yeast *Saccharomyces cerevisiae* [189, 478, 826]. A yeast cell in the G1 phase of its life cycle is spherical and grows isotropically. It then undergoes one of two fates: either it enters the mitotic phase of the life cycle and grows a bud or it forms a mating projection (shmoo) toward a cell of the opposite mating type. Both processes involve some form of symmetry breaking mechanism that switches the cell from isotropic growth to growth along a polarized axis, see Fig. 11.22. Under physiological conditions, yeast cells polarize toward an environmental spatial asymmetry. This could be a pheromone gradient in the case of mating or a bud scar deposited on the cell surface from a previous division cycle. However, yeast cells can also undergo spontaneous cell polarization in a random orientation when external asymmetries are removed. For example, in the case of budding, induced cell mutations can eliminate the recognition of bud scars. Moreover, shmoo formation can occur in the presence of a uniform pheromone concentration. The observation that cells are able to break symmetry spontaneously suggests that polarization is a consequence of internal biochemical states. Experimental studies in yeast have shown that cell polarization involves a positive feedback mechanism that enables signaling molecules already localized on the plasma membrane to recruit more signaling molecules from the cytoplasm, resulting in a polarized distribution of surface molecules. (This is analogous to the recruitment of MinD and MinE by membrane-bound MinD in bacteria.) The particular signaling molecule in budding yeast is the Rho GTPase Cdc42 (see Box 11D). As with other Rho GTPases, Cdc42 targets downstream affectors of the actin cytoskeleton. There are two main types of actin structure involved in the polarized growth of yeast cells: cables and patches. Actin patches consist of networks of branched actin filaments nucleated by the Arp2/3 complex at the plasma membrane, whereas actin cables consist of long, unbranched bundles of actin filaments. Myosin motors travel along the cables unidirectionally toward the actin barbed ends at the plasma membrane, transporting intracellular cargo such as vesicles, mRNA, and organelles. The patches act to recycle membrane-bound structures to the cytoplasm via endocytosis. During cell polarization, Cdc42-GTP positively regulates the nucleation of both types of actin structure, resulting in a polarized actin network, in which actin patches are concentrated near the site of asymmetric growth and cables are oriented toward the direction of growth.

The positive feedback mechanism that establishes cell polarity in budding yeast [9, 189, 475, 478, 934, 956] involves an actin-independent pathway, in which Bem1, an adaptor protein with multiple binding sites, forms a complex with Cdc42 that enables recruitment of more Cdc42 to the plasma membrane, see Fig. 11.23. A simple scalar RD model for positive feedback in budding yeast can be constructed as follows. Let $u(\mathbf{x},t)$ denote the concentration of some signaling molecule in the cell membrane, which is treated as a 2D bounded domain $\Sigma$. The density evolves

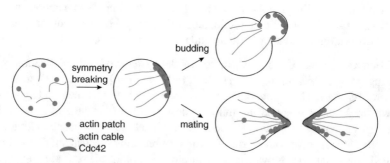

Fig. 11.22: Symmetry breaking processes in the life cycle of budding yeast. See text for details. Redrawn from [826].

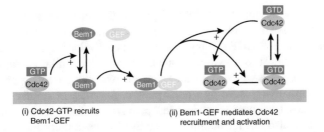

(i) Cdc42-GTP recruits
Bem1-GEF

(ii) Bem1-GEF mediates Cdc42
recruitment and activation

Fig. 11.23: Positive feedback by local activation of Cdc42. (i) Membrane-bound Cdc42-GTP recruits the activator Bem1, which subsequently binds GDP/GTP exchange factor (GEF). (ii) The BEM1-GEF complex then allows Cdc42-GTP to recruit and activate neighboring Cdc42-GDP in a positive feedback loop. This ultimately leads to the local accumulation of Cdc42-GTP and Bem1-GEF in the membrane, which is maintained by the cytosolic redistribution of these molecular components analogous to the scheme shown in Fig. 11.15.

according to the RD equation

$$\frac{\partial u}{\partial t} = D\nabla^2 u + K_+ n_c u - K_- u, \ \mathbf{x} \in \Sigma; \quad \mathbf{n} \cdot \nabla u = 0, \ \mathbf{x} \in \partial\Sigma. \tag{11.3.39}$$

Here $n_c$ is the number of molecules in the cytoplasm (which are assumed to be uniformly distributed), $D$ is the diffusivity, and $K_\pm$ represents the rates of binding/unbinding of cytosolic molecules. Binding is taken to involve positive feedback due to self-recruitment of membrane-bound molecules. Finally, we have the conservation condition

$$n_c + |\Sigma|\bar{u}(t) = M, \quad \bar{u}(t) = \frac{1}{|\Sigma|} \int_\Sigma u(\mathbf{x}, t) d\mathbf{x},$$

where $M$ is the total number of molecules in the cell. Using the Poincare inequality (Box 11B), it can be shown that such a system cannot support patterns in the long-time limit, see Ex. 11.7. This suggests that one needs to take explicit account of diffusion in the bulk. Interestingly, numerical simulations of a stochastic version of the model reveal that a localized aggregate of membrane-bound molecules can form and persist for physiologically reasonable time periods, before ultimately dispersing due to the effects of diffusion [9, 475, 546]. This stochastic-based effect has also been explored in a rigorous mathematical study [383] and by considering the spectral properties of a spatially discrete RD master equation [604], see also Sect. 16.1.

An alternative mechanism for positive feedback has been proposed [586, 827], in which spatial asymmetries are reinforced by the directed transport of Cdc42 along the actin cytoskeleton to specific locations on the plasma membrane. However, this has now been experimentally discounted due to the observation that even complete actin depolymerization has limited effects on cell polarization [951]. Moreover, mathematical modeling suggests that delivery and fusion of secretory vesicles carrying Cdc42 actually dilute polarity due to the fact that Cdc42 is less concentrated on vesicle membranes than within the polarization site [547, 794]. One of the first mass-conserving models of cell polarization in budding yeast was developed by Goryachev and Pokhilko [361]. They considered a detailed eight-variable model, based on GTP-GDP cycling of Cdc42, its activator Cdc24, and the effector Bem1, and reduced it to a two-component system with similar essential features. The model exhibits a winner-take-all behavior, whereby an emerging multi-peak pattern is eventually replaced by a polar pattern. A variety of similar Turing-based RD systems have been used to model yeast polarization, as reviewed in [363]. Some of these models also include additional negative feedback loops that result in more complex dynamical patterns, which have also been observed experimentally [432].

---

**Box 11D. Rho GTPase and cell polarization.**

---

Rho GTPase is a class of signaling molecule that plays an important role in the polarization and migration of many different cell types [393]. Rho GTPase act as intracellular molecular switches that cycle between an active GTP-bound form in the membrane and an inactive GDP-bound form in the cytosol, see Fix. 11.24(a). Guanine nucleotide exchange factors (RhoGEFs) facilitate the conversion from GDP-bound to GTP-bound form, whereas GTPase-activating proteins (RhoGAPs) enhance GTP hydrolysis and are thus negative regulators. RhoGEFs and RhoGAPs are both regulated by upstream signals. A major downstream target of the Rho GTPase signaling pathways is the actin cytoskeleton [393, 581, 667]. A variety of actin accessory proteins mediate the different components of actin dynamics within a cell and can be activated by the various signaling pathways associated with cell polarization. We describe a few of the major players; see [667] for more details. First, the actin related proteins 2 and 3 (Arp2/3) complex stimulates actin polymerization by creating new nucleation cores. The Arp2/3 complex is activated by members of the Wiskott–Aldrich syndrome protein (WASP) family protein (WAVE) complex, which localizes to lamellipodia (mesh-like actin sheets) where it facilitates actin polymerization.

The WAVE complex is activated by the small GTPase Rac1, which modulates the actin cytoskeleton dynamics by controlling the formation of lamellipodia. The formation of Filopodia (packed actin bundles), on the other hand, is regulated by another member of the small GTPases, called Cdc42. A second important accessory protein is cofilin, a member of the actin depolymerizing factor (ADF)/cofilin family, which modifies actin dynamics by increased severing and depolymerization of actin filaments via its binding to the non-barbed (pointed) ends. Cofilin is inhibited when phosphorylated by LIM-kinase, which is itself activated by p21 activated kinases (PAKs). Since the latter is a downstream target of Rac1 and Cdc42, it follows that Rac1 and Cdc42 inhibit actin depolymerization by downregulating cofilin. At first sight, one would expect the reduction of cofilin to counter the effects of Arp2/3. However, the extent to which cofilin enhances or reduces cell protrusion depends on the spatial and temporal scale over which it operates [228]. It turns out that the overall effect of Rac1 and Cdc42 is to increase actin dynamics thus promoting cytoskeletal growth. This is opposed by the action of a third type of GTPase known as RhoA, which tends to stabilize the actin network by activating ROCK-kinase, which promotes profilin (an actin-binding protein that catalyzes polymerization) and suppresses cofilin. A summary of the basic signaling pathways is given in Fig. 11.24(b).

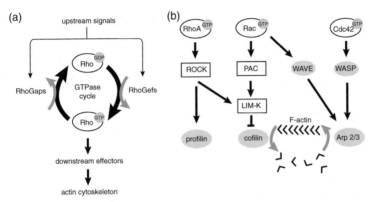

Fig. 11.24: (a) The Rho GTPase cycle. (b) Simplified signaling pathways from GTPases RhoA, Rac, and Cdc42 to actin accessory proteins cofilin and Arp 2/3 that regulate actin polymerization in the cytoskeleton. [Redrawn from Luo [581]].

**Fission yeast.** Another well-studied organism that exhibits cell polarization is the fission yeast *Schizosaccharomyces pombe*. Although it shares many of the same molecular players as budding yeast, there are some significant differences with regards to the detailed mechanisms of cell polarization (see the reviews [179, 254, 600]). Fission yeast is a rod-shaped cell consisting of two hemispheres of constant radius that cap a cylinder of increasing length. Thus growth is effectively one-dimensional (axial). There are distinct stages of cell growth that are regulated by the cell cycle, as illustrated in Fig. 11.25. Immediately following cell division, the cell initially grows at one end only, namely, the "old end" of the previous cell cycle (monopolar growth). However, during the G2 phase of the cell cycle, the cell also starts growing from the new end (bipolar growth), in a process known as "new end take off" (NETO) [625]. Cell growth then ceases during mitosis, after which the cytoskeletal growth machinery is directed toward the division site for cell separation.

Growth of fission yeast is mediated by actin filaments and microtubules (MTs). The MTs polymerize toward both tips, forming bundles of filaments fixed at the nucleus, while the dynamically unstable plus ends explore the region near the cell ends. The MTs thus form cytoskeletal tracks for molecular motors to transport land-

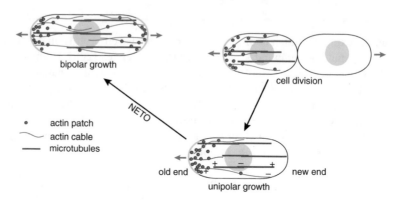

Fig. 11.25: Cytoskeleton organization of fission yeast during the cell cycle. Sites of active growth are labeled by the red lines and arrows. The microtubules (green), actin cytoskeleton (dark red), and protein-rich membrane domains (light blue) are depicted at representative cell cycle stages. [Redrawn from Chang and Martin [179].]

mark proteins to the cell tips, which then regulate cell growth. A particularly important class of regulatory protein belongs to the so-called tea system, in particular, the proteins tea1p and tea4p. These are transported along MTs to the cell tips, where they are deposited to form stable complexes. The tea1/4p complex then recruits other cell polarity factors. In particular, a major target of the tea1/4p complex is the formin for3p protein, which nucleates and polymerizes actin cables and patches.

An immediate issue that arises when comparing fission yeast to budding yeast concerns the role of the Rho GTPase Cdc42. As with other Rho GTPases, Cdc42 targets downstream affectors of the actin cytoskeleton including formin for3p protein and Arp2/3. During cell polarization in budding yeast, active Cdc42 positively regulates the nucleation of actin, resulting in a polarized actin network, in which actin patches are concentrated near the site of asymmetric growth and cables are oriented toward the direction of growth [120, 547, 586, 934]. Recent experimental evidence suggests that Cdc42 also plays an important role in regulating polarized growth in fission yeast [189, 223]. Labeling active Cdc42 within fission yeast cells using a fluorescent marker, Cdc42 oscillations were observed with an average period of 5 min. The oscillations occurred at both tips and were out of phase (anti-correlated). In the case of longer cells exhibiting bipolar growth, the mean amplitude of the oscillations was the same at both ends (symmetric, anti-correlated oscillations). On the other hand, for shorter, less mature cells exhibiting monopolar growth, the amplitude was significantly larger at the growing end (asymmetric, anti-correlated oscillations). The observed dynamics suggests that there is competition for active Cdc42 (or associated regulators) at the two ends.

**PDE-DDE model of Cdc42 oscillations in fission yeast.** The behavior of Cdc42 oscillations during cell polarization in fission yeast can be accounted for by coupling the bulk diffusion of Cdc42 in the cytoplasm with nonlinear binding/unbinding reactions at the poles of the cell [223, 963, 964]. Exploiting the rod-like geom-

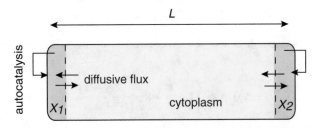

Fig. 11.26: 1D Compartmental model of Cdc42 dynamics in fission yeast.

etry of fission yeast, the cell is treated as a finite 1D domain of length $L$, see Fig. 11.26. Cdc42 diffuses within the interior of the domain, $x \in (0,L)$, and can bind to the cell membrane at the ends $x = 0, L$. Moreover, membrane-bound Cdc42 can unbind and re-enter the cytosol. The nontrivial nature of the dynamics arises from the fact that both the binding and unbinding rates at each end are taken to depend nonlinearly on the local membrane concentration. Following Ref. [223], the association rate is regulated by positive feedback and the dissociation rate is regulated by delayed negative feedback. The resulting dynamical system takes the form of a coupled PDE-DDE [963], analogous to the PDE-ODE system of Min dynamics (Sect. 11.3.3), where the bulk dynamics is described by a simple diffusion partial differential equation (PDE), and the exchange of Cdc42 between the cytosol and the end membrane compartments is modeled in terms of flux boundary conditions that involve both the positive and delayed negative feedback resulting in a delay differential equation (DDE).

Let $C(x,t)$ be the cytosolic concentration of Cdc42 at $x$ and $X_i(t)$, $i = 1,2$, the concentration of Cdc42 at the ith compartment, where $t, t > 0$, denotes time. (Concentrations are defined as the number of molecules per unit cross-section of the cell, which is assumed to be fixed.) The PDE-DDE model is taken to be

$$\frac{\partial C(x,t)}{\partial t} = D\frac{\partial^2 C(x,t)}{\partial x^2}, \quad 0 < x < L, \quad t > 0, \tag{11.3.40a}$$

with flux boundary conditions

$$-D\partial_x C(0,t) = -k^+(X_1(t))C(0,t) + k^-(X_1(t), X_1(t-\tau))X_1(t), \tag{11.3.40b}$$

$$-D\partial_x C(L,t) = k^+(X_2(t))C(L,t) - k^-(X_2(t), X_2(t-\tau))X_2(t), \tag{11.3.40c}$$

and

$$\frac{dX_1}{dt} = k^+(X_1(t))C(0,t) - k^-(X_1(t), X_1(t-\tau))X_1(t), \tag{11.3.40d}$$

$$\frac{dX_2}{dt} = k^+(X_2(t))C(L,t) - k^-(X_2(t), X_2(t-\tau))X_2(t). \tag{11.3.40e}$$

Following Ref. [223], the association rate $k^+$ is regulated by positive feedback with saturation in the form of an exponential:

$$k^+(X) = (k_0^+ + k_n^+ (X/C_{sat})^n) \exp(-X/C_{sat}), \quad n \geq 2. \tag{11.3.41}$$

Similarly, the dissociation rate $k_-$ is controlled by negative delayed feedback according to

$$k^-(X(t), X(t-\tau)) = k_0^- \left[ 1 - \frac{\varepsilon}{2} + \varepsilon \frac{X(t-\tau)^h}{X(t)^h + X(t-\tau)^h} \right], \tag{11.3.42}$$

where $\tau$ is the time delay, $k_0^-$ is the baseline dissociation rate in the absence of the delayed negative feedback, $\varepsilon$ represents the strength of the delayed negative feedback, and $h$ is the Hill coefficient. (For a detailed justification of delayed negative feedback as a mechanism for biochemical oscillations, see Ref. [690].) Equations (11.3.40a)–(11.3.40e) are supplemented by the conservation equation

$$\int_0^L C(x,t)dx + X_1(t) + X_2(t) = C_{tot}, \tag{11.3.43}$$

where $C_{tot}$ is the total number of Cdc42 molecules per unit area. We fix the units of concentration by setting $C_{sat} = 1$. The unit of time is taken to be minutes (the typical time scale of binding/unbinding and the delay $\tau$) and the baseline length is taken to be $5\mu m$ (comparable to the initial length of a fission yeast cell immediately following cell division). It follows that after non-dimensionalization a diffusion coefficient of $D = 1$ corresponds to $D \approx 0.5\mu m^2/s$ in physical units. The other parameters of the model are taken to be similar to those of Refs. [223, 963].

**Fast diffusion limit.** We begin by analyzing the original DDE model of Ref. [223], which is obtained by taking the diffusion coefficient $D \to \infty$. This provides a baseline for the analysis of the full PDE-DDE model in the case of finite $D$. Introducing the small parameter $\varepsilon = L^2 k_0^- / D$ for large $D$, the leading order terms of the diffusion equation (11.3.40a) and the boundary conditions give $C(x,t) = C_0(t)$. Using the conservation equation (11.3.43), we can rewrite $C_0(t)$ as

$$C_0(t) = \frac{C_{tot} - X_1(t) - X_2(t)}{L}. \tag{11.3.44}$$

Substituting equation (11.3.44) into (11.3.40d) and (11.3.40e) then yields the following DDE to leading order:

$$\frac{dX_1}{dt} = \frac{k^+(X_1)}{L}(C_{tot} - X_1(t) - X_2(t)) - k^-(X_1(t), X_1(t-\tau))X_1(t), \tag{11.3.45a}$$

$$\frac{dX_2}{dt} = \frac{k^+(X_2)}{L}(C_{tot} - X_1(t) - X_2(t)) - k^-(X_2(t), X_2(t-\tau))X_2(t). \tag{11.3.45b}$$

For simplicity, we take $k_0^- = 1$ and $C_{sat} = 1$ in equations (11.3.41) and (11.3.42). Following Ref. [223], suppose that we vary the cell length $L$ and take the total substrate concentration $C_{tot}$ to be a linearly increasing function of cell length $L$. (The only explicit dependence of cell length in the DDE model (11.3.45) is that it scales the rates $k_0^+$ and $k_n^+$.) In Fig. 11.27 we plot the resulting bifurcation diagram in the

Fig. 11.27: Bifurcation diagrams in the $(L, X_1)$ plane for the DDE model (11.3.45) with $C_{tot}/L = 6$ and zero delays. A stable symmetric steady state exists for sufficiently small values of $L$. As $L$ is increased the symmetric state destabilizes via a pitchfork bifurcation, leading to the emergence of a pair of stable asymmetric states. These subsequently disappear via a saddle-node bifurcation. Model parameters are $k_0^+ = 2.25$, $k_n^+ = 6.467$, $n = 4$, $k_0^- = 1$, $\varepsilon = 0.5375$, and $h = 40$.

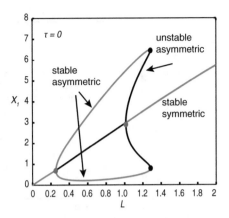

$(X_1, L)$-plane for $\tau = 0$ (zero time delay). It can be seen that for sufficiently small and large $L$ there exists a stable symmetric state $(\overline{X}_1 = \overline{X}_2)$, whereas a stable asymmetric state $(\overline{X}_1 \neq \overline{X}_2)$ exists over an intermediate range of lengths. This suggests that the cell can undergo a transition from monopolar (asymmetric) growth to bipolar (symmetric) growth as the cell length increases. Now suppose that the length is fixed, and we increase the time delay $\tau$ from zero. Using linear stability analysis it can be shown that both the symmetric and asymmetric stable steady states (assuming they exist for a given $L$) can undergo a Hopf bifurcation at some critical delay, see [963].

In Fig. 11.28(a), we plot the Hopf bifurcation curves in the $(L, \tau)$-plane for the stable symmetric and asymmetric steady states. The corresponding Hopf frequency of the Hopf curve branching from the asymmetric fixed point is plotted as a function of the cell length $L$, or equivalently the critical time delay $\tau_c$ in Fig. 11.28(b,c). It can be seen that the frequency vanishes at the Hopf bifurcation point, indicative of a Bogdanov–Takens bifurcation. The termination of the asymmetric Hopf branch at a critical cell length $L_c$ is the basic mechanism underlying the switch from an asymmetric oscillation to a symmetric oscillation. It is also consistent with the vanishing of the asymmetric steady states via a saddle-node bifurcation when $\tau = 0$, see Fig. 11.27. The same basic mechanism holds when bulk diffusion is included using the full PDE-DDE model (11.3.40). However, as shown later, the critical length where the switch occurs is sensitive to the value of the diffusion coefficient. Finally, numerical solutions of the DDE (11.3.45) show that oscillations bifurcating from the symmetric and asymmetric steady states are anti-phase in the sense that $X_1(t)$ peaks when $X_2(t)$ troughs, which is consistent with experimental observations. (In-phase oscillations can exist if the strength of the negative feedback $\varepsilon$ is sufficiently strong or the nonlinearity $h$ is sufficiently high.)

It is straightforward to use the simplified DDE model to investigate the effects of explicit time-dependent cell elongation. For the sake of illustration, suppose that both $L$ and $C_{tot}$ increase linearly with time at a rate $\gamma$. That is, equation (11.3.45) is modified by taking

$$L(t) = L_0[1 + \gamma t], \quad C_{tot}(t) = C_0[1 + \gamma t], \qquad (11.3.46)$$

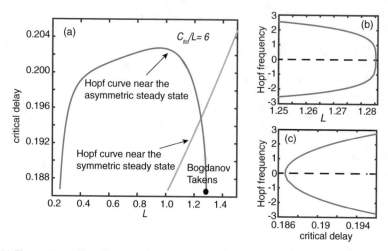

Fig. 11.28: (a) Hopf bifurcation curves for the symmetric steady state (green) and the asymmetric steady state (blue) in the $(L, \tau)$ plane with $C_{\text{tot}} = 6L$. (b,c) Plot of frequency along the Hopf curve branching from the asymmetric steady state as a function of cell length $L$ and critical delay $\tau$. Near $(\tau, L) \approx (0.18, 1.285)$, the frequency is zero indicative of a Bogdanov–Takens bifurcation. Parameters are the same as Fig. 11.27.

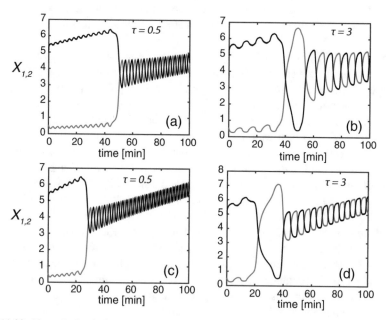

Fig. 11.29: Numerical solutions of the DDE model with $C_{\text{tot}}(t) = 6(1 + \gamma t)$ and $L(t) = 1 + \gamma t$. (a,b) Numerical solutions of $X_{1,2}(t)$ for $\gamma = 0.005$ and $\tau = 0.5, 3$, respectively. (c,d) Corresponding figures for $\gamma = 0.01$.

where $L_0, C_0$ are the initial length and total Cdc42 concentration, respectively. It is important to emphasize, however, that in the case of growing cells one can no longer treat the DDE model as the fast diffusion limit of the full PDE-DDE model (11.3.40). Indeed, integrating the diffusion equation (11.3.40) over the growing interval $[0, L(t)]$ and imposing the boundary conditions give

$$\int_0^{L(t)} \partial_t C(x,t)dx = -\frac{dX_1}{dt} - \frac{dX_2}{dt}.$$

Differentiating the conservation condition

$$C_{tot}(t) = \int_0^{L(t)} C(x,t)dx + X_1(t) + X_2(t)$$

then implies that

$$\frac{dC_{tot}(t)}{dt} = C(L(t),t)\frac{dL}{dt}.$$

This last equation holds in the large $D$ limit for which $C(L(t),t) \to C_0(t)$. However, it is not satisfied by the growth law of equation (11.3.46) for which $C_0(t)/L(t) = C_0/L_0$. Such a law is consistent if one takes proper account of diffusion in a linearly growing domain, which involves adding a convection term to the diffusion model (see below). One finds that for moderate delays (e.g., $\tau = 0.5$) the numerical solution changes from an asymmetric to a symmetric anti-phase oscillation, see Fig. 11.29. Moreover, the switch to the symmetric state occurs at a smaller value of $L$ and $C_{tot}$ (smaller times $t$) for larger $\tau$. For larger delays the oscillations become less sinusoidal, the frequency of oscillations decreases, and we observe an additional transition from large amplitude to low amplitude symmetric oscillations. Finally, the transition between asymmetric and symmetric oscillations occurs at larger cell lengths when the growth rate $\gamma$ is increased. In conclusion, the DDE model reproduces the switch from asymmetric to symmetric Cdc42 oscillations observed experimentally during NETO.

*Linear stability analysis of the full model.* Let us now return to the full PDE-DDE model given by equations (11.3.40)-(11.3.42), in order to investigate how the switch from asymmetric to symmetric oscillations depends on bulk diffusion. For the moment we take $L$ and $C_{tot}$ to be fixed with $C_{tot}$ treated as a bifurcation parameter. We also perform the rescalings $x \to \tilde{x} = x/L$, $C \to \tilde{C} = CL$, and $D \to \tilde{D} \equiv D/L^2$. The only dependence on $L$ then appears through the association rate $k_+$, as in the pure DDE model. The steady state solution for the bulk concentration $C$ satisfies (after dropping the tilde)

$$C''(x) = 0, \quad C'(0) = C'(1) = 0.$$

Hence $C(x)$ is homogeneous, i.e., $C(x) = \overline{C}$. The conservation equation of the total amount of substrate requires that

$$\int_0^1 \overline{C}dx + X_1 + X_2 = C_{tot}.$$

It follows that $\overline{C} = C_{tot} - X_1 - X_2$. The steady state solution $(\overline{X}_1, \overline{X}_2)$ satisfies

$$0 = \frac{k^+(\overline{X}_1)}{L}(C_{tot} - \overline{X}_1 - \overline{X}_2) - k^-(\overline{X}_1, \overline{X}_1)\overline{X}_1,$$

$$0 = \frac{k^+(\overline{X}_2)}{L}(C_{tot} - \overline{X}_1 - \overline{X}_2) - k_-(\overline{X}_2, \overline{X}_2)\overline{X}_2.$$

Note that the PDE-DDE has the same steady state solution as the DDE model, so that there exists a symmetric steady state solution $(\overline{X}_1 = \overline{X}_2)$ and an asymmetric steady state solution $(\overline{X}_1 \neq \overline{X}_2)$ for different choices of total substrate concentration $C_{tot}$.

Consider the perturbation near the steady state $(C_0, \overline{X}_1, \overline{X}_2)$,

$$C(x,t) = \overline{C} + e^{\lambda t}\eta(x), \quad X_i(t) = \overline{X}_i + e^{\lambda t}\phi_i.$$

Substituting into equations (11.3.40a-c) of the PDE-DDE model and linearizing about the steady state solution gives (for $k_0^- = 1$)

$$D\frac{\partial^2 \eta(x)}{\partial x^2} = \lambda \eta(x), \quad 0 < x < 1, \tag{11.3.47a}$$

$$D\partial_x \eta(0) = \frac{k^+(\overline{X}_1)}{L}\eta(0) + \left[\frac{k'_+(\overline{X}_1)}{L}\overline{C} - (1 - \frac{\varepsilon h}{4} + \frac{\varepsilon h}{4}e^{-\lambda \tau})\right]\phi_1, \tag{11.3.47b}$$

$$-D\partial_x \eta(1) = \frac{k^+(\overline{X}_2)}{L}\eta(1) + \left[\frac{k'_+(\overline{X}_2)}{L}\overline{C} - (1 - \frac{\varepsilon h}{4} + \frac{\varepsilon h}{4}e^{-\lambda \tau})\right]\phi_2. \tag{11.3.47c}$$

It then follows from equations (11.3.40d,e) that $D\partial_x \eta(0) = \lambda \phi_1$ and $-D\partial_x \eta(1) = \lambda \phi_2$. Substituting into equations (11.3.47b,c) and rearranging gives

$$\left[\lambda + 1 - \frac{\varepsilon h}{4}(1 - e^{-\lambda \tau}) - \frac{k'_+(\overline{X}_1)}{L}\overline{C}\right]\phi_1 = \frac{k^+(\overline{X}_1)}{L}\eta(0), \tag{11.3.48a}$$

$$\left[\lambda + 1 - \frac{\varepsilon h}{4}(1 - e^{-\lambda \tau}) - \frac{k'_+(\overline{X}_2)}{L}\overline{C}\right]\phi_2 = \frac{k^+(\overline{X}_2)}{L}\eta(1), \tag{11.3.48b}$$

where $k'_+(\overline{X}) = dk^+(X)/dX|_{X=\overline{X}}$. Rewriting equation (11.3.48) gives

$$\phi_1 = B_1(\lambda, \tau)\eta(0), \quad \phi_2 = B_2(\lambda, \tau)\eta(1), \tag{11.3.49}$$

with

$$B_i(\lambda, \tau) = \frac{k^+(\overline{X}_i)/L}{\lambda + 1 - \frac{\varepsilon h}{4}(1 - e^{-\lambda \tau}) - k'_+(\overline{X}_i)\overline{C}/L}.$$

Substituting equation (11.3.49) into the boundary conditions for $\eta(x)$ yields the following nonlinear boundary-value problem:

$$D\frac{\partial^2 \eta(x)}{\partial x^2} = \lambda \eta(x), \quad 0 < x < 1, t > 0,$$

$$D\partial_x \eta(0) = \lambda B_1(\lambda, \tau)\eta(0), \quad -D\partial_x \eta(1) = \lambda B_2(\lambda, \tau)\eta(1). \tag{11.3.50}$$

The solution $\eta(x)$ can be expressed in the form

$$\eta(x) = \frac{\eta_0 + \eta_1}{2}\frac{\cosh(\sqrt{\lambda/D}(x - \frac{1}{2}))}{\cosh(\frac{1}{2}\sqrt{\lambda/D})} + \frac{\eta_1 - \eta_0}{2}\frac{\sinh(\sqrt{\lambda/D}(x - \frac{1}{2}))}{\sinh(\frac{1}{2}\sqrt{\lambda/D})}, \tag{11.3.51}$$

where $\eta_0 = \eta(0)$ and $\eta_1 = \eta(1)$. The boundary conditions then require

$$\begin{pmatrix} \mathscr{A}_+(\lambda) + \lambda B_1(\lambda,\tau) & \mathscr{A}_-(\lambda) \\ \mathscr{A}_-(\lambda) & \mathscr{A}_+(\lambda) + \lambda B_2(\lambda,\tau) \end{pmatrix} \begin{pmatrix} \eta_0 \\ \eta_1 \end{pmatrix} = 0, \qquad (11.3.52)$$

where

$$\mathscr{A}_\pm(\lambda) = \sqrt{\lambda D} \frac{\tanh(\tfrac{1}{2}\sqrt{\lambda/D}) \pm \coth(\tfrac{1}{2}\sqrt{\lambda/D})}{2}. \qquad (11.3.53)$$

Setting the determinant to zero and dividing through by $\lambda D$ gives

$$\frac{\lambda}{D} B_1 B_2 + (B_1 + B_2)\sqrt{\frac{\lambda}{D}}\frac{\tanh(\tfrac{1}{2}\sqrt{\lambda/D}) + \coth(\tfrac{1}{2}\sqrt{\lambda/D})}{2} + 1 = 0. \qquad (11.3.54)$$

The presence of terms involving $\sqrt{\lambda/D}$ means that we have to introduce a branch cut in the complex $\lambda$-plane along $(-\infty,0]$. Fortunately, for finite $D,L$ this does not affect the eigenvalue relation (11.3.54) since, as $\lambda \to 0$, we have $\tanh(\sqrt{\lambda/4D}) \to \sqrt{\lambda/4D}$ and $\coth(\sqrt{\lambda/4D}) \to \sqrt{4D/\lambda}$, that is, any square roots in (11.3.54) cancel. (However, care has to be taken in the limit $D \to 0$ [963].) Note that equations (11.3.49) and (11.3.54) are well-defined provided that the denominators of the function $B_{1,2}(\lambda,\tau)$ are nonzero. Therefore, we need to check that the boundary value problem still makes sense in the singular limit.

*Symmetric case.* Let

$$A(\lambda,\tau,\overline{X}_i) = \lambda + \frac{\varepsilon h}{4} e^{-\lambda\tau} + 1 - \frac{\varepsilon h}{4} - k'_+(\overline{X}_i)\overline{C}/L,$$

and consider the eigenvalue problem associated with the symmetric steady state solution $\overline{X}_1 = \overline{X}_2 > 0$. If $A(\lambda,\tau,\overline{X}_1) = 0$, then equation (11.3.49) requires that $\eta(0) = \eta(1) = 0$. It is known that the Dirichlet boundary value problem (11.3.47a) of $\eta(x)$ only has a trivial solution, i.e., $\eta(x) = 0$. To solve for $(\phi_1,\phi_2)$, we substitute $\eta(x) = 0$ into the boundary conditions (11.3.47b) and (11.3.47c). It follows that

$$0 = \left[1 - \frac{\varepsilon h}{4} + \frac{\varepsilon h}{4} e^{-\lambda\tau} - \frac{k'_+(\overline{X}_i)}{L}\overline{C}\right]\phi_i = -\lambda\phi_i, \quad i = 1,2,$$

with the first identity holding since $A(\lambda,\tau,\overline{X}_i) = 0$. Noting that $\lambda = 0$ is not a solution of $A(\lambda,\tau,\overline{X}_i) = 0$, it follows that $\phi_1 = \phi_2 = 0$. Therefore, if $A(\lambda,\tau,\overline{X}_i) = 0$, the associated solution $(\eta(x),\phi_1,\phi_2)$ is trivial. Hence, we can assume that $A(\lambda,\tau,\overline{X}_i) \neq 0$ for the symmetric steady state, and equation (11.3.54) holds. For the symmetric steady state $\overline{X}_1 = \overline{X}_2$, we have $B_1 = B_2$ and the resulting cyclic matrix (11.3.52) has the eigenvectors $(1,1)^T, (1,-1)^T$. It follows that $\mathscr{A}_+(\lambda) \pm \mathscr{A}_-(\lambda) + \lambda B_1(\lambda,\tau) = 0$ for the in-phase $(+)$ and antiphase $(-)$ solutions, respectively, with corresponding eigenvalue equations

$$B_1 + \sqrt{\frac{D}{\lambda}}\tanh(\frac{1}{2}\sqrt{\frac{\lambda}{D}}) = 0 \text{ (in-phase)}, \qquad (11.3.55a)$$

$$B_1 + \sqrt{\frac{D}{\lambda}}\coth(\frac{1}{2}\sqrt{\frac{\lambda}{D}}) = 0 \text{ (anti-phase)}. \qquad (11.3.55b)$$

As in the case of the DDE model, it can be shown that if $\varepsilon$ and $h$ are not too large then only anti-phase oscillations can bifurcate from the symmetric steady-state, that is, the in-phase equation does not have a solution of the form $\lambda = i\omega$ [963]. Numerically solving the anti-phase equation one finds that the main effect of finite bulk diffusion is to increase the critical time delay for the onset of a Hopf bifurcation, as illustrated in Fig. 11.30.

*Asymmetric case.* Now consider the eigenvalue problem for the asymmetric steady state solution $\overline{X}_1 \neq \overline{X}_2$ where at least $B_1$ or $B_2$ is singular. Since $k^+(X)$ is monotonically increasing, it follows that $A(\lambda,\tau,\overline{X}_1)$ and $A(\lambda,\tau,\overline{X}_2)$ cannot attain zero simultaneously. Without loss of generality, we assume that $A(\lambda, \tau, \overline{X}_1) = 0$ and $A(\lambda, \tau, \overline{X}_2) \neq 0$. It follows that $\eta(0) = 0$ and $\phi_2 = B_2(\lambda, \tau)\eta(1)$. The solution for $\eta(x)$ can be rewritten as

$$\eta(x) = \frac{\eta_1}{2}\left(\frac{\cosh(\sqrt{\lambda/D}(x - \frac{1}{2}))}{\cosh(\frac{1}{2}\sqrt{\lambda/D})} + \frac{\sinh(\sqrt{\lambda/D}(x - \frac{1}{2}))}{\sinh(\frac{1}{2}\sqrt{\lambda/D})}\right).$$

The boundary conditions (11.3.47b) and (11.3.47c) require that

$$\phi_1 = \frac{\eta_1}{2}\sqrt{\frac{D}{\lambda}}[\coth(\frac{1}{2}\sqrt{\lambda/D}) - \tanh(\frac{1}{2}\sqrt{\lambda/D})],$$

$$0 = \left(\sqrt{\lambda D}\frac{\tanh(\frac{1}{2}\sqrt{\lambda/D}) + \coth(\frac{1}{2}\sqrt{\lambda/D})}{2} + \lambda B_2(\lambda, \tau)\right)\eta_1.$$

Hence, in order to have a nontrivial solution, we require

$$A(\lambda, \tau, \overline{X}_1) = 0, \quad \frac{\tanh(\frac{1}{2}\sqrt{\lambda/D}) + \coth(\frac{1}{2}\sqrt{\lambda/D})}{2} + \sqrt{\frac{\lambda}{D}}B_2(\lambda, \tau) = 0.$$

The second equation also arises from taking the limit $B_1 \to \infty$ in (11.3.54). Note that the root $\lambda$ of the second equation is dependent on the parameter $D$ while the root of the first

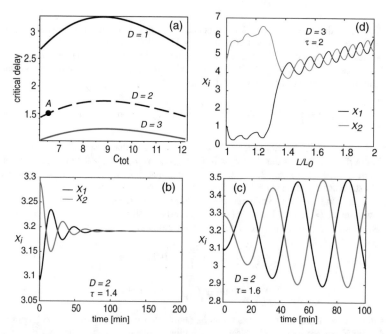

Fig. 11.30: Effects of diffusion on the Hopf bifurcations from the symmetric steady state. (a) Critical delay $\tau$ vs the total substrate concentration $C_{tot}$. (b,c) Numerical solution of $X_{1,2}$ with $(D, \tau)$ below and above the Hopf point A where $C_{tot} = 6.5$, respectively, indicating that the Hopf bifurcation at A is supercritical and the oscillation mode is anti-phase. (d) Switch from asymmetric to symmetric oscillations with $C_{tot}$ and $L$ slowly increasing functions of time $t$. Parameters: $L_0 = 1$, $C_0 = 6.5$, $\gamma = 0.01$, $k_0^+ = 2.25$, $k_n^+ = 6.467$, $n = 4$, $k_0^- = 1$, $\varepsilon = 0.5375$, and $h = 40$.

equation is independent of $D$. Again, numerically solving these equations shows that bulk diffusion increases the critical time delay [963].

Finally, note that the experimentally observed switch from asymmetric to symmetric oscillations during NETO can also be reproduced in the full PDE-DDE model [963], see Fig. 11.30(d). This requires modifying the PDE-DDE model to take into account diffusion in a growing domain $\Omega_t = [0, L(t)]$, in which the total substrate concentration $C_{\text{tot}}$ increases linearly with $L(t) = L_0(1 + \gamma t)$ according to $C_{\text{tot}} = C_0 L(t)/L_0$. Following previous studies of spontaneous pattern formation on growing domains during development [217], one can then transform the system to diffusion in a fixed domain with additional advection terms. Pattern formation on growing domains will be considered in Sect. 11.4.

## 11.4 Synaptogenesis in C. elegans and hybrid reaction–transport models

We now turn to an example of intracellular pattern formation that has been proposed as a possible mechanism for synaptogenesis in *C. elegans* [136, 137]. In contrast to cell polarization (Sect. 11.3), the separation of time scales necessary for robust pattern formation arises from different modes of transport within the same domain, rather than diffusion within separate domains, e.g., the cytoplasm and membrane. Moreover, the resulting system is not mass conserving.

During larval development of *C. elegans*, the density of ventral and dorsal cord synapses containing the glutamate receptor GLR-1 is maintained despite significant changes in neurite length [769], see Fig. 11.31. It is known that the coupling of synapse number to neurite length requires type II calcium-and calmodulin-dependent protein kinase (CaMKII) and voltage-gated calcium channels and that CaMKII regulates the active (kinesin-based) transport and delivery of GLR-1 to synapses [419, 420, 769]. However, a long outstanding problem has been identifying a possible physical mechanism involving diffusing CaMKII molecules and motor-driven GLR-1 that leads to the homeostatic control of synaptic density. The formation of a regularly spaced distribution of synaptic puncta at an early stage of development is suggestive of some form of Turing-like pattern formation, and the maintenance of synaptic density as the organism grows is suggestive of "pulse or stripe insertion" in spatially periodic patterns on growing domains, analogous to stripe insertion in patterns of skin pigmentation of the marine angelfish [217, 520]. These observations have motivated the construction of a mathematical model of synaptogenesis in *C. elegans* [136, 137], involving the interaction between a slowly diffusing species (e.g., CaMKII) and a rapidly advecting species (e.g., GLR-1) switching between anterograde and retrograde motor-driven transport (bidirectional transport). Using the classical Gierer–Meinhardt mechanism for reaction kinetics [610], it can be shown that the model generates the in-phase Turing patterns on a one-dimensional domain of fixed length. (Within the context of synaptogenesis in *C. elegans*, the periodically spaced concentration peaks can be interpreted as the loca-

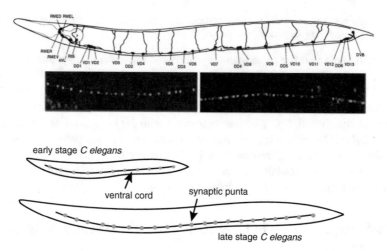

Fig. 11.31: (a) Schematic figure showing distribution of synapses of type D motor neurons in *C. elegans*. The D-type neurons include six DD and 13 VD neurons. DD forms synapses to the dorsal body muscles and VD forms synapses to the ventral muscles. Blobs, synapses to muscles; arrows, synaptic inputs to the D neurons. Below are GFP images of synaptic puncta in the dorsal and ventral cords, respectively. [Public domain figure downloaded from the WormBook/Synaptogenesis by Y. Jin: "http://www.wormbook.org/".] (b) Sketch comparing the distribution of synaptic puncta along the ventral cord of early- and late-stage *C. elegans*. New synapses are inserted during development in order to maintain synaptic density.

tions of synaptic puncta.) Moreover, when extended to the case of a slowly growing domain, new puncta are inserted so that the density of peaks is maintained. We now describe the model and its analysis in more detail.

### 11.4.1 Hybrid reaction–transport model

The dynamics of synaptic formation appears to be mediated by two distinct antagonistic effects of CaMKII, see Fig. 11.32. As the worm grows in size, the synaptic density along the ventral cord decreases, which tends to result in reduced synaptic excitation of the motor neurons. In this situation, the activation of CaMKII via voltage-gated calcium channels induces the formation of new synapses by enhancing the active transport of GLR-1. On the other hand, when the synaptic density becomes too high, the corresponding increase in excitation leads to constitutive activation of CaMKII (autophosphorylation) due to the increased calcium levels. Although constitutively active CaMKII also enhances the motor-driven transport of GLR-1 along the ventral cord, it fails to localize the receptors at synaptic sites. This is consistent with the observation that the synaptic localization of CaMKII changes in response to autophosphorylation [817].

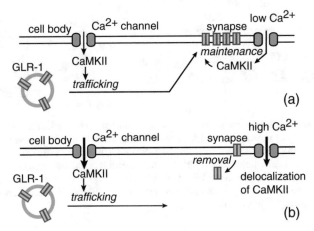

Fig. 11.32: Regulation of transport and delivery of GLR-1 to synapses by CaMKII. (a) Calcium influx through voltage-gated calcium channels activates CaMKII, which enhances the active transport and delivery of GLR-1 to synapses. (b) Under conditions of increased excitation, higher calcium levels result in constitutively active CaMKII which fails to localize at synapses, leading to the removal of GLR-1 from synapses.

To model this system, consider a one-dimensional domain of fixed length $L$, which represents a neurite in the ventral cord of *C. elegans* at a particular stage of larval development, see Fig. 11.33. Let $R(x,t)$ denote the concentration of GLR-1 receptors at position $x$ along the cell at time $t$ and let $C(x,t)$ denote the corresponding concentration of active CaMKII. For simplicity, no distinction is made between membrane-bound (synaptically localized) and cytoplasmic CaMKII. On the other, GLR-1 is partitioned into two subpopulations: those that undergo anterograde transport ($R_+$) and those that undergo retrograde transport ($R_-$) with $R(x,t) = R_+(x,t) + R_-(x,t)$. Individual receptors randomly switch between the two advective states according to a two-state Markov process

$$R_+ \underset{\alpha}{\overset{\beta}{\rightleftharpoons}} R_-,$$

with transition rates $\alpha, \beta$. We assume a symmetric process whereby $\alpha = \beta$. This yields the following system of equations [136]:

$$\frac{\partial C}{\partial t} = D_C \frac{\partial^2 C}{\partial x^2} + f(C, R_+, R_-), \tag{11.4.1a}$$

$$\frac{\partial R_+}{\partial t} = -v \frac{\partial R_+}{\partial x} + \alpha R_- - \alpha R_+ + g(C, R_+), \tag{11.4.1b}$$

$$\frac{\partial R_-}{\partial t} = v \frac{\partial R_-}{\partial x} + \alpha R_+ - \alpha R_- + g(C, R_-). \tag{11.4.1c}$$

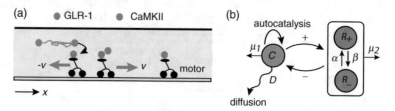

Fig. 11.33: (a) One-dimensional, three-component trafficking model. Passively transported molecules of CaMKII (green) react with motor-driven actively transported molecules of GLR-1 (red). The latter can switch between forward and backward moving states (traveling with velocities $\pm v$). Interactions between motor-driven particles such as exclusion effects (hard-core repulsion) are ignored. (b) Schematic diagram of modified Gierer and Meinhardt activator–inhibitor model with a passively diffusing activator ($C$) and an actively transported inhibitor switching between left ($R_-$) and right ($R_+$) moving states at a rate $\alpha$.

Here $D_C$ is the CaMKII diffusion coefficient and $v$ is the speed of motor-driven GLR-1. The reaction term $f(R_+, R_-, C)$ represents both the self-activation of CaMKII and the inhibition of CaMKII by GLR-1. The reaction term $g(R_\pm, C)$ represents the increase in actively transported GLR-1 due to the action of CaMKII, which is taken to be symmetric with regards to anterograde and retrograde transport. Equations (11.4.1) are supplemented by reflecting boundary conditions at the ends $x = 0, L$:

$$\left.\frac{\partial C(x,t)}{\partial x}\right|_{x=0,L} = 0, \quad vR_+(0,t) = vR_-(0,t), \quad vR_+(L,t) = vR_-(L,t). \quad (11.4.2)$$

(Note that Turing pattern formation based on advecting species has also been considered within the context of animal movement and chemotaxis [411]. However, in the latter case, all species are assumed to undergo bidirectional transport.)

It remains to specify the form of the chemical interaction functions $f$ and $g$. The precise details of the interactions between CaMKII and GLR-1 are currently unknown. This suggests considering the simplest model that can capture the activation of GLR-1 transport and delivery by CaMKII and the inhibition (synapse removal) of CaMKII by GLR-1, and the autophosphorylation properties of CaMKII. Another requirement is that the nonlinear interactions yield the Turing patterns for which the peaks of the two concentrations are in-phase. Hence, we take the classical activator–inhibitor system due to Gierer and Meinhardt (GM) [610]:

$$f(C, R_+, R_-) = \rho_1 \frac{C^2}{R_+ + R_-} - \mu_1 C, \quad g(C, R_\pm) = \rho_2 C^2 - \mu_2 R_\pm. \quad (11.4.3)$$

Here $\rho_1, \rho_2$ represent the strength of interactions, and $\mu_1$ and $\mu_2$ are the respective decay rates, see Fig. 11.33(b). It is convenient to rewrite equation (11.4.1) in terms of the variables $(C, R, J)$ where $R = R_+ + R_-$ and $J = vR_+ - vR_-$. That is, adding and subtracting the pair of equations (11.4.1b,c) yields

$$\frac{\partial C}{\partial t} = D_C \frac{\partial^2 C}{\partial x^2} + f(C,R), \tag{11.4.4a}$$

$$\frac{\partial R}{\partial t} = -\frac{\partial J}{\partial x} + h(C,R), \quad h(C,R) = 2\rho_2 C^2 - \mu_2 R, \tag{11.4.4b}$$

$$\frac{\partial J}{\partial t} = -v^2 \frac{\partial R}{\partial x} - (2\alpha + \mu_2)J. \tag{11.4.4c}$$

**Steady-state solutions.** Setting all time derivatives to zero in the system of equations (11.4.4) and eliminating the flux $J(x)$ give

$$0 = D_C \frac{d^2 C}{dx^2} + \frac{\rho_1 C^2}{R} - \mu_1 C, \tag{11.4.5a}$$

$$0 = D \frac{d^2 R}{dx^2} + 2\rho_2 C^2 - \mu_2 R, \quad D = \frac{v^2}{2\alpha + \mu_2}, \tag{11.4.5b}$$

with $D$ the effective diffusivity of the inhibitor. It follows that the three-component hybrid reaction–transport model and the classical two-component GM model have the same steady-state solutions for $(C,R)$. This includes the unique uniform solution

$$C^* = \frac{\rho_1 \mu_2}{2\rho_2 \mu_1}, \quad R_\pm^* = \frac{\rho_1^2 \mu_2}{4\rho_2 \mu_1^2}. \tag{11.4.6}$$

However, the corresponding eigenvalue problem obtained by linearizing about a given steady-state solution will differ in the two models, which means that their stability properties could also differ. A related issue is that, given a spatially varying steady-state solution for $(C,R)$ in the RD model, the corresponding hybrid system could support several different solutions for $R_+(x)$ and $R_-(x)$. One typical condition for the occurrence of pattern formation in activator–inhibitor systems such as the GM model is that the inhibitor diffuses more quickly than the activator. The above analysis suggests that $D$ is the effective diffusivity of the actively transported component. We thus obtain the condition $D_C < D$. In the particular application to synaptogenesis in C. *elegans* the following order of magnitude parameter values was used [136, 137]: $\mu_1, \mu_2, \rho_1 \sim 1/s$, $\rho_2 \sim 1/(\mu M \cdot s)$, $\alpha \sim 0.1/s$, $v \sim 1\mu$ m/s, and $D_C \sim 0.01$ $\mu m^2/s$. In addition, the typical spacing of synaptic puncta is 3-4 per 10 $\mu$m. It is clear that for this application, the effective diffusivity $D \approx 1$ $\mu m^2/s$, and thus the inequality $D_C \ll D$ is satisfied.

**Linear stability analysis.** In order to derive general conditions for a Turing instability, we proceed along analogous lines to the classical theory of diffusion-driven pattern formation (see Sect. 11.2) by linearizing about a spatially uniform fixed point and studying the spectrum of the resulting linear operator. Therefore, suppose there exists a spatially uniform fixed point $\mathbf{c}^* = (C^*, R^*, R^*)$ for which $f(C^*, R^*, R^*) = g(C^*, R^*) = 0$. In order to investigate the stability of the equilibrium solution, we introduce the perturbations

$$C(x,t) = C^* + e^{\lambda t} \Phi(x), \quad R(x,t) = 2R^* + e^{\lambda t} \Psi(x), \quad J(x,t) = e^{\lambda t} \Lambda(x). \tag{11.4.7}$$

Substituting into equation (11.4.1), after rewriting in terms of $(C, R, J)$, and linearizing give

$$\lambda \Phi(x) = D_C \frac{d^2 \Phi(x)}{dx^2} + f_c \Phi(x) + f_r \Psi(x), \tag{11.4.8a}$$

$$\lambda \Psi(x) = -\frac{d\Lambda(x)}{dx} + h_c \Phi(x) + h_r \Psi(x), \tag{11.4.8b}$$

$$\lambda \Lambda(x) = -v^2 \frac{d\Psi(x)}{dx} - (2\alpha + \mu_2)\Lambda(x). \tag{11.4.8c}$$

Here $f_c = \partial f / \partial c$ evaluated at the fixed point, etc. Setting $(\Phi(x), \Psi(x), \Lambda(x)) = \mathbf{v} e^{ik \cdot x}$ leads to the matrix eigenvalue equation

$$\lambda \mathbf{v} = \left[ -k^2 \mathbf{D} - ik\mathbf{J} + \mathbf{A} \right] \mathbf{v}, \tag{11.4.9}$$

where

$$\mathbf{D} = \begin{pmatrix} D_C & 0 & 0 \\ 0 & 0 & 0 \\ 0 & 0 & 0 \end{pmatrix}, \quad \mathbf{J} = \begin{pmatrix} 0 & 0 & 0 \\ 0 & 0 & 1 \\ 0 & v^2 & 0 \end{pmatrix}, \quad \mathbf{A} = \begin{pmatrix} f_c & f_r & 0 \\ h_c & h_r & 0 \\ 0 & 0 & -[2\alpha + \mu_2] \end{pmatrix}.$$

Hence, $\lambda$ satisfies the characteristic equation

$$\lambda^3 + b(k)\lambda^2 + c(k)\lambda + h(k) = 0, \tag{11.4.10}$$

where

$$b(k) = 2\alpha + \mu_2 + k^2 D_c - (f_c + h_r),$$

$$c(k) = (f_c h_r - f_r h_c) - (2\alpha + \mu_2)(f_c + h_r) - h_r D_C k^2 + k^2 v^2 + (2\alpha + \mu_2)D_C k^2,$$

and

$$h(k) = (2\alpha + \mu_2) \left[ (f_c h_r - f_r h_c) - k^2 [f_c D + h_r D_c] + DD_C k^4 \right].$$

In general, for each value of $k$ there will be three eigenvalues $\lambda_j(k)$, $j = 1, 2, 3$, one of which will be real and the other two either real or forming a complex conjugate pair. There are thus three solution branches or dispersion curves. Note, in particular, that in the limit $|k| \to \infty$, the three roots behave as $\lambda_1(k) \sim -k^2 \gamma$ and $\lambda_{2,3}(k) \sim \pm ik$.

Recall that a necessary condition for a Turing instability is that the steady state is stable with respect to uniform perturbation, that is, we require $\mathrm{Re}(\lambda_j(0)) < 0$, $j = 1, 2, 3$. Setting $k = 0$ in equation (11.4.10) gives the three roots

$$\lambda_1(0) = -(2\alpha + \mu_2),$$

$$\lambda_{2,3}(0) = \frac{1}{2} \left( f_c + h_r \pm \sqrt{(f_c + h_r)^2 - 4(f_c h_r - f_r h_c)} \right).$$

We thus obtain the necessary conditions

$$f_c + h_r < 0, \quad f_c h_r - f_r h_c > 0. \tag{11.4.11}$$

A stationary Turing instability will then occur if varying the bifurcation parameter $\gamma$ results in one of the real dispersion branches crossing zero from below at some critical wavenumber $k_c$, while the other pair of (possibly complex conjugate) branches have negative real parts for all $k$. A necessary condition is that there exists a wavenumber $k$ for which there is a real root $\lambda(k) = 0$. In order for this to hold, the $\lambda$-independent term in equation (11.4.10) must vanish, $h(k) = 0$, which implies

$$k^2 = \frac{1}{2DD_C}\left(f_c D + h_r D_C \pm \sqrt{(f_c D + h_r D_C)^2 - 4DD_C(f_c h_r - f_r h_c)}\right).$$

There are then two conditions for a real solution $k$: (i) the discriminant is positive, and (ii) $k^2$ is positive. The first condition implies that

$$\left(f_c + h_r\frac{D_C}{D}\right)^2 > 4\frac{D_C}{D}(f_c h_r - f_r h_c), \tag{11.4.12}$$

which certainly holds for

$$D_C \ll D = \frac{v^2}{2\alpha + \mu_2}. \tag{11.4.13}$$

Positivity of $k^2$ then requires $f_c + (D_C/D)h_r > 0$. It turns out that in order to exclude the possibility of a Turing–Hopf bifurcation, we require $h_r < 0$ [136]. This then implies that $f_c > 0$. It follows that $f_r$ and $h_c$ also have opposite sign, see second condition in (11.4.11).

In conclusion, the above analysis suggests that a Turing instability can occur when $D_C \ll D$, where $D$ is the effective diffusivity of the actively transported component GLR-1, and assuming equation (11.4.11) also holds. This is indeed found to be the case for the nonlinear reaction scheme of equation (11.4.3), see Ref. [136] and Fig. 11.34 below. Given the occurrence of the Turing patterns on a fixed 1D domain, the next issue we explore is whether or not peak insertion occurs on a growing domain, consistent with the homeostatic control of synaptic density in C. elegans.

### 11.4.2 Pattern formation on a growing domain

Let us now consider the hybrid reaction–transport model (11.4.1) on a growing domain $0 < x < L(t)$, following the analysis of Ref. [137]. Here $L(t)$ represents the length of C.elegans at a time $t$ during development, following an initial phase of synaptogenesis at time $t = 0$. This means that the model system is already operating beyond the Turing bifurcation point. If the in-phase peaks of CaMKII and GLR-1 are interpreted as synaptic sites, then the wavelength of the pattern at time $t = 0$ is equivalent to the spacing of the newly formed synapses. This suggests that under uniform growth of the domain for $t > 0$, the spacing of the synapses will increase. Therefore, in order to obtain a similar synaptic density in the adult as in the first stages of synaptogenesis, it is necessary for new synaptic puncta to be formed. From the mathematical perspective, this can be interpreted as stripe insertion of a Turing pattern on a growing domain.

Following previous studies of diffusion processes on growing domains [217], domain growth is modeled in terms of a velocity field $u$ such that $x \to x + u(x,t)\Delta t$ over the time interval $[t, t + \Delta t]$. We will assume spatially uniform growth by taking $\partial_x u = \sigma(t)$, which implies that

$$u(x,t) = \frac{x}{L(t)} \frac{dL(t)}{dt}. \tag{11.4.14}$$

Let $X \in [0, L_0]$ be the local coordinate system at the initial length $L_0$. Using a Lagrangian description, we can then represent spatial position at time $t$ as

$$x = \Lambda(X,t) \equiv \frac{XL(t)}{L_0}, \quad L(0) = L_0.$$

In order to derive the evolution equations on a growing domain, let us focus on the diffusing component $C(x,t)$; the other components can be treated in a similar fashion. As shown in Box 11E, using the particle conservation equation

$$\frac{d}{dt} \int_0^{L(t)} C(x,t)dx = \int_0^{L(t)} \left[ -\frac{\partial J(x,t)}{\partial x} + f \right] dx, \tag{11.4.15}$$

with $J(x,t) = -D\partial C(x,t)/\partial x$, and Reynold's transport theorem, we have

$$\frac{\partial C}{\partial t} + \frac{\dot{L}(t)}{L(t)} \frac{\partial [xC]}{\partial x} = D\frac{\partial^2 C}{\partial x^2} + f. \tag{11.4.16}$$

Equation (11.4.16) can then be transformed to the fixed interval $[0, L_0]$ by performing the change of variables $x \to X = (xL_0)/L(t)$. Under this transformation the advection term in equation (11.4.16) is eliminated and we obtain the modified evolution equation

$$\frac{\partial C}{\partial t} = \frac{D}{L(t)^2} \frac{\partial^2 C}{\partial x^2} - \left(\frac{\dot{L}}{L}\right) C + f(C, R_+, R_-). \tag{11.4.17a}$$

Applying a similar analysis to the advecting variables $R_\pm(x,t)$, with fluxes $J_\pm(x,t) = \mp vR_\pm$ and $f(C, R_+, R_-)$ replaced by $\pm(\alpha R_- - \alpha R_+) + g(C, R_\pm)$, we derive the following evolution equations on $[0, L_0]$:

$$\frac{\partial R_+}{\partial t} = -\frac{v}{L(t)} \frac{\partial R_+}{\partial x} - \left(\frac{\dot{L}}{L}\right) R_+ + \alpha R_- - \alpha R_+ + g(C, R_+), \tag{11.4.17b}$$

$$\frac{\partial R_-}{\partial t} = \frac{v}{L(t)} \frac{\partial R_-}{\partial x} - \left(\frac{\dot{L}}{L}\right) R_- + \alpha R_+ - \alpha R_- + g(C, R_-). \tag{11.4.17c}$$

In the above equations we have fixed the length scale by setting $L_0 = 1$.

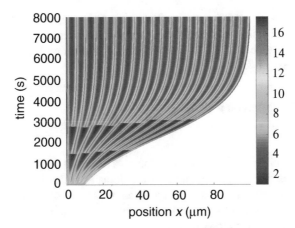

Fig. 11.34: Space–time plot showing the insertion of new potential synaptic sites as the domain representing a section of the ventral cord grows over the course of 2.5 hours. The horizontal axis represents position along the *C. elegans* ventral cord and the vertical axis represents time in seconds. Colors represent local concentration of GLR-1 (a combined total of both leftward and rightward trafficking species) in $\mu$M. Areas of high concentration represent potential synapse sites. The 1D domain grows with growth parameters $r = 0.001$ $\mu$m/s and $\Lambda_0 = 10$. Other parameters are $\rho_1 = 1$/s, $\rho_2 = 1/(\mu$M s$)$, $\mu_1 = 0.25$/s, $\mu_2 = 1$/s, $\alpha = 0.11$/s, $D = 0.01$ $\mu$m$^2$/s, and $v = 1$ $\mu$m/s.

Following Ref. [217], it is useful to make one further simplification by noting that for sufficiently slow growth, the terms involving the dilution factor $-\dot{L}(t)/L(t)$ are small compared to the remaining terms and can be neglected. It is reasonable to assume slow growth for *C. elegans*, since the larvae grow to the adult stage at an average rate of around $10^{-3}$ $\mu$m/s [769]. For the sake of illustration, consider the logistic growth

$$L(t) = e^{rt}\left[1 + \frac{1}{\Lambda_0}(e^{rt} - 1)\right]^{-1}, \qquad (11.4.18)$$

with $r = 0.001$ $\mu$m/s and $\Lambda_0 = 10$. With this choice of growth function, a section of the ventral cord grows from 10 $\mu$m to 100 $\mu$m in 2 hours. Although the logistic growth function represents the physical growth during *C. elegans* development, similar results are obtained for other choices of growth functions (such as exponential or linear). Example plots showing the evolution of the concentrations $C, R_\pm$ are shown in Fig. 11.34. It is assumed that the system operates beyond the Turing bifurcation point for pattern formation at the initial length $L_0$. The initial concentrations are chosen from a uniform random distribution near fixed point values of the spatially uniform equations. As the system evolves, we initially observe growth of the concentration $C$ of the diffusive component, and then patterns emerge as the activator is eventually tempered by increase of the advecting inhibitors $R_\pm$. In this case, the concentration of both activating and inhibiting species is in phase with each other. Once the pattern is established, it persists as the domain length increases, with areas of high concentration slowly growing farther apart. As the areas of high concentration become sufficiently separated, the pattern becomes reorganized and we see the emergence of new peaks. (Another example of diffusion on a growing domain is considered in Ex. 11.8.)

---

**Box 11E. Diffusion on a growing domain.**

---

Consider the classical diffusion equation on a finite 1D domain $0 < x < L(t)$, where $L(t)$ is the increasing length of the domain. Following previous studies of diffusion processes on growing domains [217, 824], we model domain growth in terms of a velocity field $u$ such that $x \rightarrow x + u(x,t)\Delta t$ over the time interval $[t, t + \Delta t]$. It follows that

$$\frac{dL(t)}{dt} = \int_0^{L(t)} \frac{\partial \phi}{\partial x} dx.$$

We will assume uniform growth by taking $\partial_x u = \sigma(t)$, which implies that

$$u(x,t) = x\sigma(t), \quad \sigma(t) = \frac{1}{L(t)}\frac{dL(t)}{dt}.$$

Let $X \in [0, L_0]$ be the local coordinate system at the initial length $L_0$. Using a Lagrangian description, we can then represent spatial position at time $t$ as

$$x = \Gamma(X,t) \equiv \frac{XL(t)}{L_0}, \quad L(0) = L_0.$$

In order to derive the associated diffusion equation, we consider the particle conservation equation (for the inhomogeneous Neumann boundary conditions)

$$\frac{d}{dt}\int_0^{L(t)} C(x,t)dx + \int_0^{L(t)} \frac{\partial J(x,t)}{\partial x}dx = F(t),$$

with $F(t)$ the net influx from the boundaries, and $[J(x,t) = -D\partial C(x,t)/\partial x$. Using Reynold's transport theorem (see below), the left-hand side becomes

$$\frac{d}{dt}\int_0^{L(t)} C(x,t)dx = \int_0^{L(t)} \left[\frac{\partial C(x,t)}{\partial t} + \sigma(t)\frac{\partial[xC(x,t)]}{\partial x}\right]dx.$$

We thus obtain the evolution equation

$$\frac{\partial C(x,t)}{\partial t} + \sigma(t)\frac{\partial[xC(x,t)]}{\partial x} = D\frac{\partial^2 C(x,t)}{\partial x^2} + \frac{F(t)}{L(t)}, \quad 0 < x < L(t), \quad t > 0.$$

Substituting for $\sigma(t)$ thus yields

$$\frac{\partial C(x,t)}{\partial t} + \left(\frac{\dot{L}(t)}{L(t)}\right)\left(x\frac{\partial C(x,t)}{\partial x} + C(x,t)\right) = D\frac{\partial^2 C(x,t)}{\partial x^2} + \frac{F(t)}{L(t)}.$$

Finally, we transform this equation to the fixed interval $[0, L_0]$ by performing the change of variables

$$x \to X = \frac{xL_0}{L(t)}.$$

Under this transformation the advection term is eliminated and we obtain the modified evolution equation [217]

$$\frac{\partial C(x,t)}{\partial t} = \frac{D}{L(t)^2}\frac{\partial^2 C(x,t)}{\partial x^2} - \left(\frac{\dot{L}}{L}\right)C(x,t) + \frac{F(t)}{L(t)}.$$

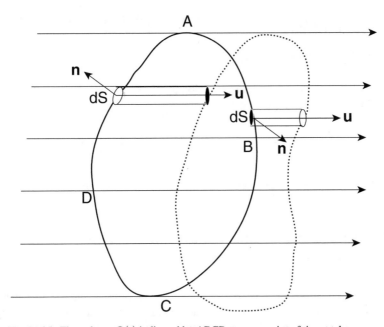

Fig. 11.35: The volume $\Omega(t)$ indicated by ABCD at some point of time t takes a new position (shown by dotted lines) after a time $\delta t$.

**Reynold's transport theorem.** We will derive the theorem for diffusion in a $d$-dimensional growing domain $\Omega(t)$ with velocity field $\mathbf{u}(\mathbf{x},t)$. From the definition of a derivative

$$\frac{d}{dt}\int_{\Omega(t)} C(\mathbf{x},t)d\mathbf{x} = \lim_{\delta t \to 0}\frac{1}{\delta t}\left[\int_{\Omega(t+\delta t)} C(\mathbf{x},t+\delta t)d\mathbf{x} - \int_{\Omega(t)} C(\mathbf{x},t)d\mathbf{x}\right].$$

Adding and subtracting $C(\mathbf{x},t+\delta)$ at $\Omega(t)$, we have

$$\frac{d}{dt}\int_{\Omega(t)} C(\mathbf{x},t)d\mathbf{x} = \lim_{\delta t \to 0}\frac{1}{\delta t}\left[\int_{\Omega(t+\delta t)} C(\mathbf{x},t+\delta t)d\mathbf{x} - \int_{\Omega(t)} C(\mathbf{x},t+\delta t)d\mathbf{x}\right]$$

$$+ \lim_{\delta t \to 0}\frac{1}{\delta t}\left[\int_{\Omega(t)} C(\mathbf{x},t+\delta t)d\mathbf{x} - \int_{\Omega(t)} C(\mathbf{x},t)d\mathbf{x}\right].$$

The second limit is simply $\int_{\Omega(t)} \partial_t C(\mathbf{x},t)d\mathbf{x}$, so it remains to determine the first limit. For small but finite $\delta$ we have to evaluate the change in an integral due to a small change in the volume, see Fig. 11.35. The surface $S(t)$ of $\Omega(t)$ moves locally in the direction of the flow $\mathbf{u}(\mathbf{x},t)$. In particular a local element $\mathbf{n}dS$ sweeps out an infinitesimal volume $\mathbf{u}\cdot\mathbf{n}dS\delta t$ in the time $\delta t$. Hence

$$\lim_{\delta t \to 0}\frac{1}{\delta t}\left[\int_{\Omega(t)} C(\mathbf{x},t+\delta t)d\mathbf{x} - \int_{\Omega(t)} C(\mathbf{x},t)d\mathbf{x}\right]$$

$$= \lim_{\delta t \to 0}\frac{1}{\delta t}\left[\int_{S(t)} C(\mathbf{x},t+\delta t)\mathbf{u}(\mathbf{x},t)\cdot\mathbf{n}dS\delta t.\right] = \int_{S(t)} C(\mathbf{x},t)\mathbf{u}(\mathbf{x},t)\cdot\mathbf{n}dS$$

$$= \int_{\Omega(t)} \nabla\cdot(C(\mathbf{x},t)\mathbf{u}(\mathbf{x},t))d\mathbf{x}.$$

The last line follows from an application of the divergence theorem. We thus obtain Reynold's transport theorem

$$\frac{d}{dt}\int_{\Omega(t)} C(\mathbf{x},t)d\mathbf{x} = \int_{\Omega(t)}\left[\frac{\partial C(\mathbf{x},t)}{\partial t} + \nabla\cdot(C(\mathbf{x},t)\mathbf{u}(\mathbf{x},t))\right]d\mathbf{x}.$$

## 11.5 Existence and stability of multi-spike solutions in the strongly nonlinear regime

Numerical simulations of a variety of classical RD systems suggest that patterns often exist in the strongly nonlinear regime, with the final steady state consisting of a spatially repeating pattern of localized spikes. Although weakly nonlinear analysis is not applicable in the large amplitude regime, it is still possible to study the existence and stability of spike patterns in certain limits. For example, in the case of classical RD systems such as the Brusselator, Gierer–Meinhardt, and Gray–Scott models, one can consider the singular limit $D_a \ll D_h$, where $D_a$ and $D_h$ are the diffusivities of a slowly diffusing activator and a fast-diffusing inhibitor, say. These systems support multi-spike solutions in which the activator is localized to narrow peak regions (spikes) within which the inhibitor is slowly varying. Asymptotic analysis can then be used to analyze the existence and linear stability of multi-spike solutions [456, 457, 518, 519, 892, 922]. We first review this approach using the example of the non-dimensionalized two-component Gierer–Meinhardt model on the domain $x \in [-1, 1]$ [456]:

$$\frac{\partial C}{\partial t} = \varepsilon^2 \frac{\partial^2 C}{\partial x^2} + \frac{C^2}{R} - C, \tag{11.5.1a}$$

$$\tau \frac{\partial R}{\partial t} = D \frac{\partial^2 R}{\partial x^2} + C^2 - R, \quad C'(\pm 1) = 0 = R'(\pm 1), \tag{11.5.1b}$$

where the slow diffusion of the activator is incorporated by taking the diffusion coefficient to be $\varepsilon^2$ with $0 < \varepsilon \ll 1$; the concentrations have also been rescaled according to $C \to \varepsilon C$, $R \to \varepsilon R$. We then consider a particular application of the theory to protein clustering in *E. coli* [657].

## 11.5.1 Existence and stability of an N-spike solution

We first derive conditions for the existence of multiple spike solutions in the limit $\varepsilon \to 0$ using the asymptotic analysis of Iron *et al.* [456, 457]. This type of solution consists of localized peaks in $C$ of width $\varepsilon$ with $C$ exponentially small outside each spike, while $R$ changes slowly within each spike. Setting time derivatives to zero in equation (11.5.1), we obtain the time-independent equations

$$0 = \varepsilon^2 \frac{d^2 C}{dx^2} + \frac{C^2}{R} - C, \tag{11.5.2a}$$

$$0 = D \frac{dR^2}{dx^2} + \frac{1}{\varepsilon} C^2 - R, \quad C'(\pm 1) = 0 = R'(\pm 1). \tag{11.5.2b}$$

Suppose that there are $N$ spikes at positions $x_1, \ldots, x_N$ with $C$ peaking at each point $x_j$ such that $C'(x_j) = 0$, $R(x_j) = U$, and $U$ independent of $j$. (In order to apply asymptotic analysis, the spikes are located away from the boundaries $x = 0, 1$.)

Inside the inner region of the $j$th spike, we transform to the stretched coordinate $y_j = \varepsilon^{-1}(x - x_j)$ and set $\widetilde{C}(y_j) = C(x_j + \varepsilon y_j)$ and $\widetilde{R}(y_j) = R(x_j + \varepsilon y_j)$. Introduce the asymptotic expansions

$$\widetilde{C} = \widetilde{C}_0 + o(1), \quad \widetilde{R} = \widetilde{R}_0 + \varepsilon \widetilde{R}_1 + O(\varepsilon),$$

with

$$\frac{d^2 \widetilde{C}_0}{dy_j^2} + \frac{\widetilde{C}_0^2}{\widetilde{R}_0} - \widetilde{C}_0 = 0, \quad -\infty < y_j < \infty; \quad \widetilde{C}_0'(0) = 0, \tag{11.5.3a}$$

$$\frac{d^2 \widetilde{R}_0}{dy_j^2} = 0, \ \widetilde{R}_0(0) = U; \quad D \frac{d^2 \widetilde{R}_1}{dy_j^2} + \widetilde{C}_0^2 = 0, \ \widetilde{R}_1(0) = 0. \tag{11.5.3b}$$

together with the far-field condition $\widetilde{C}_0 \rightarrow 0$ as $|y_j| \rightarrow \infty$. The first equation of (11.5.3b) implies that $\widetilde{R}_0(y_j) = U$ for all $-\infty < y_j < \infty$. In addition, using phase-plane analysis it can be shown that equation (11.5.3a) has the solution [456]

$$\widetilde{C}_0(y_j) = \frac{3}{2} U \operatorname{sech}^2(y_j/2) := U p(y_j). \tag{11.5.4}$$

Hence, integrating the second equation in (11.5.3b) implies that

$$\left[ \lim_{y_j \rightarrow \infty} \widetilde{R}_1' - \lim_{y_j \rightarrow -\infty} \widetilde{R}_1' \right] = -\frac{\Gamma U^2}{D}, \quad \Gamma = \int_{-\infty}^{\infty} p(y)^2 dy = 6. \tag{11.5.5}$$

Equation (11.5.5) acts as a jump condition for the first derivative of the solution for $R$ in the outer region. Since $C \sim 0$ in the outer region, we can write $R(x) = R_0(x) + O(\varepsilon)$ with

$$D \frac{d^2 R_0}{dx^2} - R_0 = 0, \quad x \in [-1, 1] \backslash \{x_1, x_2, \ldots, x_N\}, \tag{11.5.6}$$

and $R_0$ continuous across $x_j$ but $R_0'$ discontinuous according to equation (11.5.5). That is,

$$D \frac{d^2 R_0}{dx^2} - R_0 = -\Gamma U^2 \sum_{k=1}^{N} \delta(x - x_k), \quad x \in [-1, 1], \tag{11.5.7}$$

with $R_0'(x) = 0$ at $x = \pm 1$. Introducing Neumann Green's function

$$D \frac{d^2 G(x; x_k)}{dx^2} - G(x; x_k) = -\delta(x - x_k), \quad \partial_x G(\pm 1; x_k) = 0, \tag{11.5.8}$$

we can write the outer solution as

$$R_0(x) = \Gamma U^2 \sum_{j=1}^{N} G(x; x_j). \tag{11.5.9}$$

The explicit form of 1D Green's function is [456] (see also Box 6C)

$$G(x; x_k) = \begin{cases} A_k \cosh \kappa (1+x) / \cosh \kappa (1+x_k) & -1 < x < x_k \\ A_k \cosh \kappa (1-x) / \cosh \kappa (1-x_k) & x_k < x < 1 \end{cases}, \tag{11.5.10}$$

where

$$A_k = \kappa (\tanh \kappa (1 - x_k) + \tanh \kappa (1 + x_k))^{-1}, \quad \kappa = \sqrt{1/D}. \tag{11.5.11}$$

In particular,

$$\sum_{j=1}^{N} G(x; x_j) = \frac{1}{2\sqrt{D} \tanh(1/N\sqrt{D})}. \tag{11.5.12}$$

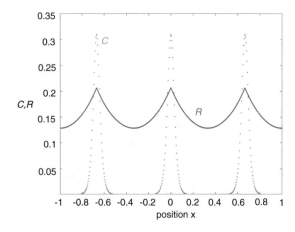

Fig. 11.36: 3-spike solution obtained from asymptotic analysis with $\varepsilon = 0.02$ and $D = 0.1$. Plots of $R(x)$ and $C(x)$.

It remains to determine the constant $U$. Setting $x = x_i$ in equation (11.5.9) yields the self-consistency condition

$$U = \Gamma U^2 \sum_{j=1}^{N} G(x_i;x_j). \tag{11.5.13}$$

It follows that a symmetric $N$-spike solution will exist provided that the positions $x_1,\ldots,x_N$ are chosen such that the sum $\mathcal{G} = \sum_{j=0}^{N-1} G(x_i;x_j)$ is independent of $i$. It can be shown that this occurs for the equally spaced solution [456]

$$x_j = -1 + \frac{2j-1}{N}, \quad j = 1,\ldots,N. \tag{11.5.14}$$

The nontrivial solution for the amplitude $U$ is then

$$U = \frac{1}{\Gamma \sum_{j=1}^{N} G(x_i;x_j)}. \tag{11.5.15}$$

In conclusion, in the limit $\varepsilon \to 0$ the leading order asymptotic solutions for the unscaled concentrations $C$, $R$, and $J$ are of the form

$$C(x) \sim U \sum_{j=1}^{N} p([x-x_j]/\varepsilon), \quad R(x) \sim \Gamma U^2 \sum_{j=1}^{N} G(x;x_j). \tag{11.5.16}$$

In Fig. 11.36 we show an example of a 3-spike solution for $\varepsilon = 0.02$ and $D = 0.1$.

In order to investigate the stability of an equilibrium spike solution $(C(x),R(x))$, we introduce the perturbations

$$C(x,t) = C(x) + e^{\lambda t}\Phi(x), \quad R(x,t) = R(x) + e^{\lambda t}\Psi(x). \tag{11.5.17}$$

Substituting into equation (11.5.1) and linearizing give

$$(1+\lambda)\Phi(x) = \varepsilon^2 \frac{d^2\Phi(x)}{dx^2} + 2\frac{C(x)}{R(x)}\Phi(x) - \frac{C^2(x)}{R^2(x)}\Psi(x), \qquad (11.5.18a)$$

$$(1+\tau\lambda)\Psi(x) = D\frac{d^2\Psi(x)}{dx^2} + \frac{2C(x)}{\varepsilon}\Phi(x), \quad x \in [-1,1]. \qquad (11.5.18b)$$

These equations are supplemented by the no-flux boundary conditions $\Phi'(\pm1) = \Psi'(\pm1) = 0$. One finds that there are two sets of eigenvalues: large or $O(1)$ eigenvalues that are bounded away from zero as $\varepsilon \to 0$ [456, 922]; small eigenvalues that approach zero as $O(\varepsilon^2)$. The source of these eigenvalues can be understood by considering the so-called shadow GM system in the limit $D \to \infty$ with $\tau = 0$. The linear operator of the latter has $N$ exponentially small eigenvalues that reflect the near translation invariance of the system and the existence of exponentially weak interactions between neighboring spikes and between spikes and the boundary. These become the small eigenvalues for finite $D$. The linear operator of the shadow system also has exactly one positive eigenvalue in the vicinity of each spike, which implies that an $N$-spike solution is unstable on an $O(1)$ time scale. However, as $D$ is decreased from infinity, the $O(1)$ eigenvalue near each spike moves into the left-half plane thus stabilizing the solution in the long-time limit. For $D < \infty$, $N \geq 2$, and $\tau \geq 0$, the following results hold (see propositions 4 and 11 of [456] and proposition 5.3 of [922]).

**Small eigenvalues.** Consider solutions of equation (11.5.18) for $\tau = 0$. The $O(\varepsilon^2)$ eigenvalues are negative only when $D < D_N^*$, where

$$D_N^* = \frac{1}{[N\ln(1+\sqrt{2})]^2}. \qquad (11.5.19)$$

There are $N-1$ small positive eigenvalues when $D > D_N^*$ and $\lambda = 0$ is an eigenvalue of algebraic multiplicity $N-1$ when $D = D_N^*$. Moreover, $D_N^*$ is a decreasing function of $N$, with $D_2^* \approx 0.32$, $D_3^* \approx 0.14$, and $D_4^* \approx 0.08$, for example. These results persist to leading order when $\tau = O(1)$.

**Large eigenvalues.** Let $\lambda_0$ be the $O(1)$ eigenvalue with the largest real part for $\tau = 0$. Then $\mathrm{Re}(\lambda_0) < 0$ when

$$D < D_N \equiv \frac{1}{\Theta_N^2}, \quad \Theta_N = \frac{N}{2}\ln\left[2 + \cos(\pi/N) + \sqrt{(2+\cos(\pi/N))^2 - 1}\right]. \qquad (11.5.20)$$

On the other hand, $\mathrm{Re}(\lambda_0) > 0$ when $D > D_N$. Note that $D_N > D_N^*$ and $D_N$ is also a decreasing function of $N$ with $D_1 = \infty$, $D_2 \approx 0.58$, $D_3 \approx 0.18$, and $D_4 \approx 0.09$, for example. It follows that there exists a parameter regime $D_N^* < D < D_N$ where an $N$-spike solution is stable on an $O(1)$ time scale but unstable in the long-time limit. Finally, if $\tau = O(1)$ then an $N$-spike solution is unstable for any $\tau \geq 0$ when $D > D_N$. For $D > D_N$ and $\tau \geq 0$, there are eigenvalues that are real and positive, and the resulting instability involves spikes being annihilated in finite time. On the other hand, for $0 < D < D_N$, an $N$-spike solution is stable with respect to the $O(1)$ eigenvalues provided that $0 \leq \tau \leq \tau_N(D)$ for some critical time constant $\tau_N$ (which can be determined numerically). When $\tau$ crosses the critical value $\tau_N(D)$, a synchronous oscillatory instability of the spike amplitudes is induced.

## 11.5.2 A nonlocal eigenvalue problem

We now describe the construction and analysis of the nonlocal eigenvalue problem developed in Ref. [922], which establishes the above results for the $O(1)$ eigenvalues. We also assume that $N \geq 2$—the analysis of single spike solutions is slightly different [922]. The first step is to look for a localized eigenfunction of $\Phi$ in the form

$$\Phi(x) \sim \sum_{k=1}^{N} c_k \phi[(x - x_k)/\varepsilon]. \tag{11.5.21}$$

Since $\Phi$ is localized to the inner region of each spike, it follows that it can be treated as a sum of the Dirac delta functions in the equation for $\Psi(x)$:

$$D\frac{d^2\Psi(x)}{dx^2} - (1 + \tau\lambda)\Psi(x) = -2U \int_{-\infty}^{\infty} p(y)\phi(y)dy \sum_{k=1}^{N} c_k \delta(x - x_k). \tag{11.5.22}$$

Here $p(y)$ is defined in equation (11.5.4). The solution for $\Psi(x)$ can be expressed in terms of Neumann Green's function

$$\frac{d^2 G_\lambda(x; x_k)}{dx^2} - \theta_\lambda G_\lambda(x; x_k) = -\delta(x - x_k), \quad \partial_x G_\lambda(\pm 1; x_k) = 0, \tag{11.5.23}$$

where $\theta_\lambda = [(1 + \lambda\tau)/D]^{1/2}$. That is,

$$\Psi(x) = \frac{\Omega}{D} \sum_{k=1}^{N} c_k G_\lambda(x; x_k), \quad \Omega = 2U \int_{-\infty}^{\infty} p(y)\phi(y)dy. \tag{11.5.24}$$

The explicit form of 1D Green's function is [456] (see also Box 6C)

$$G_\lambda(x; x_k) = \begin{cases} B_k \cosh\sqrt{\theta_\lambda}(1 + x)/\cosh\sqrt{\theta_\lambda}(1 + x_k) & -1 < x < x_k \\ B_k \cosh\sqrt{\theta_\lambda}(1 - x)/\cosh\sqrt{\theta_\lambda}(1 - x_k) & x_k < x < 1 \end{cases}, \tag{11.5.25}$$

where

$$B_k = \frac{1}{\sqrt{\theta_\lambda}}\left(\tanh\sqrt{\theta_\lambda}(1 - x_k) + \tanh\sqrt{\theta_\lambda}(1 + x_k)\right)^{-1}. \tag{11.5.26}$$

Substituting equations (11.5.21) and (11.5.24) into equation (11.5.18a), using the fact that $R(x) = U + O(\varepsilon)$ when $|x - x_j| = O(\varepsilon)$, and expressing $\Phi$ in terms of stretched variables give

$$c_j \left( \frac{d^2\phi(y)}{dy^2} - \phi(y) + 2p(y)\phi(y) \right) - p^2(y)\frac{\Omega}{D} \sum_{k=1}^{N} c_k G_\lambda(x_j; x_k) = c_j \lambda \phi(y)$$

$$(11.5.27)$$

for $-\infty < y < \infty$. The eigenvalue problem is the same for each $j$ when $c_1, \ldots, c_N$ are the components of the eigenvector $\mathbf{c}$ of the matrix equation $\mathcal{G}_\lambda \mathbf{c} = \alpha(\lambda)\mathbf{c}$, where $[\mathcal{G}_\lambda]_{ij} = G_\lambda(x_i; x_j)$. We thus obtain the nonlocal eigenvalue problem

$$\left( \frac{d^2\phi(y)}{dy^2} - \phi(y) + 2p(y)\phi(y) \right) - \frac{2\alpha(\lambda)p^2(y)}{D\sum_{j=1}^{N} G(x_i; x_j)} \frac{\int_{-\infty}^{\infty} p(y)\phi(y)dy}{\int_{-\infty}^{\infty} p^2(y)dy} = \lambda\phi(y)$$

$$(11.5.28)$$

for $-\infty < y < \infty$, and we have substituted for $\Omega$ using equation (11.5.15) and (11.5.24). We also require $\phi \to 0$ as $|y| \to 0$. The next step is to calculate the eigenvalues $\alpha(\lambda)$, which was carried out for the classical GM model in [456, 922]. The basic idea is to solve equation (11.5.18b) for $\Psi(x_n)$, $n = 1, \ldots, N$, after rewriting it as

$$D\frac{d^2\Psi(x)}{dx^2} - (1+\tau\lambda)\Psi(x) = 0, \quad x_{n-1} < x < x_n, \quad n = 1, \ldots N+1, \quad (11.5.29a)$$

$$[\Psi]_n = 0, \quad D[\Psi'] = -\Omega c_n, \quad n = 1, \ldots, N; \quad \Psi'(\pm 1) = 0, \quad (11.5.29b)$$

with $x_0 = -1$ and $x_{N+1} = 1$. One thus obtains the matrix equation

$$\mathcal{B}\mathbf{a} = \frac{\Omega}{\sqrt{(1+\tau\lambda)D}}\mathbf{c}, \quad (11.5.30)$$

where $\mathbf{a}$ is the $N$-vector with components $\Psi(x_j)$ and $\mathcal{B}$ is a tridiagonal matrix [922]

$$\mathcal{B} = \begin{pmatrix} d_\lambda & f_\lambda & 0 & \cdots & 0 & 0 & 0 \\ f_\lambda & e_\lambda & f_\lambda & \cdots & 0 & 0 & 0 \\ 0 & f_\lambda & e_\lambda & \ddots & 0 & 0 & 0 \\ \vdots & \vdots & \ddots & \ddots & \ddots & \vdots & \vdots \\ 0 & 0 & 0 & \ddots & e_\lambda & f_\lambda & 0 \\ 0 & 0 & 0 & \cdots & f_\lambda & e_\lambda & f_\lambda \\ 0 & 0 & 0 & \cdots & 0 & f_\lambda & d_\lambda \end{pmatrix}$$

$$(11.5.31)$$

with nonzero components

$$d_\lambda = \coth(2\theta_\lambda/N) + \tanh(\theta_\lambda/N), \quad e_\lambda = 2\coth(2\theta_\lambda/N), \quad f_\lambda = -\text{csch}(2\theta_\lambda/N).$$

$$(11.5.32)$$

Comparison with the solution (11.5.24) for $x = x_i$ implies that

$$\frac{1}{D}\mathcal{G} = \frac{1}{\sqrt{(1+\tau\lambda)D}}\mathcal{B}^{-1}. \quad (11.5.33)$$

Hence, $\alpha(\lambda) = D[(1 + \tau\lambda)D]^{-1/2}\kappa^{-1}(\lambda)$ where $\kappa(\lambda)$ is an eigenvalue of $\mathscr{B}$. The eigenvalues of $\mathscr{B}$ were calculated in [456, 922] for a given $\theta_\lambda$. We thus find that the eigenvalues of $\mathcal{G}$ are

$$\alpha_j(\lambda) = \frac{D}{2\sqrt{(1+\tau\lambda)D}}[\coth(2\theta_\lambda/N) - \operatorname{csch}(2\theta_\lambda/N)\cos(\pi(j-1)/N)]^{-1}.$$

(11.5.34)

In summary, the $O(1)$ eigenvalues satisfy the set of nonlocal eigenvalue problems

$$\mathbb{L}_0\phi(y) - \chi_j(\lambda)p^2(y)\frac{\int_{-\infty}^{\infty}p(y)\phi(y)dy}{\int_{-\infty}^{\infty}p^2(y)dy} = \lambda\phi(y), \quad -\infty < y < \infty; \phi \to 0 \text{ as } |y| \to \infty$$

(11.5.35)

for $j = 0, \ldots, N-1$, with the linear operator $\mathbb{L}_0$ and factor $\chi_j(y)$ given by

$$\mathbb{L}_0\phi(\lambda) = \frac{d^2\phi(y)}{dy^2} - \phi(y) + 2p(y)\phi(y),$$

(11.5.36)

$$\chi_j(y) = \frac{2}{\sqrt{1+\tau\lambda}}\frac{\tanh(1/N\sqrt{D})}{\coth(2\theta_\lambda/N) - \operatorname{csch}(2\theta_\lambda/N)\cos(\pi(j-1)/N)}.$$

(11.5.37)

*Stability conditions for the $O(1)$ eigenvalues.* It is first convenient to rewrite the nonlocal equation (11.5.35)[922]. Let $\psi(y)$ be the solution to the equation

$$\mathbb{L}_0\psi(y) \equiv \frac{d^2\psi}{dy^2} - \psi(y) + 2p(y)\psi(y) = \lambda\psi(y) + p^2(y); \quad \psi \to 0 \text{ as } |y| \to \infty. \quad (11.5.38)$$

That is, $\psi = (\mathbb{L}_0 - \lambda)^{-1}p^2$. The eigenfunction satisfying equation (11.5.35) can then be expressed as

$$\phi(y) = \chi_j(\lambda)\psi(y)\Upsilon, \quad \Upsilon = \frac{\int_{-\infty}^{\infty}p(y)\phi(y)dy}{\int_{-\infty}^{\infty}p^2(y)dy}.$$

Multiplying both sides of the equation relating $\phi$ to $\psi$ by $p(y)$ and integrating with respect to $y$ yields a transcendental equation for the eigenvalues $\lambda$:

$$g_j(\lambda) \equiv C_j(\lambda) - f(\lambda) = 0,$$

(11.5.39)

where

$$f(\lambda) \equiv \frac{\int_{-\infty}^{\infty}p(y)[\mathbb{L}_0 - \lambda]^{-1}p^2(y)dy}{\int_{-\infty}^{\infty}p^2(y)dy},$$

(11.5.40)

and

$$C_j(\lambda) = \frac{1}{\chi_j(\lambda)} = \frac{\sqrt{1+\tau\lambda}}{2\tanh(1/N\sqrt{D})}\left[\tanh(\theta_\lambda/N) + \frac{1-\cos(\pi(j-1)/N)}{\sinh(2\theta_\lambda/N)}\right].$$

(11.5.41)

We have used some identities for hyperbolic functions.

*Real eigenvalues.* The following results hold for real eigenvalues $\lambda$ [922]:

(i) The function $f(\lambda)$ has the asymptotic behavior

$$f(\lambda) \sim 1 + \frac{3}{4}\lambda_R + \kappa_c\lambda^2 + O(\lambda^3) \text{ as } \lambda \to 0; \quad f(\lambda) \to \infty \text{ as } \lambda \to v_0^-,$$

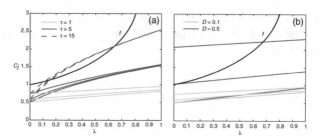

Fig. 11.37: Plots of the functions $C_j(\lambda)$, $j = 1, 2, 3$, for a 3-spike solution of the classical GM model. (a) $D = 0.1 < D_3$, and various $\tau$. (b) $\tau = 0.1$ and different diffusivities $D$. For each triplet, the lowest curve is for $C_1$ and the highest curve is for $C_3$. The black convex curve is the function $f(\lambda)$.

where $\kappa_c > 0$ and $\nu_0$ is the principal eigenvalue of the linear operator $\mathbb{L}_0$. Moreover, $f$ is a strictly monotonically increasing convex function for all $0 < \lambda < \nu_0$, that is, $f'(\lambda) > 0$ and $f''(\lambda) > 0$; $f(\lambda) < 0$ for $\lambda > \nu_0$.

(ii) For any fixed $D > 0$ and $\tau > 0$, each $C_j(\lambda)$ $j = 0, \ldots, N - 1$, is a strictly monotonically increasing concave positive function for all $\lambda \geq 0$. That is, $C_j(\lambda) > 0$, $C_j'(\lambda) > 0$, and $C_j''(\lambda) < 0$. Moreover, $C_j'(\lambda) = O(\sqrt{\tau})$ as $\tau \to \infty$.

(iii) The functions $C_j(\lambda)$ are ordered such that

$$C_N(\lambda) > C_{N-1}(\lambda) > \ldots > C_1(\lambda), \quad C_N'(\lambda) < C_{N-1}'(\lambda) > \ldots < C_1'(\lambda).$$

(iv) Set $B_j(D) = C_j(0)$ for $j = 2, \ldots, N$. Here $B_j(D)$ is a strictly monotonically increasing function of $D$ for $D > 0$ and is independent of $\tau$. In addition $B_j(D) < 1$ for $0 < D < D_N$ with $B_j(D_N) = 1$ and $D_N$ given by equation (11.5.20).

It follows from (i)-(iv) that if $D < D_N$ and $\tau$ is sufficiently small (so that $C_j'(\lambda)$ is sufficiently small), the curves $f(\lambda)$ and $C_j(\lambda)$ will not intersect for any $\lambda \geq 0$. That is, the $N$-spike solution is stable with respect to eigenmodes corresponding to real eigenvalues. The basic picture is illustrated in Fig. 11.37(a) for a 3-spike solution and $D < D_3$. One finds that there are no points of intersection for $\tau < \tau_c$ with $\tau_c = O(10)$. On the other hand, for large $\tau$, there are two real eigenvalues for each $j = 1, 2, 3$. It is clear that these eigenvalues do not appear by crossing the origin. Indeed, as highlighted in [922], instabilities induced by increasing $\tau$ for fixed $D$ can only occur via a Hopf bifurcation. On the other hand, instabilities induced by increasing $D$ for fixed $\tau$ can occur via a real eigenvalue entering the right half-plane, see Fig. 11.37(b).

*Imaginary eigenvalues.* The next step is to look for pure imaginary eigenvalues [922]. We begin by separating equation (11.5.39) into real and imaginary parts:

$$g_j = g_{j,R} + i g_{j,I}, \quad f = f_R + i f_I, \quad \lambda = \lambda_R + i \lambda_I. \tag{11.5.42}$$

If we now set $\lambda_R = 0$, the eigenvalues along the imaginary axis are the solutions of the following pair of equations

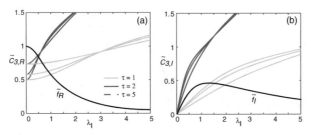

Fig. 11.38: Plot of the functions (a) $\widetilde{C}_{j,R}(\lambda_I)$ and (b) $\widetilde{C}_{j,I}(\lambda_I)$, $j = 1,2,3$, for a 3-spike solution of the classical GM model. Here $D = 0.1$ and $\varepsilon = 0.02$. The black curves in (a) and (b) are the functions $\widetilde{f}_R(\lambda_I)$ and $\widetilde{f}_I(\lambda_I)$, respectively.

$$\widetilde{g}_{j,R}(\lambda_I) \equiv \widetilde{C}_{j,R}(\lambda_I) - \widetilde{f}_R(\lambda_I), \quad \widetilde{g}_{j,I}(\lambda_I) \equiv \widetilde{C}_{j,I}(\lambda_I) - \widetilde{f}_I(\lambda_I), \tag{11.5.43}$$

with $\widetilde{C}_{j,R}(\lambda_I) = \mathrm{Re}[C_j(i\lambda_I)]$, $\widetilde{C}_{j,L}(\lambda_I) = \mathrm{Im}[C_j(i\lambda_I)]$, and

$$\widetilde{f}_R(\lambda_I) = \mathrm{Re}[f(i\lambda_I)] = \frac{\int_{-\infty}^{\infty} p(y) \mathbb{L}_0 [\mathbb{L}_0^2 + \lambda_I^2]^{-1} p^2(y)\, dy}{\int_{-\infty}^{\infty} p^2(y)\, dy},$$

$$\widetilde{f}_I(\lambda_I) = \mathrm{Im}[f(i\lambda_I)] = \frac{\lambda_I \int_{-\infty}^{\infty} p(y) [\mathbb{L}_0^2 + \lambda_I^2]^{-1} p^2(y)\, dy}{\int_{-\infty}^{\infty} p^2(y)\, dy}.$$

Without loss of generality, we can take $\lambda_I \geq 0$. The following results for $\widetilde{f}_R(\lambda_I)$ and $\widetilde{f}_I(\lambda_I)$ hold [922]:

(a) The function $\widetilde{f}_R$ has the asymptotic behavior

$$\widetilde{f}_R(\lambda_I) \sim 1 - \kappa_c \lambda_I^2 + O(\lambda_I^4) \text{ as } \lambda_I \to 0; \quad \widetilde{f}_R(\lambda_I) \to O(\lambda_I^{-2}) \text{ as } \lambda_I \to \infty,$$

where $\kappa_c > 0$. Moreover, $f$ is a strictly monotonically decreasing function for all $0 < \lambda_I < \infty$, that is, $\widetilde{f}_R'(\lambda_I) < 0$.

(b) The function $\widetilde{f}_I$ has the asymptotic behavior

$$\widetilde{f}_I(\lambda_I) \sim \frac{3}{4} \lambda_I + O(\lambda_I^3) \text{ as } \lambda_I \to 0; \quad \widetilde{f}_I(\lambda_I) \to O(\lambda_I^{-1}) \text{ as } \lambda_I \to \infty,$$

and $\widetilde{f}_I(\lambda_I) > 0$ for all $0 < \lambda_I < \infty$.

Note that $\widetilde{C}_{j,R}(0) = C_j(0) = B_j(D)$ for all $\lambda_I$. Hence, if $D < D_N$ then $\widetilde{C}_{j,R}(0) < \widetilde{f}_R(0)$. Moreover, if $\tau > 0$ then $\widetilde{C}_{j,R}(\lambda_I)$ is an increasing function of $\lambda_I$ [922], which from result (a) means that there is a single intersection point with $\widetilde{f}_R(\lambda_I)$ and thus a unique positive root $\lambda_I^*$ where $\widetilde{g}_{j,R}(\lambda_I^*) = 0$, see Fig. 11.38(a). Moreover, from result (b) we have $\widetilde{f}_I(\lambda_I) > 0$ for all $\lambda_I > 0$, whereas $\widetilde{C}_{j,I} = O(\tau)$ as $\tau \to 0$. This implies that for sufficiently small $\tau$, $\widetilde{g}_{j,I}(\lambda_I^*) < 0$ and there are no purely imaginary eigenvalues, see Fig. 11.38(b).

*Winding number argument.* The condition $\widetilde{g}_{j,I}(\lambda_I^*) < 0$ not only ensures that there are no pure imaginary eigenvalues but also implies that there are no complex-valued eigenvalues in

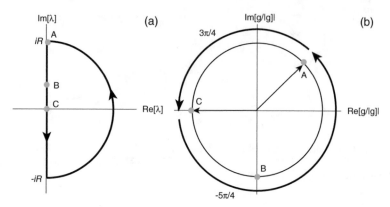

Fig. 11.39: Winding number argument of [922]. (a) Counterclockwise path along the contour $\Gamma_R \cup \Gamma_I \cup \overline{\Gamma}_I$ in the complex $\lambda$-plane, where $\Gamma_R$ is the semicircle of radius $R$, $\Gamma_I = [iR, 0]$ and $\overline{\Gamma}_I = [0, -iR]$. (b) Change in $\arg g$ along $\Gamma_I$ (from A to C) in the limit $R \to \infty$. There exists a unique point $i\lambda^*$ on the positive imaginary axis where $\widetilde{g}_{j,R}(\lambda^*) = 0$ (point B). If $\widetilde{g}_I(\lambda^*) < 0$ then $\arg g(i\lambda^*) = -\pi/2$ and $\Delta \arg g = -5\pi/4$.

the right complex half-plane. This follows from the winding number construction of [922]. The argument principle of complex analysis states that if $F(x)$ is a meromorphic function inside and on some closed contour $C$ and $F$ has no zeros or poles on $C$, then (see Box 11F)

$$\frac{1}{2\pi i} \oint_C \frac{F'(z)}{F(z)} dz = \frac{1}{2\pi i} \oint_{F(C)} \frac{dw}{w} = Z - P, \qquad (11.5.44)$$

where $w = F(z)$, and $Z, P$ denote, respectively, the number of zeros (including multiplicity) and poles (including order) of $F(z)$ inside $C$. The second integral is the winding number of $F$. Consider the contour $C$ shown in Fig. 11.39(a), which consists of the semicircle $\Gamma_R$ given by $|\lambda| = R > 0$ for $-\pi/2 < \arg\lambda \le \pi/2$, and the section $-iR \le \text{Im}[\lambda] \le iR$ of the imaginary axis. Assuming that $\alpha$ is sufficiently large so that there are no pure imaginary eigenvalues (see above), we can let $R \to \infty$ and use the argument principle to determine the number of zeros of $g_j(\lambda)$ in the right-half plane. Since $C_j(\lambda) \sim \lambda^{1/2}$ as $|\lambda| \to \infty$ while $f(\lambda) \to 0$ as $|\lambda| \to \infty$, it follows that $\arg g\lambda$ changes by an amount $\pi/2$ as $\Gamma_R$ is traversed counterclockwise. Using the fact that $g_j(\lambda)$ is analytic in the right-half plane, except at the simple pole $\lambda = \nu_0 > 0$ ($P = 1$), it follows from the argument principle that the number $M$ of zeros of $g_j(\lambda)$ in the right half-plane is

$$Z = \frac{5}{4} + \frac{1}{\pi}[\arg g_j]_{\Gamma_I}, \qquad (11.5.45)$$

where $[\arg g_j]_{\Gamma_I}$ is the change in the argument of $g$ along the semi-infinite imaginary axis $\Gamma_I = i\lambda, 0 \le \lambda < \infty$. Setting $\widetilde{C}_j(\lambda) = C_j(i\lambda)$ with $\lambda$ real, we have $\widetilde{C}_j(\lambda) \sim \sqrt{i\lambda}$ as $\lambda \to \infty$ so that $\arg g = \pi/4$ at the start of $\Gamma_I$. Moreover, $\arg g_j(0) = \pi$, since $g_j(0) = C_j(0) - f(0) < 0$. Given the unique point $i\lambda^*$ on the positive imaginary axis where $\widetilde{g}_{j,R}(\lambda^*) = 0$. If we can show that $\widetilde{g}_{j,I}(\lambda^*) < 0$, then $\arg g(i\lambda^*) = -\pi/2$ and the change of $\arg g$ along $\Gamma_I$ is $-5\pi/4$, see Fig. 11.39(b). Equation (11.5.45) then ensures that there are no eigenvalues in the right half-plane and the $N$-spike solution is stable. From result (b), we have $\widetilde{f}_I(\lambda) > 0$ for all $0 < \lambda < \infty$, whereas $\widetilde{C}_{j,I}(0) = 0$ and $\widetilde{C}_{j,I}(\lambda) = O(\tau)$ as $\tau \to 0$ for finite $\lambda$. Hence, for

$j = 1, \ldots, N$, there exists a $\tau_0 > 0$ such that $\widetilde{g}_{j,l}(\lambda^*) < 0$ and thus $Z = 0$ for all $\tau$ satisfying $0 \leq \tau < \tau_0$. This establishes the stability of an $N$-spike solution.

---

**Box 11F. Argument principle.**

---

The argument principle of complex analysis connects the winding number of a closed curve with the number of zeros and poles inside the curve. It is a consequence of Cauchy's residue theorem. The basic setup is as follows. Let $\gamma$ be a simple closed curve oriented in a counterclockwise direction. Suppose that the complex function $f(z)$ is analytic on and inside $\gamma$, except (possibly) for a set of finite order poles inside $\gamma$. Let $p_1, \ldots, p_m$ denote the poles and suppose there also exists a set of zeros $z_1, \ldots, z_n$. Take $\text{mult}(z_k)$ to be the multiplicity of the zero at $z_k$ and $\text{mult}(p_k)$ to be the order of the pole at $p_k$. Then

$$\oint_\gamma \frac{f'(z)}{f(z)} dz = 2\pi i \left( \sum_{k=1}^m \text{mult}(z_k) - \sum_{k=1}^n \text{mult}(p_k) \right). \tag{11.5.46}$$

In order to prove this result, we first note that in a neighborhood $\mathcal{U}$ of a zero $z_k$, we can write $f(z) = (z - z_k)^m g(z)$ for all $z \in \mathcal{U}$, where $m = \text{mult}(z_k)$ and $g(z)$ is a nonzero analytic function in $\mathcal{U}$. This implies

$$\frac{f'(z)}{f(z)} = \frac{m(z - z_k)^{m-1} g(z) + (z - z_k)^m g'(z)}{(z - z_k)^m g(z)} = \frac{m}{z - z_k} + \frac{g'(z)}{g(z)}, \quad z \in \mathcal{U}.$$

Since $g(z)$ is non-vanishing, the ratio $g'/g$ is analytic in $\mathcal{U}$ and

$$\text{Res}\left( \frac{f'(z)}{f(z)}, z_k \right) = m = \text{mult}(z_k).$$

Similarly, if $p_k$ is a pole of order $m$, then the Laurent series for $f(z)$ in a neighborhood $\mathcal{V}$ of $p_k$ can be written as $f(z) = (z - p_k)^{-m} h(z)$ where $h(z)$ is analytic and nonzero in $\mathcal{V}$. Hence

$$\frac{f'(z)}{f(z)} = \frac{-m(z - p_k)^{-m-1} g(z) + (z - p_k)^m g'(z)}{(z - p_k)^{-m} g(z)} = -\frac{m}{z - p_k} + \frac{g'(z)}{g(z)}, \quad z \in \mathcal{V}.$$

Again we have a simple pole of $f'(z)/f(z)$ at $p_k$ with

$$\text{Res}\left( \frac{f'(z)}{f(z)}, p_k \right) = -m = -\text{mult}(p_k).$$

Finally, the residue theorem yields the result (11.5.46), that is, the contour integral around the closed curve $\gamma$ is equal to the sum of residues of all the simple poles inside $\gamma$. The winding number (or index) of a closed curve $\gamma$ around a point $z_0$ in the complex plane is defined as

$$\mathrm{Ind}(\gamma, z_0) = \frac{1}{2\pi i} \oint_\gamma \frac{dz}{z - z_0}. \tag{11.5.47}$$

The argument principle then states that

$$\oint_\gamma \frac{f'(z)}{f(z)} dz = 2\pi i \,\mathrm{Ind}(f \circ \gamma, 0) = 2\pi i (Z - P), \tag{11.5.48}$$

where $Z$ and $P$ are the number of zeros and poles inside $\gamma$ (including multiplicities). The relationship between the contour integral and the winding number follows from performing the change of variables $w = f(z)$. This maps the contour $z = \gamma(t)$ to $w = f \circ \gamma(t)$ and $dw = f'(z)dz$ so that

$$\oint_\gamma \frac{f'(z)}{f(z)} dz = \oint_{f \circ \gamma} \frac{dw}{w}.$$

The right-hand side is the winding number.

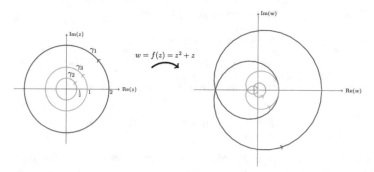

Fig. 11.40: Mapping of three circular curves in the $z$-plane to corresponding curves in the $w$-plane with $w = f(z) = z + z^2$.

**Example.** Let $f(z) = z^2 + z$ and consider the winding number of $f \circ \gamma$ around $z = 0$ for the circular curves $\gamma_r = \{z; |z| = r\}$, $r = 1/2, 1, 2$. The function $f$ has no poles and has zeros at $z = 0, -1$. The winding numbers for $r = 1/2$ and $r = 2$ are

$$\mathrm{Ind}(f \circ \gamma_{1/2}, 0) = 1, \quad \mathrm{Ind}(f \circ \gamma_2, 0) = 2,$$

since $\gamma_{1/2}$ only encloses the zero at $z = 0$, whereas $\gamma_2$ encloses both zeros. It can be checked that the image of the former curve in the $w$-plane winds around the origin once, while the image of the latter curve winds around the origin twice, see Fig. 11.40. Finally, note that the argument principle does not apply to the curve $\gamma_1$, since the zero at $z = -1$ lies on the curve. Hence, the image of the curve in the $w$-plane passes through the origin.

### 11.5.3  Spike solutions in a reaction–diffusion model of protein clustering in bacteria

We now turn to one example of intracellular pattern formation that appears to exhibit multi-spike patterns. This concerns an RD model of the self-positioning of structural maintenance of chromosomes (SMC) protein complexes in E. coli [421, 656, 657]. SMC protein complexes are required for correct chromosome condensation, organization, and segregation. MukBEF, which is the SMC in E. coli, can form a hinged loop structure to trap DNA under the action of ATP; detachment occurs following ATP hydrolysis. (Analogous structures occur in eukaryotes as illustrated in Fig. 11.41.) Experimental studies have shown that MukBEF forms protein clusters in the middle of the nucleoid in short E. coli cells, which can split into two clusters at quarter positions in longer cells [233]. (The nucleoid is an irregularly shaped region within a prokaryote cell that contains all or most of the genetic material. In contrast to the nucleus of a eukaryotic cell, it is not surrounded by a nuclear membrane.) It has been hypothesized that this self-organization could occur via Turing pattern formation. However, the classical Turing mechanism does not typically generate patterns with a fixed, predetermined phase, at least in the weakly nonlinear regime close to the Turing bifurcation point. On the other hand, far from pattern onset, RD systems can still support patterns with a well-defined periodicity, as shown above for the GM model. In large domains, the number of peaks tends to be significantly less than predicted by linear stability analysis, and the locations of the peaks are predetermined in the case of reflecting boundary conditions.

Consider the following RD system used to model the self-organization and positioning of MukBEF clusters in E. coli [421, 656, 657]:

$$\frac{\partial u}{\partial t} = D_u \frac{\partial^2 u}{\partial x^2} - \beta u (u+v)^2 + \gamma v - \sigma u + c\sigma, \tag{11.5.49a}$$

$$\frac{\partial v}{\partial t} = D_v \frac{\partial^2 v}{\partial x^2} + \beta u (u+v)^2 - \gamma v - \sigma v \tag{11.5.49b}$$

for $x \in [-L/2, L/2]$, $D_v < D_u$ and reflecting boundary conditions at $x = \pm L/2$. Note that when $\sigma = 0$, the model reduces to a mass-conserving system. Here $v$ can be interpreted as the concentration of MukBEF complexes bound to DNA, whereas $u$ represents the concentration of freely diffusing MukBEF complexes. Equations (11.5.49) are similar in form to some classical Turing models such as the Brusselator and Schnakenberg models. However, it is simpler to analyze since, in the absence of diffusion, there is a unique fixed point

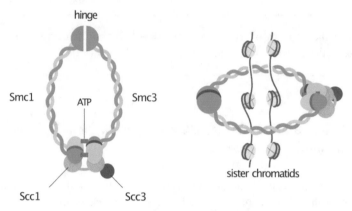

Fig. 11.41: Example of a eukaryotic SMC complex. (a) A folded SMC protein consists of a helical coiled coil with an ATPase domain at one end and a hinge domain at the other end, which interacts with the hinge domain of another SMC protein. In cohesin, this results in the formation of a Smc1-Smc3 heterodimer. (In prokaryotes, pairs of SMC proteins form homodimers.) Binding of ATP (red) promotes binding of the two ATPase domains, resulting in closure of the SMC ring. The non-SMC protein Scc1 interacts with both ATPase domains and holds them together. Cleavage of Scc1 in anaphase therefore opens the ring. (b) The cohesin complex may form a 50-nm ring around two sister chromatids. Because of its small size, however, this ring could only link nucleosomal DNA and not more complex chromatin structures. [Public domain figure uploaded to Wikimedia Commons by David O Morgan from The Cell Cycle. Principles of Control.]

$$u^* + v^* = c, \quad u^* = \frac{[1+b]c}{1+a+b}, \quad \text{with } a = \frac{\beta c^2}{\gamma}, \quad b = \frac{\sigma}{\gamma}, \tag{11.5.50}$$

which is stable for all parameter values. Linear stability analysis thus divides parameter space into two regions: one where the fixed point is stable and the other where it is Turing unstable.

We are interested in deriving conditions for the existence of multiple spike solutions in the limit $D_v \ll D_u$. This type of solution consists of localized peaks in $v$ of width $\varepsilon = O(\sqrt{D_v/\gamma})$ with $v$ exponentially small outside each spike, while $u$ changes slowly within each spike. The asymptotic analysis of spikes for the specific model of protein clustering is carried out in [657]. Here we develop the analysis using the more systematic formulation of Sect. 11.6.2 and Refs. [456, 458, 518, 519, 892], see also Ex. 11.9. First, setting $D_v = \varepsilon^2$ and $D_u = D$ (after non-dimensionalizing), the steady-state solutions are written as

$$0 = D\frac{d^2u}{dx^2} - \frac{\beta}{\varepsilon}u(\varepsilon u + v)^2 + \frac{\gamma}{\varepsilon}v - \sigma u + c\sigma, \tag{11.5.51a}$$

$$0 = \varepsilon^2\frac{d^2v}{dx^2} + \beta u(\varepsilon u + v)^2 - \gamma v - \sigma v, \tag{11.5.51b}$$

after performing the rescalings $u \to \varepsilon^{-1/2}u$, $v \to \varepsilon^{1/2}v$, and $c \to \varepsilon^{1/2}c$. Suppose that there are $N$ spikes at positions $x_1, \ldots, x_N$ with $v$ peaking at each point $x_j$ such that

$$v'(x_j) = 0, \quad u(x_j) = U, \tag{11.5.52}$$

with $U$ independent of $j$. (In order to apply asymptotic analysis, the spikes are located away from the boundaries $x = \pm L/2$.)

Inside the inner region of the $j$th spike we introduce the stretched coordinate $y_j = \varepsilon^{-1}(x - x_j)$ and set $\tilde{u}(y_j) = u(x_j + \varepsilon y_j)$ and $\tilde{v}(y_j) = v(x_j + \varepsilon y_j)$. Take $\tilde{u}$ and $\tilde{v}$ to have the asymptotic expansions $\tilde{u} = \tilde{u}_0 + \varepsilon \tilde{u}_1 + O(\varepsilon^2)$ and $\tilde{v} = \tilde{v}_0 + O(\varepsilon)$, with

$$\frac{d^2 \tilde{u}_0}{dy_j^2} = 0, \quad D \frac{d^2 \tilde{u}_1}{dy_j^2} = \beta U \tilde{v}_0^2 - \gamma \tilde{v}_0, \tag{11.5.53a}$$

$$\frac{d^2 \tilde{v}_0}{dy_j^2} + \beta U \tilde{v}_0^2 - (\gamma + \sigma)\tilde{v}_0 = 0, \quad -\infty < y_j < \infty, \tag{11.5.53b}$$

$$\tilde{v}_0'(0) = 0, \ \tilde{u}_0(0) = U, \ \tilde{u}_1(0) = 0. \tag{11.5.53c}$$

We also require $\tilde{u}_0$ to be bounded and $\tilde{v}_0 \to 0$ as $|y_j| \to \infty$. The former condition immediately implies that $\tilde{u}_0(y_j) = U$ for all $-\infty < y_j < \infty$. Next, using phase-plane analysis it can be shown that equation (11.5.53b) has the solution (Ex. 11.9)

$$\tilde{v}_0(y_j) = \frac{3}{2} \frac{\gamma + \sigma}{\beta U} \operatorname{sech}^2 \left( \frac{\sqrt{\gamma + \sigma}}{2} y_j \right). \tag{11.5.54}$$

Integrating the equation for $\tilde{u}_1$ and using the solution for $\tilde{v}_0$ yield the jump condition (Ex. 11.9)

$$D \left[ \lim_{y_j \to \infty} \tilde{u}_1' - \lim_{y_j \to -\infty} \tilde{u}_1' \right] = \frac{\Gamma}{U}, \quad \Gamma = \frac{6\sigma \sqrt{\gamma + \sigma}}{\beta}. \tag{11.5.55}$$

Equation (11.5.55) acts as a jump condition for the first derivative of the solution for $u$ in the outer region. Since $v \sim 0$ in the outer region, we can write $u(x) = u_0(x) + O(\varepsilon)$ with

$$D \frac{d^2 u_0}{dx^2} - \sigma u_0 + c\sigma = \frac{\Gamma}{U} \sum_{k=1}^{N} \delta(x - x_k), \quad x \in [-L/2, L/2], \tag{11.5.56}$$

and $u_0'(\pm L/2) = 0$. Introducing Neumann Green's function

$$D \frac{d^2 G(x; x')}{dx^2} - \sigma G(x; x') = -\delta(x - x'), \ \partial_x G(\pm L/2; x') = 0, \tag{11.5.57}$$

$$\int_{-L/2}^{L/2} G(x; x') dx = \sigma^{-1},$$

the outer solution is

$$u_0(x) = c - \frac{\Gamma}{U} \sum_{j=1}^{N} G(x; x_j). \tag{11.5.58}$$

The explicit form of 1D Green's function is (see Ex. 11.9)

$$G(x;x') = \frac{\kappa}{2} \frac{\cosh(\kappa(x+x')/L) + \cosh(\kappa(|x-x'|-L)/L)}{\sinh(\kappa)}, \quad \kappa = L\sqrt{\frac{\sigma}{D}}.$$

(11.5.59)

It remains to determine the constant $U$. Setting $x = x_i$ we have the self-consistency condition

$$U = c - \frac{\Gamma}{U} \sum_{j=1}^{N} G(x_i;x_j).$$

(11.5.60)

It follows that the $N$-spike solution will exist provided that the positions $x_1,\ldots,x_N$ are chosen such that the sum $\mathcal{G} = \sum_{j=1}^{N} G(x_i;x_j)$ is independent of $i$. It can be shown that this occurs for the equally spaced solution [657]

$$x_k = \frac{Lk}{N} - \frac{L(N+1)}{2N},$$

(11.5.61)

with

$$\mathcal{G} = \frac{\kappa}{2}\coth(\kappa/2N).$$

(11.5.62)

There are then two solutions for $U$, the smaller of which is relevant to the large amplitude regime for spikes, since $\widetilde{v}_0(y_j) \sim 1/U$:

$$U = \frac{c}{2}\left[1 - \sqrt{1 - 4\Gamma\mathcal{G}/c^2}\right].$$

(11.5.63)

Equation (11.5.58) then becomes

$$u_0(x) = c\left[1 - \rho \sum_{j=1}^{N} G(x;x_j)\right], \quad \rho = \frac{1 + \sqrt{1 - 4\Gamma\mathcal{G}/c^2}}{2\mathcal{G}}.$$

(11.5.64)

Moreover, the total amount (mass) of the component with concentration $u$ is

$$M := \int_{-L/2}^{L/2} u(x)dx = c\left[1 - \frac{\rho}{\sigma}\right],$$

(11.5.65)

where we have used $\int_{-L/2}^{L/2} G(x;x')dx = \sigma^{-1}$. Using a combination of flux-balance conditions and numerical simulations, the authors in [657] provided evidence that the particular spike pattern selected is the one that minimizes the mass $M$. First, they showed that the peaks of a displaced $N$-spike solution move toward the equally spaced solution, equation (11.5.61), which minimizes the mass for fixed $N$. Second, they found that the preferred number of peaks, $N$, in the final steady-state solution minimizes the mass with respect to $N$. In the limit $\sigma \to 0$, $N = 1$ as typically found in mass-conserving systems. Finally, note that it is also possible to carry out a more rigorous linear stability analysis of $N$-spike solutions in RD systems by solving a nonlocal eigenvalue problem along the lines outlined in Sect. 11.6.2 and Refs. [456, 458, 518, 519, 892, 922].

## 11.6 Intracellular traveling waves

There are a number of biological phenomena that have to be coordinated with incredible speed across large distances [232]. These include the propagation of nerve impulses (action potentials) along axons [496], the synchronized cell cycles of early embryogenesis [180, 231, 461, 689], and collective cell migration [206, 965]. Such forms of coordination are too fast to be achieved via passive diffusion, and in many cases the necessary machinery for active transport isn't available. Instead, fast coordination appears to be driven by traveling biochemical waves of activity that propagate information across a cell or population of cells. As in the case of Turing pattern formation, traveling waves in biological and chemical systems are typically modeled in terms of RD equations. The nature of the nonlinearities describing the underlying chemical reactions then determines the physical properties of the wave.

From a mathematical perspective, traveling waves in RD systems can be grouped into three different types: unstable waves, bistable waves, and excitable waves [114, 898]. Unstable waves tend to occur predominantly in models of population growth, dating back to the Fisher–KPP equation [311, 517], which was introduced to describe the invasion of a gene into a population. One characteristic feature of the Fisher–KPP equation and subsequent generalizations is that they support traveling fronts propagating into an unstable steady state, in which the wave speed and long-time asymptotics are determined by the dynamics in the leading edge of the wave—so–called pulled fronts [898]. In particular, a sufficiently localized initial perturbation will asymptotically approach the traveling front solution that has the minimum possible wave speed.

Bistable and excitable waves are more commonly associated with cellular and developmental systems. Recall from Chap. 5 that a common mechanism for generating bistability in a biochemical network is via nonlinear positive feedback. A threshold then exists such that values below (above) the threshold evolve to the low activity (high activity) steady state. Now suppose that bistability is incorporated into a spatial model. As soon as a transition from the low to high activity state is triggered in a local region, activity can diffuse to neighboring regions, shifting these regions above the activation threshold and therefore causing the transition from low to high activity to spread across the domain as a traveling front. An excitable medium is one that has only one stable state, but two distinct modes of returning to that state [496]). The dynamics involves a combination of positive feedback that generates a threshold-like response and negative feedback that returns the system to its steady state. Consequently, large perturbations of the resting state push the system beyond the threshold, which produces a large excursion in phase space before the system eventually returns to the resting state. A classical example is the generation of an action potential in a neuron (Chap. 3). When excitability is combined with diffusion, the resulting action potential can propagate along the neuron's axon as a traveling pulse that is characterized by a constant wave speed and amplitude. Other examples of excitability in cell biology include gene networks with both positive and negative feedbacks (Chap. 5), and $Ca^{2+}$ dynamics (see Chap. 3 and Ref. [496]).

In this section we present some examples of intracellular traveling waves, focusing on bistable (Box 11G) and unstable waves (Box 11H). (For an extensive treatment of excitable waves, see the book by Keener and Sneyd [496].) We begin by considering an example of wave propagation in a bistable RD system that has been used to model cell polarization in motile eukaryotic cells [473, 638, 639]. In this case, mass conservation leads to the stalling of the wave within the cell, leading to a stationary polarization front. We then briefly describe a second example of a bistable wave, which is found in developing eggs that are laid externally, such as those of insects, amphibians, and fish. These eggs support the fastest cell cycles seen in biology (around 8-25 min), despite their relatively large size (0.6-1.2 mm). Moreover, cell divisions within the eggs are coordinated via synchronized "mitotic waves." Such synchronization appears to be necessary for the precise execution of morphogenesis later in development. We end by considering one example of an unstable intracellular wave, which arises in a model of CaMKII translocation waves along the dendrites of neurons [112, 263, 771].

## 11.6.1 Wave-pinning model of cell polarization in motile eukaryotic cells

An important example of cell polarization occurs in a variety of motile eukaryotic cells that undergo directed motion in response to external spatial signals—eukaryotic chemotaxis [422, 553, 589]. Examples include mammalian neutrophils (white blood cells), fibroblasts (connective tissue mammalian cells responsible for wound healing), and keratocytes (fast moving epithelial cells from scales of fish). Prior to initiating movement, a given cell polarizes according to directional cues in the environment, forming nascent "front" and "back" regions, see Fig. 11.42(a). Protrusion of the front is driven by the assembly of actin cytoskeleton, whereas myosin motors at the back contract and pull up the rear, see also Sect. 14.4. The polarization of the cell is regulated by Rho GTPases and phosphoinositides (PIs), with Rac, Cdc42, and PIP3 localized at the front and Rho localized at the back. (PIs are lipids that play an important role in regulating vesicular trafficking and actin polymerization.) Although different motile eukaryotic cells utilize much of the same molecular machinery, they can exhibit significantly different forms of behavior [474]. For example, neutrophils can sense very small gradients over a large range of concentrations, polarize very quickly (in less than a minute), and do not spontaneously polarize in the absence of chemoattractant. On the other hand, fibroblasts exhibit much less sensitivity to concentration gradients, polarize and move more slowly, and spontaneously polarize after being put on an adhesive substrate. Finally, keratocytes polarize on a similar time scale to neutrophils, spontaneously polarize after being detached from surrounding cells, and react to mechanical rather than chemical stimuli. There is one further important difference between models of motiles eukaryotic cells and other examples of cell polarization such as budding yeast, which is illustrated in Fig. 11.42. Polarized eukaryotic cells such as fibroblasts and keratocytes

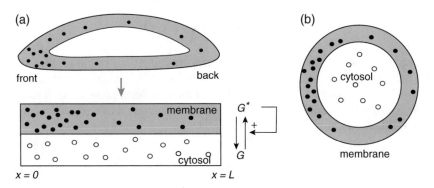

Fig. 11.42: Schematic diagram showing spatial distribution of a membrane-bound active signaling molecule $G^*$ and a cytosolic inactive form $G$. (a) A 1D bidomain model of a flattened Eukaryotic cell adhering to a substrate with no-flux boundary conditions at either end. (b) An idealized 2D or 3D cell with a spatially uniform interior and a polarized perimeter corresponding to the cell membrane.

tend to have a flattened shape when they adhere to a surface. If one imagines taking a cross-sectional slice along the polarization axis, one can treat the system as an effective 1D bidomain model with no-flux boundary conditions at either end. Both membrane-bound and cytosolic molecules diffuse along the polar axis, with a high concentration of membrane-bound molecules at the front and a low concentration at the back, see Fig. 11.42(a). On the other hand, in the case of yeast, one treats the interior of the cell as spatially uniform and determines the spatial distribution of chemicals along the cell perimeter, which in the case of a simplified two-dimensional cell consists of a circle. Cell polarization would then correspond to a localized increase in concentration somewhere on the circle.

We will focus on a RD model of cell polarity in motile eukaryotic cells developed and analyzed by Keshet *et al.* [473, 638, 639], see also [694, 699]. The simplest version of the model considers a single Rho GTPase that can transition between inactive and active forms diffusing in a bounded 1D domain of length $L$ [638], see Fig. 11.42(a). Let $a(x,t)$ and $b(x,t)$ be the concentrations of the active/inactive states. Then we have the mass-conserving system

$$\frac{\partial a}{\partial t} = D_a \frac{\partial^2 a}{\partial x^2} + f(a,b), \quad \frac{\partial b}{\partial t} = D_b \frac{\partial^2 b}{\partial x^2} - f(a,b),$$

$$\left.\frac{\partial a}{\partial x}\right|_{x=0,L} = 0 = \left.\frac{\partial b}{\partial x}\right|_{x=0,L}. \tag{11.6.1}$$

Since the rate of diffusion of the membrane-bound (active) state is significantly slower than that of the cytosolic (inactive) state, $D_a \ll D_b$. The nonlinear function $f(a,b)$ represents the difference between the rates of activation and inactivation of the Rho GTPase. Assuming there is cooperative positive feedback in the activation of the protein, which is modeled as a Hill function of index 2, then

$$f(a,b) = b \left( k_0 + \frac{\gamma a^2}{K^2 + a^2} \right) - k_- a. \tag{11.6.2}$$

It can be checked that for a range of uniform concentrations of the inactive state, $b_{min} < b < b_{max}$, the space-clamped version of the model exhibits bistability with two stable fixed points $a_{\pm}(b)$ separated by an unstable fixed point $a_0(b)$. It follows from the boundary conditions that there is mass conservation of the total amount of Rho GTPase, that is,

$$\int_0^L (a+b)dx = C. \tag{11.6.3}$$

The emergence of cell polarization in this model can be understood in terms of front propagation in a bistable RD system (see Box 11G), with the following additional features [638, 639]: (i) the inactive and active states have unequal rates of diffusion; (ii) the total amount of each GTPase is conserved. Consequently, a local stimulus induces a propagating front that decelerates as it propagates across the cell so that it becomes stationary, a process known as wave-pinning; the stationary front persists in the absence of the stimulus and represents a polarized cell. One mathematical explanation of wave-pinning proceeds as follows [638]. First, since $D_b \gg D_a$ and there are no-flux boundary conditions, one can assume that $b$ rapidly diffuses to establish a uniform concentration within the bounded domain $[0,L]$; $b$ then changes on a slower time scale as the $a$ dynamics evolves (quasi-steady-state approximation). Thus, on short time scales $b$ can be treated as a fixed global parameter of a scalar equation for $a(x,t)$ given by (11.6.1a). Suppose that initially $b_{min} < b < b_{max}$, so equation (11.6.1a) is bistable. On an infinite domain, the bistable equation supports the propagation of a traveling front linking the stable fixed point $a_+(b), a_-(b)$ (see Box 11G). That is, for $-\infty < x < \infty$ there exists a monotonically decreasing solution $a(x,t) = A(\xi)$, $\xi = x - ct$ with $\lim_{\xi \to -\infty} A(\xi) = a_+(b)$ and $\lim_{\xi \to \infty} A(\xi) = a_-(b)$. Moreover the wave speed satisfies $c = c(b)$ with

$$c(b) = \frac{\displaystyle\int_{a_-}^{a_+} f(a,b)da}{\displaystyle\int_{-\infty}^{\infty} (\partial A/\partial \xi)^2 d\xi}. \tag{11.6.4}$$

Note that the wave speed depends on the global parameter $b$. Since the denominator of (11.6.4) is always positive, the sign of $c(b)$ will depend on the sign of $I(b) \equiv \int_{a_-}^{a_+} f(a,b)da$, which has a geometrical interpretation in terms of the difference between the area of the curve $y = f(a,b)$ above the straight line $y = k_- a$ and the area below, see Fig. 11.43. In the case of a sufficiently sharp front that is away from the boundaries, these results carry over to the bounded domain $[0,L]$.

Now suppose that a transient stimulus near the edge of the cell at $x = 0$ triggers at time $t = 0$ a traveling front as described above. This implies that $b_{min} < b(0) < b_{max}$ and $I(b(0)) > 0$. As the front starts to propagate into the interior of the cell, it converts a greater fraction of the domain from $a \approx a_-(b)$ to $a \approx a_+(b)$. From the conservation condition (11.6.3), it follows that the approximately uniform concen-

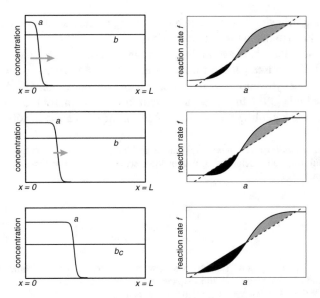

Fig. 11.43: Schematic diagram explaining the mechanism of wave-pinning developed in [638]. A sequence of snapshots of the traveling front (left column) showing that as the front advances into the domain, the background concentration $b$ of the inactive state decreases so that the front decelerates until it becomes stationary. The corresponding geometric construction of $I(b)$ (right column), which is given by the difference of the shaded areas, shows that $I(b)$ is initially positive but vanishes at the critical value $b_c$.

tration $b(t)$ of the inactive state decreases, eventually reaching a critical value $b_c$, $b_{min} < b_c < b_{max}$, for which

$$I(b_c) \equiv \int_{a_-}^{a_+} f(a, b_c)\,da = 0, \qquad (11.6.5)$$

and wave-pinning occurs. The basic steps are illustrated in Fig. 11.43. One interesting issue is to what extent the wave-pinning mechanism differs from the phenomenon of wave propagation failure due to spatial discretization. This is particularly important given that any numerical simulation of the wave-pinning model involves the introduction of a spatial grid, and the wave becomes more sensitive to discretization effects as it slows down. A careful numerical study has shown that wave-pinning and propagation failure are distinct effects [917]. In the same study, a stochastic version of the wave-pinning model was also considered, which takes into account fluctuations in the number of active and inactive molecules at low concentrations. It was found that when the total number of molecules is lowered, wave-pinning behavior is lost due to a broadening of the transition layer as well as increasing fluctuations in the pinning position.

Finally, note that the wave-pinning mechanism can also be viewed as another example of Turing-like pattern formation in a mass-conserving system (see Sect. 11.3), involving cytosolic mass redistribution and the corresponding displacement of local equilibria [391]. In particular, the uniform fixed points of the space-clamped system are local equilibria for different values of the local mass density.

*Asymptotic analysis of wave-pinning.* As shown by Mori *et al.* [639], it is possible to analyze wave-pinning in more detail using a multi time-scale analysis. The first step is to non-dimensionalize (11.6.1):

$$\varepsilon \frac{\partial a}{\partial t} = \varepsilon^2 \frac{\partial^2 a}{\partial x^2} + f(a,b), \tag{11.6.6a}$$

$$\varepsilon \frac{\partial b}{\partial t} = D \frac{\partial^2 b}{\partial x^2} - f(a,b), \tag{11.6.6b}$$

with $x \in [0,1]$ and $0 < \varepsilon \ll 1$. In order to look at the dynamics on short time-scales, set $\tau = t/\varepsilon$ and introduce the asymptotic expansions $a \sim a_0 + \varepsilon a_1 + \ldots$ and $b \sim b_0 + \varepsilon b_1 + \ldots$. This yields the following pair of equations for $a_0$ and $b_0$:

$$\frac{\partial a_0}{\partial \tau} = f(a_0,b_0), \tag{11.6.7a}$$

$$\frac{\partial b_0}{\partial \tau} = D \frac{\partial^2 b_0}{\partial x^2} - f(a_0,b_0). \tag{11.6.7b}$$

Suppose that $b_{\min} < b_0 < b_{\max}$ so that $f(a_0,b_0)$ is bistable with respect to $a_0$. Equation (11.6.7a) implies that $a_0 \to a_+(b_0)$ or $a_0 \to a_-(b_0)$ at each point $x \in [0,1]$. Assume initial conditions in which there exists a single sharp transition layer linking $a_+$ to $a_-$ (a traveling front).

Let $\phi(t)$ be the position of the transition layer with respect to the slower time-scale $t$. Divide the domain $[0,1]$ into inner and outer regions, with the latter given by $[0,\phi(t) - O(\varepsilon)) \cup (\phi(t) + O(\varepsilon),1]$. Taking the limit $\varepsilon \to 0$ in (11.6.6), with $a \sim a_0 + \varepsilon a_1 + \ldots, b \sim b_0 + \varepsilon b_1 + \ldots$, and restricting $x$ to the outer layer, we have

$$0 = f(a_0,b_0), \quad 0 = D \frac{\partial^2 b_0}{\partial x^2} - f(a_0,b_0). \tag{11.6.8}$$

Adding these two equations immediately implies that $\partial^2 b_0/\partial x^2 = 0$. Combining this with the no-flux boundary conditions, it follows that

$$b_0(x,t) = \begin{cases} b_L(t) & 0 \le x < \phi(t) - O(\varepsilon) \\ b_R(t) & \phi(t) + O(\varepsilon) < x \le 1. \end{cases} \tag{11.6.9}$$

with $b_{L,R}(t)$ independent of $x$. The corresponding solution for $a_0$ is

$$a_0(x,t) = \begin{cases} a_+(b_L) & 0 \le x < \phi(t) - O(\varepsilon), \\ a_-(b_R) & \phi(t) + O(\varepsilon) < x \le 1. \end{cases} \tag{11.6.10}$$

In order to determine the solution in the inner layer near the front, introduce the stretched coordinate $\xi = (x - \phi(t))/\varepsilon$, and set

$$A(\xi,t) = a([x - \phi(t)]/\varepsilon,t), \quad B(\xi,t) = b([x - \phi(t)]/\varepsilon,t). \tag{11.6.11}$$

Introduce the asymptotic expansion $A \sim A_0 + \varepsilon A_1 + \dots$ and similarly for $B$ and $\phi$. Substitution into (11.6.6) and expansion to lowest order in $\varepsilon$ then gives

$$\frac{\partial^2 A_0}{\partial \xi^2} + \frac{d\phi_0}{dt} \frac{\partial A_0}{\partial \xi} + f(A_0, B_0) = 0, \tag{11.6.12a}$$

$$\frac{\partial^2 B_0}{\partial \xi^2} = 0. \tag{11.6.12b}$$

The solution of (11.6.12b) is of the form

$$B_0 = \alpha_1(t)\xi + \alpha_2(t), \tag{11.6.13}$$

where $\alpha_{1,2}(t)$ are determined by matching the inner ($B_0$) and outer ($b_0$) solutions according to

$$\lim_{\xi \to -\infty} B_0(\xi) = b_L, \quad \lim_{\xi \to \infty} B_0(\xi) = b_R. \tag{11.6.14}$$

The matching conditions can only be satisfied if $B_0$ is constant in the inner layer, which means that the inactive state is uniform throughout the whole domain, $B_0 = b_0$. The next step is to determine the inner solution for $A_0$, given that $B_0$ is $\xi$-independent. That is, we have to solve the boundary value problem (11.6.12a) with matching conditions

$$\lim_{\xi \to -\infty} A_0(\xi) = a_+(b_0), \quad \lim_{\xi \to \infty} A_0(\xi) = a_-(b_0). \tag{11.6.15}$$

The inner solution thus corresponds to the standard front solution of a bistable equation with

$$\frac{d\phi_0}{dt} \equiv c(b_0) = \frac{\displaystyle\int_{a_-(b_0)}^{a_+(b_0)} f(a, b_0)\, da}{\displaystyle\int_{-\infty}^{\infty} (\partial A_0/\partial \xi)^2\, d\xi}, \tag{11.6.16}$$

see equation (11.6.4).

The final step is to incorporate the conservation condition (11.6.3), which to lowest order in $\varepsilon$ becomes

$$\int_0^1 a_0\, dx + b_0 = C. \tag{11.6.17}$$

The integral term can be approximated by substituting for $a_0$ using the outer solution:

$$\int_0^1 a_0\, dx = \int_0^{\phi(t)-O(\varepsilon)} a_0\, dx + \int_{\phi(t)+O(\varepsilon)}^1 a_0\, dx + O(\varepsilon)$$
$$= a_+(b_0)\phi_0(t) + a_-(b_0)(1 - \phi_0(t)) + O(\varepsilon).$$

Combining the various results, the analysis of a traveling front solution of (11.6.6) reduces to the problem of solving the ODE system [639]

$$\frac{d\phi_0}{dt} = c(b_0), \quad b_0 = C - a_+(b_0)\phi_0(t) - a_-(b_0)(1 - \phi_0(t)). \tag{11.6.18}$$

These equations can be used to establish that the front slows down and eventually stops to form a stable stationary front. In particular, differentiating the second relation in (11.6.18) with respect to $t$ shows that

$$\left(1 + \frac{da_+(b_0)}{db}\phi_0 + \frac{da_-(b_0)}{db}(1 - \phi_0)\right)\frac{db_0}{dt} = -[a_+(b_0) - a_-(b_0)]\frac{d\phi_0}{dt}. \tag{11.6.19}$$

Differentiating the condition $f(a_\pm(b), b) = 0$ with respect to b and imposing the condition that $f(a, b)$ is bistable, it can be established that $1 + da_\pm/db > 0$. Since $0 < \phi_0 < 1$ and $a_+(b_0) > a_-(b_0)$, it follows from (11.6.19) that $db_0/dt$ and $d\phi_0/dt$ have opposite signs. Hence, as the front advances, $b_0$ decreases until the front stalls at a critical value $b_c$ for which $c(b_c) = 0$. The corresponding front position is $\phi_c$ with

$$b_c = C - a_+(b_c)\phi_c - a_-(b_c)(1 - \phi_c). \qquad (11.6.20)$$

Explicit conditions for wave pinning can be obtained if the reaction term $f$ is taken to be a cubic

$$f(a, b) = a(1 - a)(a - 1 - b), \qquad (11.6.21)$$

rather than the Hill function (11.6.2). In this case, the speed of the front solution in the inner layer can be calculated explicitly (see §2.2), so that (11.6.18) becomes

$$\frac{d\phi_0}{dt} = \frac{b_0 - 1}{\sqrt{2}}, \quad b_0 = C - (1 + b_0)\phi_0. \qquad (11.6.22)$$

It follows that the wave stops when $b_0 = 1 \equiv b_c$ and the stall position is $\phi_c = (C - 1)/2$. Finally, note that Mori *et al.* [639] also extend the asymptotic analysis to the case of multiple transition layers, and carry out a bifurcation analysis to determine parameter regimes for which wave-pinning occurs.

---

## Box 11G. Traveling fronts in a bistable RD equation.

---

Consider a scalar bistable RD equation of the form

$$\frac{\partial u}{\partial t} = \frac{\partial^2 u}{\partial x^2} + f(u), \quad -\infty < x < \infty. \qquad (11.6.23)$$

Suppose that $f(u)$ is chosen so that the corresponding ODE, $du/dt = f(u)$, has stable equilibria at $u = u_\pm, u_+ > u_-$, separated by an unstable equilibrium at $u = u_0$. We define a traveling front solution according to

$$u(x,t) = U(x - ct) = U(\xi), \quad \xi = x - ct \qquad (11.6.24)$$

for some yet to be determined wave speed $c$, supplemented by asymptotic boundary conditions ensuring that the front links the two stable fixed points of the $x$-independent system. For concreteness, we take

$$U(\xi) \to u_+ \text{ as } \xi \to -\infty, \quad U(\xi) \to u_- \text{ as } \xi \to \infty. \qquad (11.6.25)$$

Substituting the traveling front solution into the bistable equation (11.6.23) yields the ODE

$$U_{\xi\xi} + cU_\xi + f(U) = 0, \qquad (11.6.26)$$

where $U_\xi = dU/d\xi$. Classical phase-plane analysis can now be used to find a traveling front solution by rewriting the second-order equation in the form

$$U_\xi = Z, \quad Z_\xi = -cZ - f(U). \tag{11.6.27}$$

In particular, one looks for a solution that links the excited state $(U,Z) = (u_+, 0)$ at $\xi \to -\infty$ to the state $u = u_-$ at $\xi \to \infty$—a so-called heteroclinic connection. This can be achieved using a geometric argument based on a shooting method, as illustrated in Fig. 11.44 for the cubic $f(u) = u(u-u_0)(1-u)$ with $u_- = 0, u_+ = 1$, and $0 < u_0 < 1$. Suppose that $0 < u_0 < 1/2$ so that $c > 0$ (see below). First note that irrespective of the speed $c$, the fixed points $(1,0)$ and $(0,0)$ in the phaseplane are saddles, each with one-dimensional stable and unstable manifolds. By looking at trajectories in the phase plane, it is straightforward to see that when $c \ll 1$, the unstable manifold of $(1,0)$ lies below the stable manifold of $(0,0)$ when $0 < U < 1$, whereas the opposite holds when $c$ is very large. Since these manifolds depend continuously on $c$, it follows that there must exist at least one value of $c$ for which the manifolds cross, and this corresponds to the heteroclinic connection that represents the traveling front solution. It can also be established that this front is unique.

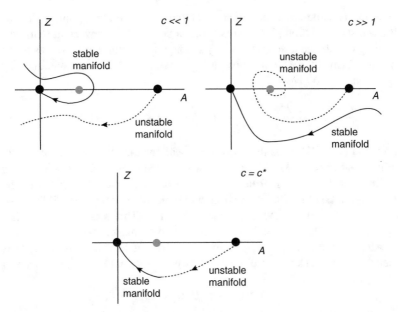

Fig. 11.44: Shooting method for constructing a front solution in the $(U,Z)$ phase plane with $Z = U_\xi$. See text for details.

A useful formula for determining the sign of the wave speed can be obtained by multiplying both sides of (11.6.26) by $U_\xi$ and integrating with respect to $\xi$:

$$c \int_{-\infty}^{\infty} (U_\xi)^2 d\xi = -\int_{-\infty}^{\infty} U_\xi f(U(\xi)) d\xi - \int_{-\infty}^{\infty} U_\xi U_{\xi\xi} d\xi = \int_{u_-}^{u_+} f(U) dU,$$

since $U(\xi)$ is monotone, and

$$\int_{-\infty}^{\infty} U_\xi U_{\xi\xi} d\xi = \int_{-\infty}^{\infty} \frac{d[U_\xi^2/2]}{d\xi} d\xi = 0.$$

As the integral on the left-hand side is positive, it follows that the sign of $c$ is determined by the sign of the area of $f$ between the two stable equilibria. In the case of the cubic, if $0 < u_0 < 1/2$ then the latter is positive and the wave moves to the right. If the negative and positive areas exactly cancel then the front is stationary or pinned, see also Sect. 11.6.1. Finally, note that for certain choices of $f$, the wave speed $c$ and wave profile $U$ can be calculated explicitly, as illustrated in Ex. 11.10.

**Stability of fronts in the bistable equation.** In order to investigate the linear stability of the above front solution, we set

$$u(x,t) = U(\xi) + \phi(\xi,t), \tag{11.6.28}$$

where $\phi$ is some small perturbation belonging to an appropriately defined Banach space $\mathscr{B}$ (complete, normed vector space). It is convenient to use the moving coordinate $\xi$ so that we may see how the perturbation evolves in the moving frame of the front. Substituting for $u$ in (11.6.23) and keeping only terms linear in $\phi$ give

$$\frac{\partial \phi}{\partial t} = \mathbb{L}\phi \equiv \frac{\partial^2 \phi}{\partial \xi^2} + c\frac{\partial \phi}{\partial \xi} + f'(U)\phi, \quad \xi \in \mathbb{R}, \quad t > 0. \tag{11.6.29}$$

Equation (11.6.29) takes the form of a linear equation with associated linear differential operator $\mathbb{L} : \mathscr{D}(\mathbb{L}) \to \mathscr{B}$ with domain $\mathscr{D}(\mathbb{L}) \subseteq \mathscr{B}$. Determining the linear stability of the front thus reduces to the problem of calculating the spectrum $\sigma(\mathbb{L})$ of $\mathbb{L}$. That is, the front will be asymptotically stable if $\|\phi\| \to 0$ as $t \to \infty$ for all $\phi \in \mathscr{B}$, with $\|\cdot\|$ the norm on $\mathscr{B}$. This is guaranteed if $\sigma(\mathbb{L})$ lies strictly in the left-hand side of the complex plane, that is, there exists $\beta > 0$ such that $Re(\lambda) \le -\beta$ for all $\lambda \in \sigma(\mathbb{L})$. The long-time asymptotics is then $\|\phi\| \sim e^{-\beta t}$. However, differentiating both sides of (11.6.26) with respect to $\xi$ gives

$$U_{\xi\xi\xi} + cU_{\xi\xi} + f'(U)U_\xi \equiv \mathbb{L}U_\xi = 0,$$

which implies that zero is an eigenvalue of $\mathbb{L}$ with associated eigenfunction $U_\xi$. This is not a major issue, once one notices that $U_\xi$ is the generator of infinitesimal translations of the front solution:

$$U(\xi + h) = U(\xi) + hU_\xi(\xi) + O(h^2).$$

Hence, such perturbations only cause a phase shift of the original front and can thus be discounted. This motivates defining stability of the solution $U$ in terms of the stability of the family of waves obtained by rigid translations of $U$. In other words, $U$ is said to be stable if and only if $u(x,t) = U(\xi) + \phi(\xi,t)$ converges to $U(\xi + h)$ for some constant, finite $h$ as $t \to \infty$. This will hold provided that zero is a simple eigenvalue of $\mathbb{L}$ and the remainder of the spectrum lies in a half-space $\{\lambda, \mathrm{Re}(\lambda) \leq -\beta\}$ for some real $\beta > 0$. It is important to note that the spectrum of $\mathbb{L}$ will depend on the choice of the Banach space $\mathscr{B}$. Restricting the class of admissible functions can push the spectrum to the left-half complex plane. However, this may exclude classes of perturbations that are physically relevant. A common choice is thus $L^2(\mathbb{R})$, which includes all normalizable, continuous functions on $\mathbb{R}$ with respect to the $L_2$ norm:

$$\|\phi\| = \int_{-\infty}^{\infty} |\phi(\xi)|^2 d\xi < \infty.$$

**Spectrum of a linear differential operator.** Before determining the stability of the traveling front solution, it is useful to briefly review various components of the spectrum of a linear differential operator. Let $\mathscr{B}$ be a Banach space and $\mathbb{L} : \mathscr{D}(\mathbb{L}) \to \mathscr{B}$ be a linear operator with domain $\mathscr{D}(\mathbb{L}) \subseteq \mathscr{B}$. For any complex number $\lambda$, introduce the new operator $\mathbb{L}_\lambda = \mathbb{L} - \lambda \mathbb{I}$, where $\mathbb{I}$ is the identity operator on $\mathscr{B}$. If $\mathbb{L}_\lambda$ has an inverse, then $R_\lambda(\mathbb{L}) = \mathbb{L}_\lambda^{-1}$ is called the resolvent of $\mathbb{L}$. Given these definitions, $\lambda$ is said to be a regular point for $\mathbb{L}$ if the following hold:

(i) $R_\lambda$ exists.
(ii) $R_\lambda$ is bounded.
(iii) $R_\lambda$ is defined on a dense subset of $\mathscr{B}$.

The spectrum $\sigma(\mathbb{L})$ is then the set of points that are not regular, which generally consists of three disjoint parts:

(a) The point or discrete spectrum is the set of values of $\lambda$ (eigenvalues) for which $R_\lambda$ does not exist.
(b) The continuous spectrum is the set of values of $\lambda$ for which $R_\lambda$ exists but is unbounded.
(c) The residual spectrum is the set of values of $\lambda$ for which $R_\lambda$ exists, is bounded, but is not defined on a dense subset of $\mathscr{B}$.

The continuous and residual spectrum belong to the essential spectrum, which is any point in $\sigma(\mathbb{L})$ that is not an isolated eigenvalue of finite multiplicity.

**Discrete spectrum.** It turns out that the essential spectrum of the linear operator defined by equation (11.6.29) lies in the left-half complex plane (see below), so that the stability of the front depends on the eigenvalues $\lambda$ of $\mathbb{L}$, where

$$\mathbb{L}\phi \equiv \phi_{\xi\xi} + c\phi_\xi + f'(U)\phi = \lambda\phi, \tag{11.6.30}$$

with $\phi \in L^2(\mathbb{R})$. Suppose that $\text{Re}(\lambda) \geq 0$ so $\phi(\xi) \sim e^{-\beta\xi}$ as $\xi \to \infty$ with $\beta \geq c$. (This follows from noting $f'(U) \to -u_0$ as $\xi \to \infty$ and analyzing the resulting constant coefficient characteristic equation.) Performing the change of variables $\psi(\xi) = \phi(\xi)e^{c\xi/2}$ yields the modified eigenvalue problem

$$\mathbb{L}_1\psi \equiv \psi_{\xi\xi} + \left(f'(U) - \frac{c^2}{4}\right)\psi = \lambda\psi, \qquad (11.6.31)$$

with $\psi \in L^2(\mathbb{R})$, since it also decays exponentially as $|\xi| \to \infty$. The useful feature of the modified operator is that it is self-adjoint, implying that any eigenvalues in the right-half complex plane are real. Multiplying both sides of the self-adjoint eigenvalue equation (11.6.31) by $\psi$ and integrating over $\mathbb{R}$, we have

$$\lambda \int_{-\infty}^{\infty} \psi^2 d\xi = -\int_{-\infty}^{\infty} \left[\psi_\xi^2 - \left(f'(U) - \frac{c^2}{4}\right)\psi^2\right] d\xi. \qquad (11.6.32)$$

Recall that $U_\xi$ is an eigenfunction of $\mathbb{L}$ with $\lambda = 0$, so that if $\Phi(\xi) = U_\xi(\xi)e^{c\xi/2}$, then $\Phi_{\xi\xi} + (f'(\psi) - c^2/4)\Phi = 0$. Hence,

$$\lambda \int_{-\infty}^{\infty} \psi^2 d\xi = -\int_{-\infty}^{\infty} \left[\psi_\xi^2 + \frac{\Phi_{\xi\xi}\psi^2}{\Phi}\right] d\xi$$

$$= -\int_{-\infty}^{\infty} \left[\psi_\xi^2 - \frac{2\psi\psi_\xi\Phi_\xi}{\Phi} + \frac{\Phi_\xi^2\psi^2}{\Phi^2}\right] d\xi$$

$$= -\int_{-\infty}^{\infty} \Phi^2 \left(\frac{d}{d\xi}(\psi/\Phi)\right)^2 d\xi.$$

This last result implies that $\lambda \leq 0$ and if $\lambda = 0$ then $\psi \sim \Phi = U_\xi$. We conclude that there are no eigenvalues in the right-half complex plane and $\lambda = 0$ is a simple eigenvalue. Thus the traveling front of the scalar bistable equation is stable.

**Essential spectrum.** We first calculate the essential spectrum of a second-order linear operator acting on $\mathcal{B} = L^2(\mathbb{R})$:

$$\mathbb{L}u = u_{xx} + pu_x - qu \qquad (11.6.33)$$

for constant positive coefficients $p, q$ and $\mathcal{D}(\mathbb{L}) = \{u : u \in L^2(\mathbb{R}), \mathbb{L}u \in L^2(\mathbb{R})\}$. We will then indicate how to generalize the analysis to the case of linear operators with $x$-dependent coefficients $p(x), q(x)$, which includes the one defined in equation (11.6.29) with $x \to \xi$. First, suppose that $\mathbb{L}_\lambda$ is not invertible for some $\lambda$. This means that there exists $\phi \in \mathcal{B}$ such that $\mathbb{L}_\lambda\phi = 0$. The latter equation is a linear second-order ODE with constant coefficients and thus has solutions of the form $e^{v_\pm x}$ with $v_\pm$ the roots of the characteristic polynomial $v^2 + pv - (q + \lambda) = 0$. Such a solution cannot decay at both $x = \pm\infty$ and so doesn't belong to $\mathscr{B}$. It follows that $\mathbb{L}$ has no eigenvalues and the resolvent $R_\lambda$ exists. We can then represent $R_\lambda$ in terms of Green's function $G$ defined according to $\mathbb{L}_\lambda^\dagger G(x - x') = \delta(x - x')$, where $\mathbb{L}^\dagger$ is the adjoint of $\mathbb{L}$ with respect to the standard inner product on $L^2(\mathbb{R})$:

$$\mathbb{L}_\lambda^\dagger u = u_{xx} - pu_x - (q + \lambda)u.$$

For any $h \in \mathscr{D}(R_\lambda) \subseteq \mathscr{B}$ we can express the solution $u = R_\lambda h$ to the inhomogeneous equation $\mathbb{L}_\lambda u = h$ as

$$u(x) = \int_{-\infty}^\infty h(y)G(y - x)dy.$$

For constant coefficients, the Green's function can be solved explicitly according to

$$G(y) = \begin{cases} \alpha e^{\mu_+ y} & y \leq 0 \\ \alpha e^{\mu_- y} & y \geq 0, \end{cases}$$

where $\mu_\pm$ are the roots of the characteristic polynomial

$$P(\mu) = \mu^2 - p\mu - (\lambda + q),$$

and $\alpha$ is chosen such that $-1 = \alpha(\mu_+ - \mu_-)$. If $P(\mu)$ has one root $\mu_+$ with positive real part and one root $\mu_-$ with negative real part, then clearly $G \in L_1(\mathbb{R})$ so that $R_\lambda$ is bounded with dense domain equal to $\mathscr{B}$. This situation holds, for example, when $\lambda$ is real and $\lambda > -p$. The roots of $P(\mu)$ vary continuously with $\lambda$ in the complex plane. Hence, the boundedness of $R_\lambda$ will break down when one of the roots crosses the imaginary axis at $ik$, say, with $\lambda = -q - k^2 - ika$. This is a parabola in the complex $\lambda$ plane $(\lambda_r, \lambda_i)$ given by $\lambda_r = -q - \lambda_i^2/p^2$. If $\lambda_r$ is to the right of this parabola,

$$\lambda_r > -q - \frac{\lambda_i^2}{p^2},$$

then $P(\mu)$ has a root on either side of the imaginary axis and $R_\lambda$ is bounded. We conclude that the essential spectrum lies to the left of the parabola,

$$\sigma(\mathbb{L}) \subseteq \{\lambda : Re(\lambda) \leq -q - Im(\lambda)^2/p^2\}. \tag{11.6.34}$$

It can be shown that the essential spectrum includes the parabola itself. It immediately follows that the essential spectrum lies in the left-half complex plane if $q > 0$.

Now note that the linear operator $\mathbb{L}$ of equation (11.6.29) is a second-order operator with constant coefficient $p = c$ and $x$-dependent coefficient $q(x) = -f'(U(x))$. Moreover, $q(x) \to q_\pm$ as $x \to \pm\infty$ with $q_+ = -f'(0) = u_0$

and $q_- = -f'(1) = 1 - u_0$. We are assuming that $f(u) = u(u - u_0)(1 - u)$. Introduce the parabolas

$$S_\pm = \{\lambda : \lambda = -q_\pm - k^2 - ikp_\pm\}. \qquad (11.6.35)$$

Let $A$ denote the union of the regions to the left of the curves $S_\pm$ that includes the curves themselves. It can then be shown that the essential spectrum of $\mathbb{L}$ lies in $A$ and includes $S_\pm$. Since $q_\pm > 0$, we deduce that the essential spectrum is bounded to the left of $Re(\lambda) = \min\{-u_0, u_0 - 1\}$ and thus does not contribute to any instabilities.

## 11.6.2 Mitotic waves

Cyclin-dependent kinase 1 (Cdk1) is the master regulator of the cell cycle, which oscillates in order to ensure that the correct sequence of downstream steps are executed [689]. (A summary of the cell cycle in eukaryotes is presented in Sect. 14.3 and Fig. 14.12.) The essential component of the oscillator is time-delayed negative feedback between Cdk1 and the Anaphase Promoting Complex (APC). The latter triggers the degradation of Cyclins, which are the regulatory subunits required for the activation of Cdk1, thus resetting Cdk1 complexes to their inactive state. The enzymatic network includes two positive feedback loops involving an inhibitor Wee1 and an activator Cdc25, respectively. These generate a bistable system, which provides robust, switch-like, and adaptable changes in Cdk1 activity that are required for the all-or-none transitions of cell cycle events. The enzymes Wee1 and Cdc25 are themselves both regulated by active Cyclin-Cdk1 complexes, namely, Wee1 is inhibited and Cdc25 is activated. Hence, active Cdk1 activates its activator and inhibits its inhibitor, generating a positive feedback loop and a double negative (positive) feedback loop, respectively, see Fig. 11.45. Following entry into mitosis and a period of high Cdk1 activity that drives cells into metaphase, Cdk1 activates the APC. Activation of the APC triggers inactivation of Cdk1 and drives anaphase and other mitotic exit processes.

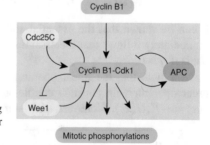

Fig. 11.45: Schematic diagram showing the Cdk1 regulatory network. See text for details.

In the mitotic regulation of *Xenopus laevis* egg extracts during the first cell cycle, inactive Cdk1 is thought to initially be activated at the centrosome, which is the organizing center of microtubules within a cell (Sect. 14.3). Activated Cdk1 then diffuses from the centrosome to neighboring regions, resulting in the catalytic activation of inactive Cdk1 through positive feedback. The combination of diffusion and catalysis leads to the propagation of a traveling wave of activated Cdk1 throughout the extract or embryo [180, 231, 461, 689]. This wave coordinates cell mitosis across the fertilized egg, ensuring for example that nuclear envelope breakdown within the interior of the egg is quickly followed by other events such as the mitotic surface contraction of the egg cortex. A one-dimensional RD system has been used to model these waves along the anterior–posterior axis of the egg [180, 231]. Let $a(x,t)$ denote the concentration of active Cdk1 and $c(x,t)$ the total concentration of Cdk1. For simplicity, the feedback between Cdk1 and Cdc25 is modeled in terms of an activating sigmoidal function, whereas the feedback between Cdk1 and Wee1 is described by an inhibiting one. The resulting equations take the form

$$\frac{\partial a}{\partial t} = D\frac{\partial^2 a}{\partial x^2} + \alpha + r_+(a)(c-a) - r_-(a)a, \tag{11.6.36a}$$

$$\frac{\partial c}{\partial t} = D\frac{\partial^2 c}{\partial x^2} + \alpha, \tag{11.6.36b}$$

where $\alpha$ is the production rate of active Cdk1, and

$$r_+(a) = \left(r_0 + r_1\frac{a^n}{K_+^n + a^n}\right), \quad r_-(a) = \left(r_2 + r_3\frac{K_-^m}{K_-^m + a^m}\right). \tag{11.6.37}$$

Here $n, m$ are the Hill coefficients and $K_+^n, K_-^m$ are corresponding dissociation constants (Chap. 3). Note that this model focuses on the activation of Cdk1 and does not explicitly include the subsequent inactivation (mitotic exit) due to the slow feedback from APC. The latter process does not affect the propagation of the front and can be modeled as a phase wave whose delays are set by the spatially distributed times of entry into mitosis generated by the Cdk1 wave. (A phase wave occurs in oscillatory media and represents a spatially varying shift in the phase of the oscillators [114].) Numerical simulations of equation (11.6.36) show that they support traveling waves of active Cdk1 consistent with experimental studies [180]. A sample set of parameter values is as follows: $D = 10\mu m^2/s$, $K_+ = 35nM$, $n = 11$, $m = 3.5$, $\alpha = 8nM/min$, $r_0 = 0.8 \ min^{-1}$, $r_1 = 4 \ min^{-1}$, $r_2 = 0.4 \ min^{-1}$, and $r_3 = 2 \ min^{-1}$.

The RD system can be reduced to a scalar equation, since the dynamics of the total Cdk1 evolves according to a linear equation that is independent of the nonlinear rates of activation and inactivation. As a first approximation, suppose that diffusion is sufficiently fast so that $c(x,t)$ can be represented by the spatially independent linear growth function $c(t) = c_0 + \alpha t$. Substitution into the equation for $a(x,t)$ then yields the non-autonomous scalar RD equation

$$\frac{\partial a}{\partial t} = D\frac{\partial^2 a}{\partial x^2} + f(a,t), \quad f(a,t) = \alpha + r_+(a)(c(t)-a) - r_-(a)a. \tag{11.6.38}$$

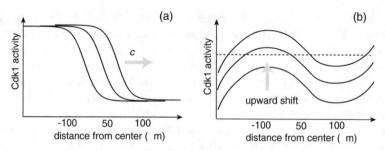

Fig. 11.46: Schematic illustration of two different mechanisms of Cdk1 waves during *Drosophila* mitosis. (a) Classical bistable wavefront in which a high activity stable steady state invades a region in a low activity metastable state. (b) Growth of a concentration gradient that preserves the spatial profile. Different spatial regions cross an activity threshold in an ordered sequence that appears as a spreading wave.

If $c(t)$ is fixed at some constant value $\bar{c}$ (adiabatic approximation), then the nonlinear function $f(a)$ exhibits bistability for a range of values of $\bar{c}$. The existence and stability of traveling fronts for such functions can then be analyzed using standard methods, as outlined in Box 11G. The non-autonomous case is less well understood. However, there has been a recent study of such a system in the case of mitotic waves in the early embryogenesis of *Drosophila* [904].

In Xenopus embryos, long-range spatial coordination mediated by a traveling wave only occurs during the first cell cycle. In later stages, dividing cells are smaller and are separated by membranes so that they act like independent oscillators. On the other hand, during the first 2 hours of development, the *Drosophila* embryo consists of a multinucleated cell with a shared cytoplasm (syncytium). Over this time interval, the syncytium undergoes 13 cycles of cell division, and each of these cycles is synchronized by waves of Cdk1 activity [231]. The waves propagate at speeds around 2-6 $\mu$m/s, with the speed slowing down as the maternal-to-zygotic transition is approached. In order to account for the slowdown of Cdk1 with cycle number, the mathematical model (11.6.36) has been modified to include another enzyme Chk1, which regulates the cell cycle by inhibiting Cdk1 activity via the modulation of the Wee1 and Cdc25 feedback loops [231, 904]. Meanwhile, Cdk1 abruptly switches off Chk1 activity at completion of the S phase of the cell cycle. The change of Cdk1 dynamics observed in the different cycles can be reproduced by simply assuming different initial levels of Chk1 activity in each cycle. One way to incorporate the effects of Chk1 is to modify the activation and inactivation rates $r_{\pm}(a)$ [904]:

$$r_+(a) = \left(r_0 + r_1 \frac{a^n}{K_+^n + a^n}\right)\left(1 - h_0 \frac{K^l}{K^l + a^l}\right), \qquad (11.6.39a)$$

$$r_-(a) = h_0 \left(r_2 + r_3 \frac{K_-^m}{K_-^m + a^m} \frac{K^l}{K^l + a^l}\right), \qquad (11.6.39b)$$

where $h_0$ specifies the initial level of Chk1 for a given cell cycle. Numerical studies of the extended model indicate an alternative way of understanding the wave-like behavior in wild type, based on the growth of a spatially varying concentration gradient [904]. Linear growth means that different regions cross a given concentration threshold at different times. This generates a spreading wave of activity that is faster than would be observed for a bistable wave. The difference between the two proposed modes of spatiotemporal Cdk1 activity is illustrated in Fig. 11.46. Note that bistable waves are observed in *Drosophila* mutants for which the feedback from active Cdk1 to Wee1 and Cdc25 is severed [904].

### 11.6.3 Reaction–diffusion model of CaMKII translocation waves

Recall from our discussion of synaptogenesis in *C. elegans*, Sect. 11.4, CaMKII ($Ca^{2+}$-calmodulin-dependent protein kinase II) is a key regulator of glutamatergic synapses and plays an essential role in many forms of synaptic plasticity. Experimental studies of hippocampal neurons have shown that chemically stimulating a local region of dendrite not only induces the local translocation of CaMKII from the dendritic shaft to synaptic targets within spines, but also initiates a wave of CaMKII translocation that spreads distally through the dendrite with an average speed of order $1\mu m/s$ [771]. In Fig. 11.47, we provide a cartoon of the mechanism for translocation waves hypothesized by Rose *et al.* [771]. Before local stimulation using a glutamate/glycine puff, the majority of CaMKII is in an inactive state and

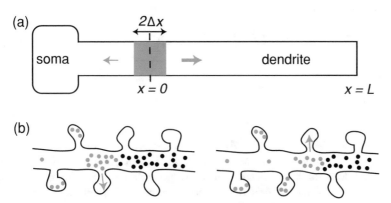

Fig. 11.47: Proposed mechanism of CaMKII translocation waves. (a) A glutamate/glycine puff activates CaMKII locally and initiates a fast $Ca^{2+}$ spike that propagates distally (indicated by larger horizontal arrow) and primes CaMKII in the remainder of the dendrite. In certain cases one also finds a second wave propagating proximally from the stimulated site to the soma (indicated by smaller horizontal arrow) (b) Activated CaMKII (gray dots) both translocates into spines and diffuses into distal regions of the dendrite where it activates primed CaMKII (black dots). The net effect is a wave of translocated CaMKII propagating along the dendrite.

distributed uniformly throughout the dendrite. Upon stimulation, all CaMKII in the region of the puff ($\sim 30\mu m$ of dendrite) is converted to an active state, probably the autonomous state of CaMKII (see Fig. 11.47(a)), and begins translocating into spines. Simultaneously, a $Ca^{2+}$ spike is initiated and rapidly travels the length of the dendrite, causing CaMKII to bind $Ca^{2+}$/CaM along the way. In this primed or partially phosphorylated state, CaMKII does not yet translocate into spines. In the meantime, a portion of the activated CaMKII from the stimulated region diffuses into the region of primed CaMKII and the two types interact, with the result that primed CaMKII is activated. Some of these newly activated holoenzymes translocate into spines while others diffuse into more distal regions of the dendrite containing primed CaMKII, and the wave proceeds in this fashion. In certain cases one also finds a second wave propagating proximally from the stimulated site to the soma [771]. A schematic diagram illustrating the progression of a translocation wave along a dendrite following the rapid priming phase is shown in Fig. 11.47(b).

A simple mathematical model of the above mechanism can be constructed using a system of RD equations for the concentrations of activated and primed CaMKII in the dendrite and spines [113, 114, 263]. (Diffusion in dendrites was also considered in Sect. 6.2.) These equations incorporate three major components of the dynamics: diffusion of CaMKII along the dendrite, activation of primed CaMKII, and translocation of activated CaMKII from the dendrite to spines. For simplicity, consider a uniform one-dimensional, nonbranching dendritic cable as shown in Fig. 11.47(a). Suppose that a region of width 30 $\mu$m is stimulated with a glutamate/glycine puff at time $t = 0$. The center of the stimulated region is taken to be at $x = 0$ and the distal end of the dendrite is at $x = L = 150\mu$m. The diffusion, activation, and translocation of CaMKII along the dendrite following stimulation is modeled according to the following system of equations:

$$\frac{\partial P}{\partial t} = D\frac{\partial^2 P}{\partial x^2} - k_0 AP, \tag{11.6.40a}$$

$$\frac{\partial A}{\partial t} = D\frac{\partial^2 A}{\partial x^2} + k_0 AP - hA, \tag{11.6.40b}$$

$$\frac{\partial S}{\partial t} = hA, \tag{11.6.40c}$$

where $D$ is the diffusivity of CaMKII within the cytosol. Here $P(x,t)$ and $A(x,t)$ denote the concentration of primed and activated CaMKII at time $t > 0$ and location $x$ along the dendrite. $S(x,t)$ denotes the corresponding concentration of CaMKII in the population of spines at the same time and distance. For simplicity, all parameters are constant in space and time. The reaction term $kAP$ represents the conversion of CaMKII from its primed to active state based on the irreversible first-order reaction scheme

$$A + P \rightarrow 2A,$$

with mass-action kinetics, where $k_0$ is the rate at which primed CaMKII is activated per unit concentration of activated CaMKII. The decay term $hA$ represents the loss of activated CaMKII from the dendrite due to translocation into a uniform distribution of spines at a rate $h$. The model assumes that translocation is irreversible over the time scale of simulations, which is reasonable given that activated CaMKII accumulation at synapses can persist for several minutes [817].

As a further simplification we will only consider the distal transport of CaMKII from the stimulated region by taking $0 \leq x \leq L$ in equation (11.6.40) and imposing closed or reflecting boundary conditions at the ends $x = 0, L$. Hence, no CaMKII can escape from the ends. In reality activated CaMKII could also diffuse in the proximal direction and trigger a second proximal translocation wave. However, the choice of boundary condition has little effect on the properties of the wave. Taking the distal half of the stimulated region to be $0 \leq x \leq 15\mu m$, consider the following initial conditions: $P(x,0) = 0$ and $A(x,0) = P_0$ for $0 \leq x \leq 15\mu m$, whereas $P(x,0) = P_0$ and $A(x,0) = 0$ for $x \geq 15\mu m$, where $P_0$ is the uniform resting concentration of CaMKII in the dendrite. Typical values of $C$ range from 0.1 to 30 $\mu M$ [771], covering two orders of magnitude. We also set $S(x,0) = 0$ everywhere. In other words, we assume that all the CaMKII is activated within the stimulated region at $t = 0$, but none has yet translocated into spines nor diffused into the nonstimulated region. We also neglect any delays associated with priming CaMKII along the dendrite. This is a reasonable approximation, since the Ca$^{2+}$ spike travels much faster than the CaMKII translocation wave [771]. Hence, by the time a significant amount of activated CaMKII has diffused into nonstimulated regions of the dendrite, any CaMKII encountered there will already be primed. The benefit of this assumption is that it eliminates the need to model the Ca$^{2+}$ spike.

Note that equation (11.6.40) is identical in form to the diffusive SI model introduced by Noble [686] to explain the spread of bubonic plague through Europe in the fourteenth century. In the latter model, $P(x,t)$ and $A(x,t)$ would represent the densities of susceptible and infective people at spatial location $x$ at time $t$, respectively, $k_0$ would be the transmission rate, and $h$ the death rate. In the absence of translocation into spines ($h = 0$), the total amount of CaMKII is conserved so that $A(x,t) + P(x,t) = P_0$ for all $x$ and $t \geq 0$. Equation (11.6.40) then reduces to the scalar Fisher–KPP equation

$$\frac{\partial A}{\partial t} = D \frac{\partial^2 A}{\partial x^2} + k_0 A (P_0 - A), \tag{11.6.41}$$

which was originally introduced to model the invasion of a gene into a population. The Fisher–KPP equation and its generalizations have been widely used to describe the spatial spread of invading species including plants, insects, genes, and diseases, see for example [655] and references therein. One characteristic feature of such equations is that they support traveling fronts propagating into an unstable steady state, in which the wave speed and long-time asymptotics are determined by the dynamics in the leading edge of the wave— so-called pulled fronts [617, 898]. In particular, a sufficiently localized initial perturbation (such as the stimulus used to generate CaMKII waves) will asymptotically approach the traveling front solution that has the minimum possible wave speed. (If we perform the change of variables $Q = P_0 - P$ in the CaMKII model, then the traveling wave solution constructed below propagates into the unstable state $A = 0, Q = 0$.) An overview of the theory of pulled fronts is presented in Box 11H and an explicit example is constructed in Ex. 11.11.

**Translocation waves for a uniform distribution of spines.** A traveling wave solution of (11.6.40) is $P(x,t) = P(\xi)$ and $A(x,t) = A(\xi)$, $\xi = x - ct$, where $c, c > 0$, is the wave speed, such that

$$P(\xi) \to P_0, \quad A(\xi) \to 0 \quad \text{as} \, \xi \to \infty,$$

and

$$P(\xi) \to P_1 < P_0, \quad A(\xi) \to 0 \quad \text{as} \, \xi \to -\infty.$$

Here $P_1$ is the residual concentration of primed CaMKII following translocation of activated CaMKII into spines. The minimum wave speed can be calculated by substituting the traveling wave solution into equation (11.6.40) and linearizing near the leading edge of the wave where $P \to P_0$ and $A \to 0$. In the traveling wave coordinate frame, (11.6.40a,b) are transformed to

$$-c\frac{dP}{d\xi} = D\frac{d^2P}{d\xi^2} - k_0 AP, \tag{11.6.42a}$$

$$-c\frac{dA}{d\xi} = D\frac{d^2A}{d\xi^2} + k_0 AP - hA. \tag{11.6.42b}$$

This is a system of two second-order ordinary differential equations in the variable $\xi$. A global view of the nature of traveling wave solutions can be obtained by identifying (11.6.42) with the equations of motion of a classical particle in two spatial dimensions undergoing damping due to "friction" and subject to an "external force". Thus we identify $A$ and $P$ with the "spatial" coordinates of the particle, $\xi$ with the corresponding "time" coordinate, and the speed $c$ as a "friction coefficient". If we ignore boundary effects by taking $-\infty < \xi < \infty$, then we can view a traveling wave solution as a particle trajectory that connects the point $(P,A) = (0,0)$ at $\xi = -\infty$ to the point $(P,A) = (P_0, 0)$ at $\xi = \infty$. A restriction on the allowed values of $c$ can now be obtained by investigating how the point $(1,0)$ is approached in the large-$\xi$ limit.

Linearizing equation (11.6.42) about the point $(P,A) = (P_0, 0)$ we obtain a pair of second-order linear equations, which have solutions of the form $(P - P_0, A) = \mathbf{V}e^{-\lambda\xi}$ where $\lambda$ and $\mathbf{V}$ satisfy the matrix equation

$$c\lambda \mathbf{V} = \begin{pmatrix} D\lambda^2 & -k \\ 0 & D\lambda^2 + k - h \end{pmatrix} \mathbf{V}, \tag{11.6.43}$$

where $k = k_0 P_0$. Solving for the eigenvalue $\lambda$ leads to the four solutions

$$\lambda = 0, \ \frac{c}{D}, \ \frac{c \pm \sqrt{c^2 - 4D(k-h)}}{2D}, \tag{11.6.44}$$

and these, along with their corresponding eigenvectors $\mathbf{V}$, determine the shape of the wave as it approaches the point $(1,0)$. Note that the last two eigenvalues have a

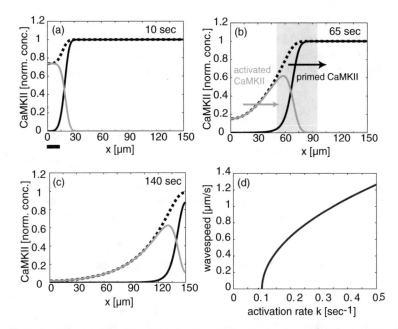

Fig. 11.48: (a-c) Three successive snapshots of a numerically simulated translocation wave propagating along a homogeneous dendrite. Solutions of (11.6.40) are plotted for parameter values consistent with experimental data on CaMKII [771, 817]. The translocation rate $h = 0.05/s$, diffusivity $D = 1\mu m^2/s$, and the activation rate $k_0 P_0 = 0.21/s$. At time $t = 0$ all of the CaMKII within the stimulated region (indicated by thick bar) is in the activated state, whereas all of the CaMKII within the nonstimulated region is in the primed state. Concentrations are normalized with respect to the initial concentration of primed CaMKII. Composite wave consists of a pulse of activated CaMKII (gray curve) moving at the same speed as a front of primed CaMKII (black curve). Also shown is the total CaMKII concentration along the dendrite (dashed black curve), which decreases with time due to translocation into spines. As indicated in the center plot, the front forms an interface between a quiescent region containing a uniform concentration of primed CaMKII and a region dominated by translocation of activated CaMKII into spines. The dynamics in the interfacial (shaded) region is dominated by diffusion–activation of primed CaMKII. (d) Plot of wave speed as a function of the activation rate $k$.

nonzero imaginary part when $c^2 < 4D(k-h)$, implying that as $\xi$ becomes large the wave oscillates about the point $(1,0)$. This cannot be allowed since it would imply that the activated CaMKII concentration $A$ takes on negative values (inspection of the corresponding eigenvectors show that their components in the $A$-direction are nonzero and so $A$ would indeed oscillate). Therefore, we must have

$$c \geq c_{\min} = 2\sqrt{D(k-h)}, \tag{11.6.45}$$

which implies that $k > h$. Note that the minimum wave speed can be identified with the linear spreading velocity of a pulled front, see appendix §3.3. This then yields a more direct method for obtaining the minimum wave speed. That is, the characteristic equation obtained from (11.6.43) yields the dispersion relation

$$c(\lambda) = D\lambda + \frac{k-h}{\lambda}. \tag{11.6.46}$$

The theory of pulled fronts shows that the minimum wave speed is obtained by minimizing $c(\lambda)$. The equation $c'(\lambda) = 0$ gives $D\lambda = (k-h)/\lambda$, which has the solution $\lambda_0 = \sqrt{(k-h)/D}$, so that $c_{min} = c(\lambda_0) = 2D\lambda_0 = 2\sqrt{D(k-h)}$. An example of a numerically determined traveling wave solution with minimal speed $c_{min}$ is shown in Fig. 11.48 for parameter values consistent with experimental studies of CaMKII$\alpha$, which is one of the two main isoforms of CaMKII. In its inactive state CaMKII$\alpha$ tends to be located in the cytosol, whereas the other isoform, CaMKII$\beta$, is weakly actin bound [817]. It can be seen in Fig. 11.48 that the wave profile of primed CaMKII is in the form of a front, whereas the co-moving wave profile of activated CaMKII is a localized pulse.

The above analysis predicts wave propagation failure when the translocation rate $h$ is greater than the effective activation rate $k$. Experimentally, $h$ is determined by globally activating CaMKII along a dendrite and determining the rate at which the level of CaMKII decays [817]. The detailed microscopic mechanism whereby CaMKII is translocated into spines is currently not known, so it is difficult to relate $h$ to individual spine properties. A simple hypothesis is that the translocation rate depends on the spine density according to $h = \rho_0 v_0$, where $v_0$ is an effective "velocity" associated with translocation into an individual spine. Since the activation rate $k = k_0 P_0$, where $P_0$ is the initial concentration of primed CaMKII in the nonstimulated region of the dendrite, the model predicts that CaMKII translocation waves will fail to propagate when $\rho_0 v_0 > k_0 P_0$. For example, this inequality predicts that dendrites with a high density of spines are less likely to exhibit translocation waves than those with a low spine density. It also predicts that dendrites with a larger initial concentration of primed CaMKII in the shaft are more likely to exhibit translocation waves than those with a smaller initial concentration. Since the initial concentration $P_0$ of primed CaMKII depends on the effectiveness of the $Ca^{2+}$ spike in both propagating along the dendrite and priming the inactive state, the model agrees with the experimental finding that translocation waves fail to propagate when L-type $Ca^{2+}$ channels are blocked [771]. One also finds that $Ca^{2+}$ spikes are less likely to propagate toward the soma, which could explain why translocation waves are more often observed propagating toward the distal end of a dendrite.

**Wave speed for a slowly modulated spine density.** It is found experimentally that there is a slow proximal to distal variation in the density of spines [41, 521]. An illustration of a typical spine density found in pyramidal neurons of mouse cortex [41] is shown in Fig. 11.49(a). Such a variation in spine density can be incorporated into equation (11.6.40) by setting $h = \bar{h} + \Delta h(\varepsilon x)$, where $\bar{h}$ denotes the translocation rate at the initiation point $x_0$ of the wave and $\Delta h(\varepsilon x)$ represents the slow modulation of the (homogenized) translocation rate over the length of a dendrite with $\varepsilon \ll 1$. The general problem of biological invasion in slowly modulated hetero-geneous environments can be analyzed using a Hamilton–Jacobi method for front velocity selection [616, 959]. This method was originally applied to homogeneous media by Freidlin using large deviation theory [320, 321, 338], see also Chap. 8, and was subsequently formulated in terms of PDEs by Evans and Sougandis [285]. We will illustrate the method by applying it to the reaction–diffusion model of CaMKII translocation waves with slow periodic modulation, see also [113]

The first step in the analysis is to rescale space and time in equation (11.6.40) according to $t \to t/\varepsilon$ and $x \to x/\varepsilon$, see [285, 321, 616]:

$$\varepsilon \frac{\partial P}{\partial t} = \varepsilon^2 D \frac{\partial^2 P}{\partial x^2} - k_0 A P, \tag{11.6.47a}$$

$$\varepsilon \frac{\partial A}{\partial t} = \varepsilon^2 D \frac{\partial^2 A}{\partial x^2} + k_0 A P - [\bar{h} + \Delta h(x)] A. \tag{11.6.47b}$$

Under the spatial rescaling the front region where $A$ $(P)$ rapidly increases (decreases) as $x$ decreases from infinity becomes a step as $\varepsilon \to 0$. This motivates the introduction of solutions of the form

$$P(x,t) \sim P_0 \left[ 1 - e^{-G_\varepsilon(x,t)/\varepsilon} \right], \quad A(x,t) \sim A_0(x) e^{-G_\varepsilon(x,t)/\varepsilon}, \tag{11.6.48}$$

with $G_\varepsilon(x,t) > 0$ for all $x > x(t)$ and $G_\varepsilon(x(t),t) = 0$. The point $x(t)$ determines the location of the front and $c = dx(t)/dt$. Substituting (11.6.48) into (11.6.47a) and (11.6.47b) gives

$$-\frac{\partial G_\varepsilon}{\partial t} = D \left[ \frac{\partial G_\varepsilon}{\partial x} \right]^2 - \varepsilon D \frac{\partial^2 G_\varepsilon}{\partial x^2} - k_0 A_0(x) \left[ 1 - e^{-G_\varepsilon(x,t)/\varepsilon} \right],$$

$$-A_0(x) \frac{\partial G_\varepsilon}{\partial t} = A_0(x) \left[ D \left[ \frac{\partial G_\varepsilon}{\partial x} \right]^2 - \varepsilon D \frac{\partial^2 G_\varepsilon}{\partial x^2} + k_0 P_0 \left[ 1 - e^{-G_\varepsilon(x,t)/\varepsilon} \right] - [\bar{h} + \Delta h(x)] \right]$$

$$+ \varepsilon^2 A_0''(x) G_\varepsilon - 2\varepsilon A_0'(x) \frac{\partial G_\varepsilon}{\partial x}.$$

Since $e^{-G_\varepsilon(x,t)/\varepsilon} \to 0$ as $\varepsilon \to 0$ for $G_\varepsilon > 0$, it follows that the limiting function $G(x,t) = \lim_{\varepsilon \to 0} G_\varepsilon(x,t)$ satisfies

$$-\frac{\partial G}{\partial t} = D \left[ \frac{\partial G}{\partial x} \right]^2 - k_0 A_0(x), \tag{11.6.49a}$$

$$-\frac{\partial G}{\partial t} = D \left[ \frac{\partial G}{\partial x} \right]^2 + k - [\bar{h} + \Delta h(x)], \tag{11.6.49b}$$

where $k = k_0 P_0$ as before. It immediately follows that

$$A_0(x) = \left[ \frac{k - \bar{h} - \Delta h(x)}{k} \right] P_0.$$  (11.6.50)

The remaining equation (11.6.49b) can be analyzed along identical lines to a previous study of the heterogeneous Fisher–KPP equation [616]. Formally speaking, comparing, equation (11.6.49b) can be rewritten as the Hamilton–Jacobi equation (see also Chap. 8)

$$\partial_t G + H(\partial_x G, x) = 0, \quad H = Dp^2 + k - [\bar{h} + \Delta h(x)],$$  (11.6.51)

where $p = \partial_x G$ is interpreted as the conjugate momentum of $x$ and $H$ is the Hamiltonian. It now follows that (11.6.49b) can be solved in terms of the Hamilton equations

$$\frac{dq}{ds} = 2Dp, \quad \frac{dp}{ds} = \frac{d\Delta h}{dq}.$$  (11.6.52)

Combining these equations yields the second-order ODE

$$\frac{d^2q}{ds^2} - 2D\Delta h(q)' = 0.$$  (11.6.53)

This takes the form of a Newtonian particle moving in a "potential" $V(q) = -2D\Delta h(q)$. Given the solution $(q(s), p(s))$, the function $G(x,t)$ is the corresponding action of the classical trajectory from $q(0) = x_0$ to $q(t) = x$, that is,

$$G(x,t) = \int_0^t \left( p(s) \frac{dq}{ds} - H(p(s), q(s)) \right) ds = \frac{1}{2D} \int_0^t \left( \frac{dq(s)}{ds} \right)^2 ds - E(x,t)t,$$  (11.6.54)

since $H$ is independent of $s$.

For certain choices of the modulation function $\Delta h(x)$, (11.6.53) can be solved explicitly [616]. In particular, suppose that the spine density curve in Fig. 11.49 is approximated by a piecewise linear function, in which the density increases linearly with distance from the soma to some intermediate location $\kappa$ along the dendrite and then decreases linearly toward the distal end. Assuming that the right-moving wave is initiated beyond the point $\kappa$, $x_0 > \kappa$, then we can simply take $\Delta h(x) = -\beta(x - x_0)$ for $\beta > 0$. Substituting into (11.6.53) and integrating twice with respect to $s$ give

$$q(s) = -D\beta s^2 + As + B.$$

The integration constants are determined by imposing the Cauchy conditions $q(0) = x_0$ and $q(t) = x$. Hence,

$$q(s) = x_0 + (x - x_0)s/t + D\beta ts - D\beta s^2.$$  (11.6.55)

The corresponding "energy" function is

$$H = \frac{1}{4D}q'(s) + k - \bar{h} + \beta(q(s) - x_0)$$

$$= \frac{(x - x_0)^2}{4Dt^2} + k - \bar{h} + \frac{\beta}{2}(x - x_0) + \frac{\beta^2}{4}Dt^2 = E(x,t).$$

Equation (11.6.54) then shows that

$$G(x,t) = \frac{(x - x_0)^2}{4Dt} - [k - \bar{h}]t - \frac{\beta}{2}(x - x_0)t - \frac{\beta^2}{12}Dt^3. \qquad (11.6.56)$$

We can now determine the wave speed $c$ by imposing the condition $G(x(t),t) = 0$. This leads to a quadratic equation with positive solution

$$x(t) = x_0 + D\beta t^2 + 2Dt\sqrt{\frac{k - \bar{h}}{D} + \frac{\beta^2}{3}t^2} = x_0 + \bar{c}t\sqrt{1 + \frac{4\beta^2 D^2 t^2}{3\bar{c}^2}} + D\beta t^2,$$

with $\bar{c} = 2\sqrt{D(k - \bar{h})}$. Finally, differentiating both sides with respect to $t$ yields

$$c \equiv \frac{dx(t)}{dt} = \bar{c}\sqrt{1 + \Gamma_0\beta^2 t^2} + \frac{\bar{c}\Gamma_0\beta^2 t^2}{\sqrt{1 + \Gamma_0\beta^2 t^2}} + 2D\beta t, \qquad (11.6.57)$$

where $\Gamma_0 = 4D^2/(3\bar{c}^2)$. For sufficiently small times such that $D\beta t \ll 1$, we have the approximation

$$c \approx \bar{c} + 2D\beta t + \frac{2(D\beta t)^2}{\bar{c}}. \qquad (11.6.58)$$

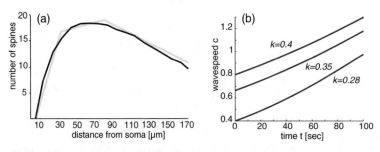

Fig. 11.49: (a) Illustrative example of the spine density variation along the basal dendrite of a pyramidal cell in mouse cortex (black curve). Density is calculated as the number of spines per $10\mu m$ segment of the dendrite from the soma to the tip of the dendrite. Abstracted from experimental data in [41]. Also shown is a simplified piecewise linear approximation of the spine density variation (gray curve). (b) Plot of time-dependent variation in wave speed $c$ given by (11.6.57) for various values of the activation rate $k$. Other parameters are $\bar{h} = 0.24s^{-1}$ and $D = 1\mu m^2/s$. At $t = 0$, $c(0) = 2\sqrt{D(k - \bar{h})}$.

Figure 11.49(b) shows example plots of the time-dependent wave speed for various choices of the activation rate $k$. It can be seen that there are significant changes in speed over a time course of 100 secs, which is comparable to the time a wave would travel along a dendrite of a few hundred microns.

---

### Box 11H. Pulled and pushed fronts

---

For a detailed account of unstable waves see the review by van Saarloos [898] and Ch. 4 of [617]. Here we highlight some of the basic result by considering a slight generalization of the Fisher–KPP equation

$$\frac{\partial u}{\partial t} = \frac{\partial^2 u}{\partial x^2} + f(u), \quad f \in C^1[0,1], \quad f(0) = f(1) = 0, \qquad (11.6.59)$$

for which the homogeneous fixed point $u = 0$ is unstable ($f'(0) > 0$) and $u = 1$ is stable ($f'(1) < 0$). We also assume that $f(u) > 0$ for all $u \in (0,1)$. We are interested in determining the long-time asymptotics of a front propagating to the right into the unstable state $u = 0$, given initial conditions for which $u(x,0) = 0$ for sufficiently large $x$. It is not possible to carry out an asymptotic analysis by simply moving to a traveling coordinate frame, since there is a continuous family of front solutions. However, considerable insight into the evolution of a localized initial condition can be obtained by linearizing about the unstable state.

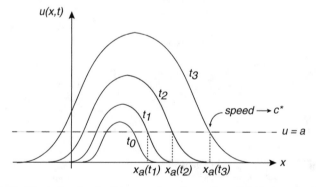

Fig. 11.50: Illustrative sketch of the growth and spreading of a solution $u(x,t)$ of the Fisher equation linearized about the unstable state $u = 0$, given a localized initial condition $u(x,t_0)$.

**The linear spreading velocity.** Linearizing (11.6.59) about $u = 0$,

$$\frac{\partial u}{\partial t} = \frac{\partial^2 u}{\partial x^2} + f'(0)u. \qquad (11.6.60)$$

Substitution of the Fourier mode $e^{-i\omega t + ikx}$ gives the dispersion relation

$$\omega(k) = i(f'(0) - k^2). \tag{11.6.61}$$

The state $u = 0$ is then said to be linearly unstable if $\text{Im}[\omega(k)] > 0$ for some range of $k$-values. In the case of the Fisher–KPP equation, after writing $k = k_r + ik_i$, this will occur when $f'(0) + k_i^2 > k_r^2$. Consider some generic initial condition $u(x,0)$ that is sufficiently localized in space (to be made precise later). Since there exists a range of unstable linear eigenmodes, we expect the localized initial condition to grow and spread out within the linear regime, as illustrated in Fig. 11.50. Tracking the evolution of a level curve $x_a(t)$ with $u(x_a(t),t) = a$, the linear spreading velocity $c^*$ is defined to be the asymptotic speed of the point $x_a(t)$ in the rightward moving edge (assuming it exists):

$$c^* = \lim_{t \to \infty} \frac{dx_a(t)}{dt}. \tag{11.6.62}$$

The linearity of the underlying evolution equation (11.6.60) means that $c^*$ is independent of the value $a$. (Note that for an isotropic medium, the leftward moving edge moves with the same asymptotic speed but in the opposite direction.) Suppose that $c^*$ is finite. If we were to move in the traveling coordinate frame $\xi = x - c^*t$, then the leading rightward edge would neither grow nor decay exponentially. Imposing this condition on the Fourier expansion of the solution $u(x,t)$ then determines $c^*$ in terms of the dispersion curve $\omega(k)$. More specifically, denoting the Fourier transform of the initial condition $u_0(x) = u(x,0)$ by $\tilde{u}_0(k)$, we have

$$u(x,t) = \int_{-\infty}^{\infty} \tilde{u}_0(k) e^{i[kx - \omega(k)t]} \frac{dk}{2\pi} = \int_{-\infty}^{\infty} \tilde{u}_0(k) e^{ik\xi} e^{-i[\omega(k) - c^*k]t} \frac{dk}{2\pi}$$

$$= \int_{-\infty}^{\infty} \tilde{u}_0(k) e^{ik\xi} e^{\psi(k)t} \frac{dk}{2\pi}, \tag{11.6.63}$$

where

$$\psi(k) = -i[\omega(k) - c^*k] = \omega_i(k) - c^*k_i - i[\omega_r(k) - c^*k_r]. \tag{11.6.64}$$

In the limit $t \to \infty$ with $\xi$ finite, we can approximate this integral using steepest descents. For the moment, we assume that $\tilde{u}(k)$ is an entire function (analytic in every finite region of the complex plane) so that we can deform the contour in the complex $k$-plane, that is, $(-\infty, \infty) \to C$, with $C$ linking points at infinity in the complex plane where $\text{Re}(\psi) < 0$.

**Method of steepest descents.** We briefly describe the method of steepest descents for a general analytic function $\psi(k)$, see Fig. 11.51. First, one would like to choose a contour $C$ so that the maximum of $\psi_r \equiv \text{Re}(\psi)$ along the contour at $k_0$, say, is as large as possible, since this point will dominate the integral. Recall, however, that one of the Cauchy–Riemann conditions on an analytic function is that $\nabla^2(\text{Re}(\psi)) = 0$, which means that $\text{Re}(\psi)$ cannot have any maxima or minima (except at singularities or branch points where $\psi$ would be non-analytic). Therefore $\nabla(\text{Re}(\psi)) = 0$ only at saddle points. Second, for a general integration contour, evaluating the integral in a neighborhood of the point $k_0$ will overestimate the value of the integral, since it doesn't take into account cancelations due to the rapidly oscillating function $e^{i\text{Im}(\psi(k))t}$. The latter issue can be eliminated by choosing the contour that is the path of steepest ascent to a saddle point and the steepest descent away from the saddle. (If there exists more than one saddle then one chooses the "highest" one.)

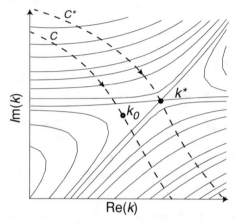

Fig. 11.51: Method of steepest descents. Sketch of $\text{Re}(\psi)$ contours in the complex $k$-plane for an analytic function in the region of a saddle point at $k^*$. The integration contour $C$ is deformed so that it passes through the saddle point.

By construction, the path is orthogonal to the contours of $\text{Re}(\psi)$. Hence, from the Cauchy–Riemann condition $\nabla(\text{Im}(\psi)) \cdot \nabla(\text{Re}(\psi)) = 0$ it follows that $\psi_i \equiv \text{Im}(\psi)$ is constant along the contour. In other words, there are no fast oscillations along the path of steepest ascent and descent and one can obtain a good estimate of the integral by Taylor expanding about the saddle point. Thus, taking $\psi_i(k) = \psi_i(k^*)$ along $C^*$, we have

$$I \equiv \int_{-\infty}^{\infty} \tilde{u}_0(k) e^{ik\xi} e^{\psi(k)t} \frac{dk}{2\pi} = \int_{C^*} \tilde{u}_0(k) e^{ik\xi} e^{\psi(k)t} \frac{dk}{2\pi}$$

$$\approx \tilde{u}_0(k^*) e^{i\psi_i(k^*)t} \int_{C^*} e^{ik\xi} e^{\psi_r(k)t} \frac{dk}{2\pi}.$$

Finally, we Taylor expand $\psi_r(k)$ to second order in $\Delta k = k - k^*$, noting that $\psi'(k^*) = 0$ and $\psi_r''(k^*) < 0$ at the saddle,

$$\psi_r(k) = \psi_r(k^*) + \frac{(\Delta k)^2}{2} \psi_r''(k^*),$$

and approximate the remaining contour integral by a Gaussian. This then gives

$$I \approx \frac{1}{\sqrt{4\pi Dt}} \tilde{u}_0(k^*) e^{i[k^*\xi + \psi_i(k^*)t]} e^{-\xi^2/4Dt} e^{\psi_r(k^*)t}, \qquad (11.6.65)$$

where $D = -\psi_r''(k^*)/2$.

**Calculation of $c^*$.** Let us now apply steepest descent to the integral (11.6.63) for $\psi(k)$ given by (11.6.64) such that $\psi_r(k) = \omega_i(k) - c^* k_i$ and $\psi_i(k) = -[\omega_r(k) - c^* k_r]$. At the (unique) saddle point $k^*$ at which $\psi'(k^*) = 0$, we have

$$c^* = \left. \frac{d\omega(k)}{dk} \right|_{k=k^*}. \qquad (11.6.66)$$

Moreover, equation (11.6.65) becomes

$$I \approx \frac{1}{\sqrt{4\pi Dt}} \tilde{u}_0(k^*) e^{i[k^*\xi - [\omega_r(k^*) - c^* k_r^*]t]} e^{-\xi^2/4Dt} e^{\psi_r(k^*)t}, \qquad (11.6.67)$$

where $D = -\frac{1}{2}\omega_i''(k^*)$. Finally, we can determine the linear spreading velocity $c^*$ by requiring that the asymptotic solution neither grows or decays with time, $\psi_r(k^*) = 0$, which implies

$$c^* = \frac{\omega_i(k^*)}{k_i^*}. \qquad (11.6.68)$$

Note that equating real and imaginary parts in (11.6.66) and combining with (11.6.68) mean that we have three equations in the three unknowns $c^*, k_r^*, k_i^*$. In the particular case of the Fisher–KPP equation (11.6.59),

$$k_r^* = 0, \quad k_i^* = \sqrt{f'(0)}, \quad c^* = 2\sqrt{f'(0)}, \quad D = 1. \qquad (11.6.69)$$

In the above analysis, it was assumed that the Fourier transform of the initial condition $u_0(x)$ was an entire function. This would apply to cases for which $u_0(x)$ is a Dirac delta function, has compact support, or decays faster than any exponential for large enough $x$ (e.g., a Gaussian). Now suppose that $u_0(x)$ falls off exponentially for large $x$, $u_0(x) \sim e^{-\lambda x}$ for some $\lambda$. Then $\tilde{u}_0(k)$ has a pole in the upper-half complex plane at $k = k'$ with $\text{Im}(k') = \lambda$. It follows that when deforming the contour $C$ in the complex $k$-plane in order to perform steepest descents, we pick up a contribution from the pole. Taking this into account, it can be shown that, within the linear regime, initial conditions whose exponential decay rate $\lambda > \lambda^*$ lead to profiles that asymptotically spread with the linear spreading velocity $c^*$. On the other hand, if $\lambda < \lambda^*$ then the profile advances at a speed faster than $c^*$ [898].

**Initial conditions.** So far we have investigated the evolution of a localized initial condition in the linear regime. It still remains to determine whether or not there are classes of initial conditions under which the full nonlinear system converges to a unique asymptotic front solution and how the speed of the front $c$ is related to the linear spreading velocity $c^*$. It turns out that for front propagation into a linearly unstable state, there are only two possibilities when starting from sufficiently steep initial conditions, that is, initial conditions that fall off faster than $e^{-\lambda^* x}$ [898]:

*Pulled front:* $c = c^*$ so that the front dynamics is determined by the behavior in the leading edge of the front where $u(x,t) \approx 0$, that is, the front is pulled along by the linear spreading of small perturbations into the linearly unstable state.

*Pushed front:* $c > c^*$ so that nonlinearities play an important role in determining the velocity of the front, pushing it into the unstable state.

In the special case of initial conditions with compact support, it can be proven that the solution evolves into a front that propagates at the minimal possible wave speed $c_{\min}$, which is bounded above and below [30]:

$$c^* = 2\sqrt{f'(0)} \le c_{\min} < 2\sqrt{\sup_u \left[\frac{f(u)}{u}\right]}. \qquad (11.6.70)$$

For any concave function, $f(u) \le uf'(0)$, the lower and upper bounds coincide and we have a pulled front; this applies to the standard Fisher–KPP equation where $f(u) = u(1-u)$. On the other, the upper and lower bounds do not coincide for concave $f(u)$. The minimal front velocity can then be larger than the linear velocity indicative of a pushed front. An example of the latter is the Ginzburg–Landau term $f(u) = u(1-u)(1+\alpha u)$ with $\alpha > 0$. One finds that for compact initial conditions, a pulled front is selected when $\alpha \le 2$, whereas a pushed front is selected when $\alpha > 2$.

## 11.7 Exercises

**Problem 11.1. Diffusion-based concentration gradient.** Consider the following system of RD equations [541, 576],

$$\frac{\partial C_1(x,t)}{\partial t} = D\frac{\partial^2 C_1}{\partial x^2} + \alpha C_0 - \beta C_1(1-C_0),$$

$$\frac{\partial C_0(x,t)}{\partial t} = -kC_0(x,t) - \alpha C_0 + \beta C_1(1-C_0)$$

for $0 < x < 1$. Here $C_0(x,t)$ and $C_1(x,t)$ denote the normalized concentrations of immobile and mobile ligand, $\alpha, \beta$ are the transitions rates for switching between the mobile and immobile states, and $k$ is the degradation rate in the immobile state.

We allow for the possibility that the binding substrate can become saturated. Units are chosen so that $D = 1$. The corresponding boundary conditions are

$$\frac{\partial C_0}{\partial t} = Q + \alpha C_0 - \beta C_1 (1 - C_0), x = 0; \quad C_0 = 0, \quad x = 1.$$

(a) Derive the following steady-state equation for $C_1$:

$$\frac{d^2 C_1}{dx^2} = \frac{kC_1}{C_1 + \Gamma}, \quad C_1(0) = \frac{Q\Gamma}{k - Q}, \quad C_1(1) = 0,$$

with

$$C_0(x) = \frac{C_1(x)}{C_1(x) + \Gamma}, \quad \Gamma = \frac{\alpha + k}{\beta}.$$

(b) Suppose that $Q \ll k$ and set $\varepsilon = Q/k$, $\psi = k/\Gamma$. Under the rescaling $C_1 = \varepsilon \Gamma C$, show that $C$ satisfies the approximate equation,

$$\frac{d^2 C}{dx^2} = \frac{\psi C}{1 + \varepsilon C}, \quad C(0) = 1, \quad C(1) = 0.$$

Carrying out a perturbation expansion in $\beta$, obtain the approximate solution

$$C_1(x) \sim \varepsilon \Gamma \frac{\sinh(\sqrt{\psi}(1 - x))}{\sinh \sqrt{\psi}} + O(\varepsilon^2).$$

Describe what happens to the concentration gradient as $\psi$ becomes large.

(c) Now suppose that $Q/k \approx 1$ and set $\varepsilon = (k - Q)/Q$, $\psi = k/\Gamma$. Under the rescaling $C_1 = \Gamma C/\varepsilon$, show that $C$ then satisfies the equation,

$$\frac{d^2 C}{dx^2} = \frac{\psi C}{1 + C/\varepsilon}, \quad C(0) = 1, \quad C(1) = 0.$$

Hence obtain the approximate solution

$$C_1(x) \sim \frac{\Gamma}{\varepsilon}(1 - x) + O(1).$$

(d) Assuming the existence of a steady-state solution $(C_0(x), C_1(x))$, linearize the RD equations by setting

$$C_n(x,t) = C_n(x) + e^{-\lambda t} c_n(x), \quad n = 0, 1.$$

Solving for $c_0(x)$ in terms of $c_1(x)$, obtain the linear equation

$$\frac{d^2 c_1(x)}{dx^2} + [\lambda - q(x, \lambda)] c_1(x) = 0,$$

where

$$q(x,\lambda) = \frac{\beta(\alpha+k)}{\beta C_1(x) + \alpha + k} \frac{\lambda - k}{\lambda - k - \alpha - \beta C_1(x)}.$$

Similarly, linearizing the boundary conditions, obtain the following boundary conditions for $c_1(x)$:

$$K(\lambda)c_1(0) = 0, \quad c_1(1) = 0,$$

where

$$K(\lambda) = \lambda - \frac{\beta(1 - Q/k)(\lambda - k)}{\lambda - [\beta C_1(0) + \alpha + k]}.$$

(e) When $K(\lambda) = 0$, there is only the single boundary condition $c_1(1) = 0$, and a nontrivial solution for $c_1(x)$ can always be constructed. Show that the equation $K(\lambda) = 0$ has two real, positive eigenvalues $\lambda_{1,2}$. The remainder of the spectrum is obtained by solving the following homogeneous Dirichlet problem:

$$\frac{d^2 c_1(x)}{dx^2} + [\lambda - q(x,\lambda)]c_1(x) = 0, \quad c_1(0) = c_1(1) = 0.$$

It can be shown that the latter eigenvalue problem also has positive eigenvalues [576], so that the steady-state solution is stable.

**Problem 11.2. Robustness of a dual protein gradient.** The concentration of Pom1p in the membrane of fission yeast evolves according to the equation

$$\frac{\partial c}{\partial t} = D\frac{\partial^2 c}{\partial x^2} - \mu c,$$

where $\mu$ is the rate of membrane dissociation, supplemented by the boundary conditions

$$D\frac{\partial c}{\partial x}\bigg|_{x=\pm L/2} = \pm J.$$

(a) Show that the steady-state solution is

$$c(x) = \frac{J}{DL_\perp} \frac{\lambda \cosh(x/\lambda)}{\sinh(L/2\lambda)},$$

where $\lambda = \sqrt{D/\mu}$ is the characteristic decay length of the spatial gradient, and $L_\perp$ is the length of the circumference.

(b) The uncertainty $\Delta x$ in the position of the center is given by

$$\Delta c = \frac{1}{2}|c''(0)|\Delta x^2.$$

Using part (a), show that

$$\Delta x = \left(\frac{4D\lambda^3 \sinh(L/2\lambda)}{Ja^2}\right)^{1/4}.$$

**Problem 11.3. Activator–inhibitor system.** Consider the activator–inhibitor system

$$\frac{\partial u}{\partial t} = \frac{\partial^2 u}{\partial x^2} + \frac{u^2}{v} - bu,$$

$$\frac{\partial v}{\partial t} = D\frac{\partial^2 v}{\partial x^2} + u^2 - v,$$

where $b, D$ are positive constants.

(a) Determine the positive steady states in the absence of diffusion and investigate their linear stability. Hence show that a necessary condition for a Turing instability is $0 < b < 1$.

(b) Determine the conditions for a Turing instability and show that the parameter region for a diffusion instability is given by

$$0 < b < 1, \quad bD > 3 + 2\sqrt{2}.$$

Sketch the corresponding region in $(b, D)$ parameter space. Also show that the critical wavenumber $k_c$ for a Turing instability is given by

$$k_c^2 = \frac{1}{2D}\left(bD - 1 + \sqrt{(bD - 1)^2 - 4Db}\right).$$

**Problem 11.4. Pattern formation on a ring.** Fill in the details of the weakly nonlinear analysis of the following RD system defined on a ring, see equations (11.2.18):

$$\frac{\partial u_1}{\partial t} = D\frac{\partial^2 u_1}{\partial \theta^2} + \kappa(u_1 + au_2 - u_1 u_2^2 - Cu_1 u_2),$$

$$\frac{\partial u_2}{\partial t} = \frac{\partial^2 u_2}{\partial \theta^2} + \kappa(-u_1 + bu_2 + u_1 u_2^2 + Cu_1 u_2),$$

with $\theta \in [0, 2\pi]$ and periodic concentrations $u_i(\theta + 2m\pi, t) = u_i(\theta, t)$ for all integers $m$.

(a) Substitute the perturbation expansion

$$u_i = \varepsilon^{1/2} u_i^{(1)} + \varepsilon u_i^{(2)} + \varepsilon^{3/2} u_i^{(3)} + \dots$$

into the RD equations and separately collect terms in powers of $\varepsilon^{1/2}, \varepsilon$, and $\varepsilon^{3/2}$, respectively. Show that this yields the linear inhomogeneous equations

$$\mathbb{L}\mathbf{u}^{(1)} = 0, \quad \mathbb{L}\mathbf{u}^{(2)} = \mathbf{h}^{(2)}, \quad \mathbb{L}\mathbf{u}^{(3)} = \mathbf{h}^{(3)},$$

where

$$\mathbb{L}\mathbf{u} = \begin{pmatrix} \kappa & \kappa a_c \\ -\kappa & b\kappa \end{pmatrix}\begin{pmatrix} u_1 \\ u_2 \end{pmatrix} + \begin{pmatrix} D & 0 \\ 0 & 1 \end{pmatrix}\begin{pmatrix} \partial^2 u_1/\partial \theta^2 \\ \partial^2 u_2/\partial \theta^2 \end{pmatrix},$$

and derive the expressions for the vectors $\mathbf{h}^{(2)}$ and $\mathbf{h}^{(3)}$ as functions of $\mathbf{u}^{(1)}$ and $\mathbf{u}^{(2)}$.

(b) Using the identity

$$\int_0^{2\pi} e^{im\theta} e^{im'\theta} \frac{d\theta}{2\pi} = \delta_{m+m',0},$$

show that $\langle e^{in\theta} | u_1^{(1)} u_2^{(1)} \rangle = 0$, and hence that the solvability condition for $\mathbf{h}^{(2)}$ is automatically satisfied.

(c) Show that $\mathbf{u}^{(2)}$ has the general form

$$\mathbf{u}^{(2)}(\theta) = \mathbf{c}^+ e^{2ni\theta} + \mathbf{c}^- e^{-2ni\theta} + \mathbf{c}^0 + \zeta \mathbf{u}^{(1)}(\theta).$$

Substituting for $\mathbf{u}^{(2)}$ into the equation $\mathbb{L}\mathbf{u}^{(2)} = \mathbf{h}^{(2)}$, derive the equations

$$\mathbf{A}(2n)\mathbf{c}^+ = \kappa C c_1 c_2 z^2 \begin{pmatrix} 1 \\ -1 \end{pmatrix}, \quad \mathbf{A}(2n)\mathbf{c}^- = \kappa C c_1 c_2 z^{*2} \begin{pmatrix} 1 \\ -1 \end{pmatrix},$$

$$\mathbf{A}(0)\mathbf{c}^0 = 2\kappa C c_1 c_2 |z|^2 \begin{pmatrix} 1 \\ -1 \end{pmatrix}.$$

**Problem 11.5. A little group theory.** (a) Show that the set of rotation and reflection matrices

$$\mathbf{M}(\theta) = \begin{pmatrix} \cos\theta & -\sin\theta \\ \sin\theta & \cos\theta \end{pmatrix}, \quad \theta \in [0, 2\pi), \quad \mathbf{M}_\kappa = \begin{pmatrix} 1 & 0 \\ 0 & -1 \end{pmatrix},$$

together with the rules of matrix multiplication, form the representation of a group acting on the linear vector space $\mathbb{R}^2$. The abstract group is $O(2)$. Setting $z = x + iy$, show that rotation by $\theta$ corresponds to the transformation $z \to e^{i\theta} z$ and reflection becomes $z \to z^*$.

(b) Suppose that there exists a matrix representation of some group. Show that a corresponding one-dimensional representation can be obtained by taking the determinants of the matrices. Calculate the representation explicitly for the dihedral group $D_3$.

(c) By considering a square centered at the origin of the $(x, y)$-plane construct the matrix representation of the dihedral group $D_4$, which consists of operations that transform the square into itself. Is the group representation irreducible? Calculate the axial isotropy subgroups of the given group representation.

**Problem 11.6. Advection–diffusion model with a reactive membrane.** Consider the following RD model defined on a semi-infinite strip, see Fig. 11.52:

$$\frac{\partial c(x,t)}{\partial t} = D_m \frac{\partial^2 c(x,t)}{\partial x^2} + k_{on} u(x,0,t) - k_{off} c(x,t),$$

$$\frac{\partial u(x,z,t)}{\partial t} = D\nabla^2 u(x,z,t) - \mathbf{v} \cdot \nabla u(x,z,t),$$

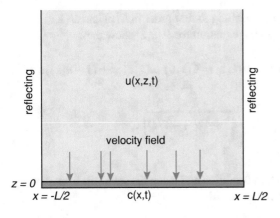

Fig. 11.52: Advection–diffusion model with a reactive membrane at $z = 0$ and reflecting boundaries at $x = \pm L/2$. The concentration in the bulk, $u(x,c,t)$ evolves according to an advection diffusion equation with a vertical velocity field $\mathbf{v} = -\alpha c(c,t)\mathbf{e}_z$, where $c(x,t)$ is the concentration in the membrane.

with $-L/2 \leq x \leq L/2$ and $0 \leq z < \infty$. Here $u(x,z,t)$ is the concentration of a signaling molecule in the bulk domain and $c(x,t)$ is the corresponding concentration in the reactive membrane. One interpretation of the advection term is that cytoplasmic transport is actively driven by molecular motors along microtubules that introduces a directional bias, see Chap. 7 and Ref. [399]. Suppose that the microtubules are aligned vertically and their density is determined by the concentration $c(x,t)$ in the reactive membrane. The velocity field $\mathbf{v}$ is then taken to be of the form

$$\mathbf{v} = -\alpha c(x,t)\mathbf{e}_z,$$

where $\alpha$ is a constant. The above equations are supplemented by reflecting boundary conditions at $x = \pm L/2$, a flux continuity equation on the membrane boundary

$$D\frac{\partial u(x,0,t)}{\partial z} + \alpha c(x,t)u(x,0,t) = k_{\text{on}}u(x,0,t) - k_{\text{off}}c(x,t) = 0,$$

and the conservation condition

$$M = \int_{-L/2}^{L/2} c(x,t)dx + \int_{-L/2}^{L/2}\int_0^\infty u(x,z,t)dz,$$

with $M$ the total number of signaling molecules.

(a) Obtain the steady-state solution

$$c_0 = \frac{k_{\text{on}}}{k_{\text{off}}}u_0(0), \quad u_0(z) = u_0(0)e^{-\xi z}, \quad \xi = \alpha c_0/D,$$

and determine $c_0$.

(b) Linearize about the steady-state solution by substituting

$$c(x,t) = c_0 + C(k)e^{ikx+\lambda t}, \quad u(x,z,t) = u_0(z) + U(k,z)e^{ikx+\lambda t}$$

into the RD equations and Taylor expanding to first order in $C(k)$ and $U(k,z)$. After
solving the resulting linear differential equation for $U(k,z)$ show that

$$C(k) = \frac{k_{\text{on}}}{(\lambda + D_m k^2 + k_{\text{off}})} C(k,0), \quad U(k,z) = C(k,0)\left[a(k)e^{-\xi z} + (1-a(k))e^{-\rho z}\right],$$

where

$$\rho = \frac{1}{2}\left[\xi \pm \sqrt{\xi^2 + 4k^2 + 4\lambda/D}\right].$$

and determine the coefficient $a(k)$.

(c) Linearizing the flux conservation condition and using the solutions from part (b),
show that $\lambda(k)$ satisfies the dispersion equation

$$(\lambda + Dk^2)\left(k_{\text{off}}(2\xi - \rho) + (\lambda + D_m k^2)(-k_{\text{on}}/D + \xi - \rho)\right) + (\xi - \rho)\xi^2 D k_{\text{off}} = 0.$$

**Problem 11.7. Positive feedback model of cell polarization.** Let $u(\mathbf{x},t)$ denote the
concentration of some signaling molecule in the cell membrane, which is treated as
a 2D bounded domain $\Sigma$. The density evolves according to the RD equation

$$\frac{\partial u}{\partial t} = D\nabla^2 u + K_+ n_c u - K_- u, \ \mathbf{x} \in \Sigma; \quad \mathbf{n} \cdot \nabla u = 0, \ \mathbf{x} \in \partial\Sigma.$$

Here $n_c$ is the number of molecules in the cytoplasm (which are assumed to be uniformly distributed), $D$ is the diffusivity, and $K_\pm$ represent the rates of binding/un-
binding of cytosolic molecules. Binding is taken to involve positive feedback due to
self-recruitment of membrane-bound molecules. Finally, we have the conservation
condition

$$n_c + |\Sigma|\bar{u}(t) = M, \quad \bar{u}(t) = \frac{1}{|\Sigma|}\int_\Sigma u(\mathbf{x},t)d\mathbf{x},$$

where $M$ is the total number of molecules in the cell. We wish to show that such
a system cannot support any patterns. (A stochastic version of the model will be
considered in Ex. 16.2 where it will be shown that spatially localized patterns may
persist for long times.)
(a) Define the mean-square deviation

$$E(t) = \frac{1}{|\Sigma|}\int_\Sigma v^2(\mathbf{x},t)d\mathbf{x}, \quad v(\mathbf{x},t) = \frac{u(\mathbf{x},t) - \bar{u}(t)}{\bar{u}(t)}.$$

Derive the inequality

$$\frac{d}{dt}\int_\Sigma v^2 d\mathbf{x} \le -2D\int_\Sigma |\nabla v|^2 d\mathbf{x}.$$

(b) Applying Poincare's inequality, equation (11.2.11) of Box 11B, and noting that
$E(t) = |\Sigma|^{-1}\int v^2 d\mathbf{x}$, show that

$$\frac{dE}{dt} \le -2\mu_1 DE(t),$$

where $\mu_1$ is the first nonzero eigenvalue of the Laplacian. Hence, $E(t) \le e^{-2\mu_1 Dt} \to 0$ as $t \to \infty$, which means that the system cannot support any patterns.

**Problem 11.8. Diffusion on a growing domain.** Consider the following diffusion equation on a growing 1D domain of length $L(t)$:

$$\frac{\partial C(x,t)}{\partial t} + \sigma(t)\frac{\partial[xC(x,t)]}{\partial x} = D\frac{\partial^2 C(x,t)}{\partial x^2} + kC(x,t)$$

for $x \in [0, L(t)]$ and $t > 0$. Here $k$ is a production rate and $\sigma(t) = \dot{L}(t)/L(t)$. Assume reflecting boundary conditions $\partial_x C = 0$ at $x = 0, L(t)$ and consider the initial condition

$$C(x,0) = \begin{cases} C_0, & 0 < x < l \\ 0, & l < x < L(t). \end{cases}$$

(a) Derive the following modified evolution equation on the domain $X \in [0, L_0]$ (see Box 11E)

$$\frac{\partial C(X,t)}{\partial t} = \frac{D}{L(t)^2}\frac{\partial^2 C(X,t)}{\partial X^2} + (k - \sigma(t))C(X,t).$$

(b) Rescale time according to

$$T = \int_0^t \frac{D}{L^2(s)}ds.$$

Assuming that the function $T = T(t)$ is invertible, derive the equation

$$\frac{\partial C(X,T)}{\partial T} = \frac{\partial^2 C(X,T)}{\partial X^2} + f(T)C(X,T),$$

where

$$f(T) = \frac{L^2(t(T))(k - \sigma(t(T)))}{D}.$$

(c) Use separation of variables to obtain the general solution

$$C(X,T) = \sum_{n=0}^{\infty} a_n \cos(n\pi X)e^{-n^2\pi^2 T} \exp\left(\int_0^T f(T')dT'\right).$$

The coefficients $a_n$ are determined from the initial condition.

(d) Determine $T$ and $f(T)$ as functions of $t$ and determine the coefficients $a_n$ for the following growth rates: (i) zero growth ($L(t) = L_0$); (ii) exponential growth ($L(t) = L_0 e^{\alpha t}$); and (iii) linear growth ($L(t) = L_0 + bt$). Plot the evolution of the solution in each case by showing successive snapshots of the profile $C(x,t)$ with $x = X/L(t) \in (0, L(t))$.

**Problem 11.9. Spike solutions in a RD model of protein clustering in bacteria.**
Consider the rescaled RD system of equations given by (11.5.51):

$$0 = D\frac{d^2u}{dx^2} - \frac{\beta}{\varepsilon}u(\varepsilon u + v)^2 + \frac{\gamma}{\varepsilon}v - \sigma u + c\sigma,$$

$$0 = \varepsilon^2\frac{d^2v}{dx^2} + \beta u(\varepsilon u + v)^2 - \gamma v - \sigma v.$$

Suppose that there are $N$ spikes at positions $x_1,\ldots,x_N$ with $v$ peaking at each point $x_j$ such that $v'(x_j) = 0$ and $u(x_j) = U$.

(a) Consider the leading order inner solution around the $j$th spike by introducing the stretched coordinate $y_j = \varepsilon^{-1}(x - x_j)$ and setting $\tilde{u}(y_j) = u(x_j + \varepsilon y_j)$, $\tilde{v}(y_j) = v(x_j + \varepsilon y_j)$. Taking $\tilde{u}$ and $\tilde{v}$ to have the asymptotic expansions

$$\tilde{u} = U + \varepsilon\tilde{u}_1 + O(\varepsilon^2), \quad \tilde{v} = \tilde{v}_0 + O(\varepsilon),$$

derive the equations

$$D\frac{d^2\tilde{u}_1}{dy_j^2} = \beta U\tilde{v}_0^2 - \gamma\tilde{v}_0, \quad \frac{d^2\tilde{v}_0}{dy_j^2} + \beta U\tilde{v}_0^2 - (\gamma+\sigma)\tilde{v}_0 = 0,$$

with $\tilde{v}_0'(0) = 0$, $\tilde{u}_0(0) = U$, $\tilde{u}_1(0) = 0$, and $\tilde{v}_0 \to 0$ as $|y_j| \to \infty$. Solve the equation for $\tilde{v}_0$ so show that

$$\tilde{v}_0(y_j) = \frac{3}{2}\frac{\gamma+\sigma}{\beta U}\operatorname{sech}^2\left(\frac{\sqrt{\gamma+\sigma}}{2}y_j\right).$$

(b) Integrating the equation for $\tilde{u}_1$ obtain the jump condition

$$D\left[\lim_{y_j\to\infty}\tilde{u}_1' - \lim_{y_j\to-\infty}\tilde{u}_1'\right] = \frac{6\sigma\sqrt{\gamma+\sigma}}{\beta}.$$

(c) Show that the leading order outer solution satisfies the equation

$$D\frac{d^2u_0}{dx^2} - \sigma u_0 + c\sigma = \frac{\Gamma}{U}\sum_{k=1}^{N}\delta(x-x_k), \quad x \in [-L/2, L/2],$$

with $u_0'(\pm L/2) = 0$. Introducing 1D Neumann Green's function

$$D\frac{d^2G(x;x')}{dx^2} - \sigma G(x;x') = -\delta(x-x'), \quad \partial_x G(\pm L/2;x') = 0,$$

$$\int_{-L/2}^{L/2}G(x;x')dx = \sigma^{-1},$$

obtain the solution

$$u_0(x) = c - \frac{\Gamma}{U} \sum_{j=1}^{N} G(x; x_j).$$

Explicitly calculate Green's function to give

$$G(x; x') = \frac{\kappa}{2} \frac{\cosh(\kappa(x+x')/L) + \cosh(\kappa(|x-x'|-L)/L)}{\sinh(\kappa)}, \quad \kappa = L\sqrt{\frac{\sigma}{D}}.$$

**Problem 11.10. Traveling fronts in a bistable equation.** Consider the scalar RD equation

$$\frac{\partial u}{\partial t} = D\frac{\partial^2 u}{\partial x^2} + f(u), \quad x \in \mathbb{R}.$$

Introducing the traveling wave solution $u(x,t) = U(\xi)$, $\xi = x - ct$, we obtain the ODE

$$\frac{d^2 U}{d\xi^2} + c\frac{dU}{d\xi} + f(U) = 0.$$

For certain choices of $f$, the wave speed $c$ and wave profile $U$ can be calculated explicitly.

(a) Suppose that $f$ is given by the cubic $f(u) = u(u-a)(1-u)$. Substitute the ansatz $U' = -AU(1-U)$ for some constant $A$ into the ODE. Collecting terms linear in $U$ and terms independent of $U$, determine $A$ and $c$. Construct the corresponding wave profile $U$.

(b) Another choice of nonlinearity for which an explicit front can be calculated is the piecewise linear function

$$f(u) = -u + H(u-a).$$

Translation symmetry of the system means that we are free to choose the solution $U$ in the moving frame to cross the threshold $a$ at $\xi = 0$ so that $U(\xi) > a$ for $\xi < 0$ and $U(\xi) < a$ for $\xi > 0$. Solve the resulting linear equation for $U$ on either side of the threshold point $\xi = 0$ to Now impose the threshold condition $V(0) = a$ and continuity of $U'(\xi)$ at $\xi = 0$ to obtain the following expression for the wave speed:

$$c = \frac{1-2a}{\sqrt{a-a^2}}.$$

**Problem 11.11. Pulled fronts in a modified Fisher–KPP equation.** Consider the scalar RD equation

$$\frac{\partial u}{\partial t} = D\frac{\partial^2 u}{\partial x^2} + f(u), \quad x \in \mathbb{R},$$

with a piecewise function

$$f(u) = \begin{cases} \mu(1-\theta)u, & 0 \le u \le \theta \\ \mu\theta(1-u), & \theta \le u \le 1 \end{cases}$$

for $\mu > 0$ and $0 < \theta < 1$. Consider a monotonically traveling front solution $u(x,t) = U(\xi)$ with $U(0) = \theta$, $U(\xi) \to 1$ as $\xi \to -\infty$ and $U(\xi) \to 0$ as $\xi \to \infty$.

(a) Show that there is a unique solution in the domain $\xi < 0$ that tends to unity as $\xi \to -\infty$, which is given by

$$U(\xi) = 1 - (1 - \theta)e^{\lambda_+ \xi}, \quad \lambda_+ = \frac{1}{2}\left(-c + \sqrt{c^2 + 4\mu\theta}\right).$$

(b) Show that the general solution for $\xi > 0$ is

$$U(\xi) = A_+ e^{\mu_+ \xi} + A_- e^{\mu_- \xi}, \quad \mu_\pm = \frac{1}{2}\left(-c \pm \sqrt{c^2 - 4\mu(1 - \theta)}\right) < 0.$$

Imposing the conditions $U(0) = \theta$ and continuity of $U'(0)$, show that there exists a unique front solution for each $c > c^*$ where

$$c_* = \sqrt{4\mu(1 - \theta)}.$$

(c) Show that there exists a unique traveling front at the minimum wave speed $c = c^*$ of the form

$$U(\xi) = \theta e^{-c^* \xi/2} + B\xi e^{-c^* \xi/2},$$

and calculate $B$. Prove that $B > 0$. [Hint: write the second-order ODE for $U(\xi)$ as a planar dynamical system.]

# Chapter 12
# Statistical mechanics and dynamics of polymers and membranes

When developing theories of nonequilibrium systems and constructing stochastic models, it is important to ensure that these are consistent with equilibrium statistical mechanics. We have encountered this issue in various contexts, including the Einstein relation and fluctuation-dissipation theorems (Chap. 2 and Chap. 10), Brownian ratchets (Chap. 4), and the law of mass action and detailed balance (Chap. 3 and Chap. 5). The essential idea is that many stochastic dynamical phenomena in cell biology involve the relaxation to an equilibrium thermodynamic state, with the relaxation occurring on relevant biological time scales. It follows that in order to understand the dynamics of cellular structures such as the cytoskeleton, plasma membrane, and nucleus, it is necessary to consider the equilibrium and near-equilibrium behavior of the physical components underlying such structures. Therefore, in this chapter, we present a basic introduction to the statistical mechanics and dynamics of polymers, membranes, and polymer networks. In order to keep the material reasonably self-contained, we use the Boltzmann-Gibbs distribution and associated partition function as the starting point of our analysis. This also provides a relatively straightforward way of defining important quantities such as free energy, entropy, and the chemical potential (Chap. 1). For a more detailed and general introduction to statistical mechanics; see Refs. [178, 489]. Extensive material on polymer physics can be found in Refs. [250, 251, 775], whereas the physics of membranes is covered in Refs. [540, 782]. Finally, Ref. [87] provides a comprehensive introduction to cell mechanics.

We begin in Sect. 12.1 with the statistical mechanics of single polymers, covering random walk models such as the freely-jointed chain, and elastic rod models or worm-like chains. The latter type of model treats a polymer as a continuous curve, whose free energy contributions arise from the stretching, bending, and twisting of the polymer. We introduce a few ideas from the continuum mechanics of elastic rods, including definitions of curvature and torsion based on the Frenet-Serret frame (Box 12A). We consider two applications of the theory. First, a random walk model is used to analyze the translocation of a flexible polymer such as DNA through a nanopore. Second, the experimentally obtained force-displacement curves for DNA

© Springer Nature Switzerland AG 2021
P. Bressloff, *Stochastic Processes in Cell Biology*, Interdisciplinary Applied Mathematics 41, https://doi.org/10.1007/978-3-030-72519-8_12

are modeled in terms of a worm-like chain. In Sect. 12.2, we explore the statistical mechanics of membranes, which are treated as thin elastic sheets. In order to construct the bending energy of the membrane, we review some basic results form membrane elasticity, including stress and strain tensors, and bending/compression moduli (Box 12B). We then construct the corresponding partition function and estimate the size of thermally driven membrane fluctuations. Since the membrane is modeled as an infinite-dimensional continuum, the partition function takes the form of a path integral (Chap. 8) whose associated free energy is a functional. The analysis of statistical properties thus requires the use of functional calculus (see Box 12C).

In Sect. 12.3, we show how to analyze the statistical dynamics of systems at or close to equilibrium. We proceed by generalizing the theory of Brownian motion introduced in Chap. 2 to more complex structures with many internal degrees of freedom. We introduce various results and concepts from classical nonequilibrium statistical physics, including Onsager's reciprocal relations, nonequilibrium forces, time correlations, and a general version of the fluctuation-dissipation theorem (see also Box 12D). We illustrate the theory by deriving Langevin equations for fluctuating polymers and membranes. Another application, namely, to liquid-liquid phase separation and the formation of biological condensates will be presented in Sect. 13.1. We then turn to polymer network models, which are used extensively to understand the rheological properties of the cytoskeleton (Sect. 12.4). We focus on the simplest classical models, analyzing the rubber elasticity of a cross-linked polymer network, swelling of a polymer gel, and the macroscopic theory of viscoelasticity in uncross-linked polymer fluids. We also briefly describe reptation theory, which is used to model the dynamics of entangled polymers.

Finally, in Sect. 12.5, we consider the dynamics of DNA within the nucleus. After describing some of the key features of nuclear organization, we introduce a classical stochastic model of a Gaussian polymer chain known as the Rouse model. The latter is used to model the subdiffusive motion of chromosomal loci, and to explore mechanisms for spontaneous DNA loop formation. The mean time to form a loop requires solving an FP equation with a nontrivial absorbing boundary condition. We introduce one method for tackling this type of problem, which is based on the Wilemski-Fixman theory of diffusion-controlled reactions (Box 12E), and involves replacing the boundary condition by a sink term in the FP equation.

## 12.1 Statistical mechanics of polymers

So far, polymers have been treated as rigid or semi-rigid rods that either act as polymerizing molecular motors (Chap. 4) or as filament tracks for the active transport of intracellular cargo (Chap. 7). Although this is a reasonable first approximation for cytoskeletal filaments such as F-actin and microtubules, other biopolymers such as DNA and many proteins tend to be more flexible. That is, thermal fluctuations induce a loss of correlations in the orientation of the polymer over length scales larger than the so-called persistence length $\xi_p$, so that the polymer can only be treated as stiff on length scales smaller than $\xi_p$. In the case of DNA, the persistence length is $\xi_p \approx 60$ nm, whereas $\xi_p \approx 4\text{–}8$ mm for microtubules. The flexible nature of a polymer has two major implications for its statistical mechanical properties. First, there are significant entropic contributions due to the large number of different spatial configurations of a flexible polymer. Second, there are energetic contributions due to elastic effects such as bending, stretching, and twisting. These various features will be explored in this section. For detailed reviews of polymer physics, see Refs. [250, 251, 775].

### 12.1.1 Random walk models of polymers

**1D random walk model.** Consider a simple 1D model of a flexible polymer. The polymer is represented as a sequence of links of length $a$ that either point in the positive or negative $x$-direction with equal probability, see Fig. 12.1. One can thus treat a given configuration or microstate of the polymer as a sample trajectory of an unbiased random walk on a 1D lattice with spacing $a$. Suppose that there are $N$ links in the chain and that one end of the chain is fixed at the origin. Let $n$ denote the number of links pointing in the positive $x$-direction. It follows that the other end of the chain (end-to-end distance) is at $x = (2n - N)a$. We will assume that $N$ is sufficiently large so that $x$ and $n$ can be treated as continuous variables. If we ignore any energy contributions from the elastic stretching, bending or twisting of the polymer, then the energy $\Phi$ of any configuration is zero. However, stretching the polymer by pulling on the free end at $x$ is resisted by an entropic force. In order to show this, we note that the number of configurations or internal microstates for fixed $n$ or $x$ is given by the combinatorial factor (see also Sect. 1.3)

$$\Omega(n) = \frac{N!}{n!(N-n)!}. \tag{12.1.1}$$

Taking logs and using Stirling's formula we have the entropy

$$S(n) = k_B[N \ln N - n \ln n - (N-n)\ln(N-n)], \tag{12.1.2}$$

which can be re-expressed in terms of $x$ according to

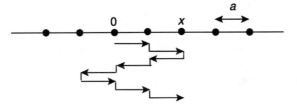

Fig. 12.1: Random walk model of a one-dimensional polymer. The links have been displaced in the vertical direction for illustrative purposes.

$$S(x) = k_B \left( N \ln N - \frac{x+Na}{2a} \ln \left[ \frac{x+Na}{2a} \right] - \frac{Na-x}{2a} \ln \left[ \frac{Na-x}{2a} \right] \right). \quad (12.1.3)$$

The entropic force is then

$$f_{ent}(x) = T \frac{dS(x)}{dx} = -\frac{k_B T}{2a} \ln \left[ \frac{1+x/Na}{1-x/Na} \right]. \quad (12.1.4)$$

Assuming that $x \ll Na$, we can Taylor expand to first order in $x$ to obtain Hooke's law for an elastic spring

$$f_{ent}(x) \approx -\frac{k_B T}{Na^2} x. \quad (12.1.5)$$

The minus sign means that the polymer resists stretching, since the force is in the negative $x$-direction. The origin of the entropic force is that when a polymer is stretched it becomes less random, in the sense that there are less configurations for larger $x$. In order to maintain the displacement $x$, the environment has to exert an opposing force $f = -f_{ent}$. Inverting equation (12.1.4), we obtain the force-displacement relation

$$z \equiv \frac{x}{Na} = \tanh \left[ \frac{fa}{k_B T} \right]. \quad (12.1.6)$$

**Freely-jointed chain model of a polymer (3D).** Let us represent a polymer in 3D as a chain of $N$ segments each of which is described by a vector $\mathbf{a}_j$ with $|\mathbf{a}_j| = a$ the length of the segment and the direction of $\mathbf{a}_j$ representing the orientation of the segment, see Fig. 12.2(a). One end of the polymer is fixed at the origin, so that the end-to-end displacement of the chain is given by $\mathbf{r} = \sum_{j=1}^N \mathbf{a}_j$. Suppose that the orientations of the segments are random in the sense that, averaging over a large population of identical polymers, we have $\langle \mathbf{a}_j \cdot \mathbf{a}_i \rangle = 0$ for all $i \neq j$. It follows that $\langle \mathbf{r} \rangle = 0$ and

$$\langle \mathbf{r} \cdot \mathbf{r} \rangle = \left\langle \left( \sum_{i=1}^N \mathbf{a}_i \right) \cdot \left( \sum_{j=1}^N \mathbf{a}_i \right) \right\rangle = \sum_{i,j=1}^N \langle \mathbf{a}_j \cdot \mathbf{a}_i \rangle = Na^2. \quad (12.1.7)$$

Let $p_N(\mathbf{r})$ denote the probability density that a chain of $N$ links has an end-to-end distance $\mathbf{r}$. It follows from the central limit theorem that for large $N$ and $|\mathbf{r}| \gg a$, $p_N(\mathbf{r})$ is given by a Gaussian

$$p_N(\mathbf{r}) = \left( \frac{2\pi Na^2}{3} \right)^{-3/2} e^{-3r^2/2Na^2}. \quad (12.1.8)$$

Here $p_N(\mathbf{r})$ corresponds to the ratio between the number of configuration $\Gamma_N(\mathbf{r})$ that take the chain from a fixed origin to $\mathbf{r}$ and the total number of configurations $\Gamma_N$. It follows that the entropy of a chain with end-to-end vector $\mathbf{r}$ is

$$S_N(\mathbf{r}) = k_B \log \Gamma_N(\mathbf{r}) = k_B \log p_N(\mathbf{r}) + k_B \log \Gamma_N = -\frac{3k_B \mathbf{r}^2}{2Na^2} + C_N, \quad (12.1.9)$$

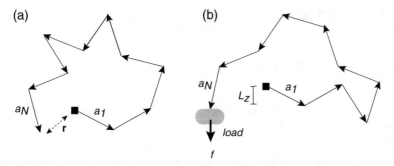

Fig. 12.2: Freely-jointed chain model of a 3D polymer. (a) End-to-end distance $\mathbf{r} = \sum_i \mathbf{a}_i$. (b) A load maintains a constant force $f$ in the positive $z$-direction.

where $C_N$ is a constant. The corresponding free energy is $F(\mathbf{r}) = -T S_N(\mathbf{r})$ (since there is no $\mathbf{r}$-dependent internal energy).

Suppose that a fixed force $f$ is now applied to the free end of the polymer in the $z$-direction using an external load, see Fig. 12.2(b). The configuration of the polymer is specified by the spherical polar angles $(\theta_j, \phi_j)$ of the $N$ links, $j = 1, \dots, N$, with $0 < \theta_j \leq \pi$ and $0 \leq \phi_j \leq 2\pi$. We identify the system as the polymer plus the applied load. Thus, when writing down the energy of the system, we have to include both the internal energy of the polymer and the potential energy of the load. The internal energy of the polymer is independent of its configuration, and can thus be set to zero. The potential energy of the load is $E = -\sum_{j=1}^{N} f a \cos(\theta_j)$. That is, whenever the polymer shortens in the $z$-direction, the load has to move against the applied force and thus gains in potential energy. The Boltzmann-Gibbs distribution with $\Theta = (\theta_1, \dots, \theta_N)$ and $\Phi = (\phi_1, \dots, \phi_N)$ is

$$p(\Theta, \Phi) = \frac{1}{Z} e^{(fa/k_B T) \sum_{j=1}^{N} \cos \theta_j} = \frac{1}{Z} \prod_{j=1}^{N} e^{(fa/k_B T) \cos \theta_j}, \qquad (12.1.10)$$

with the partition function obtained by integrating over all solid angles, $d\Omega = \prod_{j=1}^{N} \sin \theta_j d\theta_j d\phi_j$:

$$Z = \int e^{(fa/k_B T) \sum_{j=1}^{N} \cos \theta_j} d\Omega = \prod_{j=1}^{N} \left[ \int_0^{2\pi} \int_0^{\pi} e^{(fa/k_B T) \cos \theta_j} \sin \theta_j d\theta_j d\phi_j \right] = Z_1^N,$$

where

$$Z_1 = \int_0^{2\pi} \int_0^{\pi} \sin \theta e^{(fa/k_B T) \cos \theta} d\theta d\phi = 2\pi \int_0^{2\pi} e^{(fa/k_B T) \cos \theta} d\cos \theta$$

$$= 2\pi \int_{-1}^{1} e^{(fa/k_B T) x} dx = \frac{4\pi k_B T}{fa} \sinh\left(\frac{fa}{k_B T}\right).$$

It follows from the above analysis that the Boltzmann-Gibbs distribution can be factorized into a product of distributions for the $N$ independent links

$$p(\Theta, \Phi) = \prod_{j=1}^{N} p_1(\theta_j, \phi_j), \quad p_1(\theta_j, \phi_j) = Z_1^{-1} e^{(fa/k_B T) \cos \theta_j}. \tag{12.1.11}$$

In order to derive a force-displacement relation, it is necessary to determine the mean displacement of the polymer in the $z$-direction, $L_z = \mathbf{r} \cdot \mathbf{e}_z$

$$z := \langle L_z \rangle \equiv \int \left[ a \sum_{j=1}^{N} \cos \theta_j \right] p(\Theta, \Phi) d\Omega = Na \int_0^{2\pi} \int_0^\pi \cos \theta \, p_1(\theta, \phi) \sin \theta d\theta d\phi.$$

The integral can be evaluated by noting that

$$\frac{d \ln Z_1}{df} = \frac{a}{k_B T} \frac{1}{Z_1} \int_0^{2\pi} \int_0^\pi \cos \theta \sin \theta e^{(fa/k_B T) \cos \theta} d\theta d\phi,$$

which implies that

$$z = Nk_B T \frac{d \ln Z_1}{df} = Na \left[ \coth \left( \frac{fa}{k_B T} \right) - \frac{k_B T}{fa} \right]. \tag{12.1.12}$$

In the small force limit, this reduces to Hooke's law

$$f = kz, \quad k = 3k_B T / Na^2. \tag{12.1.13}$$

At large forces, equation (12.1.12) predicts that the extension saturates at $\langle L_z \rangle = Na$ (fully uncoiled polymer). The freely-jointed chain model qualitatively captures experimental data on force-extension curves for DNA at low to intermediate forces. A much better fit can be obtained using models that take into account the elastic bending energy of polymers [592]. Moreover, at large applied forces, overstretching of DNA is observed which involves a mixture of elastic strain energies and structural transitions of the DNA [829, 830, 847]. These more complicated features will be explored in Sect. 12.1.4.

> Remark 12.1. In the simple random walk model, the end-to-end distance of the polymer was treated as deterministic, whereas in the freely-jointed chain model the external force is fixed and the end-to-end distance can fluctuate. The two pictures are consistent in the case of long polymers (large $N$) as fluctuations in the end-to-end distance are negligible, which is a consequence of the law of large numbers.

**Persistence length.** In the above models, there is a fundamental length scale, namely, the length $a$ of each link, which is called the Kuhn length. Roughly speaking, one can view the Kuhn length as the length over which a polymer is essentially straight. In order to consider properties of polymers on length scales smaller than the Kuhn length, which are important for strong deformations, it is more convenient to consider a continuum model of a polymer. More precisely, suppose that a polymer is represented as a continuous curve in 3D space, parameterized by arc length $s$ with

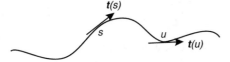

Fig. 12.3: Persistence length of a polymer represented as a continuous curve.

total contour length $L$, see Fig. 12.3. The persistence length $\xi_p$ can then be defined as the correlation length over which tangent-tangent correlations decay along the chain

$$\langle \mathbf{t}(s) \cdot \mathbf{t}(u) \rangle = e^{-|s-u|/\xi_p}. \tag{12.1.14}$$

For example, the DNA of viruses such as $\lambda$-phage has a contour length of $L \approx 16\mu$m and a persistence length of $\xi_p \approx 50$ nm at room temperature. We can relate the persistence length $\xi_p$ to the Kuhn length $a$ by calculating the mean square of the end-to-end vector $\mathbf{r} = \int_0^L \mathbf{t}(s)ds$

$$\langle \mathbf{r}^2 \rangle = \left\langle \int_0^L \mathbf{t}(s)ds \cdot \int_0^L \mathbf{t}(u)du \right\rangle = \int_0^L ds \int_0^L du \langle \mathbf{t}(s) \cdot \mathbf{t}(u) \rangle$$

$$= 2 \int_0^L ds \int_s^L du\, e^{-(u-s)/\xi_p} \approx 2 \int_0^L ds \int_0^\infty dx\, e^{-x/\xi_p} = 2L\xi_p$$

for $L \gg \xi_p$. Comparing with the analogous calculation (12.1.7) for the freely-jointed chain, we see that $a = 2\xi_p$. Note that the continuous model is the starting point for a more detailed analysis of the elastic properties of polymers using the theory of elastic beams or rods [540]; see Sect. 12.1.3. First, however, we consider a simplified model, which is equivalent to the classical Ising model of magnetic spins [178] and the model of cooperative binding considered in Chap. 3.

**The Ising model of a polymer.** Let us return to the example of a 1D polymer consisting of $N$ links of length $a$. Denote the state of each link by the binary variable $\sigma_i$ with $\sigma_i = 1$ ($\sigma = -1$) if the link points in the positive (negative) $x$-direction. Suppose that an external load maintains a constant force $f$ in the positive $x$-direction. The total extension of the polymer is $x = a\sum_{j=1}^N \sigma_j$. In contrast to the random walk model, suppose that when two neighboring links point in opposite directions, they contribute an extra $2\gamma k_B T$ of energy, where $\gamma$ is some cooperativity parameter. (This could represent an effective bending energy.) Then the total energy of the polymer plus load for a given configuration $\sigma = (\sigma_1, \ldots, \sigma_N)$ is

$$E[\sigma] = -fa \sum_{j=1}^N \sigma_j - \gamma k_B T \sum_{j=1}^{N-1} \sigma_j \sigma_{j+1}. \tag{12.1.15}$$

The corresponding Boltzmann-Gibbs distribution is

$$p(\sigma) = Z^{-1} e^{-E[\sigma]/k_B T}, \quad Z = \sum_{\sigma_1 = \pm 1} \cdots \sum_{\sigma_N = \pm 1} e^{\alpha \sum_{j=1}^N \sigma_j + \gamma \sum_{j=1}^{N-1} \sigma_j \sigma_{j+1}}, \tag{12.1.16}$$

with $\alpha = fa/k_B T$. The partition function can be treated as a generating function for the displacement $x$, that is,

$$\langle x \rangle = k_B T \frac{d}{df} \ln Z[f] = a \frac{d}{d\alpha} Z[\alpha]. \tag{12.1.17}$$

As in the Ising model of cooperative binding, one can derive an exact expression for $Z$ using transfer matrices [178]. That is,

$$Z = \text{Tr}[\mathbf{T}^N], \quad \mathbf{T} = \begin{pmatrix} e^{\alpha+\gamma} & e^{-\gamma} \\ e^{-\gamma} & e^{-\alpha+\gamma} \end{pmatrix}.$$

For large $N$, $Z \approx \lambda_+^N$, where $\lambda_+$ is the larger eigenvalue of $\mathbf{T}$,

$$\lambda_+ = e^\gamma \left[ \cosh\alpha + \sqrt{\sinh^2\alpha + e^{-4\gamma}} \right].$$

Finally, substituting the result into equation (12.1.17) shows that the mean extension is

$$\langle x \rangle = \frac{Na \sinh\alpha}{\sqrt{\sinh^2\alpha + e^{-4\gamma}}}. \tag{12.1.18}$$

Note that this reduces to the force-extension relation (12.1.6) of the random walk model in the limit $\gamma \to 0$. When $\gamma > 0$ one still recovers the maximal length $\langle x \rangle = Na$ in the limit $f \to \infty$.

## 12.1.2 Translocation of DNA through a nanopore

One application of random walk models of a polymer is the translocation of a flexible polymer such as DNA through a nanopore [658, 659, 709, 849]. A schematic illustration of the basic physical problem is shown in Fig. 12.4. The membrane is treated as an infinitesimally thin plate separating two regions I and II with electrical potentials $\Phi_1$ and $\Phi_2$, respectively. The DNA polymer is represented as a freely-jointed chain of $N$ links, each of Kuhn length $a \approx 100$ nm. For simplicity, the polymer is translocated through a pore in the membrane as a single strand. Suppose that there are $N - m$ segments in region I and $m$ segments in region II, and in each region the chain is modeled as a random walk of $m$ or $N - m$ segments pinned at the pore at one end. A final simplification of the model is that translocation is taken to be sufficiently slow so that at each step, the chains have enough time to reach thermodynamic equilibrium. (This assumption is relaxed in Ref. [195].) Note that the model differs considerably from the translocation ratchet analyzed in Sect. 4.3. The latter treats the polymer as rigid and focuses on the rectifying effects of chaperones.

For the translocation problem shown in Fig. 12.4, there are two 3D chains, each of which is restricted to lie on one side of the membrane. Now we are interested in how the total number of configurations of each chain depends on $m$ (the number of translocated segments). It can be shown that the total configurational entropy is [658, 849]

$$S(m) = -k_B(1 - \gamma)[\ln(m) + \ln(N - m)], \tag{12.1.19}$$

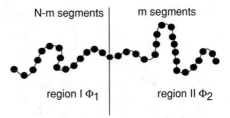

Fig. 12.4: Schematic illustration of a flexible biopolymer translating through a pore of a thin membrane. The electrical potential in each region is denoted by $\Phi_j$, $j = 1, 2$.

region I $\Phi_1$     region II $\Phi_2$

where $\gamma \approx 0.69$ is a constant that takes into account the fact that a 3D polymer cannot self-intersect (self-avoiding random walk); for a pure random walk $\gamma = 0.5$. A translocation step of the polymer leads to an increase in $m$, $m \to m+1$, which is opposed by an entropic force $f_{\text{ent}} = T dS/dm$ (assuming $m$ is large so that it can be treated as a continuous variable). There is also an electrical field acting on the (charged) polymer due to the fact that there is a potential difference $V = \Phi_1 - \Phi_2$ across the membrane. If each segment has total charge $z_s e$, where $e$ is the charge on an electron, then the change in potential energy when $m \to m+1$ is $\Delta\mu = e z_s V$. It follows that the free energy of the system for a given $m$ is

$$E(m) = k_B T (1 - \gamma) [\ln(m) + \ln(N - m)] + m\Delta\mu.$$

Setting $x = ma$ and $L = Na$, we can rewrite the free energy as

$$E(x) = k_B T (1 - \gamma) [\ln(x/a) + \ln((L - x)/a)] + x\Delta\mu/a. \tag{12.1.20}$$

Plots of $E(x)$ as a function of $x$ are shown in Fig. 12.5(a). It is now possible to reformulate translocation as Brownian motion through an energy barrier $E(x)$ with associated force $F(x) = -dE(x)/dx$ and diffusion coefficient $D$. One difference between various models is the assumed $x$-dependence of $D$ [709]. Here we will

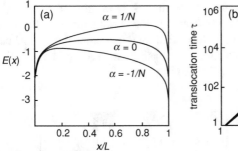

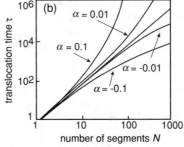

Fig. 12.5: Translocation of DNA. (a) Plot of free energy $E(x)$ (in units of $k_B T$) as a function of the length $x$ of the right-hand segment with $\gamma = 0.69$ and $L = 1$. The three curves correspond to different values of the potential energy difference with $\alpha = \Delta\mu/k_B T$. (b) Plot of MFPT $\tau$ (in units of $a^2/D$) for translocation of an $N$ unit polymer and various values of $\alpha = \Delta\mu/k_B T$. (Redrawn from Muthukumar [658].)

follow Ref. [658] and treat $D$ as a constant. The resulting FP equation is then

$$\frac{\partial p(x,t)}{\partial t} = D\frac{\partial}{\partial x}\left[\frac{p(x,t)}{k_B T}\frac{\partial E(x)}{\partial x} + \frac{\partial p(x,t)}{\partial x}\right]. \qquad (12.1.21)$$

To complete the model, absorbing boundary conditions are introduced at $x = 0, L$,

$$p(0,t) = p(L,t) = 0, \qquad (12.1.22)$$

under the assumption that if the polymer leaves the pore from either side it never returns. It follows that the conditional mean first passage time $\tau$ for successful translocation, that is, the polymer crosses the boundary at $x = L$ rather than $x = 0$, can be calculated along the lines outlined in Sect. 2.4. One finds the following asymptotic results for large $N$ [658], see also Fig. 12.5(b):

$$\frac{D\tau}{a^2} \sim \frac{k_B T}{|\Delta\mu|}N, \quad \Delta\mu \ll 0,$$

$$\frac{D\tau}{a^2} \sim \left(\frac{k_B T}{\Delta\mu}\right)^2 \exp\left(N\frac{\Delta\mu}{k_B T}\right), \quad \Delta\mu \gg 0,$$

$$\frac{D\tau}{a^2} \sim N^2, \quad \Delta\mu = 0.$$

### 12.1.3 Elastic rod model of a polymer

In order to take into account strong deformations of biopolymers and correlations on length scales smaller than the persistence length, it is necessary to consider elastic properties of the polymer, that is, energetic as well as entropic forces. This is achieved by treating a short section of polymer as an elastic rod or beam. There are then 3 types of deformation as illustrated in Fig. 12.6: stretching, bending, and twisting. In order to describe these deformations and determine the associated elastic energies, it is necessary to introduce a few basic ideas from solid mechanics. For a more detailed analysis; see Ref. [87, 540].

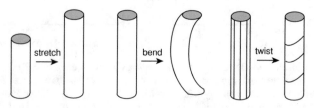

Fig. 12.6:  Three types of deformation of an elastic rod.

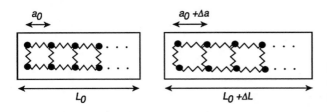

Fig. 12.7: Parallel arrays of springs in an idealized elastic rod.

**Strain energy.** Consider a simple model of a rod composed of many microscopic springs separated by a distance $a_0$ and arranged in parallel arrays or chains, see Fig. 12.7. Suppose that the number $N$ of springs in each chain is preserved during stretching, so that

$$N = \frac{L_0}{a_0} = \frac{L_0 + \Delta L}{a_0 + \Delta a} \approx \frac{L_0 + \Delta L}{a_0}\left(1 - \frac{\Delta a}{a_0}\right) \approx \frac{L_0}{a_0} + \frac{\Delta L}{a_0} - \frac{L_0}{a_0^2}\Delta a.$$

It follows that $\Delta L = (L_0/a_0)\Delta a$ to leading order. If $A$ is the cross-sectional area of the rod, then there are $A/a_0^2$ parallel chains and from Hooke's law, the force $f$ to stretch the rod is $f = k\Delta a(A/a_0^2)$, where $k$ is the spring constant of a single spring. Hence, the force per unit area is

$$\frac{f}{A} = Y\frac{\Delta L}{L_0}, \quad Y = \frac{k}{a_0}, \tag{12.1.23}$$

where $Y$ is known as Young's modulus. The strain energy of each spring is $\varepsilon_s = k(\Delta a)^2/2$. Since there is a total of $AL_0/a_0^3$ springs, the total strain energy of a uniformly stretched rod of original length $L_0$ and cross-sectional area $A$ is

$$E_s = \frac{AY}{2}\left(\frac{\Delta L}{L_0}\right)^2 L_0. \tag{12.1.24}$$

In the case of nonuniformly stretched rod, see Fig. 12.8, we partition the rod into infinitesimal sections of length $\Delta x$ such that on stretching $\Delta x \rightarrow \Delta x + \Delta v(x)$, we have the local strain energy

$$\Delta E_s = \frac{AY}{2}\left(\frac{\Delta v}{\Delta x}\right)^2 \Delta x.$$

Summing over all segments and taking the continuum limit leads to the total strain energy

$$E_s = \frac{AY}{2}\int_0^L u(x)^2 dx, \quad u(x) = \frac{dv}{dx}, \tag{12.1.25}$$

where $u(x)$ is known as the strain.

**Bending energy.** Next, we consider a bent rod that has constant curvature, see Fig. 12.9. The strain $\Delta L/L_0$ can then be calculated as a function of the distance from the

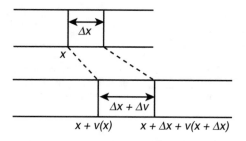

Fig. 12.8: Non-uniformly stretched rod with displacement function $v(x)$.

so-called neutral plane where there is no deformation. An infinitesimal segment at a distance $z$ from the neutral plane has arc length $L(z) = (R+z)\theta$, where $\theta = L_0/R$. Suppose that the curvature $1/R$ is sufficiently small so that $\theta \ll \pi$. It follows that such a segment has the strain

$$\frac{\Delta L(z)}{L_0} = \frac{L(z) - L_0}{L_0} = \frac{z}{R},$$

which means that the strain energy density (strain energy per unit volume) of this segment is

$$\Delta E_b = \frac{1}{2} Y \left( \frac{z}{R} \right)^2.$$

Since the volume of the rod segment is $L_0 dA(z)$, where $dA(z)$ is the infinitesimal cross-sectional area, the total bending energy is

$$E_b = \frac{K_b}{2} \frac{L_0}{R^2}, \quad K_b = Y \int_{\partial\Omega} z^2 dA(z). \tag{12.1.26}$$

(Example calculations of the so-called the cross-sectional moment of inertia $\mathscr{I} = \int_{\partial\Omega} z^2 dA(z)$ are considered in Ex. 12.1.) If the curvature of the beam varies along its arc length, then it is broken up into small elements of arc length $ds$ with curvature $\kappa(s) = 1/R(s)$, where $R(s)$ is the radius of curvature. Applying the above analysis to each small segment and summing yields the following integral formula for the bending energy:

$$E_b = \frac{1}{2} k_B T B \int_0^{L_0} \kappa(s)^2 ds, \quad B = \frac{K_b}{k_B T}, \tag{12.1.27}$$

with the constant $B$ known as the bending modulus. Using some basic differential geometry of curves (Box 12A), we can identify the curvature $\kappa(s)$ with the length of the rate of change of the tangent vector, that is, $\kappa(s) = \|\mathbf{t}'(s)\|$. Hence, we can rewrite the bending energy as

$$E_b = \frac{1}{2} k_B T B \int_0^{L_0} \left\| \frac{d\mathbf{t}(s)}{ds} \right\|^2 ds. \tag{12.1.28}$$

Another result from differential geometry is the inequality (12.1.34) of Fenchel's theorem (Box 12A). This provides a lower bound on the bending energy for a poly-

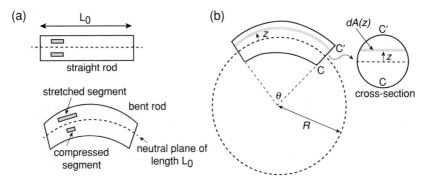

Fig. 12.9: Bending of a rod with constant curvature $R$. (a) The neutral plane divides the rod into segments that are stretched and segments that are compressed. (b) Coordinate system for calculating bending energy.

mer loop of length $L$. From the Cauchy-Schwartz inequality,

$$\oint_0^L 1 \cdot \kappa(s)ds \leq \sqrt{\oint_0^L \kappa^2(s)ds} \sqrt{\oint_0^L 1^2 ds},$$

that is

$$\oint_0^L \kappa^2(s)ds \geq \frac{4\pi^2}{L}.$$

Equation (12.1.34) then implies that for a polymer loop

$$E_b = \frac{1}{2}k_B T B \oint_0^L \kappa(s)^2 ds \geq \frac{2\pi^2 k_B T B}{L}. \tag{12.1.29}$$

---

### Box 12A Curvature, torsion and the Frenet-Serret frame.

---

**Curvature and torsion.** A curve in 3D is a vector-valued function

$$\mathbf{r}(s) = (x(s), y(s), z(s)),$$

where $s$ parameterizes the curve. It is convenient to take $s$ to be the arc length so that $s \in [0, L]$, where $L$ is the curve length. For an arbitrary parameterization, we have

$$L = \int_{\xi_0}^{\xi_1} \sqrt{d\mathbf{r}(\xi) \cdot d\mathbf{r}(\xi)} = \int_{\xi_0}^{\xi_1} \sqrt{\frac{d\mathbf{r}(\xi)}{d\xi} \cdot \frac{d\mathbf{r}(\xi)}{d\xi}} d\xi.$$

Hence, identifying $\xi$ with arc length yields

$$L = \int_0^L \left\| \frac{d\mathbf{r}(s)}{ds} \right\| ds,$$

which implies that $\|d\mathbf{r}/ds\| = 1$. Let $\mathbf{t}(s) = d\mathbf{r}/ds$ be the tangent vector to the curve. The local curvature $\kappa(s)$ of the curve is then defined according to

$$\frac{d\mathbf{t}(s)}{ds} = \kappa(s)\mathbf{n}(s), \quad \mathbf{t}(s) = \frac{d\mathbf{r}(s)}{ds}, \tag{12.1.30}$$

where $\mathbf{n}(s)$ is called the unit normal vector (chosen so that $\kappa(s)$ is always positive). Since $\mathbf{t}(s) \cdot \mathbf{t}(s) = 1$, it follows that $\mathbf{t}'(s) \cdot \mathbf{t}(s) = 0$ and thus $\mathbf{n}(s) \cdot \mathbf{t}(s) = 0$.

As $s$ varies, $\mathbf{n}(s)$ typically changes direction, which can arise from two contributions: (i) $\mathbf{n}(s)$ can rotate towards or away from $\mathbf{t}(s)$ such that the curve stays in the same flat plane. (ii) $\mathbf{n}(s)$ can rotate around the tangent vector such that there is a rotation of the plane in which the curve lies at $s$ - the osculating plane. Since $\mathbf{n}(s) \cdot \mathbf{n}(s) = 1$, we have $\mathbf{n}'(s) \cdot \mathbf{n}(s) = 0$, which means that

$$\mathbf{n}'(s) = \alpha(s)\mathbf{t}(s) + \tau(s)\mathbf{b}(s), \quad \mathbf{b}(s) = \mathbf{t}(s) \times \mathbf{n}(s),$$

with $\tau(s)$ known as the torsion. The coefficient $\alpha(s)$ can be determined by differentiating $\mathbf{t}(s) \cdot \mathbf{n}(s) = 0$ with respect to $s$:

$$0 = \mathbf{t}'(s) \cdot \mathbf{n}(s) + \mathbf{t}(s) \cdot \mathbf{n}'(s) = \kappa(s)\mathbf{n}(s) \cdot \mathbf{n}(s) + \alpha(s)\mathbf{t}(s) \cdot \mathbf{t}(s) = \alpha(s) + \kappa(s).$$

Hence

$$\mathbf{n}'(s) = -\kappa(s)\mathbf{t}(s) + \tau(s)\mathbf{b}(s). \tag{12.1.31}$$

Finally, we may calculate $\mathbf{b}'(s)$

$$\begin{aligned}
\mathbf{b}'(s) &= \mathbf{t}'(s) \times \mathbf{n}(s) + \mathbf{t}(s) \times \mathbf{n}'(s) \\
&= \kappa(s)\mathbf{n}(s) \times \mathbf{n}(s) - \kappa(s)\mathbf{t}(s) \times \mathbf{t}(s) + \tau(s)\mathbf{t}(s) \times \mathbf{b}(s) \\
&= -\tau(s)\mathbf{n}(s).
\end{aligned} \tag{12.1.32}$$

Combining equations (12.1.30), (12.1.31), and (12.1.32) yields the Frenet-Serret equations

$$\frac{d}{ds}\begin{pmatrix} \mathbf{t}(s) \\ \mathbf{n}(s) \\ \mathbf{b}(s) \end{pmatrix} = \begin{pmatrix} 0 & \kappa(s) & 0 \\ -\kappa(s) & 0 & \tau(s) \\ 0 & -\tau(s) & 0 \end{pmatrix}\begin{pmatrix} \mathbf{t}(s) \\ \mathbf{n}(s) \\ \mathbf{b}(s) \end{pmatrix}. \tag{12.1.33}$$

Note that given the curvature $\kappa(s)$ and torsion $\tau(s)$, one can reconstruct the entire curve (up to a global translation and rotation).

*Example 12.1.* Consider the following coordinate representation of a helix winding around the $z$-axis:

$$x = r\cos\theta, \quad y = r\sin\theta, \quad z = \theta/q,$$

for some constant $q$. First, we have

$$\frac{d\mathbf{r}}{d\theta} = (-r\sin\theta, r\cos\theta, 1/q),$$

which implies that

$$ds^2 \equiv d\mathbf{r}\cdot d\mathbf{r} = q^{-2}(r^2q^2+1)d\theta,$$

Hence, $s = \sqrt{1+r^2q^2}\,\theta/q$, and we can write

$$x = r\cos\left(\frac{qs}{\sqrt{1+r^2q^2}}\right), \quad y = r\sin\left(\frac{qs}{\sqrt{1+r^2q^2}}\right), \quad z = \frac{s}{\sqrt{1+r^2q^2}}.$$

Differentiating with respect to $s$ then shows that

$$\mathbf{t}(s) = \frac{rq}{\sqrt{1+r^2q^2}}\left(-\sin\left(\frac{qs}{\sqrt{1+r^2q^2}}\right), \cos\left(\frac{qs}{\sqrt{1+r^2q^2}}\right), \frac{1}{rq}\right),$$

and a second differentiation yields

$$\mathbf{t}'(s) = -\frac{rq^2}{1+r^2q^2}\left(\cos\left(\frac{qs}{\sqrt{1+r^2q^2}}\right), \sin\left(\frac{qs}{\sqrt{1+r^2q^2}}\right), 0\right).$$

Choosing the normal vector to be

$$\mathbf{n}(s) = -\left(\cos\left(\frac{qs}{\sqrt{1+r^2q^2}}\right), \sin\left(\frac{qs}{\sqrt{1+r^2q^2}}\right), 0\right)$$

and using equation (12.1.30), we obtain the curvature

$$\kappa(s) = \frac{rq^2}{1+r^2q^2}.$$

Finally,

$$\mathbf{b}(s) = \frac{rq}{\sqrt{1+r^2q^2}}\left(\frac{1}{rq}\sin\left(\frac{qs}{\sqrt{1+r^2q^2}}\right), -\frac{1}{rq}\cos\left(\frac{qs}{\sqrt{1+r^2q^2}}\right), 1\right),$$

and

$$\mathbf{b}'(s) = \frac{q}{1+r^2q^2}\left(\cos\left(\frac{qs}{\sqrt{1+r^2q^2}}\right), \sin\left(\frac{qs}{\sqrt{1+r^2q^2}}\right), 0\right),$$

so that from equation (12.1.31)

$$\tau(s) = \frac{q}{1+r^2q^2}.$$

**Fenchel's theorem.** This states that for any closed curve

$$\oint_0^L \kappa(s)ds \geq 2\pi. \tag{12.1.34}$$

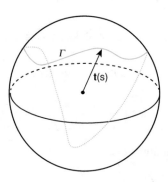

Fig. 12.10:  Tangential spherical map.

The proof use the so-called tangent spherical map, see Fig. 12.10: take the tail of the unit tangent vector $\mathbf{t}(s)$ and map it to the center of the unit sphere. The tip of $\mathbf{t}(s)$ then traces out a closed curve $\Gamma$ on the surface of the sphere as $s$ varies from 0 to $L$. If $l$ is the arc length of $\Gamma$ on the sphere, then

$$l = \oint_0^L \kappa(s)ds,$$

since $\|\mathbf{t}'(s)\| = \kappa(s)$. Moreover,

$$\mathbf{r}(s) - \mathbf{r}(0) = \int_0^s \mathbf{t}(s')ds',$$

so as the curve in $\mathbb{R}^3$ is closed, we have $\mathbf{r}(L) = \mathbf{r}(0)$, and thus

$$\oint_0^L \mathbf{t}(s')ds' = 0.$$

In order that each component of $\mathbf{t}(s')$ vanishes after integration, the curve $\Gamma$ must have support in both hemispheres. That is, $l \geq 2\pi$ (the arc length of the equator).

### 12.1.4 Worm-like chain model

Modeling a polymer as a continuous curve with elastic energy determined by continuum solid mechanics is known as a worm-like chain (WLC) [250, 251, 592, 775]. Suppose that the bending energy of the elastic rod model is dominant and consider the associated partition function

$$Z = \int \exp\left(-\frac{B}{2}\int_0^{L_0}\left\|\frac{d\mathbf{t}(s)}{ds}\right\|^2 ds\right)\int\prod_{s'}d\mathbf{t}(s'), \qquad (12.1.35)$$

where $d\mathbf{t}(s)$ denotes solid angle. This should be interpreted in terms of a path integral, since $s$ is a continuous variable (Chap. 8). We can thus proceed by discretizing the path integral into small segments of length $\Delta s$, which yields

$$Z = \int \exp\left(-\frac{B}{2\Delta s}\sum_i(\mathbf{t}_{i+1}-\mathbf{t}_i)^2\right)\prod_j d\mathbf{t}_j$$
$$= \int \exp\left(-\frac{B}{\Delta s}\sum_i(1-\cos\theta_{i+1,i})\right)\prod_j d\mathbf{t}_j, \qquad (12.1.36)$$

where $\mathbf{t}_i$ is the tangent vector of the $i$-th segment and $\mathbf{t}_{i+1}\cdot\mathbf{t}_i = \cos\theta_{i+1,i}$. Note that $\Delta s$ is much smaller than the Kuhn length $a$ of random walk models (Sect. 12.1.1).

Consider the two-point correlation $\langle\mathbf{t}_{i+1}\cdot\mathbf{t}_{i-1}\rangle$, which is evaluated using the local coordinate axes shown in Fig. 12.11:

$$\mathbf{t}_i = \mathbf{e}_1, \quad \mathbf{t}_{i-1} = \mathbf{e}_1\cos\theta_{i,i-1} + \mathbf{e}_2\sin\theta_{i,i-1},$$

and

$$\mathbf{t}_{i+1} = \mathbf{e}_1\cos\theta_{i+1,i} + \mathbf{e}_2\sin\theta_{i+1,i}\sin\phi_{i+1} + \mathbf{e}_3\sin\theta_{i+1,i}\cos\phi_{i+1}.$$

Here $\mathbf{e}_l, l = 1, 2, 3$ forms an orthonormal basis, and

$$\langle\mathbf{t}_{i+1}\cdot\mathbf{t}_{i-1}\rangle := \int \mathbf{t}_{i+1}\cdot\mathbf{t}_{i-1}\exp\left(-\frac{B}{\Delta s}\sum_i(1-\cos\theta_{i+1,i})\right)\prod_j d\mathbf{t}_j.$$

Since

$$\mathbf{t}_{i+1}\cdot\mathbf{t}_{i-1} = \cos\theta_{i+1,i}\cos\theta_{i,i-1} + \sin\theta_{i+1,i}\sin\theta_{i,i-1}\sin\phi_{i+1},$$

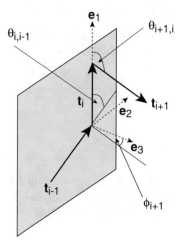

Fig. 12.11: Local coordinate axes for evaluation of the two-point correlation $\langle \mathbf{t}_{i+1} \cdot \mathbf{t}_{i-1} \rangle$.

and the bending energy is independent of the angles $\phi_i$, averaging with respect to these angles shows that

$$\langle \mathbf{t}_{i+1} \cdot \mathbf{t}_{i-1} \rangle = \langle \mathbf{t}_{i+1} \cdot \mathbf{t}_i \rangle \langle \mathbf{t}_i \cdot \mathbf{t}_{i-1} \rangle. \tag{12.1.37}$$

(For an application of this result to a freely-jointed chain with restricted bond angles; see Ex. 12.2.)

As the partition function decomposes into independent products, we can cancel out almost all the terms to obtain

$$\langle \mathbf{t}_{i+1} \cdot \mathbf{t}_i \rangle = \frac{\displaystyle\int_0^\pi \exp\left(-\frac{B}{\Delta s}(1-\cos\theta_{i+1,i})\right) \cos\theta_{i+1,i}\sin\theta_{i+1,i}d\theta_{i+1,i}}{\displaystyle\int_0^\pi \exp\left(-\frac{B}{\Delta s}(1-\cos\theta_{i+1,i})\right) \sin\theta_{i+1,i}d\theta_{i+1,i}}.$$

The right-hand side is independent of the segment label $i$, so we can write for all $i$

$$\langle \mathbf{t}_{i+1} \cdot \mathbf{t}_i \rangle = \frac{\displaystyle\int_{-1}^1 \exp\left(\frac{B}{\Delta s}\cos\theta\right)\cos\theta d(\cos\theta)}{\displaystyle\int_{-1}^1 \exp\left(\frac{B}{\Delta s}\cos\theta\right) d(\cos\theta)} = \frac{\partial}{\partial\alpha}\ln Z_1(\alpha)\bigg|_{\alpha=B/\Delta s},$$

where

$$Z_1(\alpha) = \int_{-1}^1 \exp\left(\alpha\cos\theta\right)d(\cos\theta) = 2\frac{\sinh\alpha}{\alpha}.$$

Hence

$$\langle \mathbf{t}_{i+1} \cdot \mathbf{t}_i \rangle = \coth\alpha - \frac{1}{\alpha} = \coth(B/\Delta s) - \frac{\Delta s}{B} \approx 1 - \frac{\Delta s}{B} \approx e^{-\Delta s/B} \tag{12.1.38}$$

in the limit $\Delta s \to 0$. Combining this result with the product rule (12.1.37) finally gives (in the continuum limit)

$$\langle \mathbf{t}(s) \cdot \mathbf{t}(s') \rangle = e^{-|s-s'|/B}. \tag{12.1.39}$$

Hence, the bending modulus $B$ determines the characteristic distance above which the tangent vectors decorrelate, which is precisely the persistence length. Therefore, we can make the identification $\xi_p = B$.

**Stretching under an applied force** Now suppose that there is a fixed external applied force $f$ directed along the $z$-axis. Using similar arguments to the analysis of the freely-jointed chain (Sect. 12.1.1), the continuum partition function takes the form

$$Z(f) = \exp\left( \frac{f}{k_B T} \int_0^L \mathbf{t}(s) \cdot \mathbf{e}_z ds - \frac{B}{2} \int_0^{L_0} \left\| \frac{d\mathbf{t}(s)}{ds} \right\|^2 ds \right) \int \prod_{s'} d\mathbf{t}(s'). \tag{12.1.40}$$

The corresponding partition function in the discretized version of the WLC model becomes

$$Z(f) = \int \exp\left( \sum_i \left[ \frac{f\Delta s}{k_B T} \cos\phi_i - \frac{B}{\Delta s}(1 - \cos\theta_{i+1,i}) \right] \right) \prod_j d\mathbf{t}_j, \tag{12.1.41}$$

where we have used

$$f\Delta s \sum_i \mathbf{t}_i \cdot \mathbf{e}_z = f\Delta s \sum_i \cos\phi_i.$$

We will analyze this model following along the lines of Ref. [592]; see also Chap. 9 of Ref. [663].

The first step is to assume that the polymer forms a closed loop with $N$ segments, $i = 1, \ldots, N$. This shouldn't affect the results for large $N$. We can then set

$$\sum_{i=1}^{N} \cos\phi_i = \frac{1}{2}\left[ \sum_{i=1}^{N-1} (\cos\phi_i + \cos\phi_{i+1}) + \cos\phi_1 + \cos\phi_N \right],$$

and adopt the transfer matrix approach of the Ising model (Sect. 12.1.1). The one major difference is that the discrete index $\sigma_i = \pm 1$ is replaced by the continuous index $\mathbf{t}_i$. More specifically, the partition function becomes

$$Z = \text{Tr}[M^N] := \int M(\mathbf{t}_1, \mathbf{t}_2) M(\mathbf{t}_2, \mathbf{t}_3) \ldots M(\mathbf{t}_N, \mathbf{t}_1) d\mathbf{t}_1 \ldots d\mathbf{t}_N, \tag{12.1.42}$$

where $M$ is a symmetric integral kernel

$$M(\mathbf{t}_i, \mathbf{t}_{i+1}) = \exp\left( \frac{f\Delta s}{2k_B T}(\mathbf{t}_i + \mathbf{t}_{i+1}) \cdot \mathbf{e}_z + \frac{B}{\Delta s}(\mathbf{t}_{i+1} \cdot \mathbf{t}_i - 1) \right). \tag{12.1.43}$$

The kernel $M$ defines an integral linear operator $\mathbb{M}$ acting on the Hilbert space of functions $L^2(S^2)$ with inner product

$$\langle A, B \rangle = \int_{S^2} A(\mathbf{t})B(\mathbf{t})d\mathbf{t}, \quad A, B \in L^2(S^2).$$

It is thus a compact operator[1] with a discrete spectrum consisting of the real eigenvalues $\lambda_l$ and a corresponding complete set of orthonormal eigenfunctions $\Psi_l(\mathbf{t})$. That is

$$\mathbb{M}\Psi_l(\mathbf{t}) := \int_{S^2} M(\mathbf{t}, \mathbf{t}')\Psi_l(\mathbf{t}')d\mathbf{t}' = \lambda_l \Psi_l(\mathbf{t}), \tag{12.1.44}$$

where integration is over the surface of the unit sphere. By analogy with the Ising model

$$Z = \sum_l \lambda_l^N \to \lambda_{max}^N \text{ as } N \to \infty \tag{12.1.45}$$

where $\lambda_{max}$ is the maximal eigenvalue.

As shown in Ref. [592], the Ritz variational method can be used to estimate the maximal eigenvalue $\lambda_{max}$. Completeness of the set of eigenfunctions means that an arbitrary function $V(\mathbf{t}) \in L^2(S^2)$ can be expanded as a uniformly convergent series

$$V(\mathbf{t}) = \sum_l c_l \Psi_l(\mathbf{t}) \tag{12.1.46}$$

for constant coefficients $c_l$. Define the so-called Rayleigh quotient

$$\Lambda[V] = \frac{\langle V, \mathbb{M}V \rangle}{\langle V, V \rangle}.$$

Substituting for $V$ using the eigenfunction expansion and exploiting orthonormality of the eigenfunctions shows that

$$\langle V, \mathbb{M}V \rangle = \int_{S^2} \left( \sum_l c_l \Psi_l(\mathbf{t}) \right) \mathbb{M} \left( \sum_m c_l \Psi_m(\mathbf{t}) \right) dt$$

$$= \sum_{l,m} c_l c_m \int_{S^2} \Psi_l(\mathbf{t})\mathbb{M}\Psi_m(\mathbf{t})dt = \sum_{l,m} c_l c_m \lambda_m \int_{S^2} \Psi_l(\mathbf{t})\Psi_m(\mathbf{t})dt$$

$$= \sum_{l,m} c_l c_m \lambda_m \delta_{l,m} = \sum_l \lambda_l c_l^2.$$

Performing a similar calculation for the denominator of the Rayleigh coefficient thus yields

$$\Lambda[V] = \frac{\sum_l \lambda_l c_l^2}{\sum_l c_l^2} \leq \lambda_{max}. \tag{12.1.47}$$

---

[1] A linear operator $\mathbb{L} : H \to H$ acting on a Hilbert space $H$ is said to be compact, if the image under $\mathbb{L}$ of any bounded subset of $H$ is a relatively compact subset (has compact closure). Such an operator is necessarily a bounded operator, and so continuous. Compact operators often arise in integral equations [734].

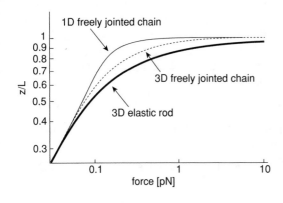

Fig. 12.12: Sketch of example force-extension curves for different polymer models. Parameters are chosen to be consistent with experimental data from DNA. The 3D elastic rod model gives the best fit, as shown elsewhere [592, 663].

The Ritz variational method consists of introducing a parameterized family of trial functions $V$ and maximizing $\Lambda[V]$ with respect to the parameter(s). Marko and Siggia [592] take $V_\omega(\mathbf{t}) = e^{\omega \mathbf{t} \cdot \mathbf{e}_z}$, and numerically look for the value $\omega^*$ that maximizes $\Lambda(\omega) = \Lambda[V_\omega]$. Setting $\Lambda^* = \Lambda(\omega^*)$, the force-displacement curve can be estimated using

$$z = Nk_B T \frac{d\ln\Lambda^*}{df}. \tag{12.1.48}$$

A reasonable functional approximation of the numerically obtained curve is [592]

$$f = f(z/L) := \frac{k_B T}{B}\left[\frac{z}{L} + \frac{1}{4(1-z/L)^2} - \frac{1}{4}\right]. \tag{12.1.49}$$

A sketch of a typical force-extension curve is shown in Fig. 12.12, where it is compared to analogous curves from the 1D and 3D freely-jointed chain models. The first gives the best match to experimental data on DNA [592, 663] for small to intermediate forces.

Finally, note that in the above analysis the external force $f$ has been fixed and the mean displacement has been determined using

$$z = -\frac{\partial F}{\partial f}, \quad F = -k_B T \ln Z,$$

with $F$ the total free energy. This can be related to the case where the displacement $z$ is fixed by performing the Legendre transformation $G(z) = F(f) + fz$ such that

$$dG(z) = dF(f) + f\,dz + z\,df = \frac{\partial F}{\partial f}df + f\,dz + z\,df = f\,dz.$$

We thus obtain the result $f(z) = G'(z)$. It follows that the original free energy can be written as

$$F = -fz + \int_0^z f(z'/L)dz'. \tag{12.1.50}$$

**Overstretching of DNA.** In Fig. 12.13, we show a sketch of a force-extension curve for DNA over a wide range of forces at room temperature. Such a curve is obtained experimentally by attaching one end of the DNA to a glass slide and the other to a micrometer, which is then pulled by optical or magnetic tweezers [829, 830]. At least five different regimes can be identified:

A  For $f < 0.01$ pN, the polymer acts like a random coil and Hooke's law holds.

B  At higher forces of around 1 pN, the relative extension levels off as it approaches unity—the polymer is nearly straight.

C  At forces beyond $f \approx 10$ pN, one finds $1 < z/L < 1 + \varepsilon$ with $\varepsilon \ll 1$ associated with elastic stretching.

D  At $f \approx 65$ pN, the DNA suddenly extends to $z \approx 1.6L$.

E  Another elastic regime occurs until the DNA breaks.

In order to account for the large force regions C and D of the force-extension curve, it is necessary to include two additional features [198]: (i) longitudinal or stretch deformations with associated elastic modulus, and (ii) each base pair can exists in two states, a regular state B and a stretched state S. Let $n_b$ and $n_s = N - n_b$ denote the number of base pairs that are in the B and S states, respectively, and let $\ell_b$ and $\ell_s$ be their corresponding molecular lengths. The chain is thus a sequence of groups of base pairs in either the B or S states. A major assumption of the model of Ref. [198] is that the bending and stretching elasticity of the chain depends only on $n_b$ and not on the details of the distribution of states along the chain. Hence, for fixed $n_b$, one can treat the polymer as a B polymer of length $n_b\ell_b$ in series with an S polymer of length $n_s\ell_s$.

The two subpolymers are modeled as WLCs with bending and elastic moduli $B_b, \kappa_b$ and $B_s, \kappa_s$, respectively. Thus, we can use a generalization of equation (12.1.50) to determine the elastic free energy of each subchain. It will be convenient to take $F$ and $z$ to be per unit length. In that case, the total elastic free energy per unit length is

$$F_{\text{elastic}} = n_b\ell_b F_b(z) + (N - n_b)\ell_s F_s(z), \qquad (12.1.51)$$

Fig. 12.13: Sketch of a typical force-extension curve for DNA at room temperature, illustrating different modes of behavior. A. Hookean regime. B. Saturation due to straightening of polymer. C. Elastic stretching of polymer. D. Sudden extension due to a structural transition. E. Another elastic regime.

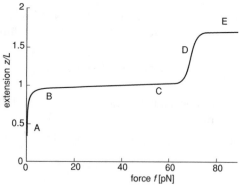
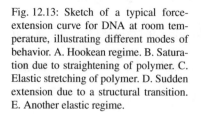

with

$$F_{b,s}(z) = -fz + \int_0^z f_{b,s}(z')dz' + \frac{f^2}{2\kappa_{b,s}}. \qquad (12.1.52)$$

The final term takes into account the additional extension due to stretching. Given $n_b$, the spatial degrees of freedom (spatial configuration) and the internal degrees of freedom (the state of each base pair) are independent. The energy of a given internal configuration can be expressed in terms of a 1D Ising model. Let $\varepsilon_S$ be the energy difference between an S state and a B state, and suppose that there is an interaction energy $\varepsilon_{bs}$ at each interface between neighboring B and S states. Introduce the Ising labels $\sigma_j$ with $\sigma_j = +1$ if the $j$-th base pair is in a B state and $\sigma_j = -1$ if it is in an S state. It follows that the energy of internal configurations is given by

$$F_{\text{internal}} = -\frac{\varepsilon_s}{2} \sum_{i=1}^N \sigma_i - \frac{\varepsilon_{bs}}{2} \sum_{i=1}^{N-1} \sigma_i \sigma_{i+1}. \qquad (12.1.53)$$

Since $F_{\text{elastic}}$ is linear in $b_b$, we can treat it as an external field of the Ising model. More specifically, setting

$$n_b = \sum_{i=1}^N (1 + \sigma_i)/2,$$

we can write the total free energy (up to a constant) as

$$F[\sigma] = -\frac{\varepsilon_{bs}}{2} \sum_{i=1}^{N-1} \sigma_i \sigma_{i+1} - h \sum_{i=1}^N \sigma_i, \qquad (12.1.54)$$

with $\sigma = (\sigma_1, \ldots, \sigma_N)$ and

$$h = \frac{1}{2}[\varepsilon_s + l_s F_s(z) - \ell_b F_b(z)]. \qquad (12.1.55)$$

The corresponding partition function is $Z = \sum_\sigma e^{-F[\sigma]/k_B T}$, which can be analyzed using transfer matrices; see Sect. 12.1.1. We thus find that the mean number of base pairs in the B state is given by

$$2\bar{n}|_b - 1 = \left\langle \sum_{i=1}^N \sigma_i \right\rangle = \frac{\partial \ln Z}{\partial h} = \frac{\sinh(h/k_B T)}{\sqrt{\sinh^2(h/k_B T) + e^{-2\varepsilon_{bs}/k_B T}}}. \qquad (12.1.56)$$

(Fluctuations about the mean for large $N$ will be negligible).

Finally, in order to determine the expected extension for a given force $f$, we imagine a B chain of extension per unit length $z_b(f) + f/\kappa_b$ in series with an S chain of extension per unit length $z_s(f) + f/\kappa_s$, where $z(f)$ is the inverse of the function $f(z)$. Thus

$$\Delta L = \bar{n}_b \ell_b[z_b(f) + f/\kappa_b] + (N - \bar{n}_b)\ell_s[z_s(f) + f/\kappa_s], \qquad (12.1.57)$$

with $\bar{n}_b$ given by equation (12.1.56), and

$$h = \frac{1}{2}[\varepsilon_s + l_s F_s(z_s(f)) - \ell_b F_b(z_b(f))].$$

One finds that the resulting force-extension curve can fit very well to experimental curves exhibiting the different regimes shown in Fig. 12.13 [198].

## 12.2 Statistical mechanics of membranes

Recall from Chap. 6 that the cell membrane of almost all organisms and many viruses are made of a lipid bilayer, as are the nuclear membrane surrounding the cell nucleus, and other membranes surrounding sub-cellular structures. Lipid bilayers are usually composed of amphiphilic phospholipids that have a hydrophilic phosphate head and a hydrophobic tail consisting of two fatty acid chains, see Fig. 12.14(a). By forming a double layer with the polar ends pointing outwards and the nonpolar tails pointing inwards, the membrane forms a physical barrier between the aqueous interior and exterior of the cell. The arrangements of lipids and various proteins, acting as receptors and channel pores in the membrane, control the entry and exit of other molecules as part of the cell's metabolism. The packing of lipids within the bilayer also affects its mechanical properties, including its resistance to stretching and bending. To a first approximation, the lipid membrane can be viewed as a two-dimensional fluid that cannot resist shear forces and whose molecular components diffuse within the membrane.

Amphiphiles such as the phospholipids of the cell membrane can self-assemble into aggregates in aqueous solution, provided that their concentration is above a critical threshold commonly known as the critical micelle concentration. There are two competing effects underlying the formation of aggregates. Clustering is favored energetically because it reduces the exposure of hydrophobic tails to contact with water. On the other hand, the uniform distribution of molecules in solution is favored

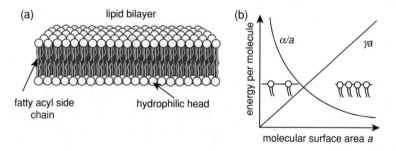

Fig. 12.14: Illustration of a lipid bilayer. [Public domain figure downloaded from Wikipedia.]

entropically. The equilibrium structure of an aggregate (e.g., number of layers, size, and shape) will depend on the details of individual lipids, such as the size of the given head group or the length of the fatty chains. We will explore the formation of molecular aggregates in Sect. 13.3. Here, we focus on the statistical mechanical properties of cell membranes. Before proceeding to the general theory, it is useful to explore some of the specific energetic properties of lipid bilayers [87].

Consider the interface formed by water and one of the layers of amphiphiles. At low densities, there is an energy cost per molecule arising from exposure of a hydrocarbon chain to water. This is taken to be of the form $\gamma a$, where $a$ is the mean surface area occupied by a molecule and $\gamma$ is the effective surface tension of the interface. On the other hand, at high densities, molecules are crowded together leading to a repulsive energy of the form $\alpha/a$, say. These two competing effects are illustrated in Fig. 12.14(b). The total energy per molecule is $E = \alpha/a + \gamma a$. The minimum free energy occurs at the equilibrium area $a_0 = \sqrt{\alpha/\gamma}$, and for small deviations about $a_0$,

$$E \approx 2\gamma a_0 + \frac{\gamma}{a_0}(a - a_0)^2.$$

Hence, up to a constant, we can take the energy density to be

$$\mathscr{E} = \frac{E}{a_0} = \gamma\left(\frac{\Delta a}{a}\right)^2. \tag{12.2.1}$$

Below we will relate the surface tension $\gamma$ to the so-called compression modulus $K_A$ of an elastic 2D sheet.

## 12.2.1 Elastic sheet model of a membrane

We begin by treating a biomembrane as a thin two-dimensional elastic sheet, just as a polymer was treated as an elastic rod (see Sect. 12.1.3). The various types of

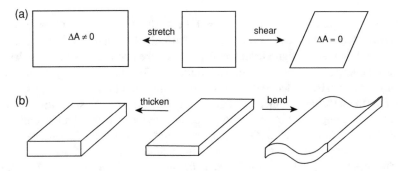

Fig. 12.15: Different types of deformation of a membrane sheet: (a) In-plane. (b) Out-of-plane.

deformations of a thin sheets are illustrated in Fig. 12.15. We will focus on bending deformations, as these are the easiest to excite via thermal fluctuations, and shear deformations tend not to be relevant in the case of fluid membranes. In the case of bending deformations, it is necessary to extend the notion of curvature from curves to surfaces. Here we give a heuristic treatment. The local curvature of a surface is found by constructing the best circle along two orthogonal directions. That is, locally one can find a coordinate system $(x, y, z)$ such that

$$z = \frac{\sigma_1}{2}x^2 + \frac{\sigma_2}{2}y^2,$$

where $\sigma_{1,2}$ are known as the principal curvatures or curvature coefficients, see Fig. 12.16. Given $\sigma_1, \sigma_2$, the energy density (per unit area) is [540]

$$\mathscr{E} = \frac{\kappa_b}{2}(\sigma_1 + \sigma_2)^2 + \kappa_G \sigma_1 \sigma_2, \tag{12.2.2}$$

where $(\sigma_1 + \sigma_2)/2$ is the mean curvature, $\sigma_1 \sigma_2$ is known as the Gaussian curvature, $\kappa_b$ is the bending modulus or rigidity, and $\kappa_G$ is the Gaussian bending rigidity or saddle-splay modulus. In the case of simple surfaces such as the sphere and cylinder, the curvature coefficients are constants. Hence, the total energy is simply obtained by multiplying the energy density by the corresponding surface area

$$E = \begin{cases} 4\pi(2\kappa_b + \kappa_G) & \text{sphere} \\ \pi\kappa_b L/R & \text{cylinder of length } L \text{ and radius } R \end{cases}. \tag{12.2.3}$$

These expressions are used to estimate the bending energy of bacterial membranes in Ex. 12.3 and the energetics of membrane tethering (extruding nanotubes) in Ex. 12.4. It turns out more generally that the contribution to the total energy from the Gaussian curvature term has a simple form that is a consequence of a basic result in differential geometry known as the Gauss-Bonet theorem. This states that for a compact manifold $\Omega$ without a boundary

$$\int_\Omega \sigma_1 \sigma_2 d^2\mathbf{r} = 4\pi(1-g), \tag{12.2.4}$$

where $g$ is the genus of the surface ($g = 0$ for a sphere, $g = 1$ for a torus, $g = 2$ for a pretzel, etc.). In the case of a compact manifold with boundary $\partial\Omega$, there is an additional boundary integral term that depends on the curvature of the boundary.

**Bending energy of a sheet.** Recall from Sect. 12.1.3 that the bending modulus $B$ of an elastic rod can be related to the Young's or elastic modulus $Y$. An analogous result holds for the bending modulus $\kappa_b$ of an elastic sheet. In order to show this, consider the gentle bending of a thin sheet into a section of cylinder with principal curvatures $(1/R, 0)$. (The corresponding Gaussian curvature is zero.) Take the surface of zero strain (the neutral surface) to run through the middle of the sheet, and assume that the thickness $d$ of the sheet is unchanged after deformation, see Figure 12.17. Since there is no compression of the neutral surface, it subtends an angle $\theta = x/R \ll 1$ on

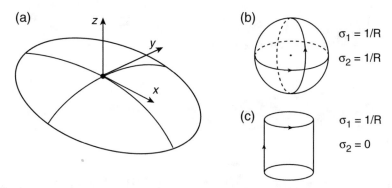

Fig. 12.16: Local curvature coefficients. (a) Local coordinate system and corresponding orthogonal circles with radii $R_1, R_2$. The local curvature coefficients are defined by $\sigma_j = 1/R_j$. (b) For a sphere of radius $R$, $\sigma_1 = \sigma_2 = 1/R$. (c) For a cylinder of radius $R$, $\sigma_1 = 1/R$ and $\sigma_2 = 0$.

the circle of curvature. The displacement in the $x$-direction is

$$\Delta x = x' - x = R\sin\theta - x \approx R(\theta - \theta^3/3!) - x \approx R\left(\frac{x}{R} - \frac{x^3}{6R^3}\right) - x = -\frac{x^3}{6R^2}.$$

Similarly, the displacement in the $z$-direction is

$$\Delta z = R\cos\theta - R \approx -\frac{R\theta^2}{2} = -\frac{x^2}{2R},$$

such that $|\Delta z| \gg |\Delta x|$. The components of the displacement vector $\mathbf{u}$ on the neutral surface are thus

$$u_x \approx 0, \quad u_z \approx \xi(x) = -\frac{x^2}{2R}.$$

In the limit $d \to 0$, we can also take $u_z \approx \xi(x)$ off the neutral surface. In order to determine the displacement $u_x$ away from the neutral surface, we exploit the fact that for a thin plate the forces perpendicular to a plate's surface are negligible compared to internal stresses. That is, to a first approximation the components of the stress tensor (see Box 12B) in the $z$-direction are zero: $\sigma_{xz} = \sigma_{zz} = 0$. This implies that the component $u_{xz}$ of the strain tensor is zero:

$$u_{xz} \approx \frac{\partial u_x}{\partial z} + \frac{\partial u_z}{\partial x} = 0 \implies u_x(x,z) = -z\frac{\partial \xi(x)}{\partial x},$$

and thus

$$u_{xx} = -z\frac{\partial^2 \xi}{\partial x^2} = \frac{z}{R}.$$

The bending energy density (see Box 12B) is thus determined by averaging with respect to $z$

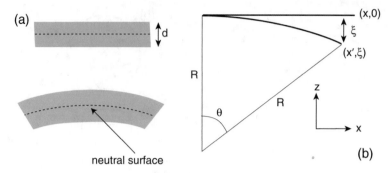

Fig. 12.17: (a) Bending of a thin sheet to form part of a cylinder of radius $R$. (b) Coordinate representation of neutral surface.

$$\mathcal{E} = \frac{K_A}{2d} \int_{-d/2}^{d/2} u_{xx}(z)^2 dz = \frac{K_A}{2dR^2} \int_{-d/2}^{d/2} z^2 dz = \frac{K_A}{2dR^2} \left[ \frac{z^3}{3} \right]_{-d/2}^{d/2} = \frac{1}{2R^2} \frac{K_A d^2}{12},$$

where $K_A$ is the so-called compression modulus. Comparison with equation (12.2.2) finally yields the following relationship between the bending modulus $\kappa_b$ and the compression modulus $K_A$:

$$\kappa_b = \frac{K_A d^2}{12}. \tag{12.2.5}$$

In the case of a lipid layer, one can estimate $K_A$ and hence $\kappa_b$ by relating the former to the surface tension $\gamma$ of the layer. The basic connection is made by noting that for a small compression of the layer, the relative change in surface can be related to the strain tensor (Box 12B) according to

$$\frac{\Delta A}{A} = u_{xx} + u_{yy}. \tag{12.2.6}$$

Hence, comparing equations (12.2.1) and (12.2.13) shows that $\gamma = K_A/2$. Taking the typical values $\gamma \sim 0.05 \text{Jm}^{-2}$ and $d \sim 4\text{nm}$, we obtain the estimate $\kappa_b \sim 10^{-19}$ $\text{J} \approx 10 k_B T$. We deduce that bending energies are comparable to thermal energies at room temperature, so the shape of a membrane is expected to fluctuate.

---

**Box 12B Membrane elasticity: strain and stress tensors.**

---

**Deformations and the strain tensor (2D).** Consider a nonuniform deformation of some domain that is specified by the displacement vector $\mathbf{u}$ such that (see Fig. 12.18(a))

$$\mathbf{r} \rightarrow \mathbf{r}' = \mathbf{r} + \mathbf{u}(\mathbf{r}).$$

(We focus on 2D but similar constructions hold in 3D.) In terms of infinitesimal displacements,

$$dx_i' = dx_i + \sum_{j=1}^{d} \frac{\partial u_i}{\partial x_j} dx_j,$$

where $d = 2$ (or $d = 3$ for corresponding deformations in 3D). Defining $d\ell^2 = \sum_{i=1}^{d} dx_i^2$ and $d\ell'^2 = \sum_{i=1}^{d} dx_i'^2$, we have

$$d\ell'^2 = d\ell^2 + 2\sum_{i,j=1}^{d} u_{ij} dx_i dx_j, \qquad (12.2.7)$$

where

$$u_{ij} = \frac{1}{2}\left[\frac{\partial u_i}{\partial x_j} + \frac{\partial u_j}{\partial x_i} + \sum_{k=1}^{d} \frac{\partial u_k}{\partial x_i}\frac{\partial u_k}{\partial x_j}\right] \qquad (12.2.8)$$

is the strain tensor.

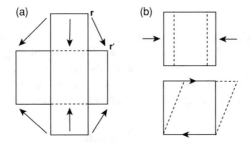

Fig. 12.18: (a) Nonuniform deformations of a membrane. (b) The type of deformation induced by a force depends on the orientation of the force with respect to the boundary it acts on.

Since $u_{ij}$ are components of a symmetric matrix, a diagonalized local coordinate system can be found such that

$$d\ell'^2 = \sum_{\alpha=1}^{d} (1 + 2U_\alpha)(dx_\alpha)^2,$$

which means $dx_\alpha' = \sqrt{1 + 2U_\alpha}\, dx_\alpha$. In the case of small deformations $\mathbf{u}$, we have to leading order

$$u_{ij} = \frac{1}{2}\left[\frac{\partial u_i}{\partial x_j} + \frac{\partial u_j}{\partial x_i}\right], \quad dx_\alpha' = (1 + U_\alpha)dx_\alpha.$$

Therefore, an infinitesimal area element ($d = 2$) changes as

$$dA' = \prod_{\alpha=1,2}(1 + U_\alpha)dA \approx (1 + \sum_{\alpha=1,2} U_\alpha),$$

Since the trace of the stress tensor is invariant under coordinate transformations, we conclude that $\Delta A/A = \mathrm{Tr}[\mathbf{u}]$. This motivates decomposing the stress tensor into the sum of a pure shear (no area change) and a compression:

$$u_{ij} = \frac{\delta_{i,j}}{d}\mathrm{Tr}[\mathbf{u}] + \left(u_{ij} - \frac{\delta_{i,j}}{d}\mathrm{Tr}[\mathbf{u}]\right). \qquad (12.2.9)$$

**Forces and the stress tensor.** A membrane deforms in response to external forces, which may have arbitrary orientations with respect to the boundaries of the membrane, see Fig. 12.18(b). Let $\mathbf{n}\,dl$ be a local element of a boundary (which is a curve for $d = 2$) with $\mathbf{n}$ the unit vector normal to the boundary. The components of the applied force $\mathbf{f}$ are related to the components of the local boundary element according to the stress tensor $\sigma_{ij}$ (see also Box 10B)

$$df_i = \sum_{j=1}^{d} \sigma_{ij} n_j dl. \qquad (12.2.10)$$

By considering moments of the forces, it can be shown that $\sigma_{ij} = \sigma_{ji}$ [540]. For a simple spring, the force is proportional to the displacement from equilibrium, which we used in applying elasticity theory to polymers. Similarly, for higher-dimensional elastic media ($d = 2,3$), the stress tensor (force) is linearly related to the strain tensor (displacement) according to the general formula

$$\sigma_{ij} = \sum_{k,l=1}^{d} c_{ijkl} u_{kl}, \qquad (12.2.11)$$

where the $c_{ijkl}$ are known as the elastic moduli. The corresponding elastic energy is then given by

$$E = \frac{1}{2}\int_\Omega \sum_{i,j,k,l=1}^{d} c_{ijkl} u_{ij} u_{kl} d\mathbf{r}. \qquad (12.2.12)$$

Since $u_{ij}$ is symmetric and $E$ is invariant under the interchange $(i,j) \leftrightarrow (k,l)$,

$$c_{ijkl} = c_{klij} = c_{ijlk} = c_{jikl}.$$

This means that there are a maximum of 6 independent elastic moduli (and 21 in 3D). However, further simplification occurs in isotropic media, where there are only two elastic moduli and the energy can be written as

$$E = \frac{1}{2}\int_\Omega \left[K(\mathrm{Tr}[\mathbf{u}])^2 + 2\mu \sum_{i,j=1}^{d}\left(u_{ij} - \frac{\delta_{i,j}}{d}\mathrm{Tr}[\mathbf{u}]\right)^2\right] d\mathbf{r}, \qquad (12.2.13)$$

> where $K$ is the bulk or compression modulus and $\mu$ is the shear modulus. Typically, one writes $K = K_A$ in 2D and $K = K_V$ in 3D. The relationship between stress and strain tensors is explored further in Ex. 12.5.

## 12.2.2 Membrane fluctuations

We now investigate the effects of thermal fluctuations, under the assumption that the dominant energy contribution is due to bending of the membrane sheet. In order to proceed, we need to introduce a particular description of a surface known as the Monge representation, which is applicable provided the fluctuations are not too large. We will also need to deal with a partition function that takes the form of a path integral (Chap. 8), and analyze statistical correlations using functional derivatives. The latter are reviewed in Box 12C. For a more detailed presentation of these methods; see Refs. [87, 782].

**Monge representation of a surface.** Suppose that we can represent a membrane surface using the so-called Monge representation $z = h(x,y)$, where $(x,y)$ parameterize the surface. The tangent plane to the surface at $\mathbf{r} = (x,y,z)$ is spanned by the two tangent vectors

$$\mathbf{r}_x = \frac{\partial \mathbf{r}}{\partial x} = (1,0,h_x), \quad \mathbf{r}_y = \frac{\partial \mathbf{r}}{\partial y} = (0,1,h_y), \tag{12.2.14}$$

where $h_x = \partial h/\partial x$, etc. The unit normal to the surface at $\mathbf{r}$ is

$$\mathbf{n} = \frac{\mathbf{r}_x \times \mathbf{r}_y}{|\mathbf{r}_x \times \mathbf{r}_y|} = \frac{1}{\sqrt{1+h_x^2+h_y^2}}(-h_x,-h_y,1). \tag{12.2.15}$$

The infinitesimal distance between two points on the surface is

$$\begin{aligned} ds^2 &= d\mathbf{r} \cdot d\mathbf{r} = (\mathbf{r}_x dx + \mathbf{r}_y dy) \cdot (\mathbf{r}_x dx + \mathbf{r}_y dy) \\ &= (1+h_x^2)dx^2 + (1+h_y^2)dy^2 + 2h_x h_y dx\,dy. \end{aligned} \tag{12.2.16}$$

and an infinitesimal area element is

$$dA = |\mathbf{r}_x \times \mathbf{r}_y| dx\,dy = \sqrt{1+h_x^2+h_y^2}dx\,dy. \tag{12.2.17}$$

Given a curve $\mathbf{r} = \mathbf{r}(s)$ on the surface, define the normal curvature at the point $s$ as (compare with the definition of curvature in Box 12A)

$$\sigma = \frac{d^2\mathbf{r}}{ds^2} \cdot \mathbf{n}. \tag{12.2.18}$$

From the chain rule of differentiation

$$\frac{d\mathbf{r}}{ds} = \mathbf{r}_x \frac{dx}{ds} + \mathbf{r}_y \frac{dy}{ds},$$

and

$$\frac{d^2\mathbf{r}}{ds^2} = \mathbf{r}_x \frac{d^2x}{ds^2} + \mathbf{r}_y \frac{d^2y}{ds^2} + \mathbf{r}_{xx} \frac{dx}{ds}\frac{dx}{ds} + 2\mathbf{r}_{xy} \frac{dx}{ds}\frac{dy}{ds} + \mathbf{r}_{yy} \frac{dy}{ds}\frac{dy}{ds}.$$

Since $\mathbf{r}_x \cdot \mathbf{n} = 0 = \mathbf{r}_y \cdot \mathbf{n}$, it follows that

$$\begin{aligned}
\sigma = \mathbf{n} \cdot \left( \mathbf{r}_{xx} \frac{dx}{ds}\frac{dx}{ds} + 2\mathbf{r}_{xy} \frac{dx}{ds}\frac{dy}{ds} + \mathbf{r}_{yy} \frac{dy}{ds}\frac{dy}{ds} \right) \\
= \frac{h_{xx}}{\Gamma} \left( \frac{dx}{ds} \right)^2 + \frac{h_{yy}}{\Gamma} \left( \frac{dy}{ds} \right)^2 + \frac{2h_{xy}}{\Gamma} \frac{dx}{ds}\frac{dy}{ds},
\end{aligned} \tag{12.2.19}$$

where $\Gamma = \sqrt{1 + h_x^2 + h_y^2}$. As a further simplification, assume that the curvature of the surface is small, so that $\Gamma \approx 1$ and $ds^2 \approx dx^2 + dy^2$.

We can now determine the principal curvatures $\sigma_1, \sigma_2$ by finding the extrema of $\sigma$ given the constraint that the length of the displacement vector $d\mathbf{r}/ds$ is unity. This can be achieved by introducing a Lagrange multiplier $\lambda$ and finding the extrema of

$$\hat{\sigma}(l, m, \lambda) = h_{xx}l^2 + h_{yy}m^2 + 2h_{xy}lm - \lambda(l^2 + m^2 - 1),$$

where $l = x'(s)$ and $m = y'(s)$. We thus obtain the equations

$$0 = \frac{1}{2}\frac{\partial \hat{\sigma}}{\partial l} = h_{xx}l + h_{xy}m - \lambda l,$$

$$0 = \frac{1}{2}\frac{\partial \hat{\sigma}}{\partial m} = h_{yy}m + h_{xy}l - \lambda m.$$

Multiplying the first equation by $l$, the second by $m$, and adding the results shows that $\lambda = \sigma$. We thus have the matrix equation

$$\begin{pmatrix} h_{xx} & h_{xy} \\ h_{xy} & h_{yy} \end{pmatrix} \begin{pmatrix} l \\ m \end{pmatrix} = \sigma \begin{pmatrix} l \\ m \end{pmatrix}. \tag{12.2.20}$$

It follows that the principal curvatures are the eigenvalues of the matrix $\mathbf{H}$ on the left-hand side. In particular, the mean and Gaussian curvatures are

$$\bar{\sigma} := \frac{\sigma_1 + \sigma_2}{2} = \mathrm{Tr}[\mathbf{H}] = \frac{h_{xx} + h_{yy}}{2}, \tag{12.2.21a}$$

$$\sigma_G := \sigma_1 \sigma_2 = \mathrm{Det}[\mathbf{H}] = h_{xx}h_{yy} - h_{xy}^2. \tag{12.2.21b}$$

Finally, note that the exact expressions for the curvatures in the Monge representation are

$$\bar{\sigma} = \frac{(1+h_y^2)h_{xx} + (1+h_x^2)h_{yy} - 2h_x h_y h_{xy}}{2(1+h_x^2+h_y^2)^{3/2}}, \tag{12.2.22a}$$

$$\sigma_G = \frac{h_{xx}h_{yy} - h_{xy}^2}{(1+h_x^2+h_y^2)^2}. \tag{12.2.22b}$$

Examples of the Monge representation are considered in Ex. 12.6 and Ex. 12.7.

**Partition function for membrane fluctuations.** When considering small thermal fluctuations of a membrane that is treated as a bending elastic sheet, one need only include contributions to the bending energy involving the mean curvature, since the contribution from the Gaussian curvature is a constant. The latter is a consequence of the Gauss-Bonet theorem and the assumption that fluctuations at the boundary of the membrane can be neglected. Using the above Monge representation, we then have the total energy

$$E[h] = \frac{\kappa_b}{2} \int_\Omega (\sigma_1 + \sigma_2)^2 d^2\mathbf{r} = \frac{\kappa_b}{2} \int_\Omega \left[\frac{\partial^2 h}{\partial x^2} + \frac{\partial^2 h}{\partial y^2}\right]^2 dx\,dy. \tag{12.2.23}$$

Introduce the path integral partition function

$$Z = \int e^{-\beta E[h]} \mathcal{D}[h], \quad \beta = \frac{1}{k_B T}. \tag{12.2.24}$$

Here $\mathcal{D}[h]$ is a formal representation of the path integral measure. Recall from Chap. 8 that one way to interpret the path integral is to discretize the spatial coordinates of the membrane to obtain a large-dimensional multiple integral, and to define the path integral as the continuum limit of the latter (assuming it exists).

For the sake of illustration, take $\Omega$ to be a square sheet of area $A = L^2$ and $0 \le x, y \le L$. Impose periodic boundary conditions and introduce the Fourier series expansions

$$h(\mathbf{r}) = \frac{1}{A} \sum_\mathbf{q} \widehat{h}(\mathbf{q}) e^{i\mathbf{q}\cdot\mathbf{r}}, \quad \widehat{h}(\mathbf{q}) = \int_\Omega \widehat{h}(\mathbf{q}) e^{-i\mathbf{q}\cdot\mathbf{r}} dx\,dy, \tag{12.2.25}$$

with

$$q_x = \frac{2\pi n}{L}, \quad q_y = \frac{2\pi m}{L}, \quad \text{for integers } n, m.$$

For simplicity, assume that

$$\widehat{h}(0) = \int_\Omega \widehat{h}(\mathbf{q}) dx\,dy = 0.$$

The partition function then becomes an infinite product of Gaussian integrals

$$Z = \int \prod_{\mathbf{q}\neq 0} \exp\left(-\frac{\beta\kappa_b}{2A}|\widehat{h}(\mathbf{q})|^2 q^4\right) \mathcal{D}[\widehat{h}] = \prod_{\mathbf{q}\neq 0} \left(\frac{2\pi A}{\beta\kappa_b}\right)^{1/2} \frac{1}{q^2}. \tag{12.2.26}$$

In order to calculate statistical quantities, we introduce the so-called generating functional (analogous to the generating function of classical probability theory)

$$\Gamma[v] = \int \exp\left(-\beta \left[E[h] - \int_\Omega h(\mathbf{r})v(\mathbf{r})dxdy\right]\right) \mathcal{D}[h]. \qquad (12.2.27)$$

Using the definition of a functional derivative (see Box 12C), we have

$$\langle h(\mathbf{r})\rangle = \frac{1}{\beta}\left.\frac{\delta \ln \Gamma}{\delta v(\mathbf{r})}\right|_{v=0}, \qquad (12.2.28)$$

and

$$G(\mathbf{r},\mathbf{r}') := \langle h(\mathbf{r})h(\mathbf{r}')\rangle - \langle h(\mathbf{r})\rangle\langle h(\mathbf{r}')\rangle = \frac{1}{\beta^2}\left.\frac{\delta^2 \ln \Gamma}{\delta v(\mathbf{r})\delta v(\mathbf{r}')}\right|_{v=0}. \qquad (12.2.29)$$

In terms of Fourier series

$$\Gamma[v] = \int \prod_{\mathbf{q}\neq 0}\exp\left(-\frac{\beta}{2A}\left[\kappa_b q^4 |\hat{h}(\mathbf{q})|^2 - \hat{h}(\mathbf{q})\hat{v}(-\mathbf{q}) - \hat{h}(-\mathbf{q})\hat{v}(\mathbf{q})\right]\right)\mathcal{D}[\hat{h}]$$

$$= \prod_{\mathbf{q}\neq 0}\left(\frac{2\pi A}{\beta \kappa_b}\right)^{1/2}\frac{1}{q^2}\exp\left(\frac{\beta}{2A\kappa_b q^4}|\hat{v}(\mathbf{q})|^2\right). \qquad (12.2.30)$$

Taking logs yields

$$\ln \Gamma[v] = \frac{\beta}{2A\kappa_b}\sum_{\mathbf{q}\neq 0}|\hat{v}(\mathbf{q})|^2 q^{-4} + \text{v-independent terms} \qquad (12.2.31)$$

$$= \frac{\beta^2}{2}\int\int v(\mathbf{r})G(\mathbf{r},\mathbf{r}')v(\mathbf{r}')d\mathbf{r}\,d\mathbf{r}' + \ldots, \qquad (12.2.32)$$

where

$$G(\mathbf{r},\mathbf{r}') = \frac{k_B T}{\kappa_b A}\sum_{\mathbf{q}\neq 0}\frac{e^{-i\mathbf{q}\cdot(\mathbf{r}-\mathbf{r}')}}{q^4}. \qquad (12.2.33)$$

It immediately follows that $\langle h\rangle = 0$ and the variance of membrane fluctuations is given by

$$\langle |\hat{h}(\mathbf{q})|^2\rangle = \frac{k_B T A}{\kappa_b q^4}$$

in Fourier space, and

$$\langle h(\mathbf{r})^2\rangle = G(\mathbf{r},\mathbf{r}) = \frac{k_B T}{\kappa_b A}\sum_{\mathbf{q}\neq 0}\frac{1}{q^4}. \qquad (12.2.34)$$

The sum on the right-hand side of (12.2.34) can be evaluated using a continuum approximation

$$\left(\frac{2\pi}{L}\right)^2 \sum_{\mathbf{q}\neq 0} \rightarrow \int d^2\mathbf{q}.$$

Introducing polar coordinates for $\mathbf{q}$ and noting that $A = L^2$,

$$\langle h(\mathbf{r})^2 \rangle = \frac{k_B T}{\kappa_b L^2}\left(\frac{L}{2\pi}\right)^2 \int_{q_{min}}^{q_{max}} \frac{2\pi q\, dq}{q^4}.$$

Here $q_{min} = 2\pi/L$, where $L$ is the size of the membrane and we have introduced an upper cut-off $q_{max} = \pi/l$, where $l$ is a typical size of a molecule forming the membrane with $l \ll L$. Evaluating the integral thus yields the approximation

$$\langle h(\mathbf{r})^2 \rangle = \frac{k_B T A}{16\kappa_b \pi^3}. \tag{12.2.35}$$

**Persistence length.** By analogy with polymers, we define the persistence length of a fluctuating membrane in terms of the two-point correlation function, see Fig. 12.19

$$g(\mathbf{r}) = \langle \mathbf{n}(\mathbf{r}) \cdot \mathbf{n}(0) \rangle. \tag{12.2.36}$$

From equation (12.2.15) and the small displacement assumption,

$$\mathbf{n} \approx (-h_x, -h_y, 1 - [h_x^2 + h_y^2]/2),$$

so that

$$g(\mathbf{r}) = \langle h_x(\mathbf{r})h_x(0) \rangle + \langle h_y(\mathbf{r})h_y(0) \rangle + 1 - \frac{1}{2}\langle h_x^2(\mathbf{r}) + h_y^2(\mathbf{r}) \rangle - \frac{1}{2}\langle h_x^2(0) + h_y^2(0) \rangle$$

$$= 1 - \frac{1}{A^2}\sum_{\mathbf{q}\neq 0} q^2 \langle |\hat{h}(\mathbf{q})|^2 \rangle [1 - e^{i\mathbf{q}\cdot\mathbf{r}}],$$

$$= 1 - \frac{1}{A^2}\sum_{\mathbf{q}\neq 0} q^2 \langle |\hat{h}(\mathbf{q})|^2 \rangle [1 - \cos(\mathbf{q}\cdot\mathbf{r})], \tag{12.2.37}$$

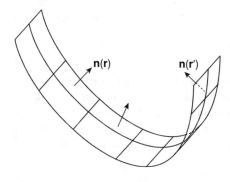

Fig. 12.19: Decorrelation of the surface normal $\mathbf{n}$ for $|\mathbf{r} - \mathbf{r}'| > \xi_p$, where $\xi_p$ is the persistence length.

since $\sin(\mathbf{q} \cdot \mathbf{r})$ is an odd function of $\mathbf{q}$. Going to a continuum limit and using equation (12.2.34),

$$
\begin{aligned}
g(\mathbf{r}) &= 1 - \frac{k_B T}{4\pi^2 \kappa_b} \int q^{-2}[1 - \cos(\mathbf{q} \cdot \mathbf{r})] d^2\mathbf{q} \\
&= 1 - \frac{k_B T}{4\pi^2 \kappa_b} \int_{q_{min}}^{q_{max}} \int_0^{2\pi} [1 - \cos(qr\cos\theta)] d\theta \frac{dq}{q} \\
&= 1 - \frac{k_B T}{2\pi \kappa_b} \int_{q_{min}}^{q_{max}} [1 - J_0(qr)] \frac{dq}{q},
\end{aligned}
\tag{12.2.38}
$$

where $J_0(z)$ is the zeroth-order Bessel function. Note that $J_0(z) \sim 1$ for small $z$ and $J_0(z) \sim 0$ for $z \geq \pi$, say. Hence, we can take the lower limit to be such that $q_{min}r = \pi$, that is, $q_{min} = \pi/r$. With $q_{max} = \pi/l$ as before

$$
g(\mathbf{r}) \approx 1 - \frac{k_B T}{2\pi \kappa_b} \int_{\pi/r}^{\pi/l} \frac{dq}{q} = 1 - \frac{k_B T}{2\pi \kappa_b} \ln\left(\frac{r}{l}\right).
\tag{12.2.39}
$$

Finally, defining the persistence length $\xi_p$ according to

$$
\langle \mathbf{n}(\mathbf{r}) \cdot \mathbf{n}(0) \rangle = e^{-r/\xi_p},
\tag{12.2.40}
$$

we set $r = \xi_p$ in the expression for $g(\mathbf{r})$ and determine $\xi_p$ by requiring $g(\xi_p) \approx 0$, which yields

$$
\xi_p \sim l \exp\left(\frac{2\pi \kappa_b}{k_B T}\right).
\tag{12.2.41}
$$

If $l = 1\,\text{nm}$ and $\kappa_b \sim 10 k_B T$ then $\xi_p \sim 10^6$ km, which implies that membranes undulate smoothly at a cellular length scale.

---

**Box 12C. Functionals and functional derivatives.**

---

In the study of continuum structures such as membranes, one encounters energies $E$ that are expressed in terms of spatial integrals of energy densities $\mathcal{E}$

$$
E[h] \equiv \int \mathcal{E}(h(\mathbf{r})) d^2\mathbf{r}.
$$

Since $E[h]$ is a real number, it can be viewed as a function of a function, which is commonly referred to as a functional. Membrane fluctuations can then be analyzed in terms of a generating functional, provided that one introduces the notion of a functional derivative. Time-dependent versions of functionals also occur in the nonequilibrium analysis of phase separation and biological condensates (see Sect. 13.1), that is,

$$F[\phi] \equiv \int \mathscr{F}(\phi(\mathbf{r},t)d^3\mathbf{r},$$

where $\phi(\mathbf{r},t)$ is the local density of a solute. Analyzing the kinetics of phase separation again requires taking functional derivatives. Functional derivatives also appear in the least-action principle of classical Newtonian mechanics, where one minimizes the classical action, which is given by an integral of a Lagrangian function with respect to time (see Box 8A),

$$S[x] = \int_0^T L(x(t),\dot{x}(t))dt.$$

In this case, minimization leads to the Euler-Lagrange equations. Generating functionals and functional derivatives also played a major role in the analysis of path integral formulations of statistical field theory (see Sect. 8.9).

Consider some functional $F[f]$ with $f : \mathbb{R} \to \mathbb{R}$. The functional derivative can be defined by considering the variation $\delta F$ that results from variation of the function $f$ by $\delta f$: $\delta F := F[f + \delta f] - F[f]$. The major step is to set $\delta f(x) = \varepsilon \eta(x)$, where $\eta$ is an arbitrary function on $\mathbb{R}$ (within some given space of allowed test functions), and then to note that $F[f + \varepsilon \eta]$ is simply a function of $\varepsilon$. This implies that the expansion in terms of powers of $\varepsilon$ is a standard Taylor expansion (assuming differentiability up to some finite or infinite order)

$$F[f+\varepsilon\eta] = F[f] + \frac{dF[f+\varepsilon\eta]}{d\varepsilon}\bigg|_{\varepsilon=0} + \frac{1}{2}\frac{d^2F[f+\varepsilon\eta]}{d\varepsilon^2}\bigg|_{\varepsilon=0} + \cdots$$

The definition of the first-order functional derivative $\delta F[f]/\delta f$ is taken to be

$$\frac{dF[f+\varepsilon\eta]}{d\varepsilon}\bigg|_{\varepsilon=0} := \int_{-\infty}^{\infty}\frac{\delta F[f]}{\delta f(x)}\eta(x)dx. \qquad (12.2.42)$$

Implicit in this definition is that the left-hand side can be expressed in the form of a linear functional with kernel $\delta F[f]/\delta f$ acting on the test function $\eta$. This is not guaranteed for arbitrary functions. A functional for which equation (12.2.42) is valid is said to be Frechet differentiable. More precisely, a functional $F[f]$ that maps an open subset of some Banach space $\mathscr{B}$ of functions $f$ onto $\mathbb{R}$, say, is called Frechet differentiable if there exists a linear continuous operator $\delta F_f : \mathscr{B} \to \mathbb{R}$ with the property

$$\lim_{\|\eta\|\to 0}\frac{|F[f+\eta] - F[f] - \delta F_f[\eta]|}{\|\eta\|} = 0.$$

Similarly, the $n$-th order functional derivative is define according to

$$\frac{d^n F[f+\varepsilon\eta]}{d\varepsilon^n}\bigg|_{\varepsilon=0} := \int_{-\infty}^{\infty} \frac{\delta^n F[f]}{\delta f(x_1)\dots\delta f(x_n)}\eta(x_1)\dots\eta(x_n)dx_1\dots dx_n.$$

As functional derivatives are constructed in terms of ordinary derivatives with respect to $\varepsilon$, most of the rules of differentiation carry over. For example, consider the product rule

$$\frac{d}{d\varepsilon}\left[F_1[f+\varepsilon\eta]F_2[f+\varepsilon\eta]\right) = \frac{dF_1[f+\varepsilon\eta]}{d\varepsilon}F_2[f+\varepsilon\eta]$$
$$+ F_1[f+\varepsilon\eta]\frac{dF_2[f+\varepsilon\eta]}{d\varepsilon}.$$

Taking the limit $\varepsilon \to 0$ and using equation (12.2.42), thus yields the product rule for functional derivatives

$$\frac{\delta(F_1 F_2)}{\delta f(x)} = \frac{\delta F_1}{\delta f(x)}F_2 + F_1\frac{\delta F_2}{\delta f(x)}. \tag{12.2.43}$$

Let us now extend the chain rule for functions to functionals. Suppose that $F[f] = G[H[f]]$. The functional derivative of $F$ with respect to $f$ is given by equation (12.2.42). Similarly,

$$\frac{dH[f+\varepsilon\eta]}{d\varepsilon}\bigg|_{\varepsilon=0} = \int_{-\infty}^{\infty} \frac{\delta H[f]}{\delta f(x)}\eta(x)dx := \widehat{\eta},$$

and

$$\frac{dG[H+\varepsilon\widehat{\eta}]}{d\varepsilon}\bigg|_{\varepsilon=0} = \int_{-\infty}^{\infty} \frac{\delta G[H]}{\delta H(y)}\widehat{\eta}(y)dy.$$

Combining these two results shows that

$$\frac{dG[H+\varepsilon\widehat{\eta}]}{d\varepsilon}\bigg|_{\varepsilon=0} = \int_{-\infty}^{\infty}\int_{-\infty}^{\infty} \frac{\delta G[H]}{\delta H(y)}\frac{\delta H[f](y)}{\delta f(x)}dxdy.$$

Finally, noting that the left-hand side is equivalent to the left-hand side of equation (12.2.42), we obtain the chain rule for functional derivatives

$$\frac{\delta F[f]}{\delta f(x)} = \int_{-\infty}^{\infty}\int_{-\infty}^{\infty} \frac{\delta G[H]}{\delta H(y)}\frac{\delta H[f](y)}{\delta f(x)}dy. \tag{12.2.44}$$

This is valid provided that the variation of $\eta$ generates all possible variations $\widehat{\eta}$ of $H$.

## 12.3 Fluctuation-dissipation theorems, Brownian motion, and nonequilibrium forces

So far, we have focused on the properties of polymers and membranes in thermodynamic equilibrium. However, their physical states fluctuate in time due to the continual bombardment of solvent molecules from the surrounding aqueous environment. In order to explore time-dependent behavior at equilibrium or close to equilibrium, it is necessary to generalize the theory of Brownian motion introduced in Chap. 2 to more complex structures with many internal degrees of freedom. This requires various results and concepts from classical nonequilibrium statistical physics, including Onsager's reciprocal relations, nonequilibrium forces, time correlations, and a general version of the fluctuation-dissipation theorem. Here we provide an introduction to these topics following closely the particular formulation of Doi [251].

### 12.3.1 Fluctuation-dissipation theorem of classical statistical mechanics

The fluctuation-dissipation (FD) theorem of statistical mechanics is based on the assumption that the response of a system in thermodynamic equilibrium to a small applied force is the same as its response to a spontaneous fluctuation. Consider a physical system with a set of discrete microstates labeled $j$ that are in thermodynamic equilibrium at a fixed temperature $T$. Suppose that a constant external force $h$ is applied to the system, let $E_j$ be the energy of a microstate in the absence of the force and let $E_j - hA_j$ be the corresponding energy of the state for $h \neq 0$, where $A_j$ is some function of state. The Boltzmann-Gibbs distribution for $h \neq 0$ is

$$p_j = Z_h^{-1} e^{-\beta(E_j - hA_j)}, \quad Z_h = \sum_j e^{-\beta(E_j - hA_j)}, \quad \beta = \frac{1}{k_B T}. \tag{12.3.1}$$

The partition function $Z_h$ can be used to generate various moments of the variable $A_j$. For example, the equilibrium mean and variance are

$$\langle A \rangle_{\text{eq,h}} = \frac{\sum_j A_j e^{-\beta(E_j - hA_j)}}{\sum_j e^{-\beta(E_j - hA_j)}} = \frac{1}{\beta} \frac{\partial \ln Z_h}{\partial h}, \tag{12.3.2}$$

and

$$\text{Var}[A]_{\text{eq,h}} = \langle A^2 \rangle_{\text{eq,h}} - \langle A \rangle_{\text{eq,h}}^2 = \frac{\sum_j A_j^2 e^{-\beta(E_j - hA_j)}}{\sum_j e^{-\beta(E_j - hA_j)}} - \left( \frac{\sum_j A_j e^{-\beta(E_j - hA_j)}}{\sum_j e^{-\beta(E_j - hA_j)}} \right)^2$$

$$= \frac{1}{\beta^2} \frac{\partial^2 \ln Z_h}{\partial h^2}. \tag{12.3.3}$$

Comparing equations (12.3.2) and (12.3.3) shows that

$$\chi_{eq} := \frac{\partial \langle A \rangle_{eq,h}}{\partial h} = \frac{1}{k_B T} \text{Var}[A]_{eq,h}. \tag{12.3.4}$$

Here, $\chi_{eq}$ is known as the equilibrium susceptibility and determines the equilibrium linear response to small perturbations in the external force $h$. It is proportional to the variance due to spontaneous fluctuations. (The equilibrium susceptibility for the classical Ising model is derived in Ex. 12.8.)

Now suppose that the applied force is present during the time interval $(-\infty, 0]$ and is then suddenly switched off at time $t = 0$. Since the system has had an infinite time to relax to equilibrium prior to $t = 0$, we know that before the force is switched off the mean $\langle A(0) \rangle = \langle A \rangle_{eq,h}$. Let $A(t|j)$ be the value of $A$ at time $t$, which evolves in the absence of the applied force from one of the initial microstates $j$ at time $t = 0$. Averaging over these initial states and using the fact that $h$ is infinitesimal, we have

$$\overline{A}(t) = \langle A(t|j) \rangle_{eq,h} = \frac{\sum_j A(t|j) e^{-\beta(E_j - hA_j)}}{\sum_j e^{-\beta(E_j - hA_j)}} \approx \frac{\sum_j A(t|j) e^{-\beta E_j}(1 + \beta h A_j(0) + \ldots)}{\sum_j e^{-\beta E_j}(1 + \beta h A_j(0) + \ldots)}$$

$$\approx \frac{\sum_j e^{-\beta E_j} A(t|j) \left(1 + \beta h A_j(0) - \beta h \frac{\sum_k e^{-\beta E_k} A_k(0)}{\sum_k e^{-\beta E_k}}\right)}{\sum_j e^{-\beta E_j}}$$

$$= \langle A \rangle_{eq,0} + \beta h \left[ \langle A(t) A(0) \rangle_{eq,0} - \langle A \rangle_{eq,0}^2 \right].$$

We have used the fact that $\langle A(t|j) \rangle_{eq,0} = \langle A \rangle_{eq,0}$ independent of $t$, that is, if the initial microstate is selected according to the Boltzmann distribution with $h = 0$ then so are subsequent microstates. It follows that the relaxation to equilibrium can be related to the autocorrelation function $C(t)$ of spontaneous fluctuations $\delta A(t) = A(t) - \langle A \rangle_{eq,0}$ in the absence of the force

$$\overline{A}(t) - \langle A \rangle_{eq,0} = \langle \delta A(t) \delta A(0) \rangle_{eq,0} = \beta h C(t). \tag{12.3.5}$$

An alternative way to characterize relaxation to equilibrium for small $h$ is in terms of the linear response function $G(t)$:

$$\overline{A}(t) - \langle A \rangle_{eq,0} = \int_{-\infty}^{t} G(t - \tau) h(\tau) d\tau.$$

Using the particular piecewise constant form for $h$, we have

$$\overline{A}(t) - \langle A \rangle_{eq,0} = h\overline{\chi}(t), \quad \overline{\chi}(t) = \int_{t}^{\infty} G(\tau) d\tau. \tag{12.3.6}$$

Comparison with equation (12.3.5) implies that $\overline{\chi}(t) = \beta C(t)$. Finally, we define the nonequilibrium susceptibility according to

$$\overline{A}(t) = \langle A \rangle_{\text{eq,h}} - h\chi(t).\tag{12.3.7}$$

(The negative sign on the right-hand side reflects the fact that the force is removed rather than added at $t = 0$.) Since $\langle A \rangle_{\text{eq,h}} \approx \langle A \rangle_{\text{eq,0}} + h\chi_{\text{eq}}$ for small $h$, we see that

$$\chi(t) = \chi_{\text{eq}} - \frac{1}{k_B T} C(t).\tag{12.3.8}$$

Equation (12.3.8) is one version of the classical FD theorem, which relates the linear response to a small change in an applied force with spontaneous fluctuations.

**FD theorem and the power spectrum.** Another well-known version of the FD theorem occurs in the case of a linear system driven by a random force $h(t)$. In terms of the causal Green's function $G(t)$

$$X(t) = \int_{-\infty}^{\infty} G(t - \tau) h(\tau) d\tau, \quad G(\tau) = 0 \text{ for } \tau < 0.$$

Equation (12.3.6) implies

$$G(t) = -\beta \frac{dC(t)}{dt}, \, t > 0,$$

so that the Fourier transform of $G$ is

$$\widetilde{G}(\omega) = -\beta \int_0^{\infty} \frac{dC(t)}{dt} e^{i\omega t} dt = -\beta + i\omega\beta \int_0^{\infty} C(t) e^{i\omega t} dt.$$

Taking the imaginary part

$$\text{Im}\,\widehat{G}(\omega) = \omega\beta \text{Re} \int_0^{\infty} C(t) e^{i\omega t} dt \quad = \frac{\omega\beta}{2} \left[ \int_0^{\infty} C(t) e^{i\omega t} + \int_0^{\infty} C(t) e^{-i\omega t} \right] dt$$

$$= \frac{\omega\beta}{2} \left[ \int_0^{\infty} C(t) e^{i\omega t} + \int_0^{\infty} C(-t) e^{-i\omega t} \right] dt = \frac{\omega\beta}{2} \int_{-\infty}^{\infty} C(t) e^{i\omega t} dt.$$

Now recall from Chap. 2 that for a stochastic process $X(t)$, the Fourier transform of the correlation function is related to the power spectrum. This then yields a second version of the FD theorem, which states that the power spectrum $S(\omega)$ is related to the Fourier transform of the linear response function $G(t)$ according to

$$S(\omega) = \frac{2k_B T}{\omega} \text{Im}[\widehat{G}(\omega)].\tag{12.3.9}$$

## 12.3.2 Brownian motion and Onsager's reciprocal relations.

The next step is to consider the Brownian motion of some molecule with a rigid shape, such as an oriented rod, see Fig. 12.20(a). Suppose that it has $M$ degrees of freedom or generalized coordinates $\mathbf{x} = (x_1, \ldots, x_M)$, which include the position of the center of mass and orientation of its body axis, for example. Assuming that the molecule moves in some external potential $U(\mathbf{x})$ and is subject to both frictional forces and random (white noise) forces, the associated Langevin equation can be written as

$$\sum_{j=1}^{M} \gamma_{ij} \frac{dX_j}{dt} = -\frac{\partial U}{\partial x_i} + \xi_i(t), \qquad (12.3.10a)$$

with

$$\langle \xi_i(t) \rangle = 0, \quad \langle \xi_i(t) \xi_j(t') \rangle = A_{ij} \delta(t - t'). \qquad (12.3.10b)$$

Here, $\gamma_{ij}$ is a generalized friction coefficient, which can be calculated using the theory of hydrodynamics at low Reynolds number (see also Box 10B). The corresponding friction matrix is generally invertible with $\eta_{ij} = (\gamma^{-1})_{ij}$ known as an Onsager coefficient, and may depend on $\mathbf{x}$. We will show that $\gamma_{ij}(\mathbf{x}) = \gamma_{ji}(\mathbf{x})$, which is Onsager's reciprocal relation, and $A_{ij} = 2\gamma_{ij} k_B T$, which is a version of Einstein's relation.

Since the results are independent of the choice of potential, consider the harmonic potential

$$U(\mathbf{x}) = \frac{1}{2} \sum_{j=1}^{M} k_j (x_j - \bar{x}_j)^2,$$

and suppose that deviations from the equilibrium configuration $\bar{\mathbf{x}}$ are small. Without loss of generality, we set $\bar{\mathbf{x}} = 0$. The Langevin equation is then linear and, using the invertibility of the friction matrix, can be written as

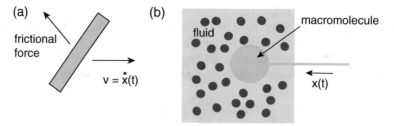

Fig. 12.20: (a) Frictional force exerted on an extended Brownian particle moving with velocity **v**. (b) Macromolecule placed in a fluid. The state variables of the macromolecule including its position $\mathbf{x}(t)$ can be treated as external parameters of the fluid motion.

$$\frac{dX_i}{dt} = -\sum_{j=1}^{M} (\gamma^{-1})_{ij} [k_j X_j + \xi_j(t)].$$

Take $X_i(0) = x_{i0}$ and determine $X_i(\Delta t)$ for small $t$:

$$X_i(\Delta t) = x_{i0} - \Delta t \sum_{j=1}^{M} (\gamma^{-1})_{ij} k_j x_{j0} + \sum_{j=1}^{M} \int_0^{\Delta t} (\gamma^{-1})_{ij} \xi_j(t) dt,$$

where $\gamma$ is evaluated at $\mathbf{x}_0$. Since $\langle \xi_i(t) \rangle = 0$, it follows that

$$\langle X_i(\Delta t) \rangle_{\mathbf{x}_0} = x_{i0} - \Delta t \sum_{j=1}^{M} (\gamma^{-1})_{ij} k_j x_{j0},$$

$$\langle X_i(\Delta t) X_l(0) \rangle = \langle x_{i0} x_{l0} \rangle - \Delta t \sum_{j=1}^{M} (\gamma^{-1})_{ij} k_j \langle x_{j0} x_{l0} \rangle.$$

Since the equilibrium distribution of $\mathbf{x}_0$ is proportional to the Boltzmann factor $\exp\left(-\sum_{i=1}^{M} k_i x_i^2 / 2k_B T\right)$, it follows that

$$\langle x_{i0} x_{l0} \rangle = \frac{\delta_{i,l} k_B T}{k_i}.$$

Therefore,

$$\langle X_i(\Delta t) X_l(0) \rangle = k_B T \left[ \delta_{i,l} k_l^{-1} - \Delta t (\gamma^{-1})_{il} \right].$$

Finally, symmetry of the time correlation function

$$\langle X_i(\Delta t) X_l(0) \rangle = \langle X_l(\Delta t) X_i(0) \rangle,$$

and arbitrariness of $\mathbf{x}_0$ implies that $(\gamma^{-1}(\mathbf{x}))_{il} = (\gamma^{-1}(\mathbf{x}))_{li}$. We thus obtain the Onsager reciprocal relation

$$\gamma_{ij}(\mathbf{x}) = \gamma_{ji}(\mathbf{x}). \tag{12.3.11}$$

The generalized Einstein relation is a consequence of the reciprocal relation. First, note that

$$\langle X_i(\Delta t) X_l(\Delta t) \rangle = \langle x_{i0} x_{l0} \rangle - \Delta t \sum_{j=1}^{M} (\gamma^{-1})_{ij} k_j \langle x_{j0} x_{l0} \rangle - \Delta t \sum_{j=1}^{M} (\gamma^{-1})_{lj} k_j \langle x_{j0} x_{i0} \rangle$$

$$+ \sum_{j,j'=1}^{M} \int_0^{\Delta t} \int_0^{\Delta t} (\gamma^{-1})_{ij} (\gamma^{-1})_{lj'} \langle \xi_j(t_1) \xi_{j'}(t_2) \rangle dt_1\, dt_2.$$

Using the equilibrium expression for $\langle X_i(\Delta t) X_l(\Delta t) \rangle = \langle x_{j0} x_{k0} \rangle$, the formula for $\langle \xi_i(t) \xi_j(t') \rangle$, and the Onsager reciprocal relation then gives

$$-2(\gamma^{-1})_{il} k_B T + \sum_{j,j'} (\gamma^{-1})_{ij} (\gamma^{-1})_{lj'} A_{jj'} = 0.$$

In other words

$$A_{ij} = 2k_B T \gamma_{ij}. \tag{12.3.12}$$

### 12.3.3 Fluctuation-dissipation theorem for a Brownian particle in a fluid

Now consider a more complicated problem, in which a macromolecule with many degrees of freedom (e.g., a flexible polymer) is moving in a fluid and affects the fluid motion, see Fig. 12.20(b). In principle, if one knew the position and momentum of every fluid molecule (represented by the symbol $\Gamma$) and the state $\mathbf{x}$ of the particle, then the time evolution of the system could be determined from the Hamiltonian $H(\Gamma;\mathbf{x})$. Assuming that the degrees of freedom of the Brownian particle evolve much more slowly than the fluid state, the generalized coordinates $\mathbf{x}$ of the Brownian particle act as external parameters of the fluid dynamics. The instantaneous forces acting on the Brownian particle when the fluid molecules are in the microstate $\Gamma$ are given by

$$\widehat{f}_i(\Gamma;\mathbf{x}) = -\frac{\partial H(\Gamma;\mathbf{x})}{\partial x_i}. \tag{12.3.13}$$

Let us define the macroscopic force at time $t$ to be the average of the instantaneous force with respect to the conditional probability density function of the fluid, $\psi(\Gamma,t;\mathbf{x})$:

$$\langle f_i(t) \rangle_{\mathbf{x}} = \int \widehat{f}_i(\Gamma;\mathbf{x}) \psi(\Gamma,t;\mathbf{x}) d\Gamma.$$

If the fluid molecules are at thermodynamic equilibrium for a given value of $\mathbf{x}$, then the average force becomes

$$\langle f_i \rangle_{\text{eq},\mathbf{x}} = \int \widehat{f}_i(\Gamma;\mathbf{x}) \psi_{\text{eq}}(\Gamma;\mathbf{x}) d\Gamma, \tag{12.3.14}$$

where $\psi_{\text{eq}}$ is the Boltzmann-Gibbs distribution

$$\psi_{\text{eq}}(\Gamma;\mathbf{x}) = \frac{e^{-H(\Gamma;\mathbf{x})/k_B T}}{\int e^{-H(\Gamma;\mathbf{x})/k_B T} d\Gamma}. \tag{12.3.15}$$

Defining the free energy according to

$$F(\mathbf{x}) = -k_B T \ln \int e^{-H(\Gamma;\mathbf{x})/k_B T} d\Gamma, \tag{12.3.16}$$

we see that

$$\langle f_i \rangle_{\text{eq},\mathbf{x}} = -\frac{\partial F(\mathbf{x})}{\partial x_i}. \tag{12.3.17}$$

Thus, the equilibrium force acting on the Brownian particle is given by the negative gradient of the free energy.

Note that $F(\mathbf{x})$ is really a constrained free energy of the particle/fluid system, since $\mathbf{x}$ is specified. For the unconstrained system in thermodynamic equilibrium, the probability density of the state $\mathbf{x}$ is given by the Boltzmann-Gibbs distribution

$$\psi_{\text{eq}}(\mathbf{x}) = \frac{1}{Z} e^{-F(\mathbf{x})/k_B T}, \quad Z = \int e^{-F(\mathbf{x})/k_B T} d\mathbf{x}. \tag{12.3.18}$$

The total free energy of the system is then

$$F_T = -k_B T \ln \left\{ \int e^{-F(\mathbf{x})/k_B T} d\mathbf{x} \right\}. \tag{12.3.19}$$

Suppose that the system $\Gamma$ is in thermodynamic equilibrium for a given $\mathbf{x}$, and that there is a sudden small jump $\mathbf{x} \to \mathbf{x} + \delta\mathbf{x}$ at time $t = 0$. (We are assuming that the state of the macromolecule can be externally controlled.) The fluid system will relax to the new equilibrium according to a time-dependent linear response function, which forms the basis of the fluctuation-dissipation theorem for the Brownian particle in a fluid. We proceed by generalizing the basic steps carried out for the simpler statistical mechanical system of Sect. 12.3.1. Now the fluid plays the role of the equilibrium system, the force $\langle f_i \rangle_{\text{eq},\mathbf{x}}$ is analogous to $\langle A \rangle_{\text{eq},h}$, and a sudden change in the position of the Brownian particle, $\mathbf{x} \to \mathbf{x} + \delta\mathbf{x}$, is analogous to the step change in the applied force $h$. Again we proceed by first calculating an equilibrium susceptibility, and then considering the time-dependent response.

First, however, we determine the change in the equilibrium forces by analogy with equation (12.3.4). Under the shift $\mathbf{x} \to \mathbf{x} + \delta\mathbf{x}$, the Hamiltonian changes as $H(\Gamma; \mathbf{x}) \to H(\Gamma, \mathbf{x} + \delta\mathbf{x})$ and the average force with respect to the new equilibrium can be written as

$$\langle f_i \rangle_{\text{eq},\mathbf{x}+\delta\mathbf{x}} \approx \langle f_i \rangle_{\text{eq},\mathbf{x}} + \sum_{j=1}^{M} \chi_{ij}^{\text{eq}} \delta x_j, \quad \chi_{ij}^{\text{eq}} = \frac{\partial \langle f_i \rangle_{\text{eq},\mathbf{x}}}{\partial x_j}. \tag{12.3.20}$$

The equilibrium susceptibility $\chi_{ij}^{\text{eq}}$ determines the equilibrium linear response of the fluid to small perturbations in the position of the Brownian particle. Using equations (12.3.13)–(12.3.15) and the chain rule

$$\chi_{ij}^{\text{eq}} = -\frac{\partial}{\partial x_j} \int \frac{\partial H(\Gamma; \mathbf{x})}{\partial x_i} \psi_{\text{eq}}(\Gamma; \mathbf{x}) d\Gamma = -\int \frac{\partial^2 H}{\partial x_i \partial x_j} \psi_{\text{eq}} d\Gamma$$

$$+ \frac{1}{k_B T} \int \frac{\partial H}{\partial x_i} \frac{\partial H}{\partial x_j} \psi_{\text{eq}} d\Gamma - \frac{1}{k_B T} \int \frac{\partial H}{\partial x_i} \psi_{\text{eq}} d\Gamma \int \frac{\partial H}{\partial x_j} \psi_{\text{eq}} d\Gamma,$$

which yields

$$\chi_{ij}^{\text{eq}} = \left\langle \frac{\partial f_i}{\partial x_j} \right\rangle_{\text{eq},\mathbf{x}} + \frac{1}{k_B T} [\langle f_i f_j \rangle_{\text{eq},\mathbf{x}} - \langle f_i \rangle_{\text{eq},\mathbf{x}} \langle f_j \rangle_{\text{eq},\mathbf{x}}]. \tag{12.3.21}$$

That is, the equilibrium susceptibility depends on the size of fluctuations in the forces that occur at equilibrium.

Let us now consider the time-dependent relaxation to the new equilibrium following a sudden perturbations $\mathbf{x} \to \mathbf{x} + \delta\mathbf{x}$. The time-dependent version of equation (12.3.20) for the average force can be written as

$$\langle f_i(t)\rangle_{\mathbf{x}+\delta\mathbf{x}} = \int \widehat{f_i}(\Gamma;\mathbf{x}+\delta\mathbf{x})\psi(\Gamma,t;\mathbf{x}+\delta\mathbf{x})d\Gamma \approx \langle f_i\rangle_{\mathrm{eq},\mathbf{x}} + \sum_{j=1}^{M} \chi_{ij}(t)\delta x_j,$$

(12.3.22a)

with the time-dependent susceptibilities $\chi_{ij}(t) \to \chi_{ij}^{\mathrm{eq}}$ as $t \to \infty$. This is the analog of equation (12.3.7). If we assume that the difference between the instantaneous force and the average force before the perturbation is a Gaussian white noise process

$$\xi_i(t) = \widehat{f_i}(\Gamma(t),\mathbf{x}) - \langle f_i\rangle_{\mathrm{eq},\mathbf{x}},$$

(12.3.22b)

with

$$\langle \xi_i(t)\rangle = 0, \quad \langle \xi_i(t)\xi_j(0)\rangle_{\mathrm{eq},\mathbf{x}} = 2k_B T \gamma_{ij}(\mathbf{x})\delta(t),$$

(12.3.22c)

then it can be shown that (see Box 12D)

$$\chi_{ij}(t) = \chi_{ij}^{\mathrm{eq}} - 2\gamma_{ij}(\mathbf{x})\delta(t).$$

(12.3.22d)

Since $\chi_{ij}$ determines the linear response to a small perturbation in $\mathbf{x}$, it follows that equation (12.3.22d) can be interpreted as a fluctuation-dissipation theorem for the generalized Brownian particle, along analogous lines to equation (12.3.8).

Next, suppose that at time $t = 0$, the system is in equilibrium for the given initial condition $\mathbf{x}(0)$, and that $\mathbf{x}(t)$ changes slowly with velocity $\dot{\mathbf{x}}(t)$ for $t > 0$. The response to a slowly varying change in $\mathbf{x}$ can be obtained by superimposing the individual responses to $d\mathbf{x}(t) = \mathbf{x}(t+dt) - \mathbf{x}(t) = \dot{\mathbf{x}}(t)dt$

$$\langle f_i(t)\rangle_{\mathbf{x}(t)} = \langle f_i\rangle_{\mathrm{eq},\mathbf{x}(0)} + \int_0^t \sum_j [\chi_{ij}^{\mathrm{eq}} - \gamma_{ij}(\mathbf{x}(t-t'))\delta(t-t')]\frac{dx_j(t')}{dt}dt'$$

$$= \langle f_i\rangle_{\mathrm{eq},\mathbf{x}(0)} + \sum_j \chi_{ij}^{\mathrm{eq}}[x_j(t) - x_j(0)] - \sum_j \gamma_{ij}(\mathbf{x})\frac{dx_j(t)}{dt}$$

$$= \langle f_i\rangle_{\mathrm{eq},\mathbf{x}(0)} + \sum_j \frac{\partial\langle f_i\rangle_{\mathrm{eq},\mathbf{x}(0)}}{\partial x_j}(x_j(t) - x_j(0)) - \sum_j \gamma_{ij}(\mathbf{x})\frac{dx_j(t)}{dt}$$

$$\approx \langle f_i\rangle_{\mathrm{eq},\mathbf{x}(t)} - \sum_j \gamma_{ij}(\mathbf{x})\frac{dx_j(t)}{dt} = -\frac{\partial F(\mathbf{x}(t))}{\partial x_i} - \sum_j \gamma_{ij}(\mathbf{x})\frac{dx_j(t)}{dt},$$

(12.3.23)

where we have used equation (12.3.20) under the assumption that $x_j(t) - x_j(0)$ is sufficiently small. The removal of a factor of 2 takes into account the fact that the lower limit of the integral on the first line is zero. Finally, Newton's law of motion implies that $\dot{p}_i = \langle f_i(t)\rangle$ where $p_i$ is the momentum conjugate to $x_i$ (see Box 8A).

However, in the low Reynold's number regime, inertial effects can be ignored, so that in the absence of external forces, we obtain the analog of equation (12.3.10), with the external potential replaced by the free energy:

$$\sum_j \gamma_{ij}(\mathbf{X})\frac{dX_j(t)}{dt} = -\frac{\partial F(\mathbf{X})}{\partial X_i} + \xi_i(t). \tag{12.3.24}$$

We have also included the effects of the fast fluctuations due to the fluid, which when averaged out recovers equation (12.3.23) with $\langle f_i(t)\rangle_{\mathbf{x}(t)} = 0$.

Although equation (12.3.24) was derived for a Brownian particle with a finite number of degrees of freedom, it forms the basis for a wide class of phenomenological models describing the dynamics of nonequilibrium systems, at least when such systems are not too far from thermodynamic equilibrium. Two examples of infinite-dimensional systems will be considered in Sect. 12.3.5, based on continuum models of a polymer and a cell membrane, respectively. Another important example arises in the kinetics of liquid-liquid phase separation; see Sect. 13.1. Now the analog of the slow variables $\mathbf{x}(t)$ of a Brownian particle is the local particle concentration $n(\mathbf{r},t)$. This connection is clearer if space is discretized so that there is a set of discrete variables $n_i(t) = n(\mathbf{r}_i,t)$. (Implicit in such a formulation is that particles are locally in equilibrium.) In more complicated systems, the coefficients $\gamma_{ij}$ are often phenomenologically-based rather than derived from hydrodynamics, for example.

**Fokker-Planck equation and the Brownian potential.** The FP equation corresponding to the Langevin equation (12.3.24) in the thermodynamic interpretation (Chap. 2) is

$$\frac{\partial \psi}{\partial t} = \sum_{i=1}^M \frac{\partial}{\partial x_i}\left(\sum_{j=1}^M \eta_{ij}(\mathbf{x})\psi(\mathbf{x},t)\frac{\partial F(\mathbf{x})}{\partial x_j}\right) + k_B T \sum_{i,j=1}^M \frac{\partial}{\partial x_i}\eta_{ij}(\mathbf{x})\frac{\partial}{\partial x_j}\psi(\mathbf{x},t), \tag{12.3.25}$$

where $\eta_{ij} = (\gamma^{-1})_{ij}$. This can be rewritten in the form of a conservation equation

$$\frac{\partial \psi}{\partial t} = -\sum_j \frac{\partial}{\partial x_j}(v_{fj}\psi), \tag{12.3.26}$$

where $\mathbf{v}_f$ is the flux velocity with components

$$v_{fi}(\mathbf{x},t) = -\sum_{j=1}^M \eta_{ij}(\mathbf{x})\frac{\partial}{\partial x_j}(k_B T \ln \psi(\mathbf{x},t) + F(\mathbf{x})). \tag{12.3.27}$$

We thus have an alternative interpretation of the FP equation, in terms of the evolution equation for a population of noninteracting Brownian particles with density $n_p\psi(\mathbf{x},t)$, where $n_p$ is the number of molecules per unit volume. The particles move deterministically in phase space according to the equation

$$\sum_j \gamma_{ij}(\mathbf{x})\frac{dx_j(t)}{dt} = -\frac{\partial[F(\mathbf{x}) + k_B T \ln \psi(\mathbf{x},t)]}{\partial x_i}, \tag{12.3.28}$$

with $\psi$ then evolving according to the particle conservation equation (12.3.26). The term $k_B T \ln \psi$ is known as the Brownian potential and plays an analogous role to the chemical potential in the case of diffusion; see Sect. 13.1. Equations (12.3.26)–(12.3.28) are the starting point for developing microscopic models of viscoelasticity in polymer solutions; see Sect. 12.4.3 and Refs. [250, 251]. The construction of the Brownian potential for a rotating rod is considered in Ex. 12.9.

Let us now introduce the so-called dynamical free energy functional

$$\mathcal{F}[\psi] = \int \psi(\mathbf{x},t)[k_B T \ln \psi(\mathbf{x},t) + F(\mathbf{x})]d\mathbf{x}. \qquad (12.3.29)$$

This functional is an effective Liapunov function, which can be used to prove irreversibility of the FP equation. That is

$$\frac{d\mathcal{F}[\psi]}{dt} = \int \frac{\delta \mathcal{F}[\psi]}{\delta \psi(\mathbf{x},t)} \frac{\partial \psi}{\partial t} d\mathbf{x}.$$

From the conservation equation (12.3.26)

$$\frac{d\mathcal{F}[\psi]}{dt} = \int \psi(\mathbf{x},t)\mathbf{v}_f(\mathbf{x},t) \cdot \nabla \frac{\delta \mathcal{F}[\psi]}{\delta \psi(\mathbf{x},t)} d\mathbf{x}$$

$$= -\int \psi \sum_{i,j=1}^{M} \eta_{ij} \frac{\partial}{\partial x_j}(k_B T \ln \psi + A)\frac{\partial}{\partial x_i}(k_B T \ln \psi + A)d\mathbf{x},$$

where we have performed an integration by parts. From Onsager's reciprocal relation, $\eta_{ij} = \eta_{ji}$, it follows that $d\mathcal{F}/dt < 0$ unless $\psi = \psi_{eq}$, with $\psi_{eq}$ given by equation (12.3.18). Thus, $\psi(\mathbf{x},t) \to \psi_{eq}(\mathbf{x})$ and $\mathcal{F}[\psi_{eq}] = F_T$; see equation (12.3.19). One of the useful features of the dynamical free energy is that it can be used to formulate the dynamics of (possibly interacting) Brownian particles in terms of a variational principle [250, 251].

---

**Box 12D. Time-dependent average forces and susceptibilities.**

---

**Liouville equation.** In order to derive equation (12.3.22), we first consider a general formulation of time correlation functions for the fluid-particle system $(\Gamma, \mathbf{x})$. Suppose that the microstate of the fluid can be expressed in terms of a set of generalized coordinates $(q_1, \ldots, q_N)$ and generalized momenta $(p_1, \ldots, p_N)$. If the Hamiltonian $H(\Gamma; \mathbf{x})$ for fixed $\mathbf{x}$ is known explicitly, then the time evolution of the microscopic state $\Gamma$ can be calculated by solving Hamilton's equations of motion (see Box 8A)

$$\frac{dq_a}{dt} = \frac{\partial H}{\partial p_a}, \quad \frac{dp_a}{dt} = -\frac{\partial H}{\partial q_a}, \quad a = 1, \ldots, N.$$

Suppose that at the initial time $t$, the microstate is $\Gamma(0) = \Gamma_0$. Let $\widehat{P}(\Gamma;\mathbf{x})$ denote the value of some quantity $P$ in the microstate $\Gamma$, given $\mathbf{x}$. For the moment, we drop the explicit dependence on $\mathbf{x}$. Since $\Gamma$ evolves in time according to Hamilton's equations, it follows that the value of $P$ at time $t$ can be written as

$$P(\Gamma_0,t) = \widehat{P}(\Gamma(t))\Big|_{\Gamma(0)=\Gamma_0},$$

and by the chain rule

$$\frac{\partial P}{\partial t} = -\sum_{a=1}^{N}\left[\dot{p}_a\frac{\partial \widehat{P}}{\partial p_a} + \dot{q}_a\frac{\partial \widehat{P}}{\partial q_a}\right] = -\sum_{a=1}^{N}\left[\frac{\partial H}{\partial p_a}\frac{\partial}{\partial q_a} - \frac{\partial H}{\partial q_a}\frac{\partial}{\partial p_a}\right]\widehat{P} \equiv -\mathbb{L}\widehat{P},$$

$$(12.3.30)$$

where $\mathbb{L}$ is known as the Liouville operator. Note that equation (12.3.30) can be interpreted as an ODE of the form $\dot{P} = -f(t)$ with $f(t) = (\mathbb{L}\widehat{P})(\Gamma(t))$ and the initial condition $P(\Gamma_0,0) = \widehat{P}(\Gamma_0)$. Its solution can be written formally as

$$P(\Gamma,t) = e^{-t\mathbb{L}}P(\Gamma,0) = e^{-t\mathbb{L}}\widehat{P}(\Gamma),\qquad(12.3.31)$$

where we have dropped the subscript on $\Gamma_0$. Suppose, in particular, that we take $P$ to be the probability density $\psi(\Gamma,t;\mathbf{x})$ for finding the system in state $\Gamma$ at time $t$, given $\mathbf{x}$ and an initial density $\psi_0(\Gamma;\mathbf{x})$. Again dropping the explicit dependence on $\mathbf{x}$, it follows from equation (12.3.31) that

$$\psi(\Gamma,t) = e^{-t\mathbb{L}}\psi_0(\Gamma).$$

Note that the equilibrium density (12.3.15) is a null vector of the operator $\mathbb{L}$, $e^{-t\mathbb{L}}\psi_{eq}(\Gamma;\mathbf{x}) = \psi_{eq}(\Gamma;\mathbf{x})$.

**Time correlation functions.** Now consider the time correlation function of two physical quantities $P$ and $Q$. Let $\widehat{P}(\Gamma;\mathbf{x})$ and $\widehat{Q}(\Gamma;\mathbf{x})$ denote the values of $P$ and $Q$ in the microstate $\Gamma$, given $\mathbf{x}$. Suppose that the system is in thermodynamic equilibrium for the given $\mathbf{x}$. We can then define the equilibrium mean according to

$$\langle P\rangle_{eq,\mathbf{x}} = \int \widehat{P}(\Gamma;\mathbf{x})\psi_{eq}(\Gamma;\mathbf{x})d\Gamma.$$

Similarly, the equilibrium time correlation $\langle P(t)Q(0)\rangle_{eq,\mathbf{x}}$ is defined as

$$\langle P(t)Q(0)\rangle_{eq,\mathbf{x}} = \int\int \widehat{P}(\Gamma;\mathbf{x})\widehat{Q}(\Gamma_0;\mathbf{x})G(\Gamma,t|\Gamma_0,0)\psi_{eq}(\Gamma_0;\mathbf{x})d\Gamma_0 d\Gamma,$$

where the propagator $G(\Gamma,t|\Gamma_0,0)$ specifies the probability that the system is in state $\Gamma$ at time $t$ given that it was in state $\Gamma_0$ at time $t = 0$. It is the solution of the Liouville equation under the initial condition $G(\Gamma,0|\Gamma_0,0) = \delta(\Gamma - \Gamma_0)$, that is, $G(\Gamma,t|\Gamma_0,0) = e^{-t\mathbb{L}}\delta(\Gamma - \Gamma_0)$. Substituting the expression for $G$ and integrating by parts with respect to $\Gamma_0$ yields

$$\langle P(t)Q(0)\rangle_{\mathrm{eq},\mathbf{x}} = \int \widehat{P}(\Gamma;\mathbf{x})e^{-t\mathbb{L}}\left[\widehat{Q}(\Gamma;\mathbf{x})\,\psi_{\mathrm{eq}}(\Gamma;\mathbf{x})\right]\delta(\Gamma - \Gamma_0)d\Gamma\,d\Gamma_0$$

$$= \int \psi_{\mathrm{eq}}(\Gamma;\mathbf{x})\widehat{P}(\Gamma;\mathbf{x})e^{-t\mathbb{L}}\widehat{Q}(\Gamma;\mathbf{x})d\Gamma. \qquad (12.3.32)$$

**Time-dependent susceptibilities.** Let us now consider the time-dependent relaxation to the new equilibrium following a sudden perturbations $\mathbf{x} \to \mathbf{x} + \delta\mathbf{x}$. In order to determine the time-dependent susceptibilities $\chi_{ij}(t)$, we consider the Liouville equation for $\psi(\Gamma,t;\mathbf{x} + \delta\mathbf{x})$

$$\frac{\partial\psi}{\partial t} = -(\mathbb{L} + \delta\mathbb{L})\psi,$$

where, from the definition of $\mathbb{L}$

$$\delta\mathbb{L} = \sum_{a=1}^{N}\left[\frac{\partial\delta H}{\partial p_a}\frac{\partial}{\partial q_a} - \frac{\partial\delta H}{\partial q_a}\frac{\partial}{\partial p_a}\right] = -\sum_{i=1}^{M}\sum_{a=1}^{N}\delta x_i\left[\frac{\partial\widehat{f_i}}{\partial p_a}\frac{\partial}{\partial q_a} - \frac{\partial\widehat{f_i}}{\partial q_a}\frac{\partial}{\partial p_a}\right].$$

We have also used equation (12.3.13), which implies $\delta H = -\sum_{i=1}^{M}\delta x_i\widehat{f_i}$. Setting $\psi(\Gamma,t;\mathbf{x} + \delta\mathbf{x}) = \psi_{\mathrm{eq}}(\Gamma;\mathbf{x}) + \delta\psi(\Gamma,t;\mathbf{x} + \delta\mathbf{x})$ gives

$$\frac{\partial\delta\psi}{\partial t} = -\mathbb{L}\delta\psi - \delta\mathbb{L}\psi_{\mathrm{eq}},$$

to leading order. The variation-of-parameters formula then yields

$$\delta\psi(\Gamma,t;\mathbf{x} + \delta\mathbf{x}) = -\int_0^t e^{-(t-t')\mathbb{L}}[\delta\mathbb{L}\psi_{\mathrm{eq}}](\Gamma;\mathbf{x})dt'.$$

From the expressions for $\mathbb{L}$, $\delta\mathbb{L}$ and $\psi_{\mathrm{eq}}$, see equation (12.3.15), it follows that

$$\delta\mathbb{L}\psi_{\mathrm{eq}} = -\sum_{i=1}^{M}\sum_{a=1}^{N}\delta x_i\left[\frac{\partial\widehat{f_i}}{\partial p_a}\frac{\partial}{\partial q_a} - \frac{\partial\widehat{f_i}}{\partial q_a}\frac{\partial}{\partial p_a}\right]\frac{e^{-H(\Gamma;\mathbf{x})/k_BT}}{\int e^{-H(\Gamma;\mathbf{x})/k_BT}d\Gamma}$$

$$= \frac{1}{k_BT}\psi_{\mathrm{eq}}\sum_{i=1}^{M}\sum_{a=1}^{N}\delta x_i\left[\frac{\partial\widehat{f_i}}{\partial p_a}\frac{\partial H}{\partial q_a} - \frac{\partial\widehat{f_i}}{\partial q_a}\frac{\partial H}{\partial p_a}\right] = \frac{1}{k_BT}\psi_{\mathrm{eq}}\sum_{i=1}^{M}\delta x_i(\mathbb{L}\widehat{f_i}).$$

Given $\psi(\Gamma,t;\mathbf{x} + \delta\mathbf{x})$, the averaged force at time $t$ is

$$\langle f_i(t)\rangle_{\mathbf{x}+\delta\mathbf{x}} = \int \widehat{f}_i(\Gamma;\mathbf{x}+\delta\mathbf{x})[\psi_{eq}(\Gamma;\mathbf{x})+\delta\psi(\Gamma,t;\mathbf{x}+\delta\mathbf{x})]d\Gamma$$

$$= \int \psi_{eq}(\Gamma;\mathbf{x})\left[\widehat{f}_i(\Gamma;\mathbf{x})+\sum_{j=1}^{M}\frac{\partial\widehat{f}_i(\Gamma;\mathbf{x})}{\partial x_j}\delta x_j\right]$$

$$\times\left[1+\frac{1}{k_BT}\sum_{j=1}^{M}\delta x_j\int_0^t dt' e^{-(t-t')\mathbb{L}}(\mathbb{L}\widehat{f}_i)(\Gamma;\mathbf{x})dt'\right]d\Gamma.$$

After dropping $O(\delta x^2)$ terms, we have

$$\langle f_i(t)\rangle_{\mathbf{x}+\delta\mathbf{x}} = \int\left[\widehat{f}_i(\Gamma;\mathbf{x})\psi_{eq}(\Gamma;\mathbf{x})+\sum_{j=1}^{M}\delta x_j\frac{\partial\widehat{f}_i(\Gamma;\mathbf{x})}{\partial x_j}\psi_{eq}(\Gamma;\mathbf{x})\right.$$

$$\left.+\psi_{eq}(\Gamma;\mathbf{x})\frac{\widehat{f}_i(\Gamma;\mathbf{x})}{k_BT}\sum_{j=1}^{M}\delta x_j\int_0^t e^{-(t-t')\mathbb{L}}(\mathbb{L}\widehat{f}_i)(\Gamma;\mathbf{x})dt'\right]d\Gamma.$$

If we now set $f_i(t)=\widehat{f}_i(\Gamma(t);\mathbf{x})$ then $\mathbb{L}\widehat{f}_i(\Gamma(t);\mathbf{x})=-\partial_t f_i(t)$. Hence, comparing the last term on the right-hand side to the general form of the time correlation function, equation (12.3.32), we obtain the compact result

$$\langle f_i(t)\rangle_{\mathbf{x}+\delta\mathbf{x}} = \langle f_i\rangle_{eq,\mathbf{x}}+\sum_{j=1}^{M}\delta x_j\left\langle\frac{\partial f_j}{\partial x_j}\right\rangle_{eq,\mathbf{x}}$$

$$+\frac{1}{k_BT}\sum_{j=1}^{M}\delta x_j\int_0^t\langle f_i(t-t')\partial_t f_j(0)\rangle_{eq,\mathbf{x}}dt'. \qquad (12.3.33)$$

From the definition of the time-dependent susceptibility in equation (12.3.22a), we see that

$$\chi_{ij}(t)=\left\langle\frac{\partial f_j}{\partial x_j}\right\rangle_{eq,\mathbf{x}}+\frac{1}{k_BT}\int_0^t\langle f_i(t-t')\partial_t f_j(0)\rangle_{eq,\mathbf{x}}dt'. \qquad (12.3.34)$$

Time shift invariance implies that

$$\int_0^t\langle f_i(t-t')\partial_t f_j(0)\rangle_{eq,\mathbf{x}}dt' = \int_0^t\langle f_i(t)\partial_t f_j(t')\rangle_{eq,\mathbf{x}}dt'$$

$$= \langle f_i(t)f_j(t)\rangle_{eq,\mathbf{x}}-\langle f_i(t)f_j(0)\rangle_{eq,\mathbf{x}}$$

$$= \langle f_i f_j\rangle_{eq,\mathbf{x}}-\langle f_i\rangle_{eq,\mathbf{x}}\langle f_j\rangle_{eq,\mathbf{x}}-\langle\xi_i(t)\xi_j(0)\rangle_{eq,\mathbf{x}},$$

where $\xi_i(t)=\widehat{f}_i(\Gamma(t),\mathbf{x})-\langle f_i\rangle_{eq,\mathbf{x}}$. Interpreting $\xi_i(t)$ as a white noise term, we take

$$\langle\xi_i(t)\xi_j(0)\rangle_{eq,\mathbf{x}}=2k_BT\gamma_{ij}(\mathbf{x})\delta(t). \qquad (12.3.35)$$

Hence, in response to a step change in $\mathbf{x}$, the time-dependent susceptibility is given by equation (12.3.22d).

$$\chi_{ij}(t) = \left\langle \frac{\partial f_j}{\partial x_j} \right\rangle_{\text{eq},\mathbf{x}} + \langle f_i f_j \rangle_{\text{eq},\mathbf{x}} - \langle f_i \rangle_{\text{eq},\mathbf{x}} \langle f_j \rangle_{\text{eq},\mathbf{x}} - 2\gamma_{ij}(\mathbf{x})\delta(t), \quad (12.3.36)$$

which yields equation (12.3.22d) on using (12.3.21).

### 12.3.4 Interacting Brownian particles

Another source of dynamical complexity for polymers and other large molecular structures in solution is that the movement of one part of the structure can affect other parts due to hydrodynamic interactions with the surrounding fluid. In order to explore this feature, consider $N$ Brownian particles that are modeled as spheres or beads [250]. These could represent different sections of a polymer, for example, see the Rouse model in Sect. 12.5.2. Denote the position of the center of mass of the $n$-th particle by $\mathbf{x}_n$, $n = 1, \ldots, N$, and suppose that it is subject to a force $\mathbf{f}_n$.[2] The force on each particle creates a local velocity flow field of the surrounding fluid, which, in turn, contributes to the motion of other particles. In order to calculate the particle velocities, it is first necessary to determine the fluid velocity field $v(\mathbf{x}, t)$. At low Reynolds number, the relevant hydrodynamical equation of motion is Stoke's equation (see also Box 10B). The latter assumes that the fluid is incompressible, $\nabla \cdot \mathbf{v} = 0$, and that the inertial forces are negligible. The Navier-Stokes equation then reduces to

$$\eta \nabla^2 v_i - \frac{\partial P}{\partial x_i} = -g_i, \quad (12.3.37)$$

where $\eta$ is the viscosity, $P$ is the pressure, and $g_i(\mathbf{x})$ is the external force acting on a unit volume of fluid. The corresponding stress tensor is

$$\sigma_{ij} = \eta \left( \frac{\partial v_j}{\partial x_i} + \frac{\partial v_j}{\partial x_i} \right) - P\delta_{i,j}. \quad (12.3.38)$$

If the force on the fluid is generated by the forces acting on the Brownian particles, then treating the particles as points, we have

$$g_i(\mathbf{x}) = \sum_{n=1}^{N} f_{ni}\delta(\mathbf{x} - \mathbf{x}_n).$$

---

[2] Non-spherical particles of finite shape will have both translational and rotational degrees of freedom that interact with the fluid. It is then necessary to take into the angular velocity of each particle and the associated external torque acting on it.

Substituting into Stokes equation yields

$$\eta \nabla^2 v_i - \frac{\partial P}{\partial x_i} = -\sum_{n=1}^{N} f_{ni} \delta(\mathbf{x} - \mathbf{x}_n). \tag{12.3.39}$$

The above equation can be solved using Fourier transforms

$$-\eta k^2 \widehat{v}_j(\mathbf{k}) + i k_j \widehat{P}(\mathbf{k}) = -\sum_{n=1}^{N} f_{nj} e^{i\mathbf{k}\cdot\mathbf{x}_n}, \tag{12.3.40}$$

$\widehat{\mathbf{v}}(\mathbf{k}) = \int \mathbf{v}(\mathbf{x}) e^{i\mathbf{k}\cdot\mathbf{x}} d\mathbf{x}$. Multiplying both sides by $k_j$, summing over $j$ and using $\mathbf{k} \cdot \widehat{\mathbf{v}}(\mathbf{k}) = 0$ (incompressible flow), shows that

$$ik^2 \widehat{P}(\mathbf{k}) = -\sum_{n=1}^{N} \mathbf{k} \cdot \mathbf{f}_n e^{i\mathbf{k}\cdot\mathbf{x}_n}.$$

Hence, we can eliminate the pressure to give

$$\widehat{v}_j(\mathbf{k}) = \frac{1}{\eta k^2} \sum_{n=1}^{N} \left( f_{nj} - \frac{k_j \mathbf{k} \cdot \mathbf{f}_n}{k^2} \right) e^{-\mathbf{k}\cdot\mathbf{x}_n} = \sum_{l=1}^{3} \frac{1}{\eta k^2} \left( \delta_{j,l} - \frac{k_j k_l}{k^2} \right) \sum_{n=1}^{N} f_{nl} e^{i\mathbf{k}\cdot\mathbf{x}_n}$$

$$\equiv \sum_{l=1}^{3} \widehat{H}_{jl}(\mathbf{k}) \sum_{n=1}^{N} f_{nl} e^{i\mathbf{k}\cdot\mathbf{x}_n}.$$

Using the convolution theorem, we deduce that

$$v_j(\mathbf{x}) = \sum_{l=1}^{3} \int H_{jl}(\mathbf{x} - \mathbf{x}') \sum_{n=1}^{N} f_{nl} \delta(\mathbf{x}' - \mathbf{x}_n) d\mathbf{x}' = \sum_{n=1}^{N} \sum_{l=1}^{3} H_{jl}(\mathbf{x} - \mathbf{x}_n) f_{nl}, \tag{12.3.41}$$

where $\mathbf{H}(\mathbf{x})$ is known as the Oseen tensor, and is given by

$$H_{jl}(\mathbf{x}) = \int \frac{1}{\eta k^2} \left( \delta_{j,l} - \frac{k_j k_l}{k^2} \right) e^{-i\mathbf{k}\cdot\mathbf{x}} \frac{d\mathbf{k}}{(2\pi)^3} = \frac{1}{8\pi\eta r} \left( \delta_{j,l} + \frac{x_j x_l}{r^2} \right), \tag{12.3.42}$$

where $r = |\mathbf{x}|$.

Since the Brownian particles move with the same velocity as the fluid

$$\frac{dr_{mj}}{dt} = v_j(\mathbf{x}_m) = \sum_{n=1}^{N} \sum_{l=1}^{3} H_{jl}(\mathbf{x}_m - \mathbf{x}_n) f_{nl}. \tag{12.3.43}$$

However, the term $n = m$ in the above sum is singular since $\mathbf{H}(\mathbf{x}) \sim 1/r$. This is a consequence of treating the Brownian particles as points. A common approximation is to take the diagonal component of the Oseen tensor to be as in the case of no hydrodynamic interactions, that is

$$\frac{dr_{mj}}{dt} = \frac{f_{mj}}{\gamma} + \sum_{n \neq m} \sum_{l=1}^{3} H_{jl}(\mathbf{x}_m - \mathbf{x}_n) f_{nl}. \tag{12.3.44}$$

The corresponding FP equation in the thermodynamic interpretation can then be written down by taking

$$f_{mj} = -\frac{\partial}{\partial x_{mj}}(k_B T \ln \psi + U).$$

This gives

$$\frac{\partial \psi}{\partial t} = \sum_{m=1}^{N} \sum_{n \neq m} \sum_{j,l=1}^{3} \frac{\partial}{\partial x_{mj}} H_{jl}(\mathbf{x}_m - \mathbf{x}_n) \left( k_B T \frac{\partial \psi}{\partial x_{nl}} + \frac{\partial U}{\partial x_{nl}} \psi \right)$$

$$+ \gamma^{-1} \sum_{m=1}^{N} \sum_{j=1}^{3} \frac{\partial}{\partial x_{mj}} \left( k_B T \frac{\partial \psi}{\partial x_{mj}} + \frac{\partial U}{\partial x_{mj}} \psi \right).$$

When studying the viscoelastic properties of a material such as a polymeric solution, see Sect. 12.4.2, the particle-fluid system is subject to a macroscopic velocity field of the form

$$\bar{v}_j(\mathbf{x}, t) = \sum_{l=1}^{3} \kappa_{jl}(t) x_j,$$

where $\kappa$ is known as the velocity gradient tensor. The corresponding microscopic velocity field $\mathbf{v}(\mathbf{x}, t)$ is obtained by requiring that (i) $\mathbf{v}$ is a solution of the Stokes equation (12.3.39) and (ii) the average of $\mathbf{v}$ with respect to the density $\psi$ is the macroscopic field, $\langle \mathbf{v}(\mathbf{x}, t) \rangle = \bar{\mathbf{v}}(\mathbf{x}, t)$. These conditions are satisfied if [250]

$$v_j(\mathbf{x}, t) = \sum_{l} \kappa_{jl}(t) x_l + \sum_{n=1}^{N} \sum_{l} H_{jl}(\mathbf{x} - \mathbf{x}_n) f_{nl}. \tag{12.3.45}$$

This follows from the observation that averaging the second term yields a spatially uniform velocity vector (ignoring boundary effects), which can only depend on the tensor $\kappa$. Since it is not possible to construct such a vector (in the absence of other tensors), the vector vanishes.

## 12.3.5 Langevin dynamics of polymers and membranes

We now analyze the Langevin dynamics of a flexible polymer and a cell membrane, both of which are modeled as continuous elastic media. First, consider a short flexible polymer of length $L < \ell_p$ so that it is reasonably straight. In that case, we can

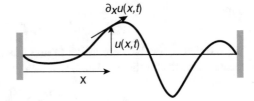

Fig. 12.21: Schematic diagram of the transverse fluctuations of a flexible polymer.

take the $x$-axis to define the average orientation of the filament and take $u(x,t)$ and $v(x,t)$ to represent the two independent transverse degrees of freedom. As a further simplification, we focus on the $u$ coordinate. As illustrated in Fig. 12.21, the local tangent to the filament is given by $\partial u/\partial x$ and the local curvature is specified by $\partial^2 u/\partial x^2$. It follows that the free energy associated with the $u$ coordinate is

$$F[u] = \frac{\kappa}{2} \int_0^L \left( \frac{\partial^2 u}{\partial x^2} \right)^2 dx, \tag{12.3.46}$$

where $\kappa = k_B T B$. The continuum analog of equation (12.3.24) is then

$$\gamma \frac{\partial u}{\partial t}(x,t) = -\frac{\delta F[u]}{\delta u(x,t)} + \xi(x,t), \tag{12.3.47}$$

where $\gamma$ is a friction coefficient and the spatiotemporal white noise term $\xi(x,t)$ represents the effects of instantaneous forces from the surrounding aqueous solution, with

$$\langle \xi(x,t) \rangle = 0, \quad \langle \xi(x,t) \xi(x',t') \rangle = 2 k_B T \gamma \delta(x-x') \delta(t-t'). \tag{12.3.48}$$

Substituting for $F[u]$ and using the definition of a functional derivative (Box 12C), we have

$$\gamma \frac{\partial u}{\partial t}(x,t) = -\kappa \frac{\partial^4 u(x,t)}{\partial x^4} + \xi(x,t). \tag{12.3.49}$$

It is simpler to consider the corresponding Langevin equation in Fourier space. Suppose that the filament is fixed at the ends $x = 0, L$, so that we have the sine series expansion

$$u(x,t) = \sum_q u_q \sin(qx), \quad q = \frac{n\pi}{L}, \quad n = 1, 2, 3, \ldots.$$

Then

$$F[u] = \frac{L}{4} \sum_q \kappa q^4 u_q^2, \tag{12.3.50}$$

and

$$\gamma \frac{\partial u_q}{\partial t} = -\kappa q^4 u_q + \xi_q(t), \tag{12.3.51}$$

with

$$\langle \xi_q(t)\xi_{q'}(t') \rangle = 2k_B T \gamma \delta(t - t')\delta_{q+q',0}.$$

We see that $u_q$ evolves according to an Ornstein-Uhlenbeck (OU) process, which means in particular that (Chap. 2)

$$\langle u_q(t)u_q(0) \rangle = \frac{2k_B T}{\kappa q^4}e^{-\omega_q t}, \quad \omega_q = \frac{\kappa q^4}{\gamma}. \tag{12.3.52}$$

Hence, we can identify $\omega_q$ as the relaxation rate of the $q$-th Fourier mode. The fact that the relaxation rate varies as $q^4$ means that there is a strong separation of time scales between the relaxation of short and long wavelength modes. An alternative Langevin equation for polymer dynamics will be introduced in Sect. 12.5.2, which ignores bending energies and treats the polymer as a Gaussian chain. This Rouse model will be used to analyze DNA dynamics within the nucleus.

We now turn to the Langevin dynamics of a membrane sheet. As a slight generalization of equation (12.2.23), let the free energy of the sheet in the Monge representation be given by the functional

$$F[h] = \int_\Omega \left[ \frac{\kappa_b}{2}[\nabla^2 h(\mathbf{r})]^2 + \frac{\sigma}{2}[\nabla h(\mathbf{r})]^2 \right] d^2\mathbf{r}, \tag{12.3.53}$$

where we have included an additional term involving the effects of stretching, with $\sigma$ the surface tension. Here $h(\mathbf{r})$ is the analog of the finite set of degrees of freedom of a Brownian particle. (This could be made more explicit by discretizing space into a set of lattice sites $\ell_j$ and setting $x_j = h(\ell_j)$.) Proceeding along analogous lines to the polymer, the continuum analog of equation (12.3.24) becomes

$$\gamma\frac{\partial h}{\partial t}(\mathbf{r},t) = -\frac{\delta F[h]}{\delta h(\mathbf{r},t)} + \xi(\mathbf{r},t), \tag{12.3.54}$$

with

$$\langle \xi(\mathbf{r},t) \rangle = 0, \quad \langle \xi(\mathbf{r},t)\xi(\mathbf{r}',t') \rangle = 2k_B T \gamma \delta(\mathbf{r} - \mathbf{r}')\delta(t - t'). \tag{12.3.55}$$

Substituting for $F[h]$ and using the definition of a functional derivative, we find that

$$\gamma\frac{\partial h}{\partial t}(\mathbf{r},t) = -\kappa_b\nabla^4 h(\mathbf{r}) + \sigma\nabla^2 h(\mathbf{r}) + \xi(\mathbf{r},t). \tag{12.3.56}$$

Transforming to Fourier space, see Sect. 12.2, gives

$$\gamma\frac{\partial \widehat{h}}{\partial t}(\mathbf{q},t) = -(\kappa_b q^4 + \sigma q^2)\widehat{h}(\mathbf{q},t) + \widehat{\xi}(\mathbf{q},t), \tag{12.3.57}$$

with

$$\langle \widehat{\xi}(\mathbf{q},t)\widehat{\xi}(\mathbf{q}',t') \rangle = 2k_B T \gamma \delta(\mathbf{q} + \mathbf{q}')\delta(t - t').$$

Here we are considering an unbounded membrane so that the vectors $\mathbf{q}$ form a continuum. Fourier transforming equation (12.3.57) with respect to time and averaging with respect to the noise then shows that the autocorrelation function for a particular mode $\mathbf{q}$ is

$$\left\langle |\widehat{h}(\mathbf{q},\omega)|^2 \right\rangle = \frac{2k_B T \gamma}{\gamma^2 \omega^2 + \omega_q^2},$$  (12.3.58)

where

$$\omega_q = \kappa_b q^4 + \sigma q^2.$$  (12.3.59)

In order to obtain the power spectral density, it is necessary to integrate over all modes, which gives

$$\left\langle |\widehat{h}(\omega)|^2 \right\rangle = \int \frac{2k_B T \gamma}{\gamma^2 \omega^2 + \omega_q^2} \frac{d^2\mathbf{q}}{(2\pi)^2} = \frac{k_B T \gamma}{\pi} \int_{q_{\min}}^{q_{\max}} \frac{q\,dq}{\gamma^2 \omega^2 + \omega_q^2}.$$  (12.3.60)

It is possible to extend the above analysis to take into account long-range hydrodynamic interactions between the membrane and the surrounding viscous aqueous environment [250]. To a first approximation, one simply convolves the right-hand side of equation (12.3.54) with the diagonal part of the Oseen tensor (12.3.42) [250]:

$$\frac{\partial h}{\partial t}(\mathbf{r},t) = \int \Lambda(\mathbf{r}-\mathbf{r}')\left\{ -\frac{\delta F[h]}{\delta h(\mathbf{r}',t)} + \xi(\mathbf{r}',t) \right\} d^2\mathbf{r}', \quad \Lambda(\mathbf{r}) = \frac{1}{8\pi\eta|\mathbf{r}|},$$  (12.3.61)

where $\eta$ is the viscosity of the surrounding fluid. It is straightforward to generalize the analysis in the Fourier domain using the convolution theorem and $\Lambda_q = 1/(4\eta q)$ with $q = |\mathbf{q}|$. In particular

$$\left\langle |\widehat{h}(\omega)|^2 \right\rangle = \int \frac{2k_B T \Lambda_q}{\omega^2 + \Lambda_q^2 \omega_q^2} \frac{d^2\mathbf{q}}{(2\pi)^2} = \frac{4\eta k_B T}{\pi} \int_{q_{\min}}^{q_{\max}} \frac{q\,dq}{(4\eta\omega)^2 + (\kappa_b q^3 + \sigma q^2)^2}.$$  (12.3.62)

Taking $q_{\min} \sim 0$ and $q_{\max} \sim \infty$, one finds that

$$\left\langle |\widehat{h}(\omega)|^2 \right\rangle \sim \omega^{-3/2} \text{ for } \omega \to \infty \qquad \left\langle |\widehat{h}(\omega)|^2 \right\rangle \sim \omega^{-1} \text{ for } \omega \to 0.$$

In Sect. 13.2.3, the dynamical equations for fluctuating membranes will extended to take into account active processes arising from interactions with the cytoskeleton.

## 12.4  Polymer networks and the cytoskeleton

The polymerized filaments of the cytoskeleton (F-actin, microtubules, and intermediate filaments) form a variety of network structures. Microtubules have a persistence length that is of the order of millimeters and so do not form contorted networks. Two common configurations are the aster-like distribution of micro-tubules radiating from the centrosome near the nucleus of a cell (see Sect. 14.3), and parallel bundles found, for example, within the axons and dendrites of neurons (see Sect. 7.1). In the former case, the filaments are not cross-linked. On the other hand, parallel bundling is mediated by microtubule associated proteins (MAPs), which are linking proteins that act like the rungs of a ladder between two neigh-boring filaments. More complicated 3D network structures can be formed by actin filaments, since they are relatively flexible, with a persistence length of around 10–20 $\mu$m. However, in common with microtubules, linking proteins are needed to form a semi-rigid network structure. Common actin-binding proteins are $\alpha$-actinin, fimbrin, and filamin. The first two link F-actin into parallel bundles. On the other hand, filamin is a hinged dimer that is approximately $V$-shaped with individual arms aligned length-wise along distinct actin filaments. Hence, this results in an irregular mesh-like polymer network, as illustrated in Fig. 12.22.

In general, many more elastic moduli are needed to describe the mechanical prop-erties of a 3D structure compared to a 2D structure such as the plasma membrane (treated as a thin sheet), see Box 12B. However, isotropic 3D materials only need two independent moduli, and a homogeneous cubic structure only needs three. The behavior of a polymer network will also depend on the density of filaments and cross-linking proteins. For sufficiently high densities, the theory of random chain networks (see below) can be used to show that the moduli are proportional to $\rho k_B T$, where $\rho$ is the density of polymer chains. On the other hand, at very low densities, the network will not be fully connected, and the shear modulus will vanish. The system is said to be below the percolation threshold. Now suppose that there are not any cross-links (or the cross-links are not permanent). The system of filaments in solution then behaves like a fluid, at least when observed over sufficiently long time scales. This means that even if the filaments are entangled, eventually thermal fluc-tuations will allow them to wiggle past each other. However, on short time scales, entanglement may result in the system responding elastically to applied forces, a phenomenon known as viscoelasticity. One thus finds that the elastic moduli of uncross-linked polymers may be time-dependent, with the effective shear modulus a decreasing function of time.

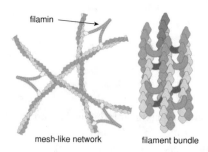

filamin

mesh-like network    filament bundle

Fig. 12.22: Two distinct configurations of actin filament networks.

The dynamics of cytoskeletal networks are much more complicated than idealized random chain networks or viscoelastic polymer fluids. It is difficult to precisely determine the characteristics of the cytoskeleton *in vivo*, so an alternative approach is to investigate the behavior of reconstituted networks of cytoskeletal proteins *in vitro* [82, 97, 745]. The latter allows a much more precise control of network parameters. Rheological measurements of these reconstituted networks have established a rich range of viscoelastic and nonlinear elastic responses that cannot be captured by traditional theories of polymers [400, 401, 564, 582]. Rheology is the general study of how materials deform and flow in response to externally applied forces. In the case of a simple elastic solid, such as rubber, the applied force is stored in the material deformation or strain, see Box 12B. Consider, for example, the experimental setup shown in Fig. 12.23(a). The nonzero component of the strain tensor is $u_{xy} = x(t)/h$, where $x(t)$ is the horizontal displacement and $h$ is the separation of the two plates. In the case of an ideal elastic solid, the corresponding stress tensor has the component

$$\sigma_{xy} = \mu u_{xy},$$

where $\mu$ is the shear modulus. The stress is $\sigma_{xy} = F/A$ where $F$ is the applied force and $A$ is the surface area of the top of the material. Hence,

$$u = u_{xy} = \frac{F}{\mu A}.$$

If the force is applied from $t = 0$ to $t = t_0$, then the strain is given by a step function, see Fig. 12.23(b). On the other hand, for a simple Newtonian fluid such as water, shear forces generate a constant flow or rate of change of strain, $\sigma = \eta \dot{u}$ where $\eta$ is the viscosity, see Fig. 12.23(c). For more complicated viscoelastic materials, such as the cytoskeleton, the stress–strain relationship takes the time-dependent form

$$\sigma(t) = \int_{-\infty}^{t} G(t - t')\dot{u}(t')dt', \tag{12.4.1}$$

where $G(t)$ is known as the relaxation modulus.

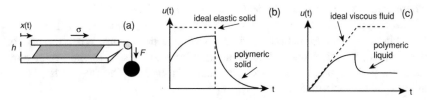

Fig. 12.23: (a) Experimental measurement of shear strain $u(t) = u_{xy}(t) = x(t)/h$. A sample is placed between two parallel plates with the upper plate displaced by the application of a force $F$ from $t = 0$ to $t = t_0$. The stress tensor is $\sigma_{xy} = F/A$, where $A$ is the upper surface area of the material. (b) Response of a solid-like material. (c) Response of a liquid-like material. [Redrawn from Doi [251].]

In the following, we briefly discuss the elasticity of cross-linked polymer networks and the macroscopic theory of viscoelasticity in uncross-linked polymer fluids. For a detailed account of microscopic theories of viscoelasticity; see Ref. [250, 251]. The more complicated response properties of cytoskeletal networks is reviewed in Ref. [331].

### 12.4.1 Random chain networks

Consider a network of flexible polymer chains that are cross-linked at many points along their length so that the network is connected, see Fig. 12.24. The theory for the elastic properties of such a network was originally developed within the context of vulcanized rubber [312]. Here we consider a simpler treatment presented in Ref. [251]. Suppose that the network is in an undeformed or reference state. The basic idea is to treat each of the polymer chains between neighboring cross-links (subchains) as flexible polymers that can be modeled as freely-jointed chains. If a subchain consists of $N$ segments and its end-to-end vector is $\mathbf{r}$, then its free energy contribution is given by $(3k_BT/2Na^2)r^2$; see Sect. 12.1.1. If interactions between subchains are ignored, then the free energy of the polymer network (per unit volume) is

$$F_0 = n_c \int \int_0^\infty \Psi_0(\mathbf{r},N) \frac{3k_BT}{2Na^2} r^2 dN d^3\mathbf{r}, \tag{12.4.2}$$

where $n_c$ is the number density of subchains and $\Psi_0(\mathbf{r},N)$ is the probability density that a subchain consists of $N$ segments and has the end-to-end vector $\mathbf{r}$. The density $\Psi_0(\mathbf{r},N)$ will depend on the particular pattern of cross-linking. We will make the simplifying assumption that the distribution of end-to-end vectors of subchains in the reference state is identical to that of a free chain (no cross-linking) at equilibrium. That is, for a chain of $N$ segments, the density $\Psi_0(\mathbf{r},N)$ is given by the Gaussian

$$\Psi_0(\mathbf{r},N) = \left(\frac{3}{2\pi Na^2}\right)^{3/2} \exp\left(-\frac{3r^2}{2Na^2}\right) \Phi_0(N), \tag{12.4.3}$$

where $\Phi_0(N)$ is the probability density of $N$ so that $\int_0^\infty \Phi_0(N)dN = 1$.

Now suppose that the polymer network is uniformly deformed with $\mathbf{x} \to \mathbf{x}' = \mathbf{Mx}$, where the transformation matrix $\mathbf{M}$ is volume preserving. When the network is deformed, we make the additional assumption that the end-to-end vector $\mathbf{r}$ of a subchain is deformed in the same way, that is, $\mathbf{r}' = \mathbf{Mr}$. The free energy change induced by the deformation is

$$F(\mathbf{M}) = n_c \int \int_0^\infty \Psi_0(\mathbf{r},N) \frac{3k_BT}{2Na^2} [(\mathbf{Mr})^2 - r^2] dN d^3\mathbf{r}. \tag{12.4.4}$$

Using the fact that the Gaussian $\Psi_0(\mathbf{r}, N)$ is isotropic, we have

$$\int \Psi_0(\mathbf{r}, N)(\mathbf{Mr})^2 d^3\mathbf{r} = \sum_{i,j,k} M_{ij}M_{ik}\int \Psi_0(\mathbf{r}, N)r_j r_k d^3\mathbf{r} = \sum_{i,j} M_{ij}M_{ij}\frac{Na^2}{3}\Phi_0(N).$$

Hence, using the unit normalization of $\Phi_0$,

$$F(\mathbf{M}) = \frac{1}{2}n_c k_B T\left(\sum_{i,j} M_{ij}M_{ij} - 3\right).$$

Let $\mathbf{D} = \mathbf{MM}^\top$ and denote the eigenvalues of $\mathbf{D}$ by $\lambda_j^2, j = 1,2,3$ (for a 3D polymer network). It immediately follows that $F(\mathbf{M}) = F(\lambda)$, where

$$F(\lambda) = \frac{1}{2}n_c k_B T\,(\mathrm{Tr}[\mathbf{D}] - 3) = \frac{1}{2}n_c k_B T\left(\sum_{j=1}^{3}\lambda_j^2 - 3\right). \tag{12.4.5}$$

For example, suppose that $x \to \Lambda_x x$, $y \to \Lambda_y y$ and $z \to \Lambda_z z$. Then

$$\mathbf{M} = \begin{pmatrix} \Lambda_x & 0 & 0 \\ 0 & \Lambda_y & 0 \\ 0 & 0 & \Lambda_z \end{pmatrix}, \tag{12.4.6}$$

and

$$F = \frac{1}{2}n_c k_B T\,(\mathrm{Tr}[\mathbf{D}] - 3) = \frac{1}{2}n_c k_B T\left(\Lambda_x^2 + \Lambda_y^2 + \Lambda_z^2 - 3\right). \tag{12.4.7}$$

In contrast to fluid membranes (Sect. 12.2), polymer networks resist shear forces. In order to determine the corresponding shear modulus $\mu$, see equation (12.2.13), consider a pure shear of the form

$$\Lambda_x = \Lambda = \frac{1}{\Lambda_y}, \quad \Lambda_z = 1.$$

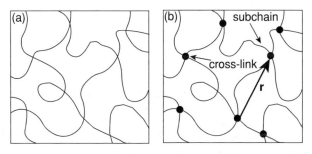

Fig. 12.24: Rubber elasticity. (a) Polymer network without cross-links. (b) Polymer network with cross-links.

In this case, equation (12.4.5) becomes

$$F = \frac{n_c k_B T}{2} \left( \Lambda^2 + \frac{1}{\Lambda^2} - 2 \right).$$

For a sufficiently small shear, $\Lambda = 1 + \delta$, $\delta \ll 1$, we have

$$F \approx 2\delta^2 n_c k_B T.$$

The corresponding stress tensor has the nonzero components $u_{xx} = \delta = -u_{yy}$, so that equation (12.2.13) reduces to $F = \mu (u_{xx}^2 + u_{yy}^2) = 2\delta^2 \mu$. It immediately follows that $\mu = n_c k_B T$. An analogous calculation for a different deformation of the polymer network is considered in Ex. 12.10

**Swelling of a polymer gel.** A polymer gel is a mixture of a cross-linked polymer network and a solvent such as water. A gel is elastic like rubber, as described above, but unlike rubber, it can change its volume by absorbing or expelling solvent. For example, swelling occurs if a dry gel is placed in a solvent, and/or if an external stretching force $f$ is applied to a gel in equilibrium with a solvent. An application to the growth of biofilms will be developed in Sect. 15.6. Here we consider a simplified model.

Consider a polymer network consisting of $N_2$ subchains each containing an average of $M$ monomers. For simplicity, take the size $v_c$ of each monomer to be the same as a solvent molecule. The undeformed volume is taken to be $V_0 = L_0^3$ with $V_0 = MN_2 v_c$. Now suppose that the dry gel is placed in a solvent and at the same time is subject to an applied force in the $z$-direction, see Fig. 12.25. In response, the gel not only elongates in the $z$-direction, but also swells in volume due to the absorption of $N_1$ solvent molecules. Assuming that the total solution is incompressible, the swollen gel has volume $V = (N_1 + MN_2)v_c$ and shape $L_z = L = \alpha L_0$, $L_x = L_y = (V/L)^{1/2}$ with $\alpha > 1$. In terms of the volume fraction

$$\phi = \frac{V_0}{V} = \frac{MN_2}{N_1 + MN_2},$$

we have the scale factors

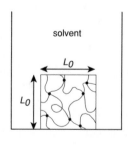

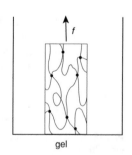

Fig. 12.25: Swelling of a polymer gel of initial volume $L_0^3$ after being placed in contact with a solvent and deformed by an applied force $f$.

$$\Lambda_z = \alpha, \quad \Lambda_x = \Lambda_y = \frac{1}{\sqrt{\alpha\phi}},$$

and we can write $\phi = 1/\Lambda_x\Lambda_y\Lambda_z$. If $F_0$ is the free energy density of the gel prior to deformation, then the free energy density $F$ in response to deformation satisfies

$$F_{gel}(\phi) := F - F_0 = \frac{V}{V_0}F_{mix} + F_{def},$$

where $F_{def}$ is given by equation (12.4.7) (for volume-preserving deformations) and $F_{mix}$ is the free energy density due to mixing $N_1$ solvent molecules with $N_2$ polymers containing $M$ monomers. The latter can be determined using the Flory-Huggins theory of gels, which is based on a generalization of the lattice model introduced in Sect. 1.3. The details will be provided in Sect. 13.1.1. Here we simply state the result based on (13.1.10) with $\phi \approx 1$

$$F_{mix} = \frac{k_B T}{v_c}[(1-\phi)\ln(1-\phi) + \chi\phi(1-\phi)], \tag{12.4.8}$$

where $\chi$ is determined by interaction energies between neighboring molecules. It follows that

$$F_{gel}(\phi) := \frac{k_B T}{v_c \phi}[(1-\phi)\ln(1-\phi) + \chi\phi(1-\phi)] + \frac{1}{2}\mu\left(\frac{2}{\alpha\phi} + \alpha^2 - 3\right).$$

The equilibrium volume fraction is determined by the condition $\partial F_{gel}/\partial\phi = 0$:

$$-\frac{\mu}{\alpha\phi^2} + \frac{\phi F'_{mix}(\phi) - F_{mix}(\phi)}{\phi^2} = 0,$$

that is,

$$\frac{\mu}{\alpha} = \phi F'_{mix}(\phi) - F_{mix}(\phi) := \Pi(\phi). \tag{12.4.9}$$

The quantity $\Pi(\phi)$ is known as the osmotic pressure, see equation (13.1.15), and here takes the particular form

$$\Pi(\phi) = \frac{k_B T}{v_c}(-\ln(1-\phi) - \phi - \chi\phi^2). \tag{12.4.10}$$

(The osmotic pressure is a measure of the additional pressure that must be applied to equilibrate the polymer solution with a pure solvent across a semipermeable membrane.) First suppose that $f = 0$ (no applied force). Placing the gel in contact with the solvent then leads to isotropic swelling, that is, $\Lambda_x = \Lambda_y = \Lambda_z$. In this case $\alpha = 1/\sqrt{\alpha\phi}$, that is, $\alpha = 1/\phi^{1/3}$. The equilibrium volume fraction is then $\phi = \phi_0$ with

$$\mu\phi_0^{1/3} = \Pi(\phi_0). \tag{12.4.11}$$

On the other hand, if $f > 0$, then $\alpha = \phi_0^{-1/3}\widehat{\alpha}$ with $\widehat{\alpha} > 1$ and the new volume fraction $\phi_1$ is given by

$$\frac{\mu\phi_0^{1/3}}{\widehat{\alpha}} = \Pi(\phi_1). \tag{12.4.12}$$

Since $\Pi'(\phi) > 0$ and $\widehat{\alpha} > 1$, it follows that $\phi_1 < \phi_0$, which means that more solvent has been absorbed and the gel is more swollen.

### 12.4.2 Viscoelasticity of polymer fluids

Let us now consider an ensemble of polymers in solution without any cross-links. A common way of characterizing the linear viscoelasticity of the resulting complex fluid is to measure the stress response to an oscillatory shear strain $u(t) = u_0 \cos \omega t$. It follows from equation (12.4.1) that

$$\sigma(t) = \int_{-\infty}^{t} G(t-t')[-u_0\omega \sin \omega t']dt' = -u_0\omega \int_0^\infty G(t')\sin \omega(t-t')dt'$$
$$= u_0[G'(\omega)\cos \omega t - G''(\omega)\sin \omega t], \tag{12.4.13}$$

where

$$G'(\omega) = \omega \int_0^\infty \sin \omega\tau G(\tau)d\tau, \quad G''(\omega) = \omega \int_0^\infty \cos \omega\tau G(\tau)d\tau. \tag{12.4.14}$$

The quantity

$$G^*(\omega) = G'(\omega) + iG''(\omega) = i\omega \int_0^\infty G(\tau)e^{-i\omega t}dt$$

is known as the complex modulus, whereas $G'(\omega)$ and $G''(\omega)$ are, respectively, called the storage modulus and loss modulus. It follows that we can rewrite the time-dependent shear stress as

$$\sigma(t) = G'(\omega)u(t) + \frac{G''(\omega)}{\omega}\frac{du}{dt}. \tag{12.4.15}$$

Hence, $G'(\omega)$ represents the elastic response, whereas $G''(\omega)$ represents the viscous response. The latter should vanish linearly with $\omega$ as $\omega \to 0$. For the sake of illustration, suppose that

$$G(t) = G_0 + G_1 e^{-t/\tau},$$

where $G_0 > 0$ $(G_0 = 0)$ for a viscoelastic solid (fluid). We then have

$$G'(\omega) = G_0 + \frac{(\omega\tau)^2}{1+(\omega\tau)^2}G_1, \quad G''(\omega) = \frac{\omega\tau}{1+(\omega\tau)^2}G_1.$$

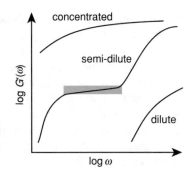

Fig. 12.26: Sketch of possible behaviors of storage modulus $G'(\omega)$ as a function of the forcing frequency $\omega$ for different concentrations. [Redrawn from Boal [87].]

Therefore, in the limit $\omega \to 0$, a viscoelastic solid behaves as $G^*(\omega) \sim G_0$, while a viscoelastic fluid behaves as $G^*(\omega) \sim i\omega G_1 \tau$.

The frequency-dependence of the storage modulus $G'(\omega)$ can be used to distinguish between different density-dependent behaviors, see Fig. 12.26. First, note that $G'(\omega)$ is a monotonically increasing function of $\omega$. Dilute polymer solutions have the smallest storage moduli with $G'(\omega) \to 0$ as $\omega \to 0$, whereas concentrated solutions can have storage moduli similar to hard plastics. The most interesting behavior occurs at intermediate concentrations, in which the system exhibits little resistance to shear at small $\omega$ (long times) but a high resistance at large $\omega$ (short times). At intermediate frequencies, storage moduli can exhibit plateau-like behavior arising from polymer entanglements. The latter are loops or sometimes knots that occur when polymers thread around each other. In this regime, the polymer solution behaves like a network of temporarily fixed cross-links. In order to develop microscopic theories of the relaxation moduli, it is necessary to determine how the stress within a polymer network is expressed microscopically at the molecular level. This can then be used to derive an equation relating the stress tensor to a given deformation history, which is known as a constitutive equation. In the unentangled regime, one needs to use various results from continuum mechanics and the behavior of objects moving in a fluid. Here we focus on the case of entangled polymers, which can be analyzed using reptation theory, which is a diffusion-based model of confined polymers. Both regimes are analyzed in some detail in refs. [250, 251].

**Reptation theory.** One way to model a solution of entangled polymers is to treat it as a network of temporarily cross-linked filaments. However, the mechanisms for creating and destroying such cross-links are not specified. An alternative approach is to represent the entanglement junctions as small rings known as slip-links, which combine pairs of polymer chains, see Fig. 12.27. The chains are assumed to be able to move freely through the slip-links via diffusion, so that a slip-link is destroyed whenever a chain end moves out of a slip-link. On the other hand, new slip-links can be created whenever polymers form new entanglement junctions. Thus, the entanglement points are constantly created and destroyed due to the 1D Brownian motion of a polymer chain along slip-links—a process known as reptation. Such constrained motion can also be represented by a tube. One important quantity is the mean distance $a$ between neighboring slip-links, which will depend on the rigidity and local density of chain segments, and is typically a few nm.

Suppose that the slip-links are distributed uniformly along a chain with equal spacing $a$, and that the molecular weight of a segment between neighboring junctions is $M_e$. The number of entanglement points per chain is $Z = M/M_e$, and the contour length of the filament is $L = Za = Ma/M_e$. A first estimate for the time-dependent relaxation of the shear stress is to determine the time-dependent probability that entanglement junctions disappear [250]. The polymer is assumed to randomly slide along the tube generated by the slip-links according to 1D Brownian motion, while keeping its length $L$ fixed. Let $s$ denote the curvilinear coordinate along the chain, with the origin $s = 0$ fixed at the position of the $n$-th slip-link $S_n$ on the chain. That is, when $t = 0$ the left and right ends of the chain are at $s = -na$ and $s = L - na$, respectively. Take $\psi_n(z,t)$ to be the probability density that the given slip-link $S_n$ still exists at time $t$ with the right end at position $s = z$. Then $\psi_n$ satisfies the FP equation

$$\frac{\partial \psi_n(z,t)}{\partial t} = D \frac{\partial^2 \psi_n(z,t)}{\partial z^2}, \tag{12.4.16}$$

where $D$ is the diffusion coefficient of the polymer along the tube. The FP equation is supplemented by the initial condition $\psi_n(z,0) = \delta(z - L + na)$ and the boundary conditions $\psi_n(0,t) = 0$, $\psi_n(L,t) = 0$. That is, $S_n$ is destroyed when either the left end ($z = L$) or the right end ($z = 0$) reaches the origin.

The solution of the boundary value problem is (Chap. 2)

$$\psi_n(z,t) = \frac{2}{L} \sum_{k=1}^{\infty} (-1)^k \sin\left(\frac{kz\pi}{L}\right) \sin\left(\frac{kna\pi}{L}\right) e^{-k^2 t/\tau_d}, \tag{12.4.17}$$

where $\tau_d = L^2/(\pi^2 D)$ is known as the reptation time, and characterizes the lifetime of an entanglement. From the Einstein relation, $D$ is related to the friction or drag coefficient $\xi$ according to $D = k_B T/\xi$. Since $\xi$ determines the force needed to pull the chain along the tube, it is proportional to the molecular weight of the chain, $M$,

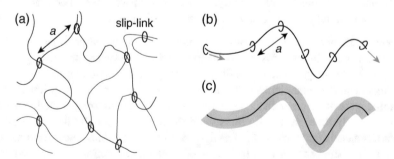

Fig. 12.27: (a) Slip-link model of entangled polymers, (b) Disappearance of a slip-link due to motion of polymer chain, and (c) Equivalent tube model.

which implies that $D \sim M^{-1}$. Given that $L \sim M$, it follows that $\tau_d \sim M^3$. We can now write down the probability density that an arbitrarily chosen entanglement point at time $t = 0$ still exists at time $t$, by averaging

$$\psi(t) = \frac{1}{Z} \sum_{n=1}^{Z} \int_0^L \psi_n(z,t)dz. \tag{12.4.18}$$

Substituting the explicit solution for $\psi_n(z,t)$ and replacing the discrete sum over $n$ by an integral (valid for large $Z$)

$$\psi(t) = \frac{8}{\pi^2} \sum_{k=1,3,5,\ldots} \frac{1}{k^2} e^{-k^2 t/\tau_d}. \tag{12.4.19}$$

Note that the dominant behavior after an initial transient phase is $\psi(t) \sim e^{-t/\tau_d}$.

Now suppose that at time $t = 0$, the polymer fluid is subjected to a shear of strain $u$ in the $x$-direction. Immediately after the introduction of the shear, all entanglement junctions act as transient cross-links. The initial stress $\sigma_{xy}(0)$ is thus given by the same expression as for rubber elasticity

$$\sigma_{xy}(0) \equiv G_0 u = n_c k_B T u = \frac{\rho N_A k_B T}{M_e} u, \tag{12.4.20}$$

where $G_0 = \mu$ is the shear modulus, $n_c$ is the number density of chains, $\rho$ is the corresponding mass density, and $N_A$ is Avagadro's number. Measurement of $G_0$ can be used to estimate the mean molecular weight $M_e$. At later times, the entanglement points start to disappear due to the Brownian motion of the chains. Since the survival probability density is given by $\psi(t)$, the shear stress at time $t$ is taken to be

$$\sigma_{xy}(t) = G(t)u, \quad G(t) = G_0 \psi(t). \tag{12.4.21}$$

Experimental measurements of the relaxation time of the shear stress shows that $\tau_d \sim M^\alpha$ with $\alpha = 3.2 - 3.4$. On the other hand, the basic reptation model predicts $\alpha = 3$. There are at least two simplifying assumptions of the model that need to be modified in order to obtain better agreement with experiments. First, the constraints represented by the distribution of slip-links are assumed to be fixed. However, in reality, these constraints are generated by other polymer chains, which are themselves undergoing reptation, which means that the equivalent confining tube has a finite lifetime. Second, the contour length of a polymer chain is assumed to be constant. However, fluctuations can also lead to contractions of the chain within the tube.

### 12.4.3 Molecular model of viscoelasticity

Following Doi [251], suppose that we represent a polymer as a dumbbell-shaped molecule consisting of two beads linked by a Hookean spring. Each bead represents

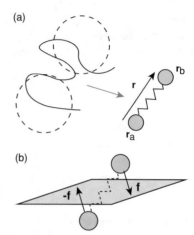

(a)

(b)

Fig. 12.28: (a) Dumbbell model of a polymer. (b) Molecular origin of the force exerted across an interior plane of the polymeric solid.

half of a polymer, as shown in Fig. 12.28(a). One can generalize the model to the case of multiple beads linked by springs (see the Rouse model of Sect. 12.5.2), but the essential ideas can be conveyed using the simpler model. Let $\mathbf{r}_a$ and $\mathbf{r}_b$ denote the positions of the two beads and write the potential energy of the spring as

$$U(\mathbf{r}) = \frac{k}{2}\mathbf{r}^2, \quad \mathbf{r} = \mathbf{r}_b - \mathbf{r}_a,$$

where $k$ is the spring constant. Suppose that the surrounding fluid is subject to a shear force that generates a flow field of the form

$$v_i(\mathbf{r},t) = \sum_{j=x,y,z} \kappa_{ij}(t)r_j,$$

where $\kappa_{ij}$ denotes the velocity gradient. The example of a shear flow in the $x$-direction is $v_x = \dot{u}(t)r_y$ and $v_y = v_z = 0$, see Fig. 12.23(a). Suppose that there are $n_p$ molecules per unit volume and denote the probability density of the different configurations by $\psi(\mathbf{r},t)$. The latter evolves according to the conservation equation (12.3.26) with the particle flux velocity $\dot{\mathbf{r}} = \mathbf{v}_f(\mathbf{r},t)$ given by a modified version of equation (12.3.28)

$$\frac{1}{2}\gamma\left(\dot{r}_i - \sum_j \kappa_{ij}r_j\right) = -k_BT\frac{\partial \ln \psi}{\partial r_i} - kr_i, \qquad (12.4.22)$$

where the frictional force now depends on the relative motion of the molecule with respect to the background fluid; see equation (12.3.45). Here $\gamma$ is an isotropic friction coefficient, and the factor of $1/2$ is due to the fact we are considering the motion of the relative position $\mathbf{r} = \mathbf{r}_b - \mathbf{r}_a$. Rearranging, we thus have

$$\dot{r}_i = v_{fi}(\mathbf{r},t) = -\frac{2}{\gamma}\left(k_BT\frac{\partial \ln \psi}{\partial r_j} + kr_j\right) + \sum_j \kappa_{ij}r_j. \qquad (12.4.23)$$

(One should really decompose the motion of the two beads in terms of the vector $\mathbf{r} = \mathbf{r}_b - \mathbf{r}_a$ and the center of mass $\mathbf{R} = (\mathbf{r}_a + \mathbf{r}_b)/2$ [251]. However, for a uniform distribution of molecules, the center of mass motions can be ignored, and we can treat $\mathbf{r}$ as the single degree of freedom per molecule.) Substituting (12.4.23) into equation (12.3.26) yields

$$\frac{\partial \psi}{\partial t} = \nabla \cdot \left( \frac{2k_B T}{\gamma} \nabla \psi + \frac{2k}{\gamma} \mathbf{r}\psi - \kappa \cdot \mathbf{r}\psi \right). \tag{12.4.24}$$

In order to calculate the stress energy tensor, we first note that the force on the bead at $\mathbf{r}_b$ has components

$$f_j = -\left( k_B T \frac{\partial \ln \psi}{\partial r_j} + kr_j \right),$$

and the force on the bead at $\mathbf{r}_a$ has components $-f_j$. Consider a plane with normal in the $z$-direction placed somewhere between the two beads, as illustrated in Fig. 12.28(b). The stress tensor $\sigma_{jz}$ is defined according to the size of the force exerted on material below the plane by material above the plane. Focusing on just a single molecule, we have $\sigma_{jz} = -f_j/A$, where $A$ is the area of the plane. The force contributes to the stress tensor provided that the plane lies between the two beads. Thus, we should write

$$\sigma_{jz} = -\frac{1}{A} f_j \Theta(z - r_{az}) \Theta(r_{bz} - z),$$

where $\Theta$ is the Heaviside function. Since the stress tensor is homogeneous, it should not depend on $z$. Therefore, averaging with respect to $z$ gives

$$\sigma_{jz} = -\frac{1}{AL} f_j (r_{bz} - r_{az}) = -\frac{1}{V} f_j r_z,$$

where $V$ is the volume. If we now sum over all molecules, and also consider forces across other planes, we see that the total stress tensor of the polymer is

$$\sigma_{jk} = n_p \int \left( k_B T \frac{\partial \ln \psi}{\partial r_j} + kr_j \right) r_k \psi d\mathbf{r}. \tag{12.4.25}$$

Integration by parts and the normalization $\int \psi(\mathbf{r}) d\mathbf{r} = 1$ gives

$$\sigma_{jk} = -n_p k_B T \delta_{j,k} + n_p k \langle r_j r_k \rangle. \tag{12.4.26}$$

Equations (12.4.23), (12.4.24), and (12.4.26) determine the so-called constitutive equations for the given model of a polymer solution. (There are also contributions to the stress tensor from the viscosity and pressure of the moving fluid, see equation (12.3.38), which we have neglected for simplicity.)

Let us now consider the particular case of the shear flow with $v_x = \dot{u} r_y$, and $v_y = v_z = 0$. The corresponding shear component of the stress tensor is $\sigma_{xy} = n_p k \langle r_x r_y \rangle$. The time-dependence of the stress tensor is determined from

$$\frac{d}{dt}\langle r_x r_y \rangle = \int r_x r_y \frac{\partial \psi}{\partial t} d\mathbf{r} = \int r_x r_y \left[ \nabla \cdot \left( \frac{2k_B T}{\gamma} \nabla \psi + \frac{2k}{\gamma} \mathbf{r} \psi \right) - \frac{\partial \dot{u} r_y \psi}{\partial r_x} \right]$$
$$= -\frac{4k}{\gamma}\langle r_x r_y \rangle + \dot{u}\langle r_y^2 \rangle, \tag{12.4.27}$$

where we have used equations (12.3.26), (12.4.23), and integration by parts. In the linear viscoelastic regime, terms of order $\dot{u}^2$ or higher are dropped. This means that we replace $\langle r_y^2 \rangle$ by its equilibrium value. Since $\psi_{\text{eq}} \sim e^{-k\mathbf{r}^2/2k_B T}$, we have $\langle r_y^2 \rangle_{\text{eq}} = k_B T/k$ and

$$\frac{d\sigma_{xy}}{dt} = -\frac{1}{\tau}\sigma_{xy} + n_p k_B T \dot{u}(t), \quad \tau = \frac{\gamma}{4k}. \tag{12.4.28}$$

We hence obtain the time-dependent stress–strain relation

$$\sigma_{xy}(t) = n_p k_B T \int_{-\infty}^{t} e^{-(t-t')/\tau} \dot{u}(t')dt'. \tag{12.4.29}$$

Comparison with the linear viscoelastic relation (12.4.1), we deduce that the relaxation modulus is

$$G(t) = G_0 e^{-t/\tau}, \quad G_0 = n_p k_B T. \tag{12.4.30}$$

The initial shear modulus $G_0$ is identical in form to that of the random chain network used to model rubber elasticity.

## 12.5 DNA dynamics in the nucleus

One of the major implications of the physics of polymers is that they typically exhibit universal behavior that is independent of their underlying composition [250, 775]. This implies that DNA should follow the same basic physical laws as commercial polymers such as polyethylene. In particular, when describing the organization of chromosomes within the nucleus, one should be able to apply classical mathematical models that determine, for example, the concentration-dependent arrangement of polymers in solution, how the end-to-end distance of a polymer varies with polymer length, and the snake-like motion of a polymer moving in a background sea of other polymer chains (reptation). Recent technological advances have resulted in high-resolution maps of the genome-wide physical contacts within and between chromosomes, which can run into the millions in the case of the human genome. Such maps have revealed a number of important organizational principles of the interphase nucleus (nucleus prior to cell mitosis), which present several challenges to reconciling classical polymer physics with the underlying biology. Three such principles are the sequestration of chromosomes into nuclear territories, the partitioning of transcriptionally active and inactive regions within the nucleus, and the organization of chromosomes into loops [13, 622, 712, 795]. In this section, we focus on the dynamics of DNA within the nucleus. After describing some of the key features of nuclear organization, we turn to a classical stochastic model of polymer dynamics, namely, the Rouse model of a Gaussian polymer chain. The latter has been used to account for the subdiffusive motion of chromosomal loci on relatively short time scales, and to explore mechanisms for spontaneous DNA loop formation.

## 12.5.1  Nuclear organization

We begin by briefly summarizing what is known about the spatial organization of DNA and associated proteins within the nucleus, which form a complex of macro-molecules known as chromatin. There are a number of important functions per-formed by chromatin. First, it packages DNA into a more compact shape so that it can fit inside the nucleus. (For example, each human cell contains around two meters of DNA, which must be tightly folded inside the cell nucleus.) Second, it controls gene expression and DNA replication by allowing RNA polymerase and transcription factors to access active regions of DNA, in spite of its tightly-packed nature. Third, it reinforces the integrity of DNA during cell mitosis and helps prevent DNA damage. Although chromatin's detailed structure is still poorly understood, it clearly exhibits organization at several different spatial scales, and its overall struc-

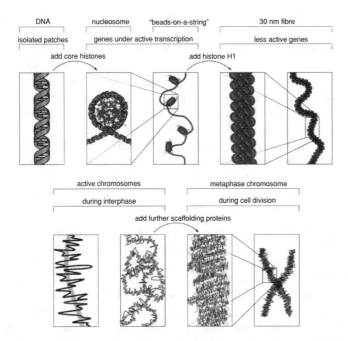

Fig. 12.29:  Schematic diagram illustrating chromatin structure at multiple spatial scales. See text for details. [Public domain figure downloaded from wikipedia commons.]

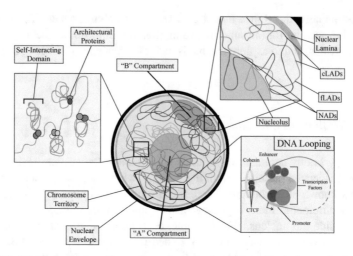

Fig. 12.30: Schematic diagram illustrating various characteristics of nuclear organization, including chromosome territories, separation of active (A compartment) and inactive (B compartment) regions, DNA loops, and associations with various substructures near the nuclear envelope (lamina, nucleolus). See text for further details. [Public domain figure downloaded from wikipedia commons.]

ture depends on the stage of the cell cycle, see Fig. 12.29. At the smallest scale, DNA wraps around so-called histone proteins forming nucleosomes, which leads to the relatively loosely packed "beads on a string" structure (euchromatin). This loose structure is exhibited by DNA coding genes that are actively transcribed ("turned on") during interphase, while inactive DNA coding genes ("turned off") are more tightly condensed by multiple histones that form 30 nm fibers (heterochromatin). Note that epigenetic chemical modification of the structural proteins in chromatin can alter the local chromatin structure, particularly via chemical modifications of histone proteins by methylation and acetylation (see Sect. 5.7). During cell mitosis, the chromatin is packed more tightly in order to facilitate segregation of the chromosomes during anaphase (see Sect. 14.3). There are a number of other levels of nuclear organization, see Fig. 12.30, which we now describe in more detail following along the lines of [795].

**Chromosome territories.** Recall from Sect. 12.1 that the configuration of a single isolated ideal polymer chain exhibits a random walk. One characteristic feature of an ideal chain is that the mean-squared end-end distance $R^2$ scales like $a^2N$, where $N$ is the number of monomers and $a$ is the effective bond length. In other words, $R \sim aN^{1/2}$. (In the case of DNA $a \sim 100$ nm.) One of the simplifications of the ideal chain model is that the spatial extent of each monomer is neglected, so that configurations include those where two monomers occupy the same space. For real chains, the excluded volume associated with monomers leads to a substantial swelling of the chain such that $R \sim aN^{3/5}$, see Fig. 12.31. On the other hand, when one con-

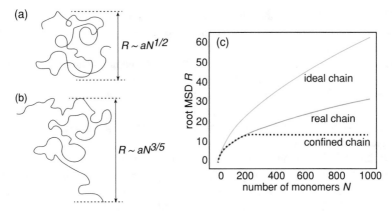

Fig. 12.31: End-to-end distance $R$ of a polymer chain consisting of $N$ monomers and bond length $a$. (a) Ideal chain, (b) Swelling of a real chain due to exclusion effects, and (c) Plots of $R$ as a function of $N$ for different cases.

siders a sufficiently concentrated polymer solution with many overlapping polymer chains, which we expect to hold in the nucleus, the excluded volume has no effect and each real polymer acts like an ideal chain. This is due to the fact that the outward pressure generated by monomers of a given polymer is canceled by the inward pressure of the other chains. Moreover, ideal chains exhibit a self-similar structure so that if two monomers are separated by $n$ links (a chemical distance $an$ along the chain) then their spatial separation is $d \sim an^{1/2}$.

In order to investigate how the theory of polymer sizes applies to chromosomes, it is necessary to visualize the 3D positions of chromosomal loci within the nucleus, and to compare their genomic and spatial distances. This can be achieved using so-called fluorescent in situ hybridization (FISH), see Fig. 12.32. Experimentally, one finds that for relatively short distances (up to around 1.8 Mb), the configurations are consistent with the random walk model. However, at longer spatial scales, the end-to-end distance tends to level off, indicative of some form of confinement

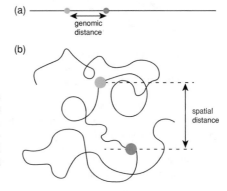

Fig. 12.32: Fluorescent in situ hybridization (FISH). (a) A stretch of chromosome whose ends a tagged with blue and red fluorochromes. The distance along the chain is called the chemical or genomic distance and (b) The physical (spatial distance) between the marked foci can be measured within the nucleus.

[896], see Fig. 12.32. In other words, chromosomes are restricted to local territories within the nucleus [218], which contradicts the expected self-similar structure of ideal chains. One possible explanation for the segregation of interphase chromosomes into territories is that, it is a consequence of the decondensation of mitotic chromosomes at the spatial positions they occupied following the last mitotic phase. The plausibility of such a mechanism has been demonstrated using a computational study [770], which also suggests that the times scale for a collection of polymers in solution to reach the equilibrium state predicted for ideal chains is much longer than biological time scales (around 500 years in human cells). The extreme slowness of equilibration is due to the fact that each polymer is effectively trapped in a tube by surrounding chains, and can only escape from the tube via reptation. Finally, note that the boundaries between chromosome territories are rather fuzzy, in the sense that there is physical contact and mixing at the edges.

Much more detailed information regarding the nonequilibrium organization of chromosomes within the nucleus has been obtained using a chromosome conformation capture technique known as Hi-C [560]. This type of experimental method quantifies the number of interactions between genomic loci that are nearby in 3D space, but may have a large genomic distance. The basic steps of chromosome conformation capture experiments are as follows: (i) covalently link chromosomes in situ with formaldehyde; (ii) cut the genome into small pieces; (iii) permanently link the cross-linked DNA fragments (ligation) and then remove the cross-links; (iv) use

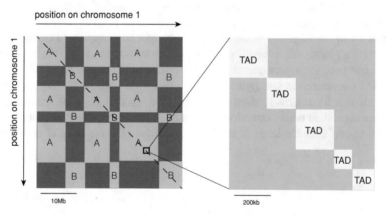

Fig. 12.33: Schematic illustration of a chromosome contact map at two levels of resolution. Hi-C data at low resolution (1Mb) typically exhibits a checkerboard pattern with regions of higher contact probability (light shade) and regions of lower contact probability (darker shade). The checkerboard pattern suggests that there are two types of compartment, A and B, which turn out to be associated with active and inactive regions of the genome, respectively. Points around the diagonal indicate contacts between adjacent regions on the genome, whereas off-diagonal points represent contacts between non-adjacent regions. When one zooms in on diagonal regions at higher resolution (100kb), one finds topologically associated domains (TADs), which have a particularly high probability of intradomain contact that are physically insulated from the remainder of the chromosome. [Redrawn from Sazer and Schiessel [795].]

massively parallel DNA sequencing to determine which DNA stretches have been ligated. The resulting data is then displayed as a genome-wide contact matrix, which specifies which parts of the genome are mutually close in space, as illustrated in Fig. 12.33. The latter is captured by the contact probability between pairs of loci, which is simply proportional to the number of ligation products between a given pair of loci. At a relatively low resolution (1Mb), the contact matrix has a checkerboard pattern with alternating regions of higher and lower contact probability. Moreover, one observes two types of compartment, A and B, which correspond to euchromatin rich and heterochromatin rich regions, respectively. At higher resolution (1kb), the diagonal of the contact matrix contains topologically associated domains (TADs), which have a particularly high probability of intradomain contact, and tend to be associated with DNA loops (see below).

**Euchromatin and heterochromatin.** Euchromatin and heterochromatin tend to be enriched in separate regions, with the former concentrated at the center of the nucleus and the latter found predominantly at the periphery, that is, in proximity to the nuclear envelope. The inner nuclear envelope contains a complex of proteins that forms a skeletal structure known as the nuclear lamina. One of the many functions of the lamina is to provide docking sites for heterochromatin at the periphery, see Fig. 12.30, and could contribute to the separation from euchromatin. It has also been suggested that nonspecific entropic forces [207] or active processes [328] could segregate chromosomes, with swollen ones enriched with euchromatin localized at the center of the nucleus and more compact ones containing heterochromatin localized at the periphery. It turns out that the A (active) and B (inactive) compartments identified using Hi-C, see Fig. 12.33, have the characteristics of euchromatin and heterochromatin, respectively. The spatial segregation of euchromatin and heterochromatin into separate A and B compartments in the nucleus is consistent with the corresponding positioning of the two chromatin types between the interior and periphery of the nucleus.

Interestingly, as noted in [795], the partitioning of chromatin into A and B compartments is analogous to a well-known phenomenon in classical polymer physics, namely, the microphase separation of the so-called block copolymers. A copolymer is a polymer that consists of a least two different types of monomer, $M_1$ and $M_2$, say. If the $M_1$ and $M_2$ monomers do not mix in solution, then they tend to form aggregates with other monomers of the same type (monomer blocks). However, in contrast to the demixing of small molecules such as oil and water, monomer demixing cannot occur at macroscopic scales, since the $M_1$ and $M_2$ blocks are connected to form a copolymer. That is, only smaller domains enriched with one type of monomer can form, which leads to the notion of microphase separation. (The phase separation of chromatin is more complicated, however, since it has not reached thermodynamic equilibrium.) Finally, note that there are examples where euchromatin is enriched at the periphery and heterochromatin at the center. One notable example is the nucleus of rod retinal cells in adult nocturnal animals [834]. Since heterochromatin has a higher refractive index than euchromatin, its central localization helps light to reach photoreceptors by acting as a collecting lens.

**DNA looping.** A semiflexible polymer in solution can spontaneously form loops (see Sect. 12.5.3). In the case of DNA, this allows DNA regions that are far apart along the linear chromosome to be brought together in three-dimensional space, for example, promoters and their regulatory elements. Consequently, DNA looping plays an important role in regulating gene expression. It also allows for additional compactification of DNA, which is important during cell mitosis. However, in order to regulate and maintain DNA loops, active mechanisms for loop formation are required.

Although the precise mechanisms underlying the formation and stabilization of DNA loops are not fully understood, a popular model for human cells is the loop extrusion mechanism [5, 323, 354, 622], see Fig. 12.34. At the simplest level, this proposes that a protein complex known as cohesin binds two sites on the chromosome, and then extrudes a growing loop of DNA between them by translocating in opposite directions along the DNA while bridging the increasingly distant chromosomal sites. Loop formation stops when cohesin encounters extrusion barriers consisting of DNA-bound proteins such as CTCF. One of the characteristic features of CTCF proteins is that the resulting barriers appear to be directional, in the sense that they only impede the traversal of complexes that extrude loops towards them from one side rather than the other. Thus, loop extrusion stops when cohesin is flanked by convergently arranged CTCFs. Once the loop-extruding complex dissociates, the loop diffusively disappears. Experimental and computational studies suggest that the interplay between loop extrusion and DNA-bound protein barriers underlies the formation of TAD patterns seen in Hi-C maps [324]. Mechanistically speaking, it is hypothesized that loop extrusion locally compactifies chromatin such that the extruding loops generate additional contacts at short genomic distances, resulting in an elevated number of contacts within a TAD. On the other hand, CTCF barriers block these extrusion-mediated contacts from propagating to neighboring TADs. (Note that an individual TAD will typically contain several loops flanked by CTCFs.) Approximately 50% of human genes are thought to be involved in long range chromatin interactions via DNA looping.

Fig. 12.34: Basic steps of the DNA loop extrusion model. (a) Cohesin ring binds to DNA, (b) Cohesin rings slide over the CTCF molecules whose binding sequence points away from the loop, which continues to grow, and (c) Loop formation stops when each of the rings has reached an inward directed CTCF sequence. [Public domain figure downloaded from wikipedia commons.]

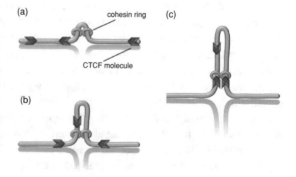

**The nucleolus.** The nucleolus, which is the most prominent structure in the cell nucleus during interphase, is the site of ribosomal RNA (rRNA) transcription, pre-rRNA processing, and ribosome subunit assembly [408, 539]. (Eukaryotic cells typically contain a single nucleolus. However, it is also possible to have several nucleoli, depending on the species of organism and the stage of the cell cycle.) Recent more detailed studies of nucleolar proteins and the analysis of genomic DNA loci associated with the nucleolus suggest that in addition to ribosome biogenesis, the nucleolus has a role as a central hub for regulating several nuclear and cellular processes [453]. For example, movement of chromatin relative to the nucleolus may trigger the activation or silencing of gene transcription, and mediate inactivation of chromatin domains. Similarly, the exchange of proteins between the nucleolus and the nucleoplasm regulates DNA repair, the stress response, and apoptosis. It is hypothesized that the membrane-less nucleolus is an example of a biological condensate that forms and maintains its structure via an active form of liquid-liquid phase separation [42, 70, 100, 293, 451]; see Sect. 13.1.

**Anomalous diffusion of chromosomal loci.** In addition to studying the spatial organization of chromosomes within the nucleus, one can also use imaging to characterize the dynamics of chromosomal loci such as telomeres. A telomere is a region of repetitive nucleotide sequences at each end of a chromosome, which protects the end of the chromosome from degradation [532]. Its state also reflects the cell's age and reproduction ability [86]. Detailed single particle tracking experiments of telomere-bound proteins have shown that their motion in mammalian cells exhibits subdiffusive behavior on short and intermediate time scales [134]. That is, the mean-square displacement (MSD) of a telomere displays power-law behavior, $\langle x^2(t) \rangle \sim t^\alpha$ with $\alpha \approx 1/3$ and $\alpha \approx 1/2$ on short and intermediate time scales, respectively. On longer time scales, one observes normal diffusion ($\alpha = 1$), after taking into account the fact that chromosomes are restricted to specific territories within the nucleus. There is also growing experimental evidence that chromosomal loci are subject to nonequilibrium active forces via interactions with cytoskeletal elements and active enzymes [135, 839, 931, 984]. For example, telomeres exhibit ATP-dependent enhanced subdiffusion and coherent motion on time scales of seconds, consistent with the action of motor-like enzymes [931, 984]. The Rouse model of classical polymer physics [250] is often used as a starting point in the development of more complicated models of chromosomal dynamics [13], as we now describe.

### 12.5.2 The Rouse model of polymer dynamics

The Rouse model of polymer dynamics, which is based on one of the simplest statistical mechanical models of an ideal polymer, namely, the Gaussian chain, provides a reasonable description of the dynamics of untangled, concentrated polymer solutions. That is, the polymer chains are assumed to be sufficiently short so that entanglements with surrounding chains, which would highly constrain the motion, can be neglected. Moreover, by considering a concentrated polymer solution, one can neglect hydrodynamic interactions mediated by the induced motion of the solvent. A characteristic feature of Rouse chains is that monomers within the chain exhibit normal diffusion on long time scales and subdiffusive motion with $\alpha = 1/2$ on short time scales. In a Gaussian chain, every bond vector $\mathbf{r}$ of the polymer is taken to be Gaussian distributed

$$p(\mathbf{r}) = \left( \frac{3}{2\pi a^2} \right)^{3/2} \exp\left( -\frac{3r^2}{2a^2} \right). \tag{12.5.1}$$

The Gaussian chain is often represented by a mechanical model consisting of $N+1$ beads connected by harmonic springs, see Fig.12.35. If $k$ is the spring constant of each link, then the total potential energy is given by

$$V(\mathbf{r}_1, \dots, \mathbf{r}_N) = \frac{k}{2} \sum_{n=1}^{N} r_n^2,$$

where $\mathbf{r}_n = \mathbf{R}_n - \mathbf{R}_{n-1}$, and $\mathbf{R}_n$ denotes the stochastic position of the $n$-th bead. It follows that if the spring constant is chosen to be $k = 3k_B T / a^2$, then the Gaussian density $p(\mathbf{r})$ can be interpreted as the Boltzmann distribution of the bond vectors. The Rouse model assumes that each bead of the chain undergoes overdamped Brownian motion, while subject to spring forces arising from nearest neighbor bonds. The resulting system of SDEs takes the form

$$\frac{d\mathbf{R}_0}{dt} = -\frac{3k_B T}{\eta a^2} (\mathbf{R}_0 - \mathbf{R}_1) + \mathbf{f}_0, \tag{12.5.2a}$$

$$\frac{d\mathbf{R}_n}{dt} = -\frac{3k_B T}{\eta a^2} (2\mathbf{R}_n - \mathbf{R}_{n-1} - \mathbf{R}_{n+1}) + \mathbf{f}_n, \quad 0 < n < N, \tag{12.5.2b}$$

$$\frac{d\mathbf{R}_N}{dt} = -\frac{3k_B T}{\eta a^2} (\mathbf{R}_N - \mathbf{R}_{N-1}) + \mathbf{f}_N, \tag{12.5.2c}$$

where $\mathbf{f}_n$ are vectors whose components $f_{n,j}$, $j = 1, 2, 3$, are independent white noise processes such that

$$\langle f_{n,j}(t) \rangle = 0, \quad \langle f_{n,j}(t) f_{m,k}(t') \rangle = 2D\delta_{m,n}\delta_{j,k}\delta(t - t').$$

We also have the Einstein relation $\eta D = k_B T$. Note that hydrodynamical interactions between the beads are neglected here; see Sect. 12.3.4.

**Fourier decomposition and Rouse modes.** Equation (12.5.2) can be analyzed by decomposing $\mathbf{R}_n$ into a set of Fourier modes $\mathbf{X}_p$, $0 \le p \le N$ according to (see Ex. 12.11)

$$\mathbf{R}_n = \mathbf{X}_0 + 2 \sum_{p=1}^{N} \mathbf{X}_p \cos\left[ \frac{p\pi}{N+1} (n + 1/2) \right]. \tag{12.5.3}$$

Using the orthogonality relation

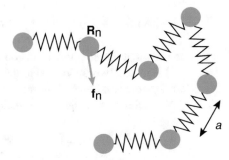

Fig. 12.35: Bean and spring representation of Rouse model.

$$\frac{1}{N+1} \sum_{n=0}^{N} \cos\left[\frac{p\pi}{N+1}(n+1/2)\right] \cos\left[\frac{q\pi}{N+1}(n+1/2)\right] = \frac{1}{2}(1+\delta_{p,0})\delta_{p,q},$$

$$(12.5.4)$$

one also has the inverse series expansion

$$\mathbf{X}_p = \frac{1}{N+1} \sum_{n=0}^{N} \mathbf{R}_n \cos\left[\frac{p\pi}{N+1}(n+1/2)\right]. \qquad (12.5.5)$$

Substitution of the cosine series expansion (12.5.3) into (12.5.2) and setting $\kappa = 3k_BT/\eta a^2 = 3D/a^2$, leads to the $N+1$ uncoupled OU processes

$$\frac{d\mathbf{X}_p}{dt} = -4\kappa \sin^2\left[\frac{p\pi}{2(N+1)}\right]\mathbf{X}_p + \mathbf{F}_p, \qquad (12.5.6)$$

with

$$\mathbf{F}_p = \frac{1}{N+1} \sum_{n=0}^{N} \mathbf{f}_n \cos\left[\frac{p\pi}{N+1}(n+1/2)\right].$$

It follows that

$$\langle F_p(t) \rangle = 0, \quad \langle F_{0,j}(t)F_{0,k}(t') \rangle = \frac{2D}{N+1}\delta_{j,k}\delta(t-t'), \qquad (12.5.7a)$$

$$\langle F_{p,j}(t)F_{q,k}(t') \rangle = \frac{D}{N+1}\delta_{p,q}\delta_{j,k}\delta(t-t'), \quad p+q>0. \qquad (12.5.7b)$$

The stochastic variables $\mathbf{X}_p(t)$ are known as the normal or Rouse modes of the Gaussian chain. The zeroth-order mode $\mathbf{X}_0$ specifies the center of mass of the beads, since $\mathbf{X}_0 = \sum_n \mathbf{R}_n/(N+1)$. Setting $p=0$ in equation (12.5.7a) and integrating with respect to time $t$ shows that

$$\mathbf{X}_0(t) = \mathbf{X}_0(0) + \int_0^t \mathbf{F}_0(\tau)d\tau,$$

which means that the mean-square displacement (MSD) of the center of mass is given by

$$\Delta_{\text{c.o.m.}} := \langle [\mathbf{X}_0(t) - \mathbf{X}_0(0)]^2 \rangle = \left\langle \int_0^t d\tau \int_0^t d\tau' \mathbf{F}_0(\tau) \cdot \mathbf{F}_0(\tau') \right\rangle = \frac{6D}{N+1} t. \quad (12.5.8)$$

It can be seen that the effective diffusion coefficient of the center of mass, $D_{\text{eff}} = D/(N+1)$ is inversely proportional to the size of the chain. The higher-order mode $\mathbf{X}_p$ determines the amplitude of vibrations whose wavelength corresponds to a subchain of $N/p$ segments. Integrating equation (12.5.7a) for $p > 0$ shows that

$$\mathbf{X}_p(t) = \mathbf{X}_p(0)e^{-t/\tau_p} + \int_0^t e^{-(t-\tau)/\tau_p} \mathbf{F}_p(\tau)d\tau, \quad (12.5.9)$$

where $\tau_p$ is the characteristic relaxation time of the $p$-th mode

$$\tau_p = \kappa^{-1} \left[ 4\sin^2\left( \frac{p\pi}{2(N+1)} \right) \right]^{-1} \approx \frac{(N+1)^2}{\kappa\pi^2 p^2}. \quad (12.5.10)$$

The final approximation holds for long wavelength modes for which $p \ll N$. It can be seen that longer wavelength modes are slower. Taking the inner product of equation (12.5.9) with $\mathbf{X}_p(0)$ and averaging then yields the two-point correlation function

$$\langle \mathbf{X}_p(t) \cdot \mathbf{X}_p(0) \rangle = \langle X_p^2 \rangle e^{-t/\tau_p}. \quad (12.5.11)$$

Here $\langle X_p^2 \rangle$ is the equilibrium expectation of the square amplitude, which is determined by the underlying Gaussian distribution. In particular, the Boltzmann-Gibbs distribution of a configuration $(\mathbf{R}_0, \ldots, \mathbf{R}_N)$ is (see Sect. 1.3)

$$p(\mathbf{R}_0, \ldots, \mathbf{R}_N) = \frac{1}{Z} \exp\left[ -\frac{3}{2a^2} \sum_{n=1}^N (\mathbf{R}_n - \mathbf{R}_{n-1})^2 \right],$$

where $Z$ is a normalization constant (partition function). Performing the linear transformation to Rouse modes then yields a product of independent Gaussians (see Ex. 12.11),

$$p(\mathbf{X}_0, \ldots, \mathbf{X}_N) = \frac{1}{Z} \exp\left[ -\frac{12}{a^2}(N+1) \sum_{p=1}^N X_p^2 \sin^2\left( \frac{p\pi}{2(N+1)} \right) \right]. \quad (12.5.12)$$

Hence

$$\langle X_p^2 \rangle = \frac{a^2}{8(N+1)\sin^2\left( \frac{p\pi}{2(N+1)} \right)} \approx \frac{(N+1)a^2}{2\pi^2 p^2} \quad (12.5.13)$$

for small $p$.

Another quantity of interest is the time correlation function of the end-to-end vector

$$\mathbf{R}_{ee} = \mathbf{R}_N - \mathbf{R}_0 = 2 \sum_{p=1}^N \mathbf{X}_p((-1)^p - 1)\cos(p\pi/2(N+1)). \quad (12.5.14)$$

Since the Rouse mode amplitudes decay as $p^{-2}$, the sum on the right-hand side will be dominated by values of $p$ that are much smaller than $N$. Therefore, to leading order, we can write

$$\mathbf{R}_{ee} = -4\sum_{p=1}^{N}{}'\mathbf{X}_p,$$

where the prime on the summation indicates that only odd $p$ terms are included. Then

$$\langle \mathbf{R}_{ee}(t)\cdot\mathbf{R}_{ee}(0)\rangle = 16\sum_{p=1}^{N}{}'\langle\mathbf{X}_p(t)\cdot\mathbf{X}_p(0)\rangle \approx \frac{8a^2}{\pi^2}(N+1)\sum_{p=1}^{N}{}'\frac{1}{p^2}e^{-t/\tau_p}. \quad (12.5.15)$$

**Mean-square displacement.** Having calculated the MSD of the center of mass, we can now determine the MSD of a typical segment or bond. First, using the fact that the different Rouse modes are statistically independent, equation (12.5.3) implies

$$\langle(\mathbf{R}_n(t)-\mathbf{R}_n(0))^2\rangle = \langle(\mathbf{X}_0(t)-\mathbf{X}_0(0))^2\rangle$$
$$+4\sum_{p=1}^{N}\langle(\mathbf{X}_p(t)-\mathbf{X}_p(0))^2\rangle\cos^2\left(\frac{p\pi}{(N+1)}(n+1/2)\right).$$

Averaging with respect to $n$, and using the result

$$\sum_{n=0}^{N}\cos^2\left(\frac{p\pi}{(N+1)}(n+1/2)\right) = \frac{1}{2}\sum_{n=0}^{N}\left[\cos\left(\frac{2p\pi}{(N+1)}(n+1/2)\right)+1\right]$$
$$= \frac{N+1}{2}(1+\delta_{p,0}),$$

we find that

$$\overline{\Delta}(t) := \frac{1}{N+1}\sum_{n=0}^{N}\langle(\mathbf{R}_n(t)-\mathbf{R}_n(0))^2\rangle$$
$$= \Delta_{\text{c.o.m.}}(t) + 2\sum_{p=1}^{N}\left(\langle X_p^2(t)\rangle + \langle X_p(0)^2\rangle - 2\langle\mathbf{X}_p(t)\cdot\mathbf{X}_p(0)\rangle\right),$$

that is

$$\overline{\Delta}(t) = \frac{6D}{N+1}t + 4\sum_{p=1}^{N}\langle X_p^2\rangle\left(1-e^{-t/\tau_p}\right). \quad (12.5.16)$$

There are then two distinguished limits. At large times $t$, $t \gg \tau_1$, the first term dominates and

$$\overline{\Delta}(t) \approx \frac{6D}{N+1}t. \quad (12.5.17)$$

On the other hand, at shorts times, $t \ll \tau_1$, the sum over $p$ dominates. Moreover, if $N \gg 1$ then the relaxation times and amplitudes can be approximated by the right-hand side of equations (12.5.10) and (12.5.13). Replacing the discrete sum by an integral, we have

$$
\begin{aligned}
\overline{\Delta}(t) &\approx \frac{2a^2}{\pi^2}(N+1)\int_0^\infty \frac{1}{p^2}\left(1-e^{-tp^2/\tau_1}\right)dp \\
&= \frac{2a^2}{\pi^2}(N+1)\int_0^\infty \left[\tau_1^{-1}\int_0^t e^{-t'p^2/\tau_1}dt'\right]dp \\
&= \frac{2a^2}{\pi^2}\frac{N+1}{2\tau_1}\sqrt{\pi\tau_1}\int_0^t \frac{dt'}{\sqrt{t'}} = \sqrt{\frac{12k_BTa^2}{\pi\eta}}t^{1/2}.
\end{aligned}
\tag{12.5.18}
$$

Thus, the short-time MSD is subdiffusive and is independent on the number of segments $N$.

**Active Rouse chains.** A number of studies have developed extensions of the classical Rouse model in order to take into account the effect of active forces on the dynamics of polymer chains. One approach is to add an extra random force on the right-hand side of equations (12.5.2a–c) that is nonthermal in origin, and thus does not obey the fluctuation-dissipation theorem [344, 695, 739]. More specifically,

$$
\mathbf{f}_n(t) \to \mathbf{f}_n(t) + \mathbf{A}_n(t),
$$

where $\mathbf{A}_n(t)$ represents some active force exerted on the $n$-th monomer. The active noise term is assumed to have zero mean, and to be a stationary process. In particular

$$
\langle \mathbf{A}_n(t)\rangle = 0, \quad \langle \mathbf{A}_n(t)\cdot\mathbf{A}_m(t')\rangle = B(n-m)C(t-t').
\tag{12.5.19}
$$

A typical choice for the time correlation function is $C(t) = C_0 e^{-t/\tau_A}$, where $\tau_A$ is the average time scale for a burst of motor activity, say. First suppose, that the active forces are monopolar, $B(n-m) = \delta_{n,m}$. When $\tau_A$ is small, the behavior is similar to the standard Rouse chain, whereas transient superdiffusive behavior ($\alpha > 1$) can be observed when $\tau_A$ is increased so that active noise is correlated over longer times [344, 695]. Very different behavior has been found in a model of a Rouse chain interacting with a population of motor-like enzymes [739]. These enzymes can temporarily bind to specific sites of the polymer and generate a dipolar force on two neighboring monomers. The active force now has correlations of the form

$$
\langle \mathbf{A}_n(t)\rangle = 0, \quad \langle \mathbf{A}_n(t)\cdot\mathbf{A}_m(t')\rangle = C_0 e^{-|t-t'|/\tau_A}[\delta_{n,m} - \delta_{n,m\pm1}],
\tag{12.5.20}
$$

and the resulting motion of monomers shares several features observed experimentally in the case of telomeres [135, 839, 931, 984]: a subdiffusive MSD and regions of correlated motion. If all sites of the polymer can be bound by the enzyme, then the active Rouse chain still exhibits subdiffusive behavior at intermediate times with $\alpha = 1/2$. However, if only a subset of sites can interact with the enzyme, and active forces dominate over thermal forces, then $\alpha < 1/2$.

Finally, note that an alternative form of extended Rouse model involves modifying the deterministic part of the dynamics, such that in normal coordinates [12]

$$\frac{d\mathbf{X}_p}{dt} = -4\kappa \sin^\beta \left[ \frac{p\pi}{2(N+1)} \right] \mathbf{X}_p + \mathbf{F}_p, \quad \beta > 1. \tag{12.5.21}$$

Transforming to real space yields an SDE with a potential containing long-range interactions. The modified Rouse chain exhibits subdiffusion at intermediate times with $\alpha = 1 - 1/\beta$.

### 12.5.3 Diffusion-controlled looping of a polymer

Consider a solution of polymer chains, in which each chain has a distribution of reactive sites along its backbone. If the intrinsic intrachain reactions are sufficiently fast, then the rate-limiting step will be the rate at which conformational changes take place that bring potentially reactive sites into close proximity. As a simple example, suppose that there are two reactive sites located at each end of the chain, respectively, so that the mutual binding of these sites signals the spontaneous formation of a polymer loop. In the diffusion-limited regime, the rate of reaction will depend on the relaxation time for the end-to-end distance $\mathbf{R}_{ee}$ to become sufficiently small. Although spontaneous loop formation is distinct from the active loop extrusion mechanism mediated by cohesin, it also plays an important role in gene regulation and DNA replication [737, 799].

Understanding the kinetics of spontaneous loop formation is a classical problem in polymer physics [250, 941, 942], which involves a nontrivial boundary value problem of a linear PDE in a high-dimensional space. A typical starting point is the FP equation corresponding to the SDE of the Rouse model (12.5.2), supplemented by an absorbing surface boundary condition for $\mathbf{R}_{ee}$. One way to estimate the reaction rate is to calculate the MFPT to reach the absorbing boundary from some initial configuration, which can be obtained using an asymptotic analysis of the principal eigenvalue of the FP differential operator in the case of a small capture radius [11, 13]. To leading order, it is found that the looping time scales as $N^{3/2}/b$, where $N$ is the number of monomers and $b$ is the capture radius. This agrees with an earlier classical study of diffusion-controlled reactions by Szabo, Schulten, and Schulten [852]. The latter authors approximated the non-Markovian dynamics of $\mathbf{R}_{ee}$, obtained from the high-dimensional Markov model, by an effective scalar SDE for $\mathbf{R}_{ee}$ moving in a radially symmetric potential. The potential was derived from the equilibrium distribution of the end-to-end distance. The first passage time problem for $|\mathbf{R}_{ee}|$ to reach $b$ could then be obtained using standard 1D methods outlined in Chap. 2.

In this section, we describe an alternative approach to calculating looping times that was originally introduced by Wilemski and Fixman using their theory of diffusion-controlled reactions [941, 942]. In contrast to the above studies, the looping time was estimated to scale as $N^2$. The basic idea of Wilemski-Fixman (WF) theory is to reduce the high-dimensional SDE to a low-dimensional non-Markovian process for the end-to-end vectors. Instead of solving the FP equation with a complicated absorbing boundary condition, a position-dependent reaction term (sink) is added to the FP equation without the absorbing boundary. This means that the solution can be expressed in terms of the known Green's function of the FP operator. One finds that the looping time is related to the time integral of a sink-sink time correlation function. Since WF theory can be applied to a wide variety of high-dimensional diffusion processes with complicated boundary conditions, we present the general theory in Box 12E, following the particular formulation of Weiss [938]. Here we focus on the particular application to polymer looping, which has been further developed by various authors [57, 562, 715, 833, 969]. These additional studies suggest that the looping time satisfies a mixed scaling law of the form $c_1 N\sqrt{N}/b + c_2 N^2$.

**FP equation for looping in the Rouse model.** Consider the Rouse model expressed in Fourier modes $\mathbf{X}_p$, equation (12.5.6), and consider only odd modes, since these contribute to the end-to-end vector $\mathbf{R}_{ee}$. We non-dimensionalize the model by expressing lengths in units of the bond length $a$ and setting $D/a^2 = 1$. The corresponding FP equation for the probability density $P(\mathbf{X},t)$ is then

$$\frac{\partial P(\mathbf{X},t)}{\partial t} = \nabla \cdot (\Gamma \mathbf{X} P(\mathbf{X},t)) + \frac{1}{2(N+1)} \nabla^2 P(\mathbf{X},t) \equiv \mathbb{L} P(\mathbf{X},t), \qquad (12.5.22)$$

where $\mathbf{X} = (\mathbf{X}'_p, p = 1, \dots, N)$ restricted to odd modes, and $\Gamma$ is a diagonal matrix with diagonal elements

$$\gamma_p = \tau_p^{-1} = 12 \sin^2[\theta_p], \quad \theta_p = \frac{p\pi}{2(N+1)}. \qquad (12.5.23)$$

In the absence of any boundary conditions, we set $P(\mathbf{X},t) = G(\mathbf{X},t|\mathbf{X}_0,0)$, where $G$ is the "free space" Green's function with stationary density given by a Gaussian Boltzmann distribution

$$g(\mathbf{X}) := \lim_{t \to \infty} G(\mathbf{X},t|\mathbf{X}_0,0) = \prod_{p \, \text{odd}} \left( \frac{(N+1)\gamma_p}{\pi} \right)^{3/2} e^{-(N+1)\gamma_p \mathbf{X}_p^2}. \qquad (12.5.24)$$

In order to investigate the looping time, it is necessary to supplement the FP equation by a nontrivial boundary condition involving the end-to-end vector $\mathbf{R}_{ee}$. Since the dynamics of the latter is a projection of a high-dimensional OU process, it is a non-Markovian Gaussian process that is determined by the correlation function $\phi(t)$ of $\mathbf{R}_{ee}$ [715, 969]. The correlation function is given by (see Sect. 12.5.2)

$$\phi(t) = \frac{\langle \mathbf{R}_{ee}(t) \cdot \mathbf{R}_{ee}(0) \rangle}{\langle \mathbf{R}_{ee}^2 \rangle} = \frac{2}{N(N+1)} \sum_{p=1}^{N} \frac{1}{\tan^2 \theta_p} e^{-\gamma_p t}. \qquad (12.5.25)$$

The conditional probability density for $\mathbf{R}_{ee}(t) = \mathbf{r}$, given that $\mathbf{R}_{ee} = \mathbf{r}_0$ is then

$$T(\mathbf{r},t|\mathbf{r}_0,0) = \left(\frac{3}{2\pi N(1-\phi^2(t))}\right)^{3/2} \exp\left(-\frac{3}{2N}\frac{(\mathbf{r}-\phi(t)\mathbf{r}_0)^2}{1-\phi^2(t)}\right). \quad (12.5.26)$$

The looping reaction is incorporated by imposing the absorbing boundary condition $P(\mathbf{X},t) = 0$ on the reaction surface

$$S_\varepsilon = \{\mathbf{X} : |\mathbf{R}_{ee}| = \varepsilon\},$$

where $\varepsilon = b/a$ and $b$ is the capture radius. (Note that it is typically assumed that the initial distribution of the polymer chain is the stationary density $g(\mathbf{X})$. However, such a distribution does not satisfy the absorbing boundary condition, that is, it should be replaced by a normalized Gaussian density outside the reactive surface but set to zero inside the surface. In practice, the difference between these initial conditions is negligible for sufficiently small $\varepsilon$.)

**Modified FP equation and WF theory.** WF theory, as outlined in Box 12E, provides an alternative approach to studying the looping problem. Instead of solving the FP equation (12.5.22) with a complicated absorbing boundary condition, the FP equation is modified by adding a reactive sink term on the right-hand side [715, 941, 942, 969]:

$$\frac{\partial Q(\mathbf{X},t)}{\partial t} = \mathbb{L}Q(\mathbf{X},t) - \kappa S(\mathbf{X})Q(\mathbf{X},t), \quad Q(\mathbf{X},0) = g(\mathbf{X}). \quad (12.5.27)$$

Following the original formulation of Wilemski and Fixman, we take the sink function $S(\mathbf{X})$ to depend on the end-to-end distance $r = |\mathbf{R}_{ee}|$ according to

$$S = S(r) = \Theta(\varepsilon - r),$$

where $\Theta$ is the Heaviside function. The limit $\kappa \to \infty$ corresponds to complete absorption by the sink, and thus connects to the more classical FPT problem with an absorbing boundary condition. For large $\kappa$, the WK approximation leads to an expression for the MFPT $\mu$ given by

$$\mu = \int_0^\infty \left(\frac{C(t)}{C(\infty)} - 1\right) dt, \quad (12.5.28)$$

where $C(t)$ is the sink-to-sink time correlation function

$$C(t) = \int_{\Omega_0}\int_{\Omega_0} S(r)T(\mathbf{r},t|\mathbf{r}',0)S(r')g(\mathbf{r}')d\mathbf{r}'d\mathbf{r}. \quad (12.5.29)$$

Unfortunately, it is not possible to obtain an exact analytical expression for $C(t)$, and hence $\mu$. However, one can use a mixture of asymptotic expansions and numerical solutions to determine how $\mu$ scales with $N$ and other parameters. In particular, one obtains the mixed scaling law [969]

$$\mu = (h_1/\varepsilon + h_2)N\sqrt{N} + h_3 N^2 \quad (12.5.30)$$

for constants $h_1, h_2, h_3$. The asymptotic analysis is found to be consistent with numerical simulations of the full Rouse model.

---

**Box 12E. Wilemski-Fixman (WF) theory of diffusion-controlled reactions.**

---

Consider the non-dimensionalized diffusion or FP equation

$$\frac{\partial p}{\partial t} = \mathbb{L}p - \kappa S(\mathbf{x})p, \quad \mathbf{x} \in \Omega \subset \mathbb{R}^n, \qquad (12.5.31)$$

where $p(\mathbf{x}, t | \mathbf{x}_0, 0)$ is the probability density for a particle to be in the position or configuration $\mathbf{x}$ at time $t$, given that it was initially at $\mathbf{x}_0$. The function $S(\mathbf{x})$ is a reactive term, $\mathbb{L}$ is a second order diffusion operator, and $\kappa$ specifies the strength of the reaction. Introduce the time-dependent Green's function $G(\mathbf{x}, t | \mathbf{x}_0, 0)$ according to

$$\frac{\partial G}{\partial t} = \mathbb{L}G, \quad G(\mathbf{x}, 0 | \mathbf{x}_0, 0) = \delta(\mathbf{x} - \mathbf{x}_0), \quad \int_\Omega G(\mathbf{x}, t | \mathbf{x}_0, 0)d\mathbf{x} = 1. \quad (12.5.32)$$

A crucial assumption of WF theory is that, $G$ has a nontrivial stationary density

$$\lim_{t \to \infty} G(\mathbf{x}, t | \mathbf{x}_0, 0) = g(\mathbf{x}). \qquad (12.5.33)$$

(We do not specify the boundary conditions on $\partial\Omega$. In practice, we assume that the domain is sufficiently large relative to the reaction domain so that we can ignore boundary effects.) The solution of equation (12.5.31) can be expressed in terms of the Green's function according to

$$p(\mathbf{x}, t | \mathbf{x}_0, 0) = G(\mathbf{x}, t | \mathbf{x}_0, 0) - \kappa \int_0^t \int_\Omega G(\mathbf{x}, t - \tau | \mathbf{y}, 0) S(\mathbf{y}) p(\mathbf{y}, \tau | \mathbf{x}_0, 0) d\mathbf{y} \, d\tau.$$

$$(12.5.34)$$

This follows from the identity

$$p(\mathbf{x}, t | \mathbf{x}_0, 0) - G(\mathbf{x}, t | \mathbf{x}_0, 0) = \int_0^t \int_\Omega \frac{d}{d\tau} [G(\mathbf{x}, t - \tau | \mathbf{y}, 0) p(\mathbf{y}, \tau | \mathbf{x}_0, 0)] d\mathbf{y} \, d\tau,$$

substituting the diffusion equations for $p$ and $G$ on the right-hand side, and using self-adjointness of $\mathbb{L}$.

**Survival probability.** Now suppose that the molecules are initially distributed according to the stationary density $g(\mathbf{x})$, and set

$$F(\mathbf{x}, t) = \int_\Omega p(\mathbf{x}, t | \mathbf{x}_0, 0) g(\mathbf{x}_0) d\mathbf{x}_0.$$

It follows from equation (12.5.34) that

$$F(\mathbf{x},t) = g(\mathbf{x}) - \kappa \int_0^t \int_\Omega G(\mathbf{x},t-\tau|\mathbf{y},0)S(\mathbf{y})F(\mathbf{y},\tau)d\mathbf{y}\,d\tau. \qquad (12.5.35)$$

We have also used the fact that

$$\int_\Omega G(\mathbf{x},t|\mathbf{x}_0,0)g(\mathbf{x}_0)d\mathbf{x}_0 = g(\mathbf{x}).$$

Introducing the survival probability $F(t)$ that no reaction occurs up to time $t$

$$F(t) = \int_\Omega F(\mathbf{x},t)d\mathbf{x}, \qquad (12.5.36)$$

we have

$$F(t) = 1 - \kappa \int_0^t \int_\Omega S(\mathbf{y})F(\mathbf{y},\tau)d\mathbf{y}\,d\tau. \qquad (12.5.37)$$

Note that if $F(\mathbf{x},t)$ is known exactly, then either of the above two equations can be used to calculate $F(t)$, and both must yield the same result. However, this is not necessarily the case for approximate solutions. The main approximation in WF theory is based on the assumption that even though a reaction can occur, the diffusion process predominates such that the spatial dependence of $F(\mathbf{x},t)$ is simply the stationary density $g(\mathbf{x})$, at least in the region around the sink. That is

$$F(\mathbf{x},t) \sim g(\mathbf{x})f(t), \qquad (12.5.38)$$

for some function $f(t)$. Integrating with respect to $\mathbf{x}$ requires

$$\int_\Omega S(\mathbf{x})F(\mathbf{x},t)d\mathbf{x} = \overline{S}f(t), \quad \overline{S} = \int_\Omega S(\mathbf{x})g(\mathbf{x})d\mathbf{x}. \qquad (12.5.39)$$

An integral equation for $f(t)$ is obtained by multiplying equation (12.5.35) by $S(\mathbf{x})$ and integrating with respect to $\mathbf{x}$. Using equation (12.5.39), we find that

$$\overline{S}f(t) = \overline{S} - \kappa \int_0^t C(t-\tau)f(\tau)d\tau, \qquad (12.5.40)$$

where $C(t)$ is the sink-sink time correlation function

$$C(t) = \int_\Omega \int_\Omega S(\mathbf{x})G(\mathbf{x},t|\mathbf{x}',0)S(\mathbf{x}')g(\mathbf{x}')d\mathbf{x}'d\mathbf{x}. \qquad (12.5.41)$$

Note that $C(\infty) = \overline{S}^2$. Since equation (12.5.40) is a convolution equation, it can be solved using Laplace transforms:

$$\widetilde{f}(s) = \frac{\overline{S}}{s[\overline{S} + \kappa\widetilde{C}(s)]}.$$

Combining equations (12.5.37) and (12.5.39) yields

$$F(t) = 1 - \kappa \overline{S} \int_0^t f(\tau) d\tau,$$

which can be Laplace transformed to obtain the result

$$\widetilde{F}(s) = \frac{1}{s} \left[ 1 - \frac{\kappa \overline{S}^2}{s[\overline{S} + \kappa \widetilde{C}(s)]} \right]. \tag{12.5.42}$$

**MFPT.** Recall that the MFPT $\mu$ for a stochastic process can be related to the survival probability according to

$$\mu = \int_0^\infty F(t) dt = \lim_{s \to 0} \widetilde{F}(s). \tag{12.5.43}$$

We thus have to determine the small $s$ behavior of $\widetilde{F}(s)$. Given the existence of a stationary distribution $g(\mathbf{x})$ of the Green's function $G(\mathbf{x}, t | \mathbf{x}_0, 0)$, it follows from equation (12.5.41) that $C(t) \sim \overline{S}^2$ for large $t$, that is, $\widetilde{C}(s) \sim \overline{S}^2 / s$ as $s \to 0$. This suggests we can write

$$\widetilde{C}(s) = \frac{\overline{S}^2}{s} + \widetilde{H}(s),$$

with $\widetilde{H}(0)$ finite. Moreover, for small $s$

$$1 - \frac{\kappa \overline{S}^2}{s[\overline{S} + \kappa \widetilde{C}(s)]} = 1 - \frac{\kappa \overline{S}^2}{\kappa \overline{S}^2 + s[\overline{S} + \kappa \widetilde{H}(s)]} = 1 - \frac{1}{1 + s[1/\kappa \overline{S} + \widetilde{H}(s)/\overline{S}^2]}$$

$$\approx 1 - \left( 1 - s[1/\kappa \overline{S} - \widetilde{H}(s)/\overline{S}^2] \right) = s \left[ \frac{1}{\kappa \overline{S}} + \frac{\widetilde{H}(s)}{\overline{S}^2} \right].$$

Hence, we can now take the limit $s \to 0$ in equation (12.5.42) to determine $\widetilde{F}(0)$ and thus $\mu$

$$\mu = \frac{1}{\kappa \overline{S}} + \frac{\widetilde{H}(0)}{\overline{S}^2}. \tag{12.5.44}$$

This is the major result of WF theory. The difficult part is specifying an appropriate sink function $S(\mathbf{x})$ and deriving an explicit expression for the corresponding quantity $\widetilde{H}(0)$. In the case of large $\kappa$ (strong reactions), we can write down an alternative expression for the MFPT. To leading order, equation (12.5.42) becomes

$$\widetilde{F}(s) = \frac{1}{s} \left[ 1 - \frac{\overline{S}^2}{s\widetilde{C}(s)]} \right] = \frac{1}{s} \frac{s\widetilde{C}(s) - \overline{S}^2}{s\widetilde{C}(s)}.$$

Introducing the function $I(t) = C(t)/C(\infty) - 1$, we have

$$\mu = \lim_{s \to 0} \widetilde{F}(s) = \lim_{s \to 0} \frac{\widetilde{I}(s)}{s\widetilde{I}(s) + 1} = \widetilde{I}(0).$$

That is

$$\mu = \int_0^\infty \left( \frac{C(t)}{C(\infty)} - 1 \right) dt. \qquad (12.5.45)$$

## 12.6 Exercises

**Problem 12.1. Bending energy of rods.** Recall that the bending energy of a rod with cross-section $\Omega$ is

$$E_b = \frac{K_b}{2} \frac{L_0}{R^2}, \quad K_b = Y \int_{\partial\Omega} z^2 dA(z),$$

where $Y$ is Young's modulus. Calculate the cross-sectional moment of inertia $\mathscr{I} = \int_{\partial\Omega} z^2 dA(z)$ for each of the cross-sections shown in Fig. 12.36. [Problem adapted from Ref. [87].]

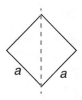
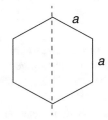

Fig. 12.36: Bending of rods with different cross-sections. The dashed line indicates the axis around which the rod is bent. All sides have the same length $a$.

**Problem 12.2. Freely-jointed chain with restricted bond angles.** Consider a polymer for which the bond angle between successive monomers is a fixed value $\alpha$, although the bonds are free to rotate around one another. The length between monomer $i$ and monomer $i+1$ defines a bond vector $\mathbf{t}_i$. Assume that all bond lengths are the same with $\mathbf{t}_i \cdot \mathbf{t}_i = b^2$ (Fig. 12.37).

(a) Show that $\langle \mathbf{t}_i \cdot \mathbf{t}_{i+k} \rangle = b^2(-\cos\alpha)^k$ with $k \geq 0$.

(b) Defining the end–to–end vector $\mathbf{r} = \sum_{i=1}^N \mathbf{t}_i$, show that in the large $N$ limit

$$\langle \mathbf{r}^2 \rangle = Nb^2 \frac{(1 - \cos\alpha)}{(1 + \cos\alpha)}.$$

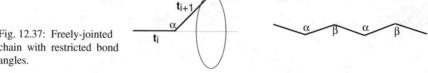

Fig. 12.37: Freely-jointed
chain with restricted bond
angles.

(c) Now consider a polymer backbone consisting of alternating bond angles $\alpha, \beta$. The bond lengths are all equal. Repeat the calculation of part (b) to show that now

$$\langle \mathbf{r}^2 \rangle = Nb^2 \frac{(1 - \cos \alpha)(1 - \cos \beta)}{(1 - \cos \alpha \cos \beta)}.$$

**Problem 12.3. Membrane of a bacterium.** Suppose that the plasma membrane of a bacterium is constructed from an initially flat sheet. How much energy (in units of $k_B T$) is required to bend the membrane into the shape shown in Fig. 12.38, comprised of a cylinder $3\,\mu\text{m}$ long and $1\,\mu\text{m}$ in diameter together with two hemispherical endcaps. Now suppose that the cell divides into two identical daughter cells such that the total volume and diameter are the same. What is the total bending energy of the daughter cells? [Problem adapted from Ref. [87].]

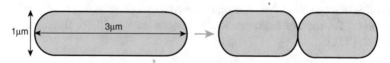

Fig. 12.38: Membrane energetics of a bacterium.

**Problem 12.4. Energetics of membrane tethering.** Membrane nanotubes play an important role in lipid and protein exchange between various organelles of the early secretory pathways such as the ER and Golgi apparatus [844, 880]. One possible mechanism of tube formation is the dynamical clustering of several molecular motors at the tip of a nanotube, which can generate the necessary force to pull the tube along cytoskeletal tracks. Such a mechanism has been established both theoretically and *in vitro*. In this problem, we determine the energetics of the extraction process. Consider a vesicle of radius R from which a tether of radius $r$ and length $L$ has been extracted, as illustrated in Fig. 12.39 (not to scale). Assume that $r \ll R, L$. Using equation (12.2.3), the energy contribution from membrane bending is

$$E_{\text{bend}} = 8\pi \kappa_b + \frac{\pi L \kappa_b}{r} + 4\pi \kappa_b,$$

where $\kappa_b$ is the bending modulus. The terms on the right-hand side correspond to the bending energy of the vesicle, cylindrical part of the tether, and the tether's end cap, respectively. Similarly, the energy contribution from membrane stretching is

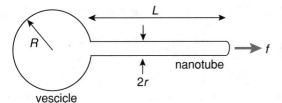

Fig. 12.39: Sketch of the extraction of a membrane tether from a vesicle by an applied force $f$ (not to scale). Here $R$ is the radius of the vesicle, $r$ is the radius of the tether and $L$ is the length of the tube. It is assumed that $r \ll R, L$.

$$E_{\text{stretch}} = \frac{K_A}{2} \frac{\Delta A^2}{A_0},$$

where $K_A$ is the bulk modulus, and $\Delta A$ is the change in surface area (neglecting the area of the end cap):

$$\Delta A = A - A_0 = 4\pi R^2 + 2\pi r L - 4\pi R_0^2 \approx 2\pi r L.$$

It follows that the total free energy of the vesicle, tether and applied load is

$$E_{\text{tot}} = 12\pi \kappa_b + \pi \kappa_b \frac{L}{r} + \frac{K_A}{2} \frac{\Delta A^2}{A_0} - \Delta p \left( \frac{4}{3} \pi R^3 + \pi L r^2 \right) - fL. \qquad (12.6.46)$$

The penultimate term on the right-hand side is the work done against the pressure difference $\Delta p$ between the inside and outside of the vesicle, and $f$ is the applied force.

(a) In order to determine the equilibrium shape of the vesicle and tube, minimize $E_{\text{tot}}$ with respect to the three variables $r, R, L$.

(b) Using the fact that the membrane tension in the elastic regime satisfies

$$\tau = K_a \frac{\Delta A}{A_0},$$

derive the Laplace-Young relation $\Delta p = 2\tau/R$ from the condition $\partial E_{\text{tot}}/\partial R = 0$. This then justifies neglecting the $\Delta p$ terms in the equations $\partial E_{\text{tot}}/\partial r = 0$ and $\partial E_{\text{tot}}/\partial L = 0$ since $r \ll R$. Hence, derive the equations

$$F = 2\pi \sqrt{2\kappa\tau}, \quad r = \sqrt{\kappa/2\tau}.$$

**Problem 12.5. Stress and strain tensors.** In the case of an isotropic material in two dimensions, the stress and strain are related according to

$$\sigma_{ij} = (K_A - \mu)\delta_{i,j}\text{Tr}[\mathbf{u}] + 2\mu u_{ij}.$$

(a) By evaluating the trace of $\sigma$, invert this relationship to show that

$$u_{ij} = \delta_{i,j}\text{Tr}[\sigma](K_A^{-1} - \mu^{-1})/4 + \sigma_{ij}/(2\mu).$$

(b) Determine $u_{xx}$ and $u_{yy}$ when $\sigma_{xx} \neq 0$ and $\sigma_{yy} = 0$. Hence show that

$$\sigma_p := -\frac{u_{yy}}{u_{xx}} = \frac{K_A - \mu}{K_A + \mu}.$$

The Poisson ratio $\sigma_p$ is a measure of how much a material contracts laterally (in the $y$-direction) when it is stretched laterally (in the $x$-direction). Note that $dy' = dy(1 + u_{yy})$.

Young's modulus in 2D can be defined according to $Y_{2D} = \sigma_{xx}/u_{xx}$. Using the above results, show that

$$Y_{2D} = \frac{4K_A}{K_A + \mu}.$$

**Problem 12.6. Curvature**  A hemispherical surface is governed by the equation $x^2 + y^2 + z^2 = R^2$, $z > 0$.
(a) Find $h(x,y)$ in the Monge representation and the normal $\mathbf{n}(x,y)$ for $x^2 + y^2 < R^2$

(b) Determine the mean and Gaussian curvatures using the exact expressions in the Monge representation.

(c) Check that your results agree with your intuition regarding properties of a sphere

**Problem 12.7. Blood cells.**  Under certain conditions, red blood cells develop dimples on their surface. Suppose that such a dimple is modeled by a Gaussian deformation

$$h(x,y) = h_0 \exp(-[x^2 + y^2]/2w^2).$$

(a) Assuming that $h_0$ is small, calculate the mean and Gaussian curvatures as a function of $(x,y)$.

(b) Show that the total bending energy

$$E = \int_{R^2} dx dy \left[ \frac{\kappa_b}{2}(\sigma_1 + \sigma_2)^2 + \kappa_G \sigma_1 \sigma_2 \right]$$

is equal to

$$E = \pi \kappa_b \left( \frac{h_0}{w} \right)^2.$$

**Problem 12.8. Equilibrium susceptibility in the Ising model.**  Consider the partition function of the Ising model

$$Z = \sum_{\sigma_1 = \pm 1} \cdots \sum_{\sigma_N = \pm 1} e^{-\beta E[\sigma] - \sum_{i=1}^{N} h_i \sigma_i}, \quad \beta = \frac{1}{k_B T},$$

with

$$E[\sigma] = -J \sum_{j=1}^{N-1} \sigma_j \sigma_{j+1}.$$

In case of polymers, $h_i$ could represent a local force on the $i$-th link of the chain. The corresponding free energy is

$$F[\mathbf{h}] = -k_B T \ln Z.$$

(a) Show that

$$-\beta \frac{\partial F}{\partial h_i} = \langle \sigma_i \rangle.$$

and

$$\chi_{ij} := -\beta \frac{\partial^2 F}{\partial h_i \partial h_j} = \langle \sigma_i \sigma_j \rangle - \langle \sigma_i \rangle \langle \sigma_j \rangle.$$

This is the analog of the equilibrium susceptibility considered in equation (12.3.21). (b) Now suppose that all links are subject to the same force $h$

$$Z(h) = \sum_{\sigma_1 = \pm 1} \cdots \sum_{\sigma_N = \pm 1} e^{-\beta E[\sigma] - h \sum_{i=1}^{N} \sigma_i}.$$

Show that

$$\chi = -\beta \frac{\partial^2 F}{\partial h^2} = \frac{1}{a^2} \left[ \langle X^2 \rangle - \langle X \rangle^2 \right],$$

where $X = a \sum_i \sigma_i$ is the total extension of the polymer. Using the approximation

$$Z(h) = \lambda_+(h)^N, \quad \lambda_+(h) = e^{\beta J} \left[ \cosh(h) + \sqrt{\sinh^2 h + e^{-4\beta J}} \right],$$

show that for small $h$,

$$\frac{\sqrt{\mathrm{Var}[X]}}{\langle X \rangle} = \frac{h}{\sqrt{N}} e^{-\beta J}.$$

Comment on the $T$ dependence of $\chi$ and the size of fluctuations.

**Problem 12.9. Rotational Brownian motion.** Consider a molecule whose shape consists of a slender rod of length $L$ with a small spherical bead attached at one end, see Fig. 12.40(a). Let $\mathbf{u}$ denote the unit axial vector of the rod and suppose that the vector $\mathbf{u}$ can change its orientation due to Brownian motion. (Rotation around $\mathbf{u}$ can be neglected for a slender rod.) The rate of change of $\mathbf{u}$ is specified by the angular velocity $\omega$ according to the vector product

$$\dot{\mathbf{u}} = \omega \times \mathbf{u}.$$

The unit vector $\omega/|\omega|$ is orthogonal to the plane of rotation, see Fig. 12.40(b). In order to derive the FP equation for the probability density $\psi(\mathbf{u}, t)$, we first need to determine the frictional torque $\mathbf{T}_f$ and external torque $\mathbf{T}_{ext}$ on the bead, analogous to the frictional and external forces associated with translational Brownian motion.

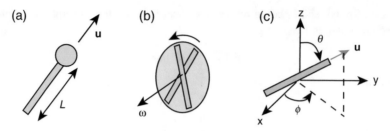

Fig. 12.40: Rotational Brownian motion. (a) Rod-shaped molecule, (b) Angular velocity, and (c) Spherical polar coordinates.

(a) The velocity of the bead is $\mathbf{v} = \omega \times L\mathbf{u}$. Use the triple vector product identity

$$\mathbf{a} \times (\mathbf{b} \times \mathbf{c}) = (\mathbf{a} \cdot \mathbf{c})\mathbf{b} - (\mathbf{a} \cdot \mathbf{b})\mathbf{c}$$

to show that

$$\mathbf{T}_f \equiv -\gamma L\mathbf{u} \times \mathbf{v} = -\gamma L^2 \omega,$$

where $\gamma$ is the friction coefficient.

(b) Suppose that the external torque is generated by a potential $U(\mathbf{u})$. Consider an infinitesimal rotation $\delta\varphi$ such that $\mathbf{u} \to \mathbf{u}' = \mathbf{u} + \delta\varphi \times \mathbf{u}$. Equating the work needed to counter the external torque, $-\mathbf{T}_{\text{ext}} \cdot \delta\varphi$ with the change in potential energy $\Delta U = U(\mathbf{u}') - U(\mathbf{u})$, and using the vector identity $\mathbf{a} \cdot (\mathbf{b} \times \mathbf{c}) = (\mathbf{a} \times \mathbf{b}) \cdot \mathbf{c}$, show that

$$\mathbf{T}_{\text{ext}} = -\mathscr{R}U, \quad \mathscr{R} = \mathbf{u} \times \nabla_{\mathbf{u}}.$$

The rotational operator $\mathscr{R}$ plays the role of $\nabla_{\mathbf{x}}$ in translational diffusion. Hence, ignoring inertial effects show that $\mathbf{T}_{\text{ext}} + \mathbf{T}_f = 0$ yields the following equation for the angular velocity:

$$\omega = -\frac{1}{\gamma_r}\mathscr{R}U, \quad \gamma_r = \gamma L^2.$$

(c) Use spherical polar coordinates, see Fig. 12.40(c), such that

$$\mathbf{u} = (\sin\theta\cos\phi, \sin\theta\sin\phi, \cos\theta), \quad \nabla_{\mathbf{u}} = \mathbf{e}_\theta \frac{\partial}{\partial\theta} + \frac{\mathbf{e}_\phi}{\sin\theta}\frac{\partial}{\partial\phi},$$

to prove the identity

$$\int_{S^2} A(\mathbf{u})[\mathscr{R}B(\mathbf{u})]d\mathbf{u} = -\int_{S^2}[\mathscr{R}A(\mathbf{u})]B(\mathbf{u})d\mathbf{u},$$

for arbitrary smooth functions $A, B$. Note that $d\mathbf{u} = \sin\theta d\phi d\theta$ and

$$\mathbf{e}_\theta = (\cos\theta\cos\phi, \cos\theta\sin\phi, -\sin\theta), \quad \mathbf{e}_\phi = (-\sin\phi, \cos\phi, 0).$$

[Hint: Apply integration by parts to each component of the above integral identity.]

(d) Introducing the Brownian potential by setting

$$\omega_f = -\frac{1}{\gamma_r}\mathscr{R}[k_B T \ln \psi + U],$$

with $\psi$ evolving according to the conservation equation

$$\frac{\partial \psi}{\partial t} = -\nabla_{\mathbf{u}} \cdot (\omega_f \times \mathbf{u})\psi,$$

obtain the FP equation for rotational Brownian motion:

$$\frac{\partial \psi}{\partial t} = \frac{1}{\gamma_r}\mathscr{R} \cdot [k_B T \mathscr{R}\psi + \psi \mathscr{R} U].$$

(e) Suppose that $U = 0$. Define the mean orientation vector according to

$$\langle \mathbf{u}(t) \rangle = \int_{S^2} \psi(\mathbf{u}, t)\mathbf{u}d\mathbf{u}.$$

Differentiating both sides with respect to $t$, substituting for $\partial_t \psi$ using the FP equation of (d), and performing integration by parts twice according to (c), show that

$$\frac{\partial}{\partial t}\langle \mathbf{u}(t) \rangle = D_r \int_{S^2} \psi \mathscr{R}^2 \mathbf{u}d\mathbf{u}, \quad D_r = \frac{k_B T}{\gamma_r}.$$

(Note that the rotational diffusion coefficient $D_r$ has units of $t^{-1}$ rather than $L^2/t$.) From the definition of $\mathscr{R}$, prove that $\mathscr{R}^2 \mathbf{u} = -2\mathbf{u}$. [Hint: Using spherical polar coordinates first establish that $\mathscr{R}_x u_y = -u_z$, $\mathscr{R}_y u_x = u_z$ and $\mathscr{R}_x u_x = 0$ plus cyclic permutations. This can be summarized as $\mathscr{R}_i u_j = -\sum_{k=1,2,3} \varepsilon_{ijk} u_k$, where $\varepsilon_{ijk}$ is the Levi Civita symbol.] Hence, show that

$$\langle \mathbf{u}(t) \rangle = \mathbf{u}_0 e^{-2D_r t}$$

for $\mathbf{u}(0) = \mathbf{u}_0$, and thus

$$\langle (\mathbf{u}(t) - \mathbf{u}(0))^2 \rangle = 2\left(1 - e^{-2D_r t}\right).$$

**Problem 12.10. Polymer network.** Consider a random chain network that is subject to a strain in the $z$-direction specified by a scaling factor $\Lambda_z = \Lambda$. The corresponding scaling factors in the $x$ and $y$-directions are equal and ensure volume conservation.

(a) Determine $\Lambda'$ where $\Lambda_x = \Lambda_y = \Lambda'$.

(b) Using equation (12.4.5) show that the free energy density of this deformation is

$$\Delta F = \frac{k_B T n_c}{2}\left(\Lambda^2 + \frac{2}{\Lambda} - 3\right).$$

(c) Setting $\Lambda = 1 + \delta$, $\delta \ll 1$, calculate $\Delta F$ to $O(\delta^2)$.

(d) Show that to leading order the strain tensor has nonzero components

$$u_{xx} = -\delta/2 = u_{yy}, \quad u_{zz} = \delta.$$

Hence, determine an alternative expression for $\Delta F$ to $O(\delta^2)$ using the formula (12.2.13).

(e) Comparing the results of (c) and (d) show that the shear modulus is $\mu = n_c k_B T$.

**Problem 12.11. The Rouse model.** Consider the Rouse model given by equations (12.5.2).

(a) First suppose $\mathbf{f}_n = 0$ for all $n$. Show that

$$\mathbf{R}_n(t) = \mathbf{X}(t)\cos(n\theta + c)$$

is a solution of equations (12.5.2a-c), with

$$\frac{d\mathbf{X}}{dt} = -4\kappa\sin^2(\theta/2)\mathbf{X},$$

provided that

$$\cos c - \cos(\theta + c) = 4\sin^2(\theta/2)\cos c$$
$$\cos(N\theta + c) - \cos([N-1]\theta + c) = 4\sin^2(\theta/2)\cos(N\theta + c).$$

The latter conditions ensure that the boundary equations ($n = 0, N$) are consistent with the bulk equations ($0 < n < N$).

(b) Using trigonometric formulas show that the pair of conditions in (a) reduce to

$$\cos c = \cos(\theta - c), \quad \cos(N\theta + c) = \cos((N+1)\theta + c),$$

which have the independent solutions

$$\theta = \frac{p\pi}{N+1}, \quad c = \theta/2$$

for integers $p$. This establishes that the general solution can be written as (12.5.3).

# Chapter 13
# Self-organization and assembly of cellular structures

There are number of highly dynamical structures within the cell nucleus, plasma membrane, and the Golgi complex that can be characterized in terms of the self-organization or self-assembly of molecular clusters [623]. For example, as described in Sect. 12.5, the mammalian cell nucleus contains a number of distinct subcompartments, the most prominent being the nucleolus, which is the site of ribosome biogenesis [408, 453, 539]. A significant feature of the nucleolus is that it is an example of a membrane-less organelle or biomolecular condensate, which is able to concentrate proteins and nucleic acids at a discrete site without a physical membrane separating it from the surrounding medium. Liquid–liquid phase separation, whereby a homogeneous solution of molecules spontaneously separates (demixes) into two coexisting liquid phases with different densities, is emerging as the key organizing principle underlying the formation of cellular condensates [42, 70, 100, 293, 451]. Related self-organizing processes are also likely to play a role in the partitioning of chromatin into different regions or territories within the nucleus [795], and the clustering of curvature-inducing proteins in active membranes. Liquid–liquid phase separation can occur either spontaneously via so-called spinodal decomposition or arise more slowly via the nucleation and growth from a metastable state. The latter type of process can also play a role in the assembly of other types of molecular clusters such as phospholipids in the plasma membrane.

In this chapter, we consider various models for the self-organization and assembly of cellular structures. We begin in Sect. 13.1 with the classical theory of liquid–liquid phase separation [251], covering both thermodynamical and kinetic aspects as developed in Chap. 12. In Sect. 13.2, we turn to more recent developments that are specific to biological condensates, such as the effects of nonequilibrium chemical reactions and protein concentration gradients. We also explore related phenomena in active membranes. In Sect. 13.3, we describe one approach to analyzing nucleation and growth, based on a discrete model of the aggregation and fragmentation of molecular clusters originally introduced by Becker and Döring. This is then used to model the self-assembly of phospholipids in the plasma membrane. Finally, in Sect. 13.4, we consider the cooperative transport of proteins between cellular organelles,

© Springer Nature Switzerland AG 2021

P. Bressloff, *Stochastic Processes in Cell Biology*, Interdisciplinary Applied Mathematics 41, https://doi.org/10.1007/978-3-030-72519-8_13

which is an self-organizing mechanism that allows organelles to maintain their distinct identities while constantly exchanging material.

## 13.1 Liquid-liquid phase separation

In this section, we review the classical theory of liquid–liquid phase separation, from both the thermodynamic and kinetic perspectives. We introduce the basic phenomena of spinodal decomposition, nucleation and growth, and Ostwald ripening. This will make use of several ideas from equilibrium and nonequilibrium statistical physics that were introduced in Chap. 12. In Sect. 13.2, we will then describe how the classical picture needs to be modified in the case of biological condensates, where phase separation can be regulated by active processes such as nonequilibrium chemical reactions.

### *13.1.1 Thermodynamics of phase separation*

Consider a homogeneous solution consisting of two molecular species, which we treat as a solute and solvent, respectively. Suppose that there are $N_p$ solute molecules with specific volume $v_p$ and $N_s$ solvent molecules with specific volume $v_s$. The total volume of the solution is given by

$$V = v_p N_p + v_s N_s,$$

and the volume fraction of solute is defined by

$$\phi = \frac{v_p N_p}{v_p N_p + v_s N_s}. \tag{13.1.1}$$

We will assume that the free energy density of the homogeneous solution is known, and can be written as $f = f(\phi, T)$, where $T$ is the temperature. (For ease of notation, we will drop the explicit $T$ dependence.)

Now suppose that a solution having volume $(V_1, \phi_1)$ is mixed with another solution having $(V_2, \phi_2)$ resulting in a homogeneous solution $(V, \phi)$. If the two solutions mix completely, then $V = V_1 + V_2$ and the volume fraction is

$$\phi = \frac{\phi_1 V_1 + \phi_2 V_2}{V_1 + V_2} = \lambda \phi_1 + (1 - \lambda)\phi_2, \quad \lambda = \frac{V_1}{V_1 + V_2}. \tag{13.1.2}$$

Prior to mixing, the free energy of the system is $V_1 f(\phi_1) + V_2 f(\phi_2)$, whereas the free energy of the final solution is $(V_1 + V_2)f(\phi)$. From the principle of free energy minimization (see Sect. 1.3), the homogeneous solution exists provided that

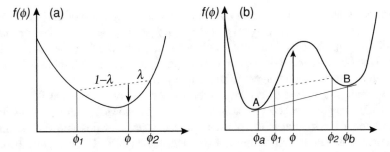

Fig. 13.1: Free energy $f$ of a homogeneous solution plotted as a function of solute volume fraction $\phi$. (a) A convex function $f$ means that the solvent and solute can mix at any composition, since mixing decreases the free energy as indicated by the arrow. (b) A concave region between $\phi_1$ and $\phi_2$ implies that mixing would raise the free energy. Therefore, phase separation occurs according to the common tangent construction.

$$(V_1 + V_2)f(\phi) < V_1 f(\phi_1) + V_2 f(\phi_2),$$

that is,

$$f(\lambda\phi_1 + (1-\lambda)\phi_2) < \lambda f(\phi_1) + (1-\lambda)f(\phi_2). \tag{13.1.3}$$

It follows that homogeneous mixing will occur for any volume ratio $V_1/V_2$ as long as equation (13.1.3) holds for all $0 \le \lambda \le 1$. This is equivalent to the condition that the free energy density is convex in the region $\phi_1 < \phi < \phi_2$, see Fig. 13.1:

$$\frac{\partial^2 f}{\partial \phi^2} > 0, \quad \text{for } \phi_1 < \phi < \phi_2. \tag{13.1.4}$$

On the other hand, if $f$ is concave in some region between $\phi_1$ and $\phi_2$, the system cannot remain homogeneous, since it can lower its free energy by separating into two solutions with concentrations $\phi_a$ and $\phi_b$. This phenomenon is known as phase separation or demixing.

The concentrations $\phi_a$ and $\phi_b$ are determined according to the common tangent construction. Suppose that we started out with a homogeneous mixture of volume fraction $\phi$ and total volume $V$, which then separates into two solutions with compositions $(V_1, \phi_1)$ and $(V_2, \phi_2)$. Given the conditions $V = V_1 + V_2$ and $\phi V = \phi_1 V_1 + \phi_2 V_2$, we have

$$V_1 = \frac{\phi_2 - \phi}{\phi_2 - \phi_1} V, \quad V_2 = \frac{\phi - \phi_1}{\phi_2 - \phi_1} V. \tag{13.1.5}$$

Hence, the free energy of the phase-separated state is

$$F = V_1 f_1(\phi_1) + V_2 f_2(\phi_2) = V\left[\frac{\phi_2 - \phi}{\phi_2 - \phi_1} f(\phi_1) + \frac{\phi - \phi_1}{\phi_2 - \phi_1} f(\phi_2)\right]. \tag{13.1.6}$$

The expression in square brackets corresponds to the height at $\phi$ of the line connecting the two points $A = (\phi_1, f(\phi_1))$ and $B = (\phi_2, f(\phi_2))$ in the plot of the free energy curve, see Fig. 13.1(b). Therefore, in order to minimize the free energy, one needs to find the points $A$ and $B$ on the curve $f(\phi)$ that minimizes the height at $\phi$. This line is given by the common tangent for the curve $f(\phi)$.

The region $\phi_a < \phi < \phi_b$ can be further divided into two regions as illustrated in Fig. 13.2. The graphical construction of Fig. 13.1(b) shows that although the global minimum of the system is the phase-separated state, it is possible for the mixed homogeneous state to be locally stable or metastable (stable to small perturbations) if $\phi$ lies in a region where $\partial^2 f / \partial \phi^2 > 0$. On the other hand, the homogeneous state is unstable when $\partial^2 f / \partial \phi^2 < 0$, since arbitrarily small fluctuations can lower the free energy. Let $\phi_a^*$ and $\phi_b^*$ be the volume fractions at which $\partial^2 f / \partial \phi^2 = 0$. Then, the solution is unstable in the region $\phi_a^* < \phi < \phi_b^*$ and metastable in the regions $\phi_a < \phi < \phi_a^*$ and $\phi_b^* < \phi < \phi_b$. Since $\phi_a, \phi_b, \phi_a^*, \phi_b^*$ are functions of temperature, they can be drawn in the $\phi - T$ plane, as shown in Fig. 13.2(b). The points $\phi_a(T)$ and $\phi_b(T)$ generate the so-called coexistence curve, whereas the curve separating the regions where the homogeneous solution is metastable from the region where it is unstable is known as the spinodal line. The peak of the spinodal is called the critical point with corresponding critical temperature $T_c$. Since $\phi_a^*(T)$ and $\phi_b^*(T)$, which both satisfy the condition $\partial^2 f / \partial \phi^2 = 0$, merge at the critical point, it follows that the critical point is determined by the pair of equations

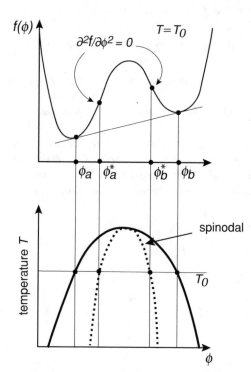

Fig. 13.2: lots of (a) free energy and (b) $\phi - T$ phase diagram of a homogeneous solution. The temperature-dependent spinodal (dashed curve) separates the regions where the solution is metastable from the region where it is unstable. See text for details.

$$\frac{\partial^2 f}{\partial \phi^2} = 0, \quad \frac{\partial^3 f}{\partial \phi^3} = 0.$$

**Lattice models.** Recall the lattice model of a solution presented in Sect. 1.3, see also Fig. 13.3(a). There we calculated the mixing entropy for $N = N_p + N_s$ boxes occupied by $n = N_p$ solute molecules and $N - n = N_s$ solvent molecules with $v_p = v_s = v_c$:

$$S_{\text{mix}} = k_B[(N+n)\ln(N+n) - n\ln n - N\ln N]$$
$$= k_B\left[-N_p\ln\left(\frac{N_p}{N_p+N_s}\right) - N_s\ln\left(\frac{N_s}{N_p+N_s}\right)\right]$$
$$= k_B N[-\phi\ln\phi - (1-\phi)\ln(1-\phi)].$$

The free energy of the solution is then $F = E - TS_{\text{mix}}$, where $E$ is the total mean energy arising from nearest neighbor interactions[1]. Let $\varepsilon_{pp}$, $\varepsilon_{ps}$ and $\varepsilon_{ss}$ denote the (negative) interaction energies of solute-solute, solute-solvent, and solvent-solvent pairs, respectively. Denote the corresponding average number of neighboring pairs by $N_{pp}, N_{ps}, N_{ss}$. The latter can be estimated as follows. Each box in the lattice has $z$ neighboring boxes, where $z$ is the coordination number. On average, $z\phi$ of these neighbors are occupied by solute molecules and the remaining $z(1-\phi)$ are occupied by solvent molecules. Therefore,

$$N_{pp} = z\phi N_p/2 = zN\phi^2/2, \quad N_{ss} = zN(1-\phi)^2/2, \quad N_{ps} = zN\phi(1-\phi).$$

We thus find that the total average energy is

$$E = \varepsilon_{pp}N_{pp} + \varepsilon_{ps}N_{ps} + \varepsilon_{ss}N_{ss} = \frac{1}{2}Nz[\varepsilon_{pp}\phi^2 + 2\varepsilon_{ps}\phi(1-\phi) + \varepsilon_{ss}(1-\phi)^2]$$
$$= \frac{1}{2}Nz\Delta\varepsilon\,\phi^2 + A_1\phi + A_0,$$

where

$$\Delta\varepsilon = \varepsilon_{pp} + \varepsilon_{ss} - 2\varepsilon_{ps},$$

and $A_0, A_1$ are constants. It turns out that determining whether or not phase separation occurs is independent of these constants. The latter are thus chosen to write the free energy in a convenient form.

Combining all of the above results, the free energy density $f(\phi) = F/V$, where $V = Nv_c$ is the total volume and $v_c$ is the volume of an individual box, is given by

$$f(\phi) = \frac{k_B T}{v_c}[\phi\ln\phi + (1-\phi)\ln(1-\phi) + \chi\phi(1-\phi)], \quad (13.1.7)$$

with

---

[1] An alternative method for deriving the free energy, starting from a Hamiltonian and using a variational method will be presented in Box 14D.

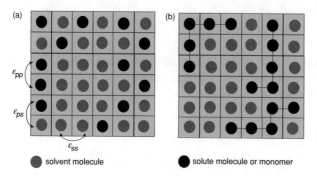

Fig. 13.3: Lattice model for (a) a simple solution and (b) a polymer solution, with each polymer consisting of $M = 5$ segments.

$$\chi = -\frac{z}{2k_BT}\Delta\varepsilon.$$

Note that $f(\phi)$ is an even function about $\phi = 1/2$. In Ex. 13.1, it is shown that $f(\phi)$ has a single minimum for sufficiently small $\chi$ but switches to two local minima as $\chi$ crosses the critical point $\chi_c = 2$. The spinodal line in the $\phi - \chi$ plane is determined by the condition $f''(\phi) = 0$, and takes the form

$$\chi = \frac{1}{2}\frac{1}{\phi(1-\phi)} \geq 2. \tag{13.1.8}$$

The symmetry of $f(\phi)$ means that the common tangent is given by the line connecting the two local minima. Consequently, the coexistence curve for $\phi(\chi)$ and $\phi_b(\chi)$ is determined by the equation $f'(\phi) = 0$, that is

$$\chi = \frac{1}{1-2\phi}\ln\left(\frac{1-\phi}{\phi}\right). \tag{13.1.9}$$

In the above analysis, the solute and solvent molecules were assumed to have the same size, that is, each occupies a single box. Now consider a polymer solution in which each polymer consists of $M$ segments, with each segment or monomer taken to be the same size as a solvent molecule, see Fig. 13.3(b). For the polymer model, the free energy density becomes

$$f(\phi) = \frac{k_BT}{v_c}\left[\frac{1}{M}\phi\ln\phi + (1-\phi)\ln(1-\phi) + \chi\phi(1-\phi)\right]. \tag{13.1.10}$$

The additional factor of $M^{-1}$ reflects the fact that the contribution to the mixing entropy due to the polymers is reduced, since $M$ segments are connected and thus cannot be placed anywhere. One finds that the spinodal line becomes (see Ex. 13.1)

$$\chi = \frac{1}{2}\left[\frac{1}{1-\phi} + \frac{1}{M\phi}\right], \tag{13.1.11}$$

with corresponding critical point $(\phi_c, \chi_c)$, where

$$\phi_c = \frac{1}{1+\sqrt{M}}, \quad \chi_c = \frac{1}{2}\left(1+\frac{1}{\sqrt{M}}\right)^2. \tag{13.1.12}$$

**Osmotic pressure.** One way to measure the thermodynamic forces that tend to mix solute and solvent is to determine the osmotic pressure. The latter is the force that occurs when a solution is brought into contact with pure solvent across a semi-permeable membrane, see Fig. 13.4. A semi-permeable membrane only permits solvent to pass through it. If the free energy of the system is lowered when solvent and solute mix, then solvent molecules tend to cross the semi-permeable membrane from the right-hand region of pure solvent into the solution. Hence, if the semi-permeable membrane could move freely, then all solvent molecules would cross into the solution region by pushing the semi-permeable membrane to the right. This would ultimately lead to a homogeneous solution. It follows that a rightward force has to be applied to the semi-permeable membrane in order to prevent it from moving. The osmotic pressure is the required force per unit area of semi-permeable membrane. Let $F_{\text{tot}}(V)$ be the free energy of the system consisting of solution of volume $V$ and pure solvent of volume $V_{\text{tot}} - V$ with $V_{\text{tot}}$ fixed. Suppose that the semi-permeable membrane is moved, resulting in a change $dV$ in the solution volume. This requires an amount of work $-\Pi dV$, which is equal to the change in the total free energy, $dF_{\text{tot}}(V)$. Therefore, the osmotic pressure is defined by the equation

$$\Pi = -\frac{\partial F_{\text{tot}}(V)}{\partial V}. \tag{13.1.13}$$

We can express the free energy $F_{\text{tot}}(V)$ in terms of the volume fraction $\phi$ according to

$$F_{\text{tot}} = V f(\phi) + (V_{\text{tot}} - V) f(0). \tag{13.1.14}$$

Since $\phi = N_p v_c / V$, we see that $\partial \phi / \partial V = -\phi/V$. Hence,

$$\Pi(\phi) = -f(\phi) + \phi f'(\phi) + f(0). \tag{13.1.15}$$

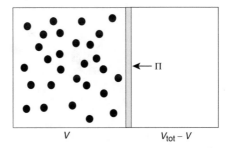

Fig. 13.4: Schematic diagram illustrating the definition of osmotic pressure.

## 13.1.2 Kinetics of phase separation and spinodal decomposition

The above thermodynamic formulation describes the equilibrium behavior of a mixture, but does not provide any information regarding the kinetics of phase separation. The latter is crucial for understanding the role of liquid–liquid phase separation in the formation of biological condensates. In particular, analyzing the kinetics is necessary in order to establish whether or not phase separation can occur over biologically relevant timescales. There are two basic dynamical mechanisms for phase separation, depending on which region of the phase diagram a solution is placed by, for example, changing the temperature [251]: (i) spinodal decomposition, which occurs when the solution is in a thermodynamically unstable state ($\phi_a^* < \phi_0 < \phi_b^*$); (ii) nucleation and growth, which occurs when the solution is in a metastable state ($\phi_a < \phi_0 < \phi_a^*$ or $\phi_b^* < \phi_0 < \phi_b$).

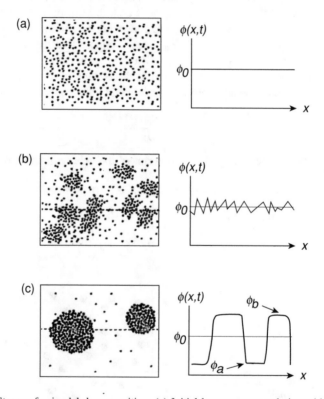

Fig. 13.5: Stages of spinodal decomposition. (a) Initial homogeneous solution with concentration $\phi_0$ lying within the spinodal region. (b) In early phase separation, solute molecules assemble to form locally concentrated regions. (c) During late phase separation, these regions grow to form macroscopic droplets. Also shown are the corresponding spatial profiles of the concentration obtaining by taking a slice through the domain. [Adapted from Doi [251].]

Spinodal decomposition involves the rapid demixing from one thermodynamic phase to two coexisting phases due to the fact that there is essentially no thermodynamic barrier to nucleation of the two phases, see Fig. 13.5. A classical example is a hot mixture of oil and water. At high temperatures, the oil and the water may mix to form a single homogeneous phase in which water molecules are interdispersed with oil molecules. If the solution is then suddenly cooled to form an unstable state, the oil molecules immediately start to cluster together to form microscopic oil-rich clusters throughout the liquid. These clusters then rapidly grow and coalesce to form macroscopic clusters or droplets, ultimately resulting in complete separation of the oil and water phases. The absence of a thermodynamic barrier means that spinodal decomposition can be treated as a diffusion problem [251].

In contrast to spinodal decomposition, nucleation and growth proceeds much more slowly as the initial formation of microscopic clusters involves a large free energy barrier [256, 701], see Fig. 13.6. During nucleation, spontaneous thermal and compositional fluctuations result in the formation of small nuclei of higher-density solute. However, such nuclei will be thermodynamically unstable due to high surface tension, unless they are larger than a critical size. The free energy cost of forming a critical nucleus, $\Delta G_{\mathrm{nucl}}$, is known as the nucleation barrier. The probability that a critical nucleus will form is proportional to the Boltzmann factor $e^{-\Delta G_{\mathrm{nucl}}/k_B T}$. Once a critical nucleus is formed, it can grow until the system reaches the stable coexisting phases, with the rate of growth determined by the transport properties of the original mixture. If $\Delta G_{\mathrm{nucl}} \gg k_B T$, then phase separation is a nucleation-limited rare event, and demixing tends to occur via the growth of a single critical cluster. On the other hand, if nucleation is relatively fast then phase separation is growth limited. In the latter regime, multiple nucleation events can occur within a single mixture. This may then lead to diffusion-mediated coarsening or Ostwald ripening, in which

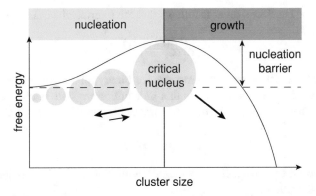

Fig. 13.6: Nucleation and growth in phase separation. Nucleation involves the formation of microscopic clusters that tend to be unstable, and thus shrink, unless they reach a critical size, after which the growth phase is initiated. The size of the nucleation barrier determines whether the process is nucleation-limited or growth-limited.

larger nuclei grow at the expense of smaller, less stable nuclei. Alternatively, smaller nuclei may collide to form a larger nucleus in a process known as coalescence. Here, we will focus on diffusion models of spinodal decomposition. Ostwald ripening will be considered in Sect. 13.1.3, and the classical theory of nucleation will be developed in Sect. 13.3 using the Becker-Döring model of aggregation-fragmentation.

**Diffusion model of spinodal decomposition.** As is clear from Fig. 13.5, in order to analyze spinodal decomposition, it is necessary to introduce a spatially varying concentration or volume fraction $\phi(\mathbf{r},t)$. Moreover, as droplets of the high density phase $\phi_b$ form in a sea of the low density phase $\phi_a$, there will be an additional contribution to the free energy due to the existence of a surface interface between the two phases. This can be taken into account by taking the total free energy to be the functional

$$F[\phi] \equiv \int \mathcal{F}(\phi(\mathbf{r},t))d^3\mathbf{r} = \int \left[ f(\phi(\mathbf{r},t)) + \frac{\alpha}{2}(\nabla\phi(\mathbf{r},t))^2 \right] d^3\mathbf{r}, \qquad (13.1.16)$$

where $f(\phi)$ is given by equation (13.1.7) and $\alpha$ is a positive constant proportional to $\chi$. Another way to interpret the free energy functional (13.1.16) is to discretize space by partitioning the cell volume into a set of small compartments labeled $\ell$ with discrete positions $\mathbf{r}_\ell$, say, and setting $\phi_\ell(t) = \phi(\mathbf{r}_\ell,t)$. Equation (13.1.16) is then replaced by a discrete sum over compartments, and one has a free energy for the set of discrete variables $\phi_\ell$. At an interface, the gradient term $\alpha(\nabla\phi)^2$ becomes very large with $\nabla\phi \approx (\phi_b - \phi_a)/\ell$, where $\ell$ is the thickness of the interface. We thus have the local estimate

$$\int (\nabla\phi)^2 d^3\mathbf{r} \approx \alpha \left( \frac{\phi_b - \phi_a}{\ell} \right)^2 \ell A,$$

where $A$ is the interfacial area. It follows that the interfacial surface tension $\tau_s$ can be estimated as

$$\tau_s \approx \alpha \frac{(\phi_b - \phi_a)^2}{\ell}. \qquad (13.1.17)$$

The thermodynamics of surface interfaces for binary mixtures are explored in Ex. 13.2 and Ex. 13.3. The basic idea is that an equilibrium state is obtained by minimizing the free energy (13.1.16) with respect to $\phi$ under the mass conservation constraint

$$\int \phi(\mathbf{r})d^3\mathbf{r} = M_0.$$

This is achieved by introducing the Lagrange multiplier $\mu$, which is the chemical potential, and setting

$$\widehat{F}[\phi] = F[\phi] + \mu \int \left( \frac{M_0}{V} - \phi(\mathbf{r}) \right) d^3\mathbf{r}.$$

The constrained minimization of the free energy leads to the functional equation (see Box 12C for a definition of functional derivatives)[2]

$$\frac{\delta \widehat{F}}{\delta \phi(\mathbf{r})} = \frac{\delta F}{\delta \phi(\mathbf{r})} - \mu = 0.$$

Conservation of solute molecules also holds during the time-dependent kinetics of phase separation, that is, $\int \phi(\mathbf{r},t)d^3\mathbf{r}$ is a constant. It follows that the concentration satisfies the conservation equation

$$\frac{\partial \phi}{\partial t} = -\nabla \cdot \mathbf{J}(\mathbf{r},t), \tag{13.1.18}$$

where $\mathbf{J}(\mathbf{r},t)$ is the solute flux. Recall from Chap. 2 that in the case of classical diffusion, the flux is given by Fick's law, $\mathbf{J} = -D\nabla\phi$. It turns out that this can be extended to more complex processes such as phase separation by noting that diffusion is really driven by the gradient of the chemical potential $\mu$. This leads to the following generalization of Fick's law (see Box 13A):

$$\mathbf{J}(\mathbf{r},t) = -\frac{1}{\gamma}\phi(\mathbf{r},t)\nabla\mu(\mathbf{r},t), \quad \mu(\mathbf{r},t) = \frac{\delta F[\phi]}{\delta\phi(\mathbf{r},t)}, \tag{13.1.19}$$

where $\gamma$ is a friction coefficient. Consistent with the equilibrium case, the time-dependent version of the chemical potential $\mu$ is given by the functional derivative of the free energy functional with respect to the concentration.

---

**Box 13A. Generalized Fick's law.**

---

In order to understand the physical motivation for equation (13.1.19), we use the fact that the only way the concentration $\phi(\mathbf{r},t)$ can change (in the absence of an external force) is via particle motion, as expressed by the conservation condition. Suppose that at each point $\mathbf{r}$ in the domain the velocity of particles is $\dot{\mathbf{r}} = \mathbf{v}(\mathbf{r},t)$ and $\mathbf{J}(\mathbf{r},t) = \phi(\mathbf{r},t)\mathbf{v}(\mathbf{r},t)$. Over a small time-interval $\delta t$, the concentration changes as (after dropping the explicit $t$ dependence)

$$\phi(\mathbf{r}) \rightarrow \phi(\mathbf{r}) - \sum_{k=1}^{n} \frac{\partial v_k(\mathbf{r})\phi(\mathbf{r})}{\partial x_k}\delta t.$$

This induces a change in the free energy according to

---

[2] More precisely, we should write $\mu = v_c \delta F/\delta\phi(\mathbf{r})$ in order to ensure the correct units, where $v_c$ is the volume of a solute molecule. For simplicity, we fix the spatial dimensions by setting $v_c = 1$ in the following.

$$F[\phi + \delta\phi] = \int \mathcal{F}\left(\phi(\mathbf{r}) - \sum_{k=1}^{n} \frac{\partial v_k(\mathbf{r})\phi(\mathbf{r})}{\partial x_k}\delta t\right)d^3\mathbf{r}$$

$$\approx F[\phi] - \int \frac{\partial\mathcal{F}}{\partial\phi(\mathbf{r})}\sum_{k=1}^{n}\frac{\partial v_k(\mathbf{r})\phi(\mathbf{r})}{\partial x_k}d^3\mathbf{r}\delta t$$

$$= F[\phi] + \int \sum_{k=1}^{n} v_k(\mathbf{r})\phi(\mathbf{r})\frac{\partial}{\partial x_k}\frac{\partial\mathcal{F}}{\partial\phi(\mathbf{r})}d^3\mathbf{r}\delta t,$$

after integration by parts. Hence, given the local changes in coordinates $\delta x_k(\mathbf{r}) = v_k(\mathbf{r})\delta t$, the corresponding change in the free energy is

$$\delta F[\phi] = \int \sum_{k=1}^{n} \delta x_k(\mathbf{r})\phi(\mathbf{r})\frac{\partial}{\partial x_k}\frac{\partial\mathcal{F}}{\partial\phi(\mathbf{r})}d^3\mathbf{r}.$$

We thus have the functional derivative equation

$$\frac{\delta F[\phi]}{\delta x_k(\mathbf{r})} = \phi(\mathbf{r})\frac{\partial}{\partial x_k}\frac{\partial\mathcal{F}}{\partial\phi(\mathbf{r})} = \phi(\mathbf{r})\frac{\partial}{\partial x_k}\frac{\delta F[\phi]}{\delta\phi(\mathbf{r})}. \qquad (13.1.20)$$

We now use a fundamental result from the theory of nonequilibrium systems operating close to equilibrium (see Box 12D and equation (12.3.24)): given a set of degrees of freedom $z_1, z_2, \ldots, z_L$ with associated free energy $F[z_1, \ldots, z_L]$, there exists a set of friction coefficients $\gamma_{ab}$ such that

$$\sum_{b=1}^{L} \gamma_{ab}\frac{dz_b}{dt} = -\frac{\partial F}{\partial z_a}.$$

If we apply an analogous result to the infinite dimensional system, then

$$\gamma(\mathbf{r})v_k(\mathbf{r}) = -\frac{\delta F[\phi]}{\delta x_k(\mathbf{r})} = -\phi(\mathbf{r})\frac{\partial}{\partial x_k}\frac{\delta F[\phi]}{\delta\phi(\mathbf{r})}.$$

(Over a small time interval $\delta t$, we can take the free energy functional to depend on the dynamical variables $\delta x_k(\mathbf{r})$, which play the role of $\delta z_a$ in the finite system.) The friction coefficient for a population of particles is $\gamma(\mathbf{r}) = \phi(\mathbf{r})\gamma$, which finally yields the generalized Fick's law (13.1.19). Before applying the theory to phase separation, it is instructive to check that Fick's law is recovered in the case of an ideal dilute solution, for which $f(\phi) = -k_B T\phi\ln\phi$. We have

$$\mu = -k_B T(\ln\phi + 1), \quad \mathbf{J} = -\gamma^{-1}k_B T\nabla\phi.$$

(The chemical potential is analogous to the Brownian potential in equation (12.3.28) for a single Brownian particle.) In particular, we recover the

Einstein relation $D\gamma = k_B T$. The inclusion of an external potential leads to the Smoluchowski equation (see Ex. 13.4).

**Cahn–Hilliard equation.** Returning to the free energy (13.1.16), the corresponding flux is

$$\mathbf{J} = -\gamma^{-1}\phi\nabla\left[f'(\phi) - \alpha\nabla^2\phi\right].$$

Substituting into the conservation equation yields the generalized diffusion equation

$$\frac{\partial\phi}{\partial t} = \gamma^{-1}\nabla\cdot\phi\nabla\left[f'(\phi) - \alpha\nabla^2\phi\right], \tag{13.1.21}$$

which is a version of the Cahn–Hilliard equation [156]. Since $\nabla f'(\phi) = f''(\phi)\nabla\phi$, it follows that the effective diffusivity is

$$D_c(\phi) = \frac{\phi}{\gamma}\frac{\partial^2 f}{\partial\phi^2}. \tag{13.1.22}$$

Recall that in the spinodal domain, $\partial^2 f/\partial\phi^2 < 0$, which means that the effective diffusivity is negative and thus solute tends to move from regions of low concentration to regions of high concentration. Consequently, if there exists a small concentration inhomogeneity, then it will grow in time as illustrated in Fig. 13.5(b).

In the early stages of phase separation, deviations from the homogeneous state are small and we can linearize the Cahn–Hilliard equation. That is, setting $\psi(\mathbf{r},t) = \phi(\mathbf{r},t) - \phi_0$, we have

$$\frac{\partial\psi}{\partial t} = \gamma^{-1}\phi_0\left[f''(\phi_0)\nabla^2\psi - \alpha(\nabla^2)^2\psi\right]. \tag{13.1.23}$$

This equation can easily be solved using Fourier transforms:

$$\widetilde{\psi}(\mathbf{k},t) = \int e^{i\mathbf{k}\cdot\mathbf{x}}\psi(\mathbf{r},t)d^3\mathbf{r}.$$

We find that

$$\frac{\partial\widetilde{\psi}}{\partial t} = -\gamma^{-1}\phi_0 k^2\left[f''(\phi_0) + \alpha k^2\right]\widetilde{\psi} \equiv \sigma(k)\widetilde{\psi}. \tag{13.1.24}$$

Since $f''(\phi_0) < 0$, we see that $\sigma(k)$ first increases as $k = |\mathbf{k}|$ increases, reaching a maximum at $k_c = \sqrt{|f''(\phi_0)|/2\alpha}$, and then decreases (see Fig. 13.7(a)). Thus, the Fourier components for all $k$ such that $\sigma(k) > 0$ grow with time, with the fast rate of growth at $k = k^*$. Hence, the structure arising from the early stages of spinodal decomposition has a characteristic length $1/k^*$.

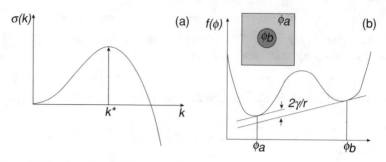

Fig. 13.7: (a) The growth rate $\sigma(k)$ of spinodal decomposition as a function of the wavenumber $k = |\mathbf{k}|$. (b) Effect of interfacial tension on a droplet.

### 13.1.3 Droplet growth and coarsening–Ostwald ripening

As phase separation proceeds, the growth of clusters ceases when $\phi(\mathbf{x},t)$ reaches $\phi_a$ or $\phi_b$, resulting in the separation of the solution into domains of low and high solute concentrations. Although $\phi(\mathbf{x},t)$ no longer changes within each domain, the size and shape of the domains evolve due to the effects of interfacial tension $\tau_s$. If the characteristic size of the domains is $L$, then the interfacial energy per unit volume is $\tau_s/L$. Structural changes of the domains can thus reduce this contribution to the free energy by effectively increasing the average size $L$, in a process known as coarsening [251]. Diffusion also plays a role in coarsening because the concentrations $\phi_a$ and $\phi_b$ in a neighborhood of a droplet deviate slightly from their values based on the common tangent construction. This is illustrated in the inset of Fig. 13.7(b), which shows a droplet of B-rich solution with solute concentration $\phi_b$ and radius $r$ surrounded by a sea of the A-rich solution with solute concentration $\phi_a$. The total free energy of the system is

$$F(r,\phi_a,\phi_b) = \frac{4\pi r^3}{3} f(\phi_b) + \left(V - \frac{4\pi r^3}{3}\right) f(\phi_a) + 4\pi r^2 \tau_s. \qquad (13.1.25)$$

Thermodynamic equilibrium is obtained by minimizing the free energy under the constraint

$$G(r,\phi_a,\phi_b) \equiv \frac{4\pi r^3}{3} \phi_b + \left(V - \frac{4\pi r^3}{3}\right) \phi_a = V_A, \qquad (13.1.26)$$

where $V_A$ is the total volume of solute. In order to carry out the constrained minimization scheme, it is necessary to introduce a Lagrange multiplier $\lambda$ and minimize the modified free energy $A = F + \lambda G$ with respect to $\lambda, r, \phi_a, \phi_b$. This yields the system of equations

$$\frac{\partial A}{\partial \lambda} = G(r, \phi_a, \phi_b) = 0,$$

$$\frac{\partial A}{\partial r} = 4\pi r^2 (f(\phi_b) - f(\phi_a) + 2\tau_s/r) + 4\pi r^2 \lambda (\phi_b - \phi_a) = 0,$$

$$\frac{\partial A}{\partial \phi_a} = -\frac{4\pi r^3}{3}(f'(\phi_a) + \lambda) = 0,$$

$$\frac{\partial A}{\partial \phi_b} = \frac{4\pi r^3}{3}(f'(\phi_b) + \lambda) = 0.$$

It immediately follows that

$$f'(\phi_b) = f'(\phi_a) = \frac{1}{\phi_b - \phi_a}\left[f(\phi_b) - f(\phi_a) + \frac{2\tau_s}{r}\right]. \qquad (13.1.27)$$

In the absence of surface tension ($\tau_s = 0$), this reduces to the common tangent construction. On the other hand, if $\tau_s > 0$ then the equilibrium concentration outside of the droplet changes according to the so-called Gibbs–Thomson law

$$\delta\phi_a = \frac{1}{(\phi_b - \phi_a)f''(\phi_a)}\frac{2\tau_s}{r}. \qquad (13.1.28)$$

As $f''(\phi_a) > 0$, it follows that $\delta\phi_a > 0$ and, hence, the concentration is higher than the one based on the common tangent construction. (There is also a shift in the concentration $\phi_b$ of the dense phase, but we will assume that this is negligible compared to $\phi_b$.)

Now suppose that there exist two droplets $\Sigma_1$ and $\Sigma_2$ with radii $r = R_1, R_2$ such that $R_2 < R_1$. From the above analysis, the concentration outside the smaller droplet will be higher than the concentration outside the larger droplet. Therefore, there will be a diffusive flux of solute from $\Sigma_2$ to $\Sigma_1$, resulting in the growth of $\Sigma_1$ at the expense of $\Sigma_2$. The same mechanism holds for multiple droplets, and results in a coarsening of the system known as Ostwald ripening [251]. The first quantitative formulation of this problem was developed by Lifshitz and Slyozov [561] and Wagner [912], and is commonly referred to as classical LSW theory. These authors derived an equation for the number density of droplets in the dilute regime (total volume fraction of droplets is small), under the crucial assumption that the interaction between droplets can be expressed solely through a common mean field. We now describe the classical formulation of LSW theory.

Consider some macroscopic domain $\Omega$ containing a collection of $N$ microscopic droplets that are well separated from each other and whose total volume fraction is small. A no-flux boundary condition on $\partial\Omega$ ensures mass conservation. Represent each droplet as a sphere of radius $R_i$ centered about $\mathbf{X}_i$, and assume that the dynamics of the droplet radii is much slower than the equilibration of the concentration profile (quasi-static approximation). The solute concentration $\phi$ exterior to the droplets then satisfies a simplified Mullins–Sekerka model:

$$\nabla^2\phi = 0, \ \mathbf{r} \in \Omega \backslash \cup_{i=1}^{N} \Omega_i; \quad \partial_n\phi = 0, \ \mathbf{r} \in \partial\Omega; \qquad (13.1.29a)$$

$$\phi = \phi_a\left(1+\frac{\ell_c}{R_i}\right) \equiv \widehat{\phi}_a(R_i) \text{ on } \partial\Omega_i, \qquad (13.1.29b)$$

where $\Omega_i = \{\mathbf{r} \in \Omega, |\mathbf{r}-\mathbf{X}_i| \leq R_i\}$. The boundary conditions (13.1.29b) are based on equation (13.1.28), rewritten in the form $\delta\phi_a/\phi_a = \ell_c/R_i$ with $\ell_c \approx 2\tau_s/[\phi_b\phi_a f''(\phi_a)]$ the capillary length constant.

The main approximation of LSW theory is to replace boundary effects and interactions between droplets by a mean field $\phi_\infty$ such that $\phi(\mathbf{x}) \approx \phi_\infty$ for $|\mathbf{r}-\mathbf{X}_i| \gg R_i$, $i = 1,\ldots,N$, and $\phi_\infty$ a constant to leading order. The quantity $\Delta = \phi_\infty - \phi_a$ is known as the supersaturation, and needs to be determined self-consistently from mass conservation. Hence, it will depend on the concentration of the original homogeneous solution and the sizes of the droplets. Let us now consider a specific droplet of radius $R$. Given the above assumptions, we can take the concentration around the droplet to satisfy the radially symmetric diffusion equation

$$0 = \frac{D}{r^2}\frac{\partial}{\partial r}r^2\frac{\partial\phi}{\partial r}, \quad r > R, \quad \phi(R) = \widehat{\phi}_a(R), \quad \phi(r) \to \phi_\infty \text{ as } r \to \infty. \quad (13.1.30)$$

The solution of the diffusion equation is given by

$$\phi(r) = \phi_\infty - \frac{R}{r}\left(\Delta - \frac{\phi_a\ell_c}{R}\right). \qquad (13.1.31)$$

The corresponding diffusive flux of solute molecules entering the droplet at its interface is

$$J_R = D\nabla\phi(R) = \frac{D}{R}\left(\Delta - \frac{\phi_a\ell_c}{R}\right). \qquad (13.1.32)$$

It follows that there exists a critical radius $R_c = \phi_a\ell_c/\Delta$ such that $J_R > 0$ when $R > R_c$ and the droplet grows due to a positive influx of solute molecules, see Fig. 13.8(a). On the other hand, small droplets with $R < R_c$ shrink as $J_R < 0$.

We can also write down a dynamical equation for the rate of change of the size of the droplet. When the radius increases by an amount $dR$, the volume increases by $dV = 4\pi R^2 dR$. Given that the expansion of the droplet involves the conversion of solute molecules from a low concentration $\phi(R)$ to a high concentration $\phi_b$, it follows that the number of molecules required to enlarge the droplet by an amount $dR$ is $(\phi_b - \phi(R))dV$. These molecules are supplied by the flux at the interface. Hence, assuming that the change in radius occurs over an infinitesimal time $dt$, we have

$$4\pi R^2(\phi_b - \phi(R))dR = 4\pi R^2 J_R dt,$$

which yields

$$\frac{dR}{dt} = \frac{D}{R(\phi_b - \phi(R))}\left(\Delta - \frac{\phi_a\ell_c}{R}\right) \equiv \frac{\Gamma}{R}\left(\frac{1}{R_c} - \frac{1}{R}\right), \qquad (13.1.33)$$

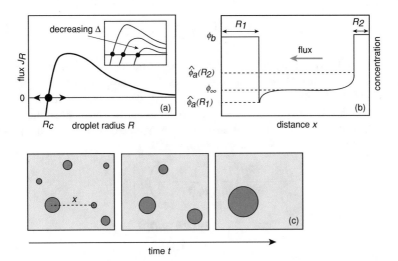

Fig. 13.8: Ostwald ripening. (a) Flux $J_R$ at the interface of a droplet as a function of its radius $R$. If $R > R_c$ then the droplet grows ($J_R > 0$), whereas the droplet shrinks when $R < R_c$ ($J_R < 0$). (b) Schematic diagram showing the concentration profile as a function of $x$ along the axis joining the centers of two well separated droplets with different radii $R_1 > R_2$. The solute concentration around the larger droplet is lower, resulting in a net diffusive flux from the small droplet to the large droplet. (c) A multi-droplet system is unstable with respect to Ostwald ripening, with large droplets growing at the expense of small droplets. As small droplets disappear, the supersaturation $\Delta$ decreases resulting in an increase in the critical radius $R_c$, see the inset of (a). This causes larger droplets to dissolve as well so that eventually only a single high concentration droplet remains. [Redrawn from Lee and Wurtz [550].]

where

$$\Gamma = \frac{D\phi_a \ell_c}{(\phi_b - \phi(R))} \approx \frac{D\phi_a \ell_c}{\phi_b}.$$

One gap in the above formulation of Ostwald ripening is how one determines the mean field $\phi_\infty$. In classical LSW theory, it is taken to be a constant in space and for each time $t$ is determined by the constraint that the volume fraction of droplets is conserved. Assuming that there are $N$ droplets at time $t$, we have

$$\frac{dR_i}{dt} = \frac{\Gamma}{R_i^2}\left(\frac{R_i}{R_c} - 1\right) \tag{13.1.34}$$

for $i = 1, \ldots, N$. Multiplying both sides by $R_i^2$, summing over $i$, and imposing conservation of the volume fraction, $4\pi \sum_i R_i^3(t)/3 = \text{constant}$, gives

$$R_c := \frac{\phi_a \ell_c}{\phi_\infty - \phi_a} = \frac{1}{N}\sum_{i=1}^{N} R_i(t).$$

After rearranging this equation we see see that

$$\phi_\infty(t) = \phi_a \left( 1 + \frac{\ell_c N}{\sum_{i=1}^{N} R_i(t)} \right). \tag{13.1.35}$$

Equation (13.1.35) implies that $\phi_\infty(t)$ decreases as the mean radius increases. The latter will occur due to the disappearance of small droplets at the expense of larger droplets. As the saturation $\Delta(t) = \phi_\infty(t) - \phi_a$ decreases the critical radius $R_c$ increases so that, ultimately, only a single droplet remains. This is illustrated in Fig. 13.8(b,c). In the case of a large number of droplets, one can introduce the number density $f(R,t)$, which satisfies the evolution equation

$$\frac{\partial f}{\partial t} + \frac{\partial}{\partial R} \left[ \frac{\Gamma}{R} \left( \frac{1}{R_c(t)} - \frac{1}{R} \right) f \right] = 0, \tag{13.1.36}$$

with

$$R_c(t) = \frac{\int f(R,t) R dR}{\int f(R,t) dR}. \tag{13.1.37}$$

It is easy to check that the evolution equation (13.1.36) has a scale invariance $R \sim t^{1/3}$. In fact it can be shown that there is a one parameter family of solutions of the form $f(R,t) = t^{-4/3} F_a(R/t)$, with $\phi_\infty = (at)^{1/3}$ and $a \in (0, 4/9]$. Moreover, all of the profiles $F_a$ have compact support, one is smooth, and the others exhibit power law behavior at the end of their support [349, 681]. LSW theory predicts that only the smooth self-similar solution is stable, and is the unique solution after a transient phase, which is independent of the initial data. However, more rigorous mathematical studies have subsequently demonstrated that the long-time behavior of the LSW model does not exhibit universal behavior, but depends sensitively on the initial data [680, 682, 683, 685]. Hence, it is necessary to go beyond mean-field theory in order to take into account higher-order effects such as screening induced fluctuations and droplet collisions.

Finally, note that one can also apply LSW theory to circular droplets in two-dimensional systems [4]. Now, however, the concentration around a droplet varies as $\ln r$ rather than $r^{-1}$, where $r$ is the distance from the center of the droplet. Thus, more care must be taken in imposing far-field conditions. One approach is to partition the region outside the droplets into a set of inner regions around each droplet together with an outer region where mean-field interactions occur. Asymptotic methods developed by Ward and collaborators [208, 729, 921] can then be used to match the inner and outer solutions along analogous lines to studies of narrow escape problems and diffusion in domains with small traps (Chap. 6), see Ref. [492]. The basic steps are outlined in Ex. 13.5.

## 13.2 Biological condensates and active processes

Membrane-less subcellular structures (biological condensates) are ubiquitous in both the cytoplasm and nucleus of cells. Examples include stress granules and processing (P) bodies in the cytoplasm, and nucleoli, Cajal bodies, and PM bodies in the nucleus. All of these structures consist of enhanced concentrations of various proteins and RNA, and proteins are continually exchanged with the surrounding medium. Major insights into the nature of biological condensates have been obtained from studies of P granules in germ cells of *C. elegans*. P granules are RNA/protein-rich bodies located in the cytosplamic region around the nucleus (perinuclear region), which play a role in asymmetric cell division. Their relatively large size (diameters of 2-4 $\mu$m) make them particularly amenable to quantitative analysis [98, 267, 549]. In particular, it has been shown that P granules fuse with one another and subsequently relax back into a spherical shape, flow freely under shear forces, and deform around surfaces of other structures. Moreover, photobleaching experiments have demonstrated that proteins are highly mobile within P granules and exchange rapidly with the surrounding cytoplasm. Taken together with subsequent studies of many other condensates, there is a growing body of evidence supporting the hypothesis that membrane-less organelles are multicomponent, viscous liquid-like structures that form via some form of liquid–liquid phase separation (see the reviews [42, 70, 100, 293, 451] and references therein). The onset of phase separation can be regulated by a number of factors: changes in protein/RNA concentration via gene expression, post-translational modifications in protein structure, and changes to salt/proton concentration and/or temperature (osmotic or pH shocks) [267].

Although intracellular biological condensates are multicomponent structures, typically containing dozens of different types of proteins and RNA, it is possible to reconstitute *in vitro* droplets that have similar features using only one or two molecular components [43, 559]. This suggests that, at least in some cases, a single protein may be necessary and sufficient to drive assembly. Some of the crucial properties of such proteins that facilitate phase separation have been identified. First, it has been found that many of the components of biological condensates contain repeats of weakly binding interaction domains. These may bind to either RNA or complementary binding partners on other proteins, and appear to be an important driver of phase separation. Moreover, since many intracellular signaling molecules exhibit such repetitive interaction domains, it is likely that the role of 'multivalent' motifs in phase separation is also important for organizing intracellular signaling networks. A second, related feature of proteins that facilitates phase separation is the existence of significant conformational heterogeneity—so-called intrinsically disordered proteins (IDPs) [267, 897, 954, 982]. More specifically, the amino acid sequences of IDPs encode an intrinsic preference for conformational disorder, and an inability to fold into stable, well-defined 3D structures under physiological conditions. This is due to the fact that IDPs have regions with weaker interchain interactions, resulting in multiple configurations separated by low energy barriers.

A major feature of biological cells is that they are driven away from equilibrium by multiple energy-consuming processes, including ATP-driven protein phosphorylation. There is growing experimental evidence that active processes also influence the phase separation of biological condensates [70, 550]. For example, various ATP-dependent disaggregases (molecules that break up molecular aggregates) and molecular motors are present in many RNA granules, and are thus in a position to control the degree of polymerization and cross-linking within condensates [466]. Indeed, depletion of ATP increases the viscosity of stress granules and nucleoli [466]. Another example is the regulation of the size distribution of nucleoli by the actin cytoskeleton, the dynamics of which is itself controlled by ATP hydrolysis [302]. The study of energy-consuming, nonequilibrium materials, also known as 'active matter,' has become a major area of interest in both the biological and physical sciences [99, 172, 389, 733, 932, 933, 990, 991].

### 13.2.1 Active regulation of biological phase separation

Classical coarsening theory predicts that only a single droplet remains following Ostwald ripening, see Sect. 13.1.3. However, in both the nucleus and cytoplasm there coexist several membrane-less organelles of the same basic composition, suggesting that there is some mechanism for suppressing Ostwald ripening. One potential candidate is the active regulation of liquid–liquid phase separation by ATP-driven enzymatic reactions that switch proteins between different conformational states (e.g., different levels of phosphorylation) [550, 958, 991]. Such a scheme has also been proposed as a mechanism for localized phase separation in *C. elegans* [549, 932] and the organization of the centrosomes prior to cell division [990]. Here we describe the suppression of Ostwald ripening, following the formulation of [550].

Consider a ternary mixture consisting of two molecular states, one phase separating ($P$) and the other soluble ($S$), together with the solvent or cytosol ($C$). It is assumed that switching between the states $P$ and $S$ occurs according to the chemical reactions

$$P \underset{h}{\overset{k}{\rightleftharpoons}} S,$$

where $k$ and $h$ are concentration-independent reaction rates. The latter reflects the nonequilibriium nature of the chemical reactions, in which detailed balance does not hold due to the action of ATP, say. (The breaking of detailed balance by ATP also plays a crucial role in allowing molecular motors to do useful work, see Chap. 4). Note that there is experimental evidence that the phase separation of intrinsically disordered proteins depends on their phosphorylation state [559].

The analysis of the model proceeds along similar lines to classical LSW theory. That is, we consider a dilute population of droplets in which the volume fraction of the droplets is small. Since the droplets are well separated, we can focus on a single droplet of radius $R$, say, together with far-field conditions that represent the effects of the other droplets. Denote the concentration of $P$ inside and outside the droplet by $\phi_{in}$ and $\phi_{out}$, respectively. The corresponding concentrations of $S$ are denoted by $\psi_{in}$ and $\psi_{out}$. Under the quasi-static approximation, the concentrations evolve according to the reaction–diffusion equations

$$D\nabla^2 \phi_{\text{in,out}} - k\phi_{\text{in,out}} + h\psi_{\text{in,out}} = 0, \tag{13.2.1a}$$

$$DV^2 \psi_{\text{in,out}} + k\phi_{\text{in,out}} - h\psi_{\text{in,out}} = 0. \tag{13.2.1b}$$

For simplicity, both solute species are assumed to have the same diffusion coefficient $D$. Equation (13.2.1) is supplemented by the following boundary conditions for the $P$ species at the interface:

$$\phi_{\text{in}}(R) = \phi_b, \quad \phi_{\text{out}}(R) = \phi_a\left(1 + \frac{l_c}{R}\right), \quad \psi_{\text{in}}(R) = \psi_{\text{out}}(R) = \Psi(R). \tag{13.2.1c}$$

Note that the $S$ species does not undergo phase separation and is thus continuous at the interface with $\Psi(R)$ determined self-consistently. Under the mean-field approximation for well separated droplets, we also have the far-field conditions

$$\phi_{\text{out}}(r) \to \phi_\infty, \quad \psi_{\text{out}}(r) \to \psi_\infty \text{ as } r \to \infty. \tag{13.2.1d}$$

Since, the concentration profiles are approximately flat in regions far from any droplet, the $P$ and $S$ concentrations will be in thermodynamic equilibrium so that

$$\phi_\infty = \frac{h}{h+k}u_\infty, \quad \psi_\infty = \frac{k}{h+k}u_\infty. \tag{13.2.2}$$

The constant $u_\infty$ can be determined using mass conservation.

Adding equations (13.2.1a,b) gives

$$DV^2(\phi_{\text{in,out}} + \psi_{\text{in,out}}) = 0. \tag{13.2.3}$$

From spherical symmetry, we have

$$\phi_{\text{in,out}}(r) + \psi_{\text{in,out}}(r) = \frac{A_{\text{in,out}}}{r} + B_{\text{in,out}}$$

for constants $A_{\text{in,out}}, B_{\text{in,out}}$. Inside the droplet ($0 \leq r < R$) the solution will diverge unless $A_{\text{in}} = 0$. Following [550], we also assume that $A_{\text{out}} = 0$ outside the droplet ($R < r$); such a condition holds in quasi-equilibrium for a population of identical droplets. Hence,

$$\phi_{\text{in,out}}(r) + \psi_{\text{in,out}}(r) = B_{\text{in,out}}. \tag{13.2.4}$$

Taking the limit $r \to \infty$ shows that $B_{\text{out}} = u_\infty \equiv \phi_\infty + \psi_\infty$. Moreover, setting $r = R$ gives

$$B_{\text{in}} = \phi_b + \Psi(R), \quad B_{\text{out}} = \phi_a\left(1 + \frac{l_c}{R}\right) + \Psi(R). \tag{13.2.5}$$

Hence, the unknown $\Psi(R)$ is determined by $u_\infty$.

It follows from the spatial uniformity of the total concentration inside and outside the droplet that equations (13.2.1a,b) decouple. In particular, the $P$ concentration satisfies

$$0 = D\nabla^2 \phi_{\text{in}} - (k+h)\phi_{\text{in}} + hB_{\text{in}}, \quad 0 \le r < R, \quad \phi_{\text{in}}(R) = \phi_b, \tag{13.2.6a}$$

and

$$0 = D\nabla^2 \phi_{\text{out}} - (k+h)\phi_{\text{out}} + hB_{\text{out}}, \quad r > R, \tag{13.2.6b}$$

with

$$\phi_{\text{out}}(R) = \phi_a \left(1 + \frac{l_c}{R}\right), \quad \phi_{\text{out}}(r) \to \phi_\infty \text{ as } r \to \infty. \tag{13.2.6c}$$

In order to solve equation (13.2.6a), set $u_{\text{in}} = r\phi_{\text{in}}$, and note that $\nabla^2 \phi = r^{-1} d^2 u/dr^2$. We thus obtain the solution (see Ex. 13.6)

$$\phi_{\text{in}}(r) = \left(\phi_b - \frac{hB_{\text{in}}}{h+k}\right) \frac{R}{r} \frac{\sinh(r/\xi)}{\sinh(R/\xi)} + \frac{hB_{\text{in}}}{h+k}, \quad \xi = \sqrt{\frac{D}{h+k}}, \tag{13.2.7}$$

where $\xi$ is a length constant that sets the scale of concentration gradients. The sinh function is required for the inner solution so that it is non-singular at $r = 0$. Similarly, the solution of (13.2.6b,c) is

$$\phi_{\text{out}}(r) = \phi_\infty - \frac{R}{r} e^{-(r-R)/\xi} \left(\Delta - \frac{\phi_a l_c}{R}\right), \quad \Delta = \phi_\infty - \phi_a. \tag{13.2.8}$$

We have used $hB_{\text{out}}/(h+k) = \phi_\infty$, and have dropped a term involving $r^{-1}e^{r/\xi}$ so that $\phi_{\text{out}}$ remains bounded as $r \to \infty$. Finally, the unknown mean field $\phi_\infty$ has to be determined self-consistently from mass conservation.

    The growth or shrinkage of the droplet will depend on the circulation of $P$ and $S$ fluxes between the droplet and the cytoplasm. Inside drops the concentration of $P$ is high so that the chemical reaction $P \to S$ dominates. There is thus an outward flux of $S$ molecules across the interface. Similarly, the reaction $S \to P$ dominates in the cytoplasm, resulting in an inward flux of $P$ molecules at the interface. The net flux of $P$ molecules into the interface at $r = R$ is thus

$$J_R = D\frac{d\phi_{\text{out}}}{dr}\bigg|_{r=R^+} - D\frac{d\phi_{\text{in}}}{dr}\bigg|_{r=R^-}. \tag{13.2.9}$$

In order to proceed further, we consider two limiting regimes [550, 958].

**Small droplet regime ($R \ll \xi$).** Suppose that the droplets have radii that are much smaller than the gradient length scale $\xi$. Also assume that $\xi$ is much larger than the typical droplet separation. These conditions will hold if the reaction rates $h, k$ are sufficiently small. Inside a droplet, the concentration is approximately a constant since $\sinh(r/\xi) \approx r/\xi$, that is, $\phi_{\text{in}}(r) \approx \phi_b$. Similarly, outside a droplet, the concentration profile is well approximated by classical LSW theory:

$$\phi_{\text{out}}(r) \approx \phi_\infty - \frac{R}{r}\left(\Delta - \frac{\phi_a l_c}{R}\right). \tag{13.2.10}$$

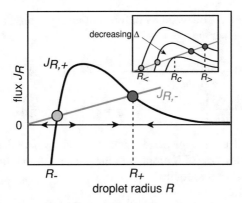

Fig. 13.9: Fluxes in small droplet regime. The net flux at a given radius $R$ and supersaturation $\Delta$ is a combination of an influx $J_{R,+}$, which is almost identical to the classical case (see Fig. 13.8(a)), and an efflux $J_{R,-}$ (straight line) that is driven by nonequilibrium chemical reactions. For a range of values of $\Delta$, there exist two steady-state radii $R_{\pm}$ for which the fluxes cancel, with $R_+$ stable and $R_-$ unstable. The inset shows the effect of varying $\Delta$ with $R_<$ and $R_>$ the smallest unstable and largest stable states, respectively. The fixed points vanish at a critical radius $R_c$ and supersaturation $\Delta_c$. [Redrawn from Lee and Wurtz [550].]

The net flux $J_R$ into the droplet is now a combination of an influx $J_{R,+}$, which is driven by the supersaturation $\Delta$ along identical lines to LSW theory, and an efflux $J_{R,-}$, which consists of the excess $S$ molecules within the droplet rapidly diffusing into the cytoplasm. This excess arises from degradation of $P$ molecules within the droplet. Hence,

$$J_{R,+} = \frac{D}{R}\left(\Delta - \frac{\phi_a \ell_c}{R}\right), \quad J_{R,-} = k\phi_b \frac{V(R)}{A(R)} = \frac{k\phi_b R}{3}, \tag{13.2.11}$$

where $V(R)$ and $A(R)$ are the volume and surface area of the droplet, respectively. In Fig. 13.9 we show how the fluxes $J_{R,\pm}$ vary with droplet radius $R$ for a fixed supersaturation $\Delta$. For this particular example, the curves of $J_{R,\pm}$ have two points of intersection $R_{\pm}$, which correspond to a pair of steady-state solutions. The state $R_-$ is unstable against Ostwald ripening so that smaller drops dissolve and large drops grow. However, in contrast to classical LSW theory, the presence of nonequilibrium chemical reactions means that there is also a stable steady-state $R_+$, around which smaller droplets grow and larger droplets shrink.

**Large droplet radius ($R \gg \xi$).** Now suppose that droplets are large compared to the gradient length scale $\xi$. In this case, the center of the droplet is far from the interface, and is thus in equilibrium, see Fig. 13.10. Indeed, from equation (13.2.7)

$$\phi_{\text{in}}(0) = \left(\phi_b - \frac{hB_{\text{in}}}{h+k}\right)\frac{R}{\xi}\frac{1}{\sinh(R/\xi)} + \frac{hB_{\text{in}}}{h+k} \approx \frac{hB_{\text{in}}}{h+k}, \quad \psi_{\text{in}}(0) \approx \frac{kB_{\text{in}}}{h+k}. \tag{13.2.12}$$

The influx and efflux are obtained by differentiating equations (13.2.8) and (13.2.7) with respect to $r$ and setting $r = R$:

$$J_{R,+} = D\frac{d\phi_{\text{out}}}{dr}\bigg|_{r=R} = \frac{D(\phi_{\infty} - \phi_{\text{out}}(R))}{\xi}\left(1 + \frac{\xi}{R}\right),$$  (13.2.13a)

$$J_{R,-} = D\frac{d\phi_{\text{in}}}{dr}\bigg|_{r=R} = \frac{D(\phi_{\text{in}}(R) - \phi_{\text{in}}(0))}{\xi}\left(1 - \frac{\xi}{R}\right).$$  (13.2.13b)

We would like to determine the steady-state radius $R^*$, where these two fluxes balance each other, in the case of $N$ droplets of the same radius $R^*$. In particular, setting $J_{R,+} = J_{R,-}$ for $\xi \ll R$, we require

$$\phi_{\text{in}}(R) - \phi_{\text{in}}(0) \approx \phi_{\infty} - \phi_{\text{out}}(R).$$  (13.2.14)

Finally, solute mass conservation implies that

$$\frac{4\pi R^3}{3}N\phi_{\text{in}}(0) + \left(V - \frac{4\pi R^3}{3}N\right)\phi_{\infty} = V\phi_{\text{tot}},$$  (13.2.15)

where $V$ is the total volume of the system and $\phi_{\text{tot}} = hu_{\text{tot}}/(h+k)$ is the global concentration of the $P$ molecules. For simplicity, the concentration gradients near the interface have been neglected since the interfacial region is small compared to the size of the droplets ($\xi \ll R$). The constraints (13.2.5), (13.2.12), (13.2.14), and (13.2.15) can now be used to determine the steady-state radius $R^*$ [550, 958]:

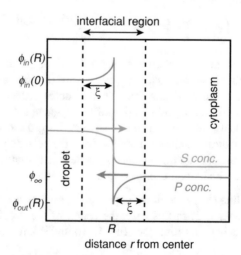

Fig. 13.10: Fluxes in large droplet regime. The concentration profiles of the $P$ and $S$ molecules are flat outside an interfacial region of half-width $\xi$. There is an influx of $P$ molecules into the droplet and an efflux of $S$ molecules out of the droplet. [Redrawn from Lee and Wurtz [550].]

$$\frac{4\pi R^{*3}}{3}N = \left(\frac{u_{\text{tot}} - \phi_a}{\phi_b} - \frac{k}{2h}\right)V. \tag{13.2.16}$$

In contrast to the small droplet regime, the steady-state $R^*$ scales with the system size $V$. It also immediately follows that there exists a maximal reaction rate $k_{\text{max}}$ above which drops dissolve ($R^* = 0$),

$$k_{\text{max}} = \frac{2(u_{\text{tot}} - \phi_a)}{\phi_b}h. \tag{13.2.17}$$

**Linear stability analysis** As previously highlighted, a crucial assumption in the above analysis of droplets is that the total concentration outside each droplet is spatially uniform ($A_{\text{out}} = 0$). This then determines the unknown S-concentration $\Psi^* = \Psi(R^*)$ at the interface according to equation (13.2.5), where $R^*$ is the steady-state radius. It also ensures continuity of the S-flux across each droplet interface. Assuming that perturbations of the steady state maintain the condition $A_{\text{out}} = 0$, one can investigate the linear stability of the steady state by considering perturbations of the form [550, 958]

$$R = R^* + \delta R, \quad \Psi = \Psi^* + \delta \Psi \text{ with } \delta \Psi = \Psi'(R^*)\delta R = \frac{\phi_a}{R^{*2}}\delta R.$$

It follows that the change in the total flux across the interface is

$$\delta J = \delta J_+ - \delta J_- = \left\{\frac{\phi_a}{R^{*2}}\left[\frac{\partial J_+}{\partial \Psi^*} - \frac{\partial J_-}{\partial \Psi^*}\right] + \left[\frac{\partial J_+}{\partial R^*} - \frac{\partial J_-}{\partial R^*}\right]\right\}\delta R. \tag{13.2.18}$$

Requiring that $\delta J$ and $\delta R$ have opposite signs at the interface then yields the stability condition

$$\frac{\phi_a}{R^{*2}}\left[\frac{\partial J_+}{\partial \Psi^*} - \frac{\partial J_-}{\partial \Psi^*}\right] + \left[\frac{\partial J_+}{\partial R^*} - \frac{\partial J_-}{\partial R^*}\right] < 0. \tag{13.2.19}$$

This can now be used to investigate the stability of the droplet state in large droplet regime, which yields the condition $k > k_c$, where [550, 958]

$$k_c(\xi \phi_b - \phi_a \ell_c) = 2h\phi_a \ell_c.$$

Substituting for $\xi$ and using $\xi \phi_b \gg \ell_c \phi_a$, $k_c \ll h$ gives

$$k_c = \frac{2\ell_c \phi_a}{\sqrt{D}\phi_b}h^{3/2}. \tag{13.2.20}$$

It is important to note that at the population level, the perturbations generate a system of non-identical droplets for which the condition $A_{\text{out}} = 0$ breaks down. In particular, this means that equation (13.2.5) can no longer be used to determine $\Psi(R)$ and hence $\delta \Psi$. Instead, one has to determine $\delta \Psi$ by imposing continuity of the S-flux at the interface [128]. To leading order, this still generates the stability condition (13.2.19), but there are higher-order corrections that involve interactions between the droplets.

## 13.2.2 Phase separation in a concentration gradient

Recall from Sect. 11.1.4 that protein concentration gradients play an important role during asymmetric cell division of the *C. elegans* zygote. Within the context of biological condensates, it has been found that RNA-protein aggregates in the form of P-granules are segregated to the posterior side of the cell, and are located in the posterior daughter cell after division. Moreover, the segregation and ripening of P-granule droplets is driven by the concentration gradient of the protein Mex-5 [98, 549, 785]. Here we describe a theoretical study of droplet ripening in a concentration gradient of a protein that regulates phase separation, which has been presented by Weber et al [932, 933].

Let $R$ denote the regulatory protein and $P$ denote the solute that undergoes phase separation in a solvent $S$. The basic regulatory mechanism is taken to be the binding of $R$ to $P$, which forms a complex $C$ that cannot phase separate:

$$R + P \underset{h}{\overset{k}{\rightleftharpoons}} C. \tag{13.2.21}$$

This has the effect of reducing the volume fraction of solute molecules that can participate in phase separation. In order to explore the consequences of this in a more quantitative fashion, consider the lattice model of Sect. 13.1.1. For simplicity, we

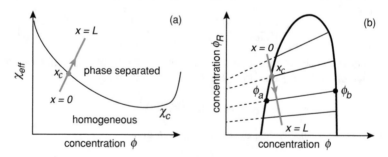

Fig. 13.11: Effect of a linearly decreasing regulator concentration $\phi_R(x)$ on phase separation, $x \in [0, L]$. (a) Sketch of phase transition curve $\chi_{\text{eff}} = \chi_c$ for phase separation as a function of $\phi$. The variation of $\chi_{\text{eff}}$ with $x$ traces a straight line in the phase diagram that crosses the critical curve at position $x_c$. (b) Coexistence curves in the $(\phi, \phi_R)$-plane for total solute concentration $\phi$ and regulator concentration $\phi_R$. The thin straight lines link points within the phase separation region to the corresponding low and high density phases $\phi_{a,b}$ on the coexistence curve. As the position $x$ varies, there is a corresponding change in $\phi_R$, which places the system at different points in the phase diagram (red curve). Droplets can only form at points that lie within the coexistence curve. The position $x_c$ marks the boundary between droplet formation and dissolution.

ignore the mixing entropy of the complex $C$ and only consider interaction energies between $P$ and $S$. It follows that the main effect of reducing $\phi_P$ is on the interaction parameter $\chi$ appearing in the free energy (13.1.7). More specifically, the interaction energy term is of the form $E = k_B T \chi \phi_P \phi_S$ with $\phi_P + \phi_C = \phi$ and $\phi$ the total volume fraction of solute. Here $\phi_S = 1 - \phi$ denotes the volume fraction of solvent and $\phi_C$ the volume fraction of complexes $C$. Assuming local equilibrium of the fast binding reactions, and taking the molecular volumes of $C$ and $P$ to be the same, we have $k\phi\phi_R/v_R = h\phi_C$, where $v_R$ is the molecular volume of $R$ molecules and $\phi_R$ is the corresponding volume fraction. Hence,

$$\phi = \frac{\Phi}{1 + K\phi_R}, \quad K = \frac{k}{h v_R},$$

and the interaction energy term can be expressed as

$$E = k_B T \chi_{\text{eff}} \phi (1 - \phi), \quad \chi_{\text{eff}} = \chi \left( 1 - \frac{K\phi_R}{1 + K\phi_R} \right). \tag{13.2.22}$$

It can be seen that increasing the regulator concentration $\phi_R$ leads to a decrease in the effective interaction parameter $\chi_{\text{eff}}$. If one now plots the coexistence curve and spinodal line in the $\phi$-$\chi_{\text{eff}}$ plane, analogous to Fig. 13.2, it follows that changing $\phi_R$ moves the system to different points in the phase diagram, which is how the regulator protein can control phase separation. In particular, the equilibrium concentrations $\phi_b$ and $\phi_a$ of the high and low-density phases will depend on $\phi_R$. This is illustrated in Fig. 13.11.

Now suppose that there exists a spatially varying concentration gradient $\phi_R(\mathbf{r})$ that, for simplicity, varies linearly in the $x$-direction, $\phi_R(\mathbf{r}) = \phi_R(0) - Ax$. (The gradient is assumed to be sufficiently shallow that changes in $\phi_R(\mathbf{r})$ are only significant on length scales comparable to the system size, which is denoted by $L$.) It immediately follows that $\chi_{\text{eff}}$ and $\phi_{a,b}$ become $x$-dependent. In order to explore the effects of a gradient on the ripening dynamics of droplets, it is necessary to extend the notion of a common far-field concentration $\phi_\infty$ to a space-dependent field $\phi_\infty(x)$. Recall from classical LSW theory that, in the absence of a regulator, the far-field represents the collective effects of surrounding droplets, under the assumption that the mean separation $d$ between droplets satisfies $d \gg R$, where $R$ is the mean droplet radius. Under the separation of length scales $R \ll d \ll L$, we can partition the system into local regions with a size corresponding to the intermediate length scale $d$. Droplets in a local domain around a spatial position $x$ will have coexisting equilibrium concentrations $\phi_a(x), \phi_b(x)$ provided that the point $(\phi, \phi_R(x))$ is located inside the phase separation region of the associated phase diagram, see Fig. 13.11. Finally, if the common far-field concentration within the local region is given by $\phi_\infty(x)$, then one can define a spatially dependent supersaturation $\Delta(x) = \phi_\infty(x) - \phi_a(x)$ and use this to analyze the dynamics of droplets by modifying classical LSW theory [932, 933].

**Dynamics of a single droplet.** Consider a single droplet of radius $R$ and center at $\mathbf{r}_0$. Introduce spherical polar coordinates $\mathbf{r} = (r, \theta, \varphi)$ with $r$ denoting radial distance from the center and $\theta, \varphi$ the azimuthal and polar angles relative to the $x$-axis. The local regulator concentration can then be written as $\phi_R(\mathbf{r}) = \phi_R(\mathbf{r}_0) - Br\cos\theta$. For simplicity the high density concentration $\phi_b$ will be taken to be $\mathbf{r}$-independent. The low density equilibrium concentration $\phi_a(\mathbf{r})$ and far-field concentration $\phi_\infty(\mathbf{r})$ have the leading order expansions

$$\phi_a(\mathbf{r}) \approx \alpha + \beta r\cos\theta, \quad \phi_\infty(\mathbf{r}) \approx \alpha_\infty + \beta_\infty r\cos\theta, \tag{13.2.23}$$

where $\alpha = \phi_a(\mathbf{r}_0), \beta = \nabla\phi_a(\mathbf{r}_0) \cdot \mathbf{e}_x$ and similarly for $\alpha_\infty, \beta_\infty$. The free solute concentration around the droplet satisfies the diffusion equation

$$\nabla^2\phi = 0, \quad R < r < \infty, \quad \phi(R) = (\alpha + \beta R\cos\theta)\left(1 + \frac{\ell_c}{R}\right), \tag{13.2.24}$$

together with the far-field condition

$$\phi(\mathbf{r}) = \alpha_\infty + \beta_\infty r\cos\theta, \quad \text{as } r \to \infty. \tag{13.2.25}$$

In contrast to the classical analysis, one can no longer assume spherically symmetric solutions. Instead, the solution of Laplace's equation is of the form

$$\phi(r, \theta) = \sum_{n=0}^{\infty} (A_n r^n + B_n r^{-n-1}) P_n(\cos\theta),$$

where the $P_n(\cos\theta)$ are Legendre polynomials. On imposing the boundary conditions, one finds that [932, 933] (see Ex. 13.7)

$$\phi(r, \theta) = \alpha_\infty\left(1 - \frac{R}{r}\right) + \beta_\infty\left(r - \frac{R^3}{r^2}\right)\cos\theta + (\alpha r/R + \beta R\cos\theta)\left(1 + \frac{\ell_c}{R}\right)\frac{R^2}{r^2}. \tag{13.2.26}$$

The droplet can grow, drift, or deform due to the normal fluxes of solute at the interface modifying the location of the interface. The local displacement $\delta R(\theta, \varphi, t)$ of the interface is determined by the normal velocity $v_n$, which for a spherical droplet is obtained by matching the influx of solute molecules into a surface patch of area $dA = R^2\sin\theta d\varphi d\theta$ with the local change in volume. Assuming that $\phi_b \gg \phi(R)$, we have

$$\phi_b v_n dA = J(R, \theta)dA,$$

where $J(r, \theta) = D\nabla\phi(r, \theta) \cdot \mathbf{e}_r$. (The flux inside the droplet vanishes as the concentration is approximately constant.) Hence,

$$v_n(\theta) = \frac{D}{\phi_b}\nabla\phi(R, \theta) \cdot \mathbf{e}_r. \tag{13.2.27}$$

We thus have $\delta R(\theta, t) = v_n(\theta)\delta t$. Assuming that the droplet maintains an approximately spherical shape, we can define the change in radius according to

$$\delta R(t) = (4\pi)^{-1} \int_0^\pi \int_0^{2\pi} \delta R(\theta,t) \sin\theta d\varphi d\theta.$$

That is,

$$\frac{dR}{dt} = \frac{D}{2\phi_b} \int_0^\pi \nabla\phi(R,\theta) \cdot \mathbf{e}_r \sin\theta d\theta = \frac{D}{\phi_b R} \left[ \alpha_\infty - \phi_a(\mathbf{r}_0)\left(1 + \frac{\ell_c}{R}\right) \right], \quad (13.2.28)$$

which is identical to the standard formula for growth of a droplet. However, in the presence of a concentration gradient there also exists a non-zero droplet drift speed. The effective change in the $x$-coordinate of the center is

$$\delta x(t) = (4\pi)^{-1} \int_0^\pi \int_0^{2\pi} \delta R(\theta,t) \cos\theta \sin\theta d\varphi d\theta,$$

which yields [933] (see Ex. 13.7)

$$\frac{dx}{dt} = \frac{D}{2\phi_b} \int_0^\pi \nabla\phi(R,\theta) \cdot \mathbf{e}_r \cos\theta \sin\theta d\theta$$

$$= \frac{D}{3\phi_b} \left[ 3\beta_\infty - 2\partial_x \phi_a(\mathbf{r}_0)\left(1 + \frac{\ell_c}{R}\right) \right]. \quad (13.2.29)$$

**Segregation of multiple droplets.** Now consider the dynamics of many droplets, labeled $i = 1,\ldots,N$, with positions $\mathbf{r}_i$ and radii $R_i$. If the droplets are well separated then

$$\frac{dR_i}{dt} = \frac{D}{\phi_b R_i} \left[ \Delta(\mathbf{r}_i) - \phi_a(\mathbf{r}_i)\frac{\ell_c}{R_i} \right], \quad (13.2.30)$$

and

$$\frac{dx_i}{dt} = \frac{D}{3\phi_b} \left[ 3\partial_x \phi_\infty(\mathbf{r}_i) - 2\partial_x \phi_a(\mathbf{r}_i)\left(1 + \frac{\ell_c}{R_i}\right) \right], \quad (13.2.31)$$

where $\Delta(\mathbf{x}) = \phi_\infty(\mathbf{x}) - \phi_a(\mathbf{x})$ is the position-dependent supersaturation. In the presence of a concentration gradient, the far field also evolves in time according to

$$\frac{\partial \phi_\infty}{\partial t} = D\frac{\partial^2 \phi_\infty}{\partial x^2} - \frac{4\pi\phi_b}{V} \sum_{i=1}^N \delta(x - x_i) R_i^2 \frac{dR_i}{dt}. \quad (13.2.32)$$

The second term on the right-hand side represent changes in the bulk concentration due to growth or shrinkage of droplets, under the assumptions that $\phi_b \gg \phi_\infty$ and the volume $V$ is much greater than the total volume occupied by the droplets. Numerical simulations of the above system of equations establishes the existence of a dissolution boundary that separates a region of droplet shrinkage from a region of droplet growth. Consequently, droplets tend to segregate towards the boundary where the supersaturation is highest [932]. This is illustrated in Fig. 13.12.

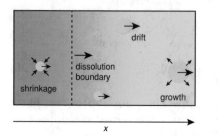

Fig. 13.12: Droplet ripening in a regulator concentration gradient. A dissolution boundary separates a region of droplet shrinkage (left of boundary) from a region of droplet growth. The boundary and droplets drift to the right, resulting in droplet segregation at the right-hand boundary. [Redrawn from Weber et al [932].]

### 13.2.3 Active membranes

We continue our discussion of phase separation by briefly considering an example of spontaneous aggregation of curvature-inducing proteins in membranes. In order to perform a variety of functions such as cell polarization and motility, eukaryotic cells dynamically reshape their membranes by controlling the local membrane lipid composition, the composition of membrane-bound proteins, the activity of membrane channels and pumps, and the recruitment of the cell's cytoskeleton [471, 525, 607, 987]. Given the difficulty of teasing apart the various molecular mechanisms responsible for shaping a cell *in vivo*, a number of *in vitro* studies have been developed that can explore mechanisms by controlling the lipid concentration and the presence of curved membrane proteins [49, 620]. However, it is more challenging to study the active cytoskeletal forces involved in reshaping a membrane even *in vitro* [166, 568]. Here we review some theoretical studies of the coupling between curved membrane components (CMCs), such as lipid domains or membrane-bound proteins, and active elements of the cytoskeleton such as F-actin or molecular motors pulling on membrane-bound filament [369, 370, 902]. Related studies consider active membranes driven by ion pumps rather than cytoskeletal elements [182, 736, 749, 750]. A common  feature of these models is the construction of a free energy functional for the membrane (see Sect. 12.3.5), which incorporates dynamically active components that have the potential to phase separate.

A schematic diagram of the basic model developed in Refs. [369, 902] is shown in Fig. 13.13(a). The membrane is assumed to be approximately flat so that its curvature can be described using the Monge representation (see Sect. 12.2). That is, its height above the $(x,y)$-plane is given by $z = h(x,y)$. (Such a situation would apply to a large vesicle or the leading edge of a motile cell, for example.) The membrane contains CMCs that promote actin polymerization and a protrusive force. The CMCs are free to diffuse laterally in the membrane with diffusivity $D$, and promote a protrusion of spontaneous negative curvature with radius of curvature $r_0$, which couples to the mean curvature of the membrane. One of the main issues of interest in the theory of active membrane is exploring mechanisms whereby the combined CMC/membrane system can undergo an instability leading to the aggregation of CMCs and the formation of regions with enhanced curvature, as illustrated in Fig. 13.13(b).

Let $\phi(\mathbf{r},t)$ denote the local concentration of CMCs in the membrane. The free energy of the composite system is taken to be

$$F[h,\phi] = \int_\Omega \left[ \frac{\kappa_b}{2}\left(\nabla^2 h + \frac{\phi}{r_0}\right)^2 + \frac{\sigma}{2}(\nabla h)^2 + f(\phi) + \frac{k_B T \chi}{2}(\nabla\phi)^2 \right] d^2\mathbf{r},$$

(13.2.33)

with $f(\phi)$ identical in form to equation (13.1.7) except that now the CMCs are solute molecules in an effective 2D domain:

$$f(\phi) = \frac{k_B T}{a^2}\left[ \phi\ln\phi + (1-\phi)\ln(1-\phi) + \chi\phi(1-\phi) \right],$$

(13.2.34)

with $a^2$ the area occupied by a CMC in the membrane. The final term is the contribution to the solute free energy from interfaces. The first two terms in the free energy (13.2.33) correspond to the standard bending and stretching energies of a membrane, see equation (12.3.53), modified to take into account the spontaneous curvature effects of the CMCs. In order to simplify the subsequent analysis, we follow Ref. [902] and assume that the CMC solution is close to the critical point for phase separation so that we can write $\phi = 1/2 + \phi'$ with $\phi'$ small. Then (after dropping the prime on $\phi'$) the free energy density $f$ can be approximated by (see Ex. 13.2)

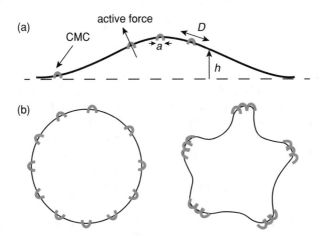

Fig. 13.13: (a) Schematic illustration of a flexible membrane subject to active protrusive forces via diffusing curved membrane complexes (CMCs) of size $a$ and spontaneous curvature $1/r_0$. The membrane is assumed to be approximately flat with height function $z = h(x,y)$ in the Monge representation. (b) The cell membrane may exist in a state with uniform curvature and a uniform distribution of CMCs. Alternatively, an instability can result in the aggregation of CMCs and associated regions of enhanced curvature or protrusions. [Redrawn from Veksler and Gov [902].]

$$f[\phi] = -\frac{b}{2}\phi^2 + \frac{c}{4}\phi^4,$$

where we have dropped constant terms and set

$$b = \frac{2k_B T}{a^2}(\chi - 2), \quad c = \frac{16k_B T}{3a^2}.$$

Note that $\chi \approx 2$ so that $b$ is small.

Following the analysis of passive membrane dynamics in Sect. 12.3.5, in particular equation (12.3.54), the equation of motion for the membrane displacement $h(\mathbf{r},t)$ is

$$\gamma_h \frac{\partial h}{\partial t} = -\frac{\delta F}{\delta h(\mathbf{r},t)} + \Gamma(\phi(\mathbf{r},t) - \langle\phi\rangle), \tag{13.2.35}$$

where $\gamma_h$ is the friction coefficient for membrane fluctuations, which is related to the viscosity $\eta$ of the surrounding fluid. Indeed, taking the Oseen Green's function $\Lambda(\mathbf{r})$ in equation (12.3.61) to have support over a domain of size $a$, one can take $\gamma_h^{-1} = a/8\eta$ [902]. In equation (13.2.35) we are neglecting spontaneous random forces, but include an additional active force, which acts to deform the membrane whenever the local CMC concentration deviates from its mean value $\langle\phi\rangle$. This force is produced by the cytoskeleton and arises from the consumption of energy via ATP hydrolysis, for example. The direction of the active force is determined by the sign of the constant $\Gamma$ with $\Gamma > 0$ indicating a protrusive force. It remains to specify the dynamics of $\phi(\mathbf{r},t)$. This is obtained by following the kinetics of phase separation in Sect. 13.1. In particular, combining equations (13.1.18) and (13.1.19), we have

$$\frac{\partial\phi}{\partial t} = -\nabla\cdot\mathbf{J} = \frac{a^2}{\gamma_c}\nabla\cdot\left(\phi\nabla\frac{\delta F}{\delta\phi}\right). \tag{13.2.36}$$

where $\gamma_c$ is a friction coefficient for the CMCs and the factor $a^2$ ensures that the units of variables and parameters are consistent. (It can be derived from first principles by starting from a lattice model along the lines of Sect. 13.1.1.) Note that the model assumes that the cytoskeletal forces are predominantly normal to the membrane surface and thus do not contribute to the motion of the CMCs.

Substituting for $F$ using equation (13.2.33) and evaluating the various functional derivatives gives the following closed system of equations:

$$\frac{\partial h}{\partial t} = \frac{a}{8\eta}\left\{-\kappa_b\nabla^2[\nabla^2 h + \phi] + \sigma\nabla^2 h + \Gamma(\phi(\mathbf{r},t) - \langle\phi\rangle)\right\}, \tag{13.2.37a}$$

$$\frac{\partial\phi}{\partial t} = \frac{Da^2}{k_B T}\nabla\cdot\left([\phi + 1/2]\nabla\left(\left[\frac{\kappa_b}{r_0^2} - b\right]\phi + c\phi^3 + \frac{\kappa_b}{2r_0^2} + \frac{\kappa_b}{r_0}\nabla^2 h - k_B T\chi\nabla^2\phi\right)\right). \tag{13.2.37b}$$

From equation (13.2.37b), a variety of distinct currents contribute to the distribution of CMCs in the membrane. We focus on the most significant currents identified in Refs. [370, 482]. First,

$$\mathbf{J}_{\text{diff}} = -2D(\chi - 2)\nabla\phi, \quad \mathbf{J}_{\text{disp}} = -\frac{\kappa_b a^2}{\gamma_c r_0^2}\phi\nabla\phi.$$

These both tend to disperse the CMCs. On the other hand, there are two currents that tend to aggregate the CMCs via direct attractive interactions and due to the effects of spontaneous curvature:

$$\mathbf{J}_{\text{agg}} = D\phi\left[2(\chi - 2)\nabla\phi + a^2\chi\nabla\nabla^2\phi\right], \quad \mathbf{J}_{\text{curv}} = \frac{\kappa_b a^2}{\gamma_c r_0}\phi\nabla\nabla^2 h.$$

In order to develop intuition about the opposing effects of aggregation and dispersal, suppose that we focus on the current $\mathbf{J} = \mathbf{J}_{\text{diff}} + \mathbf{J}_{\text{curv}}$. The evolution equation for $\phi$ is then

$$\frac{\partial\phi}{\partial t} = D\nabla \cdot \left(2(\chi - 2)\nabla\phi - \frac{\kappa_b a^2}{k_B T r_0}\phi\nabla\nabla^2 h\right).$$

Given a steady-state membrane curvature with Monge function $h(\mathbf{r})$, the steady-state distribution of CMCs takes the form

$$\phi(\mathbf{r}) \sim \exp\left[\frac{\kappa_b a^2}{2k_B T r_0(\chi - 2)}\nabla^2 h(\mathbf{r})\right]. \tag{13.2.38}$$

This establishes that CMCs tend to accumulate in regions of membrane curvature that match their spontaneous curvature. Such behavior has been observed in bacteria, where curved membrane proteins are found to aggregate near the poles [776].

The question remains whether or not CMCs can undergo spontaneous aggregation and phase separation, resulting in the formation of curved membrane shapes. This can be investigated by carrying out a linear stability analysis of the coupled system given by equations (13.2.37) [482, 902]. Several results emerge from such an analysis. First, in the absence of active forces $f$, the aggregation of CMCs is prevented in the limit of small deformations $h$ due to the entropic costs. However, these can be overcome in the present of direct interactions that generate $\mathbf{J}_{\text{agg}}$. In the presence of active forces, the resulting deformations are greatly amplified resulting a variety of behaviors, including static protrusions, cortical waves, and oscillations [370].

## 13.3 Nucleation and growth of molecular clusters

The analysis of liquid-liquid phase separation from a metastable solution via the nucleation and growth of solute clusters is much harder to analyze than spinodal decomposition. Understanding nucleation and growth processes is not only of interest within the context of biological condensates such as the nucleolus, but is also relevant to the self assembly of other cellular structures, including lipid bilayers, spherical and cylindrical vesicles, and protein aggregates such as amyloid fibrils.

Classical nucleation theory was originally developed within the context of physical processes such as liquid condensation in vapors [256, 701]. The main assumptions of the classical approach are that (i) the system is below the temperature of the gas/liquid phase transition and (ii) the density of liquid droplets (clusters or nuclei of vapor molecules) is dilute. One can then subdivide the system into an ensemble of nuclei that are separated from the pure vapor by distinct interfacial boundaries with associated macroscopic surface energies. Moreover, interactions between nuclei can be neglected. The system is said to be in a weakly supersaturated metastable state, in the sense that there is an energy barrier between the given state and a lower energy state where there is an irreversible conversion of monomers to supercritical clusters that grow ever larger, resulting in the separation of the liquid and vapor phases. This is analogous to liquid-liquid phase separation in the metastable state. In this section we describe one approach to analyzing nucleation and growth, based on a discrete model of the aggregation and fragmentation of molecular clusters originally introduced by Becker and Döring [52].

### 13.3.1 Becker-Döring model of aggregation-fragmentation

Becker and Döring [52] introduced a system of kinetic equations, in which clusters form by monomers colliding with each other and then grow via subsequent collisions between clusters and monomers, see Fig. 13.14. A cluster can also partially fragment by losing a monomer. (A more general class of aggregation-fragmentation models allow clusters of various sizes to coalesce with each other, and for larger clusters to fragment into pairs of smaller clusters [929].) The main simplifying assumption is that interactions between clusters are ignored, which is reasonable when the cluster density is relatively small. If $u_n, n \geq 2$, denotes the concentration of clusters of size $n$ and $u_1$ denotes the concentration of monomers then the Becker-Döring (BD) equations take the form

$$\frac{du_n}{dt} = J_{n-1} - J_n, \quad n \geq 2, \tag{13.3.1a}$$

$$\frac{du_1}{dt} = -J_1 - \sum_{n \geq 1} J_n, \tag{13.3.1b}$$

where we have introduced the particle fluxes

$$J_n = a_n u_1 u_n - b_{n+1} u_{n+1}. \tag{13.3.2}$$

Here $a_n$ and $b_n$ denote the rates of aggregation and partial fragmentation of a cluster of size $n$. More precisely, equations (13.3.1a) and (13.3.1b) are a slightly modified version of the original BD equations, whereby the total mass of the system is conserved [40, 718]. The original model took the monomer concentration $u_1$ to be fixed [52], which was also assumed in the model of actin polymerization (see Sect. 4.1).

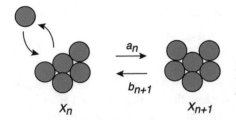

Fig. 13.14: Aggregation/fragmentation steps of the Becker-Döring model. Here $X_n$ denotes a cluster of size $n$.

In recent years the BD equations (13.3.1a) and (13.3.1b) have been applied to a wide range of chemical and biological processes including micelle and vesicle formation [212, 927], protein aggregation [252, 972], viral capsid assembly [434], the formation of robust protein concentration gradients [793], and vesicular transport in axons [117]. The final application is developed further in Ex. 13.8. There have also been several mathematical studies of the existence and uniqueness of steady-state solutions and large-time asymptotics [40, 718, 928]. See Refs [416, 929] for mathematical reviews of the BD equations.

Here we will focus on deterministic versions of the BD model under the assumption that the number of monomers and the cluster size are both large. However, there are a number of biological aggregation processes where there is a fixed maximum cluster size, which could be of the order of ten or a hundred, after which the clustering process stops or switches to some other dynamical regime. In the case of finite systems, where the maximum cluster size is capped, it is necessary to go beyond the mass-action kinetics of the BD equations by considering a discrete stochastic model [75, 252, 810, 972]. The associated master equation can then be analyzed using various asymptotic methods.

**Constant monomer concentration.** Let us begin the analysis of the BD equations in the simpler case that the monomer concentration is fixed, $u_1(t) = u$ for all $t$, and the forward and backward rate coefficients are size-independent, $a_n = a, b_n = b$. This model was briefly studied in Sect 4.1 within the context of actin polymerization. We will follow closely the analysis of Wattis and King [928]. We first look for an equilibrium solution that satisfies detailed balance. That is, setting $J_n = 0$ for all $n \geq 1$ yields the following solution for the concentrations:

$$u_n^{\mathrm{eq}} = \left(\frac{au}{b}\right)^{n-1} u \equiv \theta^{n-1} u. \qquad (13.3.3)$$

Requiring that $u_n \to 0$ as $n \to \infty$ implies that an equilibrium solution only exists for $\theta < 1$ or $u < b/a$; the corresponding mass density is

$$\rho_0 \equiv \sum_{n \geq 1} n u_n = \frac{u}{(1-\theta)^2}. \qquad (13.3.4)$$

If $\theta \geq 1$ then one has to consider a more general class of (nonequilibrium) steady-state solutions for which $J_n = J \neq 0$. There is then a constant flux through the system and detailed balance no longer holds. Iteratively solving the constant flux conditions

$auu_n - bu_{n+1} = J$ yields the family of steady-state solutions

$$u_n^{ss} = u \left[ \theta^{n-1} - \frac{J(\theta^n - \theta)}{au^2(\theta - 1)} \right].$$  (13.3.5)

If $\theta < 1$ then $u_n^{ss} \to -J/b(1 - \theta)$ as $n \to \infty$. It follows that $u_n^{ss}$ becomes negative at large $n$ for $J > 0$, whereas for $J < 0$ the concentrations remain bounded away from zero, implying that the mass density is infinite. Hence, the only steady-state solution is the equilibrium solution when $\theta < 1$. In the case $\theta > 1$, there exists a steady-state solution $u_n^{ss} = u$ with $J = au^2(1 - 1/\theta) = u(au - b)$. This is the least divergent solution in the sense that all other values of $J$ produce solutions with $u_n \to \pm\infty$ as $n \to \infty$. It turns out that in the large $t$ limit, the system approaches the least divergent solution when $\theta > 1$ and the equilibrium solution when $\theta < 1$. Finally, for $\theta = 1$ one finds that

$$u_n^{ss} = u \left( 1 - \frac{J(n-1)}{bu} \right).$$  (13.3.6)

The least divergent solution is $u_n = u$ with $J = 0$, which is a mixture of the two other cases.

Another way to differentiate between the two cases $\theta > 1$ and $\theta < 1$ is in terms of the behavior of the function

$$V = \sum_{n=1}^{\infty} u_n \left( \log(u_n/u\theta^{n-1}) - 1 \right).$$  (13.3.7)

It can be shown that

$$\frac{dV}{dt} = -\sum_{n \geq 1} (auu_n - bu_{n+1})(\log(auu_n) - \log(bu_{n+1})) \leq 0,$$

with equality only holding at equilibrium. (Since log is a monotonic function, it follows that $(A - B)(\log A - \log B) \geq 0$.) For $\theta < 1$, the equilibrium is a minimizer of $V$, whereas $V$ is not bounded below when $\theta > 1$. Hence, only in the former case is $V$ a Lyapunov function that guarantees convergence to the equilibrium solution. For $\theta > 1$ the steady-state mass density is infinite, suggesting that the approach to steady state is not uniform in $n$. However, asymptotic approximation methods can be used to determine the large-time behavior [928]. (In fact, in the case of monodisperse initial data for which $u_n(0) = 0$ for all $n > 1$, one can write down an exact solution for $u_n(t)$ [928]. However, exact solutions for more general rate coefficients are not known.)

*Large-time kinetics for* $\theta < 1$. Since the system is expected to converge to the equilibrium solution $u_n^{eq} = u\theta^{n-1}$, we introduce the rescaling $\psi_n(t) = u_n(t)/u_n^{eq}$ so that $\psi_n(t) \to 1$ as $t \to \infty$. The BD equations reduce to

$$\frac{1}{b} \frac{d\psi_n}{dt} = \psi_{n-1}(t) - \psi_n(t) - \theta\psi_n(t) + \theta\psi_{n+1}(t), \quad n \geq 2,$$  (13.3.8)

with $\psi_1(t) = 1$. We also assume that at $t = 0$ there are no clusters of infinite size so that $\psi_n(0) \to 0$ and $n \to 0$. In order to generate an approximate solution for $\psi_n(t)$, we suppose that $\psi_n(t)$ is a slowly varying function of $n$ at large times; this assumption will be verified below. Taking the continuum limit of equation (13.3.8) by setting $\Psi_n(t) = \psi(n,t)$ and Taylor expanding $\Psi(n\pm 1,t)$ to second order about $n$ yields the advection–diffusion equation

$$\frac{\partial \Psi}{\partial t} = \frac{b}{2}(1+\theta)\frac{\partial^2 \Psi}{\partial n^2} - b(1-\theta)\frac{\partial \Psi}{\partial n}, \qquad (13.3.9)$$

with effective diffusivity $D = b(1+\theta)/2$ and drift velocity $V = b(1-\theta)$. The boundary conditions are $\Psi(0,t) = 1$ and $\Psi(n,t) = 0$ for $n \to \infty$. Since we are dealing with the large $n,t$ regime, we expect the solution to be of the form of a traveling wave $\Psi(n,t) = f(n-Vt)$ with $f(x) \to 1$ as $x \to -\infty$ and $f(x) \to 0$ as $x \to \infty$; this satisfies the boundary conditions. It follows that we have the asymptotic solution

$$\Psi(n,t) = \frac{1}{2}\mathrm{erfc}\left(\frac{n-Vt}{\sqrt{4Dt}}\right), \qquad \mathrm{erfc}(x) = \frac{2}{\sqrt{\pi}}\int_x^\infty e^{-y^2}\,dy,$$

where $\mathrm{erfc}(x)$ is the complementary error function. Hence,

$$u_n(t) \sim \frac{b}{2a}\left(\frac{au}{b}\right)^n \mathrm{erfc}\left(\frac{n-[b-au]t}{\sqrt{2[b+au]t}}\right), \qquad t \to \infty. \qquad (13.3.10)$$

For large times $t$, the cluster distribution function $u_n(t)$ has three regions as illustrated in Fig. 13.15(a): (i) For $n \ll Vt$ the system has effectively reached equilibrium; (ii) For $n = Vt + O(t^{1/2})$ there is a transition region; (iii) For $n \gg Vt$ the concentration is still asymptotically small. Note that as time increases the transition region moves to larger $n$ and widens.

*Large-time kinetics for $\theta \geq 1$.* The case $\theta > 1$ can be analyzed along similar lines to $\theta < 1$. Now one rescales the concentrations according to $\psi_n(t) = u_n(t)/u$ since the

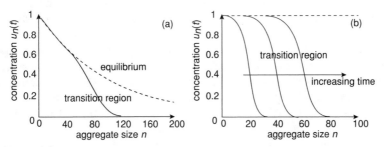

Fig. 13.15: Large-time asymptotics of cluster distribution function $u_n(t)$ for fixed $t$. (a) Case $\theta < 1$. (b) Case $\theta > 1$ shown at at times $t = 200, 400, 600$ with $a = u = 1, b = 0.9$. Speed of the front is $V = 0.1$ and the width of the front grows with $t$. [Redrawn from Wattis [929].]

system converges to the steady-state solution $u_n(t) = u$. Taking the continuum limit in $n$ with $\Psi(n,t) = \psi_n(t)$ results in the advection–diffusion equation

$$\frac{\partial \Psi}{\partial t} = \frac{b}{2}(1+\theta)\frac{\partial^2 \Psi}{\partial n^2} - b(\theta - 1)\frac{\partial \Psi}{\partial n}. \qquad (13.3.11)$$

We thus obtain the asymptotic solution

$$u_n(t) \sim \frac{u}{2}\mathrm{erfc}\left(\frac{n - [au - b]t}{\sqrt{2[b + au]t}}\right), \quad t \to \infty. \qquad (13.3.12)$$

The evolution of this traveling front solution is illustrated in Fig. 13.15(b). Finally, in the special case $\theta = 1$ the continuum limit yields a pure diffusion equation, and the approach to the equilibrium solution $u_n = u$ occurs via a stationary diffusive wave according to

$$u_n(t) \sim u\,\mathrm{erfc}\left(\frac{n}{2\sqrt{bt}}\right), \quad t \to \infty. \qquad (13.3.13)$$

**Constant mass formulation.** The analysis of the full BD equations (13.3.1a) and (13.3.1b) is much more involved due to the presence of nonlinearities. Formally speaking, multiplying equation (13.3.1a) by $n$, summing with respect to $n$ and adding equation (13.3.1b) gives

$$\frac{d}{dt}\sum_{n\geq 0} n u_n = -J_1 - \sum_{n\geq 1} J_n + \sum_{n\geq 2} n(J_{n-1} - J_n)$$

$$= -J_1 - \sum_{n\geq 1} J_n + 2(J_1 - J_2) + 3(J_2 - J_3) + \ldots = 0.$$

This implies that the total mass is conserved:

$$\sum_{n\geq 1} n u_n(t) = \sum_{n\geq 1} n u_n(0) \equiv \rho_0. \qquad (13.3.14)$$

One subtlety regarding the above derivation of the conservation equation is that it has been assumed one can reverse the order of infinite summation and differentiation. It turns out that for certain choices of the $n$-dependent transition rates $a_n, b_n$, commutativity breaks down, reflecting the fact that the total mass of the equilibrium solution is less than the initial mass [40, 718], see below. The physical interpretation of such behavior is that an equilibrium solution no longer exists, since there is an irreversible transfer of monomers to ever larger clusters. The nucleation of these clusters can ultimately lead to crystallization of a solid from solution, for example.

A first step in the analysis of the BD equations (13.3.1a) and (13.3.1b) with constant mass is to look for steady-state solutions $J_n(t) = J$ for all $n \geq 0$. The only physical solution is the equilibrium solution $J = 0$, otherwise $du_1/dt \to -\infty$. Hence

$$b_{n+1}u_{n+1} = a_n u_1 u_n, \quad n \geq 1,$$

which on rearranging and iterating gives $u_1 = u$ with

$$u_n = Q_n u^n, \quad Q_n = \frac{a_{n-1} a_{n-2} \dots a_1}{b_n b_{n-1} \dots b_2}. \tag{13.3.15}$$

In contrast to the constant concentration case, $u$ has to be determined self-consistently from the mass conservation condition

$$\rho_0 = F_0(u) \equiv \sum_{n \geq 1} n Q_n u^n. \tag{13.3.16}$$

We will assume that for a given choice of $a_n$ and $b_n$ the infinite series defining $F_0(u)$ has a finite radius of convergence $u = R$. There then exists a steady-state solution provided that equation (13.3.16) has a solution for which $u \leq R$, see below.

Approach to equilibrium can be established by constructing an appropriate Liapunov function [718]. That is, consider the function

$$L = \sum_{n=1}^{\infty} u_n \left[ \log(u_n/Q_n) - 1 \right]. \tag{13.3.17}$$

Differentiating with respect to $t$ gives

$$\frac{dL}{dt} = \sum_{n \geq 1} \frac{du_n}{dt} \log(u_n/Q_n) = (-J_1 - \sum_{n \geq 2} J_n) \log(u/Q_1) + \sum_{n \geq 2} (J_{n-1} - J_n) \log(u_n/Q_n)$$

$$= J_1 \log(u_2 Q_1^2 / Q_2 c_1^2) + J_2 \log(u_3 Q_2 / u u_2 Q_3) + J_3 \log(u_4 Q_3 / u_3 u Q_4) + \dots$$

$$= J_1 \log(u_2 Q_1^2 / Q_2 c_1^2) + \sum_{n \geq 2} J_n \log(u_{n+1} b_{n+1} / u u_n a_n) + \dots.$$

Substituting for $J_n$ then gives

$$\frac{dL}{dt} = (a_1 u^2 - b_2 u_2) \log(u_2 b_2 / a_1 u^2) + \sum_{n \geq 2} (a_n u u_n - b_{n+1} u_{n+1}) \log(b_{n+1} u_{n+1} / a_n u u_n)$$

$$\leq 0, \tag{13.3.18}$$

Moreover, $L$ is bounded from below. First note that each term

$$f(u_n) \equiv u_n \left[ \log(u_n/Q_n) - 1 \right]$$

is a convex function of $u_n$ so that at an arbitrary value $u_n^*$,

$$f(u_n) - f(u_n^*) \geq f'(u_n^*)[u_n - u_n^*],$$

which implies that

$$L \geq \sum_{n \geq 0} \left( u_n^* [\log(u_n^*/Q_n) - 1] + (u_n - u_n^*) \log(u_n^*/Q_n) \right).$$

Let us choose $u_n^* = Q_n z^n$ so that

$$L \quad \geq \sum_{n \geq 0} (n u_n \log z - Q_n z^n) \quad = (\rho_0 - c) \log z - \sum_{n \geq 0} Q_n z^n$$

$$= (\rho_0 - c) \log z - F_0(z) > -\infty.$$

It follows that $L$ must approach a limit as $t \to \infty$ such that $dL/dt \to 0$. Since every term on the right-hand side of the penultimate line of Eq. (13.3.18) is non-positive, it follows that the individual terms approach zero:

$$J_n \equiv a_n u u_n - b_{n+1} u_{n+1} \to 0 \quad \text{as } t \to \infty, \quad n \geq 0.$$

Therefore,

$$u_n - Q_n u^n \to 0 \quad \text{as } t \to \infty, \quad n \geq 1. \tag{13.3.19}$$

However, we still need to determine how $u$ behaves as $t \to \infty$. From the conservation equation, we have

$$\lim_{t \to \infty} \sum_{n \geq 1} n u_n(t) = \rho_0. \tag{13.3.20}$$

On the other hand, Eq. (13.3.19) tells us that

$$\sum_{n \geq 1} \lim_{t \to \infty} n u_n(t) = \sum_{n \geq 1} n Q_n \left[ \lim_{t \to \infty} u(t)^n \right]. \tag{13.3.21}$$

If we can interchange the two limit operations $t \to \infty$ and $n \to \infty$, then we obtain the asymptotic results

$$\lim_{t \to \infty} u(t) = \bar{u}, \quad \rho_0 = F_0(\bar{u}). \tag{13.3.22}$$

Ball et al [40] carried out a rigorous analysis of the BD equations, in which they investigated the existence of steady-state solutions and mass conservation under various conditions on the rate constants $a_n, b_n$ and the initial states $u_n(0)$. One of their major findings was that in certain regimes, the system exhibits a form of metastability. We distinguish three different cases (see also [718, 929]).

(i) Case A: $R = \infty$. In this case the function $F_0(u)$ has an infinite radius of convergence and so $0 \leq F_0(u) < \infty$ for all $0 \leq u < \infty$ with $F_0(u) \to \infty$ as $u \to \infty$. Given any finite initial mass $\rho_0$, there exists a unique equilibrium solution $u_1 = u$ for which $F_0(u) = \rho_0$. A simple example is obtained by taking aggregation rates $a_n = 1$ and fragmentation rates $b_n = n$, which implies that fragmentation dominates in large clusters. Then $Q_n = 1/n!$ and

$$F_0(u) = \sum_{n \geq 1} n \frac{u^n}{n!} = u e^u.$$

This is a monotonically function of $u$ with $F_0(0) = 0$ and $F_0(\infty) = \infty$. Thus there exists a unique solution to the equation $F_0(u) = \rho_0$.

(ii) Case B: $0 < R < \infty$ and $F_0(R) = \infty$. Now the function $F_0(u)$ has a finite radius of convergence and as $u \to R^-$, we have $F_0(u) \to \infty$. For any initial mass $\rho_0$, the equation $F_0(u) = \rho_0$ has a unique solution for $u \in [0, R)$. In particular, for large initial masses, there is an upper limit to the equilibrium monomer concentration. A simple example occurs for $n$-independent aggregation-fragmentation rates, $a_n = a, b_n = b$, as in the case of linear polymerization and the formation of cylindrical micelles (see Sect. 13.3.3). Then $Q_n = (a/b)^{n-1}$ and

$$F_0(u) = \sum_{n \geq 1} n \left(\frac{au}{b}\right)^{n-1} = \frac{u}{(1 - au/b)^2}.$$

The equilibrium equation $\rho_0 = F_0(u)$ then has the unique solution

$$u = \frac{2\rho_0}{1 + 2a\rho_0/b + \sqrt{1 + 4a\rho_0/b}}.$$

Note that $u \to 0$ as $\rho_0 \to 0$ and $u \to (b/a)^-$ as $\rho_0 \to \infty$. The radius of convergence is $R = (b/a)^-$.

(iii) Case C: $0 < R < \infty$ and $F_0(R) < \infty$. Again the function $F_0(u)$ has a finite radius of convergence $R$, but there now exists a critical mass $\rho_c$ at which the monomer concentration takes its maximal value, $\rho_c = F_0(R)$. If $\rho_0 < \rho_c$ then there exists a unique equilibrium solution with the same mass $\rho_0$. As in Case B, convergence to equilibrium is strong in the sense that

$$\lim_{t \to \infty} \sum_{n \geq 1} n |u_n(t) - Q_n u^n| = 0, \quad F_0(u) = \rho_0. \tag{13.3.23}$$

This means that the two limits $t \to \infty$ and $n \to \infty$ are interchangeable. On the other hand, if $\rho_0 > \rho_c$ then there is no equilibrium solution with the same mass as the initial data. One finds that

$$\lim_{t \to \infty} u_n(t) = Q_n R^n,$$

where $R$ is the radius of convergence of the series representation of $F_0$. Now convergence is weak, that is, individual terms of the series (13.3.23) converge to zero, but their sum does not[3]. It follows that the two limits are not interchangeable:

$$\sum_{n \geq 1} n u_n(t) = \rho, \quad \sum_{n \geq 1} n Q_n R^n = F_0(R) = \rho_c < \rho_0.$$

---

[3] A more familiar example of weak convergence was highlighted in [929]. Consider the diffusion equation $u_t = u_{xx}$ on $\mathbb{R}$ with initial condition $u(x, 0) = e^{-x^2}$. One expects the integral $I[u] = \int_{-\infty}^{\infty} u(x) dx$ to be conserved, that is, $I[u] = 2\sqrt{\pi}$ for all $t \geq 0$. However, the solution $u(x, t) - e^{-x^2/4(t+1)}/\sqrt{t+1}$ implies that pointwise $u(x, t) \to 0$ as $t \to 0$. Since $I[0] = 0$, we see that the limit $t \to \infty$ does not commute with evaluating the integral, which is analogous to performing an infinite summation.

The excess mass $\rho_0 - \rho_c$ reflects the fact that there is an irreversible transfer of monomers to ever larger clusters (nucleation and growth), which ultimately leads to phase separation (formation of macroscopic clusters), see Sect. 13.3.2. One biological example of Case C is the formation of spherical micelles, see Sect. 13.3.3

### 13.3.2 Nucleation in the BD model

We begin by briefly describing the classical theory of nucleation as illustrated in Fig. 13.16, see also the discussion of phase separation in Sect. 13.1. Suppose that one starts with a pure monomer solute and rapidly increases the concentration of the solute beyond the critical concentration $u_c = R$ (stage I). The solute is then said to be supersaturated and enters a nucleation stage in which monomers are rapidly converted into small clusters, some of which survive possible conversion back to monomers and grow into larger stable nuclei or particles (stage II). Let $n_c$ denote the critical cluster size for nucleation. Since small clusters have relatively large free energies, the formation of small clusters is difficult and this generates an energy barrier $\Delta G_c$ to the nucleation of stable large clusters. A quasi-steady state is then reached, in which stable nuclei are produced at a constant rate $J$. A classical formula for the nucleation rate is

$$J \sim u a(n_c) e^{-\Delta G_c / k_B T},$$

where $u$ is the monomer concentration, $a(n_c)$ is the rate at which monomers bind to clusters of critical size $n_c$. However, this formula is based on the assumption that every cluster that grows beyond $n_c$ continues to grow without decaying back to a smaller size. However the actual nucleation rate is smaller, since growth is not irreversible, and so one must scale $J$ by a correction factor $Z, 0 < Z < 1$, known as the Zeldovich factor. Eventually, the monomer concentration becomes so depleted that the number of stable clusters becomes approximately constant, and the system enters stage III. In this growth and coarsening stage larger nucleated particles grow at the expense of smaller particles, which causes the total number of particles to decrease.

The BD equations were originally introduced as a model of nucleation in which the monomer concentration is effectively fixed [52]. In order to understand how this relates to the constant-mass version of these equations considered in Sect. 13.3, suppose the rates $a_n, b_n$ are chosen so that Case C holds. If the initial monomer concentration is just above the radius of convergence $R$, $u_1(0) = u = R + \varepsilon$, and there are initially no clusters, then the nucleation rate for the formation of clusters is exponentially small, and $u_1(t) \approx u$ for exponentially large times. Since $u > R$, there does not exist an equilibrium solution, that is, $n Q_n u^n$ diverges (albeit slowly) as $n \to \infty$. This suggests looking for a nonequilibrium steady-state solution for which $J_n(t) = J$ for all $n \geq 1$, and identifying $J$ as the nucleation rate. Such a state is metastable, since eventually the pool of monomers will be depleted due to the irreversible growth

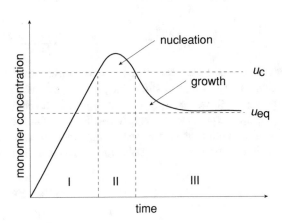

Fig. 13.16: Illustration of the classical mechanism for nucleation and growth of molecular aggregates in solution. (I) A rapid increase in the concentration of free monomers in solution. (II) Once the monomer concentration $u$ exceeds a critical value $u_c$, it becomes supersaturated and nucleation occurs, which significantly reduces the concentration of free monomers in solution. (III) Following nucleation, growth and cluster coarsening occurs under the control of the diffusion of the monomers through the solution.

of arbitrarily large clusters. Rigorous results on metastable states in the BD equations have been developed by Penrose [718]. Here we will follow a more physical approach [256, 701].

First, note that we can use statistical mechanics to relate $Q_n$ to the free energy change $G_n$ for the reaction

$$nX_1 \leftrightarrow X_n,$$

where $X_n$ denotes a cluster of size $n$. That is, the ratio of the forward and backward rates of $X + X_{n-1} \leftrightarrow X_n$ is given by (Chap. 1)

$$\frac{a_{n-1}u}{b_n} = e^{-\Delta g_{n-1}/k_B T},$$

where $\Delta g_{n-1}$ is the change in free energy for adding one monomer to the cluster $X_{n-1}$. It follows that

$$b_n = u a_{n-1} e^{[G_n - G_{n-1}]/k_B T}, \tag{13.3.24}$$

and

$$u^{n-1} Q_n \equiv \prod_{j=2}^{n} \frac{a_{j-1}u}{b_j} = e^{-G_n/k_B T}, \qquad G_n = \sum_{j=2}^{n} \Delta g_{j-1}. \tag{13.3.25}$$

In cases where the equilibrium solution exists, we have the Boltzmann distribution

$$u_n^{\mathrm{eq}} = u^n Q_n = u e^{-G_n/k_B T}. \tag{13.3.26}$$

Consider a steady-state solution for which $J_n(t) = J > 0$ for all $n$. Then

$$J = a_n u u_n - b_{n+1} u_{n+1}, \tag{13.3.27}$$

which on dividing by $a_n u^{n+1} Q_n$ becomes

$$\frac{J}{a_n u^{n+1} Q_n} = \frac{u_n}{u^n Q_n} - \frac{u_{n+1}}{u^{n+1} Q_{n+1}}.$$

Summing this equation from $n = 1$ to some maximum value $n = N$ (with $Q_1 = 1$) gives

$$J \sum_{n=1}^{N} \frac{1}{a_n u^{n+1} Q_n} = 1 - \frac{u_{N+1}}{u^{N+1} Q_{N+1}}. \tag{13.3.28}$$

In the metastable state of nucleation processes, the condensed (liquid) phase is thermodynamically stable, which means that clusters of sufficiently large size have a faster growth rate then shrinkage rate, that is, $u a_{N-1} > b_N$ (or $\Delta g_N < 0$) for sufficiently large $N$. Thus, $u^{N-1} Q_N$ grows exponentially with $N$ for large $N$. On the other hand, the assumption of small depletion of monomers requires that $u_N \ll u$. Therefore, taking the limit $N \to \infty$ in equation (13.3.28) yields the steady-state flux

$$J = u \left( \sum_{n=1}^{\infty} \frac{1}{a_n u^n Q_n} \right)^{-1}. \tag{13.3.29}$$

This determines the nucleation rate. Also setting $N = r - 1$ in equation (13.3.28) gives

$$u_r = Q_r u^r \left( 1 - J \sum_{n=1}^{r-1} \frac{1}{a_n u^n Q_n} \right) = u_r^{eq} \frac{\sum_{n=r}^{\infty} \frac{1}{a_n u^n Q_n}}{\sum_{n=1}^{\infty} \frac{1}{a_n u^n Q_n}} < u_r^{eq}. \tag{13.3.30}$$

In classical nucleation theory, the flux $J$ is estimated using steepest descents (see Box 11H for a discussion of the method). Assuming that nucleation is dominated by large clusters, we can treat the cluster size $n$ as a continuous variable and approximate the sum in equation (13.3.29) by an integral. Let $n_c$ denote the critical cluster size for which the term $\Gamma_n = a_n u^n Q_n$ is minimized,

$$\frac{d}{dn} (a_n u^n Q_n) = 0 \quad \text{at } n = n_c.$$

Assuming that the sum in (13.3.29) is dominated by values in a neighborhood of $n_c$ and $a_n$ is a slowly varying function of $n$, we can then use the approximation

$$\sum_{j=1}^{\infty} \frac{1}{a_n u^n Q_n} \approx \frac{1}{a(n_c)} \int_{-\infty}^{\infty} e^{G(n)/k_B T} \, dn \approx \frac{1}{a(n_c)} e^{G(n_c)/k_B T} \int_{-\infty}^{\infty} e^{-|G''(n_c)|(n-n_c)^2/k_B T} \, dn$$

$$= \sqrt{\frac{2\pi k_B T}{|G''(n_c)|}} \frac{e^{G(n_c)/k_B T}}{a(n_c)}.$$

We thus obtain the classical Zeldovich formula for the nucleation rate:

$$J = u \sqrt{\frac{|G''(n_c)|}{2\pi k_B T}} a(n_c) e^{-G(n_c)/k_B T}. \tag{13.3.31}$$

### *13.3.3 Self-assembly of phospholipids*

Amphiphiles, such as the phospholipids of the plasma membrane, can spontaneously form aggregates in aqueous solution [463, 664]. Aggregation is driven by the fact that it is energetically favorable to form clusters, since they shield the hydrophobic regions of individual lipids from contact with water. However, formation of clusters lowers the number of objects in the system, thus reducing the entropy. Competition between these two factors is taken into account by the free energy of the system. The resulting geometrical arrangement of the lipids depends on the amphiphile concentration, the molecular volume, the length of the hydrocarbon chain, and properties of the solvent. Examples include monolayer cylindrical or spherical aggregates (micelles), planar and spherical bilayers (Sect. 12.2), as illustrated in Fig. 13.17. Here we will focus on spherical and cylindrical micelles.

Let $u_n$ be the concentration of a micelle of size $n$. A standard calculation of the entropy in the dilute limit (see also Sect. 1.4) shows that the total free energy of the system is

$$F = - \sum_{n=1}^{\infty} u_n \varepsilon_n + k_B T \sum_{n=1}^{\infty} u_n (\ln u_n - 1), \qquad (13.3.32)$$

with the monomer density $u_1 = u$ given by

$$u = \rho_0 - \sum_{n=2}^{\infty} n u_n. \qquad (13.3.33)$$

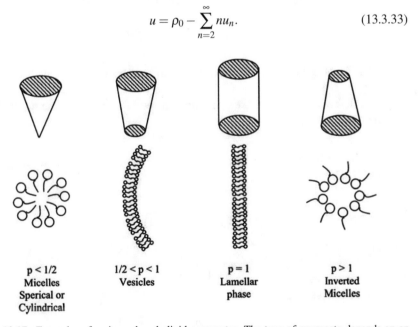

| p < 1/2 | 1/2 < p < 1 | p = 1 | p > 1 |
|---|---|---|---|
| Micelles | Vesicles | Lamellar | Inverted |
| Sperical or | | phase | Micelles |
| Cylindrical | | | |

Fig. 13.17: Examples of various phospholipid aggregates. The type of aggregate depends on an effective packing parameter $p = v/la$, which describes the geometry of the volume occupied by an individual molecule (e.g., cone, truncated cone, or cylinder). Here $a$ is the polar head surface, $v$ is the tail volume, and $l$ is the hydrophobic tail length.

Here $\rho_0$ is the total density and $\varepsilon_n = ne_1 - e_n$ where $e_n$ is the energy of an $n$-cluster, that is, $-\varepsilon_n$ is the total binding energy of the cluster for $n \geq 2$ and $\varepsilon_1 = 0$. The equilibrium density of $n$-clusters ($n \geq 2$), assuming it exists, can be found by minimizing the free energy with respect to $u_n, n \geq 2$, under the constraint (13.3.33). Introducing a Lagrange multiplier $\mu$ (which is the chemical potential), we minimize the modified free energy

$$\widehat{F} = F - \mu \left( u + \sum_{n=2}^{\infty} n u_n - \rho_0 \right).$$

That is,

$$\frac{\partial F}{\partial u_n} = n\mu, \quad n \geq 1,$$

which gives $-\varepsilon_n + k_B T \ln u_n = n\mu$ or

$$u_n^{\text{eq}} = e^{(\varepsilon_n + n\mu)/k_B T} = u^n e^{\varepsilon_n / k_B T}, \quad n \geq 2.$$

Comparison with equation (13.3.26) shows that the activation free energy used in the BD model is

$$G_n = -\varepsilon_n + k_B T (n-1) \ln(1/u). \tag{13.3.34}$$

Finally, we obtain a self-consistency condition for the monomer density in terms of the total density $\rho_0$ (see equations (13.3.16) and (13.3.33)):

$$\rho_0 = F_0(u) \equiv \sum_{n=1}^{\infty} n u^n e^{\varepsilon_n / k_B T}. \tag{13.3.35}$$

The behavior of the system will then be determined by the $n$-dependence of the energy $\varepsilon_n$.

Let us first consider a rod-like cylindrical micelle for which the energy is typically modeled as

$$\varepsilon_n = (n-1)\alpha k_B T, \tag{13.3.36}$$

where $\alpha k_B T$ is the monomer-monomer bonding energy [463]. Substituting into equation (13.3.35) and using

$$\sum_{n=1}^{\infty} n x^n = x \frac{d}{dx} \sum_{n=1}^{\infty} x^n = \frac{x}{(1-x)^2}, \quad x < 1$$

we find that

$$\rho_0 = \frac{u}{(1-ue^\alpha)^2}.$$

This has the unique solution

$$u = u(\rho_0) = \frac{1 + 2\rho_0 e^\alpha - \sqrt{1 + 4\rho_0 e^\alpha}}{2\rho_0 e^{2\alpha}}, \tag{13.3.37}$$

with $u(\rho_0) < e^{-\alpha}$ for all $0 < \rho_0 < \infty$ and $u(\rho_0) \to e^{-\alpha}$ as $\rho_0 \to \infty$. It follows that the system is an example of Case B within the context of the BD model (Sect. 13.3), since the radius of convergence of $F_0(u)$ is $R = e^{-\alpha}$ and $F_0(u) \to \infty$ as $u \to R^-$. Hence, cylindrical micelles form an equilibrium, polydisperse distribution of sizes in a process known as micellization. The average cluster size in equilibrium is then

$$\langle n \rangle \equiv \left( \sum_{n \geq 1} u_n^{eq} \right)^{-1} \sum_{n \geq 1} n u_n^{eq} = \rho_0 \frac{1 - u e^\alpha}{u} = \frac{\sqrt{1 + 4\rho_0 e^\alpha} - 1}{2u(\rho_0)e^\alpha}. \qquad (13.3.38)$$

The behavior of spherical micelles is very different. A spherical aggregate with $n$ molecules will have a radius $r = (3nv_0/4\pi)^{1/3}$, where $v_0$ is molecular volume. There will be an energy $4\pi r^2 \gamma$ associated with surface tension, where $\gamma$ is the interfacial energy per unit area. Thus, we can write the total energy of a spherical micelle for $n \gg 1$ as [463]

$$\varepsilon_n \sim (n-1)\alpha k_B T - \frac{3}{2}\sigma n^{2/3}, \qquad (13.3.39)$$

where $\sigma = 2\gamma(4\pi v_0^2/3)^{1/3}$. In this case, equation (13.3.35) yields the self-consistency condition

$$\rho_0 = u \sum_{n=1}^{\infty} n(u e^\alpha)^{n-1} \exp\left( -\frac{3\sigma n^{2/3}}{2k_B T} \right). \qquad (13.3.40)$$

The radius of convergence is finite, $R = e^{-\alpha}$, but $F_0(R) < \infty$, corresponding to Case III of the BD model. Hence, there is a critical micellar concentration $u = e^{-\alpha}$ beyond which nucleation occurs and spherical micelles grow indefinitely, resulting in phase separation.

**Multi-scale analysis of cylindrical micelles.** Numerically solving the BD equations for cylindrical micelles shows that if the initial monomer concentration $u(0) \gg e^{-\alpha}$ and $u_n(0) = 0$ for $n \geq 2$, then the approach to equilibrium (micellization ) can be divided into three stages [664], see Fig. 13.18: (1) the monomer concentration decreases rapidly due to the formation of many small size clusters; (2) aggregates steadily increase in size until their distribution becomes a self-similar solution of the diffusion equation; (3) the final approach to equilibrium, which can be modeled using an FP equation. We shall describe a recent multi-scale analysis of the BD equations by Neu et al [664], which captures each of these stages. (An alternative multi-scale analysis is carried out by Wattis and King [928].)

First, combining equations (13.3.24), (13.3.34) and (13.3.36) shows that for cylindrical micelles the off and on rates $b_n, a_{n-1}$ are related according to $b_n = a_{n-1}e^{-\alpha}$ so that

$$J_n = b_{n+1}\left( e^\alpha u u_n - u_{n+1} \right).$$

Setting $b_{n+1} = 1$ and substituting into the BD equation (13.3.1a) and rearranging yields the system of ODEs (see Ex. 13.9)

$$\frac{du_n}{dt} + (e^\alpha u - 1)(u_n - u_{n-1}) = u_{n+1} - 2u_n + u_{n-1}, \quad n \geq 2, \qquad (13.3.41)$$

which is supplemented by the conservation condition $\sum_{n=1}^{\infty} n u_n = \rho_0$. Suppose that the initial condition is $u_n(0) = \rho_0 \delta_{n,1}$ with $\rho_0 \gg e^{-\alpha}$. We can eliminate one of the

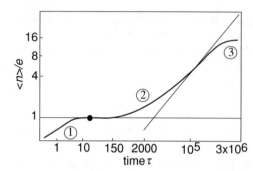

Fig. 13.18: Log-Log plot of the mean cluster size $\langle n \rangle / e$ as a function of the scaled time $\tau$ (thick solid line), which shows three distinct stages. The black dot represents the beginning of the second phase of growth, which asymptotes to the straight line of slope 1/2, indicating self-similar diffusive growth. [Redrawn from Neu et al [664].]

parameters $\alpha, \rho_0$ by rescaling the cluster densities and time:

$$u_n = \rho_0 r_n, \quad \tau = e^{\alpha} \rho_0 t = \frac{t}{\varepsilon},$$

with $0 \ll \varepsilon < 1$. Then

$$\frac{dr_n}{dt} + (r_1 - \varepsilon)(r_n - r_{n-1}) = \varepsilon(r_{n+1} - 2r_n + r_{n-1}), \quad n \geq 2, \quad (13.3.42a)$$

with $1 = \sum_{n=1}^{\infty} n r_n$ and initial conditions $r_n(0) = \delta_{n,1}$. Multiplying equation (13.3.42a) by $n$ and summing with respect to $n$ gives (see Ex. 13.9)

$$\frac{dr_1}{d\tau} + r_1(r_1 + r_c) + \varepsilon(r_1 - r_2 - r_c) = 0, \quad (13.3.42b)$$

where $r_c = \sum_{n \geq 1} r_n$ is the total (rescaled) density of clusters. Similarly, directly summing equation (13.3.42a) with respect to $n$, and combining with equation (13.3.42b) yields

$$\frac{dr_c}{d\tau} + r_1 r_c + \varepsilon(r_1 - r_c) = 0. \quad (13.3.42c)$$

Note that the average cluster size is $\langle n \rangle = 1/r_c$.

**Initial transient regime.** Since $r_1(0) = 1$ and $r_n(0) = 0$ for $n > 1$, it follows that there will be an initial regime in which we can take the limit $\varepsilon \to 0$ to obtain the planar system [664]

$$\frac{dr_1}{ds} = -(r_1 + r_c), \quad \frac{dr_c}{ds} = -r_c, \quad (13.3.43)$$

with renormalized time $s = \int_0^{\tau} r_1(y) dy$ such that $df/d\tau = r_1 df/ds$ for arbitrary $f$. (Although $r_n$ for $n \geq 2$ is negligible in this regime, the sum over all aggregate sizes need not be.) The planar system has the solution

$$r_1(s) = (1 - s)e^{-s}, \quad r_c = e^{-s},$$

which implies that

$$\tau = \int_0^s \frac{e^z}{1 - z} dz. \quad (13.3.44)$$

It follows that $\tau \to \infty$ as $s \to 1^-$ with $r_1(1) = 0$ and $r_c(1) = 1/e$. These are the limiting values of $r_1, r_c$ at the end of the initial stage. Finally, setting $\varepsilon = 0$ in equation (13.3.42a) shows that

$$\frac{d}{ds}(r_n e^s) = r_{n-1} e^s,$$

which can be solved recursively starting from $n = 2$ to yield

$$r_n(s) = \left(\frac{s^{n-1}}{(n-1)!} - \frac{s^n}{n!}\right) e^{-s}, \quad n \geq 2. \tag{13.3.45}$$

As $\tau \to \infty$, $r_n \to (n-1)e^{-1}/n!$, and numerically speaking there are negligible numbers of clusters with more than five monomers. The average cluster size is $\langle n \rangle = 1/r_c = e$, which is much smaller than the equilibrium value $\langle n \rangle \sim \sqrt{\rho_0 e^\alpha} \gg 1$ (take $\rho_0 e^\alpha \gg 1$ in equation (13.3.38)). Hence, there must be additional growth on time scales larger than $t = O(\varepsilon)$.

**Intermediate transient regime.** Returning to the exact equation (13.3.42a), we see that when $r_1 = O(\varepsilon)$ and $r_n = O(1)$ for $n \geq 2$, all terms on the right-hand side are $O(\varepsilon)$. This suggests rescaling $r_1 = \varepsilon R_1$ and using the original time $t = \varepsilon \tau$ in equation (13.3.42a) [664]:

$$\frac{dr_n}{dt} = -(R_1 - 1)(r_n - r_{n-1}) + (r_{n+1} - 2r_n + r_{n-1}), \quad n \geq 3, \tag{13.3.46a}$$

and

$$\frac{dr_2}{dt} = -(R_1 - 1)(r_2 - \varepsilon R_2) + (r_3 - 2r_2 + \varepsilon R_1). \tag{13.3.46b}$$

Moreover, equations (13.3.42b) and (13.3.42c) become

$$(R_1 - 1)r_c - r_2 + \varepsilon\left(\frac{dR_1}{dt} + R_1^2 + R_1\right) = 0, \tag{13.3.46c}$$

and

$$\frac{dr_c}{dt} + (R_1 - 1)r_c + \varepsilon R_1 = 0, \tag{13.3.46d}$$

with $r_c = \varepsilon R_1 + \sum_{n \geq 2} r_n \approx \sum_{n \geq 2} r_n$. In the limit $\varepsilon \to 0$, we have $R_1 - 1 = r_2/r_c$ and we obtain the closed system of equations

$$\frac{dr_n}{dt} = -\frac{r_2(r_n - r_{n-1})}{r_c} + (r_{n+1} - 2r_n + r_{n-1}), \quad n \geq 2, \tag{13.3.47}$$

with $r_1 \equiv 0$ and initial conditions $r_n(0) = (n-1)e^{-1}/n!$. At large times the difference $r_n - r_{n-1}$ becomes small, suggesting that one can approximate $r_n$ by a continuum limit. Therefore, set [664]

$$r_n(t) \sim \delta^a r(x, T), \quad x = \delta n, \quad T = \delta^b t. \tag{13.3.48}$$

Here the scale factor $\delta \to 0$ is introduced in order to ensure that we are working in the large-$n$ and large-$t$ regime when $x, T = O(1)$. The index $a$ can be found from the conservation condition $\sum_{k \geq 2} n r_n = 1$:

$$1 = \delta^{a-2} \sum_{n \geq 2} (n\delta) r(n\delta, T) \delta \sim \int_0^\infty x r(x, T) dx,$$

provided $a = 2$. Similarly, $r_c \sim \delta \int_0^\infty r(x, T) dx \equiv \delta R_c$. Rescaling equation (13.3.47) yields

$$\delta^b \frac{\partial r}{\partial T} = -\frac{\delta^2 r(2\delta, T)[r(x, T) - r(x - \delta, T)]}{\delta R_c}$$
$$+ r(x + \delta, T) - 2r(x, T) + r(x - \delta, T).$$

If we choose $b = 2$, then we can divide through by $\delta^2$ and take the limit $\delta \to 0$ to obtain the PDE

$$\frac{\partial r(x, T)}{\partial T} = -\frac{r(0, T)}{R_c(T)} \frac{\partial r(x, T)}{\partial x} + \frac{\partial^2 r(x, T)}{\partial x^2}. \tag{13.3.49}$$

Since $r_1 = 0$ in equation (13.3.47), we have the boundary condition $r(0, T) = 0$, which leads to the diffusion equation

$$\frac{\partial r(x, T)}{\partial T} = \frac{\partial^2 r(x, T)}{\partial x^2}, \quad r(0, T) = 0. \tag{13.3.50}$$

In order to solve the diffusion equation, we need to specify the initial condition. Numerically speaking, one finds that in the intermediate transient regime the distribution of cluster sizes is concentrated about zero in the limit $T \to 0^+$. This suggests the solution

$$r(x, T) = -\frac{\partial}{\partial x} \left( \frac{e^{-x^2/4T}}{\sqrt{\pi T}} \right) = \frac{x}{2\sqrt{\pi} T^{3/2}} e^{-x^2/4T}. \tag{13.3.51}$$

The normalization is chosen so that the mass conservation condition $\int_0^\infty x r(x, T) dx = 1$ is satisfied. In terms of the original variables, we have

$$r_n(t) \sim \frac{n}{2\sqrt{\pi} t^{3/2}} e^{-n^2/4t}, \tag{13.3.52}$$

and the mean cluster size is $\langle n \rangle \sim \sqrt{\pi t}$. This is consistent with the straight-line of slope $1/2$ in the log–log plot of Fig. 13.18.

**Late transient regime.** The large-time limit of equation (13.3.52) does not match the equilibrium distribution

$$u_n^{\text{eq}} = u^n e^{(n-1)\alpha}. \tag{13.3.53}$$

Since we are working in the regime $\rho_0 e^\alpha = 1/\varepsilon \gg 1$, we have from equation (13.3.37) that $u \sim \rho_0 \varepsilon [1 - \sqrt{\varepsilon}]$ so that

$$r_n^{eq} = \rho_0^{-1} u_n^{eq} \sim \varepsilon \left[1 - \sqrt{\varepsilon}\right]^n \sim \varepsilon e^{-n\sqrt{\varepsilon}}.$$

Also $\langle n \rangle \sim 1/\sqrt{\varepsilon}$, whereas at the end of stage II we have $\langle n \rangle \sim \sqrt{\pi t}$. This suggests that the final stage occurs at times $t = O(\varepsilon^{-1/2})$. Therefore, try the same scaling (13.3.48) as the intermediate regime with $a = b = 2$ and $\delta = \sqrt{\varepsilon}$:

$$r_n(t) = \varepsilon r(x,t), \quad x = \sqrt{\varepsilon} n, \quad T = \varepsilon t,$$

such that

$$r_c \sim \varepsilon^{1/2} \int_0^\infty r(x,T) dx \equiv \varepsilon^{1/2} R_c, \quad 1 = \int_0^\infty r(x,T) dx.$$

Equations (13.3.42a) and (13.3.42c) with $r_1 = \varepsilon R_1$ now become

$$\varepsilon^3 \frac{\partial r}{\partial T} = -\varepsilon^2 (R_1 - 1)[r(x,T) - r(x - \sqrt{\varepsilon}, T)]$$
$$+ \varepsilon^2 [r(x + \sqrt{\varepsilon}, T) - 2r(x,T) + r(x - \sqrt{\varepsilon}, T)], \quad (13.3.54a)$$

and

$$\varepsilon \frac{dR_c}{dt} + (R_1 - 1)R_c + \varepsilon^{1/2} R_1 = 0. \tag{13.3.54b}$$

The second equation implies that $R_1 - 1 = -\varepsilon^{1/2}/R_c + O(\varepsilon)$. Substituting into equation (13.3.54a), dividing through by $\varepsilon^3$, and taking the limit $\varepsilon \to 0$ then gives [664]

$$\frac{\partial r(x,T)}{\partial T} = -\frac{1}{R_c(T)} \frac{\partial r(x,T)}{\partial x} + \frac{\partial^2 r(x,T)}{\partial x^2}, \tag{13.3.55}$$

with $R_c(T) = \int_0^\infty r(x,T) dx$ and boundary condition $r(0,T) = 1$. The latter follows from equation (13.3.54b). Finally, it can be shown that the solution to this PDE matches the solution of the intermediate stage diffusion equation in the limit $T \to 0^+$ and matches the equilibrium solution in the limit $T \to \infty$, see Neu et al [664] for details.

## 13.4 Cooperative transport of proteins between cellular organelles

The extensive secretory pathway of eukaryotic cells is a critical system for the maturation and transportation of newly synthesized lipids and proteins to specific target sites within the cell membrane. The first major organelle of the secretory pathway is the endoplasmic reticulum (ER) [565], see Fig. 13.19. Proteins and lipids destined for the plasma membrane enter the ER from the nucleus as they are translated by ER-associated ribosomes, where they fold into their proper 3D structure. One

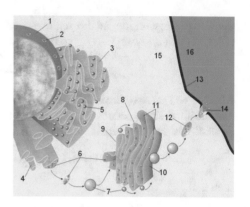

Fig. 13.19: Diagram of secretory pathway including nucleus, ER, and Golgi apparatus. 1. Nuclear membrane; 2. Nuclear pore; 3. RER; 4. SER; 5. Ribosome; 6. Protein; 7. Transport vesicles; 8. Golgi apparatus; 9. *Cis* face of Golgi apparatus; *Trans* face of Golgi apparatus ; 11 Cisternae of Golgi apparatus; 12. Secretory vesicle; 13. Plasma membrane; 14. Exocytosis; 15 Cytoplasm; 16. Extracellular domain. [Public domain image from WikiMedia Commons].

important aspect of the secretory pathway is that it is tightly regulated [565]. Proteins accumulate at specific exit sites and leave the ER in vesicles that transfer the cargo to organelles forming the Golgi network, where final packaging and sorting for target delivery is carried out. In most eukaryotic cells, the Golgi network is confined to a region around the nucleus known as the Golgi apparatus. Another the significant features of the secretory pathway is that there is a constant active exchange of molecules between organelles such as the ER and Golgi apparatus, which have different lipid and protein compositions. Such an exchange is mediated by motor-driven vesicular transport. Vesicles bud from one compartment or organelle, carrying various lipids and proteins, and subsequently fuse with another compartment. Transport in the anterograde direction from the ER to Golgi has to be counterbalanced by retrograde transport in order to maintain the size of the compartments and to reuse components of the transport machinery. Since bidirectional transport would be expected to equalize the composition of both compartments, there has been considerable interest in understanding the self-organizing mechanisms that allow such organelles to maintain their distinct identities while constantly exchanging material [623]. One transport model for generating stable, non-identical compartments has been proposed by Heinrich and Rapoport [406], and developed further by a number of groups [84, 245, 357]. These models primarily focus on two of the essential steps of vesicular transport, namely budding and fusion, both of which involve a complex network of molecular interactions between vesicles, transported molecules and recipient organelles [512, 566]. Such nonlinear cooperative interactions provide the basic mechanism for the self-organization of distinct organelles and, hence, we refer to the whole process as an example of cooperative vesicular transport.

Budding from a donor compartment is mediated by cytosolic protein coats that bind to the membrane, induce curvature, and eventually pinch off a vesicle. Protein coats, which are recruited to the membrane by the hydrolysis of G-proteins, are also involved in the selective concentration of specific proteins within the budding vesicle. Following budding and detachment from the donor compartment, vesicles are transported to the acceptor compartment where they undergo fusion. The latter is mediated by the pair-wise interaction of vesicle ($v-$) and target ($t-$)

soluble N-ethylmaleimide-sensitive factor attachment protein receptors (SNARES). Although $v-$snares and $t-$snares occur in multiple homologous variants, they tend to form high-affinity pairs that provide a recognition mechanism for membrane fusion between specific vesicles and a compartment. Since each type of protein coat preferentially loads particular high-affinity pairs of SNARES during budding, it follows that a given vesicle fuses preferentially with a compartment that contains higher levels of these SNARES, thus further increasing their concentration within the compartment. In the case of a single type of protein coat mediating vesicular transport between two compartments, this mechanism would lead to a steady state in which many vesicles with a low SNARE content move in one direction (from the first to the second compartment, say), whereas a few vesicles with a large SNARE content move in the opposite direction. The total protein fluxes would then be balanced, while maintaining a higher concentration of SNARES in the second compartment. However, lipid balance would not be maintained because there would be a net flux of vesicles in the anterograde direction. The simultaneous balance of lipid fluxes could be achieved by having a second type of protein coat that preferentially loads a different set of SNARES that are concentrated in the first compartment, resulting in a net flux of vesicles in the retrograde direction. The role of protein coats and SNARES in vesicular budding and fusion forms the basis of the model introduced by Heinrich and Rapoport [406]. Note that in the case of the early secretory pathway coat protein complex II (COPII) vesicles mediate transport from the ER to the Golgi, whereas coat protein complex I (COPI) vesicles mediate transport in the opposite direction.

We will describe a simplified version of the model by Rapoport and Heinrich [406], see [111, 245]. Suppose that there are two compartments $j = 1, 2$ that contain two types of surface protein $X = U, V$ whose concentrations at time $t$ are denoted by $U_j(t), V_j(t)$, respectively, see Fig. 13.20. For simplicity, we do not explicitly model the interactions between complementary SNARE proteins, that is, we take each protein $U, V$ to be self-interacting. These proteins are exchanged between the compart-

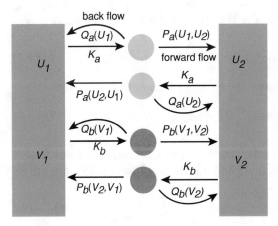

Fig. 13.20: Schematic diagram of the exchange of vesicles between two organelles. Shown is the limiting case in which one class of vesicle (shaded blue, labeled $\alpha = a$) only carries protein U and the other type of vesicle (shaded red, labeled $\alpha = b$) only carries protein V. When a vesicle of type $\alpha$ buds from a given compartment, it can immediately re-fuse with the same compartment at a rate $Q_\alpha$ or fuse with the other compartment at a rate $P_\alpha$. Both compartments produce vesicles of type $\alpha$ at a rate $K_\alpha$.

ments via two types of protein-coated vesicles labeled by $\alpha = a, b$ with $a$-vesicles transporting protein $U$ and $b$-vesicles transporting protein $V$. As a further simplification, we neglect the actual physical process of vesicular transport and the associated transport delays. Thus, we assume vesicles that bud from the $j$th compartment are immediately available for fusion with either compartment. Let $Q_a(U_j)$ denote the backward rate per unit area of fusion of an $a$-vesicle with its source compartment, whereas $P_a(U_j, U_{\bar{j}})$ is the forward rate per unit area at which an $a$-vesicle from compartment $j = 1, 2$ fuses with compartment $\bar{j} = 2, 1$. The backward and forward rates for $a$-vesicles are taken to have the explicit form [406]

$$Q_a(U_j) = \gamma + \kappa f_a(U_j)U_j, \tag{13.4.1}$$

$$P_a(U_j, U_{\bar{j}}) = \gamma + \kappa f_a(U_j)U_{\bar{j}}. \tag{13.4.2}$$

Here $f_a(U)$ is the protein and coat specific concentration within an $a$-vesicle generated from a compartment with concentration $U$. It is taken to have the Michaelis–Menten form

$$f_a(U) = W_a \frac{U}{U + C_a}. \tag{13.4.3}$$

(In a more general model, vesicles would transport both proteins in a competitive manner [111, 245, 357, 406].) Note that $\gamma$ can be interpreted as a background fusion rate, whereas $\kappa$ determines the increase in the reaction rate due to binary interactions between surface proteins of the vesicle and compartment, respectively, based on the law of mass action. For simplicity, $\kappa$ is taken to be independent of the particular protein. Analogous equations hold for $b$-vesicles with $a \to b$ and $U \to V$.

Let $N_j^\alpha(t)$ denote the number density of $\alpha$-vesicles produced by the $j$th compartment at time $t$, $\alpha = a, b$. The number of vesicles evolves according to the equation

$$\frac{dN_j^a}{dt} = A_j K_a - A_j Q_a(U_j)N_j^a - A_{\bar{j}} P_\alpha(U_j, U_{\bar{j}})N_j^a. \tag{13.4.4}$$

Here $A_j(t)$ is the surface area of the $j$th compartment and $K_\alpha$ is the rate of production of $\alpha$-vesicles. Assuming that all vesicles have the same surface area $\Delta A$ with $\Delta A \ll A_j$, the rate of change of the number of U proteins within the $j$th compartment is

$$\frac{d[A_j U_j]}{dt} = A_j \Delta A \left[ -K_a + Q_a(U_j)N_j^a + P_a(U_{\bar{j}}, U_j)N_{\bar{j}}^a \right] f_a(U_j). \tag{13.4.5}$$

We also have a conservation equation for the total number $M_U$ of U proteins:

$$M_U = \sum_{j=1,2} \left( A_j(t)U_j(t) + \Delta A f_a(U_j)N_j^a(t) \right). \tag{13.4.6}$$

Identical equations hold for $b$-vesicles and $V$ proteins by taking $a \to b$ and $U \to V$ in equations (13.4.4), (13.4.5), and (13.4.6). In this simplified model, competition between the proteins arises via the transport of membrane. The rate of change of membrane surface area of the $j$th compartment is

$$\frac{d[A_j]}{dt} = A_j \Delta A \left[ -K_a + Q_a(U_j)N_j^a + P_a(U_{\bar{j}}, U_j)N_{\bar{j}}^a \right] \tag{13.4.7}$$

$$+ A_j \Delta A \left[ -K_b + Q_b(V_j)N_j^b + P_b(V_{\bar{j}}, V_j)N_{\bar{j}}^b \right], \tag{13.4.8}$$

which is supplemented by the conservation condition for the total amount of membrane $A_{\text{tot}}$

$$A_{\text{tot}} = \sum_{j=1,2} \left( A_j(t) + \Delta A[N_j^a(t) + N_j^b(t)] \right). \tag{13.4.9}$$

We will focus on steady-state solutions. Setting to zero all time-derivatives in equation (13.4.4) with $U_j, V_j$ fixed, immediately implies that at steady state

$$N_j^a = \frac{A_j K_a}{A_j Q_a(U_j) + A_{\bar{j}} P_a(U_j, U_{\bar{j}})}, \quad N_j^b = \frac{A_j K_b}{A_j Q_b(V_j) + A_{\bar{j}} P_b(V_j, V_{\bar{j}})}. \tag{13.4.10}$$

Setting $d[A_j X_j]/dt = 0$ in equation (13.4.5) and substituting the steady-state expressions for $N_j^a, N_j^b$ gives

$$G_a(U_1, U_2) f_a(U_1) = \widehat{G}_a(U_1, U_2) f_a(U_2), \tag{13.4.11a}$$

$$G_b(V_1, V_2) f_b(V_1) = \widehat{G}_b(V_1, V_2) f_b(V_2), \tag{13.4.11b}$$

where

$$G_a(U_1, U_2) = \frac{P_a(U_1, U_2)}{A_1 Q_a(U_1) + A_2 P_a(U_1, U_2)}, \quad \widehat{G}_a(U_1, U_2) = \frac{P_a(U_2, U_1)}{A_2 Q_a(U_2) + A_1 P_a(U_2, U_1)}, \tag{13.4.12}$$

and similarly for $a \to b$, $U \to V$. The steady-state version of (13.4.7) yields

$$K_a G_a(U_1, U_2) + K_b G_b(V_1, V_2) = K_a \widehat{G}_a(U_1, U_2) + K_b \widehat{G}_b(V_1, V_2). \tag{13.4.13}$$

It turns out that the system undergoes a symmetry breaking bifurcation even if the kinetics of both proteins are the same, that is, $W_a = W_b = W$, $C_a = C_b$, and $K_a = K_b = K$. In this case the functions $f_\alpha, P_\alpha, Q_\alpha, G_\alpha, \widehat{G}_\alpha$ become independent of the label $\alpha = a, b$. In this simplified case, one solution to equations (13.4.11a)–(13.4.13) is the symmetric solution

$$(U_1, V_1, A_1) = (U_2, V_2, A_2) = (U^*, V^*, A^*),$$

with $(U^*, V^*, A^*)$ determined from the conservation conditions (13.4.9) and (13.4.6). In order to search for symmetry-breaking solutions (non-identical protein concentrations in the two compartments), suppose that $A_1 = A_2 = A_{\text{tot}}/2$ (equal surface areas) and

$$M_U = M_V \equiv M. \tag{13.4.14}$$

Assuming that $\varepsilon = \Delta A/A_{\mathrm{tot}} \ll 1$, the conservation conditions (13.4.9) and (13.4.6) imply that

$$A_2 \approx A_{\mathrm{tot}} - A_1, \quad A_2 X_2 \approx M - A_1 X_1 \tag{13.4.15}$$

for $X = U, V$, and the approximate solution of equations (13.4.11a), (13.4.11b) and (13.4.13) is

$$U_1 = C, \quad U_2 = C_{\mathrm{tot}} - C, \quad V_1 = C_{\mathrm{tot}} - C, \quad V_2 = C, \tag{13.4.16}$$

with $C_{\mathrm{tot}} = 2M/A_{\mathrm{tot}}$ and $C$ satisfying the equation

$$\frac{P(C, C_{\mathrm{tot}} - C)f(C)}{Q(C) + P(C, C_{\mathrm{tot}} - C)} = \frac{P(C_{\mathrm{tot}} - C, C)f(C_{\mathrm{tot}} - C)}{Q(C_{\mathrm{tot}} - C) + P(C_{\mathrm{tot}} - C, C)}. \tag{13.4.17}$$

A symmetry-breaking solution exists if equation (13.4.17) has solutions for which $C \neq C_{\mathrm{tot}}/2$. Setting

$$G(C) = \frac{P(C, C_{\mathrm{tot}} - C)f(C)}{Q(C) + P(C, C_{\mathrm{tot}} - C)}, \tag{13.4.18}$$

we plot $G(C)$ and $G(C_{\mathrm{tot}} - C)$ in Fig. 13.21(a) for various values of $\kappa$. For each $\kappa$, the intersections of the solid curve $G(C)$ with the dashed curve $G(C_{\mathrm{tot}} - C)$ determines the possible steady-state values for $C$. As $\kappa$ increases, the symmetric solution at $C = C_{\mathrm{tot}}/2$ bifurcates into two additional non-symmetric solutions at a critical value $\kappa_c$, as illustrated in Fig. 13.21(b). Note that $G(C)$ is a measure of the steady-state protein flux of $U$ in the anterograde direction, which is balanced by the corresponding flux $G(C_{\mathrm{tot}} - C)$ in the opposite direction. (In the given parameter regime, the steady-state fluxes of proteins $U$ and $V$ have the same magnitude.) Fig. 13.21(b) shows that the stable flux is independent of $\kappa$ for $\kappa < \kappa_c$ and a decreasing function of $\kappa$ for $\kappa > \kappa_c$. A similar bifurcation scenario occurs when other parameters are varied such as the baseline flux rate $\gamma$.

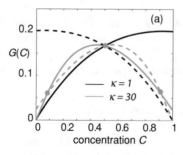 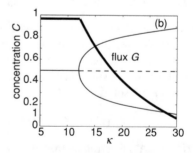

Fig. 13.21: Symmetry breaking in two compartment model. (a) Plot of functions $G(C)$ (solid curves) and $G(C_{\mathrm{tot}} - C)$ (dashed curves) for various values of $\kappa$. Points of intersection of $G(C)$ and $G(C_{\mathrm{tot}})$ for fixed $\kappa$ determine steady-states. Other parameter values are $\gamma = 1, W = 1, C_S = 1$, $A_{\mathrm{tot}} = 1$, and $C_{\mathrm{tot}} = 1$. (b) Bifurcation diagram for steady-state concentration $C$ in compartment 1 as a function of the parameter $\kappa$. Thick curve shows the variation of the steady-state flux $G(C)$ along a stable branch.

Finally, note that in the case of small cellular organelles, protein sorting can be affected by stochastic fluctuations in the number of budding and fusing vesicles. This has been explored by Vagne and Sens [895], who treat the perfect sorting of two membrane components that are initially mixed in a single compartment as a first passage time problem. They show that the mean sorting time displays two distinct regimes that depend on the ratio of vesicle fusion to budding rates. In the case of relatively fast budding, sorting is fast but leads to a broad size distribution of sorted compartments. On the other hand, relatively fast fusion results in two well-defined sorted compartments, but sorting is exponentially slow.

## 13.5 Exercises

**Problem 13.1. Phase separation in a lattice model.** The free energy density of a solution obtained from a simple lattice model is, see Sect. 13.1.1

$$f(\phi) = \frac{k_B T}{v_c} [\phi \ln \phi + (1 - \phi) \ln(1 - \phi) + \chi \phi (1 - \phi)].$$

Note that $f(\phi)$ is an even function about $\phi = 1/2$.

(a) Plot $f(\phi)$ for various values of $\chi$ (with $k_B T / v_c = 1$) in order to establish that $f(\phi)$ has a single minimum for sufficiently small $\chi$ but switches to two local minima as $\chi$ crosses $\chi = \chi_c = 2$.

(b) Show that the spinodal line with $f''(\phi) = 0$ is given by

$$\chi = \frac{1}{2} \frac{1}{\phi(1 - \phi)} \geq 2.$$

(c) Use the common tangent construction to show that the coexistence curve is determined by the equation $f'(\phi) = 0$ and thus

$$\chi = \frac{1}{1 - 2\phi} \ln \left( \frac{1 - \phi}{\phi} \right).$$

(d) Construct the phase diagram in the $\phi - \chi$ plane.

(e) Determine the spinodal line and critical point for a solution of polymers of length $M$.

**Problem 13.2. Surface tension of a 1D interface.** The free energy of a binary mixture with space-dependent concentration $\phi(\mathbf{r})$ is (see Sect. 13.1)

$$F[\phi] = \int \left[ \frac{\alpha}{2} (\nabla \phi)^2 + f(\phi) \right] d^3 \mathbf{r},$$

where $f$ is the local free energy density (13.1.7).

(a) Suppose that the solution is close to the critical point for phase separation so that we can write $\phi = 1/2 + \psi$ with $\psi$ small. Show that to $O(\psi^4)$ the free energy can be approximated by carrying out a Ginsburg–Landau expansion in $\psi$:

$$F[\psi] = \int \left[ -\frac{\varepsilon}{2} \psi(\mathbf{r})^2 + \frac{c}{4} \psi(\mathbf{r})^4 + \frac{\alpha}{2} (\nabla \psi)^2 \right] d^3\mathbf{r},$$

where we have dropped constant terms and set

$$b = \frac{2k_B T}{v_c} (\chi - 2), \quad c = \frac{16 k_B T}{3 v_c}.$$

Note that $\chi \approx 2$ so that $b$ is small.

(b) The equilibrium state is determined by minimizing the free energy under the constraint $\int \psi(\mathbf{r}) d^3\mathbf{r} = \text{constant}$:

$$\frac{\delta F[\psi]}{\delta \psi(\mathbf{r})} = \mu,$$

where $\mu$ is the chemical potential. Taking $\int \psi(\mathbf{r}) d^3\mathbf{r} = 0$ so that $\mu = 0$ derive the equation

$$-\varepsilon \psi(\mathbf{r}) + a \psi(\mathbf{r})^3 - b \nabla^2 \psi(\mathbf{r}) = 0.$$

(c) Consider the 1D equation

$$-b\psi(x) + c\psi(x)^3 - \alpha \frac{d^2}{dx^2} \psi(x) = 0.$$

Interpret the solution as an interface separating the two binary phases by imposing the boundary conditions

$$\lim_{x \to \pm\infty} \psi(x) = \pm\sqrt{\frac{b}{c}}.$$

Obtain the interface solution

$$\psi(x) = \sqrt{\frac{b}{c}} \tanh(x/\xi), \quad \xi = \sqrt{2\alpha/b},$$

and interpret $\xi$.

(d) Set

$$f(\psi) = -\frac{b}{2} \psi^2 + \frac{c}{4} \psi^4.$$

Multiply the ODE in part (c) by $\psi'(x)$ and show that

$$f(\psi(x)) - \frac{\alpha}{2} (\psi'(x))^2 = \text{constant} = f(\psi_0),$$

that is, the free energy density is the same as that of a uniform bulk solution $\psi_0$. Defining the surface tension (interfacial free energy) by $\Gamma = F[\psi] - F[\psi_0]$, show that

$$\Gamma = \alpha \int_{-\infty}^{\infty} \psi'(x)^2 dx.$$

**Problem 13.3. Fluctuations of a curved interface.** Let us return to the free energy of Ex. 13.2:

$$F[\phi] = \int \left[ \frac{\alpha}{2} (\nabla \phi)^2 + f(\phi) \right] d^3 \mathbf{r},$$

where $f$ is the local free energy density (13.1.7).

(a) Consider a curved surface interface solution $\phi(\mathbf{r}) = \Phi(x, y, z - h(x, y))$. Substitute this solution into the free energy functional and derive a PDE for $\Phi$ by minimizing the free energy according to

$$\frac{\delta F[\Phi]}{\delta \Phi(\mathbf{r})} = 0.$$

Show that the PDE reduces to the form

$$\frac{\partial f(\Phi)}{\partial \Phi} = \alpha \left[ 1 + h_x^2 + h_y^2 \right] \Phi_{zz},$$

assuming that spatial variations of $\Phi$ with respect to $x, y$ are much smaller than with respect to $z$ and dropping terms involving $h_{xx}, h_{yy}$. It follows that $\Phi$ can be approximated by

$$\Phi(\mathbf{r}) = \Psi \left( \frac{z - h}{\sqrt{1 + h_x^2 + h_y^2}} \right),$$

where

$$\alpha \Psi_{zz} = \frac{\partial f(\Psi)}{\partial \Psi}.$$

(b) Multiply both sides of the above PDE for $\Phi$ by $\Phi_z$ and integrate with respect to $z$ using the boundary conditions $\Phi(z) \to \phi_\pm$ for $z \to \pm\infty$, where $\phi_\pm$ are the bulk compositions that minimize $f(\phi)$. Use this to show that the surface free energy is

$$F_S[h] \equiv F[\Phi] - F[\phi_-] = \Gamma \int \sqrt{1 + h_x^2 + h_y^2} \, dx dy,$$

where

$$\Gamma = \alpha \int \Psi_z(z)^2 dz.$$

Hint: Note that

$$\int (\nabla \phi)^2 d^3 \mathbf{r} = -\int \nabla^2 \phi \, d^3 \mathbf{r} \approx -\int (1 + h_x^2 + h_y^2) \Phi_z^2 d^3 \mathbf{r}.$$

(c) For small curvature, the surface free energy can be approximated by

$$F_S[h] = \frac{\Gamma}{2} \int_\Lambda \left( h_x^2 + h_y^2 \right) dxdy.$$

Taking $\Lambda$ to be a square of sides $L$, introduce the partition function

$$Z = \int \prod_{\mathbf{r}} dh(\mathbf{r}) e^{-F_s[h]/k_B T}.$$

Impose periodic boundary conditions and expand $h(\mathbf{r})$, $\mathbf{r} = (x,y)$, as a Fourier series with $\widehat{h}(0) = 0$,

$$h(\mathbf{r}) = \frac{1}{A} \sum_{\mathbf{q}} \widehat{h}(\mathbf{q}) e^{i\mathbf{q}\cdot\mathbf{r}}, \quad \widehat{h}(\mathbf{q}) = \int_\Omega \widehat{h}(\mathbf{q}) e^{-i\mathbf{q}\cdot\mathbf{r}} dxdy,$$

with

$$q_x = \frac{2\pi n}{L}, \quad q_y = \frac{2\pi m}{L}, \quad \text{for integers } n, m.$$

Hence, show that

$$Z = \prod_{\mathbf{q} \neq 0} \left( \frac{2\pi L^2}{\beta \Gamma} \right)^{1/2} \frac{1}{q}.$$

(d) Introduce the generating functional

$$\Gamma[v] = \int \exp\left( -\beta \left[ F_S[h] - \int_\Omega h(\mathbf{r}) v(\mathbf{r}) dxdy \right] \right) D[h]$$

such that

$$\langle h(\mathbf{r}) \rangle = \frac{1}{\beta} \frac{\delta \ln \Gamma}{\delta v(\mathbf{r})} \bigg|_{v=0},$$

and

$$G(\mathbf{r},\mathbf{r}') := \langle h(\mathbf{r}) h(\mathbf{r}') \rangle - \langle h(\mathbf{r}) \rangle \langle h(\mathbf{r}') \rangle = \frac{1}{\beta^2} \frac{\delta^2 \ln \Gamma}{\delta v(\mathbf{r}) \delta v(\mathbf{r}')} \bigg|_{v=0}.$$

Rewriting $\Gamma[v]$ in terms of Fourier series, show that

$$\langle h(\mathbf{r})^2 \rangle \equiv G(\mathbf{r},\mathbf{r}) = \frac{k_B T}{\Gamma L^2} \sum_{q=q_{min}}^{q_{max}} q^{-2}.$$

Evaluate the sum on the right-hand side in the continuum limit. (See the analysis of membrane fluctuations in Sect. 12.2.)

**Problem 13.4. The Smoluchowski equation.** Consider particles with dilute concentration $\phi(x,t)$ diffusing in a 1D domain. Let $U(x)$ denote an external potential. The free energy functional is given by

$$F[\phi] = \int_{-\infty}^{\infty} [k_B T \phi \ln \phi + \phi U(x)] dx.$$

(a) Using particle conservation and the generalized Fick's law (13.1.19), derive the Smoluchowski equation

$$\frac{\partial \phi}{\partial t} = D \frac{\partial}{\partial x} \left[ \frac{\partial \phi}{\partial x} + \frac{\phi}{k_B T} \frac{\partial U}{\partial x} \right],$$

where $D = k_B T / \eta$ and $\eta$ is the friction coefficient.

(b) In the case of a gravitational potential $U(x) = mgx$, where $m$ is mass of a single particle and $g$ is the gravitational acceleration, shows that the equilibrium solution is given by

$$\phi_{eq}(x) \propto \exp\left( -\frac{mgx}{k_B T} \right).$$

**Problem 13.5. Ostwald ripening in two dimensions.** Consider $N$ droplets of radii $R_i = \varepsilon \rho_i$. Let $\Omega_i = \{ \mathbf{x} \in \Omega; |\mathbf{x} - \mathbf{x}_i| \le \varepsilon \rho_i \}$ where $\mathbf{x}_i \in \mathbb{R}^2$ is the center of the droplet. Suppose that the concentration $u(\mathbf{x})$ outside the droplets satisfies the quasi-static diffusion equation

$$\nabla^2 u = 0, \quad \mathbf{x} \in \Omega \backslash \cup_{i=1}^N \Omega_i, \quad \partial_n u = 0 \text{ on } \partial\Omega,$$

$$u = \frac{1}{\rho_i} \text{ on } \partial\Omega_i,$$

and $u \sim u_\infty$ far from any droplet (in the dilute droplet regime). (A factor of $\varepsilon$ has been absorbed into $u$.) Divide the domain $\Omega \backslash \cup_{i=1}^N \Omega_i$ into $N$ inner regions around the droplets and an outer region, and use matched asymptotics along analogous lines to Ex. [6.6].

(a) Introducing the stretched coordinates $\mathbf{y} = \varepsilon^{-1}(\mathbf{x} - \mathbf{x}_i)$, show that the inner solution around the $i$th droplet is

$$w(\mathbf{x}) = \frac{1}{\rho_i} + v A_i(v) \ln(|\mathbf{x} - \mathbf{x}_i|/\varepsilon\rho_i),$$

where $v = -1/\ln \varepsilon$. The coefficients $A_i(v)$ are determined by matching with the outer solution.

(b) Show that the outer solution can be written as

$$u(\mathbf{x}) = u_\infty - 2\pi v \sum_{i=1}^N A_i(v) G(\mathbf{x}, \mathbf{x}_i), \quad \mathbf{x} \notin \{ \mathbf{x}_j, j = 1, \dots, N \},$$

where $G(\mathbf{x}, \mathbf{x}_i)$ is Neumann Green's function,

$$\nabla^2 G = \frac{1}{|\Omega|} - \delta(\mathbf{x} - \mathbf{x}_0), \quad \mathbf{x} \in \Omega, \quad \partial_n G = 0 \text{ on } \partial\Omega, \quad \int_\Omega G d\mathbf{x} = 0.$$

Note that as $\mathbf{x} \to \mathbf{x}_0$, $G$ can be decomposed as

$$G(\mathbf{x}, \mathbf{x}_0) = -\frac{\ln |\mathbf{x} - \mathbf{x}_0|}{2\pi} + R(\mathbf{x}, \mathbf{x}_0),$$

where $R$ is the regular part of Green's function. Deduce the self consistency condition

$$\sum_{i=1}^{N} A_i(v) = 0.$$

By matching inner and outer solutions show that

$$-(1 - v \ln \rho_j + 2\pi v R(\mathbf{x}_j, \mathbf{x}_j)) A_j - 2\pi v \sum_{i \neq j} A_i(v) G(\mathbf{x}_j, \mathbf{x}_i) = \frac{1}{\rho_j} - u_\infty, \quad j = 1, \ldots, N.$$

We thus obtain $N + 1$ equations for the $N + 1$ unknowns $A_1, \ldots, A_N, u_\infty$.
(d) Given that to leading order,

$$A_j = u_\infty - \frac{1}{\rho_j},$$

show that the mean-field $u_\infty$ is given by the harmonic mean $\rho_{\text{harm}}$ of the droplet radii, and write down the leading order expressions for the inner and outer solutions. Hence, show that

$$\frac{d\rho_j}{dt} = \frac{1}{\varepsilon^3} \frac{\partial w}{\partial \rho}\bigg|_{\rho = \rho_j} \sim \frac{v}{\varepsilon^3} \left[ \frac{1}{\rho_{\text{harm}}} - \frac{1}{\rho_j} \right] \frac{1}{\rho_j}.$$

(Note that the physical lengths are $\varepsilon \rho$, etc. and we have taken $w \to \varepsilon w$.)

**Problem 13.6. Phase separation driven by nonequilibrium chemical reactions.**
Consider the system of equations for the $P$ solution in a ternary mixture, see (13.2.6):

$$0 = D\nabla^2 \phi_{\text{in}} - (k + h)\phi_{\text{in}} + h B_{\text{in}}, \quad 0 \leq r < R, \quad \phi_{\text{in}}(R) = \phi_b,$$

and

$$0 = D\nabla^2 \phi_{\text{out}} - (k + h)\phi_{\text{out}} + h B_{\text{out}}, \quad r > R,$$

with

$$\phi_{\text{out}}(R) = \phi_a \left(1 + \frac{l_c}{R}\right), \quad \phi_{\text{out}}(r) \to \phi_\infty \text{ as } r \to \infty.$$

Show that the solution to the boundary value problem is

$$\phi_{\text{in}}(r) = \left(\phi_b - \frac{h B_{\text{in}}}{h + k}\right) \frac{R}{r} \frac{\sinh(r/\xi)}{\sinh(R/\xi)} + \frac{h B_{\text{in}}}{h + k},$$

and

$$\phi_{\text{out}}(r) = \phi_\infty - \frac{R}{r}e^{-(r-R)/\xi}\left(\Delta - \frac{\phi_a \ell_c}{R}\right),$$

where $\Delta = \phi_\infty - \phi_a$, and $\xi = \sqrt{D/(h+k)}$.

**Problem 13.7. Phase separation in a concentration gradient.** Suppose that the free solute concentration around a single droplet of radius $R$ satisfies the diffusion equation

$$\nabla^2 \phi = 0, \quad R < r < \infty, \quad \phi(R) = (\alpha + \beta R\cos\theta)\left(1 + \frac{\ell_c}{R}\right),$$

together with the far-field condition

$$\phi(\mathbf{r}) = \alpha_\infty + \beta_\infty r\cos\theta, \quad \text{as } r \to \infty.$$

(a) Show that the solution to this boundary value problem is

$$\phi(r,\theta) = \alpha_\infty\left(1 - \frac{R}{r}\right) + \beta_\infty\left(r - \frac{R^3}{r^2}\right)\cos\theta + (\alpha r/R + \beta R\cos\theta)\left(1 + \frac{\ell_c}{R}\right)\frac{R^2}{r^2}.$$

(b) Using the solution in part (a), evaluate the integral on the right-hand side of the equations

$$\frac{dR}{dt} = \frac{D}{2\phi_b}\int_0^\pi \nabla\phi(R,\theta)\cdot\mathbf{e}_r\sin\theta d\theta,$$

and

$$\frac{dx}{dt} = \frac{D}{2\phi_b}\int_0^\pi \nabla\phi(R,\theta)\cdot\mathbf{e}_r\cos\theta\sin\theta d\theta.$$

**Problem 13.8. Modified Becker-Döring model for reversible vesicular transport.** Consider the modified BD model (see Fig. 13.22)

$$\frac{du_n}{dt} = J_{n-1} - J_n, \quad n \geq 1, \quad \frac{du_0}{dt} = -J_0, \quad \frac{dc}{dt} = -\sum_{n\geq 0} J_n,$$

with fluxes

$$J_n = a_n c u_n - b_{n+1} u_{n+1}.$$

These equations have been used to model a non-spatial version of reversible vesicular transport along an axon (see Chap. 7 and Ref. [118]). In particular, $u_n, n = 0, 1, ..,$ denotes the number of motor complexes at time $t$ that are carrying $n$ vesicles, whereas $c(t)$ is the number of vesicles in the axonal membrane. (It is assumed that

Fig. 13.22: Schematic diagram of reversible vesicular transport model. Each motor-complex can reversibly exchange a vesicle with a synaptic target, and there is clustering of vesicles bound to motors and bound to targets.

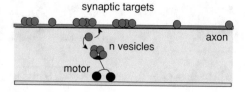

there is no upper bound for the carrying capacity of a motor complex—this is not a major issue, since the steady-state solution satisfies $\lim_{n\to\infty} u_n = 0$.) Moreover, it is assumed that motor complexes can only exchange one vesicle at a time with synaptic targets in the membrane. We thus have the following reaction scheme

$$X + U_n \xrightarrow{a_n} U_{n+1}, \quad U_n \xrightarrow{b_n} X + U_{n-1}, \quad n \geq 1, \quad X + U_0 \xrightarrow{a_0} U_1.$$

Here $X$ denotes a membrane-bound vesicle, $U_n$ denotes a motor-complex with $n$ vesicles, $b_n$ is the rate at which a vesicle is transferred from the complex to a synaptic target, and $a_n$ is the rate of the reverse process. Hence, in the transport model, membrane-bound vesicles play the role of monomers and motor complexes play the role of clusters, with $n$ now labeling the number of motor-bound vesicles rather than cluster size. (Another major difference between the transport model and cluster formation models is that the fastest diffusing element in the latter is a monomer, whereas in the transport model the "monomer" is membrane bound and does not diffuse.)

(a) By summing the equations for $\dot{u}_n$ with respect to $n$ and assuming uniform convergence, show that the total number of motor complexes is conserved, $\sum_{n\geq 0} u_n(t) = M$. Hence show that the total number of vesicles is also $c(t) + \sum_{n\geq 1} n u_n(t) = V$.

(b) Using the results of part (a), derive the following conditions for the existence of a steady-state solution $J_n = 0$:

$$M = u_0 \left( 1 + \sum_{n\geq 1} Q_n c^n \right) \equiv u_0 (1 + F_0(c)),$$

and

$$V = c + \left( \sum_{n\geq 1} n Q_n c^n \right) u_0 = c + \frac{F_1(c)}{1 + F_0(c)} M,$$

where

$$F_0(c) = \sum_{n\geq 1} Q_n c^n, \quad F_1(c) \equiv c F_0'(c) = \sum_{n\geq 1} n Q_n c^n.$$

Give a necessary condition for the existence of a steady state in terms of $c$ and the convergence properties of $F_0, F_1$.

(c) Suppose that $a_n = a$ and $b_n = b$ for all $n$. Calculate the functions $F_0(c)$ and $F_1(c)$ assuming $ac/b < 1$, and thus obtain the result

$$V = c + \frac{ac/b}{1-(ac/b)^2}M.$$

(d) Show that the function

$$\widehat{L} = \sum_{n=0}^{\infty} u_n\left[\log(u_n/Q_n c^n) - 1\right],$$

with $Q_0 = 1$, satisfies

$$\frac{d\widehat{L}}{dt} = \sum_{n\geq 0}(a_n c u_n - b_{n+1}u_{n+1})\log(b_{n+1}u_{n+1}/a_n c u_n) - \frac{d}{dt}\left[\rho\log c - c\right].$$

Hence, introducing the modified function $L = \widehat{L} + \rho\log c - c$, show that $dL/dt \leq 0$. Also establish that $L_0$, and hence $L$, are bounded from below, which means that $L$ is a Liapunov function. Deduce that

$$u_n - Q_n c^n u_0 \to 0 \quad \text{as } t \to \infty, \quad n \geq 1.$$

**Problem 13.9. Self-assembly of cylindrical micelles.** Consider the BD equations (13.3.1a) and (13.3.1b):

$$\frac{du_n}{dt} = J_{n-1} - J_n, \quad n \geq 2,$$
$$\frac{du_1}{dt} = -J_1 - \sum_{n\geq 1}J_n,$$

with particle fluxes

$$J_n = a_n u_1 u_n - b_{n+1}u_{n+1},$$

where $u_n$ is the concentration of an $n$-micelle, and set $u_1 = u$. In the case of cylindrical micelles, take

$$b_n = ua_{n-1}e^{[G_n - G_{n-1}]/k_B T}, \quad G_n = -\varepsilon_n - k_B T(n-1)\ln u, \quad \varepsilon_n = (n-1)\alpha k_B T.$$

(a) Show that for cylindrical micelles, if $b_k = 1$ for all $k$ then the BD equations can be written as

$$\frac{du_n}{dt} + (e^\alpha u - 1)(u_n - u_{n-1}) = u_{n+1} - 2u_n + u_{n-1}, \quad n \geq 2,$$

where $u_n$ is the concentration of an $n$-micelle, which is supplemented by the mass conservation condition $\sum_{n=1}^{\infty} nu_n = \rho_0$.

(b) By rescaling the cluster densities and time according to

$$u_n = \rho_0 r_n, \quad \tau = e^{\alpha} \rho_0 t = \frac{t}{\varepsilon},$$

with $0 \ll \varepsilon < 1$, show that the BD equations can be rewritten as

$$\frac{dr_n}{d\tau} + (r_1 - \varepsilon)(r_n - r_{n-1}) = \varepsilon(r_{n+1} - 2r_n + r_{n-1}), \quad n \geq 2,$$

with $1 = \sum_{n=1}^{\infty} n r_n$.

(c) Multiplying the above equation by $n$ and summing with respect to $n$, derive the equation

$$\frac{dr_1}{d\tau} + r_1(r_1 + r_c) + \varepsilon(r_1 - r_2 - r_c) = 0,$$

where $r_c = \sum_{n \geq 1} r_n$. Similarly, directly summing equation with respect to $n$, show that

$$\frac{dr_c}{d\tau} + r_1 r_c + \varepsilon(r_1 - r_c) = 0.$$

# Chapter 14
# Dynamics and regulation of the cytoskeleton

Another system where the principles of self-organization are widely applied is the cytoskeleton [427, 442, 627, 662, 787, 967]. As described in Chap. 4 and Sect. 12.4, the cytoskeleton is made up of polymers such as microtubules (MTs) and F-actin, which determine cell size, shape and polarity, maintain the structural integrity of the cell, and form the mitotic spindle apparatus during cell division. F-actin and MTs are polarized filaments that are intrinsically unstable, undergoing continuous turnover of sub-units by addition at their plus end and depolymerization at their minus end. The continuous dynamical exchange of the sub-units and their interactions with filament-associated proteins provides the basis for the self-organization of different cytoskeletal structures. This has been demonstrated *in vitro* for MT networks, which form by simply combining tubulin, MT motors, and ATP in solution [662]. By varying the relative concentrations of motors and tubulin, one can generate different network patterns, including random networks, vortices, and asters.

In this chapter, we focus on the dynamical role of polymerization in the function of the cytoskeleton, and how it is regulated. We begin in Sect. 14.1 by considering several distinct mechanisms for filament length regulation, including interactions between polymerizing filaments and molecular motors or protein concentration gradients, and a diffusion-secretion model of flagellar length control in *Salmonella*. We then develop a detailed stochastic model of eukaryotic flagellar length control (Sect. 14.2), which is analyzed in terms of a doubly stochastic Poisson process (Box 14A). The dynamics of the mitotic spindle during various stages of cell mitosis is then considered in Sect. 14.3. We describe the search-and-capture model for the interactions between chromosomes and MTs of the mitotic spindle, dynamical instabilities in the positioning of the chromosomes, and spindle length control. In Sect. 14.4 we discuss the important role of actin polymerization and Brownian ratchets in cell motility. This also requires an introduction to the theory of cell adhesion. Finally, in Sect. 14.5 we develop the theory of cytoneme-based morphogenesis. Cytonemes are thin, actin-rich filaments that can dynamically extend up to several hundred microns to form direct cell-to-cell contacts.

© Springer Nature Switzerland AG 2021
P. Bressloff, *Stochastic Processes in Cell Biology*, Interdisciplinary Applied
Mathematics 41, https://doi.org/10.1007/978-3-030-72519-8_14

## 14.1 Filament length regulation

At least three distinct classes of filament length-control mechanisms have been identified [747].

1. *Molecular rulers.* In the case of linear structures such as filaments, size control can be achieved by a molecular ruler protein, whose length is equal to the desired length of the growing structure. One classical example is the length of the $\lambda$-phage tail, which is determined by the size of the gene H product (gpH) [491]. During assembly of the tail, gpH is attached to the growing end in a folded state, and protects the growing end from the terminator gene product U (gpU). As the tail elongates, gpH stretches such that when it is fully extended, further growth exposes the tail to the action of gpU, see Fig. 14.1.

2. *Quantal synthesis.* Size could be controlled by synthesizing exactly enough material to build a structure of the appropriate size—a process known as quantal synthesis. For example, precursor protein levels are known to affect the length of flagella in the unicellular green alga *Chlamydomonas reinhardtii* [551], and the length of sea urchin cilia is correlated with the concentration of the protein tektin [843]. One prediction of the quantal synthesis model is that doubling the number of flagella should halve their length. However, studies of *Chlamydomonas* mutants indicate a much weaker dependence of length on the number of flagella, suggesting that there is an additional length-control mechanism involving dynamic balance [595], see below.

3. *Dynamic balance.* Dynamic structures are constantly turning over so that in order for them to maintain a fixed size, there must be a balance between the rates of assembly and disassembly. If these rates depend on the size in an appropriate way, then there will be a unique balance point that stabilizes the size of the organelle. For example, eukaryotic flagellar MTs undergo continuous assembly and disassembly at their tips, in which a constant rate of disassembly is balanced by a length-dependent rate of assembly due to a fixed number of molecular motors transporting tubulin dimers from the cell body, leading to a fixed flagellar length [593, 595]. The transport of tubulin is also thought to be one possible mechanism for controlling axonal elongation. In this case diffusion, possibly combined with an active component, transports newly synthesized tubulin at the somatic end of the axon to the + end of MTs at the axonal tip. As the axon grows, the concentration of tubulin at the tip decreases until there is a bal-

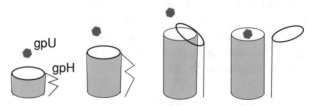

Fig. 14.1: Mechanism of cell size control: Molecular ruler in bacteriophage tail.

ance between polymerization and depolymerization [606], see Ex. 14.1. A different balance mechanism appears to control the length of MTs in yeast, where kinesin motors move processively to the MT tips where they catalyze disassembly. Longer MTs recruit more kinesin motors from the cytoplasm, which results in a length-dependent rate of disassembly. When this is combined with a length-independent rate of assembly, a unique steady-state MT length is obtained [899]. In more complex cytoskeletal structures such as the mitotic spindle (Sect. 14.3), it is likely that multiple length-dependent mechanisms play a role [150]. Dynamic balance mechanisms also control other filament-like structures. For example, actin filaments within stereocilia of the inner ear [778], constantly treadmill back toward the cell body, with disassembly at the base balanced by assembly at the tip. The latter depends on the diffusion of actin monomers to the tip, which results in a length-dependent rate of assembly. Yet another example is the control of the hook length in bacterial flagella [495].

In this section we consider several dynamic balance mechanisms for filament length-control. We begin by describing a simple model of regulation based on filament-motor interactions, and then turn to an alternative mechanism based on the action of a protein concentration gradient. Finally, we consider a diffusion-based secretion model of filament assembly in *Salmonella*. The role of intraflagellar transport in the length control of Eukaryotic flagella will be considered in Sect. 14.2

### 14.1.1 Filament length regulation by motor-tip interactions

There have been a number of theoretical models that combine the dynamics of molecular motors with the dynamics of MT assembly and disassembly. Several studies have focused on motor regulation of depolymerization at the tip combined with constant growth [371, 426, 613, 756] or treadmilling [477], whereas others have considered the effects of motors on dynamic instabilities, that is, the frequency of catastrophes [336, 337, 381, 529, 878]. Here we describe a simplified version of the first mechanism along the lines of [426, 529]. The regulation of the catastrophe rate is considered in Ex. 14.2.

Let $\rho(x,t)$ denote the density of molecular motors along a single filament, which is assumed to evolve according to the mean-field model [529, 713]

$$\frac{\partial \rho}{\partial t} = -v\frac{\partial}{\partial x}[\rho(1-\rho/\rho_{max})] + k_{on}c(1-\rho/\rho_{max}) - k_{off}\rho. \qquad (14.1.1)$$

Here $v$ is the speed of the motors in the absence of crowding, $\rho_{max}$ is the maximum possible motor density, and the factors $(1-\rho/\rho_{max})$ take into account the effects of molecular crowding on the processivity of the motors and on their rate of binding to the filament (see also Sect. 7.6). The latter is proportional to the bulk motor concentration $c$ in the cytoplasm, which is assumed fixed. As a further simplification, the effects of crowding on the drift term are neglected so that the PDE is linear. Introducing the fractional motor density $p(x,t) = \rho(x,t)/\rho_{max}$, we have

$$\frac{\partial p}{\partial t} = -v\frac{\partial p}{\partial x} + \bar{k}_{on}(1-p) - k_{off}p, \quad \bar{k}_{on} = \frac{k_{on}C}{\rho_{max}}. \tag{14.1.2}$$

Equation (14.1.2) is supplemented by the boundary condition $p(0,t) = 0$ (absorbing boundary condition at the base of the filament) and the initial condition $p(x,0) = 0$. Away from the tip of the MT, it is assumed that the motor density reaches a quasi-steady state $p_s$ given by the solution to the equation

$$\frac{dp_s}{\partial x} + \frac{1}{\lambda}p_s = \frac{\bar{k}_{on}}{v}, \quad \lambda = \frac{v}{k_{off} + \bar{k}_{on}}.$$

Hence,

$$p_s(x) = p_0\left[1 - e^{-x/\lambda}\right], \quad p_0 = \frac{\bar{k}_{on}\lambda}{v} = \frac{\bar{k}_{on}}{k_{off} + \bar{k}_{on}}.$$

Following [426], suppose that the depolymerization rate at the tip is $\gamma p_e(t)$, where $p_e(t)$ is the fractional motor density at the tip and $\gamma$ is a constant. Taking the rate of growth to be a constant $\alpha$, the filament length $L(t)$ evolves as

$$\frac{dL}{dt} = \alpha - \gamma p_e. \tag{14.1.3}$$

Let $a$ be the lattice spacing of a single sub-unit of the filament. The rate at which molecular motors enter the tip (assuming the site is unoccupied) is then $(v - L')/a$ with $L' = dL/dt$. It follows that $p_e$ satisfies the equation

$$\frac{dp_e}{dt} = \frac{v - L'}{a}p(L-a,t)(1 - p_e) - k_{e,off}p_e, \tag{14.1.4}$$

where $k_{e,off}$ is the rate of unbinding at the tip. The quasi-static assumption implies that

$$p(L-a,t) = p_s(L-a) \approx p_0(1 - e^{-L/\lambda}).$$

Therefore, at steady-state ($p_e' = L' = 0$), equation (14.1.4) implies that $p_e = p_e(L)$ with

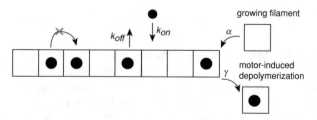

growing filament

motor-induced depolymerization

Fig. 14.2: Schematic diagram of filament length regulation by molecular motor-based depolymerization.

$$p_e(L) = \frac{p_0(1 - e^{-L/\lambda})}{ak_{e,\text{off}}/v + p_0(1 - e^{-L/\lambda})},$$

which can be rearranged to give

$$1 - e^{-L/\lambda} = \frac{ak_{e,\text{off}}}{p_0 v} \frac{p_e}{1 - p_e}.$$

However, we also have $p_e = \alpha/\gamma$, which then determines the filament length to be

$$L_s = -\lambda \ln\left[1 - \frac{ak_{e,\text{off}}}{p_0 v} \frac{\alpha}{\gamma - \alpha}\right]. \tag{14.1.5}$$

The existence of a steady-state length requires the argument of the logarithm to be positive, which means that

$$p_0 > p_{0c} = \frac{\alpha k_{e,\text{off}} a}{v(\gamma - \alpha)}.$$

Note that $p_0$ is a monotonically increasing function of the bulk motor concentration $c$. Hence, for a steady-state length to occur the bulk motor concentration must exceed a minimal concentration $c_0$ with

$$\frac{k_{\text{on}} c_0}{k_{\text{off}} \rho_{\max} + k_{\text{on}} c_0} = \frac{ak_{e,\text{off}}}{v(\gamma/\alpha - 1)},$$

which yields

$$c_0 = \frac{k_{\text{off}} k_{e,\text{off}} \rho_{\max} a}{k_{\text{on}}} \frac{1}{v(\gamma/\alpha - 1) - ak_{e,\text{off}}}. \tag{14.1.6}$$

Note that $L_s$ is a monotonically decreasing function of $c$ with $L_s \to \infty$ as $c \to c_0$ from above.

The above model ignores statistical fluctuations. However, Monte Carlo simulations of a stochastic version of the model show that there is a unimodal distribution of filament lengths that is peaked around the steady-state value of mean-field theory [426, 529]. The stochastic model takes the form of a totally asymmetric exclusion process (TASEP), see Sect. 7.6, in which a bound motor can hop to a neighboring site if it is unoccupied, see Fig. 14.2. Similarly a motor in the bulk can only bind to a site if it is unoccupied. A monomer can be removed from the tip of the filament when it is occupied by a motor. Following removal of the monomer, the motor steps backwards if the penultimate site is unoccupied, otherwise it is also removed. The tip motor can also spontaneously unbind without removal of the end monomer. The results from computer simulations are shown in Fig. 14.3 for both the density-controlled model and a flux-controlled model. The latter takes the depolymerization rate to depend on the flux of motors at the tip, rather than the motor density. That is [529],

$$L' = \alpha - a\rho_{\max} p_0 \left[1 - e^{-L/\lambda}\right](v - L'),$$

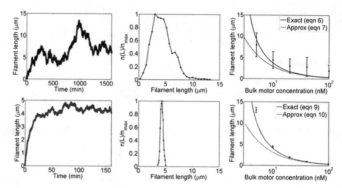

Fig. 14.3: Filament dynamics and steady-state filament length for length regulation by depolymerization. Top row: density-controlled depolymerization; Bottom row: flux-controlled depolymerization. Left: example trace of filament length versus time from a simulation of TASEP model. Middle: normalized filament length distribution averaged with respect to ten stochastic simulations (after removal of initial transients). Right: comparison of steady-state filament length based on mean-field theory with stochastic simulation (error bars are standard deviations of steady-state length distributions). Parameters of the mean-field model are $v = 3\,\mu$m/min, $k_{on} = 2$ nM$^{-1}$ $\mu$m$^{-1}$ min$^{-1}$, $k_{off} = 0.25$ min$^{-1}$, $k_{e,off} = 1.45$ min$^{-1}$, $\gamma = 1.025\,\mu$m/min, $a = 8$ nm, $\delta = 8$ nm, and $\rho_{max} = 125\,\mu$m$^{-1}$. For the density-controlled model $\alpha = 1.0\,\mu$m/min, while for the flux-controlled model $\alpha = 0.5\,\mu$m/min. The stochastic simulations use the same parameters except $\gamma = 1.5\,\mu$m/min and $k_{e,off} = 1$ min$^{-1}$ for the density-controlled model. [Adapted from Kuan and Betterton [529].]

and the steady-state length is

$$L_s = -\lambda \ln \left[ 1 - \frac{\alpha}{p_0 v} \delta \rho_{max} \right]. \qquad (14.1.7)$$

One finds that that statistical fluctuations are reduced in the flux-controlled model.

## 14.1.2 Microtubule regulation by stathmin protein gradients

An alternative mechanism for regulating filament growth is a signaling mechanism involving a protein concentration gradient (see Sect. 11.1). In order to illustrate this, we consider the regulation of MT growth by the signaling proteins Rac1 and stathmin [974]. Suppose that MTs switch between growing and shrinking phases within a bounded domain of length $L$, see Fig. 14.4. Growth is inhibited by active stathmin within the cytoplasm whose concentration increases away from the leading edge of the cell at $x = L$. This gradient is maintained by membrane-bound active Rac1 at the leading edge, which locally deactivates stathmin via phosphorylation. For the moment, we will assume that the concentration of active Rac1 is fixed, and thus independent of the number of MTs attached to the leading edge. The basic assumptions of the stathmin-regulated MT growth model are as follows:

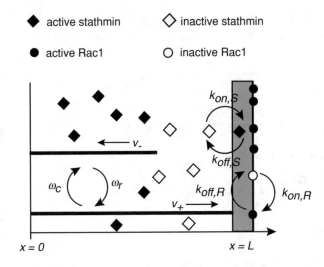

Fig. 14.4: Rac1/Stathmin regulation model of MT growth. Active Rac1 in the leading edge ($x = L$) generates a gradient of phosphorylated stathmin. As $x$ decreases, the concentration of active (dephosphorylated) stathmin becomes larger, thus increasing the likelihood that an MT undergoes catastrophe. An MT bound to the leading edge can activate Rac1, resulting in a positive feedback loop.

1. The fraction of active Rac1 in the leading edge is given by the constant $r_{on}$, $0 \leq r_{on} \leq 1$.
2. Stathmin in both the active (dephosphorylated) and inactive (phosphorylated) states diffuse in the cytosol with the same diffusion coefficient $D$. Activation of stathmin takes place in the cytosol with a constant rate $k_{on,S}$ while deactivation only occurs at the leading edge under the regulation of the active Rac1 at a rate $k_{off,S}$.
3. MTs stochastically switch between a growth state and a shrinkage state at a catastrophe rate $\omega_c$ and a rescue rate $\omega_r$. MTs polymerize in the positive $x$-direction at an average velocity $v_+$ in the growth state and depolymerize at an average velocity $-v_-$ in the shrinkage state with $v_\pm > 0$.
4. The growth of MTs is regulated by the local stathmin concentration either by directly increasing the catastrophe rate or by sequestering tubulin (see below).

Since we are initially ignoring the feedback loop associated with the activation of Rac1 by MTs at the leading edge, we can consider the dynamics of a single MT. Let $p_\pm(x,t)$ denote the probability that the MT has length $x$ at time $t$ and is in the growth (+) or shrinkage (−) phase. The densities $p_\pm$ evolve according to the extended Dogterom–Leibler model [249]

$$\frac{\partial p_+}{\partial t}(x,t) = -\frac{\partial [v_+(x,t)p_+(x,t)]}{\partial x} - \omega_c(x,t)p_+(x,t) + \omega_r p_-(x,t), \qquad (14.1.8a)$$

$$\frac{\partial p_-}{\partial t}(x,t) = v_- \frac{\partial p_-(x,t)}{\partial x} + \omega_c(x,t)p_+(x,t) - \omega_r p_-(x,t). \qquad (14.1.8b)$$

where the space-time dependence of the catastrophe rate $w_c$ and growth velocity $v_+$ arises from their dependence on the stathmin concentration, see below. We also impose the reflecting boundary condition $v_+(x,t)p_+(x,t) - v_- p_-(x,t)$ at $x = 0, L$.

Let $S_{on}(x,t)$ and $S_{off}(x,t)$ denote the concentration of active and inactive stathmin, respectively, at position $x$ at time $t$. The stathmin concentrations are taken to satisfy the diffusion equations

$$\frac{\partial S_{off}}{\partial t}(x,t) = D\frac{\partial^2 S_{off}(x,t)}{\partial x^2} - k_{on,S}S_{off}(x,t), \qquad (14.1.9a)$$

$$\frac{\partial S_{on}}{\partial t}(x,t) = D\frac{\partial^2 S_{on}(x,t)}{\partial x^2} + k_{on,S}S_{off}(x,t), \qquad (14.1.9b)$$

supplemented by the boundary conditions

$$\frac{\partial S_{on}}{\partial t}(L,t) = -\frac{D}{\delta}\frac{\partial S_{on}}{\partial x}\bigg|_{x=L} + k_{on,S}S_{off}(L,t) - r_{on}k_{off,S}S_{on}(L,t),$$

$$\frac{\partial S_{off,on}}{\partial x}\bigg|_{x=0} = 0. \qquad (14.1.10)$$

Following Zeitz and Kierfeld [974], we are assuming that there exists a boundary layer of width $\delta$ at the leading edge $x = L$, within which stathmin molecules deactivate (phosphorylate) at a rate $r_{on}k_{off,S}$ and activate (dephosphorylate) at a rate $k_{on,S}$. Outside this boundary layer, only dephosphorylation occurs. The stathmin model is coupled to the MT growth model by taking the catastrophe rate, and possibly, the growth velocity, to depend on the local concentration of active stathmin. We consider two forms of coupling [974]:

(A) One suggested pathway for stathmin to suppress MT growth is by direct interaction with an MT filaments, resulting in an increase in the catastrophe rate. Experimental data suggests a linear increase of the catastrophe rate with the concentration of active stathmin so we take

$$\omega_c(x,z,t) = \omega_c^0 + k_c S_{on}(x,z,t), \qquad (14.1.11)$$

with $k_c = 0.005 s^{-1}\mu M^{-1}$ and $\omega_c^0 = 7 \times 10^{-4} s^{-1}$ [171, 384]

(B) Another possible pathway is via sequestering of tubulin. It turns out that a single active stathmin protein sequesters two tubulin proteins [171, 384],

$$2T + S \rightleftharpoons ST_2.$$

If this is combined with the kinetics of activation/deactivation of stathmin, then at chemical equilibrium, the normalized concentration of free tubulin $t \equiv [T]/[T_0]$, where $[T_0]$ is the total tubulin concentration, can be expressed as a nonlinear function of the normalized active stathmin concentration $s_{on} = S_{on}/[T_0]$ [974]:

$$t(s_{on}) = \frac{1}{3}\left[1 - 2s_{on} + \frac{k(1 - 2s_{on})^2 - 3}{k\alpha(s_{on})} + \alpha(s_{on})\right],$$

with $k \equiv K_0[T_0]^2$, where $K_0$ is the equilibrium constant for the stathmin activation reaction,

$$\alpha(s) = \left[(1-2s)^3 + \frac{9}{k}(1+s) + \beta(s)\right]^{1/3},$$

and

$$\beta(s) = 3\sqrt{\frac{3}{k^3}[1+k^2(1-2s)^3 + k(2+10s-s^2)]}.$$

Since the MT growth velocity $v_+$ depends on the local tubulin concentration, it follows that a spatial variation in active stathmin concentration leads to a spatial variation in the growth velocity. That is, $[T](x) = [T_0]t(s_{on}(x))$ and

$$v_+(x) = (\kappa_{on}[T](x) - \kappa_{off})d, \qquad (14.1.12)$$

where $d \approx 0.6nm$ is the effective tubulin dimer size and $\kappa_{on}, \kappa_{off}$ are binding and unbinding rates. Following Zeitz and Kierfeld [974], we take $\omega_{on} \equiv \kappa_{on}[T_0] = 143$ s$^{-1}$, $\kappa_{off}d = 3.6$nm s$^{-1}$. Note that the growth velocity is effectively a function of the stathmin concentration under the assumption that the tubulin-stathmin reactions are fast relative to other relevant processes. Experimentally one finds that the average time spent in the growing state, $\langle \tau_+ \rangle = 1/\omega_c$ is a linear function of the growth velocity, so that the catastrophe rate also becomes space-dependent:

$$\omega_c(x,z) = \frac{1}{a + bv_+(x,z)} \qquad (14.1.13)$$

for constant coefficients $a = 20$ s and $b = 1.38 \times 10^{10}$ s$^2$ m$^{-1}$ [469].

**Steady-state solution with no feedback.** Next we turn to steady-state solutions of the 2D MT/stathmin model given by equations (14.1.8), (14.1.9) and (14.1.10), see also Sect. 4.2. First, setting $\omega_c = \omega_c(x)$ and $v_+ = v_+(x)$, adding equations (14.1.8a) and (14.1.8b), and setting time derivatives to zero yields the $x$-independent steady-state equation $\partial_x[v_+(x)p_+(x) - v_-p_-(x)] = 0$. This together with the reflecting boundary conditions implies that $v_+(x)p_+(x) - v_-p_-(x) = 0$. It follows that $p_-(x) = v_+(x)p_+(x)/v_-$. Substituting into equation (14.1.8a) gives

$$\frac{\partial v_+(x)p_+(x)}{\partial x} - \left[\frac{\omega_r}{v_-} - \frac{\omega_c(x)}{v_+(x)}\right]v_+(x)p_+(x) = 0.$$

Hence,

$$v_+(x)p_+(x) = v_+(0)p_+(0)\exp\left(\int_0^x \lambda(x')dx'\right), \quad \lambda(x) = \frac{\omega_r}{v_-} - \frac{\omega_c(x)}{v_+(x)}.$$

It follows that the steady-state probability density for the MT length $x$ is

$$p(x) \equiv p_+(x) + p_-(x) = \left(1 + \frac{v_+(x)}{v_-}\right)p_+(x)$$

$$= \mathcal{N}\left(1 + \frac{v_-}{v_+(x)}\right)\exp\left(\int_0^x \lambda(x')dx'\right). \qquad (14.1.14)$$

The factor $\mathcal{N}$ is determined by the normalization condition $\int_0^L p(x)dx = 1$.

It remains to determine the $x$-dependence of the functions $v_+(x)$ and $\omega_c(x)$ by finding the steady-state solution of equation (14.1.9). In the case of a fixed Rac1 concentration, the steady-state solution for $S_{\text{off}}$ is of the form

$$S_{\text{off}}(x) = \Lambda_0 \cosh(v_0 x), \quad v_0 = \sqrt{\frac{k_{\text{on,S}}}{D}}. \tag{14.1.15}$$

The coefficient $\Lambda_0$ depends on the steady-state boundary condition at the leading edge $(x = L)$, where deactivation of stathmin takes place:

$$\frac{D}{\delta}\frac{\partial S_{\text{off}}}{\partial x}\bigg|_{x=L} = k_{\text{off,S}} r_{\text{on}} S_{\text{on}}(L) - k_{\text{on,S}} S_{\text{off}}(L). \tag{14.1.16}$$

Let $S(x) = S_{\text{on}}(x) + S_{\text{off}}(x)$ be the total stathmin concentration at $x$. Since $S(x)$ evolves according to the 1D steady-state diffusion equation with reflecting boundaries, it follows that $S(x) = S_{\text{tot}} = \text{constant}$. Hence

$$S_{\text{on}}(x) = S_{\text{tot}} - S_{\text{off}}(x) = S_{\text{tot}} - \Lambda_0 \cosh(v_0 x). \tag{14.1.17}$$

Substituting equation (14.1.17) into the boundary condition (14.1.16) gives

$$\Lambda_0 = \Lambda_0(r_{\text{on}}) = \frac{S_{\text{tot}} k_{\text{off,S}} r_{\text{on}}}{(D/\delta) v_0 \sinh(v_0 L) + (r_{\text{on}} k_{\text{off,S}} + k_{\text{on,S}}) \cosh(v_0 L)}. \tag{14.1.18}$$

Finally, one can determine the average MT length by substituting for $S_{\text{on}}$ into either model of stathmin-MT coupling: equation (14.1.11) (direct interactions) or equations (14.1.12) and (14.1.13) (indirect interactions via tubulin sequestering). For example, in the former case, we have

$$\omega_c(x) = \omega_c^0 + k_c \left[S_{\text{tot}} - \Lambda_0 \cosh(v_0 x)\right]. \tag{14.1.19}$$

Substituting equation (14.1.19) into the steady-state probability density of equation (14.1.14) gives

$$p(x) = \mathcal{N}\left(1 + \frac{v_+}{v_-}\right) \exp\left[\gamma x + \frac{k_c}{v_+}\frac{\Lambda_0}{v_0}\sinh(v_0 x)\right], \tag{14.1.20}$$

where

$$\gamma = \frac{\omega_r}{v_-} - \frac{\omega_c^0 + k_c S_{\text{tot}}}{v_+}, \tag{14.1.21}$$

and

$$\mathcal{N}\left(1 + \frac{v_+}{v_-}\right) = \frac{N}{L}\left[\int_0^L \exp\left[\gamma x + \frac{k_c}{v_+}\frac{\Lambda_0}{v_0}\sinh(v_0 x)\right]dx\right]^{-1}.$$

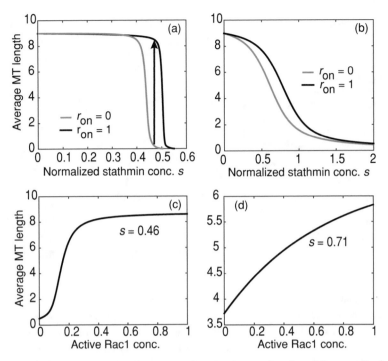

Fig. 14.5: Steady-state solutions for the average MT length $\bar{x}$ as a function of the normalized stathmin concentration $s = S_{\text{tot}}/[T_0]$, where $[T_0]$ is the total tubulin concentration, and the active Rac1 concentration $r_{\text{on}}$ for fixed $s$. (a,c) Tubulin-sequestering stathmin. (b,d) Catastrophe-promoting stathmin. For sufficiently large $s$, the model acts like a switch, jumping from a small $\bar{x}$ in the absence of active Rac1 ($r_{\text{on}} = 0$) to a large $\bar{x}$ for constitutively active Rac1 ($r_{\text{on}} = 1$). Parameters values are taken from [974], where one can find a table listing the various parameters and references to supporting experimental literature: $v_+ = 0.06\,\mu\text{m/s}$, $v_- = 0.18\,\mu\text{m/s}$, $w_r = 0.18\,\text{s}^{-1}$, $k_c = 0.005\,\text{s}^{-1}\,\mu\text{M}^{-1}$, $D = 15\,\mu\text{m}^2\text{/s}$, $k_{\text{on,S}} = 1\,\text{s}^{-1}$, $k_{\text{off,S}} = 300\,\text{s}^{-1}$, $\delta = 0.02\,\mu\text{m}$.

We conclude that for a fixed concentration of active Rac1 within the membrane, the average MT length is

$$\bar{x} = \int_0^L x p(x)\,dx.$$

In Fig. 14.5 we plot the mean MT length distribution $\bar{x}$ as a function of $S_{\text{tot}}$ for both forms of MT-stathmin interactions, based on similar parameter values to those of Zeitz and Kierfeld [974].

A number of observations can be made [974]. First, the tubulin-sequestering mechanism generates a switch-like variation of mean MT length as a function of the normalized stathmin concentration $s = S_{\text{tot}}/[T_0]$, where $[T_0]$ is the total tubulin concentration. Second, comparison of the steady-state probability density $p(x)$ between the cases $r_{\text{on}} = 1$ (active Rac1 present) and $r_{\text{on}} = 0$ (active Rac1 absent) shows that the former can lead to a bimodal probability density. This can be understood by noting that the monotonically decreasing concentration $S_{\text{on}}(x)$ of active

Fig. 14.6: Sketch of average MT length $\bar{x}$ as a function of the normalized stathmin concentration $s = S_{\text{tot}}/[T_0]$ for tubulin-sequestering stathmin, and feedback regulation of Rac1 by MT contacts at the leading edge. Dashed lines are for no feedback (fixed fraction of active Rac1) for $r_{\text{on}} = 0$ and $r_{\text{on}} = 1$. The solid blue and green curves correspond to the case of feedback with either $N = 1$ or $N = 10$ MTs. [Redrawn from Zeitz and Kierfeld [974].]

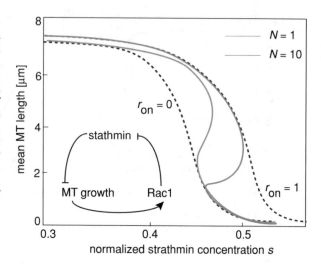

stathmin results in $\lambda(x)$ being an increasing function of x for both stathmin models. In a certain parameter regime, one finds that $\lambda(x)$ changes sign on $0 < x < L$. That is, $\lambda(x) < 0$ for small $x$ (MTs shrink on average) and $\lambda(x) > 0$ for large $x$ (MTs grow on average). Thus active stathmin promotes the growth of long MTs, and shrinkage of short MTs resulting in a bimodal MT length distribution. Interestingly, the subpopulation of long fast-growing MTs is suggestive of pioneering microtubules that have been observed in the leading edge of migrating cells [946].

**Bifurcation analysis with feedback.** Now suppose that we include the feedback loop, where the activation of Rac1 is regulated by MTs in contact with the leading edge, and thus the level of active Rac1 depends on the total number $N$ of MTs in the system. The net result is a positive feedback loop, as illustrated in the inset of Fig. 14.6. The fraction of active Rac1 at the leading edge is now a dynamical variable $r_{\text{on}}(t)$, which evolves according to the equation

$$\frac{dr_{\text{on}}}{dt} = p_{\text{MT}}(t)k_{\text{on,R}}(1 - r_{\text{on}}(t)) - k_{\text{off,R}}r_{\text{on}}(t), \qquad (14.1.22)$$

where the mean number of MTs within a contact domain of width $\delta$ is

$$p_{\text{MT}}(t) = N \int_{L-\delta}^{L} p_+(x,t)dx. \qquad (14.1.23)$$

In steady state, we have the implicit equation

$$r_{\text{on}} = \frac{1}{1 + k_{\text{off,R}}/[k_{\text{on,R}}p_{\text{MT}}(r_{\text{on}})]}, \qquad (14.1.24)$$

where $p_{MT}(r_{on})$ is obtained by extracting $p_+(x)$ from equation (14.1.20), substituting into equation (14.1.23), evaluating the integral, and noting that $p_+(x)$ depends on the parameter $\Lambda_0(r_{on})$ given by equation (14.1.18). Numerically solving for $r_0$ and then determining the corresponding mean MT length establishes that the system can act as a bistable switch [974].

For small stathmin concentrations, there is only a single fixed point at a high level of active Rac1, resulting in a large mean MT length and a bimodal length distribution, i.e., pioneering MTs, see Fig. 14.6. There also exists a single fixed point at high stathmin concentrations, corresponding to a low level of Rac1 and short MTs. However, at intermediate stathmin concentrations there are two stable fixed points separated by a single unstable fixed point. Hence, the system exhibits bistability and hysteresis over a range of stathmin concentrations that widens as the number of MTs increases.

### 14.1.3 Secretion models of flagellar length control in Salmonella

The flagellar motor of Salmonella Typhimurium (*S. Typhimurium*) consists of three major parts: the basal body, the hook, and the filament, see Fig. 14.7(a). The basal body is fixed in the inner and outer membranes, whereas the filament and hook are located externally to the cell. The largest component of the motor is the filament, which extends more than 10 $\mu$m, and propels the bacterium forward when rotated by the motor at the base. The filament is linked to the basal body via the hook, which acts as a universal joint. Assembly of the flagellum is initiated by the insertion of a ring structure within the cytoplasmic membrane, followed by the construction of a secretory apparatus. The latter is needed to export flagellar structural sub-units beyond the membrane so that they can self-assemble into the hook and then the filament. The sub-units are secreted from the cytoplasm via the hydrolysis of ATP by the ATPase protein FliI. Since, the hook, hook-filament junction, and filament are constructed from distinct sub-units, the underlying gene regulatory networks have to switch the secretion of one type of sub-unit to the next at appropriate times during assembly.

The synthesis and assembly of the flagellar motor involves the expression of more than 60 genes, which are organized into a transcriptional hierarchy that is based on three promoter classes that are temporally ordered [16, 185, 447], see Fig. 14.7(b). The master operon of the early (class 1) genes is *flhDC*, which lies at the top of the hierarchy and determines whether or not to proceed with the construction of flagella. It produces the proteins FlhD and FlhC, which form a heteromultimeric unit that initiates transcription of intermediate (class 2) genes and auto-represses *flhDC* transcription (switches off class 1). This results in the secretion of the hook constituent protein FlgE, which continues until the hook is around 55 nm long. Secretion then switches to proteins FlgK and FlgL (additional class 2 factors), which form the hook-filament junction, followed by the production and secretion of FliC (class 3 factor), which is the sub-unit of the filament. Two major regulatory components

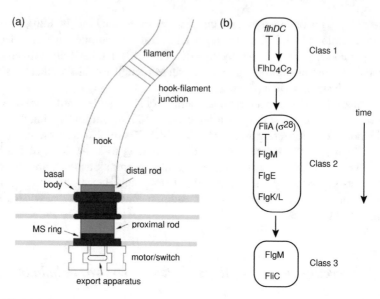

Fig. 14.7: (a) Schematic diagram of the major components of the *Salmonella* flagellar motor. (b) Three stages of gene regulation and secretion during assembly of the flagellar motor. See text for details.

of flagellar assembly are the transcription factor $\sigma^{28}$ and its inhibitor FlgM. Both proteins are produced during the second stage of construction (hook assembly), with FlgM also expressed during the third stage. $\sigma^{28}$ is an activator of RNA polymerase that is required for transcription of all class 3 factors. However, activity of $\sigma^{28}$ is repressed by FlgM during stage 2, thus preventing class 3 transcription. Upon completion of hook assembly, the specificity of the ATPase is modified so that secretion switches from FlgE, FlK, and FlgL (hook and hook-filament junction monomers) to FliC (filament substrate). The crucial point is that FlgM is also now secreted, which releases $\sigma^{28}$ from its inactivation, resulting in the production of the class 3 factors FlgM and FliC. Since FlgM represses $\sigma^{28}$, there is a negative feedback loop.

**Secretion model of filament assembly.** We will focus on a mathematical model of the final stage of flagellar motor assembly, namely, construction of the filament [495]. A crucial aspect of the model is determining how the cell "knows" when the filament has reached the correct length. This requires that one develops a model of length-dependent secretion in a narrow tube. Since the latter has an inner diameter of around 2 nm, movement of molecules along the tube is going to be single file. We will assume that the secretion apparatus is at $x = 0$ and that the growing tip is at $x = L(t)$. Let $p(x,t)$ denote the expectation that at time $t$ the position $x$ is occupied by an unfolded monomer of length $l$ with $l \ll L(t)$. If there are $N$ unfolded monomers within the tube at time $t$, then

$$\int_0^L p(x,t)\,dx = Nl.$$

As we showed in Sect. 6.7, although single-file diffusion is anomalous at the level of a single-tagged particle, the average flux of particles obeys normal diffusion. Hence, neglecting exclusion effects within the bulk, we can take

$$\frac{\partial p}{\partial t} = -\frac{\partial J}{\partial x}, \quad J = -D\frac{\partial p}{\partial x}, \tag{14.1.25}$$

with $J/l$ the number flux of monomers. Since the latter must match the rate of polymerization at the growing tip,

$$\frac{J(L,t)}{l} = k_+ p(L,t), \tag{14.1.26}$$

with $k_+$ the polymerization rate constant. Note that at the end of the filament there is a cap constructed of FliD that prevents the escape of FliC and facilitates polymerization.

The next step is to model the translocation of monomers (substrate) into the tube via the secretion apparatus. Following Keener [495], this is modeled by assuming that the ATPase can be in two states corresponding to whether or not it is bound by its cognate substrate. Moreover, unbinding only occurs via translocation (injection) of the monomer into the tube. Let $P(t)$ be the probability that the APTase is in the bound state. Then

$$\frac{dP}{dt} = K_{on}(1 - P(t)) - k_{off}(1 - p(0,t))P(t), \tag{14.1.27}$$

where $p(0,t)$ is the probability that the $x = 0$ end of the hook is occupied by a monomer at time $t$, and $(1 - p(0,t))$ is an exclusion factor. (In terms of a discrete exclusion process, we are treating $l$ as the lattice size so that if $x = 0$ is unoccupied then the whole monomer can be translocated into the tube.) It follows that the total unbinding (translocation) rate must be the same as the flux of monomers at $x = 0$,

$$\frac{J(0,t)}{l} = k_{off}(1 - p(0,t))P(t). \tag{14.1.28}$$

Finally, the length of the filament evolves according to

$$\frac{dL}{dt} = \frac{J(L(t),t)}{\beta l}, \tag{14.1.29}$$

where $\beta$ is the number of monomers per unit length of filament. (This assumes that only FliC is secreted—the equation will have to be modified when secretion of FlgM is also taken into account, see below.)

Equations (14.1.25)–(14.1.29) define a free boundary value problem, see also Ex. 14.1. We transform this to a problem on a fixed domain by introducing the dimensionless variables [495]

$$t = \frac{l^2}{D}\tau, \quad x = L(\tau)y.$$

In the new coordinate system, we have

$$\frac{\partial p}{\partial \tau} = \frac{s'(\tau)y}{s(\tau)}\frac{\partial p}{\partial y} + \frac{1}{s^2}\frac{\partial^2 p}{\partial y^2},$$

where $s$ is the scaled filament length $s(\tau) = L(\tau)/l$, and the growth of the filament is determined by

$$s' = -\frac{1}{\beta l s}\frac{\partial p}{\partial y}\bigg|_{y=1}. \tag{14.1.30}$$

Now taking $u = s^2$ gives

$$u\frac{\partial p}{\partial \tau} = \frac{u'(\tau)y}{2}\frac{\partial p}{\partial y} + \frac{\partial^2 p}{\partial y^2}, \quad u' = -\varepsilon\frac{\partial p}{\partial y}\bigg|_{y=1}, \tag{14.1.31}$$

where $\varepsilon = 2/\beta l$. Taking biophysical reasonable parameter values $\beta = 2.4/\text{nm}$ and $l = 75$ nm, it follows that $\varepsilon \approx 0.01 \ll 1$. It remains to specify the boundary conditions in the new coordinates. We simplify the problem by assuming that the ATPase is in quasi-steady state, so that

$$P(t) = \frac{K_{\text{on}}}{k_{\text{off}}(1 - p(0,t)) + K_{\text{on}}}.$$

The boundary condition at $y = 0$ thus takes the form, see equation (14.1.28),

$$\frac{\partial p}{\partial y}\bigg|_{y=0} = -\sqrt{u}\frac{l^2}{D}\frac{k_{\text{off}}K_{\text{on}}(1 - p(0,t))}{k_{\text{off}}(1 - p(0,t)) + K_{\text{on}}}. \tag{14.1.32}$$

Similarly, the boundary condition at $y = 1$ is

$$\frac{\partial p}{\partial y}\bigg|_{y=1} = -\sqrt{u}\frac{l^2}{D}k_+ p(1,t). \tag{14.1.33}$$

Numerically one finds that $p$ rapidly reaches a quasi-steady state over time scales for which the effective time constant $u$ and boundary conditions are slowly varying (due to small $\varepsilon$). The quasi-steady state of equation (14.1.31) with $u'$ treated as a constant is obtained by integrating with respect to $y$ twice:

$$p(y) = A\int_0^y \exp(-r^2 u'/4)dr + B,$$

where $A, B$ are integration constants that are determined by the boundary conditions (14.1.32) and (14.1.33). Defining

$$E(a) = \int_0^1 \exp(-ay^2/4)dy,$$

the boundary conditions can be expressed as

$$A = -sK_D \frac{1-B}{K_a(1-B)+1}, \quad K_b A \exp(-u'/4) = -sK_D(AE(u')+B),$$

where

$$K_D = \frac{k_{\text{off}}l^2}{D}, \quad K_a = \frac{k_{\text{off}}}{K_{\text{on}}}, \quad K_b = \frac{k_{\text{off}}}{k_+}.$$

We now make the further simplifications: $u' = 0$ to leading order in $\varepsilon$ and $K_a \ll 1$. The boundary conditions then simplify as

$$A + sK_D = sK_D B = -(K_b + sK_D)A,$$

which yields

$$A = -\frac{sK_D}{sK_D + K_b + 1}.$$

Substituting the solution for $p(y)$ into equation (14.1.30) and reintroducing physical variables and parameters, we obtain the following equation for the rate of growth of the filament tip [495]:

$$\frac{dL}{dt} = \frac{1}{\beta} \frac{k_{\text{off}}}{k_{\text{off}}lL/D + 1 + k_{\text{off}}/k_+}. \tag{14.1.34}$$

The secretion model thus predicts that the rate of growth is an inverse function of the filament length, at least for long filaments.

Although the results of [495] are consistent with early observations of filament growth [454], a more recent experimental study based on advances in fluorescent imaging suggested that flagellar filaments grow at a constant rate of around 13nm/min [891]. A possible mechanism for constant growth, involving interactions between secreted sub-units has also been proposed [287], although convincing evidence for such a mechanism has not been obtained [447]. Moreover, one of the possible limitations of the study by Turner et al. [891] is that it measured filament regrowth after shearing, rather than the growth of nascent filaments. Experimental studies of the latter [763] provide further evidence supporting the secretion model of Keener [495], with the energy necessary for the direct injection of proteins into the secretion apparatus provided by coupling to a proton-driven motor.

Finally, note that the secretion model has been extended to provide a possible mechanism for length regulation, based on the coupling between the length-dependent rate of secretion and the negative feedback that regulates the production of FlgM [495]. The basic idea is that for short filaments FlgM is secreted rapidly, so that its concentration within the basal body remains low, allowing active $\sigma^{28}$ to produce FliC and more FlgM. On the other hand, as the filament grows, the rate of secretion slows down, the concentration of FlgM builds up again, inhibiting $\sigma^{28}$

activity and ultimately switching off the production of FliC. An immediate consequence of this built in length sensing mechanism is that if the filament is severed, then secretion of FlgM increases significantly due to the length-dependence of
the secretion rate, signaling that more FliC is needed for regrowth. The latter thus
accounts for the experimentally observed regrowth of damaged filaments [454].

## 14.2 Intraflagellar transport in Eukaryotes

Radioactive pulse labeling has been used to measure protein turnover in eukaryotic
flagella. Such measurements have established that turnover of tubulin occurs at the
+ end of flagellar MTs, and that the assembly (rather than disassembly) part of
the turnover is mediated by intraflagellar transport (IFT). This is a motor-assisted
motility within flagella in which large protein complexes move from one end of
the flagellum to the other [802]. Particles of various sizes travel to the flagellar tip
(anterograde transport) at $2.0\,\mu$m/s, and smaller particles return from the tip (retrograde transport) at $3.5\,\mu$m/s, after dropping off their cargo of assembly proteins at
the + end. A schematic diagram of IFT transport is shown in Fig. 14.8. Immunoflu-
orescence analysis indicates that the number of IFT particles (estimated to be in the
range 1–10) is independent of length [593, 595]. If a fixed number $N$ of transport
complexes move at a fixed mean speed $\bar{v}$, then the rate of transport and assembly
should decrease inversely with the flagellar length $L$. On the other hand, measurements of the rate of flagellar shrinkage when IFT is blocked indicate that the rate of
disassembly is length-independent. This has motivated the following simple deterministic model for length control [593]:

$$\frac{dL}{dt} = \frac{a\bar{v}N}{2L} - V,\tag{14.2.1}$$

where $a$ is the size of the precursor protein transported by each IFT particle and $V$
is the speed of disassembly. Equation (14.2.1) has a unique stable equilibrium given
by $L^* = a\bar{v}N/2V$. Using the experimentally based values $N = 10$, $\bar{v} = 2.5\,\mu$m/s,

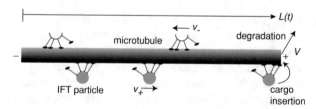

Fig. 14.8: Schematic diagram of intraflagellar transport (IFT), in which IFT particles travel with
speed $v_\pm$ to the $\pm$ end of a flagellum. When an IFT particle reaches the + end it releases its
cargo of protein precursors that contribute to the assembly of the flagellum. Disassembly occurs
independently of IFT transport at a speed $V$.

$L^* = 10\,\mu\text{m}$ and $V = 0.01\,\mu\text{m/s}$, the effective precursor protein size is estimated to be $a \approx 10$ nm. (A stochastic version of a model for intraflagellar transport has also been developed using the theory of continuous-time random walks [109], see Ex. 14.3.)

## 14.2.1 IFT regulation at the basal body

However, the above model is oversimplified. In particular, there is growing experimental evidence that the flux of IFT particles into the flagellum is regulated by the amount of accumulated IFT particles at the base of the flagellum [578, 579, 953]. Moreover, recent photobleaching studies have shown that there is constant turnover of IFT particles within flagella, presumably through the exchange of IFT particles between the basal body and the cytoplasm [579]. An emerging picture is that IFT particles enter the flagellum through the flagellar pore, a membrane-spanning structure at the base of the flagellum that may be homologous to a nuclear pore. The latter regulates the exchange of macromolecules between a cell's cytoplasm and the nucleus (Chap. 6). There is also a microtubule-organizing center known as the basal body, which anchors the flagellar MTs at the plasma membrane and integrates them with the cytoplasmic MTs. IFT proteins dock around the basal body and assemble into IFT particles or trains of particles prior to entering the flagellum [227]. It appears that the rate at which IFT particles enter the flagellum depends on the amount of docked IFT particles in the basal body, with faster growing flagellar having more localized IFT particles [578, 953]. This suggests that there is some length-dependent mechanism for regulating the accumulation of IFT particles at the basal body (and possibly the loading of cargo to docked IFTs [953]).

Following Ludington et al. [579], consider a one-dimensional flagellum of length $L$ with the basal body at $x = 0$ and the tip at $x = L$, as shown in Fig. 14.9. Suppose that there are $M_0$ binding sites for IFT particles in the basal body, and the concentration of IFTs within the cytoplasm is $B$. Denote the binding and unbinding rates by $k_+$ and $k_-$, respectively. Assuming that $M_0$ is sufficiently large, the kinetic equation for the number $m(t)$ of bound IFTs at time $t$ is

$$\frac{dm}{dt} = k_+ B[M_0 - m(t)] - k_- m(t),  \tag{14.2.2}$$

which has the steady-state solution

$$m^* = M_0 X^*, \quad X^* = \frac{k_+ B}{k_+ B + k_-}.  \tag{14.2.3}$$

Fits with experimental data suggest that $k_+ B/k_- \sim 10\,\mu\text{m}$ [578, 579]. Now suppose that there is some signaling mechanism within the flagellum such that the binding rate is a decreasing function of length $L$, and set $k_+ = \bar{k}_+ C_0(L)$. We will give one example of such a signaling mechanism below, see also Ref. [579]. Under the adi-

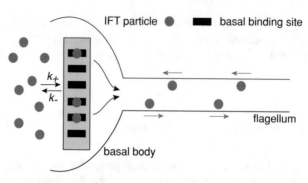

Fig. 14.9: Schematic diagram of IFT flux regulation in the basal body. IFT particles (filled circles) can undergo binding/unbinding reactions with $M_0$ sites (filled rectangles) in the basal body at rates $k_{\pm}$. The number of bound IFTs determines the rate at which IFTs are injected into the flagellum. Once in the flagellum, IFTs are actively transported to the tip, where they deliver their cargo and are then transported back to the basal body along the lines shown in Fig. 14.1. Some signaling mechanism within the flagellum (not shown) results in the binding rate $k_+$ being dependent on the flagellar length $L$, resulting in a length-dependent IFT flux regulation.

abatic approximation that the growth-rate of the flagellum is much slower than the various kinetic processes, we can still treat $m^*$ as a constant with

$$m^* = m^*(L) \equiv \frac{\bar{k}_+ C_0(L) B}{\bar{k}_+ C_0(L) B + k_-} M_0. \tag{14.2.4}$$

The rate of injection of IFTs into the flagellum is then taken to be $\lambda_0 = \eta m^*(L)$, which means that the influx is a monotonically decreasing function of $L$.

The critical flagellar length is determined by the balance between the influx and the length-independent rate of disassembly, along analogous lines to equation (14.2.1). That is, suppose each injected particle remains in the flagellum a time $T = 2L/v + \tau$ before being removed, where $v$ is the harmonic mean of the antero-grade and retrograde speeds of each IFT particle, and $\tau$ is the time spent at the tip. We will take $\tau = 1$ s and $v = 2$ $\mu$m/s. It follows that the steady-state number $N^*$ of particles in the flagellum is

$$N^* = \lambda_0 T = \eta \frac{\bar{k}_+ C_0(L) B}{\bar{k}_+ C_0(L) B + k_-} M_0 (2L/v + \tau). \tag{14.2.5}$$

Hence, setting $N = N^*(L)$ in equation (14.2.1) we deduce that the critical length is determined by the intercept of the monotonically decreasing function $N^*(L)/L$ with a constant $\zeta = 2V/av$.

The one remaining component of the model is the specification of the length-dependent function $C_0(L)$ of the IFT binding rate $k_+$. Ludington et al. [579] considered several different signaling mechanisms for generating this length dependence, and compared the models with experimental data on the length-dependence of IFT. They applied two distinct ways of quantifying IFT. First, they used live imaging to

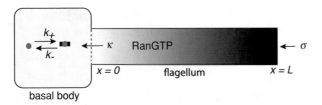

Fig. 14.10: Schematic diagram of RanGTP concentration gradient model of IFT flux regulation. A source of RanGTP at the tip of the flagellum sets up a concentration gradient along the flagellum resulting a length-dependent concentration of RanGTP in the basal body. This in turn regulates the binding rate of IFTs to sites in the basal body.

measure the rate at which individual IFT particles were injected into flagella using GFP-tagged kinesin motors. Second, they fixed the cells and imaged IFT proteins that had accumulated at the basal bodies. The various deterministic models were then fitted with experimental curves obtained by averaging with respect to the flagellar population. For the sake of illustration, we will consider one of the models that fit particularly well with photobleaching data. It is a diffusion-based model of RanGTP concentration gradient formation. RanGTP is a small enzyme that is known to play an important role in regulating nuclear transport through the nuclear pore complex, and it is hypothesized that RanGTP plays an analogous role in regulating IFT particle influx. In particular, a decrease in RanGTP concentration at the basal body as cell length increases leads to a reduction in IFT particle influx.

Suppose that RanGTP is produced at a rate $\sigma$ at the tip ($x = L$), resulting in a concentration gradient, see Fig. 14.10. Assume that cytoplasmic RanGTP concentration is negligible and $\kappa$ is the flow rate through the pore at $x = 0$. Then the RanGTP concentration per unit volume $C(x,t)$ evolves as

$$\frac{\partial C}{\partial t} = D\frac{\partial^2 C}{\partial x^2} - \gamma C, \quad x \in [0,L], \tag{14.2.6}$$

where $\gamma$ is a degradation rate. The boundary conditions are

$$D\frac{\partial C}{\partial x} = \kappa C, \quad x = 0; \quad D\frac{\partial C}{\partial x} = \sigma, \quad x = L. \tag{14.2.7}$$

Integrating equation (14.2.6) with respect to $x$ and using the boundary conditions gives

$$\frac{dR}{dt} = \sigma - \kappa C(0,t) - \gamma R,$$

where $R(t)$ is the total number of RanGTP molecules per unit area,

$$R(t) = \int_0^L C(x,t)dx. \tag{14.2.8}$$

If we assume that diffusion is fast so that the characteristic length $\sqrt{D/\gamma} \gg L$, then $C(x,t)$ is approximately uniform and we can take $C(0,t) \approx R(t)/L$. Therefore,

$$\frac{dR}{dt} = \sigma - \kappa\frac{R}{L} - \gamma R. \tag{14.2.9}$$

Equation (14.2.9) has the steady-state solution $R = \sigma L/(\gamma L + \kappa)$, so that the concentration at the basal pore is

$$C_0 = C_0(L) = \frac{\sigma}{\gamma L + \kappa}. \tag{14.2.10}$$

Typical values of the parameters are [579] $\sigma \sim 5-20$ s$^{-1}$, $\kappa \sim$ 5-25 $\mu$m/s, $\gamma \sim$ 10 $-$ 400 s$^{-1}$.

### 14.2.2 Doubly stochastic model of IFT transport

One limitation of the above model is that it cannot account for the statistical features of the time series of IFT injections [578]. The sequence of time intervals between consecutive injections exhibits transient periodicity, bursting activity, power-law dependencies, memory effects, and non-exponential interval statistics. Moreover, there are correlations between the frequency and size of injected IFT particles, with larger sizes (more IFT proteins within a particle) tending to occur less frequently. Based on these observations, Ludington et al [578] suggested that the process of IFT injection is stochastic and exhibits avalanche-like behavior. The length-dependent binding of IFTs to the basal body then regulates the mean rate and size of IFT injections. Ludington et al [578] also developed a computational model of the avalanche-like behavior, based on a cellular analog of a sandpile model. More specifically, they introduced a trafficking model for the passage of bound IFTs through the flagellar pore at the distal end of the basal body. The build up of IFTs at the opening of the flagellum due to jamming effects then generated avalanche-like events, which were fitted to the experimental data.

An alternative approach to modeling the statistics of IFT injection events has been developed in terms of a stochastic version of the model shown in Fig. 14.9 [121]. This model combines two sources of stochasticity. The first involves fluctuations in the number of bound IFTs $M(t)$ due to the finite number $M_0$ of binding sites in the basal body ($M_0 \sim 100-1000$). That is, $M(t)$ is a discrete stochastic variable evolving according to a birth–death master equation. Under a system-size expansion with respect to $M_0$, see Sect. 3.3, the fraction $X(t) = M(t)/M_0$ of bound IFTs evolves according to the SDE

$$dX = A(X)dt + \sqrt{\frac{B(X)}{M_0}}dW(t), \tag{14.2.11}$$

where $W(t)$ is an independent Wiener process, and

$$A(x) = (1-x)k_+B - k_-x, \quad B(x) = (1-x)k_+B + k_-x.$$

The rate of injection into the flagellum is then itself a stochastic variable, since $\lambda(t) = \eta M_0 X(t)$. (Certain care must be taken, however, since there is a nonzero probability that $X(t)$ becomes negative. We will assume that this does not cause problems for sufficiently large $M_0$.)

A second level of stochasticity is introduced by taking the number $N(t)$ of IFT particles injected into the flagellum in the time interval $[0,t]$ to be a Poisson process with rate $\lambda(t)$ [121]. Since the rate $\lambda(t)$ is itself stochastic, it follows that the process of IFT injection is described by a doubly stochastic Poisson process (DSPP). That is, for a given realization of the continuous stochastic process up to time $t$, $\{X(s), 0 \leq s < t\}$, the conditional probability distribution $P_n(t) \equiv \mathbb{P}[N(t) = n | \{X(s), 0 \leq s < t\}]$ is given by

$$P_n(t) = \frac{\Lambda(t)^n}{n!} e^{-\Lambda(t)}, \quad \Lambda(t) := \int_0^t \lambda(\tau)d\tau = \eta M_0 \int_0^t X(t')dt'. \quad (14.2.12)$$

The analysis of DSPPs shows that determining the first-order and second-order statistics of the number $N(t)$ of injected IFT particles requires calculating the corresponding statistics of $X(t)$. Assuming that $X(0) = X^*$ with $X^*$ given by equation (14.2.3), then the Gaussian process $X(t)$ is stationary. It follows that $\langle X(t) \rangle = X^*$ and

$$R_X(t_1, t_2) \equiv \langle [X(t_1) - \langle X(t_1) \rangle][X(t_2) - \langle X(t_2) \rangle] \rangle = \frac{X^*(1+\Theta)}{2M} e^{-\Gamma|t_2 - t_1|},$$

where

$$\Gamma = k_+B + k_-, \quad \Theta = \frac{k_- - k_+B}{k_- + k_+B}.$$

It can be shown that (see Box 14A)

$$\mathbb{E}[N(t)] = \mathbb{E}_X[\Lambda(t)], \quad R_N(t_1, t_2) = R_\Lambda(t_1, t_2) + \mathbb{E}_X[\Lambda(t_1)], \quad (14.2.13)$$

where expectation is taken with respect to the stochastic process $X$. Hence, we find that

$$\mathbb{E}[N(t)] = \eta M_0 \int_0^t \langle X(\tau) \rangle d\tau = \eta M_0 X^* t, \quad (14.2.14)$$

and (see Ex. 14.4):

$$\text{Var}[N(t)] = M_0 \eta^2 \frac{X^*(1+\Theta)}{2} \left\{ \frac{2t}{\Gamma} - \frac{2}{\Gamma^2} \left[ 1 - e^{-\Gamma t} \right] \right\} + \mathbb{E}[N(t)]. \quad (14.2.15)$$

Note that $\text{Var}[N(t)] > \mathbb{E}[N(t)]$. The latter is a basic property of DSPPs, namely, that the variance is greater than a Poisson process with intensity given by the mean of the stochastic intensity—a feature known as over-dispersion. Ignoring transient

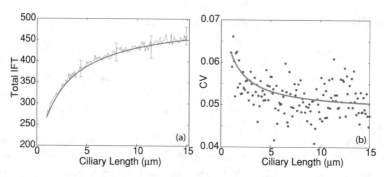

Fig. 14.11: Simulation of the DSPP with $X(t)$ evolving according to the SDE (14.2.11) averaged over 100 trials. (a) Plot of $\langle F(t) \rangle$ for large $t$. Simulations with error bars shown in blue. Analytical curve shown in red. (b) Plot of coefficient of variation versus ciliary length. Blue points are simulation results. Analytical curve shown in red. Other parameter values are $M = 100$, $B\bar{k}_+/k_- = 10$, $\eta = 1\ \mathrm{s}^{-1}$, $\kappa/\sigma = 1\ \mu\mathrm{m}$, $\gamma/\sigma = 4$, $\tau = 1\ \mathrm{s}$, and $v = 2\ \mu\mathrm{m/s}$.

statistics we conclude that for large $t$,

$$\mathbb{E}[N(t)] \sim \lambda_0 t, \quad \lambda_0 = \eta M_0 X^*, \tag{14.2.16}$$

and

$$\mathrm{Var}[N(t)] \sim \lambda_1 t, \quad \lambda_1 = \lambda_0 + \eta^2 M X^* \frac{1 + \Theta}{\Gamma}. \tag{14.2.17}$$

It follows from equations (14.2.14) and (14.2.15), that the corresponding Fano factor is

$$F_N(t) := \frac{\mathrm{Var}[N(t)]}{\mathbb{E}[N(t)]} = 1 + \left( \frac{\lambda_1}{\lambda_0} - 1 \right) \left( 1 - \frac{1 - e^{-\Gamma t}}{\Gamma t} \right). \tag{14.2.18}$$

It is clear that $F_N(t)$ is a monotonically increasing function of time with $F_N(t) = 1$ and $F_N(t) \to \lambda_1/\lambda_0 > 1$ as $t \to \infty$. Recall that a homogeneous Poisson process has a Fano factor of one. Hence, at larger times the DSPP exhibits non-Poisson-like behavior with a Fano factor greater than one, which is consistent with bursting. This is also consistent with what is observed experimentally [578]. One discrepancy between the stochastic model and the experimental data is that at small times the experimentally determined Fano factor dips below one, indicative of behavior more regular than Poisson (e.g., transient periodicity). The latter could be due to some form of refractoriness.

One method for simulating a DSPP is to use a thinning algorithm [557]. Consider an inhomogeneous Poisson process on the time interval $[0, T]$ with rate function $\lambda(t)$, and assume there exists a constant $\lambda^*$ such that $\lambda^* \geq \lambda(t)$ on $[0, T]$. To simulate the inhomogeneous Poisson process, first consider the homogeneous Poisson process with rate $\lambda^*$. We now generate a sequence of times $T_1, T_2, ..., T_m$, for $m \in \mathbb{N}$ with $0 < T_1 < T_2 < ... < T_m \leq T$, with $T_i$, $i = 1, ..., m$ corresponding to the time of the $i$th injection of IFTs docked at the basal body into the flagellum. To obtain the sequence of injection times for the inhomogeneous Poisson process with rate $\lambda(t)$, we accept each $T_i$ generated from the homogeneous Poisson process with probability $\lambda(T_i)/\lambda^*$. The resulting sequence of injection times correspond to the inhomogeneous Poisson process with rate function $\lambda(t)$. For a rigorous proof, see [557]. The thinning algorithm can be applied to the stochastic IFT model by utilizing the following procedure:

- Generate a stochastic trajectory $X(t)$ according to equation (14.2.11) on the interval $[0,T]$.
- Compute $\lambda(X(t)) = \eta M X(t)$ and let $\lambda^* = \max(\lambda(X(t)))$.
- Generate a sequence of times $T_1, T_2, ..., T_m$ with $0 < T_1 < T_2 < ... < T_m \leq T$ from an exponential distribution with parameter $\lambda^*$.
- For each $T_i$, $i = 1, .., m$, generate a random number $U_i$ distributed uniformly on the interval $[0,1]$.
- If $\lambda(X(T_i))/\lambda^* \geq U_i$ accept $T_i$ as a firing time generated by the inhomogeneous Poisson process with rate $\lambda(X(t))$. Otherwise, do not include $T_i$ as a firing time generated by the inhomogeneous Poisson process.

Figure 14.11 shows numerical plots of $\langle F(T) \rangle$ and the coefficient of variation (CV), $\sqrt{\text{Var}[F(T)]}/\langle F(T) \rangle$ versus flagellar length, with stochastic trajectories generated by the aforementioned numerical procedure, together with analytical curves determined by equations (14.2.14) and (14.2.15).

---

**Box 14A. Doubly stochastic Poisson (Cox) processes.**

---

Doubly stochastic Poisson processes (DSPPs) were first introduced by Cox [213] as a generalization of an inhomogeneous Poisson process, in which the time-dependent transition rate depends on a second, independent stochastic process. The general theory of DSPPs was subsequently developed by Grandell [373]. Example applications include photon and electron detection [788], see also Sect. 10.2.3, occurrences of credit events in finance [544], neural coding [72, 527, 819], and intraflagellar transport [121].

Let $\{N(t), t \geq 0\}$ be a counting process with positive intensity $\lambda(x(t))$, which depends on a second, independent stochastic process $\{X(t), t \geq 0\}$; for simplicity, we take $X(t) \in \mathbb{R}$. A major quantity of interest is the averaged distribution

$$\mathcal{P}_n(t) = \mathbb{E}_X \left[ \frac{1}{n!} \left( \int_0^t \lambda(X(s))ds \right)^n \exp \left( -\int_0^t \lambda(X(s))ds \right) \right],$$

where $\mathbb{E}_X$ denotes expectations with respect to the stochastic process $X$. Introducing the characteristic function

$$G_{\Lambda(t)}(z) = \mathbb{E}_X \left[ e^{iz\Lambda(t)} \right], \quad \Lambda(t) = \int_0^t \lambda(X(s))ds,$$

it immediately follows that $\mathcal{P}_n(t)$ is related to the $n$th derivative of $G_{\Lambda(t)}(z)$:

$$\mathcal{P}_n(t) = \frac{(-i)^n}{n!}G_\Lambda^{(n)}(i).$$

We can express the characteristic function for $N(t)$ in terms of $G_{\Lambda(t)}$:

$$G_{N(t)}(z) \equiv \mathbb{E}\left[e^{izN(t)}\right] = \sum_{n\geq0}e^{izn}\mathcal{P}_n(t) = \sum_{n\geq0}\mathbb{E}_X\left[\frac{1}{n!}\left(e^{iz}\Lambda(t)\right)^n e^{-\Lambda(t)}\right]$$

$$= \mathbb{E}_X\left[\sum_{n\geq0}\left(\frac{1}{n!}\left(e^{iz}\Lambda(t)\right)^n\right)e^{-\Lambda(t)}\right] = \mathbb{E}_X\left[e^{e^{iz}\Lambda(t)}e^{-\Lambda(t)}\right]$$

$$= G_{\Lambda(t)}(i - ie^{iz}).$$

It immediately follows that

$$\mathbb{E}[N(t)] \equiv -i\frac{dG_{N(t)}(z)}{dz}\bigg|_{z=0} = -i\frac{dG_{\Lambda(t)}(i-ie^{iz})}{dz}\bigg|_{z=0} = \mathbb{E}_X[\Lambda(t)].$$

**Characteristic functional and correlations.** In order to determine more general statistics of the DSPP such as the covariance we need to determine the joint characteristic function of a finite set of variables $\{N(t_1),\ldots,N(t_m)\}$. This can be achieved using the notion of a characteristic functional [92]. (Functionals were introduced in Box 12C.) The latter is defined according to

$$\Phi_N[v] \equiv \mathbb{E}\left[\exp\left(i\int_0^T v(\sigma)dN(\sigma)\right)\right], \qquad (14.2.19)$$

for fixed $T$, where $v$ is a real-valued function and the counting integral is

$$\int_0^T v(\sigma)dN(\sigma) = \sum_{i=1}^{N(T)} v(\omega_i),$$

with $\omega_i$ denoting the occurrence times of the DSPP. Expectation is taken with respect to both stochastic processes $N(t),X(t)$. In order to evaluate the characteristic functional, we first condition on a particular realization $\{x(t), 0 \leq t \leq T\}$ of the stochastic process $X(t)$ over the time interval $[0,T]$. We write the corresponding conditioned characteristic functional as

$$\Phi_N[v|x] \sim \mathbb{E}\left\{\exp\left(i\sum_{k=1}^M v(\sigma_k)\Delta N(\sigma_k)\right)\right\}.$$

Following along analogous lines to the analysis of path integrals (Chap. 8), we have discretized time into $M$ intervals of size $\Delta\sigma$, and expectation is taken with respect to the inhomogeneous Poisson process with intensity $\lambda(t) = \lambda(x(t))$. (We are assuming that the limit $M \to \infty, \Delta\sigma \to 0$ with $M\Delta\sigma = T$ is well-defined.) It follows that

$$\Phi_N[v|x] \sim \prod_{k=1}^{M} \left[ (1 - \lambda(\sigma_k)\Delta\sigma) + \lambda(\sigma_k)\exp(iv(\sigma_k))\Delta\sigma \right]$$

$$\sim \prod_{k=1}^{M} \exp\left( \left[ e^{iv(\sigma_k)} - 1 \right] \lambda(\sigma_k)\Delta\sigma \right)$$

$$\sim \exp\left( \sum_{k=1}^{M} \left[ e^{iv(\sigma_k)} - 1 \right] \lambda(\sigma_k)\Delta\sigma \right).$$

If we now retake the continuum limit, we see that

$$\Phi_N[v|x] = \exp\left( \int_0^T \left[ e^{iv(\sigma)} - 1 \right] \lambda(\sigma)d\sigma \right).$$

Finally, taking expectation with respect to the stochastic process $X(t)$ yields

$$\Phi_N[v] = \mathbb{E}_X \left\{ \exp\left( \int_0^T \left[ e^{iv(\sigma)} - 1 \right] \lambda(\sigma)d\sigma \right) \right\}. \qquad (14.2.20)$$

Now take $v(\sigma)$ to be the following piecewise function [92]:

$$v(\sigma) = \begin{cases} \sum_{i=1}^{m} \alpha_i; & 0 \leq \sigma < t_1 \\ \sum_{i=2}^{m} \alpha_i; & t_1 \leq \sigma < t_2 \\ \vdots & \vdots \\ \alpha_m; & t_{m-1} \leq \sigma < t_m \\ 0; & t_m \leq \sigma < T, \end{cases}$$

where $0 < t_1 < t_2 < \cdots < t_m < T$. From equation (14.2.19), the corresponding characteristic functional is

$$\Phi_N[v] = \mathbb{E}\{\exp[i(\alpha_1 + \cdots + \alpha_m)N(t_1) + i(\alpha_2 + \cdots + \alpha_m)(N(t_2) - N(t_1)) + \cdots + i\alpha_m(N(t_m) - N(t_{m-1}))]\}$$

$$= \mathbb{E}\{\exp[i(\alpha_1 N(t_1) + \cdots + \alpha_m N(t_m))]\} = G_{N(t_1),\ldots,N(t_m)}(\alpha_1,\ldots,\alpha_m),$$

where $G_{N(t_1),\ldots,N(t_m)}$ is the joint characteristic function of $(N(t_1),\ldots,N(t_m))$. On the other hand, from equation (14.2.20) we have

$$\Phi_N[v]$$

$$= \mathbb{E}_X \left\{ \exp\left[ \left( e^{i(\alpha_1 + \cdots + \alpha_m)} - 1 \right) \int_0^{t_1} \lambda(\sigma)d\sigma + \cdots + \left( e^{i\alpha_m} - 1 \right) \int_{t_{m-1}}^{t_m} \lambda(\sigma)d\sigma \right] \right\}$$

$$= \mathbb{E}_X \left\{ \exp \left[ \left( e^{i(\alpha_1 + \cdots + \alpha_m)} - e^{i(\alpha_2 + \cdots + \alpha_m)} \right) \Lambda(t_1) + \cdots + \left( e^{i\alpha_m} - 1 \right) \Lambda(t_m) \right] \right\}$$

$$= G_{\Lambda(t_1),\ldots,\Lambda(t_m)} \left( -ie^{i(\alpha_1 + \cdots + \alpha_m)} + ie^{i(\alpha_2 + \cdots + \alpha_m)}, \ldots, i - ie^{i\alpha_m} \right),$$

where $G_{\Lambda(t_1),\ldots,\Lambda(t_m)}$ is the joint characteristic function of $(\Lambda(t_1),\ldots,\Lambda(t_m))$. For the sake of illustration, consider the case $m = 2$ and the covariance function

$$R_N(t_1, t_2) = \mathbb{E}[N(t_1)N(t_2)] - \mathbb{E}[N(t_1)]\mathbb{E}[N(t_2)]$$

$$= -\left. \frac{\partial^2 G_{N(t_1),N(t_2)}(\alpha_1, \alpha_2)}{\partial \alpha_1 \partial \alpha_2} \right|_{\alpha_1 = \alpha_2 = 0} - \mathbb{E}_X[\Lambda(t_1)]\mathbb{E}_X[\Lambda(t_2)]$$

$$= -\left. \frac{\partial^2 G_{\Lambda(t_1),\Lambda(t_2)}(-ie^{i(\alpha_1 + \alpha_2)} + ie^{i\alpha_2}, i - ie^{i\alpha_2})}{\partial \alpha_1 \partial \alpha_2} \right|_{\alpha_1 = \alpha_2 = 0}$$

$$\qquad - \mathbb{E}_X[\Lambda(t_1)]\mathbb{E}_X[\Lambda(t_2)]$$

$$= \mathbb{E}_X[\Lambda(t_1)\Lambda(t_2)] + \mathbb{E}_X[\Lambda(t_1)] - \mathbb{E}_X[\Lambda(t_1)]\mathbb{E}_X[\Lambda(t_2)]$$

$$= R_\Lambda(t_1, t_2) + \mathbb{E}_X[\Lambda(t_1)].$$

Expressing $\Lambda(t)$ in terms of the intensity finally gives

$$R_N(t_1, t_2) = \int_0^{t_1} \int_0^{t_2} R_\lambda(\tau, \tau') d\tau d\tau' + \int_0^{t_1} \mathbb{E}_X[\lambda(\tau)] d\tau, \quad t_1 < t_2,$$

where

$$R_\lambda(\tau, \tau') = \mathbb{E}_X[\lambda(\tau)\lambda(\tau')] - \mathbb{E}_X[\lambda(\tau)]\mathbb{E}_X[\lambda(\tau')].$$

**Inter-event interval density.** Consider a counting process with stationary increments. Let $N(t_1, t_2)$ denote the number of events in the interval $(t_1, t_2]$ so that $N(t) = N(0, t)$. Introduce the survivor function for the time $X$ between consecutive events:

$$F_X(x) = \mathbb{P}[X > x] = \lim_{\delta \to 0^+} \mathbb{P}[N(0, x) = 0 | N(-\delta, 0) > 0]. \qquad (14.2.21)$$

This is the probability that following an event at $t = 0^-$, there is not another event up to time $t = x$. By stationarity,

$$\mathbb{P}[\{N(0, x) = 0\} \cup \{N(-\delta, 0) > 0\}] = \mathbb{P}[N(0, x) = 0] - \mathbb{P}[N(-\delta, x) = 0]$$
$$= \mathbb{P}[N(x) = 0] - \mathbb{P}[N(x + \delta) = 0].$$

Hence

$$\mathbb{P}[N(0, x) = 0 | N(-\delta, 0) > 0]\mathbb{P}[N(\delta) > 0] = \mathbb{P}[N(x) = 0] - \mathbb{P}[N(x + \delta) = 0].$$

Dividing both sides by $\delta$ and taking the limit $\delta \to 0$ with

$$v = \lim_{\delta \to 0^+} \mathbb{P}[N(\delta) > 0],$$

(assuming $v$ exists) establishes that

$$F_X(x) = -\frac{1}{v}\frac{dP_0(x)}{dx},$$

where $P_n(x) = \mathbb{P}[N(x) = k]$, $k = 0, 1, \ldots$. Finally, the inter-event interval density is

$$\rho(\tau) = -\left.\frac{dF_X(x)}{dx}\right|_{x=\tau} = \frac{1}{v}\frac{d^2 P_0(\tau)}{d\tau^2}.$$

In the case of a Poisson process, $v$ can be identified with the Poisson rate $\lambda_0$, whereas in the case of a DSPP $v = \mathbb{E}[\lambda]$ and $P_0(\tau) \to \mathcal{P}_0(\tau) = \mathbb{E}[P_0(\tau)]$:

$$\rho(\tau) = \frac{1}{\mathbb{E}[\lambda]}\frac{d^2}{d\tau^2}\mathbb{E}\left[\exp\left(-\int_0^\tau \lambda(s)ds\right)\right].$$

**Ornstein–Uhlenbeck process.** Suppose that $\lambda(t) = X(t)$ with $X(t)$ given by the OU process (Sect. 2.2)

$$dX = k(X^* - X)dt + \sigma dW(t),$$

where $W(t)$ is a Wiener process, and $X(0) = X_0$. We will assume that $\sigma \ll X^*$ so that the chance of meaningless negative values for $\lambda$ can be ignored. We will assume that the Gaussian process is stationary, which corresponds to taking the initial condition $X(0) = X^*$. It follows that $\langle X(t)\rangle = X^*$ and

$$R_X(t_1, t_2) \equiv \langle [X(t_1) - \langle X(t_1)\rangle][X(t_2) - \langle X(t_2)\rangle]\rangle = \frac{\sigma^2}{2k}e^{-k|t_2 - t_1|}.$$

Hence,

$$\mathbb{E}[N(t)] = \int_0^t \langle X(\tau)\rangle d\tau = X^* t,$$

and

$$R_N(t_1, t_2) = \frac{\sigma^2}{2k}\int_0^{t_1}\int_0^{t_2} e^{-k|\tau' - \tau|}d\tau d\tau' + \mathbb{E}[N(t_1)] = \frac{\sigma^2}{2k}\mathcal{A}(t_1, t_2) + \mathbb{E}[N(t_1)],$$

where for $t_1 \leq t_2$

$$\mathcal{A}(t_1, t_2) = \frac{2t_1}{k} - \frac{2}{k^2}\left[1 - e^{-kt_1}\right] + \frac{1}{k^2}\left[1 - e^{-k(t_2 - t_1)} - e^{-kt_1} + e^{-kt_2}\right].$$

In particular, setting $t_1 = t_2 = t$ yields the variance

$$\text{Var}[N(t)] = \frac{\sigma^2}{k} \left[ \frac{t}{k} - \frac{1}{k^2} \left[ 1 - e^{-kt} \right] \right] + \mathbb{E}[N(t)].$$

Note that for all $t > 0$ we have $\text{Var}[N(t)] > \mathbb{E}[N(t)]$. The latter is a basic property of DSPPs, namely, that the variance is greater than a Poisson process with intensity given by the mean of the stochastic intensity.

## 14.3 Cell mitosis in eukaryotes

Mitosis is a phase in a eukaryotic cell's life cycle, during which it segregates its already-duplicated chromosomes in preparation for cell division, or cytokinesis. (Recall from Chap. 6 and Chap. 12 that chromosomes are thread-like structures consisting of DNA tightly coiled around proteins called histones, which allow the extensive genome to be packed inside the nucleus.) Following duplication of its DNA, a cell synthesizes many additional macromolecules, so at the end of the period between cell divisions (interphase), all of the materials needed to form two viable cells are present. Mitosis and cytokinesis then separate this biochemically doubled cell into two essentially identical objects, each equipped to grow and divide again. The major molecular machinery responsible for organizing and segregating the duplicated chromosomes is known as the mitotic spindle, which is an assembly of MTs spreading radially from two poles. Mitosis consists of several distinct phases as illustrated in Fig. 14.12 [603]. An image of a mitotic spindle obtained using fluorescent microscopy is shown in Fig. 14.13.

(i) *Prophase.* The first physical step known as prophase involves restructuring the chromosomes within a dividing cell (chromosome condensation), so that each is sufficiently compact to be separable within a space no bigger than a single cell. At the start of prophase there are two identical copies of each chromosome in the cell due to replication in interphase. These copies are referred to as sister chromatids and are attached to each other by a DNA element called the centromere. Such coupling is crucial for the proper functioning of mitosis, since accurate segregation of sister chromatids depends on their being attached until the moment when all chromatids simultaneously begin segregation. The centromere is also the domain where each chromosome develops specializations such as the kinetochore, which is the protein structure on chromatids where the mitotic spindle fibers attach during cell division to pull sister chromatids apart.

(ii) *Prometaphase.* Spindle formation initiates the process of chromosome organization during prometaphase. The main step is the attachment of all chromosomes to spindle MTs in such a way that each chromatid of every chromosome is associated with MTs that are in turn associated with one and only one end of the mitotic spindle. A second step in prometaphase is the migration of all chromosomes to the spindle mid-plane or equator, a process called congression.

(iii) *Metaphase.* Once the chromosomes are positioned at the equator, the cell is said to be in metaphase. Normal cells include quality control processes that determine whether each chromosome is properly attached to the spindle before segregation is allowed to begin; this is the spindle assembly checkpoint (SAC). Shortly after this checkpoint has been satisfied, the cohesins that have been holding sister chromatids together are cleaved by a protease.

(iv) *Anaphase.* MT-generated forces acting on the now-independent sister chromatids move them to opposite ends of the cell in a process called anaphase. If the nuclear envelope dispersed during spindle formation, then it now reforms on the still-condensed chromosomes by the application of vesicles derived largely, if not entirely, from the previously dissociated nuclear envelope. As these membranes are fusing to define the two nuclear compartments, the cell initiates cytokinesis, the process that divides the cytoplasm into two approximately equal parts, each of which contains its own nucleus. At the same time, the chromosomes de-condense, and the daughter cells return to interphase.

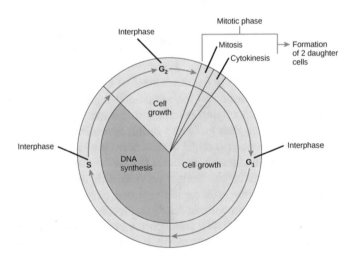

Fig. 14.12: The cell cycle of eukaryotes consists of four distinct phases or stages G1-S-G2-M. Here G1 and G2 are gap phases, while S is the synthesis stage where DNA replication occurs. The M phase consists of two tightly coupled processes; mitosis in which the cell's chromosomes are divided into two, and cytokinesis where the cell's cytoplasm divides in half to form distinct cells. Activation of each phase is dependent on the proper progression and completion of the previous one. Cells that have temporarily stopped dividing are said to have entered a state of quiescence called the G0 phase. The mitotic phase is itself subdivided into four distinct phases, see text for details. [Public domain figure downloaded from Boundless.]

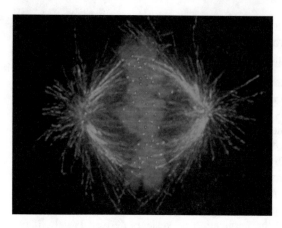

Fig. 14.13: Image of the mitotic spindle in a human cell showing MTs in green, chromosomes (DNA) in blue, and kinetochores in red. [Public domain figure downloaded from Wikipedia.]

A number of mathematical and computational models have been developed in order to gain an understanding of how the spindle machinery performs its complex functions. Typically, a model focuses on one or two aspects of mitosis such as spindle assembly, positioning, maintenance and elongation, chromosomal capture and congression, and the spindle assembly checkpoint. In the following we describe some of these models; see also the reviews [402, 634, 636, 717].

### 14.3.1 Search-and-capture model of MT/kinetochore attachment

A crucial step in prometaphase is the attachment of each chromosome to an MT of the mitotic spindle. According to the search-and-capture model of Kirschner and Mitchison [511], the underlying mechanism involves the nucleation of MTs in random directions, which then grow and shrink dynamically in order to search space and eventually encounter a target kinetochore, see Fig. 14.14. Analysis of MTs searching for a single kinetochore shows that MT dynamic instability (Chap. 4) provides an effective search mechanism provided it is regulated appropriately, that is, MTs do not waste time growing in the wrong direction and don't undergo premature catastrophe when growing in the right direction [410, 424]. However, as highlighted by Wollman et al. [948], although the estimated capture time is consistent with the duration of the mitotic phase, it does not take into account realistic geometries nor the capture of multiple chromosomes. Using a combination of mathematical analysis and computer simulations, Wollman et al show that unbiased search-and-capture for multiple chromosomes is not efficient enough to account for the duration of the prometaphase. On the other hand, if there exists a spatial gradient in some stabilizing factor that biases MT dynamics toward the chromosomes, then one obtains more realistic capture times [948]. One candidate molecule for acting as a stabilizing factor is Ran-GTP [162].

   We now develop the analysis of unbiased search-and-capture, following along the lines of Wollman et al. [948], under the simplifying assumption that the rescue rate following each catastrophe is zero ($k_{rescue} = 0$). (This simplification avoids the need to deal with sticky boundaries, see below.) The authors consider MTs nucleating from two centrosomes, which could be located at the two focal points of an ellipsoid representing the cell shape, see Fig. 14.14(c). Pairs of chromosomes are linked together by kinetochores, which are the fixed targets of searching MTs, and are distributed randomly around the equatorial plane. Each centrosome has hundreds of nucleating sites from which newly formed MTs grow and shrink according to the Dogterom–Leibler model [249]. We begin by considering MTs from a single nucleation site searching for a single kinetochore. Let $P(n)$ be the probability that $n$ sequentially nucleated MTs fail to capture the kinetochore but the $(n + 1)$-th MT is

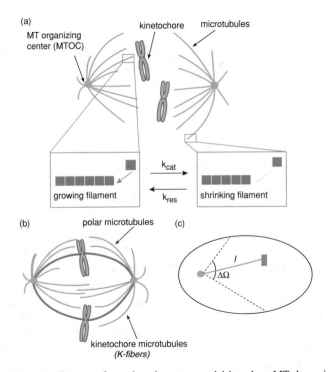

Fig. 14.14: Schematic diagram of search-and-capture model based on MT dynamic instability. (a) During prometaphase MTs randomly probe the cellular domain by alternating between growth and shrinkage phases until they capture the kinetochores. This process of dynamic instability can be quantified by four parameters: the rates of growth ($V_g$) and shortening ($V_S$) and the rates of catastrophe ($k_{cat}$) and rescue ($k_{res}$). To capture all kinetochores in a reasonable time frame, the dynamic instability parameters have to be optimized, but this is not sufficient to make the process fast enough. One possible mechanism for accelerating the search-and-capture is the presence of a RanGTP gradient around the chromosomes that biases the MT dynamics [948]. (b) At the end of prometaphase, all the kinetochores are attached to MTs, one from each pole of the mitotic spindle, and are co-aligned along the mid-plane. (c) Illustration of the search cone of a nucleation site on one of the centrosomes of a cell, with the cell treated as an ellipsoid. A target falls within the search cone at a distance $l$ from the site.

successful. Let $P(\tau|n)$ denote the conditional probability that given $n$ cycles of failure, the time to capture is less than $\tau$. The total probability of capture before time $\tau$ is then

$$P(\tau) = \sum_{n=0}^{\infty} P(\tau|n)P(n), \quad P(n) = p(1-p)^n, \tag{14.3.1}$$

where $p$ is the probability of an MT nucleating in the right direction and reaching the kinetochore before undergoing catastrophe. Suppose that the kinetochore is at a radial distance $l$ from the nearest pole of the mitotic spindle, and has an effective target radius of $r$. Assuming that the MTs are nucleated in random directions, it follows that the probability $P_1$ of an MT nucleating in the right direction is given by the solid angle subtended by the target:

$$P_1 = \frac{\pi r^2}{4\pi l^2} = \frac{r^2}{4l^2}.$$

Suppose that the time from nucleation of an MT to its catastrophe is exponentially distributed. In the absence of rescue, the probability $P_2$ of reaching the kinetochore before catastrophe is

$$P_2 = \int_{T_s}^{\infty} k_{cat}e^{-k_{cat}t}\,dt = e^{-k_{cat}l/V_g},$$

where $T_s = l/V_g$ is the time to reach the kinetochore given a constant speed of growth $V_g$. Therefore, the probability of reaching the kinetochore is

$$p = P_1 P_2 = \frac{r^2}{4l^2}e^{-lk_{cat}/V_g}. \tag{14.3.2}$$

It remains to calculate the conditional probability $P(\tau|n)$. First, note that $P(\tau|n) = Q(\tau - \Delta\tau|n)$ where $\Delta\tau = l/V_g$ is the time for the $(n+1)$-th MT to reach the kinetochore, and $Q(\tau|n)$ is the conditional probability that given $n$ cycles of failure ($n \geq 1$), the total time taken up by these cycles is less than $\tau$. In order to calculate the latter, we need to determine the average lifetime $T_{cycle}$ of an unsuccessful cycle, starting from nucleation through catastrophe to complete depolymerization. The mean time to a catastrophe is $T_c = 1/k_{cat}$ and the subsequent time for the MT to shrink is given by the mean length at the start of catastrophe divided by the speed of shrinkage, $V_g/(V_s k_{cat})$. Therefore,

$$T_{cycle} = \frac{V_g + V_s}{V_s k_{cat}}. \tag{14.3.3}$$

The duration of each nucleation cycle, in the absence of rescue, is an exponential random variable with mean $T_{cycle}$. We now use the basic result that the sum of $n$ exponential random variables is a Gamma random variable (see Box 14B), that is,

$$Q(\tau|n) = \frac{1}{T_{cycle}^n (n-1)!} \int_0^{\tau} s^{n-1}e^{-s/T_{cycle}}\,ds. \tag{14.3.4}$$

We can now evaluate $P(\tau)$ according to (see Ex. 14.5)

$$P(\tau) = p + \sum_{n=1}^{\infty} (1-p)^n p Q(\tau - \Delta\tau | n) = p + (1-p)(1 - e^{-p(\tau-\Delta\tau)/T_{cycle}})$$
(14.3.5)

for $\tau > \Delta\tau$. If $p \ll 1$ then the characteristic number of unsuccessful searches is $n \gg 1$ and the typical search time is $\tau \gg \Delta t$. Hence

$$P(\tau) \approx 1 - e^{-p\tau/T_{cycle}},$$

which implies that the average time to capture is

$$T_{capture} = \frac{T_{cycle}}{p} = \frac{V_g + V_s}{V_s k_{cat}} \frac{4l^2}{r^2} e^{lk_{cat}/V_g}.$$
(14.3.6)

It follows that the optimal catastrophe frequency is $k_{cat} = V_g/l$ [424, 948]. Extensions to the case of multiple MTs and multiple kinetochores are considered in Ex 14.5. For example, one finds that for $N$ nucleating MTs and a single kinetochore the mean time to capture is $T_{capture}/N$. Moreover, for $N$ MTs and $M$ kinetochores $(N \geq M)$ the mean time to capture all of the kinetochores is approximately

$$T_{N,M} = (T_{capture}/N) \ln M.$$

It turns out that the time to capture $M = 46$ chromosomes using up to 1000 searching MTs is substantially greater than experimental measurements of 20–30 min. This was shown by Wollman et al. [948] using Monte Carlo computer simulations of their model based on the following algorithm:

1. Chromosome positions are generated randomly within a sphere of radius 10 $\mu m$ representing the nucleus, and for each chromosome the two-pole kinetochore distances are calculated.

2. For each of these distances, the probability of a successful search $p$ is calculated using equation (14.3.2).

3. The number $n$ of unsuccessful searches is generated randomly from the geometric probability distribution $P(n) = p(1-p)^n$.

4. The duration of each unsuccessful search is generated randomly using the exponential probability distribution $P(\tau) \sim e^{-t/T_{cycle}}$ with $T_{cycle}$ given by equation (14.3.3). The sum of the $n$ random duration times is added to the successful search time $l/V_g$ to determine the total search time for the given kinetochore.

5. The above four steps are repeated $N$ times (for $N$ MTs) and the smallest search time is chosen. This is then repeated for each of the kinetochores and the largest of the $M$ search times is identified as the total capture time. (Since $N \gg M$, steric effects are ignored, that is, one neglects the fact that once an MT has found a kinetochore, the number of searching MTs is reduced.)

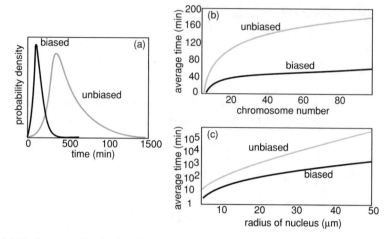

Fig. 14.15: Summary sketch of results from model simulations of Wollman et al. [948], comparing capture times for biased and unbiased search, which are shown by black and gray curves, respectively. (a) Distribution of capture times in the case of 250 MTs searching for 46 chromosomes. (b) Variation of mean capture time with the number of chromosomes for 1000 searching MTs (unbiased search) and 250 MTs (biased search), respectively. (c) The unbiased model exhibits an exponential increase in capture time as a function of nucleus radius, whereas the capture time of the biased model varies as a power law (approximately cubic).

6. The average capture time is determined by repeating the above five steps over multiple trials.

The results obtained by Wollman et al. [948] for the above-unbiased search scenario are sketched in Fig. 14.15, which shows the distribution of capture times and the variation of the mean with respect to nucleus size and number of chromosomes. Also shown are the corresponding results for a biased search, in which the catastrophe rate of MTs is modulated by a RhoGTPase concentration gradient, resulting in a significant reduction in capture time, see [948] for further details.

Finally, note that another possible mechanism for reducing the capture time would be to have some form of cooperative effect between chromosomes so that once several central chromosomes have been captured by random search, the remainder could be captured more quickly. Cooperative effects might be mediated by molecular motors and bundles of MTs nucleated from the chromosomes [402, 636].

---

**Box 14B. Gamma distribution.**

---

Let $T = \sum_{j=1}^{n} \tau_j$, where the $\tau_j$ are independent, exponential random variables with mean $\bar{\tau}$, that is, $\tau_j$ has the probability density

$$p(\tau_j) = \frac{1}{\bar{\tau}} e^{-\tau_j/\bar{\tau}}.$$

The probability density for the random variable $T$ is

$$\rho(T) = \int_0^\infty \cdots \int_0^\infty \delta\left(T - \sum_{l=1}^n \tau_l\right) \prod_{k=1}^n p(\tau_k) d\tau_k$$

$$= \frac{1}{\bar{\tau}^n} \int_0^\infty \cdots \int_0^\infty \delta\left(T - \sum_{l=1}^n \tau_l\right) e^{-\sum_{j=1}^n \tau_j/\bar{\tau}} \prod_{k=1}^n d\tau_k.$$

Introduce the Fourier representation of the Dirac delta function,

$$\delta\left(T - \sum_{l=1}^n \tau_l\right) = \int_{-\infty}^\infty e^{iz(T - \sum_{l=1}^n \tau_l)} \frac{dz}{2\pi}.$$

Substituting into the integral expression for $\rho(T)$ and reordering the multiple integral yields

$$\rho(T) = \frac{1}{\bar{\tau}^n} \int_{-\infty}^\infty e^{izT} \prod_{l=1}^n \left[ \int_0^\infty e^{-\tau_l(iz+1/\bar{\tau})} d\tau_l \right] \frac{dz}{2\pi}$$

$$= \frac{1}{\bar{\tau}^n} \int_{-\infty}^\infty e^{izT} \frac{1}{(iz + \bar{\tau}^{-1})^n} \frac{dz}{2\pi}$$

$$= \frac{(-i)^n}{\bar{\tau}^n} \int_{-\infty}^\infty e^{izT} \frac{1}{(z - i\bar{\tau}^{-1})^n} \frac{dz}{2\pi}.$$

The remaining integral can be calculated using the calculus of residues. That is, treat $z$ as a complex variable and close the contour in the lower-half complex plane. Recall that for any analytic function $f(z)$, the integral around a closed contour $C$ with $\omega$ an $n$th order pole within the interior of $C$ is given by

$$\frac{1}{2\pi i} \int_C f(z) \frac{dz}{(z - \omega)^n} = \frac{1}{(n-1)!} \frac{d^{n-1}}{dz^{n-1}} f(z) \bigg|_{z=\omega}.$$

Taking $C$ to be the semi-circle in the lower half-plane, $\omega = i/\bar{\tau}$, and $f(z) = e^{izT}$, we see that

$$\rho(T) = \frac{1}{\bar{\tau}^n} \frac{T^{n-1}}{(n-1)!} e^{-T/\bar{\tau}}.$$

**First passage time theory for search-and-capture with rescue.** A more detailed mathematical analysis of first passage time problems in the search-and-capture model has been developed by Gopalakrishnan and Govindan [358]. They allow

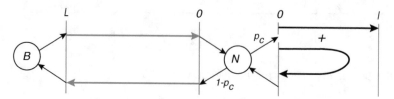

Fig. 14.16: Schematic illustration of the search-and-capture model analyzed in Ref. [358].

for MT rescue, which means that one has to keep track of both nucleation events and collisions of a growing MT with the cell wall. This involves two separate sticky boundary conditions. Suppose that an MT is nucleated at a rate $r_n$ from a centrosome in an arbitrary direction that lies in a cone subtending a solid angle $\Delta\Omega$. If the kinetochore is located at a distance $l$ from the centrosome and has a cross-sectional area $a$, then it subtends a solid angle $\Delta_c = a/l^2$ with respect to the centrosome. Hence, the probability of being nucleated in a direction that finds the target is $p_c = \Delta\Omega_c/\Delta\Omega$. Furthermore, suppose that if the MT nucleates outside the target cone, which occurs with probability $1 - p_c$, then it can potentially grow until it hits a cell boundary at a distance $L$ from the centrosome. (For simplicity, the search cone solid angle $\Delta\Omega$ is taken to be sufficiently small so that the relevant region of the cell wall is approximately equidistant from the nucleation site.) Finally, whenever the MT hits the boundary at $x = L$, its growth velocity $v_+$ drops to zero and it sticks to the wall until it transitions to a shrinkage state at a rate $r_b$.

The total MT state space $\Sigma$ can thus be decomposed as $\Sigma = N \cup A_b \cup B \cup A_c$, where $N$ is the nucleation state, $B$ is the state of being attached to the cell boundary, $A_b$ are the active states that the MT is outside the target cone and has length $X(t) \in (0,L)$, and $A_c$ are the active states that the MT is inside the target cone and has length $X(t) \in (0,l)$. Let $S(t)$ denote the state of the MT at time $t$. If $S(t) \in A_b$ then $X(t)$ evolves according to the Dogtorem–Leibler model with sticky boundary conditions at $x = 0, L$. On the other hand, if $S(t) \in A_c$ then $X(t)$ evolves according to the Dogtorem–Leibler model with a sticky boundary condition at $x = 0$ and an absorbing boundary condition at $x = l$. If $S(t) = N$ then the MT transitions to a growing state, which either belongs to $A_b$ with probability $1 - p_c$ or belongs to $A_c$ with probability $p_c$. The time $\tilde{\tau}_n$ spent in state $N$ is exponentially distributed with mean time $r_n^{-1}$. Similarly, the time $\tilde{\tau}_b$ spent in state $B$ is exponentially distributed with mean time $r_b^{-1}$. A schematic diagram of the search and capture model is shown in Fig. 14.16. The MFPT to capture a target was analyzed by the authors using the forward CK equation and Green's function methods [358]. Simplified methods of analysis have subsequently been developed in terms of either the backward CK equation [646] or probabilistic methods [126], as outlined in Sect. 4.2 and Box 4A. An analogous probabilistic approach will be used to analyze cytoneme-based search-and-capture processes in Sect. 14.5.

## *14.3.2  Chromosome movements and directional instability*

It is found that the movements of chromosomes during prometaphase and metaphase are characterized by periods of ballistic motion or "runs" at approximately constant speed, separated by abrupt reversals in direction of movement [170, 377, 761, 825]. These oscillations arise from interactions between attached kinetochore MTs (kMTs) and the corresponding chromosomes. When a chromosome becomes bioriented, that is, the sister chromatids are attached to kMTs emanating from opposite poles, the duration of movements toward and away from the nearest pole are balanced so that the chromosome is aligned with the spindle equator at metaphase.

**Hill sleeve model of kMT-kinetochore interactions.** One of the first models of kMT/chromosome interactions was developed by Hill [410], who treated the outer region of a kinetochore as a "sleeve" containing a sequence of tubulin-binding sites that holds a kMT within the sleeve, see Fig. 14.17(a). The sleeve is assumed to be around 40 nm thick, which means that it can accommodate up to $M = 65$ tubulin sub-units of the kMT. The model keeps track of the position $n$ of the kMT tip within the sleeve, $n = 1,\ldots,M$, with limits $n = 1$ (fully inserted) and $n = M$ (almost unattached). Motion of the kMT tip is modeled as a random walk in the free energy landscape sketched in Fig. 14.17(b). Let $-a$, $a > 0$, be the binding energy of a single tubulin site. This effect tends to pull the kMT into the sleeve. However, the attractive force is opposed by two frictional forces—an external frictional force $F$ that opposes movement of the whole chromosome and a surface roughness acting at the interface between the kMT and the kinetochore. The latter increases linearly with the length of the interface, which is proportional to $M - n + 1$. If $l$ is the length of a single step, then the forward transition $n \to n + 1$ (withdrawal) requires climbing a barrier height of size $[\Delta E_+(n)$, whereas the reverse transition $n + 1 \to n$ (insertion) requires climbing a barrier height of size $\Delta E_-(n)$, where

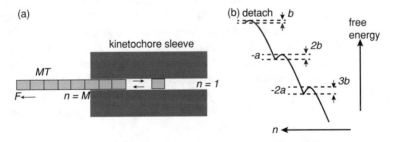

Fig. 14.17: Hill "sleeve model." (a) Schematic diagram of a kinetochore sleeve interacting with a kMT. The position of the kMT tip changes either due to thermal motion of the sleeve or by addition or loss of tubulin sub-units at the kMT tip. (b) Free energy diagram for a kinetochore sleeve interacting with a kMT in the absence of external friction ($F = 0$).

$$\Delta E_+(n) = a + (M-n+1)b - \frac{Fl}{2}, \quad \Delta E_-(n) = (M-n+1)b + \frac{Fl}{2}. \quad (14.3.7)$$

(For convenience, the external frictional force is divided equally between the two transitions.) From the theory of chemical kinetics, the forward and backward transition rates take the form

$$k_\pm(n) = \kappa e^{-\Delta E_\pm(n)/k_B T} \quad (14.3.8)$$

for a background hopping rate $\kappa$. The position of the tip can also change due to polymerization or depolymerization. We will assume that the rate of adding a tubulin sub-unit at the tip is $\alpha c$, where $c$ is the background concentration of tubulin, and the rate of removing a sub-unit is $\beta e^{-a/k_B T}$. Here $\alpha, \beta$ are constants and the Boltzmann factor takes into account the binding energy of the sub-unit to the kinetochore sleeve. We can now write down a kinetic equation for the mean tip location $n(t)$:

$$\frac{dn}{dt} = k_+(n) - k_-(n) + \beta e^{-a/k_B T} - \alpha c. \quad (14.3.9)$$

Hence, at equilibrium the mean tip location $n^*$ is obtained by setting $\dot{n} = 0$:

$$\kappa e^{-(M-n+1)b/k_B T} \left[ e^{-Fl/2k_B T} - e^{Fl/2k_B T} e^{-a/k_B T} \right] = \beta e^{-a/k_B T} - \alpha c,$$

which gives

$$n^* = M + 1 - \frac{k_B T}{b} \ln \Gamma, \quad \Gamma = \kappa^{-1} \frac{\beta e^{-a/k_B T} - \alpha c}{e^{-Fl/2k_B T} - e^{Fl/2k_B T} e^{-a/k_B T}}. \quad (14.3.10)$$

Suppose that depolymerization dominates over polymerization so that the tendency of the kinetochore is to move toward the nearest pole. An important feature of Hill's sleeve model is that over a wide range of external forces $F$, the speed of depolymerization-coupled kinetochore movements remains constant. This is a consequence of the observation that for a given $F$, the steady-state position $n^*(F)$ of the kMT tip within the sleeve is fixed, and thus the sleeve moves at an average speed equal to the rate of kMT shortening. In other words, the sleeve keeps up with the tip of the depolymerizing MT. If the external force changes, $F \to F'$, the sleeve will shift to a new steady-state position $n^*(F')$, where it will continue on at the rate of kMT shortening. This approximate load independence is consistent with experimental observations [825]. Finally, note that a more mathematical treatment of the Hill sleeve has been developed by Shtylla and Keener based on a jump-diffusion process [821, 822].

**Chromosomal oscillations.** As it stands, the sleeve model of single kMT-kinetochore interactions cannot explain the oscillatory switching of chromosome motion toward and away from a spindle pole, see Fig. 14.18. In order to account for such dynamics, it is necessary to consider some combination of the following mechanisms: tensional coupling between sister chromatids, polar ejection forces due to the pushing action of polar (non-kinetochore) MTs on the arms of a chromosome, the space-dependent modulation of kMT catastrophe and rescue frequencies, and the possible action of depolymerizing and polymerizing molecular motors such as kinesin and dynein [196, 334, 335, 476, 567, 821]. Joglekar and Hunt [476] considered a simple extension of the Hill sleeve model by including tensional coupling and polar ejection forces and allowing detached MTs to switch from catastrophe to rescue. Following the Hill model, kMTs are assumed to be in catastrophe so that in order to maintain an average kMT tip location within the kinetochore, the associated chromatid tends to move toward the pole of the attached kMT. It follows that there is essentially a tug-of-war between the sister chromatids moving in opposite directions. As the sister chromatids start to separate, tensional forces increase until one of the chromatids loses all of its depolymerizing kMTs, at which point it follows the motion of the other chromatid. Consequently the chromosome moves toward the pole of the winning chromatid. However, as it approaches the pole, the density of polar MTs increases, resulting in an increase in the polar ejection force. The closer the kinetochore moves toward the pole, the more strongly the polar ejection force opposes its advancement, and eventually the last depolymerizing kMT detaches from the kinetochore as the load exceeds the detachment force. Meanwhile growing MTs from the opposite pole can be recruited by the sister chromatid resulting in an abrupt reversal in direction and the cycle repeats. An immediate consequence of the interplay between polar ejection forces, MT dynamic instabilities, and the "Hill sleeve" is that the chromosomes tend to position themselves toward the spindle equator.

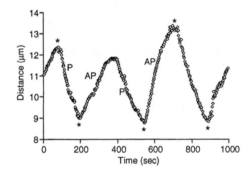

Fig. 14.18: Sawtooth-like *in vivo* oscillations of a non-oriented chromosome (attached to one spindle-pole) observed in a newt mitotic cell. The chromosome switches between poleward (P) and antipoleward (AP) motion. [Adapted from Inoue et al. [455].]

One of the limitations of the Joglekar and Hunt model is that it neglects the experimental observation that there is a tension-dependent increase in the rescue frequency of kMTs. This particular mechanism has been explored in a number of computational models by Gardner et al [334, 335, 837]. A schematic illustration of the basic model is shown in Fig. 14.19; only one chromatid pair is shown for simplicity. The minus end of each kMT is fixed at its spindle pole and the plus end of the kMT is fixed at the corresponding kinetochore; thus the details of the kMT-kinetochore interface, as specified in the Hill sleeve model, are neglected. For convenience, distance from the spindle equator towards the right (left) pole is denoted by the positive coordinate $x$ ($\bar{x}$), and the direction of all forces $F$ ($\bar{F}$) acting on the right-half (left-half) of the spindle are defined with respect to the poleward direction. Let $X_K$ and $\bar{X}_K$ denote the current position of the right and left sister kinetochore plates

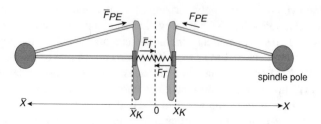

Fig. 14.19: Tension-mediated regulation of kMT/kinetochore dynamics. Here $F_T$ and $F_{PE}$ denote the tension force and polar ejection force acting on the rightward kMT whose tip is at position $X_K$ relative to the equator. Similarly, $\overline{F}_T$ and $\overline{F}_{PE}$ denote the corresponding forces on the leftward kMT whose tip is at $\overline{X}_K$. Note the sign conventions.

with respect to the spindle equator, and let $V_K$ and $\overline{V}_K$ denote the corresponding velocities. For concreteness, let us focus on a kMT attached to the right-hand pole; a similar formulation holds for those attached to the other pole. Each kMT is modeled in terms of the Dogterom–Leibler model of MT catastrophe and rescue [249], see Sect. 4.2, so that the mean velocity is

$$\langle V_K \rangle = \frac{k_- v_+ - k_+ v_-}{k_+ + k_-},$$

where $k_+$ $(k_-)$ is the catastrophe (rescue) frequency, and $v_+$ $(v_-)$ is the rate of growth (shrinkage). The catastrophe and rescue frequencies $k_\pm$ are assumed to be regulated by mechanical forces acting on the kMT-kinetochore system, perhaps in combination with a chemical gradient in some kMT catastrophe promoter, and are thus $x$-dependent. The basic components of the model are as follows [837]:

1. *Tension force between the kinetochores* $F_T$. The tension force is modeled as a linear spring, so that the magnitude of the force depends on the distance between the sister kinetochores according to

$$F_T = \kappa(X_K + \overline{X}_K - d_0), \qquad (14.3.11)$$

where $\kappa$ is the spring constant and $d_0$ is the equilibrium length of the spring. Note that, given our sign convention, $F_T = \overline{F}_T$. The tension force provides the coupling between the two kinetochores.

2. *Polar ejection force* $F_{PE}$. This is directed toward the spindle equator and is due to the interaction between the chromosome arms and the plus ends of polar MTs. Hence, it is proportional to the density of polar MTs emanating from the pole. From geometric arguments one can take

$$F_{PE} = \rho X_K^2 \qquad (14.3.12)$$

for some constant $\rho$.

3. *Rescue frequency.* The rescue frequency is taken to be a function of the total force $F = F_T + F_{PE}$ according to

$$k_- = k_{-,0}e^{F/k_B T}.$$

4. *Chemical gradient.* It is assumed that a kMT catastrophe promoter forms a spatial gradient due to spatial segregation of a kinase/phosphatase system that regulates the promoter; the kinase phosphorylates (deactivates) the promoter, whereas the phosphatase dephosphorylates (activates) the promoter. The spatial gradient is modeled by taking the kinase to be localized to the surface of the spindle poles, whereas the phosphatase is distributed homogeneously throughout the cell volume. Let $c_A$ and $c_B$ denote the concentration of deactivated and activated catastrophe promoter, respectively, such that $c_A(x,t) + c_B(x,t) = c_T$ with $c_T$ a constant. The reaction–diffusion equation for $c_A$ is

$$\frac{\partial c_A}{\partial t} = \frac{\partial^2 c_A}{\partial x^2} - kc_A,$$

supplemented by the boundary conditions

$$-D\frac{\partial c_A}{\partial x}\bigg|_{x=0} = k^* c_B(0), \quad -D\frac{\partial c_A}{\partial x}\bigg|_{x=L/2} = 0.$$

Here $k$ and $k^*$ are the rates of activation and deactivation due to the action of kinds/phosphatase, and $L$ is the spatial separation of the poles. The analysis of intracellular spatial gradients was discussed more fully in Sect. 11.1. One finds that the steady-state concentration is of the form

$$c_A(x) = c_T\left(Ae^{-\gamma x/L} + Be^{\gamma x/L}\right), \quad \gamma = 2\sqrt{k/D}.$$

5. *Catastrophe frequency.* Given the steady-state spatial gradient $c_B(x)$ of catastrophe promoter, the catastrophe frequency of the kMT tip is taken to be

$$k_+ = k_{+,0} + \beta c_B(X_K).$$

Computer simulations of the above model, involving different combinations of the various components, suggest that in order to match experimental data of chromosomal dynamics during metaphase of budding yeast, including the tendency to cluster at the cell equator, it is necessary to combine MT dynamic instabilities with tension-dependent rescue and either polar ejection forces or a spatial gradient in some catastrophe promoter [335, 837]. Finally, note that some more recent theoretical studies have considered explicit models of the mechanobiochemical feedback mechanism, in which tension-dependent molecular sensor molecules enzymatically regulate a phosphorylation cascade that alters the dephosphorylation rate at the kMT tip [567, 821]. Other recent models have emphasized the role of chromokinesin

motors on polar MTs that exert a polar ejection force on the chromosomal arms
[160, 850].

## 14.3.3 Force balance and spindle length control

It is hypothesized that the interplay between kMT dynamic instability, kMT–kinetochore
interactions, various populations of molecular motors, and elastic or viscoelastic
forces play an important role in later stages of mitosis. This includes the maintenance of spindle length (distance between the poles) in metaphase, the separation
of sister chromatids in anaphase A, and the increase in spindle length in anaphase
B [197, 365, 603]. For example, during anaphase A, there is a net poleward flux
of kMTs due to depolymerization at their pole-associated minus ends, which may
be supplemented by a so-called "pacman" mechanism in which the kinetochores
actively "chew" their way toward the poles by depolymerization of kMTs at their
plus ends. On the other hand, during anaphase B, a new subset of motors known
as ipMTs drive spindle elongation. Examples of changes in spindle length during
different stages of mitosis are shown in Fig. 14.20 for several organisms. A number of computational models have been developed in order to study one or more of
these stages by considering the various forces acting on kMTs and the chromosomes
[150, 196, 220, 301, 364]. Here we will illustrate the flavor of force-balance mod-

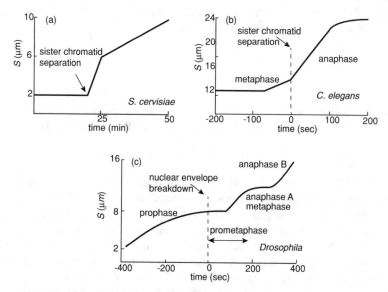

Fig. 14.20: Changes in spindle length $S(t)$ with time $t$ during formation of mitotic spindles for
three cell types: (a) budding yeast (*Saccharomyces cerevisiae*) from metaphase through anaphase
B—the latter is biphasic; (b) *Caenorhabditis elegans* single-cell embryos; and (c) *Drosophila
melanogaster* embryos. Redrawn from Goshima and Scholey [365].

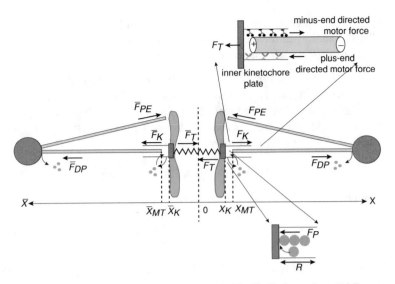

Fig. 14.21:  Force-balance model. See text for details. Redrawn from [196].

els by considering in more detail the model of chromosome motility in *drosophila* embryos [196]. Fig. 14.21 is a schematic diagram indicating the various forces acting on the kinetochores and kMTs of the mitotic spindle. For simplicity, we only consider one kMT per kinetochore, although it is possible to extend the model to multiple kMTs [196]. The sign convention of spatial coordinates and forces are as in Fig. 14.19. In contrast to the models of Gardner et al [335, 837], the tip of a kMT can move relative to its kinetochore, so that one must now consider separate force-balance equations for the velocities of the tips and kinetochore. Moreover, polar ejection and tension forces are supplemented by several additional forces including the action of kinesin and dynein motors at the kMT tip.

Following the formulation of the Gardner et al models, let $X_K$ and $\overline{X}_K$ denote the current position of the right and left sister kinetochore plates with respect to the spindle equator, and let $V_K$ and $\overline{V}_K$ denote the corresponding velocities. Similarly, let $X_{MT}$ and $\overline{X}_{MT}$ be the current position of the plus ends of the right and left kMTs with respect to the spindle equator, and denote the corresponding poleward sliding rates by $V_{MT}$ and $\overline{V}_{MT}$. (Note that $V_K = \dot{X}_K$, whereas the relationship between $X_{MT}$ and $V_{MT}$ is more complicated due to dynamic instability at the plus end of a kMT, see below.) The force-balance equations for the right and left kinetochores (in a low Reynolds number regime) are given by

$$\mu V_K = F_K - F_P - F_T - F_{PE}, \qquad (14.3.13a)$$

$$\mu \overline{V}_K = \overline{F}_K - \overline{F}_P - \overline{F}_T - \overline{F}_{PE}, \qquad (14.3.13b)$$

where $\mu$ is the drag coefficient, $F_T$ and $F_{PE}$ are given by equations (14.3.11) and (14.3.12). The additional forces on the right kinetochore are as follows (with analogous definitions for the left kinetochore):

1. *Net kinetochore motor force $F_K$.* Let $n_\pm$ denote the density of bound plus-end (minus-end) moving kinetochore motors. These motors act on the region of the MT tip that is inserted into the kinetochore; this region has length $R - [X_{MT} - X_K]$, where $R$ is the length of the kinetochore sleeve. If $f_\pm$ denotes the force generated by each plus and minus end-directed motor, then

$$F_K = (X_K + R - X_{MT})(n_- f_- - n_+ f_+). \qquad (14.3.14)$$

The motors are assumed to obey linear force–velocity relations (see Sect. 14.1.1),

$$f_\pm = F_\pm \left(1 - \frac{v_\pm}{V_\pm}\right), \qquad (14.3.15)$$

where $F_\pm$ is the stall force, $V_\pm$ is the motor velocity without a load, and $v_\pm$ is the current velocity of the motors with

$$v_- = -v_+ = V_K - V_{MT}. \qquad (14.3.16)$$

2. *Plus end polymerization force $F_P$.* This will only arise if the MT impinges on the kinetochore plate, that is, $X_K > X_{MT}$, in which case it is taken to have the linear form

$$F_P = \varepsilon(X_K - X_{MT}), \qquad (14.3.17)$$

where $\varepsilon$ is the elastic modulus of the plate.

In order to obtain a closed set of equations, it is necessary to determine the kMT sliding velocity $V_{MT}$ and the position of the tip $X_{MT}$. The former is determined by considering the force-balance equation for the kMTs. It turns out that the viscous drag on the kMT is negligible compared to other forces. The other forces acting on a kMT are the counter motor force $-F_K$, the counter polymerization force $-F_P$ and a depolymerization force $F_{DP}$ at the minus end. The latter is assumed to be generated by a set of depolymerization motors with a linear force–velocity relationship. At steady state the depolymerization velocity at the minus end of the kMT due to the action of the motors is equal to the sliding velocity $V_{MT}$. It follows that

$$n_0 F_0(1 - V_{MT}/V_0) = F_K - F_P, \qquad (14.3.18)$$

where $n_0$ is the number of active depolymerization motors, $F_0$ is the stall force, and $V_0$ is their maximum velocity. The position $X_{MT}$ will spend both on the sliding velocity and the removal or addition of tubulin sub-units at the tip. As in the Gardner et al model [335, 837], it is assumed that the plus end undergoes dynamic instability, characterized by the stochastic switching of MTs between growing and shrinking phases, that is between rescue and catastrophe, see Sect. 4.2. The dynamic stability is specified by four parameters given by the growth and shrinkage velocities $v_g, v_s$ and the rates of rescue and catastrophe $k_{cat}$ and $k_{res}$. The standard catastrophe model of MT dynamics is supplemented by the action of depolymerase enzymes. When tension on the kMT is low, the depolymerase acts freely on the MT plus end by suppressing the rescue frequency by some factor $\gamma > 1$. On the other hand, when the tension is high, the action of the depolymerase is blocked by some form of structural change of the kMT so that the rescue frequency recovers proportionally to the tension force. Finally, if the MT plus end is in contact with the kinetochore plate, then the catastrophe frequency is scaled up, whereas the rescue frequency returns to its low tension value.

Extensive numerical simulations of the above model have shown that it provides a quantitative description of the rapid, highly dynamic properties of metaphase and anaphase A that have been observed experimentally during *drosophila* embryo mitosis [196], see Fig. 14.22. It also provides a possible scenario for the switch from metaphase to anaphase A based on the degradation of the cohesive bonds between the sister kinetochores and the removal of PE forces. One prediction of the model is that increasing the level of dynein activity suppresses chromosome oscillations in metaphase. The basic idea is that the increased minus end-directed force pulls kMTs further into the kinetochores, which increases the distance between sister chromatids and thus increases the tension on the kMTs. This promotes kMT rescue and stabilizes the kMTs within the kinetochores—recall that the basic mechanism for chromosome oscillations is depolymerization of kMT tips within the kinetochore. One limitation of the model is that it assumes that the spindle poles are fixed, and thus does not provide an explanation of spindle length maintenance during metaphase. Another limitation is that it neglects the possible role of external factors such as morphogen gradients and the dynamics of a postulated viscoelastic spindle matrix.

In order to develop a model of spindle length control, it is necessary to consider the balance of forces acting on the spindle poles. Here we describe a simple force-balance model due to Goshima et al [364, 365], see Fig. 14.23. Let $S(t)$ and $L(t)$ denote the spindle length and length of the overlapping region of interpolar MTs (ipMTs) where kinesin-5 motors act to separate the poles. The main components of the model are the net forces acting on the poles and the polymerization/depoly-

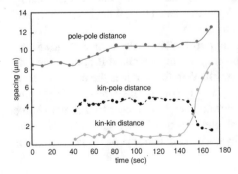

Fig. 14.22: Dynamics of spindle poles and chromatids in Drosophila embryos. During metaphase (80-135 s), the chromatids remain at the spindle equator, and do not exhibit oscillations between the spindle poles as observed in some other organisms. During anaphase A (135-175 s), chromatids rapidly move toward the spindle poles, which are held at an approximately constant spacing of 10 $\mu$m. Redrawn from Civelekoglu-Scholey et al. [196].

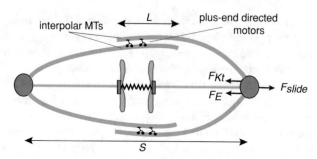

Fig. 14.23: Schematic diagram of force-balance model of spindle length control. Kinesin-5 motors slide apart antiparallel inter parallel Mts (ipMTs) generating a net sliding force $F_{\text{slide}}$ on a spindle pole that acts to separate the centrosomes. This is counteracted by a kinetochore reaction force $F_{\text{Kt}}$ that tends to pull the kinetochore and pole together due to a Hill sleeve mechanism, for example, and a hypothesized elastic restoring force $F_{\text{E}}$, which could be due to MT elasticity or a spindle matrix. The spatial separation of the poles is denoted by $S$, whereas the length of ipMT overlap is given by $L$.

merization of ipMTs. Let $V_{\text{poly}}$ be the rate of polymerization of the ipMT plus ends, $V_{\text{depoly}}$ be the rate of depolymerization of all MTs at the centrosome, and $V_{\text{slide}}$ the rate at which antiparallel ipMTs slide apart. It follows that

$$\frac{dS}{dt} = 2(V_{\text{slide}} - V_{\text{depol}}), \quad \frac{dL}{dt} = 2(V_{\text{poly}} - V_{\text{slide}}). \tag{14.3.19}$$

The factors of two reflect the fact that pairs of ipMTs overlap. These equations are coupled to a force-balance equation, see Fig. 14.23,

$$\mu \frac{dS}{dt} = 2\left(F_{\text{slide}} - F_{\text{Kt}} - F_{\text{E}}\right), \tag{14.3.20}$$

where

$$F_{\text{slide}} = \alpha L \left(1 - \frac{V_{\text{slide}}}{V_{\text{max}}}\right), \quad F_{\text{E}} = \beta(S - S_0). \tag{14.3.21}$$

Here $F_{\text{slide}}$ is the kinesin-5-dependent force that slides apart the ipMTs and is proportional to the concentration of motors $\alpha$, whereas $F_{\text{E}}$ is the elastic restoring force that is taken to behave like a Hookean spring with effective spring constant $\beta$. Finally, $F_{\text{Kt}}$ is taken to be a constant kinetochore force that tends to pull the kinetochore and pole together via some Hill sleeve mechanism, say.

As is stands, such a model does not support a stable steady-state length $S$, but can be used to model spindle length separation during anaphase B [144]. One way to achieve stability is to introduce some mechanism that couples the rate of depolymerization $V_{\text{depot}}$ to the sliding force $F_{\text{slide}}$. For example, Goshima et al [364] propose that increases in $F_{\text{slide}}$ push the minus ends of MTs closer to the centrosome where depolymerizing motors act. Using the theory of Brownian ratchets (Sect. 4.3), the probability that an MT is within a critical distance $\delta$ of the centrosome is

$P(x < \delta) = 1 - \exp(-F_{\text{slide}}\delta/Nk_BT)$. It is assumed that the sliding force is divided equally by the $N$ MTs attached to the centrosome. Assuming that depolymeriztion by motors only occurs if $x < \delta$ then

$$V_{\text{depol}} = V_0 + V_1\left(1 - e^{-F_{\text{slide}}\delta/Nk_BT}\right). \tag{14.3.22}$$

Combining these various equations we obtain the pair of equations

$$\frac{dL}{dt} = 2(V_{\text{poly}} - V_{\text{slide}}) \tag{14.3.23a}$$

$$\mu\frac{dS}{dt} = 2\left(\alpha L\left(1 - \frac{V_{\text{slide}}}{V_{\text{max}}}\right) - F_{\text{Kt}} - \beta(S - S_0)\right), \tag{14.3.23b}$$

with

$$V_{\text{slide}} = \frac{1}{2}\frac{dS}{dt} + V_0 + V_1\left(1 - e^{-\alpha L\left(1 - \frac{V_{\text{slide}}}{V_{\text{max}}}\right)\delta/NK_bT}\right).$$

The steady-state solution for $S$ and $L$ are then

$$L = \frac{\Gamma_0 k_B T N}{\alpha\delta(V_{\text{poly}}/V_{\text{max}} - 1)}, \quad S = S_0 - \frac{F_{\text{Kt}}}{\beta} - \frac{\Gamma_0 N k_B T N}{\beta\delta}, \tag{14.3.24}$$

where

$$\Gamma_0 = \ln\left(1 - \frac{V_{\text{poly}} - V_0}{V_1}\right).$$

## 14.4 Cell motility and focal adhesions

Just as the polymerization and depolymerization of microtubules plays an essential role in cell mitosis, the growth and shrinkage of actin polymers plays a major role in generating the forces necessary for various forms of cell motility [243, 632, 731, 746]. For example, the movement of crawling cells such as amoeba, keratocytes, fibroblasts, and migrating neurons involves the protrusion of lamellipodia and filopodia at the leading edge of the cell, see Fig. 14.24(a), which requires actin polymerization at the cell membrane boundary [104]. On the other hand, intracellular pathogens such as *Listeria* propel themselves within a host cell by assembling the host cell's actin into a comet-like tail. The tail consists of oriented cross-linked networks of actin filaments whose growing ends orient toward the bacterial surface, thus thrusting the pathogen forward [872, 876], see Fig. 14.24(b). A major challenge is linking the complex biochemical processes regulating actin polymerization with mechanical properties of the cell and the associated forces. In the case of crawling cells, there are contractile forces on the actin cytoskeleton due to the action of myosin motors, traction forces from the drag between the cytoskeleton and surface adhesion complexes, membrane tension resisting the actin polymerization force, viscoelastic stresses arising from deformations of the actin network, and viscous drag

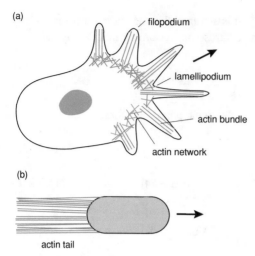

Fig. 14.24: Examples of actin-based cell motility (a) Crawling eukaryotic cell. (b) Pathogen such as *Listeria* propelled by an actin tail assembled from the cytoskeleton of the host cell.

between actin filaments and cytosolic fluid flows [222, 635]. As in the case of cell mitosis, cell motility is a vast subject in its own right and we cannot hope to do it justice here. Instead, we will focus on some aspects that relate most closely to the themes of this book. For an excellent introduction to cell motility see the book by Bray [104].

### 14.4.1 Tethered ratchet model

We begin by considering a microscopic model of cell protrusion based on an extension of the polymerization ratchet model [724]. This was developed by Mogilner and Oster within the context of the simpler problem of *Listeria* propulsion [631, 633], and has subsequently been incorporated into more complex models of cell crawling [423, 502, 632, 635]. Recall from Sect. 4.3 that the speed of growth of a polymerization ratchet depends on the diffusion coefficient of membrane Brownian motion. Within the context of bacterial motion, this would imply that the bacterial velocity depends on its diffusion coefficient, and thus on its size. However, such size-dependence has not been observed experimentally, which led Mogilner and Oster to propose an elastic ratchet model, in which thermal bending fluctuations of a semi-stiff actin filament, rather then bacterial diffusion [631], generates the gap necessary for insertion of an additional monomer, with the resulting growth generating the force to propel the bacterium forward. The elastic ratchet model was itself superseded by the tethered ratchet model [633], in order to account for a number of additional experimental observations [158]. In particular, Brownian fluctuations are almost completely suppressed during *Listeria* propulsion due to the fact that the bacterium is tightly bound to its actin tail. One thus observes smooth particle trajectories that are persistent in both direction and curvature. It is known that the surface of

*Listeria* is coated with nucleation promotion factor ActA, which transiently binds the Arp2/3 complex on the sides of attached actin filaments; Arp2/3 is known to mediate nucleation of side-branched filaments. The tethered ratchet model is one way to resolve the dilemma of how the actin tail can be attached to the bacterium, and yet propel the bacterium forward via growth of unattached active filaments. More specifically, it proposes that there are two classes of filament: some are attached, under tension and nucleating rather than growing, while others are unattached and pushing via an elastic ratchet mechanism. In the following, we will describe the tethered ratchet model in more detail.

Suppose that there are $n_u(t)$ unattached filaments and $n_a(t)$ attached filaments at time $t$ and that the bacterium is moving at speed $v$. There are three forces acting on the bacterium, neglecting any elastic recoil forces of the actin tail, see Fig 14.25: a load force $F_L = \gamma v + F_{ext}$, where $\gamma$ is a viscous drag coefficient and $F_{ext}$ represents any experimentally imposed external forces; a tensional force $F_a = n_a f_a$ due to attached filaments, with $f_a$ the force per filament; a pushing force $F_u = n_u f_u$ due to unattached filaments, with $f_u$ the force exerted by a single unattached filament via an elastic ratchet mechanism. The corresponding force-balance equation is

$$F_L + n_a f_a = n_u f_u. \tag{14.4.1}$$

It is assumed that the two filament populations evolve according to the simple kinetic equations

$$\frac{dn_a}{dt} = \sigma - kn_a, \quad \frac{dn_u}{dt} = kn_a - \kappa n_u, \tag{14.4.2}$$

where $\sigma$ is the nucleation rate of side branches, $k$ is the rate of detachment, and $\kappa$ is the rate of capping of unattached filaments which can then no longer polymerize and push on the bacterium's surface. It remains to specify the dependence of the forces

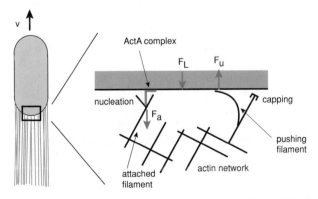

Fig. 14.25: Schematic diagram of tethered ratchet model, redrawn from [633]. There are three main forces acting on the Listeria: a load force $F_L$ due to viscous drag and any externally imposed forces (in an experiment), a polymerization ratchet force $F_u$ due to unattached filaments, and a tensional force $F_a$ due to attached filaments with $F_L + F_a = F_u$.

$f_a, f_u$ and the detachment rate $k$ on the bacterium velocity $v$. The force–velocity relation for a single polymerizing filament is taken from the polymerization ratchet model [631]:

$$v = v_+ e^{-f_u l / k_B T} - v_-, \tag{14.4.3}$$

where $v_+ = k_{on} l M$ is the polymerization velocity and $v_- = k_{off} l$ is the depolymerization velocity. Here $k_{on}$ and $k_{off}$ are the rates of monomer assembly and disassembly, $M$ is the concentration of monomers available for polymerization, and $l$ is the effective increase in filament length due to addition of one monomer.

In order to estimate the average attachment force $f_a$, it is assumed that an attached filament acts like a Hookean spring. Suppose that the filament binds to an ActA complex at time $t = 0$. The force acting on the resulting bond is given by $f(t) = \eta v t$, where $\eta$ is the effective spring constant. Using the basic theory of chemical bond breaking [286], the rate of detachment takes the velocity-dependent form

$$k(v,t) = k_0 e^{\eta v t / f_b}, \tag{14.4.4}$$

where $x_b = f_b / \eta$ can be interpreted as bond length at which the bond breaks sharply. The probability $p(t)dt$ of the bond first breaking in the time interval $(t, t + dt)$ is given by the product of no failure in the interval $(0,t)$ times the probability of subsequent failure within the interval $(t, t + dt)$. Hence,

$$p(t) = k(v,t) e^{-\int_0^t k(v,s)ds}.$$

Set $v_0 = f_b k_0 / \eta$, which can be interpreted as the velocity at which the bond stretches to the critical length $x_b$ over the characteristic bond lifetime $1/k_0$. Rescaling velocity and time according to $\mu = v/v_0$ and $\tau = k_0 t$, we have

$$p(\tau) = \exp\left( \mu\tau + \frac{1}{\mu}(1 - e^{-\mu\tau}) \right).$$

It follows that the mean attachment time of a filament (for constant $v$) is

$$\langle t \rangle = \frac{1}{k_0} \int_0^\infty \tau p(\tau) d\tau = \frac{1}{k_0} w(\mu), \tag{14.4.5}$$

with

$$w(\mu) = \int_0^\infty \tau \exp\left( \mu\tau + \frac{1 - e^{-\mu\tau}}{\mu} \right) d\tau.$$

We now identify the mean detachment rate as $k = 1/\langle t \rangle$ and take the average force $f_a$ exerted by a single attached filament to be $f_a = \eta v \langle t \rangle$. Thus,

$$k(\mu) = \frac{k_0}{w(\mu)}, \quad f_a = f_b \mu w(\mu). \tag{14.4.6}$$

Note that the function $w(\mu)$ has the following properties:

1. If $\mu \ll 1$ then $w(\mu) \approx 1$, which implies that for sufficiently slow movement ($v \ll v_0$) the effective detachment rate is equal to the force-free rate ($k \approx k_0$) and $f_a \approx f_b v/v_0$.

2. If $\mu \gg 1$ then $w(\mu) \approx \mu^{-1} \ln \mu$.

Consider the case of constant propulsion speed $v$. The steady-state numbers of attached and detached filaments are then

$$n_a = \sigma/k, \quad n_u = \sigma/\kappa.$$

Substituting the force-balance equation (14.4.1) into the velocity equation (14.4.3) gives

$$v = v_+ \exp\left[-l(n_a f_a/n_u + F_L/n_u)/k_B T\right] - v_-.$$

Since $n_a/n_u = \kappa/k = w(\mu)\kappa/k_0$ and $f_a = f_b\mu w(\mu)$, the velocity satisfies the implicit equation

$$v = v_+ \exp\left[-l\left(\frac{f_b\kappa}{k_0 v_0}vw^2(v/v_0) + \frac{F_L\kappa}{\sigma}\right)/k_B T\right] - v_-. \tag{14.4.7}$$

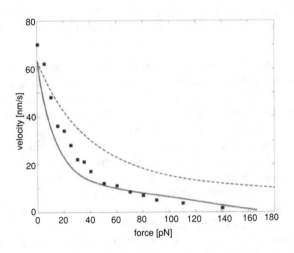

Fig. 14.26: The load-velocity curve for the tethered ratchet model of Mogilner and Oster [633]. The solid curve is generated from equation (14.4.7) using the following parameter values: monomer size $l = 2.2$ nm, polymerization velocity $v_+ = 500$ nm/s, depolymerization velocity $v_- = 2.2$ nm/s, nucleation rate $\sigma = 10$ s$^{-1}$, capping rate $\kappa = 0.5$ s$^{-1}$, free detachment rate $k_0 = 0.5$ s$^{-1}$, thermal energy $k_B T = 4.1$ pN nm, effective length of bond $x_b = 0.4$ nm, effective strength of bond $f_b = 10$ pN, and spring coefficient $\eta = 1$ pN/nm. The dashed curve is obtained by introducing a threefold increase in the nucleation rate $\sigma$, and illustrates the effect of filament density on the load-velocity behavior. Finally, the squares represent data from stochastic model simulations with a reduced polymerization velocity $v_+ = 240$ nm/s. (Adapted from Mogilner and Oster [633].)

Using biophysically based estimates for the various parameters, Mogilner and Oster [633] numerically solved the equation for $v$ and obtained speeds of the order 10 nm /s, which is consistent with experimental data, see Fig. 14.26. They also showed that the load-velocity relation exhibits biphasic behavior, whereby the velocity decreases rapidly with $F_L$ at low load forces and decreases more slowly at high load forces. This is a consequence of the modeling assumptions regarding chemical bond breaking. At high velocities increasing the external load helps the attached filaments to hold on longer, thus increasing the resistive force $F_a$ which itself slows the bacterium further. On the other hand, at sufficiently slow velocities the external load has a minor effect on the resistive force $F_a$ and the velocity decreases more slowly.

The tethered ratchet model has been refined over the years in order to include more details regarding nucleation and capping mechanisms, for example, and to try and account for an ever increasing amount of new biophysical data [635]. Several other propulsion mechanisms have also been proposed, which can match some types of data better than ratchet models. One alternative model assumes that all filaments are attached to the surface of the bacterium, with the pushing (barbed) ends clamped to an end-tracking protein at the surface, which processively tracks the growing filament tip via phosphorylation of ATP [242, 243]. Another class of model is based on the analysis of elastic deformations of an actin gel near the surface of the bacterium, which provides a mesoscopic description of cell motility at a length scale larger than individual proteins [342, 587]. The gel is approximated as a continuous elastic medium with stress generated at the surface interface by growing actin filaments. We will briefly describe a simple 1D version of this model.

**Elastic gel model of *Listeria* propulsion.** An alternative approach to modeling *Listeria* propulsion is to treat the actin tail as an elastic gel. Following Ref. [342], the bacterium is modeled as a cylinder with a circular cross-section of area $A_b = \pi r_b^2$, see Fig. 14.27. The main simplification of the 1D model is to assume that the actin gel is produced only at the back of the bacterium. (A more detailed 3D model also allows for production from the curved surface of the bacterium [342].) The elastic deformations of the gel involved in propulsion occur on length scales much smaller than those where depolymerization becomes relevant. Therefore, the tail is modeled as a semi-infinite tube of cross-sectional area $A_t$, consisting of homogeneous elastic material. The latter is characterized by a compression modulus $Y$ and a longitudinal elastic strain $u_{zz}$ with associated axial stress $\sigma_{zz} = Yu_{zz}$ (see Box 12B). Radial and shear components of the stress vanish on the cylindrical surface of the tail, since viscous forces arising from friction against the outer medium are negligible. The compression of the actin gel is in reaction to the external force $F_{ext}$ acting on the bacterium, and also provides the propulsive force on the bacterium. The bacterium moves forward at a speed $v$ due to polymerization at the interface between the bacterium and tail.

Let $v_p$ be the rate of polymerization. The rate at which filaments are generated across the back surface of the bacterium is equal to the flux of new material through the cross section of the tail which extends at the speed $v$. In other words, $v_p A_b = v A_t$. Assuming that the gel is incompressible, the addition of a volume $A_b \Delta z$ of new material is redistributed over a region of width $\Delta z'$ with $A_t \Delta z' = A_b \Delta z$. (Over a time interval $\Delta t$, we have $\Delta z' = v\Delta t$ and $\Delta z = v_p\Delta t$.) Hence the axial stress and strain are given by

$$\frac{\sigma_{zz}}{Y} = u_{zz} = \frac{\Delta z' - \Delta z}{\Delta z} = \frac{A_b}{A_t} - 1. \tag{14.4.8}$$

Since the elastic stress balances the external force, we have

$$\sigma_{zz} = -\frac{F_{ext}}{A_t} = -\frac{F_{ext}}{A_b}\left(1 + \frac{\sigma_{zz}}{Y}\right),$$

which on rearranging gives

$$\frac{\sigma_{zz}}{Y} = \frac{1}{1 + F_{ext}/A_b Y} - 1. \tag{14.4.9}$$

Hence, the force–velocity relation is

$$v = \frac{1}{1 + F_{ext}/A_b Y} v_p(F_{ext}), \tag{14.4.10}$$

where we now make explicit the fact that the polymerization rate will also depend on the applied force. The latter is taken to be

$$v_p(f) = v_0 e^{-F_{ext}a\Delta A/[k_B T A_b]} = v_0 e^{-\varepsilon F_{ext}/Y A_b}, \tag{14.4.11}$$

where $\Delta A$ is the typical cross-sectional area of an actin filament, $a$ is the size of a monomer, and $\varepsilon = Ya\Delta A/k_B T$. It turns out that $\varepsilon \approx 1$ and $F_{ext}/A_b Y \ll 1$ so that [342]

$$v = \frac{v_0}{1 + F_{ext}/A_b Y} e^{-F_{ext}/A_b Y} \approx v_0\left(1 - \frac{2F_{ext}}{A_b Y}\right). \tag{14.4.12}$$

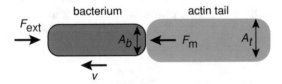

Fig. 14.27: Schematic diagram of a 1D elasticity model of Listeria propulsion. The actin tail exerts a force $F_m$ on the bacterium, which moves at speed $v$ against the external force $F_{ext}$ due to polymerization at the bacterium/tail interface. The external force also generates an axial compression of the actin tail resulting in an axial stress $\sigma_{zz}$, where $z$ is the axial spatial direction. The cross-sectional areas of the bacterium and tail are $A_b$ and $A_t$, respectively.

## 14.4.2 Crawling cells and the motor-clutch mechanism

The majority of migratory cells that crawl along some cellular substrate such as the extracellular matrix (ECM) rely on lamellipodial motility. A prerequisite for directed motion is that the cell is polarized, i.e., it has a well-defined front and rear. (Mechanisms of cell polarization for a range of cellular processes were considered in Sect. 11.3.) The leading edge of a crawling cell consists of a lamellipod, which is a flat leaf-like extension that is filled with a dense actin network. Roughly speaking, following cell polarization, the migration of a cell consists of three components driven by interconnected but distinct processes [222, 635], see also Fig. 14.28:

1. *Cell protrusion:* Models of cell protrusion combine ratchet models of actin polymerization with the so-called dendritic-nucleation hypothesis. The latter posits that nascent actin filaments branch from the sides of existing filaments such that there is approximately a 70° angle between "mother" and "daughter" filaments, and all leading edge filaments have their barbed ends oriented toward the direction of protrusion at an angle of around 35° [165, 583, 651, 797]. One finds that the growing barbed ends grow at a rate around 0.1 $\mu$m/s, thus pushing out the membrane of the leading edge. However, capping of the growing ends restricts the length of individual filaments to $0.1 - 1$ $\mu$m. These features make sense, since intermediate-length filaments that subtend the leading edge at an angle are neither too rigid nor too flexible, and are protected from immediate capping. The net protrusion of the leading edge is determined by actin polymerization minus the centripetal rearward flow of the actin network. This retrograde flow is a dissipative mechanism for reducing stresses on the actin network.

2. *Cell contraction:* Contraction of the rear of the cell is thought to be mediated by myosin II motors acting on actin fibers in an analogous fashion to muscle contraction. A gradient of adhesion forces that is high at the leading edge and low at the trailing edge (see below) means that contraction will lead to preferential forward movement of the rear, provided that it is not too strong that it cancels protrusion by imposing a rearward stress on the actin network.

Fig. 14.28: Main components of cell crawling. Side view shows protrusion at the leading edge due to polymerization of the actin network followed by contraction or retraction of the rear due to action of myosin II motors. Net displacement occurs provided that there is a gradient of cell adhesion from front to rear. Adapted from Mogilner [635].

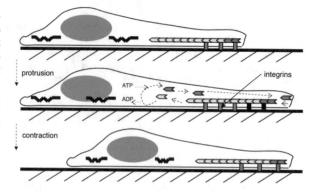

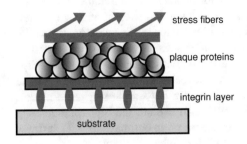

Fig. 14.29: Schematic diagram of a focal adhesion (FA) consisting of a lower layer of integrin receptors attached to a substrate, and a submembrane plaque consisting of multiple proteins that bind to intracellular integrin domains. Actin polymerization and the actomyosin contractile machinery generate forces that affect mechanosensitive proteins in the various components of the FA, including the integrin receptors.

3. *Cell adhesion:* Recall from our discussion of the mitotic spindle in Sect. 14.3 that inertial effects can be ignored in the low Reynolds number regime, so there has to be force balance. In order to balance propulsive forces at the leading edge and contractile forces at the rear, adhesion forces between the cell and substrate are required [714]. The latter forces are mediated by a layer of transmembrane receptors called integrins [906]. The integrin layer is part of a large multiprotein complex known as a focal adhesion (FA), which links the contractile machinery of the actin cytoskeleton to the substrate ECM, see Fig. 14.29. More specifically, integrins are heterodimeric proteins whose extracellular domains attach to the substrate, while their intracellular domains act as binding sites for various submembrane proteins, resulting in the formation of the plaque. The plaque consists of more than 50 different types of protein, which play a role in intracellular signaling, actin linkage and force transduction, and actin polymerization. (There is growing experimental evidence that integrins/FAs also act as biochemical sensors of the local microenvironment (e.g., the rigidity and composition of the ECM), see [494, 538] and references therein. In particular, FAs self-assemble and grow under the application of pulling forces (stresses) transmitted by the actin cytoskeleton, see Sect. 14.4.4.)

There are a wide variety of mathematical and computational models of cell migration, including a number of integrative models of the whole cell that can make predictions about the changes in cell shape during migration [222, 423, 502, 589, 635]. Mathematically speaking, the modeling of cell shape involves a difficult free boundary value problem, in which the cell is represented as a time-dependent domain $\Omega(t)$ whose boundary $\partial\Omega(t)$ evolves according to boundary conditions that themselves depend on various densities that are defined by dynamical equations on $\Omega(t)$. We refer the reader to the given citations for more details. Here we focus on the interplay between retrograde acton flow, cell protrusion, and adhesion at the leading edge of a cell.

Experimental studies of migratory cells suggest that at the leading edge the FA acts like a molecular clutch [345, 380, 481, 563, 626]. That is, for high adhesion or drag, the retrograde flow of actin is slow and polymerization results in a net protrusion—the clutch is in "drive." On the other hand, if adhesion is weak then retrograde flow can cancel the polymerization, and the actin network treadmills, i.e., the clutch is in "neutral." The dynamical interplay between retrograde flow and the assembly and disassembly of focal adhesions leads to a number of behaviors that are characteristic of physical systems involving friction at moving interfaces [309, 796, 838]. These include biphasic behavior in the velocity-stress relation and stick-slip motion. The latter is a form of jerky motion, whereby a system spends most of its time in the "stuck" state and a relatively short time in the "slip" state. Various insights into the molecular clutch mechanism and its role in substrate stiffness-dependent migration have been obtained using simple stochastic models [177, 226, 558, 691, 780, 812]. Such models can capture the biphasic stick-slip force–velocity relation, and establish the existence of an optimal substrate stiffness that is sensitive to the operating parameters of the molecular clutch.

Consider a model for the motor-clutch mechanism first introduced by Chan and Odde [177], see Fig. 14.30. (A mean-field analysis of a simplified version of the model will be considered in Sect. 14.4.3.) The model considers opposing forces on an F-actin bundle, consisting of myosin motors that transport F-actin retrogradely from the leading edge, and adhesion-based molecular clutches that exert an anterograde traction force. The clutches stochastically bind to the F-actin at a constant rate and unbind according to a rate that is tension-dependent. Suppose that there are $N$ clutches, labeled $n = 1, \ldots, N$. Let $q_n(t) \in \{0, 1\}$ specify whether the clutch is bound to the F-actin ($q_n = 1$) or unbound ($q_n = 0$) at time $t$. The force on the $n$th clutch when it is engaged is given by Hooke's law $F_n = \kappa_c(x_n - x_s)$, where $\kappa_c$ is the clutch spring constant, $x_n$ is the position of the clutch bound to actin, and $x_s$ is the position of the substrate. (More precisely, if $X_n$ is the position of the base of the $n$th spring when the substrate is relaxed ($x_s = 0$), then the end points of an engaged clutch are at $x_s + X_n$ and $x_n + X_n$, respectively. The clutch force is thus independent of $X_n$ and $x_n \to x_s$ when the clutch disengages.) When the clutch unbinds it immediately relaxes such that $x_n = x_s$ and there is no force. The substrate position is determined by a force-balance equation between the total clutch force on the substrate and the substrate spring force:

$$\kappa_s x_s = \kappa_c \sum_{n=1}^{N} q_n(x_n - x_s),$$

where $\kappa_s$ is the substrate spring constant. It follows that

$$x_s = \frac{\kappa_c \sum_{n=1}^{N} q_n x_n}{\kappa_s + N \kappa_c}. \tag{14.4.13a}$$

Denote the constant binding rate by $k_{\mathrm{on}}$ and the force-dependent unbinding rate by $k_{\mathrm{off},n}$ with the latter given by the so-called Bell–Evans formula [55, 286, 288, 808]

$$k_{\mathrm{off},n} = k_{\mathrm{off}} e^{r(x_n - x_s)}, \quad r = \frac{\kappa_c x_b}{k_B T}. \tag{14.4.13b}$$

Here $x_b$ is the characteristic bond rupture displacement [286]. A similar exponential dependence of unbinding rates on the applied force was assumed in the tug-of-war model of collective motor transport, see Sect. 4.5, and the tethered ratchet model of Sect. 14.4.1. Although most adhesion bonds exhibit this type of behavior, and are referred to as slip bonds, there also exist so-called catch bonds whose strength

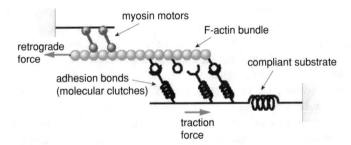

Fig. 14.30: Schematic illustration of the motor-clutch model introduced by Chan and Odde [177]. Myosin motors retract F-actin in the retrograde direction, while the molecular clutches and compliant substrate, each modeled as Hookean springs resist the motion, exerting an anterograde traction force. Clutches are formed by the binding of adhesion molecules (integrins) to the F-actin bundle. Binding occurs at a constant rate, whereas the rate of unbinding increases with tension. [Redrawn from Danuser et al. [222].]

can actually increase with applied force, see Box 14C. The classical theory of cell adhesion with slip bonds is explored further in Ex. 14.6.

In between binding or unbinding events, the clutch positions evolve according to the equations

$$\frac{dx_n}{dt} = G_n(\mathbf{x}, \mathbf{q}) \equiv (1 - q_n)\frac{dx_s}{dt} + q_n V, \qquad (14.4.13c)$$

where $V$ is the actin retrograde flow velocity. The latter is obtained using a linear force–velocity relationship for molecular motors that accounts for the slowing down of the motors when acting against an elastically loaded substrate (Chap. 4):

$$V(t) = v\left(1 - \frac{\kappa_s x_s(t)}{M F_m}\right), \qquad (14.4.13d)$$

where $M$ is the number of motors, $\kappa_s x_s(t)/M$ is the force acting on each motor, $F_m$ is the single motor stall force, and $v$ is the motor velocity of a motor without any load. Whenever a clutch is bound to the F-actin it moves at the retrograde speed $V$, whereas when it is unbound it moves with the substrate. Multiplying both sides of equation (14.4.13c) by $q_n$, summing over $n$ with $q_n(1 - q_n) = 0$ and $q_n^2 = q_n$, and substituting into the time derivative of equation (14.4.13b) shows that

$$\frac{dx_s(t)}{dt} = \frac{\kappa_c f_b(t)}{\kappa_s + N\kappa_c} V(t), \quad f_b(t) = \sum_{n=1}^{N} q_n(t), \qquad (14.4.13e)$$

where $f_b(t)$ is the number of bound clutches at time $t$. Note that the right-hand side of equation (14.4.13c) depends on the positions of all the clutches due to equations (14.4.13d,e). We thus have a stochastic hybrid system (Chap. 5) on $\mathbb{R}^N \times 2^N$ with an additional reset rule, namely, whenever $q_n$ switches from 1 to 0, $x_n(t) \to x_s(t)$.

Numerical simulations of the model, combined with experimental studies of embryonic chick neurons, identified different dynamical regimes that depend on the stiffness of the substrate [177]. On stiff substrates, clutches bind to the F-actin bundle but abruptly unbind, since the lack of compliance in the substrate results in a rapid building of tension within engaged clutches. This significantly shortens F-actin/clutch interaction lifetimes so that only a few clutches are engaged at any one time, and the F-actin bundle exhibits "frictional slippage." On the other hand, the compliance of soft substrates slows the rate at which tension builds along individually engaged clutches. Hence, myosin motors initially work near their unloaded sliding velocity, leading to high rates of F-actin retrograde flow. However, as the tension builds, clutches largely remain engaged due to sharing of the mechanical load. This then provides considerable resistance to the motor force, substantially slowing retrograde flow. Eventually, the load becomes so great that the stochastic loss of one clutch leads to a cascading failure event, where all clutches rapidly disengage and the substrate snaps back to its initial rest position. One thus finds an oscillatory "load-and-fail" traction force dynamics with periods of around 10–100 s. In summary, the motor-clutch model predicts higher retrograde flow rates with lower traction forces on stiff substrates, and lower retrograde flow rates with higher traction forces on soft substrates. Moreover, since clutches cannot resist retrograde flow if the substrate is too compliant, this suggests that there exists an optimal stiffness where one finds a maximum traction force. Adding some form of feedback signal could then allow cells to tune their behavior by shifting the value of the optimal stiffness.

---

### Box 14C. Catch bonds.

---

The concept of a catch bond was originally introduced in a theoretical study [229], but was subsequently observed experimentally in a variety of molecular systems [314, 379, 594, 873, 983]. A catch bond is defined as a bond whose lifetime can increase when it is stretched by a mechanical force. This contrasts with the more familiar slip bond, whose lifetime always decreases. A simple way to understand the effect of an applied force on bond lifetime is to consider the energy potential of a receptor-ligand interaction, as illustrated in Fig. 14.31. From the theory of noise-induced escape (Chap. 2 and Chap. 8), the bond lifetime increases with the barrier height $\Delta E_0$ separating the bound state from the dissociated state. In the classical theory of bond adhesion [55, 286], an applied force $f$ induces a linear change in barrier height according to

$$\Delta E(f) = \Delta E_0 \pm f \Delta x, \qquad (14.4.14)$$

where $\Delta x$ is the displacement of the ligand. In the case of slip bonds the sign is negative, which means that the force pulls out the ligand and reduces the barrier height. On the other hand, if the sign is positive, then the free energy landscape is such that the force pushes the ligand deeper into the receptor, resulting in a catch bond. Of course, given sufficient force, a catch bond transitions to a slip bond. (A useful analogy is the physical connection between two ropes with hooks on their ends. Pulling on the ropes forces the hooks to be more interlocked, unless the force is so strong that it deforms the hooks and breaks the link.)

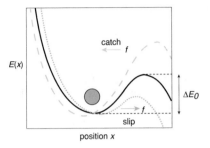

Fig. 14.31: Effect of an applied force $f$ on the interaction potential of a receptor-ligand bond.

There have been several different models of catch bonds and their transitions to slip bonds [735]. Here we will focus on one of the simplest, known as the two-pathway model [720]. The latter assumes that the receptor-ligand interaction potential has a single bound state at $x_0$ and two barriers at $x_c$ (catch barrier) and $x_s$ (slip barrier). Crossing either barrier leads to an unbound sate, and an applied force raises the catch barrier but lowers the slip barrier, see Fig. 14.32(a). The force-dependence of the inverse lifetime or unbinding rate is taken to be of the form

$$\alpha \equiv \frac{K_c x_c}{K_s x_s} > 1,$$

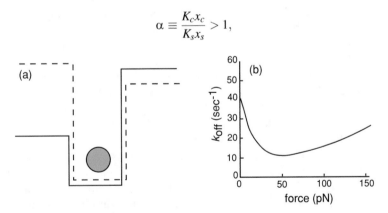

Fig. 14.32: Two-pathway model of a catch bond. (a) Affect of an applied force on the catch and slip barriers. Solid lines (no force), dashed lines (with force). (b) Sketch of a typical curve showing variation of bond lifetime with the strength of the applied force.

$$k_{\text{off}}(f) = K_c e^{-x_c f/k_B T} + K_s e^{x_s f/k_B T}, \qquad (14.4.15)$$

with $x_{c,s} > 0$. Assuming the relationship

$$\alpha \equiv \frac{K_c x_c}{K_s x_s} > 1,$$

one finds that $k_{\text{off}}$ has a minimum at the critical force

$$f_c = \frac{k_B T}{x_s + x_c} \ell \alpha.$$

A sketch of $k_{\text{off}}(f)$ as a function of $f$ for parameters fitted to the behavior of the adhesion molecule L-selectin is shown 14.32(b). For a comparison with experimental data, see Ref. [720].

### 14.4.3 Mean-field analysis of a stochastic motor-clutch model

We now describe a simpler example of a molecular clutch model, which is more amenable to analysis [780]. A schematic diagram of the system is shown in Fig. 14.33. The actin cytoskeleton is treated as a rigid slider that moves over the substrate (retrograde flow) under the action of a constant driving force $F$. The latter represents the combined effects of contractile forces exerted by myosin motors at the trailing edge of the cell and actin polymerization at the leading edge. The force $F$ is balanced by two time-dependent forces: an elastic force due to integrins that are stochastically bound at the interface, and a viscous friction force $\xi v$, where $v$ is the sliding velocity and $\xi$ is a friction coefficient. When a bond is attached to the cytoskeleton it stretches at the velocity $v$, but its extension $x$ is immediately reset to zero whenever it unbinds. The latter occurs at an $x$-dependent rate $k_{\text{off}}(x)$. The bond subsequently rebinds at a constant binding rate $k_{\text{on}}$. The bonds are coupled due to the fact that the sliding velocity depends on the sum of the elastic forces generated by the closed bonds. However, using a mean-field approximation, one can derive an effective single-bond dynamics with a constant sliding velocity that is determined self-consistently from the single-bond statistics [780].

Suppose that the FA is of constant size so that there is a fixed maximal number of $N$ assembled bonds. Each bond, $i = 1, \ldots, N$, can be in one of two states denoted by $q_i(t) \in \{0, 1\}$, with the bond closed (open) if $q_i(t) = 1$ ($q_i(t) = 0$). The closed bonds are modeled as Hookean springs with spring constant $\kappa$ and time-dependent extension $x_i(t)$ that stretches at the sliding velocity $v$. On the other hand, an open

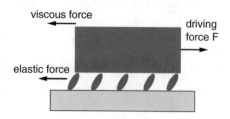

Fig. 14.33: Simplified model of forces acting on the actin cytoskeleton. The driving force $F$ for sliding is balanced by elastic adhesive forces and velocity-dependent viscous forces.

bond is assumed to immediately reset to its equilibrium state with $x_i = 0$. The force-balance equation, which holds in the low Reynolds number regime where inertial forces can be ignored, takes the form [780]

$$v = v(\mathbf{x}) := \frac{1}{\xi}\left(F - \kappa\sum_{i=1}^{N}q_i(t)x_i(t)\right), \qquad (14.4.16)$$

with $\mathbf{x} = (x_1,\ldots,x_N)^{\top}$. Here $F$ is the constant driving force and $\xi$ is a friction coefficient. The extensions $x_i$ thus evolve according to the system of equations

$$\frac{dx_i}{dt} = v(\mathbf{x}) \text{ if } q_i(t) = 1, \quad x_i = 0 \text{ if } q_i(t) = 0, \qquad (14.4.17)$$

with $i = 1,\ldots,N$. Finally, the stochastic switching of the discrete state $q_i(t)$ between the values $0,1$ is determined by a constant binding rate $k_{on}$ and a stretch-dependent off-rate given by the Bell–Evans formula $k_{off}(x) = k_{off}e^{bx}$, see equation (14.4.13b).

The above dynamical system is an example of a stochastic hybrid system, since it couples a set of continuous variable $\mathbf{x}(t)$ with a Markov chain for the discrete states $\mathbf{q} = (q_1,\ldots,q_N)$. Let $\rho(\mathbf{x},t,\mathbf{q})$ denote the probability density that at time $t$ the bonds are in the configurational state $\mathbf{q}$ and have extensions $\mathbf{x}$. (Note that $x_j = 0$ for all $j$ such that $q_j = 0$.) As shown in Ref. [780], $\rho(\mathbf{x},t,\mathbf{q})$ evolves according to a differential Chapman-Kolmogorov (CK) equation, which takes the form

$$\frac{\partial\rho(\mathbf{x},t,\mathbf{q})}{\partial t} = -\sum_{i=1}^{N}q_i\frac{\partial J(\mathbf{x},t,\mathbf{q})}{\partial x_i} - \sum_{i=1}^{N}q_ik_{off}(x_i)\rho(\mathbf{x},t,\mathbf{q}) - \sum_{i=1}^{N}(1-q_i)k_{on}\rho(\mathbf{x},t,\mathbf{q})$$

$$+ \sum_{i=1}^{N}\delta(x_i)\left\{(1-q_i)\int_{0}^{\infty}k_{off}(x_i')\,\rho(\mathbf{x},t,\mathbf{q})|_{(x_i,q_i)=(x_i',1)}\,dx_i'\right.$$

$$\left. + q_i[k_{on}\rho(\mathbf{x},t,\mathbf{q})|_{q_i=0} - J(\mathbf{x},t,\mathbf{q})]\right\}, \qquad (14.4.18)$$

where $J(\mathbf{x},t,\mathbf{q}) = v(\mathbf{x})\rho(\mathbf{x},t,\mathbf{q})$. The first summation on the right-hand side of equation (14.4.18) represents the stretching of closed bonds in between switching events. The next two summations represent reaction terms associated with the unbinding and binding of bonds. Finally the terms in the bracket $\{\cdot\}$ represent the various fluxes into or out of the state $x_i = 0$ due to the following events: unbinding of the $i$th bond in the closed state followed by resetting, binding of the $i$th bond in the open state, and exit from zero extension due to stretching of the $i$th bond. The last term is required in order to ensure that there is no flux entering the system from $x_i < 0$. Total probability is then conserved, as can be shown by integrating both sides of equation (14.4.18) with respect to $\mathbf{x}$ and summing over all configurations $\mathbf{q}$:

$$\frac{d}{dt}\left\{\sum_{\mathbf{q}}\left[\prod_{j=1}^{N}\int_{0}^{\infty}dx_j\right]\rho(\mathbf{x},t,\mathbf{q})\right\} = 0. \qquad (14.4.19)$$

First, consider the case of a single bond ($N = 1$). Setting $x_1 = x$, $\rho(x,t,1) = p(x,t)$ and $\rho(0,t,0) = P_0(t)$, equation (14.4.18) reduces to the pair of equations

$$\frac{\partial p(x,t)}{\partial t} = -\frac{\partial v(x)p(x,t)}{\partial x} - k_{\text{off}}(x)p(x,t) + \delta(x)[k_{\text{on}}P_0(t) - J(0,t)], \quad (14.4.20a)$$

$$\frac{dP_0(t)}{dt} = -k_{\text{on}}P_0(t) + \int_0^\infty k_{\text{off}}(x')p(x',t)dx', \quad (14.4.20b)$$

with

$$J(x,t) = v(x)p(x,t), \quad v(x) = \frac{1}{\xi}(F - \kappa x). \quad (14.4.21)$$

Rather than including the Dirac delta function terms on the right-hand side of (14.4.20a), we can equivalently impose the boundary condition $J(0,t) = k_{\text{on}}P_0(t)$. The boundary condition reflects the fact that the dynamics is on the half-line, and ensures that there is probability conservation. In particular,

$$\int_0^\infty p(x,t)dx + P_0(t) = 1. \quad (14.4.22)$$

The model of a single clutch is an example of dynamical process with spatially-dependent stochastic resetting and a refractory period [289] (Chap. 7). That is, $k_{\text{off}}(x)$ is the resetting rate, the reset location $X_r = 0$, and the refractory period following resetting is determined by an exponential waiting time density with rate $k_{\text{on}}$.

In Ref. [780], the steady-state probability density for multiple bonds is calculated using a mean-field approximation. This involves taking the large-$N$ limit and replacing $v(\mathbf{x})$ by a space-independent mean velocity $\langle v \rangle$. The resulting probability density then factorizes into the product of $N$ single-bond densities, which can be used to derive a self-consistency condition for $\langle v \rangle$. Consider a particular bond $k$, and introduce the marginal densities and fluxes

$$\rho_{k,1}(x,t) = \sum_{\mathbf{q}} \delta_{q_k,1} \left[ \prod_{j=1}^N \int dx_j \right] \delta(x_k - x)\rho(\mathbf{x},t,\mathbf{q}), \quad (14.4.23a)$$

$$\rho_{k,0}(t) = \sum_{\mathbf{q}} \delta_{q_k,0} \left[ \prod_{j=1}^N \int dx_j \right] \rho(\mathbf{x},t,\mathbf{q}), \quad (14.4.23b)$$

$$J_k(x,t) = \sum_{\mathbf{q}} \delta_{q_k,1} \left[ \prod_{j=1}^N \int dx_j \right] \delta(x_k - x)J(\mathbf{x},t,\mathbf{q}). \quad (14.4.23c)$$

Summing the full CK equation (14.4.18) with respect to $q_i$ and integrating with respect to $x_i$ for all $i \neq k$ gives the following pair of equations:

$$\frac{\partial \rho_{k,1}(x,t)}{\partial t} = -\frac{\partial J_k(x,t)}{\partial x_k} - k_{\text{off}}(x)\rho_{k,1}(x,t) + \delta(x)[k_{\text{on}}\rho_{k,0}(t) - J_k(0,t)],$$

$$\frac{\partial \rho_{k,0}(t)}{\partial t} = -k_{\text{on}}\rho_{k,0}(t) + \int_0^\infty k_{\text{off}}(x')\rho_{k,1}(x',t)dx'.$$

Note that all off-diagonal terms $i \neq k$ in the various summations of equation (14.4.18) cancel. As it stands, we don't have a closed single-bond equation because $\rho_{k,1}(x,t)$, $\rho_{k,0}(t)$ and $J_k(x,t)$ all depend on the full probability density. The mean-field approximation is to assume that for large $N$, the individual bonds are statistically independent so that we can factorize the multi-bond density into the product of $N$ single-bond densities:

$$\rho(\mathbf{x},t,\mathbf{q}) = \prod_{j=1}^{N} [q_j p(x_j,t) + (1-q_j)P_0(t)]. \tag{14.4.24}$$

Substituting into equations (14.4.23a-c) gives $\rho_{k,1}(x.t) = p(x,t)$, $\rho_{k,0}(t) = P_0(t)$ and $J_k(x,t) = N\langle v(t) \rangle$ for all $k = 1,\ldots,N$, where

$$\langle v(t) \rangle = \frac{1}{\xi}(F - \kappa N \langle X(t) \rangle), \tag{14.4.25}$$

with $X(t) = \int_0^\infty x p(x,t)dx$.

We thus obtain the effective single-bond CK equation

$$\frac{\partial p(x,t)}{\partial t} = -\langle v(t) \rangle \frac{\partial p(x,t)}{\partial x} - k_{\text{off}}(x)p(x,t) \tag{14.4.26a}$$

$$\frac{dP_0(t)}{dt} = -k_{\text{on}}P_0(t) + \int_0^\infty k_{\text{off}}(x')p(x',t)dx', \tag{14.4.26b}$$

together with the boundary condition $k_{\text{on}}P_0(t) = \langle v(t) \rangle p(0,t)$. A transcendental equation for $\langle v \rangle$ can be obtained in the steady state, see Ex. 14.7. First, equation (14.4.26a) becomes

$$\langle v \rangle \frac{dp^*(x)}{dx} = -k_{\text{off}}e^{bx}p^*(x), \tag{14.4.27}$$

with boundary condition $p^*(0) = k_{\text{on}}P_0*/\langle v \rangle$. The solution is

$$p^*(x) = \frac{P_0^* k_{\text{on}}}{\langle v \rangle} \exp\left(\frac{k_{\text{off}}}{r\langle v \rangle}[1 - e^{rx}]\right). \tag{14.4.28}$$

The constant $P_0^*$ is then determined using the normalization condition (14.4.22),

$$P_0^* = \left[1 + \frac{k_{\text{on}}}{b\langle v \rangle}e^{k_{\text{off}}/b\langle v \rangle}\Gamma(0;k_{\text{off}}/b\langle v \rangle)\right]^{-1}, \tag{14.4.29}$$

where $\Gamma$ is the incomplete Gamma function

$$\Gamma(z;k) = \int_1^\infty y^{z-1}e^{-ky}dy. \tag{14.4.30}$$

Finally, the self-consistent equation for $\langle v \rangle$ is obtained by substituting for $p^*(x)$ in the mean-field equation (14.4.25):

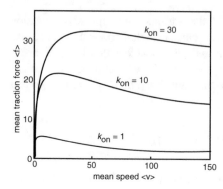

Fig. 14.34: Sketch of curves showing the biphasic relationship between the mean traction force $\langle f \rangle$ and the mean velocity $\langle v \rangle$ for different values of the binding rate $k_{on}$ and $\xi = 1$. Dimensionless quantities are used by taking $k_{off}^{-1}$, $b^{-1}$ and $\kappa/b$ to be the units of time, length and force, respectively. Typical values for various parameters are $k_{off} = 1$ s$^{-1}$, $b = 0.5$ nm$^{-1}$, $\kappa = 10$ pN nm$^{-1}$, $F_D = 20$ pN, and $\xi = 10$ pN s nm$^{-1}$. [Redrawn from Sabass and Schwarz [780].]

$$\langle v \rangle = \frac{F_D}{\xi} - \frac{N\kappa P_0^* k_{on}}{\xi} \frac{1}{\langle v \rangle} \int_0^{\infty} x \exp\left(\frac{k_{off}}{b\langle v \rangle}\left[1 - e^{bx}\right]\right) dx. \quad (14.4.31)$$

Note that the second term on the right-hand side, which takes the form $N\langle f \rangle/\xi$ for the mean traction force $\langle f \rangle$, vanishes in the limit $\langle v \rangle \to \infty$. In the case of fast unbinding, $k_{off}/b\langle v \rangle \gg 1$, one can carry out an asymptotic expansion of the incomplete Gamma function to show that (see Ex. 14.7) $p_0 \approx k_{off}/(k_{on} + k_{off})$. Similarly, carrying out an asymptotic expansion of the mean traction force gives

$$\langle f \rangle \approx \frac{k_{on}}{k_{on} + k_{off}} \frac{\kappa\langle v \rangle}{k_{off}}.$$

In Fig. 14.34 we sketch some example curves of the mean-field model, showing how the mean traction force $\langle f \rangle$ varies with the mean velocity $\langle v \rangle$. It can be seen that there is a maximum in the mean traction force at intermediate velocities. This can be understood within the mean-field framework by noting that the average transmitted force of an engaged clutch increases monotonically with speed, while the probability of the clutch being bound decreases at higher speeds. In Ref. [780] it is also shown that for the given parameter values, the mean-field model agrees well with numerical simulations of the full model when $N = 25$. However, the mean-field approximation breaks down for a sufficiently weak friction coefficient $\xi$. In this regime, stochastic bond dynamics leads to irregular frictional slippage that is not accounted for in the mean-field model; this effect is suppressed at higher values of $\xi$.

### 14.4.4 Force-induced growth of focal adhesions

The above motor-clutch models include details regarding the state of individual adhesion molecules. Other models simply keep track of the total population of bound adhesions, but incorporate other features such as the coupling between adhesion and protrusion at the leading edge of a cell [939, 985] or the spatial variation of the traction force and rate of retrograde flow as one moves away from the leading edge [216, 816]. These models are motivated by experimental evidence that in addition to their mechanical role of providing a traction force that opposes retrograde flow, integrins couple to various signaling pathways such as Rac GTPases and Rho/ROCK, which modulate actin polymerization and activate myosin motors [89, 332]. Moreover, protrusion can activate the formation of nascent integrins at the leading edge [192], which subsequently disassemble toward the rear resulting in an adhesion gradient. (The existence of an adhesion gradient was assumed in one of the earliest whole-cell mechanical models of cell motility [244].) The coupling between cell protrusion and the formation of nascent integrins is one example of the force-driven adsorption and growth of FAs. Such a phenomenon also occurs in mature FAs and is thought to help protect adhesions from bond rupture [538, 808]. There have been a number of theoretical studies of this process, based on the statistical physics of adsorption and aggregation [71, 238, 677–679, 693, 815, 913, 914]. For the sake of illustration, we will focus on the detailed analytical study of Ref. [71], since this incorporates many of the features considered in other models, and builds upon the theory of nonequilibrium systems developed in Chap. 12 and Chap. 13.

The main components of the model are based on the schematic diagram of Fig. 14.29. The FA is divided into two layers. The lower layer is attached to the substrate, and represents the mechanosensitive proteins that change their configuration and binding energies on the application of a shear stress. These include both the integrins and force-sensitive plaque proteins. The upper layer consists of plaque proteins that are insensitive to external forces, but which can bind to sites in the lower layer. Although these upper layer proteins are passive with respect to applied forces, their association with proteins in the lower layer depends on the force-induced configuration states of the latter. Moreover, the proteins in the upper layer are directly connected to the actin stress fibers and thus transmit the force to the lower layer. For simplicity the substrate is taken to have infinite stiffness; the case of soft substrates is considered in Ref. [678].

**Elasticity of focal adhesions.** We begin by considering the elastic forces in the lower layer. The integrins are represented as particles in a 1D lattice coupled by springs of stiffness $k$, see Fig. 14.35. They are grafted to the substrate by a second set of springs with stiffness $k_b$. Let $u_n$ denote the displacement of the $n$th integrin particle from equilibrium. From Hooke's law the force-balance equation takes the form

$$k[u_{n+1} + u_{n-1} - 2u_n] - k_b u_n + f_n = 0, \qquad (14.4.32)$$

where $f_n$ is the actin force transmitted to the $n$th particle via the upper layer. Setting $x = na$, $u(x) = u_n$, $f(x) = f_n$ and taking the continuum limit leads to the ODE

$$ka^2\frac{d^2u}{dx^2} - k_b u(x) + f(x) = 0, \quad f(x) = g\phi(x). \qquad (14.4.33)$$

The force distribution $f(x)$ is taken to depend on the local volume fraction $\phi(x)$ of plaque proteins in the upper layer, with $g$ the average force when $\phi = 1$. This is based on the observation that integrins in the lower layer will only be connected to the actin stress fibers via their association with plaque proteins in the upper layer. The

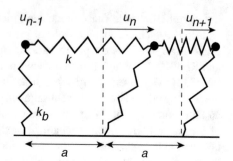

Fig. 14.35: Lower layer of integrins is modeled as a 1D elastic chain that is grafted to the substrate via springs of stiffness $k_b$. [Redrawn from Besser and Safran [71].]

probability of such an association is taken to depend on the local volume fraction of plaque proteins. (This will be explored in more detail below.) Solving equation (14.4.33) using a 1D Green's function (See Box 6C) yields

$$u(x) = \frac{\kappa g}{2k_b} \int_{-\infty}^{\infty} e^{-\kappa|x-x'|} \phi(x') dx' \qquad (14.4.34)$$

for $\kappa^2 = k_b a^2 k$. A reasonable assumption is to take $k_b \gg k$ so that $\kappa \gg 1$. It follows that the integral solution will be dominated by contributions in a small neighborhood of $x' = x$. Hence, Taylor expanding $\phi(x')$ about $x$ gives

$$u(x) = \frac{\kappa g}{2k_b} \int_{-\infty}^{\infty} e^{-\kappa|x-x'|} \left[ \phi(x) + \phi'(x)(x-x') + \ldots \right] dx'$$

$$= \frac{g}{k_b} \left( \phi(x) + \frac{1}{\kappa^2} \phi''(x) + \ldots \right). \qquad (14.4.35)$$

**Adsorption kinetics.**

The model of Besser and Safran [71] assumes that the growth of the FA is generated by the adsorption of plaque proteins from the cytoplasm into the upper layer via the association with mechanosensitive integrin proteins in the lower layer. This can be formulated in terms of the classical theory of surfactant adsorption [240]. In general, the adsorption process can be partitioned into two steps: (i) diffusion of free plaque proteins dissolved in the cytoplasm to adhesion sites, and (ii) binding to integrins that have been activated by direct stretching or by in-plane compression (see below). In Chap. 6 we considered various examples of binding processes that were diffusion-limited, that is, binding was much faster than diffusion. In the current case, it turns out that diffusion is relatively fast so that surfactant adsorption operates in the kinetic limited regime. This means that we can take the local volume fraction of plaque proteins in the bulk to be a constant $\phi_b$. The kinetics of adsorption is then driven by the difference between the chemical potential of plaque proteins in the cytoplasm, $\mu_b$, and the corresponding chemical potential in the upper layer, see Fig. 14.36.

Recall from Chap. 13 the general formula (13.1.19) for Fickian flux, which in 1D takes the form

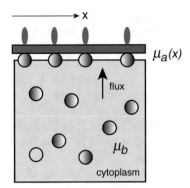

Fig. 14.36: Schematic diagram of surfactant adsorption driven by a difference in chemical potentials.

$$J = -\frac{1}{\gamma}\phi\frac{\partial\mu}{\partial x},$$

where $\gamma = D/k_BT$ is a friction coefficient. It follows that the flux into the upper layer
interface is given by

$$J = \frac{D\phi_b}{k_BT}\frac{\mu_b - \mu_a(x,t)}{a},$$

where the integrin spacing $a$ is used as a length scale for the layer width. From
particle conservation and the fact that $\phi$ is defined per unit volume rather than per
unit area, we have

$$\frac{\partial\phi(x,t)}{\partial t} = \frac{D\phi_b}{a^2 k_BT}[\mu_b - \mu_a(x,t)]. \tag{14.4.36}$$

One contribution to the spatial dependence of the chemical potential $\mu_a$ is the distribution of activated integrin sites, which depends on $\phi$ and the spatial variation of
the stresses from the actin cytoskeleton. The latter dependence leads to anisotropic
growth. In order to specify the chemical potential $\mu_a$ we now need to determine the
free energy of plaque proteins comprising the upper layer.

**Free energy of the adsorbed plaque proteins.** Let us return to the 1D lattice model
and take $t_n \in \{0,1\}$ to be the site-variable for plaque proteins, with $t_n = 1$ indicating
that a plaque protein is located near the $n$-th binding site and $t_n = 0$ otherwise.
However, the plaque protein can only actually bind if the corresponding integrin
receptor is activated by the applied forces. For the moment we will not specify the
details of mechanotransduction, and simply denote the probability of activation by
the function $p_n(t_n)$. The energy function or Hamiltonian for the lattice of plaques is
then taken to be of the form [71]

$$H = k_BT\sum_n \alpha_n(t_n)t_n + \frac{1}{2}\sum_{m,n} J_{nm}t_n(1-t_m), \quad \alpha_n = -\frac{\varepsilon_b}{k_BT}p_n, \tag{14.4.37}$$

where $\varepsilon_b$ is the binding energy gained when a plaque protein associates with an
activated integrin. Consistent with experimental observations of condensed protein
plaques, neighboring plaque proteins are assumed to have an effective attraction.

Thus $J_{nm} = JC_{nm}$ where $C_{nm}$ is the connectivity matrix for nearest-neighbor coupling and $J > 0$. The presence of the interaction term means that it is difficult to evaluate the partition function $Z$ and associated free energy $F$, where

$$Z = \left( \prod_n \sum_{t_n=0,1} \right) e^{-H(\mathbf{t})/k_B T}, \quad F = -k_B T \ln Z. \tag{14.4.38}$$

Therefore, following Ref. [71], the free energy is approximated using the variational method outlined in Ref. [782] and Box 14D. The analysis is very similar to that of a binary mixture.

More specifically, introduce the simplified Hamiltonian

$$H_0 = k_B T \sum_n [\alpha_n(t_n) - \sigma_n] t_n, \tag{14.4.39}$$

where the $\sigma_n$ parameterize $H_0$. The corresponding partition function is

$$Z_0 = \prod_n \left( \sum_{s_n=0,1} \right) e^{-H_0(\mathbf{s})/k_B T} = \prod_n \frac{1}{1 - \phi_n}, \quad \phi_n = \frac{1}{1 + e^{\alpha_n(1) - \sigma_n}}.$$

Since $\langle t_n \rangle_0 = \phi_n$, it follows that

$$\langle H - H_0 \rangle_0 = \frac{1}{2} \sum_{m,n} J_{mn} \phi_m (1 - \phi_n) + k_B T \sum_n \sigma_n \phi_n.$$

The approximate free energy defined by equation (14.4.60) or (14.4.62) of Box 14D thus takes the form

$$\widetilde{F} = k_B T \sum_n [(1 - \phi_n) \ln(1 - \phi_n) + \phi_n \ln \phi_n] + k_B T \sum_n \alpha_n(1) \phi_n + \frac{1}{2} \sum_{m,n} J_{mn} \phi_m (1 - \phi_n). \tag{14.4.40}$$

Finally, the variational method establishes that (see equation (14.4.63)) $F \leq \widetilde{F}$ for all $\{\phi_n\}$.

As in the analysis of binary mixtures (Sect. 13.1), one can take a continuum limit of the above lattice free energy by assuming nearest-neighbor interactions with coupling strength $J$ and lattice spacing $a$. This yields the free energy functional

$$F[\phi] = \frac{1}{a} \int \left[ f(\phi) + k_B T \alpha(1,x) \phi + \frac{Ja^2}{4} \left( \frac{\partial \phi}{\partial x} \right)^2 \right] dx, \tag{14.4.41}$$

with

$$f(\phi) = k_B T \left( (1 - \phi) \ln(1 - \phi) + \phi \ln \phi \right) + \frac{J}{2} \phi(1 - \phi). \tag{14.4.42}$$

The final component of the model is to determine $\alpha(1,x)$ by considering the process of mechanotransduction. As we will see below, $\alpha(1,x)$ depends on the displacement field $u(x)$ of the elastic integrin layer.

**Mechanotransduction of integrins.** The detailed molecular mechanisms underlying integrin-protein mechanotransduction are not yet well understood. In Ref. [71], it is assumed that forces arising from the actin cytoskeleton increase the probability of a conformational switch from an inactivated to an activated state. This is modeled by taking the inactivated integrin-protein complex (mechanosensor) to have a spherical shape, while the activated state is more elongated resulting in the exposure of potential binding sites for plaque proteins, see Fig. 14.37. The deformation required to achieve a switch from a spheroid to an ellipsoid shape can be achieved either by stretching the complex or by compressing it. In order to determine the statistics of this process, it is necessary to consider the energetics associated with changes in conformational state.

For simplicity, suppose that the $n$th integrin-protein complex has two conformational states; an activated state $s_n = 1$ and an inactivated state $s_n = 0$. In order to switch from the latter to the former, it is necessary to overcome a free energy barrier $\Delta G$. This energy can be supplied by external forces that either stretch or compress the mechanosensor. Let $\mathcal{U}_n$ denote the relative amount of in-plane compression ($\mathcal{U}_n < 0$) or in-plane expansion ($\mathcal{U}_n > 0$). In terms of the continuum model $\mathcal{U}_n \to u'(x)$. Since compression is energetically favored by the activated state, it contributes a term $\tau \mathcal{U}_n s_n$ to the total energy for some constant $\tau$. The term is non-local, since it involves the relative displacement of neighboring integrins, see Fig. 14.35. Actin forces can also directly stretch the mechanosensors. Although stretching does not contribute to the displacements $u_n$ described by equation (14.4.33), it does contribute a term to the energetics of the activated state. That is, if $d$ is the length scale of molecular deformation during activation, then the change in energy is $-df_n s_n$, where $f_n$ is the local force. Recall that the force $f_n$ can only act on the mechanosensor if there is a plaque protein close to the integrin-binding site ($t_n = 1$). That is, we can write $f_n = g t_n$. Combining all of these results, the Hamiltonian that accounts for the energetics of the activated state is

$$H_{\text{int}} = \sum_n s_n (\Delta G + \tau \mathcal{U}_n - dg t_n) \equiv \sum_n h_n(t_n) s_n. \qquad (14.4.43)$$

In terms of the associated Boltzmann-Gibbs distribution, we have

Fig. 14.37: Model of mechanotransduction, in which an inactivated spherical mechanosensor is deformed by compression or stretching into an activated state. In-plane compression arises from elastic interactions between neighboring sensors, whereas stretching is due to direct coupling of the sensors to the actin stress fibers. [Redrawn from Besser and Safran [71].]

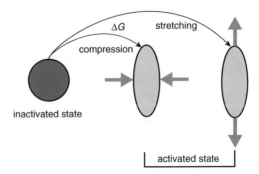

$$\langle s_n \rangle = \frac{e^{-h_n/k_B T}}{1 + e^{-h_n/k_B T}} = \frac{1}{2}\left(1 - \tanh(h_n/2k_B T)\right).$$

Substituting for $h_n$ and using the fact that $\mathcal{U}_n$ is a small quantity, we can Taylor expand the tanh function to first order in $\mathcal{U}_n$:

$$\langle s_n \rangle = \frac{1}{2}\left(1 - \tanh([\Delta G - dgt_n]/2k_B T) - \mathrm{sech}^2([\Delta G - dgt_n]/2k_B T)\frac{\tau \mathcal{U}_n}{2k_B T}\right)$$
$$+ O(\mathcal{U}_n^2) \equiv p_n(t_n). \tag{14.4.44}$$

Hence, the function $\alpha_n(t_n)$ appearing in equation (14.4.37) has the explicit form for $t_n = 1$

$$\alpha_n(1) = \frac{\varepsilon_b}{2k_B T}\left[A_0(g) + A_1(g)\mathcal{U}_n\right], \tag{14.4.45}$$

with

$$A_0(g) = \tanh([\Delta G - dg]/2k_B T) - 1, \quad A_1(g) = \frac{\tau}{2k_B T}\mathrm{sech}^2([\Delta G - dg]/2k_B T). \tag{14.4.46}$$

Finally, in the continuum limit $\alpha_n(1) \to \alpha(1, x)$ with $\mathcal{U}_n \to u'(x)$, we have

$$\alpha(1, x) = \frac{\varepsilon_b}{2k_B T}\left[A_0(g) + A_1(g)\frac{\partial u}{\partial x}\right]. \tag{14.4.47}$$

**Analysis of model: traveling wave solution.** We wish to solve the kinetic equation (14.4.36) with the chemical potential $\mu_a(x, t)$ determined from the free energy (14.4.41) according to

$$\mu_a(x, t) = \frac{\delta F[\phi]}{\delta \phi(x, t)}. \tag{14.4.48}$$

Following Ref. [71], it is convenient to use the Ginsburg-Landau expansion of the free energy about the concentration at the interface where $\phi \approx 1/2$. That is, setting $\psi = \phi - 1/2$, we have from Ex. 13.2

$$F[\psi] = \frac{1}{a}\int\left[k_B T\alpha(1, x)\psi - \frac{\varepsilon}{2}\psi^2 + \frac{c}{4}\psi^4 + \frac{B}{2}\left(\frac{\partial \psi}{\partial x}\right)^2\right]dx, \tag{14.4.49}$$

where

$$\varepsilon = (J - 4k_B T), \quad c = \frac{16k_B T}{3}, \quad B = \frac{Ja^2}{2}.$$

and $\alpha(1, x)$ is given by equation (14.4.47). It then follows that

$$\mu_a = \frac{\delta F[\psi]}{\delta \psi} = \frac{\varepsilon_b}{2}\left[A_0(g) + A_1(g)\frac{\partial u}{\partial x}\right] - \varepsilon\psi + c\psi^3 - B\frac{\partial^2\psi}{\partial x^2}. \tag{14.4.50}$$

We can now write down a closed kinetic equation for $\psi$ by substituting for $\mu_a$ in equation (14.4.36) and noting from equation (14.4.35) that $u \approx g\phi/k_b$:

$$\frac{\partial \psi}{\partial t} = \Gamma\left\{\Delta\mu_0(g) - \frac{\varepsilon_b g}{2k_b}A_1(g)\frac{\partial \psi}{\partial x} + \varepsilon\psi - c\psi^3 + B\frac{\partial^2\psi}{\partial x^2}\right\}, \tag{14.4.51}$$

and

$$\Delta\mu_0(g) = \mu_b - \frac{\varepsilon_b}{2}A_0(g), \quad \Gamma = \frac{D\phi_b}{a^2 k_B T}.$$

Equation (14.4.51) can be analyzed in a similar fashion to the 1D interface equation of Ex. 13.2, except that now we have a traveling front solution [71]. That is, set $\psi(x,t) = \Psi(x - vt) = \Psi(\xi)$ with $v$ the wave speed,

$$B\frac{\partial^2\Psi}{\partial \xi^2} + V(g)\frac{\partial \Psi}{\partial \xi} + \varepsilon\Psi - c\Psi^3 + \Delta\mu(g) = 0, \tag{14.4.52}$$

and

$$V(g) = \frac{v}{\Gamma} - \frac{\varepsilon_b g}{2k_b}A_1(g). \tag{14.4.53}$$

Equation (14.4.52) has a solution of the form, see Ex. 14.8,

$$\Psi(\xi) = \alpha\tanh(\gamma\xi) + \beta. \tag{14.4.54}$$

That is, substituting such a solution into (14.4.52) yields a cubic in $\tanh(\gamma\xi)$. Requiring that the coefficients of the cubic vanish yields four algebraic equations for the four unknowns $\alpha, \beta, \gamma, v$. Note that $|\gamma|^{-1}$ determines the width of the front, whereas $\alpha$ and $\beta$ are related to the far-field values $\Psi(\pm\infty)$. In particular, assuming $\gamma < 0$, we have

$$\alpha + \beta = \Psi(-\infty), \quad \beta - \alpha = \Psi(\infty). \tag{14.4.55}$$

After some algebra, one finds a simple expression for the function $V(g)$, see Ex. 14.8, namely

$$V = 3\sqrt{2Bc}\beta\,\text{sign}(\gamma), \tag{14.4.56}$$

with $\beta$ related to $\Delta\mu_0(g)$ according to the cubic

$$\beta - \frac{4c}{\varepsilon}\beta^3 = \frac{\Delta\mu_0(g)}{2\varepsilon}. \tag{14.4.57}$$

Fig. 14.38: Diagram illustrating how the direction of the applied force determines the front and back of the growing FA.

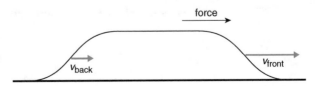

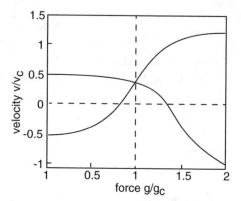

Fig. 14.39: Sketch of front and back velocities as a function of the applied force $g$.
The velocity is expressed in units of $v_0 = 3/sqrt2Bc\Gamma|\mu_b|/\varepsilon$ and the force is in units
of the critical force $g_c$ at which $v_{back} = v_{front}$
and $\Delta\mu_0 = 0$. [Redrawn from Besser and
Safran [71].]

Assuming that $\Delta\mu_0$ is small, this can be inverted to give [71]

$$V = V(\Delta\mu_0)\text{sign}(\gamma), \quad V(\Delta\mu_0) \approx \frac{3\sqrt{Bc}}{\sqrt{2}\varepsilon}\Delta\mu_0 + \frac{3\sqrt{Bc^3}}{\sqrt{2}\varepsilon^4}\Delta\mu_0^3 \geq 0. \quad (14.4.58)$$

Based on the above solution, the model predicts that there is a force-induced
anisotropy in the growth of the FA. First, note that the direction of the force determines the front and back of the growing region. The former can be treated as a front
with $\gamma < 0$ and the other as a front with $\gamma > 0$, see Fig. 14.38. It then follows from
equations (14.4.53) and (14.4.58) that the front interface has a higher speed than the
back interface:

$$v_{front} = \Gamma V(\Delta\mu_0) + \frac{\varepsilon_b g \Gamma}{2k_b} A_1(g), \quad (14.4.59a)$$

$$v_{front} = -\Gamma V(\Delta\mu_0) + \frac{\varepsilon_b g \Gamma}{2k_b} A_1(g). \quad (14.4.59b)$$

The second term on the right-hand side of these equations arises from the elastic
compression of the lower integrin layer. Such a term is absent in classical domain
growth of adsorbed surfactants, which means that the back and front edges grow
symmetrically into the dilute region. This symmetry is broken by the elastic compression term. At the front end the compression of the FA stimulates growth due to
activation of integrins so that it enhances the rightward shift, whereas dilation of the
FA at the back end means that plaque proteins tend to desorb due to inactivation of
integrins. This depletion and shrinking of the back edge has the effect of moving
it to the right as well. In Fig. 14.39 we sketch example plots of the front and back
velocities as a function of the non-dimensionalized force. At a critical force $v_c$ both
ends move at the same velocity and there is no growth (treadmilling).

**Box 14D Variational method for approximating free energies.**

Consider some statistical mechanical system characterized by an exact Hamiltonian $H$ defined on some state space $\Omega$. The Boltzmann-Gibbs distribution is (see Chap. 1)

$$P(\omega) = Z^{-1} e^{-H(\omega)/k_B T}, \quad Z = \int_\Omega e^{-H(\omega)/k_B T} d\omega,$$

where $\omega$ represents the set of state-space variables and $d\omega$ is an appropriate integral measure. Introduce the following free energy functional $F$ that is parameterized by a probability density $P$:

$$F[P] = k_B T \int_\Omega P(\omega) \ln P(\omega) d\omega + \int_\Omega P(\omega) H(\omega) d\omega. \qquad (14.4.60)$$

The Boltzmann–Gibbs distribution is recovered by minimizing the functional $F[P]$ with respect to $P$:

$$\frac{\delta F[P]}{\delta P(\omega)} = 0 \implies P(\omega) \sim e^{-H(\omega)/k_B T}.$$

The correct normalization can be enforced by adding the term

$$\lambda \left[ \int_\Omega P(\omega) d\omega - 1 \right]$$

to the right-hand side of equation (14.4.60), where $\lambda$ is a Lagrange multiplier. At the minimum one sees that the free energy takes the expected thermodynamic form

$$F = -k_B T \ln Z.$$

The basic idea of the variational method is that one replaces the general problem of minimizing $F$ with respect to all admissible probability distributions $P$, by the simpler problem of minimizing with respect to the set of parameters of a prescribed distribution $P_0$. (This is analogous to the Ritz variational method introduced in the analysis of the worm-like chain model of a polymer, see Sect. 12.1.4.) That is, $F$ is approximated by

$$\widetilde{F} = k_B T \int_\Omega P_0(\omega) \ln P_0(\omega) d\omega + \int_\Omega P_0(\omega) H(\omega) d\omega, \qquad (14.4.61)$$

with $P_0 \sim e^{-H_0/k_B T}$ for a simplified Hamiltonian $H_0$. Setting

$$Z_0 = \int_\Omega e^{-H_0(\omega)/k_B T}, \quad F_0 = -k_B T \ln Z_0,$$

it follows that

$$\widetilde{F} = F_0 + \langle H - H_0 \rangle_0, \qquad (14.4.62)$$

where

$$\langle H - H_0 \rangle_0 = \int_\Omega [H(\omega) - H_0(\omega)] P_0(\omega) d\omega.$$

For any $P_0$, the approximate free energy $\widetilde{F}$ is an upper bound for the exact free energy $F$. In order to show this, observe that

$$\int_\Omega P_0(\omega) \ln \left[ \frac{P_0(\omega)}{P(\omega)} \right] d\omega = \int_\Omega P(\omega) \ln \left[ V(\omega) \ln V(\omega) - V(\omega) + 1 \right] d\omega \geq 0,$$

where $V = P_0/P$. The final inequality follows from positivity of $P$ and the inequality $x \ln x \geq x - 1$ for all $x > 0$. Hence,

$$\int_\Omega P_0(\omega) \ln P_0(\omega) d\omega \geq \int_\Omega P_0(\omega) \ln P(\omega) d\omega.$$

Substituting for $P_0$ and $P$, rewritten as

$$P_0 = e^{[F_0 - H_0]/k_B T}, \quad P = e^{[F - H]/k_B T},$$

then gives

$$\int_\Omega P_0(\omega) [F_0 - H_0] d\omega \geq \int_\Omega P_0(\omega) [F - H] d\omega,$$

which on rearranging yields the result

$$F \leq F_0 + \langle H - H_0 \rangle_0. \qquad (14.4.63)$$

Given a parameterized form for $P_0$, one tries to get as close as possible to the exact free energy by minimizing the right-hand side with respect to the set of parameters.

**Free energy of mixing.** The variational method can be used to derive the free energy functional (13.1.16) used in the analysis of phase separation of a binary mixture. The starting point is the Hamiltonian $H$ of a lattice model where the state at lattice site $i$ is given by $s_i \in \{0, 1\}$. Here $s_i = 1$ means that the site is occupied by species A (solute), whereas $s_i = 0$ means that it is occupied by species B (solvent). The Hamiltonian takes the form

$$H = \frac{1}{2} \sum_{i,j} J_{ij} s_i (1 - s_j), \qquad (14.4.64)$$

with $J_{ij}$ the net pairwise interaction between the two components. The coupling between different sites means that it is difficult to analyze the corresponding partition function. Therefore, we introduce the simpler Hamiltonian

$$H_0 = k_B T \sum_i \sigma_i s_i, \tag{14.4.65}$$

with parameters $\sigma_i$ scaled by the factor $k_B T$ for convenience. The next step is to determine the approximate free energy $\widetilde{F}$, equation (14.4.62), which can then be minimized with respect to the $\sigma_j$.

First, we have

$$Z_0 = \prod_i \left( \sum_{s_i=0,1} e^{-\sigma_i s_i} \right) = \prod_i \frac{1}{1-\phi_i}, \quad \phi_i = \frac{1}{1+e^{-\sigma_i}}.$$

It immediately follows that

$$\langle s_i \rangle_0 = -\frac{\partial \ln Z_0}{\partial \sigma_i} = \phi_i.$$

Similarly,

$$\langle H - H_0 \rangle_0 = \left\langle \frac{1}{2} \sum_{i,j} J_{ij} s_i (1-s_j) - k_B T \sum_i \sigma_i s_i \right\rangle_0$$
$$= \frac{1}{2} \sum_{i,j} J_{ij} \phi_i (1-\phi_j) - k_B T \sum_i \sigma_i \phi_i,$$

since $s_i$ and $s_j$ for $i \neq j$ are statistically independent with respect to the distribution $P_0$. Combining these results with equation (14.4.62), we find that an upper bound of the exact free energy is

$$\widetilde{F} = k_B T \sum_i [(1-\phi_i) \ln(1-\phi_i) + \phi_i \ln \phi_i] + \frac{1}{2} \sum_{i,j} J_{ij} \phi_i (1-\phi_j). \tag{14.4.66}$$

This is precisely the spatially discretized version of the continuum free energy given by equation (13.1.16). The latter is recovered by assuming nearest-neighbor interactions and taking the continuum limit.

## 14.5 Cytoneme-based morphogen gradients

Cytonemes are thin, actin-rich filaments that can dynamically extend up to several hundred microns to form direct cell-to-cell contacts, see Fig. 14.40. There is increasing experimental evidence that these direct contacts allow the active transport of morphogen to embryonic cells during development [153, 203, 387, 388, 522, 523, 900, 979]. Cytonemes were first characterized in the wing imaginal disc

Fig. 14.40: Micrograph showing cytonemes extending from tracheal cells of a *Drosophila* larva, which are marked with membrane-tethered mCherry fluorescent protein. Some of the cytonemes contact the underlying wing imaginal disc and transport the Dpp morphogen protein (marked with Green Fluorescent Protein) to the tracheal cells. [Creative commons figure originally generated by Thomas Korenberg.]

of *Drosophila* [752], and have been associated with the transport of both morphogenetic protein Decapentaplegic (Dpp) and Hedgehog (Hg) [85, 372, 773]. Many cytonemes in *Drosophila* are found to extend from morphogen-producing cells to target cells. Morphogens are actively transported along the cytonemes in a bidirectional fashion, probably via myosin motors that actively "walk" along the actin filaments of a cytoneme. The amount of morphogen delivered to a cell will then depend on the flux of particles along a cytoneme and the number of cytonemes that form a stable contact with the target cell. Cytonemes can also emanate from receptor-bearing target cells, transporting their receptors to the vicinity of source cells. Increasing experimental evidence indicates that cytonemes also mediate morphogen transport in vertebrates [292, 783]. Examples include sonic hedgehog (Shh) cell-to-cell signaling in chicken limb buds [790] and Wnt signaling in the neural plate of zebrafish [840, 841]. The latter involves a different morphogen transport mechanism, in which Wnt is clustered at the membrane tip of growing signaling filopodia that make temporary contacts with target cells and deliver a burst of morphogen. In this case, the amount of morphogen within a target cell will depend on the temporal frequency of contacts by cytonemes. Another complicating factor is the expansion of the neural plate during development, which means that cells are continuously moving out of the cytonemal area of influence [841].

The precise biochemical and physical mechanisms underlying how cytonemes find their targets, form stable contacts and deliver their cargo to target cells are currently unknown. However, it has been hypothesized that cytonemes find their targets via a random search process based on alternating periods of retraction and growth [522]. Indeed, imaging studies in *Drosophila* [85] and chick [790] show that cytonemes actively expand and contract. In contrast to diffusion-based morphogenesis, there have been a relatively small number of mathematical modeling studies of cytoneme-based morphogenesis. These have mainly focused on one-dimensional (1D) models, in which cytonemes from a source cell transport morphogen to a 1D array of target cells. Transport occurs via two distinct mechanisms. The first involves active motor-driven transport of morphogen packets (vesicles) along static cytonemes with fixed contacts between a source cell and a target cell [124, 508, 509, 865, 866]. The second is based on nucleating cytonemes from a

source cell that dynamically grow and shrink until making contact with one of the
target cells [126, 133] and delivering a morphogen burst, as in vertebrate Wnt signaling. The delivery of a single burst can be analyzed in terms of a first passage
time (FPT) problem with a sticky boundary at the source cell. The latter takes into
account the exponentially distributed waiting time required for nucleation of a new
growing filament, following any return to the source cell. After delivery of a morphogen burst, the cytoneme retracts and a new search-and-capture process is initiated. This then leads to a sequence of search-and-capture events, whereby morphogen accumulates in the target cells. Assuming that the build up of resources
within each target is counterbalanced by degradation, there will exist a steady-state
morphogen distribution in the long-time limit, which takes the form of a morphogen
gradient. One way to calculate the statistics of resource accumulation is to formulate multiple search-and-capture events as a $G/M/\infty$ queue [126] along the lines
outlined in Sect. 7.4.

   One interesting issue is whether or not cytoneme-based morphogenesis has any
possible advantages over diffusion-based mechanisms outlined in Sect. 11.1. In the
case of the 1D models considered in Refs. [124, 508, 509, 865, 866], it would
appear that both mechanisms exhibit similar behaviors, although the details may
differ. Moreover, diffusion can support longer-range morphogen gradients, since
cytonemes only extend up to a few hundred microns. On the other hand, the more
specific and directional nature of the cytoneme-based contact between source and
target cells, suggests that there is more flexibility in generating different types of
morphogen fields in higher spatial directions.

### 14.5.1 Bidirectional transport model of cytoneme-based morphogen gradients

Teimouri and Kolomeisky [865, 866] first introduced a compartmental model of
cytoneme-based morphogenesis in invertebrates, which consists of a discrete set of
$N+1$ cells arranged on a line. A source cell at one end makes direct contact with
each of the $N$ target cells via a single cytoneme per cell. Assuming that the rate
$w_n$ of morphogen transport decreases with distance $L_n$ between target and source
cells, they showed how a steady-state morphogen gradient could be established.
Here we describe an extension of their model, in which the motor-driven transport
of morphogens along cytonemes is modeled explicitly [124, 508].

   Consider a single cytoneme of length $L$ linking a source cell to a single target cell,
see Fig. 14.41. Let $u(x,t)$ denote the density of motor-cargo complexes at $x \in [0,L]$
along the cytoneme at time $t$. Suppose that the complexes can be partitioned into
anterograde $(+)$ and retrograde $(-)$ subpopulations labeled by $u_+(x,t)$ and $u_-(x,t)$,
respectively. (In the case of myosin transport along actin filaments, the retrograde
flow could be due to treadmilling [661, 970]. For recent evidence of myosin motor-
based transport of puncta along cytonemes within the *Drosophila* wing imaginal
disc, see [184]).) The densities evolve according to the standard velocity jump pro-

cess also used to model microtubule catastrophes (Chap. 4) and 1D bacterial run-and-tumble (Chap.10):

$$\frac{\partial u_+}{\partial t} = -v_+ \frac{\partial u_+}{\partial x} + \alpha u_- - \beta u_+, \tag{14.5.1a}$$

$$\frac{\partial u_-}{\partial t} = v_- \frac{\partial u_-}{\partial x} - \alpha u_- + \beta u_+, \tag{14.5.1b}$$

where $v_\pm$ are the speeds of the $\pm$ states, $\alpha$ is the rate of switching from the retrograde to the anterograde state, and $\beta$ is the switching rate from anterograde to retrograde. Equation (14.5.1) is supplemented by the boundary conditions

$$u_+(0,t) = \kappa C_0(t), \quad u_-(L,t) = 0, \tag{14.5.2}$$

where $C_0(t)$ is the density of vesicles in the source cell and $\kappa$ is an injection rate. We assume that initially there are no particles within the cytonemes so $u_\pm(x,0) = 0$ for all $0 \le x \le L$. The transport component of the model couples to the number of vesicles in the source and target cells according to

$$\frac{dC_0}{dt} = Q - J(0,t), \quad \frac{dC_1}{dt} = J(L,t) - kC_1(t), \tag{14.5.3}$$

where $Q$ is the particle production rate in the source cell, $k$ is a morphogen degradation rate in the target cell, and $J(x,t)$ is particle flux at position $x$ at time $t$,

$$J(x,t) = v_+ u_+(x,t) - v_- u_-(x,t). \tag{14.5.4}$$

We now calculate the steady-state solution $C_0^*, C_1^*$ as a function of cytoneme length. (An analogous advection–diffusion model is analyzed in Ex. 14.9.) Setting time derivatives to zero and adding equations (14.5.1a) and (14.5.1b) gives

$$\frac{d}{dx}\left(v_+ u_+(x) - v_- u_-(x)\right) = 0.$$

It follows that there is a stationary flux $J(x) = J_0$. Setting $dC_n/dt = 0$ in equation (14.5.3) implies that $J_0 = J(0) = Q$ and $C_1^* = Q/k$. Using these results to eliminate $u_-$ from equation (14.5.1a), we have

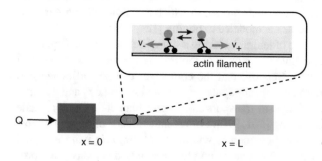

Fig. 14.41: Bidirectional transport model of a single cytoneme.

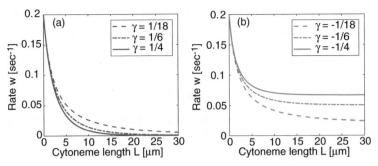

Fig. 14.42: Transport rate $w(L)$ plotted as a function of cytoneme length $L$ for various values of $\gamma$. (a) $\gamma > 0$ so that $\bar{v} < 0$ and the flux decays asymptotically to zero. (b) $\gamma < 0$ so that $\bar{v} > 0$ and the asymptote is nonzero. Parameter values are as follows: $v_+ = 0.2\mu m\ s^{-1}$, $\alpha = 0.1\ s^{-1}$, $\beta = 0.1$ $s^{-1}$, $Q = 0.1\ s^{-1}$, $\kappa = 0.01\ s^{-1}$, $k = 0.05 s^{-1}$, $L = 10\ \mu m\ s^{-1}$. The parameter $\gamma$ (with units $\mu m^{-1}$) is varied by varying $v_-$.

$$\frac{du_+}{dx} + \gamma u_+ = -\frac{\alpha Q}{v_+ v_-}, \quad \gamma = \frac{\beta v_- - \alpha v_+}{v_+ v_-}.$$

This has the solution (after imposing the boundary condition at $x = 0$)

$$u_+(x) = \kappa P_0^* e^{-\gamma x} - \frac{\alpha Q}{\gamma v_+ v_-}\left[1 - e^{-\gamma x}\right]. \tag{14.5.5}$$

(In the non-generic case that $\gamma = 0$, the concentration decreases linearly with $x$.) Finally, imposing the absorbing boundary condition at $x = L$ determines $C_0^*$:

$$C_0^* = \frac{Q}{w(L)}, \quad C_1^* = \frac{Q}{k}, \tag{14.5.6}$$

where

$$w(L) = \frac{\kappa v_+ e^{-\gamma L}}{1 + \alpha\left[1 - e^{-\gamma L}\right]/\gamma v_-}. \tag{14.5.7}$$

Note that $w(L) > 0$, since $e^{-\gamma L} - 1$ has the same sign as $\gamma$. An important quantity, which generalizes to the multi-cell case, is the ratio of the target and source densities,

$$\frac{C_1^*}{C_0^*} = \frac{w(L)}{k}. \tag{14.5.8}$$

The length-dependence of this ratio is determined by the function $w(L)$, which can be identified as an effective transport coefficient [865].

The above analysis establishes that bidirectional transport leads to a stationary flux that is an exponentially decaying function of cytoneme length, see Fig. 14.42. However, the asymptotic value $\lim_{L\to\infty} w(L)$ depends on the sign of $\gamma$ and thus the sign of the mean velocity $\bar{v}$:

$$\bar{v} = \frac{\alpha v_+ - \beta v_-}{\alpha + \beta}, \tag{14.5.9}$$

where $\bar{v}$ is the mean velocity of a motor-cargo complex. If $\bar{v} < 0$ ($\gamma > 0$), then $w(L)$ decays to zero as cytoneme length tends to infinity. On the other hand, if $\bar{v} > 0$ ($\gamma < 0$), then the anterograde transport state is dominant and one finds that

$$\lim_{L\to\infty} w(L) = \kappa v_+ \left(1 - \frac{\beta v_-}{\alpha v_+}\right). \tag{14.5.10}$$

Interestingly, an experimental study of cytoneme-based transport in the wing imaginal disc of *Drosophila* provides evidence for a significant retrograde component of motor-based transport [184]. The authors observed puncta moving at similar anterograde and retrograde speeds of around $0.4\mu m/s$. This does not necessarily imply that $\gamma = 0$, since the switching rates $\alpha$ and $\beta$ could differ. Thus, rather than taking $v_- < v_+$ and $\alpha = \beta$, we could obtain a nonzero $\gamma$ by taking $v_+ = v_-$ and $\alpha \neq \beta$.

**Multiple target cells.** It is straightforward to extend the single cytoneme model to multiple cytonemes of length $L_n$ linking a source cell to multiple target cells, $n = 1,\dots,N$, see Fig. 14.43. Let $u_+^n(x,t)$ and $u_-^n(x,t)$ denote the anterograde and retrograde subpopulations in the cytoneme contacting the $n$th target cell. Then the bidirectional model for $u_\pm^n$ takes the same form as equation (14.5.1) on $x \in (0,L_n)$,

$$\frac{\partial u_+^n}{\partial t} = -v_+ \frac{\partial u_+^n}{\partial x} + \alpha u_-^n - \beta u_+^n, \tag{14.5.11a}$$

$$\frac{\partial u_-^n}{\partial t} = v_- \frac{\partial u_-^n}{\partial x} - \alpha u_-^n + \beta u_+^n, \tag{14.5.11b}$$

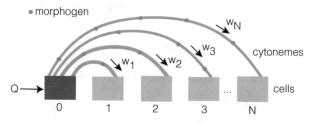

Fig. 14.43: Schematic diagram of the cytoneme-based transport model introduced in Ref. [866]. A source cell (dark blue) generates morphogens at a rate $Q$, which are then delivered to other cells (light blue) via a set of tubular cytonemes. It is assumed that a single cytoneme links the source cell (labeled $n = 0$) to each target cell (labeled $n = 1,\dots,N$), and transports morphogens (red dots) at an effective rate $w_n$.

with the boundary conditions

$$u_+^n(0,t) = \kappa P_0(t), \quad u_-^n(L_n,t) = 0. \tag{14.5.12}$$

(For simplicity, we assume that vesicles are distributed uniformly between the cytonemes. However, one could take a non-uniform distribution with $u_+^n(0,t) = \kappa f_n C_0(t)$ and $\sum_n f_n = 1$ [124].) Extending (14.5.3) to the case of multiple target cells yields

$$\frac{dC_0}{dt} = Q - \sum_{m=1}^{N} J_m(0,t), \quad \frac{dC_n}{dt} = J_n(L_n,t) - kC_n, \tag{14.5.13}$$

where

$$J_n(x,t) = v_+ u_+^n(x,t) - v_- u_-^n(x,t).$$

Solving the steady-state equations shows that

$$v_+ u_+^n(x) - v_- u_-^n(x) = J_n^*, \tag{14.5.14}$$

where $J_n^*$ is the stationary flux reaching the $n$th target cell, and

$$u_+^n(x) = \kappa C_0^* e^{-\gamma x} - \frac{\alpha J_n^*}{\gamma v_+ v_-} \left[1 - e^{-\gamma x}\right]. \tag{14.5.15}$$

Imposing the absorbing boundary condition at $x = L_n$ implies that $J_n^* = C_0^* w(L_n)$, with $w(L)$ given by equation (14.5.7). Finally, the stationary versions of equations (14.5.12) show that

$$C_0^* = \frac{Q}{\sum_{m=1}^{N} w_m}, \quad C_n^* = \frac{Q}{k} \frac{w_n}{\sum_{m=1}^{N} w_m} = \frac{w_n}{k} C_0^*, \tag{14.5.16}$$

where $w_n = w(L_n)$. This is identical in form the stationary solution of the compartmental model introduced in Ref. [866], except that here $w_n$ is derived explicitly from a transport model.

**Robustness of cytoneme-based morphogen gradient.** Following the analysis of diffused-based concentrations in Sect. 11.1, we will characterize the sensitivity of the cytoneme-based morphogen gradient to fluctuations in $Q$ by considering the corresponding induced spatial shift in morphogen concentration. For analytical convenience, we replace the discrete set of target cells $n = 1,\ldots,N$ by a continuum of target cells distributed uniformly on the domain $y \in [0,L]$ such that $C_n(t) \to C(y,t)$ and $J_n(t) \to J(y,t)$. The source cell is treated as a single compartment with particle density $C_0(t)$, and the cytoneme linking the target cell at $y$ is taken to have length $y$. Equation (14.5.1) still holds with $u_\pm^n(x,t) \to u_\pm(x,y,t)$, while the boundary conditions (14.5.2) become

$$u_+(0,y,t) = \kappa C_0(t), \quad u_-(y,y,t) = 0. \tag{14.5.17}$$

Finally, the transport component of the model couples to the distributions of particles in the source and target cells according to

$$\frac{dC_0}{dt} = Q - \int_0^L J(0,y,t)dy, \quad \frac{\partial C(y,t)}{\partial t} = J(y,y,t) - kC(y,t), \quad (14.5.18)$$

where $J(x,y,t)$ is the particle flux at position $x$ at time $t$ through the cytonemes linking the source cell to cells at $y$:

$$J(x,y,t) = v_+u_+(x,y,t) - v_-u_-(x,y,t). \quad (14.5.19)$$

For fixed $y$, the steady-state solution of the transport equation (14.5.1) can be solved as before. We thus obtain the steady-state solutions

$$C_0^* = \frac{Q}{\int_0^L w(y)dy}, \quad C^*(y) = \frac{Q}{k}\frac{w(y)}{\int_0^L w(y)dy}. \quad (14.5.20)$$

Consider some threshold morphogen concentration $C_0$ and denote the cellular position where this threshold occurs by $y_0$, that is, $C^*(y_0) = C_0$. We wish to determine the shift in threshold position $y_0 \to y_0 + \Delta y_0$ in response to a shift in the production rate, $Q \to Q + \Delta Q$. Since $\Gamma := \int_0^L w(y)dy$ is fixed, we have

$$Qw(y) = (Q + \Delta Q)w(y + \Delta y) = (Q + \Delta Q)w(y) + Qw'(y))\Delta y.$$

Rearranging and taking the limits $\Delta Q, \Delta y \to 0$ yields the sensitivity

$$\left.\frac{dy}{dQ}\right|_{y=y_0} = -\frac{w(y_0)}{Qw'(y_0)}. \quad (14.5.21)$$

It follows from equation (14.5.7) that

$$\left.\frac{dy}{dQ}\right|_{y=y_0} = \frac{1}{Q\gamma}\frac{v_-\gamma + \alpha(1 - e^{-\gamma y_0})}{v_-\gamma + \alpha}. \quad (14.5.22)$$

The sensitivity with respect to $Q$ has different behavior depending on the sign of $\gamma$. This is illustrated in Fig. 14.44. If $\gamma > 0$, so that the mean velocity of motor-cargo complex satisfies $\bar{v} < 0$, then $0 < e^{-\gamma y_c} \leq 1$. Hence,

$$0 < \left.\frac{dy}{dQ}\right|_{y=y_0} < \frac{1}{Q\gamma}, \quad (14.5.23)$$

and the sensitivity with respect to fluctuations in $Q$ is bounded regardless of the size of $y_c$. On the other hand, if $\gamma < 0$ with $\bar{v} > 0$, then $e^{-\gamma y_c} \geq 1$. It follows that

$$\left.\frac{dy}{dQ}\right|_{y=y_0} = \frac{1}{Q|\gamma|}\left[\frac{1}{1 - |\gamma|v_-/\alpha}e^{|\gamma|y_c} - 1\right] > 0. \quad (14.5.24)$$

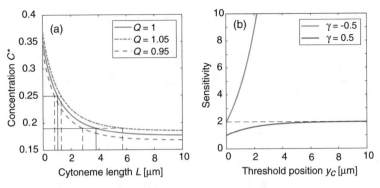

Fig. 14.44: Sensitivity of the threshold position $y_0$ with respect to fluctuations in $Q$. (a) A perturbation in $Q$ also perturbs the threshold position $y_c$. (b) Plot of sensitivity $dy/dQ$ as a function of the threshold position $y_0$. If $\gamma > 0$ then the sensitivity is bounded, whereas if $\gamma < 0$ then the sensitivity increases exponentially. Other parameters as in Fig. 14.42.

Therefore, the sensitivity is always positive and increases exponentially with respect to $y_c$. The analysis strongly suggests that in order to ensure that the morphogen gradient is not exponentially sensitive to fluctuations in the production rate, it is necessary that $\bar{v} < 0$. This reinforces the observation of Ref. [865] that robustness comes at the expense of an energy cost, in this case a strong retrograde flow. Note, however, there is also an energy cost in the case of diffusion-based morphogenesis, since it is necessary to produce a sufficient number of proteins at the local source in order to generate a robust gradient. The more targeted cytoneme-based mechanism allows less protein to be produced.

**Accumulation time.** So far we have only considered the steady-state solution of the cytoneme-based model of morphogen gradient formation. As in the case of diffusion-based models of gradient formation, Sect. 11.1, it is also important to determine the accumulation time to steady state. Analogous to equations (11.1.30) and (11.1.31), the accumulation time for a single target cell of length $L$ is defined according to

$$\tau_1 = \int_0^\infty R_1(t)dt, \quad R_1(t) = 1 - \frac{C_1(t)}{C_1^*}. \tag{14.5.25}$$

In terms of the Laplace transform

$$\widetilde{R}_1(s) = \int_0^\infty R_1(t)e^{-st}dt,$$

we have $\tau_1 = \lim_{s \to 0} \widetilde{R}(s)$. Integration by parts proves that $\lim_{s \to 0} s\widetilde{C}_1(s) = C_1^*$. Hence,

$$\tau_1 = \lim_{s \to 0} \frac{1}{s}\left(1 - \frac{s\widetilde{C}_1(s)}{C_1^*}\right) = -\frac{1}{C_1^*}\frac{d}{ds}s\widetilde{C}_1(s)\Big|_{s=0}. \tag{14.5.26}$$

The next step is to evaluate $\widetilde{C}_1(s)$. Taking Laplace transforms of the second equation in (14.5.3) yields

$$s\widetilde{C}_1(s) = \frac{v+s}{s+k}\widetilde{u}_+(L,s),$$

where we have used the initial condition $u_\pm(L,0) = 0$. Substituting this and equation (14.5.16) into (14.5.26), and using

$$\lim_{s\to 0} s\widetilde{u}_+(L,s) = u_+(L,\infty) = \frac{1}{v_+}J(L,\infty) = \frac{k}{v_+}C_1^*,$$

we have

$$\tau_1 = \frac{1}{k} - \frac{v_+}{Q}\frac{d}{ds}\left. s\widetilde{u}_+(L,s)\right|_{s=0}. \tag{14.5.27}$$

In order to calculate $\widetilde{u}_+(x,s)$, we introduce the operators

$$\mathbb{L}_+ = \frac{1}{\alpha}(\partial_t + v_+\partial_x + \beta), \quad \mathbb{L}_- = \frac{1}{\beta}(\partial_t - v_-\partial_x + \alpha)u_-,$$

and set $\mathbb{L} = \mathbb{L}_-\mathbb{L}_+$. Equations (14.5.11a,b) for $u_\pm$ on $x \in [0,L]$ can be rewritten as

$$\mathbb{L}_+u_+ = u_-, \quad \mathbb{L}_-u_- = u_+. \tag{14.5.28}$$

Since $\mathbb{L}_+$ and $\mathbb{L}_-$ commute, it follows that $\mathbb{L}u_\pm = u_\pm$, which is a version of Telegrapher's equation. Imposing the initial conditions $u_\pm(x,0) = 0$ for $x \in [0,L]$, and taking Laplace transforms yields

$$\partial_x^2\widetilde{u}_\pm(x,s) - 2\left(s\delta - \frac{\gamma}{2}\right)\partial_x\widetilde{u}_\pm(x,s) - \frac{s(s+\alpha+\beta)}{v_+v_-}\widetilde{u}_\pm(x,s) = 0,$$

where $2\delta = 1/v_- - 1/v_+$. The corresponding general solution is given by

$$\widetilde{u}_\pm(x,s) = \left[A_\pm(s)e^{f(s)x} + B_\pm(s)e^{-f(s)x}\right]e^{(s\delta-\gamma/2)x}, \tag{14.5.29}$$

where

$$4f^2(s) = \left(\frac{1}{v_+} + \frac{1}{v_-}\right)^2 s^2 + 2\left(\frac{1}{v_+} + \frac{1}{v_-}\right)\left(\frac{\beta}{v_+} + \frac{\alpha}{v_-}\right)s + \gamma^2.$$

The coefficients $A_\pm(s), B_\pm(s)$ for fixed $s$ are calculated as follows. First, taking the time derivative of the boundary condition (14.5.12) at $x = 0$, substituting for $\dot{C}_0(t)$ using (14.5.13), and then Laplace transforming we find that

$$\left(\frac{s}{\kappa} + v_+\right)(A_+(s) + B_+(s)) = v_-(A_-(s) + B_-(s)) + \frac{Q}{s}.$$

Similarly, Laplace transforming the remaining boundary condition at $x = L$ gives

$$A_-(s)e^{Lf(s)} + B_-(s)e^{-Lf(s)} = 0.$$

Finally, Laplace transforming the first equation of (14.5.28), substituting for $\tilde{u}_\pm$ using (14.5.29), and comparing coefficients yields

$$g(s)A_+(s) = \alpha A_-(s), \quad h(s)B_+(s) = \alpha B_-(s),$$

where

$$g(s) = (1 + v_+\delta)s + \left(\beta - \frac{v_+\gamma}{2}\right) + v_+ f(s),$$

$$h(s) = (1 + v_+\delta)s + \left(\beta - \frac{v_+\gamma}{2}\right) - v_+ f(s).$$

We now have four conditions and four unknowns. After some algebra, we can solve for $A_+(s)$ and $B_+(s)$ and thus determine $u_+(L,s)$. Substituting the result into equation (14.5.26) thus yields the following expression for the accumulation time [124]:

$$\tau_1 = \frac{1}{k} + 2\kappa v_+^2 e^{-\gamma L/2} \frac{d}{ds} \left.\frac{e^{s\delta L} f(s)}{a(s)q(s)}\right|_{s=0}, \tag{14.5.30}$$

where

$$a(s) = h(s)e^{-Lf(s)} - g(s)e^{Lf(s)},$$

and

$$q(s) = s + \kappa v_+ - \frac{\kappa v_-}{\alpha} g(s)h(s)\left[e^{-Lf(s)} - e^{Lf(s)}\right]\frac{1}{a(s)}.$$

Example plots of $\tau_1$ as a function of cytoneme length $L$ are depicted in Fig. 14.45 for different $\gamma$. The corresponding concentration profile $C_1(t)$ in the case $\gamma = -0.5$ is also shown. It can be seen that $\tau_1$ is an increasing function of cytoneme length and also increases with $\gamma$. The latter means that increasing the level of retrograde flow increases the accumulation time.

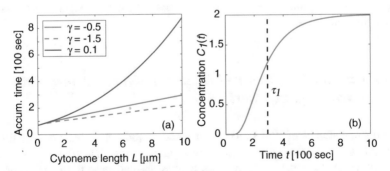

Fig. 14.45: Accumulation time $\tau_1$ for a single target cell. (a) Plot of the accumulation time as a function of cytoneme length for various values of $\gamma$. (b) Plot of the approach of $C_1(t)$ to the steady state $C_1^*$ for $\gamma = -0.5$. Other parameter values are as in Fig. 14.42.

Fig. 14.46: Two forms of contact between a cytoneme tip and a target cell. (a) Direct coupling. (b) Indirect synaptic coupling.

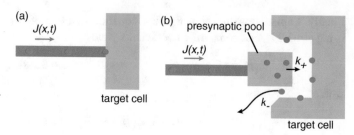

**Synaptic coupling model.** It is also possible to modify the boundary conditions of the transport model in order to take into account other possible forms of contact between the cytoneme tip and a target cell [508]. For example, in Fig. 14.46 we compare direct contact, as modeled above, with an indirect form of synaptic contact. In the latter case, it is assumed that a cytoneme delivers vesicles of morphogen to a presynaptic vesicular pool. The vesicles are then released into the synaptic cleft via exocytosis, and subsequently internalized by the target cell via endocytosis. This requires modifying the boundary condition at the cytoneme tip. Let $B_1(t)$ and $C_1(t)$ be the concentrations of vesicles in the presynaptic and postsynaptic domains, respectively. Equation (14.5.3) becomes

$$\frac{dC_0}{dt} = Q - J(0,t), \quad \frac{dB_1}{dt} = J(L,t) - k_+ B_1(t), \quad \frac{dC_1}{dt} = -k_- C_1(t) + k_+ B_1(t),$$
(14.5.31)

where $k_+$ is the rate of release of presynaptic vesicles into the synaptic cleft, and $k_-$ is an effective degradation rate of postsynaptic vesicles, which could be due to failure to be endocytosed. One way to modify equation (14.5.2) is to take

$$u(0,t) = \kappa C_0(t), \quad u(L,t) = \widehat{\kappa} B_1(t).$$
(14.5.32)

That is, the likelihood of vesicles returning to the source cell is assumed to depend on the concentration of vesicles in the presynaptic pool. A comparison of direct and synaptic forms of contact suggests that the latter can support longer range concentration gradients [508].

### 14.5.2 Directional search-and-capture model of cytoneme-based morphogen transport

Experimental studies of Wnt signaling in zebrafish [840, 841] indicate that Wnt is clustered at the membrane tip of growing signaling filopodia. When the filopodia make contact with target cells, the morphogens are delivered to the cells and the filopodia are pruned off within 10 minutes of making contact. Hence, the amount of morphogen delivered to a cell will depend on the rate of filopodia growth, the concentration of morphogen at the tips, and the frequency of contacts between source and target cells. One way to model this process is to adapt the so-called "search-and-capture" model of cell mitosis, which was described in detail in Sect. 14.3. In particular, consider a single cytoneme, which nucleates from a source cell and dynamically grows and shrinks until making contact with a target cell [126]. Suppose that morphogen is localized at the tip of the growing cytoneme, which is delivered as a "morphogen burst" when the cytoneme makes temporary contact with the target cell before subsequently retracting. The mean time to capture by a target cell can be calculated using the renewal approach of Sect. 7.5. Moreover, it can be shown how multiple rounds of search-and-capture, morphogen delivery, cytoneme retraction and nucleation events lead to the formation of a morphogen gradient. This is achieved by formulating the morphogen bursting model as a queuing process [126], analogous to the study of transcriptional bursting in gene networks [531] (Sect. 5.4), and active search processes (Sect. 7.4). Here we consider a simpler version of the search-and-capture model of cytoneme-based morphogenesis, which involves stochastic resetting with delays [133].

Consider a source cell with a single nucleation site from which a cytoneme grows toward one of $N$ target cells, $i = 1, \ldots, N$, that are distributed in some bounded domain $\Omega \subset \mathbb{R}^d$, see Fig. 14.47. We will assume that the cytoneme nucleates in a random direction such that the probability of being oriented toward the $i$th target cell at position $\mathbf{x}_i \in \Omega \subset \mathbb{R}^d$ is given by a fixed value $p_i$. (The source cell is taken to be at the origin.) The specific value of $p_i$ will depend on the angle subtended by the $i$th target cell with respect to the source cell, which itself will depend on the cell size and its distance from the source cell. Given that there is a nonzero probability of the cytoneme searching in a "wrong" direction, that is, in a direction without a target, we take $p_{\text{tot}} = \sum_{i=1}^{N} p_i < 1$. Once a cytoneme has nucleated in a particular direction, it grows according to a dynamical process with stochastic resetting. (The general theory of stochastic resetting is developed in Sect. 7.5.) Taking one end of the cytoneme to be fixed at $x = 0$ (the nucleation site), the other end is represented by a stochastic variable $X(t)$, which can also be identified as the length of the cytoneme (ignoring curvature). Suppose that the cytoneme can exist in one of two discrete states: a right-moving (anterograde) state with speed $v_+$ or a left-moving (retrograde) state with speed $v_-$. The cytoneme undergoes the state transition $v_+ \rightarrow v_-$ at a resetting rate $r$,

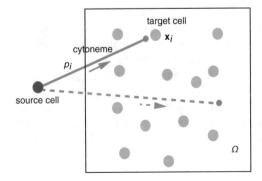

Fig. 14.47: Schematic diagram of a single cytoneme nucleating from a source cell and penetrating a domain $\Omega \subset \mathbb{R}^d$ consisting of $N$ target cells labeled $i = 1, \ldots, N$. The center of the $i$-th target cell is at $\mathbf{x}_i \in \Omega$. Each time a new cytoneme nucleates, it grows in a random direction such that the probability of being oriented towards the $i$-th target cell is $p_i$ with $p_{\text{tot}} = \sum_{i=1}^{N} p_i < 1$. The dashed line represents an orientation in which the cytoneme fails to find a target cell.

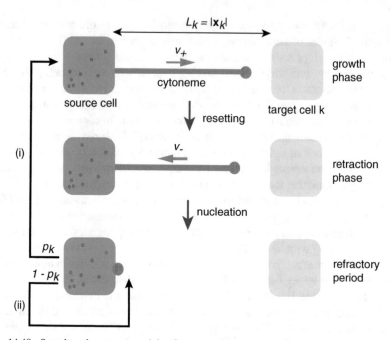

Fig. 14.48: Search-and-capture model of cytoneme-based morphogen transport. A single cytoneme nucleates from a source cell and grows at a speed $v_+$. Each time it nucleates, the cytoneme is oriented towards the $k$th target cell with probability $p_k$ (pathway (i)) or oriented towards another target cell with probability $1 - p_k$ (pathway (ii)). Prior to finding a target, the cytoneme may randomly switch to a retraction phase (resetting) and return to the origin at speed $v_-$. After a refractory period (nucleation waiting time), a new cytoneme starts to grow and the process repeats. The distance of the target cell from the source cell is $L_k = |\mathbf{x}_k|$.

after which it returns to the origin. At the origin the particle enters a refractory state for an exponentially distributed waiting time with rate $\eta$, prior to re-entering the anterograde state at a new, randomly selected orientation. The different stages of the search process for the $k$th target cell are shown in Fig. 14.48.

We are interested in how the above search process generates a morphogen gradient in $\Omega$, under the assumption that each time the cytoneme reaches a target cell, it delivers a package of morphogen and then retracts back to the origin. A new cytoneme loaded with morphogen then nucleates from the source cell and another round of search-and-capture is initiated. If we assume that morphogen degrades at a fixed rate within each target cell, then the distribution of morphogen across the set of target cells will reach a steady state that determines the morphogen gradient. The latter will depend on the resetting rate $r$, the speeds $v_\pm$, the positions $\mathbf{x}_i$ and the corresponding probabilities $p_i$. We will develop the analysis in stages, starting with a single target, then a single search-and-capture event for multiple targets, and finally multiple search-and-capture events.

**Mean first passage time for a single target.** Let us begin by considering a single target cell $(N = 1)$ at a distance $L$ from the source cell. Suppose that each time the cytoneme nucleates it grows toward the target cell with probability $p < 1$ and grows in the wrong direction with probability $1 - p$. (This is a simplified representation of the effects of other targets or failure to find a target.) In both cases, the cytoneme may switch to a shrinking phase at a resetting rate $r$ and return to the origin (nucleation site) at a speed $v_-$. It is convenient to partition the set of cytoneme states $\Sigma$ according to $\Sigma = \mathcal{N} \cup A \cup \overline{A}$, where $\mathcal{N}$ is the nucleation state, $A$ are the growing/shrinking states oriented in the direction of the target, and $\overline{A}$ are the corresponding states when the cytoneme is oriented in another direction. We denote the state of the cytoneme at time $t$ by $K(t)$. During each growth/shrinkage phase, let $p_n(x,t)$ be the probability density that at time $t$ the cytoneme tip is at a distance $X(t) = x$ from the nucleation site and in either the anterograde state $(n = +)$ or the retrograde state $(n = -)$. Similarly, let $P_0(t)$ denote the probability that the particle is in the refractory state at time $t$. If $K(t) \in A$ then $X(t) \in (0, L)$ evolves according to the Chapman-Kolmogorov (CK) equation

$$\frac{\partial p_+}{\partial t} = -v_+ \frac{\partial p_+}{\partial x} - r p_+ \quad x \in (0, L), \tag{14.5.33a}$$

$$\frac{\partial p_-}{\partial t} = v_- \frac{\partial p_-}{\partial x} + r p_+, \tag{14.5.33b}$$

$$\frac{dP_0}{dt} = v_- p_-(0,t) - \eta P_0(t), \tag{14.5.33c}$$

together with the boundary conditions

$$v_+ p_+(0,t) = \eta P_0(t), \quad p_-(L,t) = 0. \tag{14.5.33d}$$

If the cytoneme hits $x = 0$ $(x = L)$ first then it enters the state $\mathcal{N}$ (is captured by the target cell). On the other hand, if $K(t) \in \overline{A}$ then one simply waits a time $\sigma$ before the cytoneme returns to the state $\mathcal{N}$. The waiting time $\sigma$ is a random variable with mean $\mathbb{E}[\sigma] = \bar{\sigma}$. We will assume that the cytoneme keeps growing until it resets so that

$$\bar{\sigma} = \frac{1}{r}\left(1 + \frac{v_+}{v_-}\right). \tag{14.5.34}$$

(For simplicity, we ignore the possibility that the cytoneme hits another target and then retracts.) Finally, if $K(t) = \mathcal{N}$ then the cytoneme transitions to a growing state, which either belongs to $A$ with probability $p$ or belongs to $\overline{A}$ with probability $1 - p$. The time $\tau$ spent in state $\mathcal{N}$ is exponentially distributed with mean time $\eta^{-1}$. A schematic diagram of the different states is shown in Fig. 14.49.

Note that the nucleation boundary condition at $x = 0$ is mathematically identical to the so-called sticky boundary condition used in models of bidirectional transport and microtubular catastrophes [126, 358, 646], see also Sect. 4.2 and Sect. 10.4. The absorbing boundary condition at $x = L$ means that once the cytoneme reaches its target, it delivers its packet of morphogen and the search process ends. We also impose the initial conditions $P_0(0) = 0$, $p_n(x,0) = \delta_{n,+}\delta(x)$. That is, the cytoneme

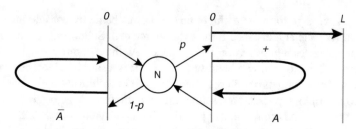

Fig. 14.49: Schematic illustration of different cytoneme states when $p < 1$.

starts in the growth phase at $x = 0$. Summing equations (14.5.33a) and (14.5.33b) and then integrating with respect to $x$ over the interval $[0, L]$ shows that

$$\frac{d}{dt} \int_0^L p(x,t)dx = -[v_+ p_+(x,t) - v_- p_-(x,t)]\big|_0^L,$$

where $p = p_+ + p_-$. Given the boundary conditions (14.5.33d), it follows that

$$\frac{d}{dt} \int_0^L p(x,t)dx + \frac{dP_0}{dt} = -v_+ p_+(L,t) \equiv -J(t),$$

where $J(t)$ is the probability flux into the target. Introducing the survival probability

$$Q(t) = \int_0^L p(x,t)dx + P_0(t), \qquad (14.5.35)$$

we see that the first passage time density $f(t)$ is

$$f(t) = -\frac{dQ(t)}{dt} = J(t).$$

Hence, the mean first passage time $T$ for the cytoneme to be captured by the target, assuming that it starts at $x = 0$ is given by

$$T = -\int_0^\infty t \frac{dQ(t)}{dt} dt = \int_0^\infty Q(t)dt. \qquad (14.5.36)$$

We now calculate the MFPT to find the target in the two cases $p = 1$ and $p < 1$.

**Case $p = 1$.** When $p = 1$ (zero probability of failure) the above model can be mapped onto a stochastic resetting process for a particle with a refractory (nucleation) period and a finite return time [133]. This is illustrated in Fig. 14.50, which shows a sample particle trajectory prior to capture by the target at $x = L$. We can thus use renewal theory to determine the MFPT $T$ in terms of the Laplace transform of the survival probability without reset, which we denote by $Q_0(t)$. For the given system, the latter is defined according to $Q_0(t) = \int_0^L p_+(x,t)dx$, with

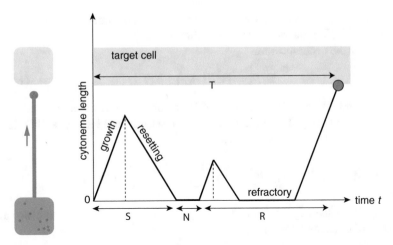

Fig. 14.50: Mapping of cytoneme search-and-capture to a first passage time problem for a particle under stochastic resetting. Resetting events are Poissonian with a rate $r$. The nucleation intervals are generated by an exponential waiting time density $\psi(\tau) = \eta e^{-\eta \tau}$ with rate $\eta$. There is an absorbing boundary at $x = L$. Also shown is the decomposition of the first passage time according to the sum $\mathcal{T} = \mathcal{S} + \mathcal{N} + \mathcal{R}$.

$$\frac{\partial p_+}{\partial t} = -v_+ \frac{\partial p_+}{\partial x} \quad x \in (0, L). \tag{14.5.37}$$

This has the solution $p_+(x,t) = \delta(x - v_+ t)$, which means that $Q_0(t) = H(L/v_+ - t)$ where $H$ is the Heaviside function. Moreover, the Laplace transform is

$$\tilde{Q}_0(s) = \int_0^\infty e^{-st} H(L/v_+ - t) dt = \int_0^{L/v_+} e^{-st} dt = \frac{1}{s}\left(1 - e^{-sL/v_+}\right). \tag{14.5.38}$$

We will use the particular renewal method of Sect 7.5 and Refs. [53, 130, 704, 705]. This exploits the fact that resetting eliminates any memory of previous search stages. Consider the following set of first passage times;

$$\begin{aligned}
\mathcal{T} &= \inf\{t > 0; X(t) = L\}, \\
\mathcal{S} &= \inf\{t > 0; X(t) = 0\}, \\
\mathcal{R} &= \inf\{t > 0; X(t + \mathcal{S} + \tau) = L\}.
\end{aligned} \tag{14.5.39}$$

Here $\mathcal{T}$ is the FPT for finding the target irrespective of the number of resettings, $\mathcal{S}$ is the FPT for the first resetting and return to the origin, $\tau$ is the first refractory period, and $\mathcal{R}$ is the FPT for finding the target given that at least one resetting has occurred. Next we introduce the sets $\Omega = \{\mathcal{T} < \infty\}$ and $\Gamma = \{\mathcal{S} < \mathcal{T} < \infty\} \subset \Omega$. That is, $\Omega$ is the set of all events for which the particle is eventually absorbed by the target (which has measure one), and $\Gamma$ is the subset of events in $\Omega$ for which the particle resets at least once. It immediately follows that $\Omega \backslash \Gamma = \{\mathcal{T} < \mathcal{S} = \infty\}$. In other words, $\Omega \backslash \Gamma$ is the set of all events for which the particle is captured by the target without any resetting. We now use a probabilistic argument to calculate the MFPT $T = \mathbb{E}[\mathcal{T}]$ in the presence of resetting $(r > 0)$.

Consider the decomposition

$$\mathbb{E}[\mathcal{T}] = \mathbb{E}[\mathcal{T}1_{\Omega\backslash\Gamma}] + \mathbb{E}[\mathcal{T}1_{\Gamma}]. \tag{14.5.40}$$

The first expectation on the right-hand side can be evaluated by noting that it is the MFPT for capture by the target without any resetting, and the probability density for such an event is $-e^{-rt}\partial_t Q_0(t)$. Hence,

$$\mathbb{E}[\mathcal{T}1_{\Omega\backslash\Gamma}] = -\int_0^\infty te^{-rt}\frac{dQ_0(t)}{dt}dt = \left(1 + r\frac{d}{dr}\right)\tilde{Q}_0(r), \tag{14.5.41}$$

where $\tilde{Q}_0(r)$ is the Laplace transform of $Q_0(t)$ with $s = r$. The second expectation can be further decomposed as

$$\mathbb{E}[\mathcal{T}1_{\Gamma}] = \mathbb{E}[(\mathcal{S} + \tau + \mathcal{R})1_{\Gamma}] = \mathbb{E}[\mathcal{S}1_{\Gamma}] + \overline{\tau}\mathbb{P}[\Gamma] + \mathbb{E}[\mathcal{R}1_{\Gamma}]$$
$$= \mathbb{E}[\mathcal{S}1_{\Gamma}] + (\overline{\tau} + T)\mathbb{P}[\Gamma]. \tag{14.5.42}$$

Here $\mathbb{E}[\tau] = \overline{\tau} = \eta^{-1}$ is the mean refractory period, and we have used the result $\mathbb{E}[\mathcal{R}1_{\Gamma}] = T P[\Gamma]$. The latter follows from the fact that return to the origin restarts the stochastic process without any memory.

In order to calculate $\mathbb{E}[\mathcal{S}1_{\Gamma}]$, it is necessary to incorporate the time to return to the origin following the first return event. The first resetting occurs with probability $re^{-rt}Q_0(t)dt$ in the interval $[t, t + dt]$. At time $t$ the particle is at position $v_+t$ and thus takes an additional time $v_+t/v_-$ to return to $x = 0$. We thus find

$$\mathbb{E}[\mathcal{S}1_{\Gamma}] = \int_0^\infty re^{-rt}\left(1 + \frac{v_+}{v_-}\right)Q_0(t)dt = -r\left(1 + \frac{v_+}{v_-}\right)\frac{d}{dr}\tilde{Q}_0(r). \tag{14.5.43}$$

Moreover, from the definitions of the first passage times and the effect of resetting,

$$\mathbb{P}[\Gamma] = \mathbb{P}[\mathcal{S} < \infty]\mathbb{P}[\mathcal{R} < \infty], \tag{14.5.44}$$

with $\mathbb{P}[\mathcal{R} < \infty] = 1$ and

$$\mathbb{P}[\mathcal{S} < \infty] = \int_0^\infty re^{-rt}Q_0(t)dt = r\tilde{Q}_0(r). \tag{14.5.45}$$

Combining equations (14.5.41)–(14.5.45) yields the implicit equation

$$T = \left(1 + r\frac{d}{dr}\right)\tilde{Q}_0(r) + r\overline{\tau}\tilde{Q}_0(r) - r\left(1 + \frac{v_+}{v_-}\right)\frac{d}{dr}\tilde{Q}_0(r) + r\tilde{Q}_0(r)T. \tag{14.5.46}$$

Rearranging this equation yields the result

$$T = \frac{\widetilde{Q}_0(r) + r\overline{\tau}\widetilde{Q}_0(r) - r\frac{v_+}{v_-}\widetilde{Q}'_0(r),}{1 - r\widetilde{Q}_0(r)}. \tag{14.5.47}$$

Note that this is a general formula for a dynamical process with stochastic resetting, finite return times, and refractory periods [53, 130, 704, 705]. For our particular model, we substitute equation (14.5.38) into equation (14.5.47) to give

$$T = T(L) := \frac{1}{r}\left[(e^{rL/v_+} - 1)(1 + r\overline{\tau} + v_+/v_-) - \frac{rL}{v_-}\right]. \tag{14.5.48}$$

In the limit $r \to 0$, $T(L) \to L/v_+$, which is simply the deterministic time for the cytoneme tip to travel the distance $L$.

**Case $p < 1$.** In order to include the effects of failure in the absence of resetting, it is necessary to generalize the FPTs defined by equation (14.5.39). Let $I(t)$ denote the number of resettings in the interval $[0,t]$. Assuming that the cytoneme starts out in the growing state, we set

$$\begin{aligned}
\mathcal{T} &= \inf\{t \geq 0; X(t) = L,\, I(t) \geq 0\}, \\
\mathcal{T}_A &= \inf\{t \geq 0; X(t) = L,\, I(t) \geq 0 | K(0) \in A\}, \\
\mathcal{S}_A &= \inf\{t \geq 0; X(t) = 0,\, I(t) = 1 | K(0) \in A\}, \\
\mathcal{R}_{\overline{A}} &= \inf\{t \geq 0; X(t + \tau + \sigma) = L,\, I(t + \tau + \sigma) \geq 1 | K(0) \in \overline{A}\}, \\
\mathcal{R}_A &= \inf\{t \geq 0; X(t + \tau + \mathcal{S}_A) = L,\, I(t + \tau + \mathcal{S}_A) \geq 1 | K(0) \in A\}.
\end{aligned} \tag{14.5.49}$$

These are the natural extensions of the FPTs defined in equation (14.5.39), which keep track of whether or not the nucleating cytoneme is oriented toward the given target. Next we define the sets

$$\begin{aligned}
\Omega &= \{\mathcal{T} < \infty\}, \quad \Omega_A = \{\mathcal{T} < \infty\} \cap \{K(0) = A\} \subset \Omega, \\
\Gamma &= \{\mathcal{S}_A < \mathcal{T}_A < \infty\} \subset \Omega_A, \quad \overline{\Gamma} = \{\mathcal{T} < \infty\} \cap \{K(0) = \overline{A}\} \subset \Omega, \tag{14.5.50}
\end{aligned}$$

where $\Omega$ is the set of all events for which the cytoneme is eventually absorbed by the target ($\mathbb{P}[\Omega] = 1$), $\Omega_A$ is the subset of events conditioned on starting in the state $A$, and $\Gamma$ ($\overline{\Gamma}$) is the subset of events in $\Omega_A$ ($\Omega$) conditioned on starting in the state $A$ ($\overline{A}$) and having at least one resetting event. It follows that $\Omega = \Omega_A \cup \overline{\Gamma}$, and

$$\Omega_A \backslash \Gamma = \{\mathcal{T} < \mathcal{S}_A = \infty\},$$

where $\Omega_A \backslash \Gamma$ is the set of all events for which the cytoneme is captured by the target without any resettings. Also note that

$$\mathbb{P}[\Omega] = p\mathbb{P}[\Omega_A] + (1-p)\mathbb{P}[\overline{\Gamma}] = p + 1 - p = 1,$$

that is, $\mathbb{P}[\Omega_A] = 1 = \mathbb{P}[\overline{\Gamma}]$.

For $p < 1$ we have

$$
\begin{aligned}
T_p := \mathbb{E}[\mathcal{T}] &= p\mathbb{E}[\mathcal{T}_A 1_{\Omega_A}] + (1-p)\mathbb{E}[(\mathcal{R}_{\overline{A}} + \tau + \sigma)1_{\overline{\Gamma}}] \\
&= p\mathbb{E}[\mathcal{T}_A 1_{\Omega_A}] + (1-p)(\bar{\tau} + \bar{\sigma} + \mathbb{E}[\mathcal{R}_{\overline{A}} 1_{\overline{\Gamma}}]) \\
&= (1-p)(\bar{\sigma} + \bar{\tau}) + p\mathbb{E}[\mathcal{T}_A 1_{\Omega_A}] + (1-p)\mathbb{E}[\mathcal{T}],
\end{aligned}
\tag{14.5.51}
$$

where $\bar{\sigma} = r^{-1}(1 + v_+/v_-)$, and we have again used the fact that return to the origin restarts the stochastic process without any memory so $\mathbb{E}[\mathcal{R}_{\overline{A}}] = \mathbb{E}[\mathcal{T}]$. Rearranging,

$$
T_p = \frac{(1-p)[\bar{\tau} + \bar{\sigma}]}{p} + \mathbb{E}[\mathcal{T}_A 1_{\Omega_A}].
\tag{14.5.52}
$$

The analysis of $\mathbb{E}[\mathcal{T}_A 1_{\Omega_A}]$ proceeds along similar lines to the case $p = 1$ by performing the decomposition

$$
\begin{aligned}
\mathbb{E}[\mathcal{T}_A 1_{\Omega_A}] &= \mathbb{E}[\mathcal{T}_A 1_{\Omega_A \setminus \Gamma}] + \mathbb{E}[\mathcal{T}_A 1_\Gamma] = \mathbb{E}[\mathcal{T}_A 1_{\Omega_A \setminus \Gamma}] + \mathbb{E}[(\mathcal{S}_A + \tau + \mathcal{R})1_\Gamma] \\
&= \mathbb{E}[\mathcal{T}_A 1_{\Omega_A \setminus \Gamma}] + \mathbb{E}[\mathcal{S}_A 1_\Gamma] + (\bar{\tau} + T_p)\mathbb{P}[\Gamma],
\end{aligned}
\tag{14.5.53}
$$

with $\mathbb{P}[\Gamma] = r\widetilde{Q}_0(r)$, see equations (14.5.44) and (14.5.45). The term $\mathbb{E}[\mathcal{T}_A 1_{\Omega_A \setminus \Gamma}]$ is the MFPT for capture by the target without any resetting, given that the cytoneme is oriented in the correct direction, and is thus given by equation (14.5.41):

$$
\mathbb{E}[\mathcal{T}_A 1_{\Omega_A \setminus \Gamma}] = \left(1 + r\frac{d}{dr}\right)\widetilde{Q}_0(r).
\tag{14.5.54}
$$

Similarly, $\mathbb{E}[\mathcal{S}_A 1_\Gamma]$ is given by equation (14.5.43):

$$
\mathbb{E}[\mathcal{S}_A 1_\Gamma] = -r\left(1 + \frac{v_+}{v_-}\right)\frac{d}{dr}\widetilde{Q}_0(r).
\tag{14.5.55}
$$

Finally, combining equations (14.5.52)–(14.5.55) yields the implicit equation

$$
\begin{aligned}
T_p &= \frac{(1-p)[\bar{\tau} + \bar{\sigma}]}{p} + \left(1 + r\frac{d}{dr}\right)\widetilde{Q}_0(r) + r\widetilde{Q}_0(r)(\bar{\tau} + T_p)) \\
&\quad - r\left(1 + \frac{v_+}{v_-}\right)\frac{d}{dr}\widetilde{Q}_0(r).
\end{aligned}
\tag{14.5.56}
$$

Rearranging this equation yields the result

$$
T_p = \frac{1}{1 - r\widetilde{Q}_0(r)}\left\{\frac{(1-p)[\bar{\tau} + \bar{\sigma}]}{p} + \widetilde{Q}_0(r) + r\bar{\tau}\widetilde{Q}_0(r) - r\frac{v_+}{v_-}\widetilde{Q}_0'(r)\right\}.
\tag{14.5.57}
$$

The first term in $\{\cdot\}$ has a simple interpretation, as can be seen by noting that

$$
\frac{(1-p)[\bar{\tau} + \bar{\sigma}]}{p} = \frac{(1-p)[\bar{\tau} + \bar{\sigma}]}{1 - (1-p)} = [\bar{\tau} + \bar{\sigma}]\sum_{m=1}^{\infty}(1-p)^m.
$$

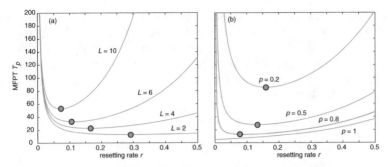

Fig. 14.51: Plots of MFPT $T_p(L)$ as a function of the resetting rate $r$. (a) Various cytoneme lengths $L$ and $p = 0.5$. (b) Various probabilities $p$ and $L = 5$. Dimensionless units with $v_+ = 1, \bar{\tau} = 1, v_- = 5$. Optimal resetting rates are indicated by the filled circles.

That is, each time there is an excursion in the wrong direction, which occurs with probability $(1 - p)$, an additional mean-time penalty of $\bar{\tau} + \bar{\sigma}$ is incurred. Finally, substituting for $\tilde{Q}_0(r)$ using equation (14.5.38) and setting $\bar{\sigma} = r^{-1}(1 + v_+/v_-)$, we obtain the result

$$T_p = T_p(L) := \frac{(1 - p)[\bar{\tau} + r^{-1}(1 + v_+/v_-)]}{p} e^{rL/v_+} + T(L), \qquad (14.5.58)$$

with $T(L)$ given by equation (14.5.47).

It is clear from equation (14.5.58) that $T_p(L) \to \infty$ as $r \to 0$, since $\bar{\sigma} \to \infty$, which is a consequence of the possible failure of the search process in the absence of resetting. This suggests that there exists an optimal resetting rate at which the MFPT is minimized, which is indeed found to be the case. In Fig. 14.51 we show plots of $T_p$ as a function of the resetting rate for various distances $L$ and probabilities $p$. It can be seen that the optimal resetting rate $r_{\text{opt}}$ at which each curve has a minimum is an increasing function of $L$ and a decreasing function of $p$, with $r_{\text{opt}} \to 0$ as $p \to 1$.

### 14.5.3 Splitting probabilities and conditional MFPTs for multiple targets

We now extend the analysis of Sect. 16.3.2 in order to calculate the splitting probability $\pi_j$ and conditional MFPT $T_j$ to be captured by the $j$th target cell, $j = 1, \ldots, N$, without previously being captured by any other target. The set of cytoneme states is partitioned as $\Sigma = \mathcal{N} \cup \overline{A} \cup_{j=1}^{N} A_j$, where $\mathcal{N}$ is the nucleation state, $A_j$ are the growing/shrinking states oriented in the direction of the $j$th target for $j = 1, \ldots, N$, and $\overline{A}$ is any state oriented away from all of the targets. Let $K(t)$ denote the state of the cytoneme at time $t$. If $K(t) \in A_j$ then $X(t) \in (0, L_j)$ evolves according to equation (14.5.33) with $L = L_j$. If the cytoneme hits $x = 0$ first then it enters the state

$\mathcal{N}$, otherwise it is captured by the $j$th target cell. If $K(t) = \mathcal{N}$ then the cytoneme transitions to a new growing state in one of the states $A_k$ with probability $p_k$ or the state $\overline{A}$ with probability $\bar{p} = 1 - \sum_{i=1}^{N} p_i$, after a waiting time $\tau$. Again let $I(t)$ denote the number of resettings in the interval $[0, t]$. Assuming that the cytoneme starts out in the growing state, we introduce a set of FPTs, which are the multi-target analogs of equation (14.5.49):

$$
\begin{aligned}
\mathcal{T}_j &= \inf\{t \geq 0; X(t) = L_j, \, I(t) \geq 0\}, \\
\widehat{\mathcal{T}}_j &= \inf\{t \geq 0; X(t) = L_j, \, I(t) \geq 0 | K(0) = A_j\}, \\
\mathcal{S}_j &= \inf\{t \geq 0; X(t) = 0, \, I(t) = 1 | K(0) \in A_j\}, \\
\mathcal{R}_{ji} &= \inf\{t \geq 0; X(t + \tau + \mathcal{S}_i) = L_j, \, I(t + \tau + \mathcal{S}_i) \geq 1 | K(0) \in A_i\}, \\
\overline{\mathcal{R}}_j &= \inf\{t \geq 0; X(t + \tau + \sigma) = L_j, \, I(t + \tau + \sigma) \geq 1 | K(0) \in \overline{A}\}.
\end{aligned}
\tag{14.5.59}
$$

Here $\mathcal{T}_j$ is the FPT for finding the $j$th target irrespective of the number of resettings, $\widehat{\mathcal{T}}_j$ is the corresponding FPT conditioned on starting in the state $A_j$, $\mathcal{S}_j$ is the FPT for the first resetting and return to the origin starting from the state $A_j$, $\mathcal{R}_{ji}$ is the FPT for finding the $j$th target after at least one resetting conditioned on starting from the state $A_i$, and $\overline{\mathcal{R}}_j$ is the analogous FPT starting from the failure state $\overline{A}$. Next we define the sets

$$
\begin{aligned}
\Omega_j &= \{\mathcal{T}_j < \infty\}, \quad \Omega_{ji} = \{\mathcal{T}_j < \infty\} \cap \{K(0) = A_i\} \subset \Omega_j, \\
\Gamma_j &= \{\mathcal{S}_j < \mathcal{T}_j < \infty\} \subset \Omega_{jj}, \quad \overline{\Gamma}_j = \{\mathcal{T}_j < \infty\} \cap \{K(0) = \overline{A}\} \subset \Omega_j,
\end{aligned}
\tag{14.5.60}
$$

where $\Omega_j$ is the set of all events for which the cytoneme is eventually absorbed by the $j$th target cell without being absorbed by any other target, $\Omega_{ji}$ ($\overline{\Gamma}_j$) is the subset of events in $\Omega_j$ conditioned on starting in the state $A_i$ ($\overline{A}$), and $\Gamma_j$ is the subset of events in $\Omega_{jj}$ that reset at least once. It follows that $\Omega_j = \cup_{i=1}^{N} \Omega_{ji} \cup \overline{\Gamma}_j$, and

$$
\Omega_{jj} \backslash \Gamma_j = \{\mathcal{T}_j < \mathcal{S}_j = \infty\},
$$

where $\Omega_{jj} \backslash \Gamma_j$ is the set of all events for which the cytoneme is captured by the $j$th target without any resettings. During each search phase directed towards the $j$th target, we denote the survival probability without resetting by $Q_j(t)$, whose Laplace transform is

$$
\widetilde{Q}_j(s) = \frac{1}{r}\left(1 - e^{-sL_j/v_+}\right).
\tag{14.5.61}
$$

The splitting probability $\pi_j$ can be decomposed as

$$
\pi_j := \mathbb{P}[\Omega_j] = p_j \mathbb{P}[\Omega_{jj}] + \sum_{i \neq j} p_i \mathbb{P}[\Omega_{ji}] + \overline{p}\mathbb{P}[\overline{\Gamma}_j],
\tag{14.5.62}
$$

We have the further decomposition

$$
\mathbb{P}[\Omega_{jj}] = \mathbb{P}[\Omega_{jj} \backslash \Gamma_j] + \mathbb{P}[\Gamma_j].
\tag{14.5.63}
$$

Let us consider the latter decomposition first. The probability that the cytoneme is captured by the $j$th target in the interval $[t, t+dt]$ without any resettings is

$$\mathbb{P}[\Omega_{jj}\backslash\Gamma_j] = -\int_0^\infty e^{-rt}\frac{dQ_j(t)}{dt}dt = \left(1 - r\widetilde{Q}_j(r)\right). \tag{14.5.64a}$$

Next, from the definitions of the first passage times, we have

$$\mathbb{P}[\Gamma_j] = \mathbb{P}[\mathcal{S}_j < \infty]\mathbb{P}[\mathcal{R}_{jj} < \infty] = \mathbb{P}[\mathcal{S}_j < \infty]\pi_j. \tag{14.5.64b}$$

We have used the renewal property of resetting to set $\mathbb{P}[\mathcal{R}_{jj} < \infty] = \pi_j$. The probability $\mathbb{P}[\mathcal{S}_j < \infty]$ is determined by noting that during a growth phase in the $j$th direction, we require that the cytoneme returns to the origin before reaching the target at $L_j$. The probability of first switching to the return phase in the time interval $[t, t+dt]$ is equal to the product of the reset probability $re^{-rt}dt$ and the survival probability $Q_j(t)$. Hence,

$$\mathbb{P}[\mathcal{S}_j < \infty] = \int_0^\infty re^{-rt}Q_j(t)dt = r\widetilde{Q}_j(r). \tag{14.5.64c}$$

Finally, turning to the decomposition (14.5.62), we have

$$\mathbb{P}[\Omega_{ji}] = \mathbb{P}[\mathcal{S}_i < \infty]\mathbb{P}[\mathcal{R}_{ji} < \infty] = r\widetilde{Q}_i(r)\pi_j, \tag{14.5.64d}$$

and

$$\mathbb{P}[\overline{\Gamma}_j] = \mathbb{P}[\overline{\mathcal{R}}_j < \infty] = \pi_j. \tag{14.5.64e}$$

Combining equations (14.5.64a)–(14.5.64e) yields the implicit equation

$$\pi_j = p_j\left(1 - r\widetilde{Q}_j(r)\right) + r\pi_j\sum_{l=1}^N p_l\widetilde{Q}_l(r) + \overline{p}\pi_j,$$

which on rearranging gives

$$\pi_j = \frac{p_j\left(1 - r\widetilde{Q}_j(r)\right)}{1 - r\sum_{l=1}^N p_l\widetilde{Q}_l(r) - \overline{p}} = \frac{p_j\left(1 - r\widetilde{Q}_j(r)\right)}{\sum_{l=1}^N p_l\left(1 - r\widetilde{Q}_l(r)\right)}. \tag{14.5.65}$$

Summing both sides of equation (14.5.65) with respect to $j$ implies that $\sum_{j=1}^N \pi_j = 1$. That is, in the presence of resetting, the probability of eventually finding a target is unity.

Similarly, we decompose the MFPT $\mathbb{E}[\mathcal{T}_j 1_{\Omega_j}] := \pi_j T_j$ as

$$\mathbb{E}[\mathcal{T}_j 1_{\Omega_j}] = p_j\mathbb{E}[\widehat{\mathcal{T}}_j 1_{\Omega_{jj}\backslash\Gamma_j}] + p_j\mathbb{E}[\widehat{\mathcal{T}}_j 1_{\Gamma_j}] + \sum_{i\neq j} p_i\mathbb{E}[\mathcal{T}_j 1_{\Omega_{ji}}] + \overline{p}\mathbb{E}[\mathcal{T}_j 1_{\overline{\Gamma}_j}].$$

$$\tag{14.5.66}$$

The first expectation can be evaluated by noting that it is the MFPT for capture by the $j$th target without any resetting. From equation (14.5.41), we thus have

$$\mathbb{E}[\widehat{\mathcal{T}}_j 1_{\Omega_{jj}\setminus\Gamma_j}] = \left[1 + r\frac{d}{dr}\right]\widetilde{Q}_j(r). \tag{14.5.67a}$$

The second expectation is further decomposed as

$$\mathbb{E}[\widehat{\mathcal{T}}_j 1_{\Gamma_j}] = \mathbb{E}[(\mathcal{S}_j + \tau + \mathcal{R}_{jj})1_{\Gamma_j}] = \mathbb{E}[\mathcal{S}_j 1_{\Gamma_j}] + \bar{\tau}\mathbb{P}[\Gamma_j] + \mathbb{E}[\mathcal{R}_{jj} 1_{\Gamma_j}]. \tag{14.5.67b}$$

In order to calculate $\mathbb{E}[\mathcal{S}_j 1_{\Gamma_j}]$, we need to calculate the mean time that the cytoneme returns to the origin before reaching the target at $L_j$. Following along analogous lines to equation (14.5.43), we have

$$\mathbb{E}[\mathcal{S}_j 1_{\Gamma_j}] = \mathbb{P}[\mathcal{R}_{jj} < \infty]\int_0^\infty re^{-rt}t\left(1 + \frac{v_+}{v_-}\right)Q_j(t)dt$$
$$= -r\pi_j\left(1 + \frac{v_+}{v_-}\right)\frac{d}{dr}\widetilde{Q}_j(r). \tag{14.5.67c}$$

Hence,

$$\mathbb{E}[\widehat{\mathcal{T}}_j 1_{\Gamma_j}] = -r\pi_j\left(1 + \frac{v_+}{v_-}\right)\frac{d}{dr}\widetilde{Q}_j(r) + r\widetilde{Q}_j(r)\pi_j(\bar{\tau} + T_j). \tag{14.5.67d}$$

In addition,

$$\mathbb{E}[\mathcal{T}_j 1_{\Omega_{ji}}] = \mathbb{E}[(\mathcal{S}_i + \tau + \mathcal{R}_{ji})1_{\Omega_{ji}}] = \mathbb{E}[\mathcal{S}_i 1_{\Omega_{ji}}] + \bar{\tau}\mathbb{P}[\Omega_{ji}] + \mathbb{E}[\mathcal{R}_{jj} 1_{\Omega_{ji}}] \tag{14.5.67e}$$

$$= -r\pi_i\left(1 + \frac{v_+}{v_-}\right)\frac{d}{dr}\widetilde{Q}_i(r) + r\widetilde{Q}_i(r)\pi_j(\bar{\tau} + T_i).$$

Finally,

$$\mathbb{E}[\mathcal{T}_j 1_{\overline{\Gamma}_j}] = \mathbb{E}[(\sigma + \tau + \overline{\mathcal{R}}_j)1_{\overline{\Gamma}_j}] = (\bar{\sigma} + \bar{\tau})\mathbb{P}[\overline{\Gamma}_j] + \mathbb{E}[\overline{\mathcal{R}}_j 1_{\overline{\Gamma}_j}] \tag{14.5.67f}$$
$$= \pi_j(\bar{\sigma} + \bar{\tau} + T_j).$$

Combining equations (14.5.67a)–(14.5.67f) yields the implicit equation

$$\pi_j T_j = p_j\left[1 + r\frac{d}{dr}\right]\widetilde{Q}_j(r) - r\pi_j\left(1 + \frac{v_+}{v_-}\right)\left[\sum_{l=1}^N p_l\frac{d}{dr}\widetilde{Q}_l(r)\right]$$
$$+ r(\bar{\tau} + T_j)\pi_j\left[\sum_{l=1}^N p_l\widetilde{Q}_l(r)\right] + \bar{p}\pi_j(\bar{\sigma} + \bar{\tau} + T_j), \tag{14.5.68}$$

which can be arranged to give the general result

$$\pi_j T_j = \left\{ \frac{1}{1 - r\sum_{l=1}^{N} p_l \widetilde{Q}_l(r) - \bar{p}} \right\} \left\{ p_j \left[ 1 + r\frac{d}{dr} \right] \widetilde{Q}_j(r) \right. \tag{14.5.69}$$

$$\left. + r\pi_j \sum_{l=1}^{N} p_l \left[ \bar{\tau}\widetilde{Q}_l(r) - \left(1 + \frac{v_+}{v_-}\right)\frac{d}{dr}\widetilde{Q}_l(r) \right] + \bar{p}\pi_j(\bar{\sigma} + \bar{\tau}) \right\}.$$

Substituting for $\widetilde{Q}_j(r)$ using equation (14.5.61), equations (14.5.65) and (14.5.69) become

$$\pi_j = \frac{p_j e^{-rL_j/v_+}}{\sum_{l=1}^{N} p_l e^{-rL_l/v_+}}, \tag{14.5.70}$$

and

$$\pi_j T_j = \left\{ \frac{1}{\sum_{l=1}^{N} p_l e^{-rL_l/v_+}} \right\} \left\{ \frac{p_j L_j}{v_+} e^{-rL_j/v_+} + \bar{p}\pi_j(\bar{\sigma} + \bar{\tau}) \right. \tag{14.5.71}$$

$$\left. + \pi_j \sum_{l=1}^{N} p_l \left[ \left( \bar{\tau} + \frac{1}{r}\left(1 + \frac{v_+}{v_-}\right) \right) \left(1 - e^{-rL_l/v_+}\right) - \frac{L_l}{v_+}\left(1 + \frac{v_+}{v_-}\right)e^{-rL_l/v_+} \right] \right\}.$$

As in the single target case, $T_j \to \infty$ in the limit $r \to 0$ due to the presence of the term $\bar{\sigma}$. Finally, summing both sides of equation (14.5.71) with respect to $j$ yields the unconditional MFPT

$$T = \sum_{j=1}^{N} \pi_j T_j = \left\{ \frac{1}{\sum_{l=1}^{N} p_l e^{-rL_l/v_+}} \right\} \left\{ \sum_{j=1}^{N} \frac{p_j L_j}{v_+} e^{-rL_j/v_+} + \bar{\tau}\left(1 - \sum_{l=1}^{N} p_l e^{-rL_l/v_+}\right) \right.$$

$$\left. + \frac{1}{r}\left(1 + \frac{v_+}{v_-}\right)\left[ \bar{p} + \sum_{l=1}^{N} p_l e^{-rL_l/v_+}\left( e^{rL_l/v_+} - \left[1 + \frac{rL_l}{v_+}\right] \right) \right] \right\}. \tag{14.5.72}$$

### 14.5.4 Multiple search-and-capture events and morphogen gradient formation

In order to determine the morphogen gradient, we now consider the statistics of morphogen accumulation in the target cells in response to multiple rounds of search-and-capture events. We assume that the build up of resources within each target is counterbalanced by degradation at a rate $\gamma$, so that there is a steady-state amount of morphogen in the long-time limit. The various stages are illustrated in Fig. 14.52, where morphogen localized at the tip of a growing cytoneme is delivered as a "morphogen burst" whenever the cytoneme makes temporary contact with a target cell before subsequently retracting. Furthermore, suppose that the total time for the particle to unload its cargo, return to the nucleation site and start a new search process is given by the random variable $\tau$ with waiting time density $\rho(\tau)$, which for simplic-

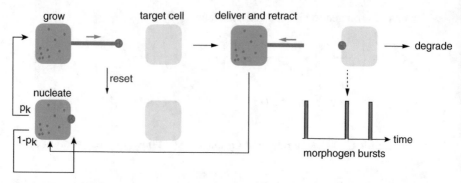

Fig. 14.52: Multiple search-and-capture events. Alternating periods of growth, shrinkage, nucle-ation and target capture generates a sequence of morphogen bursts in a given target cell that is analogous to the arrival of customers in a queuing model. This results in the accumulation of mor-phogen within the cell, which is the analog of a queue. Degradation corresponds to exiting of customers after being serviced by an infinite number of servers.

ity is taken to be independent of the location of the targets (under the assumption of instantaneous reset). Let $M_k(t)$ be the number of resource packets in the $k$th target that have not yet degraded at time $t$. As shown in Sect. 7.4 and Refs. [126, 133], the accumulation of resources within the targets can be analyzed by reformulating the multiple search-and-capture model as a G/M/$\infty$ queuing process. Here we simply quote the results for the steady-state mean and variance of $M_k$.

First, the mean is given by

$$\overline{M}_k = \frac{\pi_k}{\gamma[T + \tau_{\text{cap}}]}, \qquad (14.5.73)$$

where $\pi$ and $T$ are the splitting probabilities and unconditional MFPT of a sin-gle search-and-capture event, and $\tau_{\text{cap}} = \int_0^\infty \tau \rho(\tau) d\tau$ is the mean loading/unloading time. Hence, the relative distribution of resources across the population of compet-ing targets is determined by the splitting probabilities $\pi_k$, whereas the total number of resources depends on the unconditional MFPT $T$, that is,

$$\overline{M}_{\text{tot}} = \sum_{k=1}^{N} \overline{M}_k = \frac{1}{\gamma(T + \tau_{\text{cap}})}.$$

On the other hand, the variance (and higher-order moments of the resource distri-bution) depend on the full conditional FPT densities in Laplace space that were calculated in section 3. In particular, the variance takes the form

$$\text{Var}[M_k] = \overline{M}_k \left[ \frac{\pi_k \widetilde{\rho}(\gamma) \widetilde{f}_k(\gamma)}{1 - \sum_{j=1}^{N} \pi_j \widetilde{\rho}(\gamma) \widetilde{f}_j(\gamma)} + 1 - \overline{M}_k \right], \qquad (14.5.74)$$

where $\widetilde{f}_k(\gamma)$ are the conditional FPT densities.

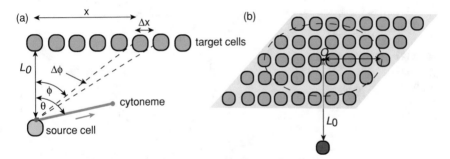

Fig. 14.53: (a) Single 1D layer of target cells and a single source cell that extends a cytoneme at a random orientation $\theta \in [0, \pi/2]$. (b) Single 2D layer of target cells.

**Example: single-layer of target cells.** Consider a single layer of target cells as shown in Fig. 14.53(a). We assume that there is a single source cell that extends a cytoneme in a random direction $\theta \in [0, \pi/2]$ in the plane as illustrated in the diagram. (For the sake of illustration, we take the maximum orientation to be $\pi/2$; however, this is not a necessary condition.) If the layer subtends an angle $\Theta < \pi/2$ with respect to the source cell, then the cytoneme will extend in the wrong direction whenever $\theta > \Theta$. We also take the target cells to be sufficiently close together so that there are no "gaps" between them. A simple trigonometric calculation can be used to estimate the probability that the cytoneme extends towards the $k$th target. Suppose that the $k$th target cell subtends an angle $\Delta \phi_k$ and that the distribution of cytoneme directions is uniform on $[0, \pi/2]$. It follows that $p_k \approx 2\Delta \phi_k / \pi$. Moreover, $\Delta \phi_{k+1} \approx \phi(x_k + \Delta x) - \phi(x_k)$, where $\tan \phi(x) = x/L_0$, $L_0$ is the perpendicular distance of the source cell from the target layer, and $x_k = k \Delta x$. That is,

$$\Delta \phi_{k+1} \approx \sec^{-2}(\phi_k) \frac{\Delta x}{L_0} = \left(1 + \frac{x_k^2}{L_0^2}\right)^{-1} \frac{\Delta x}{L_0}.$$

Hence

$$p_{k+1} \approx \frac{2\Delta x L_0}{\pi L_k^2}, \quad L_k^2 = L_0^2 + x_k^2, \quad k = 0, \ldots, N-1. \tag{14.5.75}$$

In the following we will fix the length scale by taking the total size of the target array to be $L_{\text{target}} = 1$ and set $\Delta x = L_{\text{target}}/N$. Similarly, the time scale is fixed by setting $v_+ = 1$. Experimental studies of cytoneme-mediated transport of Wnt morphogen in zebrafish [841] and Shh in chicken [790] indicate that the growth rate of a cytoneme is of the order $v_+ \sim 0.1 \ \mu\text{m/s}$. Cytoneme lengths vary from $10 - 100 \ \mu\text{m}$ so if we take $L_{\text{target}} = 1$ to correspond to a length of $25 \ \mu\text{m}$ (around 20 cells), then the fundamental time scale is 250 s (around 4 minutes). We will mainly focus on the dependence of the morphogen distribution on $L_0, N, r$ by taking $v_- \gg v_+$ (fast return speed). Finally, a new cytoneme can be formed approximately twice every

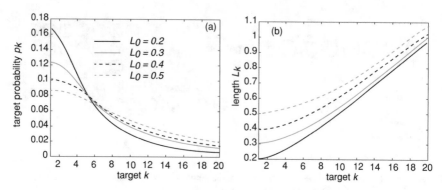

Fig. 14.54: Plots of (a) probability $p_k$ and (b) distance $L_k$ as a function of $k$ for $N = 20$ and various lengths $L_0$.

minute. Therefore, we take $\tau_{\mathrm{ref}} = \tau_{\mathrm{cap}} = 0.1$ and assume that degradation occurs on a time scale of hours by setting $\gamma^{-1} = 100$.

In Fig. 14.54 we plot the target probability $p_k$ and target-source separation $L_k$ as a function of $k$ for various choices of $L_0$ and $N = 20$. As expected, reducing the perpendicular distance of the source cell from a target array of fixed length leads to a greater variation of $p_k$ and $L_k$ with $k$. In addition, the total probability $p_{\mathrm{tot}} = \sum_{k=1}^{N} p_k$ increases as $L_0$ decreases. In Fig. 14.55 we show plots of the splitting probability $\pi_k$ as a function of $k$, which is obtained by substituting equation (14.5.75) into (14.5.70). It can be seen that reducing $L_0$ for fixed resetting rate sharpens the spatial ($k$-dependent) variation of $\pi_k$ along the target array, as does increasing $r$ for fixed $L_0$. Note in particular that fast resetting significantly amplifies the spatial variation of $\pi_k$ compared to $p_k$, see Figs. 14.55(a) and 14.54(a).

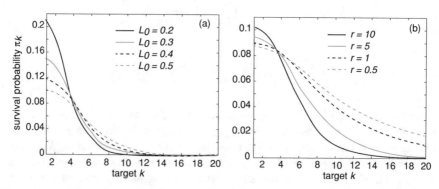

Fig. 14.55: Plots of splitting probability $\pi_k$ as a function of $k$ for $N = 20$: (a) various $L_0$ and $r = 1$; (b) various $L_0$ and $r = 10$; (c) various resetting rates $r$ and $L_0 = 1$.

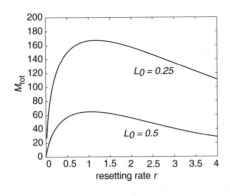

Fig. 14.56: Steady-state mean number of resources $M_{\text{tot}}$ delivered to all of the targets as a function of the resetting rate $r$ for $L_0 = 0.25, 0.5$. Other parameters are $\tau_{\text{ref}} = \tau_{\text{cap}} = 0.1$, $\gamma = 0.01$, $v_+ = 1$ and $v_- = 100v_+$.

It follows from equation (14.5.73) that $\pi_k$ determines the corresponding steady-state mean distribution of resources $\overline{M}_k$ up to the normalization factor

$$\overline{M}_{\text{tot}} := \sum_{k=1}^{N} \overline{M}_k = \frac{1}{\gamma(T + \tau_{\text{cap}})}, \qquad (14.5.76)$$

where $T$ is the unconditional MFPT. The latter is determined by substituting equation (14.5.75) into (14.5.72). Hence, while $\pi_k$ specifies the relative distribution of resources to the target cells, that is the shape and steepness of the morphogen gradient, $T$ fixes the total amount delivered to all the target cells. As in the case of a single target, $T$ has a minimum at an optimal resetting rate $r_{\text{tot}}$, which implies that $M_{\text{tot}}$ has a maximum at the same value of $r$. This is illustrated in Fig. 14.56. Our analysis suggests that although varying the rate of resetting controls the steepness of the gradient, $r$ should lie in an interval around $r_{\text{opt}}$ in order to ensure sufficient resources are delivered to the targets. Finally, note that a similar construction can be applied to a 2D layer of target cells as illustrated in Fig. 14.53(b). To a first approximation, the probability $p_k$ and distance $L_k$ of the $k$th target cell will depend on the in-plane radial distance of the target cell from the point $O$ where the perpendicular projection from the source cell intersects the layer.

**Multiple nucleating sites and noise reduction.** It is straightforward to extend the model to include multiple nucleation sites on the source cell or on a local cluster of source cells, provided that each nucleation site is independent and described by the same statistics. In particular, suppose that there is a total of $Z$ independently nucleating cytonemes with the same mean nucleation time. Let $M_k^{\mu}(t)$ be the number of morphogens present in the $k$th target cell at time $t$ that were delivered by the $\mu$-th cytoneme and set $M_k^{\Sigma}(t) = \sum_{\mu=1}^{Z} M_k^{\mu}(t)$. Since the $M_k^{\mu}(t)$ are independent identically distributed random variables, we have the steady-state mean and variance

$$\langle M_k^{\Sigma} \rangle = Z \langle M_k \rangle = \frac{Z\lambda_k}{\gamma}, \quad \text{Var}[M_k^{\Sigma}] = Z\text{Var}[M_k].$$

Thus multiple independent cytonemes scale up the morphogen concentration gradient by a factor of $Z$, and reduce fluctuations according to $Z^{-1/2}$. This scaling provides one way to compensate for the effects of the factor $\Gamma = \gamma \sum_{j=1}^{N} \pi_j (T_j + \langle \widehat{\tau} \rangle)$ in equation (14.5.73). Another interesting extension would be to consider morphogenesis in a growing domain such as the neural plate of zebrafish [841].

## 14.6 Exercises

**Problem 14.1. Advection–diffusion model of axonal length control.** Consider a 1D continuum model of the diffusive transport of tubulin along an axon [606]. Let $c(x,t)$ denote the concentration of tubulin at position $x$ along the axon at time $t$. Suppose that at time $t$ the axon has length $L(t)$ so that $x \in [0, L(t)]$. The transport of tubulin is modeled macroscopically in terms of an advection–diffusion equation with an additional decay term representing degradation at a rate $\gamma$:

$$\frac{\partial c}{\partial t} = D \frac{\partial^2 c}{\partial x^2} - V \frac{\partial c}{\partial x} - \gamma c.$$

Such a model can be derived from a more detailed stochastic model of active transport with $V$ the effective drift due to motor-driven transport and $D$ the effective diffusivity. It is assumed that there is a constant flux of newly synthesized tubulin from the cell body at $x = 0$ so that

$$-D \frac{\partial c}{\partial x} \bigg|_{x=0} = \sigma.$$

The flux at the growing end $x = L(t)$ is equal to the difference between the fluxes associated with MT assembly and disassembly:

$$-D \frac{\partial c}{\partial x} \bigg|_{x=L} = \varepsilon_l c(L) - \gamma_l,$$

where $\varepsilon_l$ and $\gamma_l$ are the rates of polymerization and depolymerization. Finally, the rate of growth is also taken to be proportional to the difference between these two fluxes according to

$$\frac{dL}{dt} = a \left[ \varepsilon_l c(L(t), t) - \gamma_l \right].$$

The constant $a$ depends on the size of each tubulin dimer, the number of MTs at the tip, and the cross-sectional area of the axon.

(a) Determine the steady-state solution $c(x)$ in terms of the steady-state length $L$ using the boundary conditions at $x = L$. Hence, derive the following transcendental equation for $L$ by imposing the boundary condition at $x = 0$:

$$F(L) \equiv e^{-\lambda_- L} - e^{-\lambda_+ L} = \frac{D\sigma}{\gamma} \frac{1}{c_L}(\lambda_+ - \lambda_-),$$

where $c_L = \varepsilon_L/\gamma_L$ and

$$\lambda_\pm = \frac{V}{2D}\left[1 \pm \sqrt{1 + 4D\gamma/V^2}\right].$$

Determine an approximation for $L$ in (i) the small $L$ regime and (ii) the large $L$ and large $V$ regime.

(b) Now suppose that diffusion is dominant, the concentration rather than the flux at $x = 0$ is fixed, and the rate of polymerization is infinitely fast. Then

$$\frac{\partial c}{\partial t} = \frac{\partial^2 c}{\partial x^2}, \quad 0 < x < L(t),$$

with

$$c(0,t) = c_0, \quad c(L(t),t) = 0, \quad -\frac{\partial c}{\partial x}\bigg|_{x=L(t)} = \beta\frac{dL(t)}{dt}.$$

We have set $D = 1$ for simplicity. Trying a solution of the form $c(x,t) = 1 + A\,\mathrm{erf}(x/2\sqrt{t})$, show that

$$u(x,t) = c_0\left[1 - \frac{\mathrm{erf}[x/(2\sqrt{t})]}{\mathrm{erf}(\lambda)}\right],$$

where the error function $\mathrm{erf}(x) = (2/\sqrt{\pi})\int_0^x e^{-t^2}\,dt$, and $\lambda$ satisfies the transcendental equation

$$\sqrt{\pi}\beta\,\mathrm{erf}(\lambda)e^{\lambda^2} = 1.$$

This is known as the Neumann solution to the one-phase Stefan problem, and has been used to model the growth of the acrosome of the sea cucumber *Thyone* during fertilization [719].

**Problem 14.2. Filament length control via regulation of the MT catastrophe rate.** Some experimental and modeling studies have suggested that, rather than directly regulating depolymerization, kinesin motors promote the transition from growing to shrinking phases (catastrophes) in dynamic MTs [336, 337, 381, 529, 878], see Fig. 14.57. Kuan *et al* Consider a modified version of the Dogterom–Leibler model of MT catastrophes [249] [529]:

$$\frac{\partial n_+}{\partial t} = -v_+\frac{\partial n_+}{\partial L} - k_+ n_+ + k_- n_-,$$

$$\frac{\partial n_-}{\partial t} = v_-\frac{\partial n_-}{\partial L} - k_- n_- + k_+ n_+,$$

in which the transition rate $k_+$ from the growing to shrinking phase is taken to depend on the fractional density $p_e$ of motors at the tip: $k_+ = \bar{k}_+ + \alpha p_e$. Here

$n_\pm(L,t)$ represent the number density of filaments of length $L$ in the growing and shrinking phase, respectively. In steady-state (assuming it exists), the transition rate $k_+$ will be $L$-dependent due to the $L$-dependence of $p_e$.

(a) Show that the steady-state number density of filaments of length $L$ is given by $n_\pm(L) = n(L)/v_\pm$ with

$$n(L) = n(0)e^{k_- L/v_-} \exp\left(-\frac{1}{v_+} \int^L k_+(L')dL'\right).$$

(b) Obtain the length dependence of $k_+$ by solving equation (14.1.4) with $\dot{p}_e = 0$ and $\dot{L} = v_+$, the speed of the growing phase:

$$p_e = p_e(L) = \frac{p_0(1 - e^{-L/\lambda})}{ak_{e,\text{off}}/(v - v_+) + p_0(1 - e^{-L/\lambda})},$$

and $k_+(L) = \bar{k}_+ + \alpha p_e(L)$.

(c) Assuming that $e^{-L/\lambda} \ll 1$ obtain the approximation

$$n(L) \approx n(0)e^{-[(\bar{k}_+ + \Delta k_+)/v_+ - k_-/v_-]L}, \qquad (14.6.77)$$

with

$$\Delta k_+ = \frac{\alpha(v - v_+)p_0}{ak_{e,\text{off}} + (v - v_+)p_0}.$$

It follows that increasing $p_0$ by increasing the bulk motor concentration results in a higher catastrophe rate and, hence, shorter filament lengths.

**Problem 14.3. Stochastic model of intraflagellar transport.** Consider a particle undergoing a unidirectional random walk along a single filament track as shown in Fig. 14.58. The track is modeled as a finite 1D lattice with lattice spacing $\ell$. Suppose that at time $t$ there are $N(t) + 1$ lattice sites labeled $n = 0, \ldots, N$, with $n = 0$ corresponding to the $-$ end and $n = N(t)$ to the $+$ end. Suppose that $N(0) = 0$ and

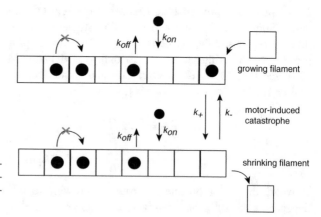

Fig. 14.57: Schematic diagram of filament length regulation by altering the frequency of catastrophes.

the particle starts at the minus end. During the $j$th cycle of the dynamics with $N(t) = j$, the particle walks from the minus to the plus end. The times $\tau$ between successive steps are taken to be independent, identically distributed random variables with a common waiting time density $\psi(\tau)$. When the particle reaches the current $+$ end, the length of the filament is increased by one lattice site to form the new $+$ end. Once the particle has reached this new lattice site, the hopping process reverses direction. After returning to the $-$ end the particle reverses direction again immediately, and the process continues iteratively. For simplicity the waiting time density is taken to be the same in both directions, and we are neglecting depolymerization at the plus end.

(a) Let $f_j(t)$ be the FPT density for the particle to travel from one end to the other when the length is $L_j = (j+1)\ell$. Similarly, let $g_n(t)$ be the probability density that the particle has just completed the $n$th visit to the $+$ end at time $t$ and filament length has increased by one unit. Explain the meaning of the following iterative equations

$$f_j(t) = \int_0^t \psi(\tau) f_{j-1}(t-\tau) d\tau,$$

with $f_0(t) = \psi(t)$ and

$$g_n(t) = \int_0^t \int_0^{t'} f_n(t-t') f_{n-1}(t'-t'') g_{n-1}(t'') dt'' dt'$$

for $n \geq 2$ with $g_1(t) = f_1(t)$.

(b) Using Laplace transforms show that for $n \geq 2$

$$\tilde{g}_n(s) = \tilde{f}_j(s) \prod_{k=1}^{n-1} \left[\tilde{f}_k(s)\right]^2 = \{\tilde{\psi}(s)\}^{(n+1)^2 - 2}.$$

(c) Let $L(t)$ be the length of the flagellum at time $t$ and introduce the length probability $P_j(t) = \text{Prob}[L(t) = L_j]$. Explain the iterative equation

$$P_j(t) = \int_0^t F_{j+1}(t-t') g_j(t') dt'$$

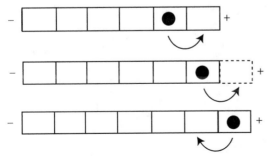

Fig. 14.58: Particle hopping along a one-dimensional filament that is modeled as a discrete lattice. When the particle reaches the $+$ end of the filament, a lattice site is added to form the new $+$ end and the particle reverses its direction. The particle also reverses direction at the $-$ end.

for $j \geq 1$, where $F_{j+1}(t - t')$ is the probability that the particle has not completed the $(j+1)$-th trip to the plus end at time $t$, starting from the plus end at time $t'$. Show that

$$F_{j+1}(\tau) = \int_\tau^\infty \left[ \int_0^{t'} f_{j+1}(t' - t'') f_j(t'') dt'' \right] dt'$$

for $j \geq 1$. Determine $\widetilde{F}_j(s)$ and hence show that

$$\widetilde{P}_j(s) = \frac{\widetilde{g}_j(s) - \widetilde{g}_{j+1}(s)}{s},$$

with $g_0(t) = \delta(t)$. Determine $\sum_{j=0}^\infty \widetilde{P}_j(s)$ and interpret the result.

(d) A useful way to characterize the stochastic growth of the filament is in terms of the mean and variance of the length $L(t)$. Let $\eta_n(t) = \sum_{j=0}^\infty j^n P_j(t)$ denote the $n$th moment of the distribution $P_j(t)$. Then

$$\langle L \rangle = 1 + \ell \eta_1(t), \quad \langle \Delta L^2 \rangle = \ell^2 [\eta_2(t) - \eta_1(t)^2].$$

Using part (c) show that

$$\widetilde{\eta}_1(s) = \frac{1}{s} \sum_{j=1}^\infty \widetilde{g}_j(s), \quad \widetilde{\eta}_2(s) = \frac{1}{s} \sum_{j=1}^\infty (2j - 1) \widetilde{g}_j(s).$$

(e) Suppose that the waiting density has finite first and second moments so that for small $s$, $\widetilde{\psi}(s) \sim 1 - \tau s$ and $\widetilde{g}_j(s) \approx e^{-(j+1)^2 \tau s}$. The sums in part (d) can then be approximated for small $s$ as

$$\widetilde{\eta}_1(s) \sim \frac{1}{2s} \sqrt{\frac{\pi}{\tau s}}, \quad \widetilde{\eta}_2(s) \sim \frac{1}{\tau s^2}.$$

Using a Tauberian theorem, derive the large-$t$ behavior

$$\eta_1(t) \sim \sqrt{t/\tau}, \quad \eta_2(t) \sim t/\tau.$$

**Problem 14.4. Doubly stochastic model of intraflagellar transport.** Consider the doubly stochastic model of Sect. 14.2. Suppose that the intensity of the DSPP is given by $\lambda(X(t)) = \eta M X(t)$, where $X(t)$ is the fraction of bound binding sites in the basal body. Performing the Ito integral shows that

$$X(t) = X^* \left[1 - e^{-\Gamma t}\right] + X_0 e^{-\Gamma t} + \sqrt{\frac{\Gamma}{M_0}} \int_0^t e^{-\Gamma(t-t')} \sqrt{X^* + \Theta X(t')} dW(t')$$

for a fixed initial condition $X(0) = X_0$ with $X^*$ given by equation (14.2.3) and

$$\Gamma = k_+ B + k_-, \quad \Theta = \frac{k_- - k_+ B}{k_- + k_+ B}.$$

(a) Determine the variance

$$\text{Var}[X] \equiv \langle [X(t) - \langle X(t) \rangle][X(t) - \langle X(t) \rangle] \rangle.$$

(b) Hence, using equation (14.2.13) show that

$$\text{Var}[N(t)] = M\eta^2 \left[ \frac{X^*}{2} \mathcal{A}(t) + \frac{X^*}{2} \Theta[\mathcal{A}(t) - 2\mathcal{B}(t)] + \Theta X_0 \mathcal{B}(t) \right] + \mathbb{E}[N(t)],$$

with

$$\mathcal{A}(t) = \frac{2t}{\Gamma} - \frac{2}{\Gamma^2} \left[ 1 - e^{-\Gamma t} \right] - \frac{1}{\Gamma^2} \left[ 1 - e^{-\Gamma t} \right]^2,$$

and

$$\mathcal{B}(t) = \frac{2}{\Gamma^2} \left[ 1 - e^{-\Gamma t} \right] - \frac{2t}{\Gamma} e^{-\Gamma t} - \frac{1}{\Gamma^2} \left[ 1 - e^{-\Gamma t} \right]^2.$$

**Problem 14.5. Search-and-capture model.** Let $P(\tau)$ be the probability that a single MT finds a single kinetochore before time $\tau$. From equation (14.3.1),

$$P(\tau) = p + \sum_{n=1}^{\infty} Q(\tau - \Delta\tau|n) P(n),$$

with $P(n) = p(1-p)^n$, $\Delta\tau = V_g/x$ and $Q(\tau|n)$ given by equation (14.3.4).
(a) Show that the probability of capture before time $\tau$, $\tau > \Delta\tau$, is

$$P(\tau) = p + (1-p)(1 - e^{-p(\tau - \Delta\tau)/T_{\text{cycle}}}).$$

(b) If $p \ll 1$ and $\Delta\tau \ll \tau$, then

$$P(\tau) \approx 1 - e^{-p\tau/T_{\text{cycle}}}.$$

Using the expressions for $p$ and $T_{\text{cycle}}$, see equations (14.3.2) and (14.3.3), show that the optimal catastrophe frequency is $k_{\text{cat}} = V_g/x$.

(c) Now suppose that there are $N$ independent searching MTs, and denote the probability that a single kinetochore is found before time $\tau$ by $P^{(N)}(\tau)$. Explain the formula

$$P^{(N)}(\tau) = 1 - (1 - P(\tau))^N.$$

Using the expression for $P(\tau)$ in part (b), show that the average time to capture is $T_{\text{capture}}/N$, where $T_{\text{capture}}$ is the result for $N = 1$.

(d) Finally, consider $N$ MTs and $M$ kinetochores. Since the attachment of a kinetochore is independent of all other attachment events, the probability that all kinetochores will be attached to an MT before time $\tau$ is

$$P^{(N,M)}(\tau) = P^{(N)}(\tau)^M = \left(1 - e^{-pN\tau/T_{\text{cycle}}}\right)^M.$$

The corresponding density function is $f(t) = dP^{(N,M)}(t)/dt$. The most likely time $t_0$ when the last kinetochore is captured is obtained by finding the maximum of the function $f(t)$. Show that

$$t_0 = \frac{T_{\text{cycle}}}{pN} \ln M.$$

It turns out from numerical studies that $t_0$ is a reasonable estimate of the average time to capture all the kinetochores.

**Problem 14.6. Classical theory of cell adhesion.** Consider a collection of $N_T$ molecules forming a cluster of adhesion molecules attached to an upper surface, which can bind to corresponding sites on a lower surface, see Fig. 14.59. The geometrical arrangement means that bound adhesions share the load. Let $n(t)$ denote the number of bound adhesions (closed bonds) at time $t$. Each bond can break at a rapture rate $k_{\text{off}} = k_0 e^{F/nF_b}$, where $F$ is the applied force on the upper surface, and $F_b$ specifies the characteristic bond breaking force. The binding rate is a constant $k_{\text{on}}$. For large $N_T$, the number of closed bonds evolves according to the kinetic equation

$$\frac{dn}{dt} = \gamma(N_T - n) - ne^{f/n},$$

where we have fixed the time units by setting $t \to k_0 t$, and taken $f = F/F_b$ and $\gamma = k_{\text{on}}/k_0$.

(a) Show that the ODE exhibits a saddle-node bifurcation such that no steady-state solution exists when $f \geq f_c$, where

$$f_c = N_T g(\gamma/e),$$

and $g(a)$ is defined to be the solution of $xe^x = a$. [Hint: perform the change of variables $y = f/n$ and consider zeros of the function $H(y) = \gamma N_T y/f - \gamma - e^y$.] Hence adhesion characterized by a finite number of bonded molecules is stable only up to a critical force $f_c$. The critical force can be increased by increasing the relative binding rate $\gamma$ or the total number $N_T$ of adhesion molecules.

Fig. 14.59: Schematic diagram of an adhesion cluster under a force $F$. At time $t$, $n(t)$ adhesion molecules form closed bonds, whereas the remaining $N_t - n(t)$ bonds are open. Adhesions bind at a constant rate $k_{\text{on}}$ and unbind at a force dependent rate $k_{\text{off}}(F/n(t))$.

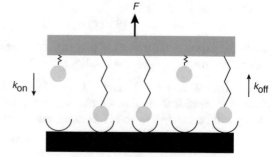

(b) Now consider a stochastic version of the model. Let $p_n(t)$ be the probability that $n$ bonds are formed at time $t$. We then have the birth–death master equation (see Chap. 3)

$$\frac{dp_n}{dt} = \omega_+(n-1)p_{n-1} + \omega_-(n+1)p_{n+1} - [\omega_+(n) + \omega_-(n)]p_n,$$

with

$$\omega_-(n) = ne^{f/n}, \quad \omega_+(n) = \gamma(N_T - n).$$

In contrast to the deterministic equation, which predicts infinitely long cluster lifetimes for $f < f_c$, the stochastic model predicts finite lifetimes for any value of $f$. The average lifetime under the initial condition $n(0) = N_T$ corresponds to the MFPT $\tau$ for the cluster to reach the state $n = 0$. Using the analysis in Sect. 3.3.3, show that

$$\tau = \sum_{n=1}^{N_T} \frac{1}{\omega_-(n)} + \sum_{n=1}^{N_T-1} \sum_{m=n+1}^{N_T} \frac{1}{\omega_+(m)} \prod_{l=m-n}^{m} \frac{\omega_+(l)}{\omega_-(l)}.$$

Plot $\tau$ as a function of $f/N_T$ for $N_t = 1, 2, 5, 10, 15, 25$ and $\gamma = 1$.

**Problem 14.7. Stochastic motor-clutch model.** Consider the mean-field approximation of the second motor-clutch model analyzed in Sect. 14.4.2. The steady-state equations for the probabilities are

$$\int_0^\infty p^*(x)dx + P_0^* = 1,$$

and

$$\langle v \rangle \frac{dp^*(x)}{dx} = -k_{\mathrm{off}} e^{bx} p^*(x),$$

with boundary condition $p^*(0) = k_{\mathrm{on}} P_0^* / \langle v \rangle$. The mean velocity is determined self-consistently from the force-balance equation (14.4.25).

(a) Obtain the solution

$$p^*(x) = \frac{P_0^* k_{\mathrm{on}}}{\langle v \rangle} \exp\left( \frac{k_{\mathrm{off}}}{b\langle v \rangle} \left[ 1 - e^{bx} \right] \right),$$

with

$$P_0^* = \left[ 1 + \frac{k_{\mathrm{on}}}{b\langle v \rangle} e^{k_{\mathrm{off}}/b\langle v \rangle} \Gamma(0; k_{\mathrm{off}}/b\langle v \rangle) \right]^{-1},$$

and

$$\Gamma(z;k) = \int_1^\infty y^{z-1} e^{-ky} dy.$$

(b) Using asymptotic expansions of the integral expression in the case of fast unbinding, $k_{\mathrm{off}}/b\langle v \rangle \gg 1$, obtain the approximation

$$P_0^* \approx \frac{k_{\mathrm{off}}}{k_{\mathrm{on}} + k_{\mathrm{off}}}.$$

(c) Similarly, carry out an asymptotic expansion of the mean traction force

$$\langle f \rangle = \kappa \frac{P_0^* k_{\mathrm{on}}}{\langle v \rangle} \int_0^\infty x \exp\left(\frac{k_{\mathrm{off}}}{b\langle v \rangle}\left[1 - e^{bx}\right]\right) dx$$

to obtain the approximation

$$\langle f \rangle \approx \frac{k_{\mathrm{on}}}{k_{\mathrm{on}} + k_{\mathrm{off}}} \frac{\kappa \langle v \rangle}{k_{\mathrm{off}}}.$$

**Problem 14.8. Traveling wave solution of focal adhesion model.** Consider the model of force-induced growth of a focal adhesion analyzed in Sect. 14.4.4. A traveling wave solution $\psi(x,t) = \Psi(\xi)$, $\xi = x - vt$, for the concentration of plaque proteins in the upper layer satisfies equation (14.4.52):

$$B\frac{\partial^2 \Psi}{\partial x^2} + V(g)\frac{\partial \Psi}{\partial x} + \varepsilon\Psi - c\Psi^3 + \Delta\mu(g) = 0,$$

with

$$V(g) = \frac{v}{\Gamma} - \frac{\varepsilon_b g}{2k_b} A_1(g).$$

(a) Consider the trial solution

$$\Psi(\xi) = \alpha \tanh(\gamma\xi) + \beta.$$

Substituting into the ODE for $\Psi$, derive a cubic equation for $\tanh(\gamma\xi)$.

(b) By requiring that the four coefficients of the cubic vanish derive four algebraic equations for the four unknowns $\alpha, \beta, \gamma, v$. Here $|\gamma|^{-1}$ determines the width of the front, whereas $\alpha$ and $\beta$ are related to the far-field values $\Psi(\pm\infty)$.

(c) Use the equations of part (b) to derive the growth velocity formula

$$V = 3\sqrt{2Bc}\beta\,\mathrm{sign}(\gamma),$$

with $\beta$ related to $\Delta\mu_0(g)$ according to the cubic

$$\beta - \frac{4c}{\varepsilon}\beta^3 = \frac{\Delta\mu_0(g)}{2\varepsilon}.$$

**Problem 14.9. Advection–diffusion model of cytoneme-based morphogenesis.** Suppose that, instead of equation (14.5.1), the transport of morphogens along a cytoneme of length $L$ is described by the advection–diffusion equation

$$\frac{\partial u}{\partial t} = -v\frac{\partial u}{\partial x} + D\frac{\partial^2 u}{\partial x^2}, \quad x \in (0, L),$$

where $v$ is the average speed of motor-cargo transport and $D$ is a diffusion coefficient. The corresponding boundary conditions are

$$u(0,t) = \kappa C_0(t), \quad u(L,t) = 0,$$

where $C_0(t)$ is the density of vesicles in the source cell. Equation (14.5.3) still holds except that now the flux is

$$J(x,t) = vu(x,t) - D\frac{\partial u(x,t)}{\partial x}.$$

(a) Show that the ratio of the target and source densities at steady state is

$$\frac{C_1^*}{C_0^*} = \frac{w(L)}{k}, \quad w(L) := \frac{\kappa v e^{-\gamma L}}{e^{-\gamma L} - 1}, \quad \gamma = -\frac{v}{D}.$$

Compare plots of $w(L)$ for positive and negative values of $\gamma$.

(b) Define the accumulation time $\tau_1$ according to equation (14.5.25). Use Laplace transforms to calculate $\tau_1$. First, Laplace transform the second equation of (14.5.3) and use equation (14.5.26) to show that

$$\tau_1 = \frac{1}{k} - \frac{1}{Q}\frac{d}{ds}\, s\hat{J}(L,s)\Big|_{s=0}.$$

Second, Laplace transform the advection–diffusion equation to obtain the general solution

$$\hat{u}(x,s) = e^{-\frac{1}{2}\gamma x}[A(s)\sinh(\gamma_1(s)x) + B(s)\cosh(\gamma_1(s)x)],$$

where

$$\gamma_1(s) = \sqrt{\left(\frac{\gamma}{2}\right)^2 + \frac{s}{D}}.$$

Finally, determine $A(s), B(s)$ using the boundary conditions. Hence, derive the following expression for $\tau_1$:

$$\tau_1 = \frac{1}{k} + \frac{D}{\kappa v^2}\left[\kappa(e^{\gamma L} - 1 - \gamma L) + \gamma(e^{\gamma L} - 1)\right].$$

# Chapter 15
# Bacterial population growth and collective behavior

In this chapter, we consider various topics related to bacterial population growth and collective behavior. We begin in Sect. 15.1 by describing a continuum model of bacterial population growth based on an age-structured evolution equation. Such an equation supplements the continuously varying observational time by a second time variable that specifies the age of an individual cell since the last division. Whenever a cell divides, the age of the daughter cells is reset to zero. Although the total number of cells grows exponentially with time, the normalized age distribution approaches a steady state. The latter determines the effective population growth rate via a self-consistency condition. The age-structured model is extended in Sect. 15.2 in order to keep track of both the age and volume distribution of cells. This is then used to explore various forms of cell length regulation, including timer, sizer, and adder mechanisms. In Sect. 15.3, we consider to what extent single-cell molecular variation plays a role in population-level function. This is explored within the context of phenotypic switching in switching environments, which is thought to be an important factor in the phenomenon of persistent bacterial infections following treatment with antibiotics. At the population level, phenotypic switching is modeled in terms of a stochastic hybrid system. We show how WKB methods can be used to analyze population extinction. We then turn to a discussion of bacterial quorum sensing (QS) in Sect. 15.4. This is a form of collective cell behavior that is triggered by the population density reaching a critical threshold, which requires that individual cells sense their local environment. In Sect. 15.5, we investigate synchronization in a population of synthetic gene oscillators that are dynamically coupled to an external medium via a QS mechanism. In Sect. 15.6, we describe some mathematical models for the growth of bacterial biofilms, which consist of a large number of cells living within a self-secreted polymer-rich matrix. Finally, in Sect. 15.7, we briefly describe an example of motility-based phase separation on populations of motile bacteria.

© Springer Nature Switzerland AG 2021
P. Bressloff, *Stochastic Processes in Cell Biology*, Interdisciplinary Applied
Mathematics 41, https://doi.org/10.1007/978-3-030-72519-8_15

## 15.1 Stochastic models of bacterial population growth

When bacteria are placed in a medium that provides all of the nutrients that are necessary for their growth, the population exhibits four phases of growth that are representative of a typical bacterial growth curve, see Fig. 15.1.

*Lag phase.* Upon inoculation into the new medium, bacteria do not immediately reproduce, and the population size remains constant. During this period, called the lag phase, the cells are metabolically active and increase only in cell size. They are also synthesizing the enzymes and factors needed for cell division and population growth under their new environmental conditions.

*Exponential growth phase.* The population then enters the exponential growth phase in which each cell generation occurs in the same time interval as the preceding ones, resulting in a balanced increase in the constituents of each cell (balanced population growth). The growth phase continues until nutrients are depleted or toxic products accumulate, at which time the cell growth rate slows, and some cells may begin to die. Under optimum conditions, the maximum population for some bacterial species at the end of the growth phase can reach a density of 10 to 30 billion cells per mL.

*Stationary phase.* Bacterial growth is followed by the stationary phase, in which the size of a population of bacteria remains constant, even though some cells continue to divide and others begin to die.

*Death phase.* The stationary phase is followed by the death phase, in which the death of cells in the population exceeds the formation of new cells.

In the case of balanced population growth, the quantities of cellular components double on average during each cell cycle and then halve at cell division. However, individual cells can deviate significantly from the average due to a combination of molecular noise and stochastic growth and cell division. A variety of experimental and theoretical studies have investigated the distribution of cell sizes and synthesized proteins across a population of cells under well-controlled conditions, with the goal of understanding the relative contributions of intracellular noise, the random partitioning of proteins at cell division, and the stochastic exponential growth of cells [10, 67, 101, 105–107, 322, 356, 418, 448, 449, 479, 515, 516, 585, 807].

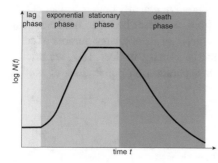

Fig. 15.1: Schematic of a bacterial growth curve, showing four distinct stages.

Roughly speaking, models can be distinguished according to (i) the quantity whose distribution is tracked over multiple cycles of cell division (e.g., protein expression, cell size), (ii) whether cells divide synchronously or asynchronously, (iii) whether cell division is symmetric or asymmetric, and (iv) the particular model of cell growth between cell divisions. (See Refs. [181, 479] for reviews of different approaches to modeling cell-population dynamics.) In this section, we focus on cell growth, and then explore its implications for cell size regulation at the population level in Sect. 15.2.

### 15.1.1 Age-structured evolution equations

In a typical cell culture, the number of cells per mL is very large ($O(10^8)$) so that the total number $N(t)$ of cells at time $t$ can be treated as a deterministic continuous variable. In the balanced growth phase, we have $N(t) = N_0 e^{kt}$, where $N_0$ is the initial cell number and $k$ is the population growth rate; the latter can be determined experimentally from standard bulk-culture measurements. However, the observation of population growth is not sufficient to determine the time scales of growth and division at the single-cell level, fluctuations in division times, and cell-age distributions. On the other hand, the relationship between bulk and single-cell behavior can be investigated using so-called age-structured models and compared to more precise measurements of the stochastic growth of individual cells, as recently demonstrated for *Caulobacter cresentus* bacterial cells [465, 470]. Age-structured models are probably best known within the context of birth-death processes in ecology, where the birth and death rates depend on the age of the underlying populations [194, 374, 605, 911]. These could be cells undergoing differentiation or proliferation [772, 848, 986], or whole organisms undergoing reproduction [503]. There are also a growing number of applications of age-structured models within cell biology, including cell motility [298, 299], microtubule catastrophes [472], and stochastically-gated diffusion [125]. Here we will follow the formulation of Ref. [470]. For a more detailed analysis of age-structured models within the context of ecology; see the book by Iannelli and Milner [452].

The first step is to supplement the continuously varying observational time $t$ by a second time variable $\tau$ that denotes the age of an individual cell since the last division. Whenever a cell divides, the age of the daughter cells is reset to zero. Let $p(\tau)$ denote the density of cell ages, that is, $p(\tau)d\tau$ is the probability that a cell lives to age $\tau$ and then divides in the interval $[\tau, \tau + d\tau]$. Let $\alpha(\tau)$ be the rate at which cells of age $\tau$ divide. It follows that

$$p(\tau)d\tau = \left[1 - \int_0^\tau p(\tau')d\tau'\right]\alpha(\tau)d\tau,$$

where the factor in square brackets is the probability that a cell does not divide before the age $\tau$. Rearranging, we have

$$\alpha(\tau) = \frac{p(\tau)}{1 - \int_0^\tau p(\tau')d\tau'}. \tag{15.1.1}$$

Note that $p(\tau)$ being independent of the time $t$ is a consequence of the balanced growth condition.

Next we define $n(t, \tau)$ to be the age-dependent number density: $n(t, \tau)d\tau$ is the number of cells present at time $t$ with ages between $\tau$ and $\tau + d\tau$. Introducing the normalized density

$$G(t, \tau) = \frac{n(t, \tau)}{N(t)}, \quad \int_0^\infty G(t, \tau)d\tau = 1, \tag{15.1.2}$$

the balanced growth condition means that

$$\lim_{t \to \infty} G(t, \tau) = G^*(\tau),$$

where $G^*(\tau)$ denotes a steady-state age distribution. Suppose that each cell divides into $v$ daughter cells. The question we wish to explore, following Ref. [470], is how $G^*(\tau)$ and the growth rate $k$ are related, given $p(\tau)$ and $v$. In order to proceed, one has to find the steady-state solution of an age-structured evolution equation for $G(t, \tau)$, which is itself constructed from a corresponding evolution equation for $n(t, \tau)$.

Combining the aging of cells with cell-division events, we have

$$n(t + dt, \tau)d\tau = n(t, \tau - dt)d\tau - [\alpha(\tau)dt]n(t, \tau - dt)d\tau.$$

After rearranging and dividing through by $d\tau$,

$$n(t + dt, \tau) - n(t, \tau) = -n(t, \tau) + n(t, \tau - dt) - [\alpha(\tau)dt]n(t, \tau - dt).$$

Now dividing both sides by $dt$ and taking the limit $dt \to 0$, we obtain the evolution equation

$$\frac{\partial n}{\partial t} + \frac{\partial n}{\partial \tau} = -\alpha(\tau)n(t, \tau). \tag{15.1.3}$$

The loss of each cell of age $\tau$ due to cell division simultaneously results in $v$ daughter cells of zero age. This is represented mathematically by the boundary condition

$$n(t, 0) = v\left[\int_0^\infty \alpha(\tau)n(t, \tau)d\tau\right] := v\rho(t)N(t), \tag{15.1.4}$$

where $\rho(t)$ is defined as the cell-averaged rate at which newborn cells arise across the whole cell population. The evolution equation is known as the McKendrick-Foerster equation in classical ecology [605, 911]. For the moment, we are assuming that cell division is symmetric, that is, all daughter cells are identical. The case of asymmetric cell division will be considered later.

Although the number of cells $n(t, \tau)$ will grow exponentially with time $t$, we expect the normalized age distribution $G(t, \tau)$ to approach a steady state with respect to $t$. Using equations (15.1.3) and (15.1.4),

$$\frac{dN(t)}{dt} = \int_0^\infty \frac{\partial n(t,\tau)}{\partial t} d\tau = \int_0^\infty \left[ -\frac{\partial n}{\partial \tau} - \alpha(\tau)n(t,\tau) \right] d\tau$$

$$= n(t,0) - n(t,\infty) - \int_0^\infty \alpha(\tau)n(t,\tau)d\tau = (v-1)\rho(t)N(t). \qquad (15.1.5)$$

This can be used to derive an evolution equation for $G(t,\tau)$ by differentiating both sides of the identity $n(t,\tau) = G(t,\tau)N(t)$:

$$\frac{\partial n(t,\tau)}{\partial t} = N(t)\frac{\partial G(t,\tau)}{\partial t} + \frac{dN(t)}{dt}G(t,\tau)$$

$$= N(t)\frac{\partial G(t,\tau)}{\partial t} + G(t,\tau)(v-1)\rho(t)N(t).$$

Substituting for $\partial_t n$ using equation (15.1.3) and dividing through by $N(t)$ yields the evolution equation

$$\frac{\partial G}{\partial t} + \frac{\partial G}{\partial \tau} = -\alpha(\tau)G(t,\tau) - (v-1)\rho(t)G(t,\tau). \qquad (15.1.6)$$

The corresponding boundary condition is obtained by dividing both sides of (15.1.4) by $N(t)$,

$$G(t,0) = v\left[ \int_0^\infty \alpha(\tau)G(t,\tau)d\tau \right] := v\rho(t). \qquad (15.1.7)$$

**Steady-state solution.** In the steady state (assuming it exists), we take $G(t,\tau) = G^*(\tau)$ and $\rho(t) = \rho^*$. Setting $\rho(t) = \rho^*$ in equation (15.1.5) gives

$$\frac{dN(t)}{dt} = (v-1)\rho^* N(t),$$

which has the solution

$$N(t) = N_0 e^{kt}, \qquad (15.1.8)$$

with $k = (v-1)\rho^*$. Moreover, equations (15.1.6) and (15.1.7) become

$$\frac{dG^*}{d\tau} + [\alpha(\tau) + k]G^*(\tau) = 0, \qquad (15.1.9a)$$

$$G^*(0) = v\left[ \int_0^\infty \alpha(\tau)G^*(\tau)d\tau \right] := v\rho^*. \qquad (15.1.9b)$$

Equation (15.1.9a) can be rewritten as

$$\frac{d}{d\tau}\left[ G^*(\tau)e^{k\tau + A(\tau)} \right] = 0, \quad A(\tau) = \int_0^\tau \alpha(\tau')d\tau'.$$

Integrating with respect to $\tau$ and using equation (15.1.9b) leads to the result (for $v > 1$)

$$G^*(\tau) = \frac{kv}{v-1} e^{-k\tau} e^{-A(\tau)}. \tag{15.1.10}$$

It remains to determine the growth rate $k$ (or equivalently $\rho^*$). This can be achieved using the normalization of the probability density $G^*(\tau)$:

$$1 = \int_0^\infty G^*(\tau) d\tau = \frac{kv}{v-1} \int_0^\infty e^{-k\tau} e^{-A(\tau)} d\tau$$

$$= -\frac{v}{v-1} \int_0^\infty e^{-A(\tau)} \frac{d}{d\tau} e^{-k\tau} d\tau = \frac{v}{v-1} \left[ 1 - \int_0^\infty \alpha(\tau) e^{-k\tau} e^{-A(\tau)} d\tau \right],$$

after integrating by parts. Hence,

$$\int_0^\infty \alpha(\tau) e^{-k\tau} e^{-A(\tau)} d\tau = \frac{1}{v}. \tag{15.1.11}$$

The left-hand side can be reduced further as follows. Setting $Q(\tau) = \int_0^\tau p(\tau') d\tau'$ and using equation (15.1.1) implies that

$$\frac{dQ(\tau)}{d\tau} = p(\tau) = \alpha(\tau)[1 - Q(\tau)].$$

This equation can be rewritten in the form

$$\frac{d}{d\tau} \left[ e^{A(\tau)} Q(\tau) \right] = \alpha(\tau) e^{A(\tau)} = \frac{d}{d\tau} e^{A(\tau)}.$$

Integrating both sides with respect to $\tau$ and multiplying both sides by $e^{-A(\tau)}$, yields

$$Q(\tau) = 1 - e^{-A(\tau)}.$$

Finally, differentiating both sides with respect to $\tau$, we deduce that

$$p(\tau) = \alpha(\tau) e^{-A(\tau)}.$$

Hence, the condition (15.1.11) determining $k$ can be rewritten as

$$\int_0^\infty e^{-k\tau} p(\tau) d\tau = \frac{1}{v}. \tag{15.1.12}$$

Moreover, the equation (15.1.10) can be rewritten as

$$G^*(\tau) = \frac{kv}{v-1} e^{-k\tau} \left[ 1 - \int_0^\tau p(\tau') d\tau' \right]. \tag{15.1.13}$$

Equations (15.1.8), (15.1.12) and (15.1.13) specify the complete analytic steady-state solution for symmetric cell division; the particular case of binary division ($v = 2$) has also been studied in detail within the context of classical ecology [703]. It immediately follows from equation (15.1.13) that $G^*(\tau)$ is a monotonically decreasing function of $\tau$. Moreover, in the case of cell division, the division time density $p(\tau)$ is unimodal. Since $p(\tau) \sim d[e^{k\tau}G^*(\tau)]/d\tau$, the most probable division time corresponds to the inflection point of the function $e^{k\tau}G^*(\tau)$; the slope of the function provides an estimate of the division time density [470]. Finally, for a given density $\rho(\tau)$, a larger $v$ increases the growth rate $k$ and skews the age density $G^*(\tau)$ towards smaller ages. Examples of the functions $G^*(\tau)$ and $p(\tau)$ are given in Fig. 15.2.

Further simplification can be achieved in the case $v = 1$, which is applicable to *Caulobacter cresentus* bacteria under specific experimental conditions [470]. For these, cells divide into two morphologically and functionally distinct daughter cells: a nonmotile stalked cell, which can itself subsequently divide, and a motile swarmer cell that cannot undergo further division but can differentiate into a stalked cell. In microfluidic experiments, one can keep stalker cells and remove swarmer cells after each division, so that the stalked-cell dynamics is equivalent to cells simply being renewed after each division, that is, $v = 1$ and $N(t)$ is a constant $N^*$ (zero growth rate). Equation (15.1.9a) can be integrated directly to show that $G^*(\tau) = c[1 - Q(\tau)]$, with the multiplicative constant $c$ determined by the normalization condition

$$
\begin{aligned}
1 &= \int_0^\infty G^*(\tau')d\tau' = c \int_0^\infty [1 - Q(\tau')]d\tau' \\
&= c \int_0^\infty d\tau' \int_{\tau'}^\infty p(\tau'')d\tau'' = c \int_0^\infty d\tau'' p(\tau'') \int_0^{\tau''} d\tau' \\
&= c \int_0^\infty p(\tau)\tau d\tau := c\mu_\tau.
\end{aligned}
$$

Hence

$$
G^*(\tau) = \frac{1}{\mu_\tau} \int_\tau^\infty p(\tau')d\tau', \tag{15.1.14}
$$

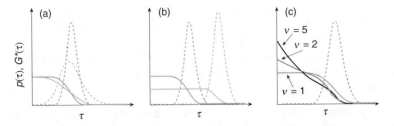

Fig. 15.2: Comparison of the division time density $p(\tau)$ and the steady-state age density $G^*(\tau)$ for a given progeny number $v$. (a) Cell-age densities (bold orange and blue curves) for $v = 1$ corresponding to two division time densities $p(\tau)$ with the same mean but different variances. Note that the inflection points of the $G^*(\tau)$ curves coincide, but have different slopes. (b) Same as (a) except the division time densities now have the same variances but different means. (c) Cell-age density $G^*(\tau)$ corresponding to the same division time density but different $v = 1,2,5$. [Redrawn from Jafarpour et al. [470].]

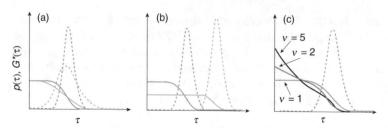

Fig. 15.3: Sketch of steady-state cell-age densities for various temperatures. Under the rescaling $\widehat{G}^*(\tau/\mu_a) = \mu_a G^*(\tau)$, all the curves collapse on to a single curve [Adapted from Jafarpour et al. [470].]

where $\mu_\tau$ is the mean division time. The mean age of the population can now be determined as follows:

$$
\begin{aligned}
\mu_a &:= \int_0^\infty \tau G^*(\tau)d\tau = \frac{1}{\mu_\tau}\int_0^\infty \tau d\tau \int_\tau^\infty p(\tau')d\tau' \\
&= \frac{1}{\mu_\tau}\int_0^\infty p(\tau')d\tau' \int_0^{\tau'} \tau d\tau \\
&= \frac{1}{2\mu_\tau}\int_0^\infty \tau'^2 p(\tau')d\tau' = \frac{\sigma_\tau^2 + \mu_\tau^2}{2\mu_\tau},
\end{aligned}
\tag{15.1.15}
$$

where $\sigma_\tau^2$ is the variance of division times.

Jafarpour et al. [470] applied the above theory to experimental measurements of *Caulobacter cresentus* bacterial growth. That is, they constructed histograms of the measured cell-age distributions and division time distributions, and used the latter to construct predicted cell-age distributions. They found excellent agreement between the measured and predicted cell-age distributions spanning a physiological range of temperatures. One important feature of these curves is that, under the rescaling $\widehat{G}^*(\tau/\mu_a) = \mu_a G^*(\tau)$, all the probability densities collapsed on to a single curve, indicative of some form of universal behavior, see Fig. 15.3. This universality was explored in previous work by considering a biophysical model of cell growth [464, 465], which is based on a stochastic version of a classical kinetic model of an autocatalytic cycle introduced by Hinshelwood [417]. The analysis is described in more detail in Sect. 15.1.2.

**Time-dependent solutions along characteristics.** A well-known feature of an age-structured evolution equation of the form (15.1.3) is that time-dependent solutions can be obtained using the method of characteristics [452]; see also Box 2C. First, suppose $t > \tau$ and consider straight line characteristics $t = \tau + t_0$ originating from the $t$-axis with $\tau$ the characteristic time and $t_0$, $t_0 > 0$, the Cauchy data parameter,

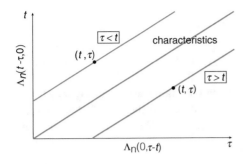

Fig. 15.4: Characteristics used to determine $\Lambda_n(t, \tau)$ in terms of the initial data $\Lambda_n(t - \tau, 0)$ for $t > \tau$ and $\Lambda_n(0, \tau - t)$ for $\tau > t$.

see Fig. 15.4. It follows that for fixed $t$

$$\frac{dn(t - \tau + \tau', \tau')}{d\tau'} = -\alpha(\tau')n(t - \tau + \tau', \tau'),$$

with $n(t - \tau, 0)) = v\rho(t - \tau)N(t - \tau)$. Hence

$$n(t, \tau) = v\rho(t - \tau)N(t - \tau)e^{-\int_0^t \alpha(\tau')d\tau'}, \quad t > \tau. \tag{15.1.16}$$

Similarly, when $t < \tau$ we consider straight line characteristics $\tau = t + \tau_0$ originating from the $\tau$-axis with $t$ now the characteristic time and $\tau_0$, $\tau_0 > 0$, the Cauchy data parameter, see Fig. 15.4. For fixed $\tau$

$$\frac{dn(t', \tau - t + t')}{dt'} = -\alpha(\tau - t + t')n(t', \tau - t + t'),$$

with $n(0, \tau - t) = n_0(\tau - t)$. Therefore

$$n(t, \tau) = n_0(\tau - t)e^{-\int_0^\tau \alpha(\tau - t + t')dt'}. \tag{15.1.17}$$

**Asymmetric cell division.** Motivated by the distinction between stalked and swarming *Caulobacter cresentus* cells, we follow Jafarpour et al [470] and now consider a cell population with two cell types: normal reproducing cells and quiescent cells that transition to normal cells before dividing. As before, normal cells have a division rate $\alpha(\tau)$ and a division time density $\rho(\tau) = \alpha(\tau)e^{-A(\tau)}$ with $A(\tau) = \int_0^\tau \alpha(\tau')d\tau'$. The age-dependent densities of normal cells are again denoted by $n(t, \tau)$ and $G(t, \tau) = n(t, \tau)/N(t)$ with $N(t) = \int_0^\infty n(t, \tau)d\tau$. Each normal cell divides to produce $v$ normal cells and $v_q$ quiescent cells. The quiescent cells transition to normal cells at a rate $\alpha_q(\tau_q)$ with a corresponding waiting time density $\rho_q(\tau_q) = \alpha_q(\tau_q)e^{-A_q(\tau_q)}$ and $A_q(\tau_q) = \int_0^{\tau_q} \alpha_q(\tau_q')d\tau_q'$. Finally, the corresponding age-dependent densities of quiescent cells are denoted by $n_q(t, \tau_q)$ and $G_q(t, \tau_q) = n_q(t, \tau_q)/N_q(t)$ with $N_q(t) = \int_0^\infty n_q(t, \tau_q)d\tau_q$.

The age-structured evolution equations are given by

$$\frac{\partial n}{\partial t} + \frac{\partial n}{\partial \tau} = -\alpha(\tau)n(t,\tau), \tag{15.1.18a}$$

$$\frac{\partial n_q}{\partial t} + \frac{\partial n_q}{\partial \tau_q} = -\alpha_q(\tau_q)n_q(t,\tau_q). \tag{15.1.18b}$$

The loss of each normal cell of age $\tau$ due to cell division simultaneously results in $v$ normal cells and $v_q$ quiescent cells of zero age. Moreover, the loss of each quiescent cell of age $\tau_q$ results in a single normal cell of zero age. We thus have the boundary conditions

$$n(t,0) = v\rho(t)N(t) + \rho_q(t)N_q(t), \quad n_q(t,0) = v_q\rho(t)N(t), \tag{15.1.19}$$

where

$$\rho(t) = \int_0^\infty \alpha(\tau)G(t,\tau)d\tau, \quad \rho_q(t) = \int_0^\infty \alpha_q(\tau_q)G_q(t,\tau_q)d\tau_q. \tag{15.1.20}$$

Finally, following along similar lines to the derivation of equation (15.1.5), we have (see Ex. 15.1)

$$\frac{dN(t)}{dt} = (v-1)\rho(t)N(t) + \rho_q(t)N_q(t), \tag{15.1.21a}$$

$$\frac{dN_q(t)}{dt} = v_q\rho(t)N(t) - \rho_q(t)N_q(t). \tag{15.1.21b}$$

Suppose that there exists a steady-state solution $G(t,\tau) = G^*(\tau), \rho(t) = \rho^*$ and $G_q(t,\tau_q) = G_q^*(\tau_q), \rho_q(t) = \rho_q^*$. Also assume that both cell populations grow exponentially at the same rate $k$ so that $\phi = N_q(t)/N(t)$ is a constant. It then follows from equation (15.1.21a) that (see Ex. 15.1)

$$k = (v-1)\rho^* + \rho_q^*\phi = \rho^*(v-1+\phi\gamma), \tag{15.1.22a}$$

$$k = v_q\frac{\rho^*}{\phi} - \rho_q^* = \rho^*\left(\frac{v_q - \phi\gamma}{\phi}\right), \tag{15.1.22b}$$

where $\gamma = \rho_q^*/\rho^*$. Repeating the steady-state analysis of the symmetric case, we obtain the steady-state equations

$$\frac{dG^*}{d\tau} + [\alpha(\tau) + k]G^*(\tau) = 0, \tag{15.1.23a}$$

$$\frac{dG_q^*}{d\tau_q} + [\alpha_q(\tau_q) + k]G_q^*(\tau_q) = 0, \tag{15.1.23b}$$

with

$$G^*(0) = \rho^*(v + \phi\gamma), \quad G_q^*(0) = \rho^*v_q/\phi. \tag{15.1.24}$$

These have the solutions

$$G^*(\tau) = \rho^*(v + \phi\gamma)e^{-k\tau}e^{-A(\tau)}, \quad G_q^*(\tau_q) = \frac{\rho^* v_q}{\phi}e^{-k\tau_q}e^{-A_q(\tau_q)}. \quad (15.1.25)$$

Finally, imposing the normalization conditions $\int_0^\infty G^*(\tau)d\tau = 1$, $\int_0^\infty G_q^*(\tau)d\tau = 1$ and using equation (15.1.22), we find (see Ex. 15.1)

$$\left\langle e^{-k\tau} \right\rangle_p = 1 - \frac{k}{\rho^*(v + \phi\gamma)} = \frac{1}{v + \phi\gamma}, \quad (15.1.26a)$$

$$\left\langle e^{-k\tau_q} \right\rangle_{pq} = 1 - \frac{k\phi}{\rho^* v_q} = \frac{\gamma\phi}{v_q}. \quad (15.1.26b)$$

Eliminating $\gamma\phi$ from the pair of equations (15.1.26a,b) yields an equation that determines the growth rate $k$ (analogous to equation (15.1.12))

$$\left\langle e^{-k\tau} \right\rangle_p = \frac{1}{v + v_q \left\langle e^{-k\tau} \right\rangle_{pq}}. \quad (15.1.27)$$

Substituting this back into equation (15.1.26) determines the product $\phi\gamma$, and equation (15.1.22) then gives the individual values of $\rho^*, \gamma, \phi$. This completely specifies the steady state solution.

## 15.1.2 Stochastic Hinshelwood model for population growth

The universal behavior of fluctuations during the growth and division of *Caulobacter cresentus* bacterial cells has been accounted for theoretically by Iyer-Biswas et al. [464] using a minimal biophysical model. The latter assumes that the cell components controlling cell growth are linked via a Hinshelwood cycle of autocatalytic reactions [417], whereby each chemical species catalyzes the production of the next, see Fig. 15.5. The stochastic Hinshelwood cycle (SHC) consists of $N$ species $\{X_1, X_2, \ldots, X_N\}$. The mean rate of production of $X_j$ is taken to be $k_j x_{j-1}$, $1 \le j \le N$, where $x_j$ is the copy number of $X_j$, and $X_0 \equiv X_N$. The reaction scheme is

$$X_{j-1} \overset{k_j x_{j-1}}{\longrightarrow} X_{j-1} + X_j. \quad (15.1.28)$$

(We will use $X_i$ to denote the chemical species and the stochastic copy number of the given reactant.) The corresponding master equation for the probability distribution $P(\mathbf{x}, t)$, $\mathbf{x} = (x_1, x_2, \ldots x_N)$ is

$$\frac{\partial P}{\partial t} = \sum_{j=1}^N k_j x_{j-1} [P(x_1, \ldots, x_j - 1, \ldots x_N) - P(x_1, \ldots, x_j, \ldots x_N)], \quad (15.1.29)$$

with $P(x_1, \ldots, x_j - 1, \ldots x_N) = 0$ if $x_j = 0$. Multiplying both sides by $x_i$ and integrating with respect to $\mathbf{x}$, we see that all terms on the right-hand side vanish except when $j = i$, for which

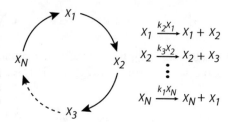

$$X_1 \xrightarrow{k_2 X_1} X_1 + X_2$$

$$X_2 \xrightarrow{k_3 X_2} X_2 + X_3$$

$$\vdots$$

$$X_N \xrightarrow{k_1 X_N} X_N + X_1$$

Fig. 15.5: Hinshelwood model for exponential growth, showing the autocatalytic cycle, in which each species activates production of the next, and the corresponding reactions.

$$\frac{d\langle X_i \rangle}{dt} = k_i \int x_i x_{i-1} \left[ P(x_1, \ldots, x_i - 1, \ldots x_N) - P(x_1, \ldots, x_i, \ldots x_N) \right] d\mathbf{x} = k_i \langle X_{i-1} \rangle.$$

In vector form with $\mu_j(t) = \langle X_j(t) \rangle$, we have

$$\frac{d\mu(t)}{dt} = \mathbb{K} \mu(t), \qquad (15.1.30)$$

where $\mathbb{K}$ is a cyclic matrix of period $N$, that is, $\mathbb{K}^N = k_1 k_2 \ldots k_N \mathbb{I}_N$, and $\mathbb{I}_N$ is the $N \times N$ identity matrix.

Consider the eigenvalue equation $\mathbb{K} \mathbf{u} = \lambda \mathbf{u}$, which can be iterated to give $\mathbb{K}^N \mathbf{u} = \kappa^N \mathbf{u} = \lambda^N \mathbf{u}$, where $\kappa$ is the geometric mean of the reaction rates $k_j$. It follows that the $m$-th eigenvalue is

$$\lambda_m = \kappa e^{2\pi i m/N}, \qquad \kappa^N = k_1 k_2 \ldots k_N,$$

and the $q$-th component of the corresponding eigenvector $\mathbf{u}_m$ is

$$u_m^{(q)} = \lambda_m^{-q} \prod_{p=1}^{q} k_p.$$

The eigenvalue with the largest real part is $\lambda_N = \kappa$, and this will dominate the asymptotic dynamics with relaxation time scale $\kappa^{-1}$. In particular, suppose that we expand the first moment vectors in terms of the eigenvectors $\mathbf{u}_m$:

$$\mu(t) = \sum_{m=1}^{N} c_m(t) \mathbf{u}_m,$$

so that

$$\sum_{m=1}^{N} \dot{c}_m(t) U_{qn} = \sum_{n=1}^{N} \lambda_n c_n(t) U_{nq}, \qquad U_{qm} = u_m^{(q)}.$$

Since the eigenvalues are distinct, the matrix of eigenvectors $\mathbb{U}$ is invertible so that

$$\dot{c}_m(t) = \lambda_m c_m(t), \qquad c_m(t) = e^{\lambda_m t} c_m(0),$$

and the eigenvalue expansion takes the form

$$\mu_j(t) = \sum_{m=1}^{N} e^{\lambda_m t} U_{jm} c_m(0) = \sum_{m,i=1}^{N} e^{\lambda_m t} U_{jm} U_{mi}^{-1} \mu_i(0).$$

Hence, in the asymptotic limit $(t \gg \kappa^{-1})$

$$\mu_j(t) \sim \left( \sum_{i=1}^{N} U_{Ni}^{-1} \mu_i(0) \right) e^{\kappa t} U_{jN}, \qquad (15.1.31)$$

which implies that the mean copy number of all the reactants evolve asymptotically as the single exponential $e^{\kappa t}$. It also follows that

$$\frac{\mu_j(t)}{\mu_r(t)} = \frac{U_{jN}}{U_{rN}},$$

which is independent of initial conditions.

An interesting interpretation of the geometric mean $\kappa$ can be obtained by assuming that each individual reaction rate is given by the Arrhenius form (see Chap. 2):

$$k_i = A_i \exp(-\Delta E_i / k_B T),$$

where $\Delta E_i$ is the activation energy of reaction $i$. Then

$$\kappa = (k_1 k_2 \ldots k_N)^{1/N}$$
$$= (A_1 A_2 \ldots A_N)^{1/N} \exp \left( -\frac{\Delta E_1 + \ldots + \Delta E_N}{N k_B T} \right)$$
$$\equiv A \exp(-\Delta E / k_B T),$$

where $\Delta E$ is the arithmetic mean of the elementary activation energies. Hence, assuming that each reaction step has an energy barrier that is of the order of a typical enzyme reaction, then so does the effective growth rate. This is consistent with experimental measurements of growth rates in single bacterial cells [465].

The next step is to determine the asymptotic behavior of the covariance matrix $\mathbb{C}(t)$ with

$$\mathbb{C}_{ij} = \langle X_i X_j \rangle - \langle X_i \rangle \langle X_j \rangle.$$

A basic result from the theory of chemical master equations with transition rates that are linear in the copy number is that the covariance matrix satisfies the same Ricatti equation as the corresponding OU process obtained using a system size expansion (see Chap. 5). That is

$$\frac{d\mathbb{C}(t)}{dt} = \mathbb{K}\mathbb{C}(t) + \mathbb{C}(t)\mathbb{K}^{\top} + \Theta^{\top}, \qquad \Theta_{ij} = \delta_{ij} \mu_j(t). \qquad (15.1.32)$$

It can be shown that in the asymptotic limit [464]

$$\mathbb{C}_{ij}(t) \sim U_{iN} U_{jN} e^{2\kappa t} \sum_{r=1}^{N} b_r \mu_r(0),  \tag{15.1.33}$$

with the coefficients $b_r$ dependent on the rates but not the initial conditions. Comparing equations (15.1.31) and (15.1.33), one finds that the rescaled covariance

$$\widehat{C}_{ij}(t) = \langle Y_i(t) Y_j(t) \rangle - \langle Y_i(t) \rangle \langle Y_j(t) \rangle, \quad Y_i(t) = \frac{X_i(t)}{\mu_i(t)},$$

takes the time-independent asymptotic form

$$\widehat{C}_{ij}(t) \sim \left( \sum_{l=1}^{N} U_{Nl}^{-1} \mu_l(0) \right)^{-2} \sum_{r=1}^{N} b_r \mu_r(0).  \tag{15.1.34}$$

In particular, $\mathrm{Var}[X_i(t)/\mu_i(t)] \sim$ constant. Since the rescaled covariance is independent of $i, j$ and time, it follows that the random variables $Y_i(t)$ are perfectly correlated. This only occurs if the random variables $Y_i(t)$ are linearly related. We conclude that although the means $\mu_i(t)$ grow exponentially in time, the rescaled random variables $X_i(t)/\mu_i(t)$ have the same time-independent probability density in the asymptotic limit, which is also observed experimentally [465]. This, in turn, implies that the $n$-th order moment of $X_i$ varies as $e^{n\kappa t}$.

## 15.2 Cell size regulation

Micro-organisms such as bacteria regulate their cell size, which is clear from the fact that their cell size distributions have a small coefficient of variation (CV), reaching as low as $CV = 0.1$. How cells control their size and maintain cell homeostasis is still an open question [188, 418, 943]. Classical studies of the cell cycle and cell size homeostasis have focused on two basic regulatory mechanisms, see Fig. 15.6(a,b). The first is a "sizer" mechanism, whereby cell division is triggered when a critical size is reached. Hence, the mean cell size can be recovered in a single generation. Sizer regulation implies that cells that are born smaller than average divide at the common average size but have a longer-than-average inter-division time until they reach a certain critical size. The second form of regulation is a "timer" mechanism, in which a fixed time interval elapses between cell cycles. Timer regulation means that cells that are born small typically divide when they are smaller than the average size. The efficacy of timer regulation for cell size control is thus dependent on the single-cell growth kinetics. If cells grow at a rate that is proportional to their size (exponential growth), then a stable size distribution may not be achieved. A simple argument for this is as follows [418]. Suppose that a cell grows exponentially at a rate $\kappa$ over a fixed time $\tau$. Let $v$ be the size at birth and $\hat{v}$ the size at the inter-division time $\tau$ so that $\hat{v} = v e^{\kappa \tau}$. Introducing a mean size $v_0$ and taking logs implies that

$$\hat{x} = x + \kappa\tau - \ln 2, \quad x = \ln(v/v_0), \quad \hat{x} = \ln(\hat{v}/v_0).$$

Size homeostasis could be achieved in the absence of noise provided that $\tau = \kappa^{-1}\ln 2$. On the other hand, if $\tau$ fluctuates according to a Wiener process, then the log sizes will have fluctuations that grow as the square root of the number of divisions. This suggests that the cell size distribution will not reach stationarity via a timer strategy. However, the argument breaks down if deviations from exponential growth occur at small cell sizes, for example. Hence, more quantitative studies of cell growth and division are necessary in order to determine the type of regulatory control mechanism [767].

In fact, recent advances in single-cell imaging have identified a third regulatory mechanism known as "adder" regulation, in which a fixed increment of material is added over the cell cycle [854], see Fig. 15.6(c,d). Deviations from the average size

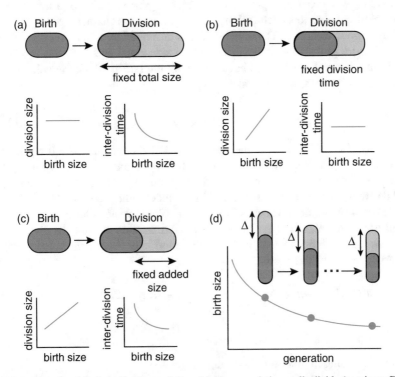

Fig. 15.6: Three models for cell size regulation. (a) Sizer regulation: cells divide (or trigger DNA replication) upon reaching a total critical size, (b) Timer regulation: cells divide after a fixed time period from birth, and (c) Adder regulation: cells divide after the addition of a fixed critical size. The three different mechanism can be distinguished by determining the correlations between birth size and division size, as well as between birth size and inter-division time. (d) The adder rule leads to progressive regression to the mean cell size over several generations. Only the case of a larger than average cell is shown; a similar picture holds for a smaller then average cell. [Redrawn from Willis et al (2017) [943].]

are inherited at a reduced level over multiple generations, rather than eliminated within a single generation as for sizer regulation. Although the adder mechanism has been identified in an increasing number of bacterial species, the jury is still out regarding how universal it is, and to what extent it depends on environmental conditions [943]. In the following, we will describe two different types of model for investigating the distribution of cell sizes, one based on a generalization of age-structured models [54, 744, 767] and the other on discrete stochastic maps [585].

### 15.2.1 PDE model of cell size regulation

The age-structured model introduced in Sect. 15.1 only kept track of the age of a cell following cell division, and thus neglected cell size. Moreover, the probability of dividing was taken to depend on the age $\tau$ alone, indicative of a timer mechanism. Now suppose that each cell in a culture at time $t$ is characterized by its age $\tau$ and its current volume $v$. The division rate is described by a function $\alpha(\tau, v)$ such that $\alpha(\tau, v)dt$ is the probability that a cell of size $v$ and age $\tau$ divides in the next small time interval $dt$. We also have to specify the rate at which the volume of an individual cell grows, which is taken to be

$$\frac{dv}{dt} = f(v). \tag{15.2.1}$$

(Typically, $f$ is independent of the age $\tau$.) We will assume that there is a minimum volume $v_0$ for which a cell is viable and take $v \geq v_0$ and $\alpha(\tau, v) = 0$ for $v \leq 2v_0$. Let $n(t, \tau, v)$ determine the number of cells in $[v, v+dv]$ and $[\tau, \tau+d\tau]$ at time $t$. We will derive a generalized age-structured PDE model, following [54, 744, 854].

Consider the number density of cells with volume $v$ such that $v_1 \leq v \leq v_2$. In a small time interval, this will change due to (i) a fraction of the cells dividing and thus resetting their age to zero, and (ii) cells entering or leaving the volume interval due to cell growth. Therefore

$$\int_{v_1}^{v_2} n(t+dt, \tau+dt, v)dv = \int_{v_1}^{v_2} n(t, \tau, v)[1 - \alpha(\tau, v)dt]dv$$
$$+ \int_{v_1-f(v_1)dt}^{v_1} n(t, \tau, v)dv - \int_{v_2-f(v_2)dt}^{v_2} n(t, \tau, v)dv.$$

Taylor expanding to first order in $dt$ and $d\tau$ gives

$$\int_{v_1}^{v_2} [n(t, \tau, v) + \partial_\tau n(t, \tau, v)d\tau + \partial_t n(t, \tau, v)dt] dv$$
$$= \int_{v_1}^{v_2} n(t, \tau, v)[1 - \alpha(\tau, v,)dt]dv - [n(t, \tau, v_2)f(v_2) - n(t, \tau, v_1)f(v_1)] dt.$$

Noting that

$$n(t,\tau,v_2)f(v_2) - n(t,\tau,v_1)f(v_1) = \int_{v_1}^{v_2} \frac{\partial}{\partial v}[n(t,\tau,v)f(v)]dv,$$

we have

$$\int_{v_1}^{v_2} [(\partial_t + \partial_\tau + \alpha(\tau,v))n(t,\tau,v) + \partial_v[n(t,\tau,v)f(v)]]\,dv = 0.$$

Since this equation holds for arbitrary values of $v_1, v_2$ such that $v_0 < v_1 < v_2 < \infty$, we finally obtain the PDE

$$\frac{\partial n}{\partial t} + \frac{\partial n}{\partial \tau} + \frac{\partial f(v)n}{\partial v} + \alpha(\tau,v)n = 0. \tag{15.2.2}$$

The loss of each cell of age $\tau$ due to cell division simultaneously results in $v$ daughter cells of zero age. Consider the case of symmetric binary division for which $v = 2$ and the daughter cells are identical. Hence, if the volume of the mother cell lies in the interval $[2v, 2v + 2dv]$, then each daughter cell will have volume in the interval $[v, v + dv]$. This is represented mathematically by the boundary condition

$$n(t,0,v) = 4\left[\int_0^\infty \alpha(\tau,2v)n(t,\tau,2v)d\tau\right]. \tag{15.2.3}$$

The total number of cells at time $t$ evolves according to

$$\begin{aligned}
\frac{dN(t)}{dt} &= \int_0^\infty d\tau \int_{v_0}^\infty dv\, \frac{\partial n(t,\tau)}{\partial t} \\
&= \int_0^\infty d\tau \int_{v_0}^\infty dv\left[-\frac{\partial n}{\partial \tau} - \frac{\partial f(v)n}{\partial v} - \alpha(\tau,v)n(t,\tau,v)\right] \\
&= \int_{v_0}^\infty n(t,0,v)dv - \int_0^\infty d\tau \int_{v_0}^\infty dv\,\alpha(\tau,v)n(t,\tau,v) \\
&= \rho(t)N(t), \tag{15.2.4}
\end{aligned}$$

where we have used the boundary condition (15.2.3), noted that $\alpha(\tau,v) = 0$ for $v \le 2v_0$, and defined

$$\rho(t) = \int_0^\infty d\tau \int_{v_0}^\infty dv\,\alpha(\tau,v)G(t,\tau,v),$$

with $n(t,\tau,v) = G(t,\tau,v)N(t)$. Since

$$\frac{\partial n(t,\tau,v)}{\partial t} = N(t)\frac{\partial G(t,\tau,v)}{\partial t} + G(t,\tau,v)\rho(t)N(t),$$

it follows that $G$ evolves as

$$\frac{\partial G}{\partial t} + \frac{\partial G}{\partial \tau} + \frac{\partial f(v)G}{\partial v} + (\rho(t) + \alpha(\tau,v))G = 0. \tag{15.2.5}$$

Note that in the case of a timer mechanism, $\alpha = \alpha(\tau)$, equation (15.2.2) reduces to the age-structured evolution equation (15.1.3) for $n = n(t, \tau)$, after integrating with respect to $v$. On the other hand, in the case of a sizer mechanism, $\alpha = \alpha(v)$ and $f = f(v)$, integration with respect to $\tau$ leads to the evolution equation

$$\frac{\partial n(t,v)}{\partial t} + \frac{\partial f(v)n(t,v)}{\partial v} + \alpha(v)n(t,v) = 4\alpha(2v)n(t,2v). \tag{15.2.6}$$

Moreover, we can set $\alpha(v) = f(v)\gamma(v)$, where $\gamma(v)dv$ is the probability that a cell divides when its volume lies in the interval $[v, v + dv]$.

**Stationary solution.** Assume exponential growth so that $N(t) \sim e^{kt}$ and hence $\partial_t \log N = k$. If a stationary solution $G^*(\tau, v)$ exists, then it satisfies the equation

$$\frac{\partial G^*}{\partial \tau} + \frac{\partial f(v)G^*}{\partial v} + (k + \alpha(\tau, v))G^* = 0. \tag{15.2.7}$$

The corresponding boundary condition is

$$G^*(0,v) = 4\left[\int_0^\infty \alpha(\tau, 2v)G^*(\tau, 2v)d\tau\right]. \tag{15.2.8}$$

Motivated by the sizer mechanism, suppose that the division rate is independent of cell age, $\alpha = \alpha(v)$. Then we have the quasi-linear PDE

$$\frac{\partial G^*}{\partial \tau} + f(v)\frac{\partial G^*}{\partial v} + \Gamma(v)G^* = 0, \tag{15.2.9}$$

with $\Gamma(v) = k + f'(v) + \alpha(v)$, and the boundary condition

$$G^*(0,v) = B(v) \equiv 4\alpha(2v)\left[\int_0^\infty G^*(\tau, 2v)d\tau\right]. \tag{15.2.10}$$

This can be solved using the method of characteristics (Box 2C). Let $\sigma$ denote the parameter of a characteristic curve such that

$$\frac{d\tau}{d\sigma} = 1, \quad \frac{dv}{d\sigma} = f(v), \quad \frac{dG^*}{d\sigma} + \Gamma(v(\sigma))G^* = 0.$$

The first two characteristic equations imply that $\sigma = \tau$ and

$$\tau = \int_{\bar{v}}^v f(v')^{-1}dv' \equiv F(v) - F(\bar{v}), \quad \bar{v} = v(0).$$

Solving the characteristic equation for $G^*$ with Cauchy data $G^*(0, v(0)) = B(\bar{v})$ gives

$$G^*(\tau, v(\tau)) = B(\bar{v})\exp\left(-\int_0^\tau \Gamma(v(\tau'))d\tau'\right).$$

We now want to eliminate $\bar{v}$ by expressing the solution in terms of a given $(\tau, v)$. Noting that

$$\bar{v} = F^{-1}(F(v) - \tau), \quad v(\tau') = F^{-1}(F(v) - \tau + \tau'),$$

we have

$$G^*(\tau, v) = B(F^{-1}(F(v) - \tau)) \exp\left(-\int_0^\tau \Gamma(F^{-1}(F(v) - \tau + \tau'))d\tau'\right).$$

Since we are interested in the stationary size distribution, we eliminate the age $\tau$ by introducing the marginal density

$$\Phi(v) = \int_0^\infty G^*(\tau, v)d\tau.$$

Substituting for $G^*$ and performing the change of variables $z = F^{-1}(F(v) - \tau)$ gives

$$\Phi(v) = \int_{v_0}^v \frac{B(z)}{f(z)} \exp\left(-\int_0^{F(v)-F(z)} \Gamma(F^{-1}(F(z) + \tau'))d\tau'\right) dz.$$

We have used $F^{-1}(F(v) - \tau) = v$ when $\tau = 0$ and $B(z) = 0$ for $z \le v_0$. After the second change of variables $y = F^{-1}(F(z) + \tau')$,

$$\Phi(v) = \int_{v_0}^v \frac{B(z)}{f(z)} \exp\left(-\int_z^v \frac{\Gamma(y)}{f(y)}dy\right) dz. \tag{15.2.11}$$

This is an implicit equation for $\Phi(v)$, since the boundary condition (15.2.10) can be expressed as

$$B(v) = 4\alpha(2v)\Phi(2v). \tag{15.2.12}$$

Using $\int_z^v dy = \int_{v_0}^v dy - \int_{v_0}^z dy$ and defining a new variable $\phi(v)$ according to

$$\Phi(v) = \phi(v)\exp\left(-\int_{v_0}^v \frac{\Gamma(y)}{f(y)}dy\right), \tag{15.2.13}$$

the implicit equation for $\Phi$ reduces to the following implicit equation for $\phi$

$$\phi(v) = \int_{v_0}^v K(z)\phi(2z)dz, \quad K(v) = \frac{4\alpha(2v)}{f(v)}\exp\left(-\int_v^{2v} \frac{\Gamma(y)}{f(y)}dy\right). \tag{15.2.14}$$

Note that $\phi(v_0) = 0$.

In order to solve equation (15.2.14), we first rewrite it in the form

$$\phi(v) = \phi_\infty - \int_v^\infty K(z)\phi(2z)dz,$$

where $\phi_\infty = \lim_{v\to\infty} \phi(v)$ exists. Iterating once gives

$$\phi(v) = \phi_\infty - \phi_\infty \int_v^\infty K(z_1)dz_1 + \int_v^\infty K(z_1)\left[\int_{2z_1}^\infty K(z_2)\phi(2z_2)dz_2\right]dz_1.$$

Iterating $M$ times and taking the limit $M \to \infty$ (assuming it exists) yields the formal solution

$$\phi(v) = \phi_\infty \lim_{M\to\infty}\left(1 + \sum_{m=1}^M (-1)^m \int_v^\infty dz_1 K(z_1)\ldots\int_{2z_{m-1}}^\infty dz_m K(z_m)\right) \equiv \phi_\infty R(v)$$

(15.2.15)

for $v \geq v_0$. A self-consistency condition is that $R(v_0) = 0$. The unknown constant $\phi_\infty$ is then determined by substituting for $\phi(v)$ in equation (15.2.13) and imposing normalization of the probability density $\Phi(v)$. It remains to determine the growth rate $k$. Following along similar lines to Sect. 15.1, an implicit equation for $k$ can be obtained by noting from equation (15.2.4)

$$k = \int_{2v_0}^\infty \alpha(v)\Phi(v)dv,$$

(15.2.16)

which is implicit, since $\Phi(v)$ involves integrals of $\Gamma(z)$, and in general can only be solved numerically.

**Example 15.1.** As a simple example, suppose that $v_0 = 0$, $f = 1\mu m^3/h$ and $\alpha = 1h^{-1}$. This corresponds to linear growth of individual cells and no volume/age regulation of cell division. It immediately follows from normalization that

$$k = \int_0^\infty \Phi(v)dv = 1.$$

From the definition of $K(v)$ in equation (15.2.14),

$$K(v) = 4e^{-(k+1)v},$$

which on substituting into equation (15.2.15) yields [744]

$$R(v) = 1 - \frac{4}{k+1}\sum_{m=0}^\infty (-1)^m \left(\frac{4}{k+1}\right)^m \pi_m^{-1}e^{-(k+1)\sigma_m v},$$

(15.2.17)

with

$$\sigma_m = \sum_{j=0}^m 2^j, \quad \pi_m = \prod_{j=0}^m \sigma_j.$$

It can be seen that $R(v) \to 1$ as $v \to \infty$ and one finds numerically that $R(0) = 0$ provided that $k = 1$ as required. Finally,

$$\Phi(v) = \phi_\infty R(v)e^{-(k+1)v},$$

with $\phi_\infty \sim 6.9 \text{ h}^{-1}$ from normalization. A sketch of the resulting size distribution is shown in Fig. 15.7.

Finally, note that a generalized age-structured model of the form (15.2.2) has been fitted numerically with experimental data taken from E. coli growing in a nutrient-rich medium [767]. This particular study indicates that the data provides

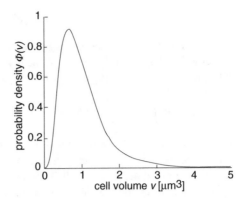

Fig. 15.7: Sketch of probability density $\Phi$ as function of cell size $v$ for a simple division rule with $v_0 = 0$, $f = 1\,\mu\mathrm{m}^3/\mathrm{h}$ and $\alpha = 1$ $\mathrm{h}^{-1}$.

support for a sizer mechanism rather than a timer mechanism, that is, $\alpha = \alpha(v)$ rather than $\alpha = \alpha(\tau)$.

## 15.2.2 Discrete-time stochastic model for cell size regulation

One limitation of age-structured PDE models is that they tend to be analytically intractable, except in a few simple cases. This, together with advances in cell imaging, has led to an alternative modeling approach that describes single cells over many divisions [10, 585, 630, 875]. The latter is more amenable to analysis as it considers only a single individual described by a discrete-time stochastic process or stochastic map. Here we describe the particular formulation of Ref. [585]. For simplicity, we focus on symmetric binary division; the more general case of asymmetric division can be found in Ref. [585].

**Stochastic growth model.** At the $n$-th generation a cell born with volume $v_n$ grows exponentially in time to a final size $\hat{v}_n$, after which it divides into two daughter cells of volume $\hat{v}_n/2$ (assuming symmetric cell division). Let $\phi$ specify the relationship between birth size and division size according to $\hat{v}_n = \phi(v_n)$. We will focus on the affine linear function $\phi(v) = \Delta + cv$. Given exponential growth of cell size prior to cell division, the division time $\tau(v_n)$ then satisfies

$$\phi(v_n) = v_n e^{\kappa \tau(v_n)} \implies \tau(v_n) = \frac{1}{\kappa} \ln\left(\frac{\phi(v_n)}{v_n}\right).$$

Note that the timer mechanism corresponds to the case $\Delta = 0$, since $\tau(v_n) = \kappa^{-1} \ln c$ is a constant. On the other hand, $c = 0$ is a sizer policy, since the final size before division is fixed. Finally, the adder policy corresponds to the case $f(v) = \Delta + v$. A stochastic version of the model is obtained by adding errors in the growth time (multiplicative noise) and the growth model (additive noise)

$$\hat{v}_n = \phi(v_n)e^{\kappa \Delta \tau_n} + \Delta v_n, \tag{15.2.18}$$

where $\Delta \tau_n$ and $\Delta v_n$ are random variables.

Let $P_n(v)$ describe the density of birth volumes for the $n$-th generation, and let $\widehat{P}_n(v')$ be the corresponding density of final volumes. Then

$$\widehat{P}_n(\hat{v}) = \int P_n(v)\delta[\hat{v} - (f(v)e^{\kappa \Delta \tau} + \Delta v)]\rho(\Delta \tau, \Delta v)d\Delta \tau d\Delta v dv, \tag{15.2.19}$$

where $\rho(\Delta \tau, \Delta v)$ is the joint probability density for the noisy growth model. For simplicity, the noise is independent across generations and independent of the birth volume. It immediately follows that for symmetric division into two daughter cells, $P_{n+1}(v) = 2\widehat{P}_n(2v)$, the size density evolves according to the recursive formula [418]

$$P_{n+1}(v) = \int P_n(v')\delta[v - (f(v')e^{\kappa \Delta \tau} + \Delta v)/2]\rho(\Delta \tau, \Delta v)d\Delta \tau d\Delta v dv'$$

$$\equiv \int_0^\infty K(v, v')P_n(v')dv'. \tag{15.2.20}$$

A stationary solution $P_{n+1}(v) = P_n(v) = P(v)$ (if it exists) satisfies

$$P(v) = \int_0^\infty K(v, v')P(v')dv', \tag{15.2.21}$$

which is known as a homogeneous Fredholm integral equation of the second kind [734]. It is analogous to the matrix eigenvalue problem $P_j = \sum K_{jk}P_k$ in finite-dimensional linear algebra.

Introduce the notation

$$\mathbb{K}a(v) = \int_0^\infty K(v, v')a(v')dv'$$

for any function $a \in L^2([0, \infty))$. Mathematically speaking, $K(v, v')$ is then referred to as the integral kernel and $P(v)$ is an eigenfunction of the corresponding integral operator $\mathbb{K}$ with eigenvalue 1. We then note that

$$\langle b, \mathbb{K}a \rangle = \int_0^\infty b(v)\left(\int_0^\infty (K(v, v')a(v')dv'\right)dv = \int_0^\infty \left(\int_0^\infty b(v)K(v, v')dv\right)a(v')dv'$$

$$= \langle \mathbb{K}^\dagger b, a \rangle,$$

after reversing the order of integration, where $\mathbb{K}^\dagger$ is the adjoint integral operator. It immediately follows that the adjoint integral operator has the kernel $K^\dagger(v, v') = K(v', v)$. From the definition of $K(v, v')$ in equation (15.2.20) and properties of the Dirac delta function, we have

$$\int_0^\infty K(v, v')dv = \int \rho(\Delta \tau, \Delta v)d\Delta \tau d\Delta v = 1.$$

That is, the adjoint integral operator $\mathbb{K}^\dagger$ has an eigenfunction $a^\dagger(v) = 1$ with eigenvalue 1. This implies that $\mathbb{K}$ has at least one eigenfunction of eigenvalue 1. However, there is no guarantee that one of these eigenfunctions $q(v)$, say, is a probability density. The latter requires that $q(v) \geq 0$ for all $v \in [0,\infty)$ and $\int_0^\infty q(v)dv = 1$ (after rescaling)).

One example where an eigenfunction exists but is not normalizable occurs in the case of the timer mechanism with exponential temporal growth [585]. Suppose that there is no additive noise ($\Delta v \equiv 0$) and $\Delta \tau$ is generated from a zero mean Gaussian with variance $\sigma^2$. Setting $f(v) = cv$ then gives

$$K(v,v') = \int \delta[v - cv'e^{\kappa\Delta\tau}/2] \frac{1}{\sqrt{2\pi\sigma^2}} e^{-\Delta\tau^2/2\sigma^2} d\Delta\tau$$

$$= \frac{1}{\kappa v\sqrt{2\pi\sigma^2}} \exp\left[-\frac{1}{2\kappa^2\sigma^2}\left(\ln\left[\frac{2v}{cv'}\right]\right)^2\right].$$

Multiplying both sides by $1/v'$ leads to a log-normal distribution:

$$K(v,v')\frac{1}{v'} = \frac{1}{v}\ln\mathcal{N}_{v'}(\ln(2v/c), \kappa^2\sigma^2).$$

Since the log-normal distribution has unit normalization, it follows that

$$\int_0^\infty K(v,v')\frac{1}{v'}dv' = \frac{1}{v}.$$

In other words, the non-normalizable function $1/v$ is an eigenfunction of $\mathbb{K}$ with eigenvalue 1. In the absence of any other natural volume scales in the system, dimensional analysis shows that this eigenfunction is unique.

Although it is not possible to rigorously prove existence of a stationary probability density $P(v)$, one can show that if $P(v)$ exists, then it is unique and the system converges to $P(v)$ [585]. First, given the conditions $P_n(v) \geq 0$ for all $v \in [0,\infty)$ and $\int_0^\infty P_n(v)dv = 1$, it follows that for any $\varepsilon > 0$, there exists a cut-off volume $\mathcal{V}$ such that $P_n(v > \mathcal{V}) < \varepsilon$ for all $n$. It is then possible to discretize the integral equation (15.2.20). First, setting $v_j = j\delta v$ for $j = 0,\ldots,M$ and $\delta v = \mathcal{V}/M$, we take

$$p_{n,j} = P_n(j\delta v)\delta v, \quad j = 0,\ldots,M-1, \quad p_{n,M} = \text{Prob}[v \geq \mathcal{V}|\text{generation } n],$$

$$K_{ij} = \frac{M}{\mathcal{V}} \int_{(i-1)\delta v}^{i\delta v} dv \int_{(j-1)\delta v}^{j\delta v} dv' K(v,v'), \quad 0 \leq i,j \leq M-1,$$

and

$$K_{M+1,j} = \frac{M}{\mathcal{V}} \int_{\mathcal{V}}^\infty dv \int_{(j-1)\delta v}^{j\delta v} dv' K(v,v').$$

The last definition ensures that $\sum_{i=0}^M K_{ij} = 1$ for $j = 0,1,\ldots,M-1$, which is necessary for $\mathbf{K}$ to be a stochastic matrix. Under the assumption that a stationary density exists, it can be shown that for sufficiently large cut-off $\mathcal{V}$ the contribution from the final term $K_{iM}p_{n,M}$ in the discretized sum $p_{n+1,i} = \sum_{j=0}^{M} K_{ij}p_{n,j}$ can be made arbitrarily small. Thus, $K_{iM}$ can be chosen such that $\mathbf{K}$ is an irreducible stochastic matrix. It then follows from the Perron-Frobenius theorem (see Box 3A) that the largest eigenvalue of the matrix $\mathbf{K}$ is equal to 1, and the system converges to the corresponding unique positive eigenvector as $n \to \infty$.

**Moments of the stationary distribution.** Suppose that the system converges to a unique stationary density $P(v)$. The stochastic growth model (15.2.18) implies that immediately after cell division the birth volume of a daughter cell is

$$v_{n+1} = \frac{1}{2}f(v_n)e^{\kappa \Delta \tau_n} + \Delta v_n. \tag{15.2.22}$$

Taking moments of both sides with respect to the stationary density for $v$ and the additive and multiplicative noise distributions for $\Delta \tau$ and $\Delta v$ gives

$$\langle v \rangle = \frac{1}{2}\langle f(v) \rangle \langle e^{\kappa \Delta \tau} \rangle.$$

For simplicity, we assume that $\langle \Delta v \rangle = 0$. In the case of a linear growth function, $f(v) = cv + \Delta$,

$$\langle v \rangle = \frac{1}{2}(\Delta + c\langle v \rangle \langle e^{\kappa \Delta \tau} \rangle),$$

which can be rearranged to yield

$$\langle v \rangle = \frac{\Delta \langle e^{\kappa \Delta \tau} \rangle}{2 - c\langle e^{\kappa \Delta \tau} \rangle}. \tag{15.2.23}$$

Similarly, for the linear growth function it can be shown that the variance of $v$ is

$$\sigma_v^2 = \frac{\Delta^2 \langle e^{2\kappa \Delta \tau} \rangle}{4 - c^2 \langle e^{2\kappa \Delta \tau} \rangle} \left( \frac{\langle \Delta v^2 \rangle}{\Delta^2 \langle e^{2\kappa \Delta \tau} \rangle} + \frac{4\langle e^{2\kappa \Delta \tau} \rangle - 4\langle e^{\kappa \Delta \tau} \rangle^2}{(2 - c\langle e^{\kappa \Delta \tau} \rangle)^2 \langle e^{2\kappa \Delta \tau} \rangle} \right). \tag{15.2.24}$$

Note that the mean is only positive when $c < 2/\langle e^{\kappa \Delta \tau} \rangle$ and the variance is only positive when $c < 2/\sqrt{\langle e^{2\kappa \Delta \tau} \rangle}$. Non-existence of these moments for sufficiently large $c$ does not necessarily mean that the stationary density does not exist, since it could simply be heavy-tailed. Indeed, an asymptotic analysis of the stationary probability density (if it exists) indicates that for multiplicative noise ($\Delta v \equiv 0$) [585]

$$P(v) \sim \frac{1}{v^{1+\beta}} \quad \text{for } v \to \infty,$$

with $\beta$ related to the coefficient $c$ according to

$$\beta = \frac{2(\ln 2 - \ln c)}{\kappa^2 \sigma^2}.$$

## 15.3 Bacterial persistence and phenotypic switching

Advances in high-resolution measurements of gene expression in single cells have revealed that there is significant variation in gene expression across individual cells of the same cell type [145, 234, 258, 588, 730]. It is thought that such variation is a consequence of the stochastic nature of gene expression, involving both gene regulation (Chap. 5) and cell growth. A major issue is to what extent single-cell molecular variation might play a role in population-level function. That is, does cell variation indicate a diversity of hidden functional processes within an ensemble of identical cells, and can this diversity facilitate collective behavior that would be absent in a homogeneous population [259, 265, 971]? It is important to note that variation or variability of a trait refers to the entire characteristics of the distribution of the trait over the ensemble of cells. This should be distinguished from statistical quantities such as variance. In particular, as illustrated in Fig. 15.8, two different single-cell measurements can have the same variance but a different variation at the population level. Moreover, the two different populations can respond very differently when there are thresholding effects, such as resistance to an antibiotic or a chemotherapy agent.

One example where it may be advantageous for a population to maintain a diversity of cell phenotypes is in fluctuating, unpredictable environments. Since each phenotype is better suited to a particular environmental condition, there is a corresponding fitness tradeoff [575, 688]. Such a strategy may be preferable to one in which individual cells sense and respond to the environment, which comes at a metabolic cost, particularly when there is insufficient time for signal transduction of environmental signals, or where the environment is favorable most of the time but occasionally causes a catastrophe that wipes out most of the population. The potential advantages of phenotypic variation in fluctuating environments has been known for a long time in ecology and population genetics, where it is known as bet hedging [811]. More recently, there have been a number of experimental [1, 38, 575] and modeling [1, 168, 169, 446, 536, 575, 601, 649, 716, 759, 871, 909, 937] studies of

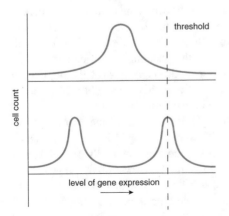

Fig. 15.8: Schematic illustration of two different single-cell distribution with the same variance in gene expression but different variation.

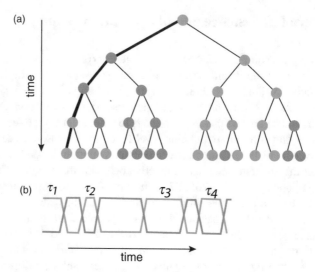

Fig. 15.9: Population-level phenotypic switching. (a) An individual mother cell grows and divides into two daughter cells, producing a family tree over multiple generations. Switching between two phenotypes is shown using two different colors. At a given generation, the relative abundance of each of the two phenotypes can be determined by measuring across the corresponding row. Alternatively, one can keep track of phenotypic switching along one lineage of the expanding tree (indicated by the thicker branches). If this single lineage is observed for a sufficiently long time then one should recover the steady-state population incidence of the two phenotypes, assuming the system is ergodic. (b) Keeping track of the durations $\tau_l$ of each phenotype generates the residence time distribution.

bet hedging in microbial populations. Note that by observing the evolution of successive generations, one can obtain statistics regarding the distribution of switching or residence times, the relative abundance of different phenotypes and the fitness of the population, see Fig. 15.9.

A classic example of phenotypic variation occurs in *E. coli* populations, which maintain a subset of cells in a quiescent or persistent phenotypic state [38, 536]. The resulting bimodal population distribution, reflecting subpopulations of persistent and proliferating cells, is generated by individual *E. coli* cells stochastically switching into and out of persistence. Switching between phenotypic states with different growth rates is an important factor in the phenomenon of persistent bacterial infections after treatment with antibiotics. Although most of the population is rapidly killed by the treatment, a small genetically identical subset of dormant persister cells can survive an extended period of exposure. When the drug treatment is removed, the surviving persisters randomly transition out of the dormant state, causing the infection to reemerge. This is illustrated in Fig. 15.10(a). Persistence of *E. coli* in response to antibiotics has been studied experimentally by Balaban et al. [38] using microfluidic devices. These allow direct observation of the growth of individual bacteria under normal conditions, and the analysis of survivors following antibiotic treatment. The authors found that a wild type population can be divided

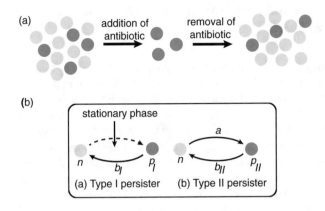

Fig. 15.10: Phenotypic heterogeneity in a bacterial population. (a) Dynamics of a heterogeneous population consisting of normal (light) and persister (dark) cells. The persisters survive the addition of an antibiotic, which allows the population to recover after the removal of the antibiotic and (b) Reaction schemes for type I and type II bacterial persisters.

into three subpopulations: normal cells ($n$); continuously generated type II persisters ($p_{II}$); and stationary phase type I persisters ($p_I$), see Fig. 15.10(b). Type I persisters constitute a preexisting population of essentially non-growing (dormant) cells that are generated by trigger events during the stationary phase of cell growth. There is negligible switching $n \to p_I$ during exponential growth of the normal population. On the other hand, there are transitions $p_I \to n$ at some rate $b_I \sim 0.05$ hours$^{-1}$. Type II persisters constitute a subpopulation of slowly growing cells with growth rate $\mu_p$ that is an order of magnitude slower than the growth rate $\mu_n$ of the normal population. They arise at a switching rate $a \sim 10^{-3}$ hours$^{-1}$ from the normal population, and switch back at a rate $b_{II} \sim 10^{-7} - 10^{-4}$ hours$^{-1}$.

Another example of phenotypic variation due to switching occurs in the lactose metabolism of *E. coli*: if an inducer that is not metabolized binds to the repressor of the lac operon, then only a part of the bacterial population expresses lac [832]. This allows a more robust response to changes in lactose densities in the environment. Similarly, *E. coli* infected by $\lambda$-phage are stochastically driven toward one of two possible fates [28]: either they are killed (lysed) or survive as lysogens, whereby the phage DNA in incorporated into their chromosomes. (See Chap. 5 for more details.) A fourth example is the stochastic switching between expression and non-expression of cell surface pili during infection of the urinary tract by *E. coli* [577, 947]. A final example occurs in the quorum sensing system of *V. harveyi* (Sect. 15.4.1), where it has been found that only a fraction of the cells respond to an autoinducer [19]. Phenotypic switching thus appears to be a common mechanism for generating cell variation within microbial populations [575, 688]. As far as we are aware, there are no known examples of bet hedging in healthy mammalian tissues, perhaps due to the interdependence of cells in multicellular organisms. However, it is possible that mammalian cancer cells exhibit phenotypic switching in order to survive chemotherapies [382, 981].

### 15.3.1 Modeling phenotypic switching in random environments as a stochastic hybrid system

Consider a bacterial colony that is in the exponential growth phase of its evolution, see Fig. 15.1. Suppose that individual bacterial cells can switch between two different phenotypes labeled by $A, B$ and that the environment switches between two states labeled $\sigma \in \{0, 1\}$. The switching times of the phenotypes and the environmental states are taken to be exponentially distributed. One can then model the evolution of the phenotypic subpopulations in terms of a stochastic hybrid system [445, 446]; see Box 5D. (There have have also been studies that treat the environment as a continuous stochastic process or include the effects of intrinsic noise due to finite system size [445, 937].) Let $x_j(t)$ denote the number of bacteria in a large population that has phenotype $j$ at time $t$, and let $\sigma(t)$ be the current state of the environment. The population equations take the form

$$\frac{dx_A}{dt} = \gamma_\sigma^A x_A - k_A x_A + k_B x_B, \tag{15.3.1a}$$

$$\frac{dx_B}{dt} = \gamma_\sigma^B x_B + k_A x_A - k_B x_B. \tag{15.3.1b}$$

Here the growth rates of the two subpopulations $A$ and $B$, $\gamma_\sigma^A, \gamma_\sigma^B$, are indexed by the current state $\sigma(t) = \sigma \in \{0, 1\}$ of the environment, and $k_A, k_B$ are the phenotypic switching rates. The environment is also assumed to switch according to a two-state Markov chain

$$\{\sigma = 0\} \underset{\alpha}{\overset{\beta}{\rightleftharpoons}} \{\sigma = 1\}.$$

Finally, we will assume that state $\{0\}$ favors phenotype $A$ and state $\{1\}$ favors phenotype $B$, that is, $\gamma_0^A > \gamma_0^B$ and $\gamma_1^A < \gamma_1^B$.

Following [871], define the population fitness $f$ when $\sigma = 0$ as the fraction of the total population that is in state $A$:

$$f(t) = \frac{x_A(t)}{N(t)}, \quad N(t) = x_A(t) + x_B(t).$$

A nonlinear dynamical equation for $f$ can be derived by first noting that

$$\frac{dN}{dt} = \gamma_\sigma^A x_A + \gamma_\sigma^B x_B = \gamma_\sigma(t) N, \tag{15.3.2}$$

with

$$\gamma_\sigma(t) = \gamma_\sigma^B + \Delta \gamma_\sigma f(t), \quad \Delta \gamma_\sigma = \gamma_\sigma^A - \gamma_\sigma^B.$$

Here we have $\Delta \gamma_1 < 0 < \Delta \gamma_0$. Differentiating the expression for $f$ with respect to time then shows that

$$\frac{df}{dt} = \frac{1}{N}\frac{dx_A}{dt} - \frac{x_A}{N^2}\frac{dN}{dt} = (\gamma_\sigma^A - k_A)f + k_B(1 - f) - f(\gamma_\sigma^B + \Delta \gamma_\sigma f).$$

This can be rewritten as

$$\frac{df}{dt} = -\Delta\gamma_\sigma(f - f_\sigma^+)(f - f_\sigma^-),\tag{15.3.3}$$

where $f_\sigma^\pm$ are the roots of the quadratic equation

$$f^2 - \left(1 - \frac{k_A + k_B}{\Delta\gamma_\sigma}\right)f - \frac{k_B}{\Delta\gamma_\sigma} = 0.$$

That is,

$$f_\sigma^\pm = \frac{\Gamma_\sigma \pm \sqrt{\Gamma_\sigma^2 + 4k_B\Delta\gamma_\sigma}}{2\Delta\gamma_\sigma}, \quad \Gamma_\sigma = \Delta\gamma_\sigma - k_A - k_B.$$

Equation (15.3.3) implies that in the environmental state $\sigma$, $f(t)$ will converge towards the state $f_\sigma^+$. Since the zero state favors phenotype $A$, we have $f_1^+ < f_0^+$ and thus after a transient, $f(t) \in [f_1^+, f_0^+]$. We now note that the stochastic hybrid dynamics of $f(t)$ is independent of $N(t)$. Equation (15.3.3) is thus a one-dimensional stochastic hybrid system, whose corresponding CK equation can be solved along similar lines to the model of an unregulated gene network with promoter noise; see Chap. 5 and Ref. [490]. One finds that the stationary density for $f$ is [445, 446]

$$\Pi_0^*(f) = \frac{\mathcal{N}}{\Delta\gamma_0}(f_0^+ - f)^{g-1}(f - f_0^-)^{-g-1}(f - f_1^+)^h(f_1^- - f)^{-h},\tag{15.3.4a}$$

$$\Pi_1^*(f) = -\frac{\mathcal{N}}{\Delta\gamma_1}(f_0^+ - f)^g(f - f_0^-)^{-g}(f - f_1^+)^{h-1}(f_1^- - f)^{-h-1},\tag{15.3.4b}$$

where

$$g = \frac{\beta}{\Delta\gamma_0(f_0^+ - f_0^-)}, \quad h = \frac{\alpha}{\Delta\gamma_1(f_1^+ - f_1^-)},$$

and $\mathcal{N}$ is a normalization factor that ensures

$$\int_{f_1^+}^{f_0^+} [\Pi_0^*(f) + \Pi_1^*(f)]df = 1.$$

It follows from equation (15.3.2) that the average growth rate is given by

$$\gamma_{av} = \lim_{T\to\infty} \frac{1}{T}\int_0^T \gamma_\sigma(t)dt.$$

Hufton *et al.* assume ergodicity and replace the time average by an ensemble average. That is, the fraction of time the environment is in the state $\sigma$ is given by the stationary distribution $\rho_\sigma$ with

$$\rho_0 = \frac{\alpha}{\alpha+\beta}, \quad \rho_1 = \frac{\beta}{\alpha+\beta}.$$

Hence, the ensemble averaged growth rate can be decomposed as

$$\gamma_{\text{av}} = \rho_0 \mathbb{E}[\gamma | \sigma = 0] + \rho_1 \mathbb{E}[\gamma | \sigma = 1], \tag{15.3.5a}$$

with

$$\mathbb{E}[\gamma | \sigma] = \gamma_\sigma^B + \Delta \gamma_\sigma \mathbb{E}[f | \sigma], \tag{15.3.5b}$$

and

$$\mathbb{E}[f | \sigma] \rho_\sigma = \int_{f_1^+}^{f_0^+} f \Pi_\sigma^*(f) df. \tag{15.3.5c}$$

Hufton et al. [446] use equation (15.3.5) to calculate the average growth rate as a function of model parameters. They obtain results that are consistent with previous studies: (i) In slow switching environments the optimal phenotypic switching rates are equal to the phenotypic switching rates ($\beta = k_A, \alpha = k_B$), (ii) No phenotypic switching (homogeneity) is preferred in the regime of fast environmental switching. The preferred phenotype is the one that grows faster on average: always phenotype $A$ if $\rho_0 \gamma_0^A + \rho_1 \gamma_1^A > \rho_0 \gamma_0^B + \rho_1 \gamma_1^B$, and always phenotype $B$ otherwise, and (iii) At intermediate switching rates the optimal phenotypic switching rates are slower than the environmental switching rates.

As a modification of the previous model, suppose that the environment is almost always in the state $\sigma = 0$, which is favorable to population $A$, but that there are occasional catastrophes where the number in state A is suddenly decreased. Moreover, suppose that there is feedback from the microbial population and the environment: the probability of a catastrophe (such as an antibiotic treatment) depends on the state of the population. This could mimic, for example, the host immune response. Using mathematical modeling, one finds that that there are two favored tactics for microbial populations in environments with a given feedback function: keep all the population in the fast growing state, regardless of the environmental response, or alternatively, use switching to maintain a population balance that reduces the likelihood of an environmental response. Which of these strategies is optimal depends on the parameters of the model. In the absence of any feedback between the population and environment, phenotypic switching is always unfavorable. However, as the environment becomes more responsive, switching can be advantageous. This is explored further in Ex. 15.2 using a model due to Visco et al. [909].

### 15.3.2 Stochastic model of population extinction

We now turn to a recent WKB analysis of extinction in a population model consisting of two distinct phenotypes [569], which can be interpreted as normal and persister cells, respectively. As in the previous model, the populations are assumed to be well-mixed so that spatial effects can be ignored. Assuming a maximum carrying capacity of $K$, the normals grow at a rate $B(1 - x_A/K)$ and die at a rate that is set to unity, while the persisters remain static. The mean-field kinetic equations (for large populations) are

$$\frac{dx_A}{dt} = B\left(1 - \frac{x_A}{K}\right)x_A - x_A - \alpha x_A + \beta x_B, \tag{15.3.6a}$$

$$\frac{dx_B}{dt} = \alpha x_A - \beta x_B. \tag{15.3.6b}$$

The rate equations have a trivial fixed point at $x_A = x_B = 0$, which represents population extinction, and a nontrivial fixed point $Q$ at $x_A^* = K(1 - 1/B)$, $x_B^* = \alpha x_A^*/\beta$. The latter fixed point is stable provided that $B > 1$, whereas the zero state is a saddle point. Hence, in the absence of noise and $x_A(0) > 0$, the system converges to the nontrivial fixed point $Q$, resulting in a viable steady-state population of phenotypes.

As with biochemical processes, one can write down a stochastic version of the population model based on the discrete nature of individuals. The resulting master equation for the probability distribution $P_{nm}(t) = \mathbb{P}[X_A(t) = n, X_B(t) = m]$ is

$$\frac{dP_{n,m}}{dt} = \widehat{A}P_{n,m}, \quad \frac{dP_{0,0}}{dt} = P_{1,0}, \tag{15.3.7}$$

with

$$\widehat{A}P_{n,m} = B(n-1)\left(1 - \frac{n-1}{K}\right)P_{n-1,m} + (n+1)P_{n+1,m} - Bn\left(1 - \frac{n}{K}\right)P_{n,m} - nP_{n,m}$$

$$+ \alpha(n+1)P_{n+1,m-1} + \beta(m+1)P_{n-1,m+1} - \alpha nP_{n,m} - \beta(1 - \delta_{n,K})mP_{n,m}.$$

Here $(n,m) \in [0,K] \times [0,\infty)$ with $P_{n<0,m} = P_{n,m<0} = P_{n>K,m} = 0$. In the stochastic model the state $n = m = 0$ is now an absorbing extinction state. The extinction state reflects the existence of a simple zero eigenvalue of the generator $-\widehat{A}$ of the Markov chain with corresponding eigensolution $P_{n,m} = \delta_{n,0}\delta_{m,0}$. Since all other eigenvalues are positive definite, it follows that all other eigensolutions decay to zero, and hence the population goes extinct in the limit $t \to \infty$. We thus have a situation analogous to the stochastic genetic switch considered in Chap. 5. Now the metastable state is the nontrivial fixed point $Q$, and there is an absorbing boundary at the zero fixed point. We can decompose the probability density for all nonzero states as

$$P_{n,m}(t) = \sum_r \phi_{n,m}^{(r)} e^{-\lambda^{(r)}t},$$

where the sum is over all nonzero eigenvalues, and

$$P_{0,0}(t) = 1 - \sum_{n,m}' P_{n,m}(t).$$

Here $'$ indicates that the sum excludes $(0,0)$. If the system is initially close to $Q$ and $K$ is sufficiently large, then we expect extinction to be a rare event. Following from the analysis of Chap. 8, we assume that the smallest nonzero eigenvalue is exponentially small with respect to the system size $K$, in contrast to all other nonzero eigenvalues. We thus have the quasi-stationary approximation for $(n,m) \neq (0,0)$,

$$P_{n,m}(t) \approx \Pi_{n,m} e^{-\lambda^{(0)}t},$$

where $\Pi_{n,m} = \phi_{n,m}^{(0)}$. Choosing $\Pi_{n,m}$ to have unit normalization (with $\Pi_{0,0} = 0$), it follows that

$$P_{0,0} \approx 1 - e^{-\lambda^{(0)}t},$$

and we can identify $1/\lambda^{(0)}$ as the mean time to extinction [33]. Moreover, equation (15.3.7) implies that $\lambda^{(0)} = \Pi_{1,0}$.

The quasi-stationary density can be calculated using a WKB approximation along similar lines to Chap. 8. Consider the WKB ansatz

$$P_{n,m}(t) \approx e^{-K\Phi(x,y)} e^{-\lambda^{(0)}t}, \quad (n,m) \neq (0,0); \quad \lambda^{(0)} \approx e^{-K\Phi(0,0)},$$

with $x = n/K, y = m/K$ treated as continuous variables. Substituting the WKB ansatz into the quasi-stationary equation $\widehat{A}\Pi_{n,m} = 0$ gives, to leading order in $1/K$, the zero-energy Hamilton-Jacobi equation [569]

$$H(x,y,\partial_x\Phi,\partial_y\Phi) = 0, \tag{15.3.8}$$

with effective Hamiltonian

$$H(x,y,q_x,q_y) = Bx(1-x)(e^{q_x} - 1) + x(e^{-q_x} - 1) + \alpha x(e^{-q_x+q_y} - 1)$$
$$+ \beta y(e^{q_x-q_y} - 1). \tag{15.3.9}$$

Hamilton's equation are given by

$$\dot{x} = Bx(1-x)e^{q_x} - xe^{-q_x} - \alpha xe^{-q_x+q_y} + \beta ye^{q_x-q_y}, \tag{15.3.10a}$$
$$\dot{y} = \alpha xe^{-q_x+q_y} - \beta ye^{q_x-q_y}, \tag{15.3.10b}$$
$$\dot{q}_x = -B(1-2x)(e^{q_x} - 1) - (e^{-q_x} - 1) - \alpha(e^{-q_x+q_y} - 1), \tag{15.3.10c}$$
$$\dot{q}_y = -\beta(e^{q_x-q_y} - 1). \tag{15.3.10d}$$

The zero-energy dynamics in the invariant plane $q_x = q_y = 0$ recovers the rate equations (15.3.6a) and (15.3.6b), after rescaling by $K$. Thus, the two saddles $O = (0,0,0,0)$ and $R = (n_*/K, m_*/K, 0, 0)$ in the four-dimensional phase plane originate from the fixed points of the two-dimensional deterministic system. There is an additional saddle at $S = (0,0,-\ln B, -\ln B)$, which corresponds to the fluctuational extinction point [33, 269, 569]. We now have to determine the appropriate zero-energy path along which to evaluate the action of the Hamiltonian system, which we then identify with the quasi-potential. Given an established population at $t \to -\infty$, the trajectory starts at the saddle $R$. The issue is then whether it terminates at $O$ or $S$, both of which represent extinction of the two populations. It can be shown that the correct choice is the fluctuational extinction point $S$ [262, 504], so that the optimal fluctuational path is the heteroclinic connection $\mathcal{C}$ from $R$ to $S$. Hence[1],

---

[1] It is important to note that the assumption of large system size ($n, m \gg 1$) implicit in the WKB analysis breaks down around $x = y = 0$. This means that one has to introduce a boundary layer of

$$\Phi(0,0) = \int_{\mathscr{C}} [q_x dx + q_y dy]. \qquad (15.3.11)$$

The extinction model has two degrees of freedom (population densities of two phenotypes) and a four-dimensional phase-space. Since there is only one independent integral of motion (the energy), it follows that in general the optimal fluctuational path has to be calculated numerically. It is also possible to carry out a multiscale perturbation analysis when $B \approx 1$, where all of the fixed points approach each other (due to the fact that for $B < 1$ the deterministic system also becomes extinct). Analytical results can be obtained in both the fast and slow switching limits. Moreover, it is possible to analyze the effects of a catastrophe (temporary reduction in the growth rate $B$) on the population extinction risk. This issue was previously explored in Ref. [34] for a single population of normals, which is obtained in the fast switching limit $\alpha, \beta \to \infty$ (see below). The main finding of the two-population model is that, the presence of a persister subpopulation dramatically reduces the increase in extinction probability due to the same type of catastrophe [569]. Moreover, the reduction is significantly greater in the slow phenotypic switching regime.

**One-population model in the fast switching regime.** In the fast switching limit $\alpha, \beta \to \infty$, we can eliminate the variables $y, q_y$ by taking $y \to \alpha x/\beta$ and $q_y \to q_x$. Hamilton's equations for $x, p_x$ become

$$\dot{x} = Bx(1-x)e^{q_x} - xe^{-q_x}, \qquad (15.3.12a)$$

$$\dot{q}_x = -B(1-2x)(e^{q_x}-1) - (e^{-q_x}-1), \qquad (15.3.12b)$$

and the associated Hamiltonian is

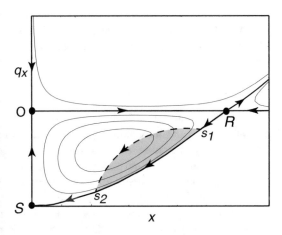

Fig. 15.11: Schematic diagram indicating the phase plane $(x, q_x)$ for the one-population model of extinction. The fixed points are the trivial state $O$, the metastable state $R$ and the fluctuational fixed point $S$. The zero-energy trajectories are indicated by thick solid lines, including the optimal fluctuational path given by the heteroclinic connection $R \to S$. The dashed curve indicates a fluctuational path segment with nonzero energy $E_c$. This arises in response to a catastrophe consisting of the instantaneous reduction of the carrying capacity to zero at $t = 0$ and full recovery at $t = T$ (see text for details).

width $\sim 1/K$ around the origin. However, since this yields a subleading contribution in $K^{-1}$ to the mean extinction time [35], it can be ignored (see also [415]).

$$H(x,q_x) = Bx(1-x)(e^{q_x}-1) + x(e^{-q_x}-1). \tag{15.3.13}$$

The corresponding phase plane is shown in Fig. 15.11. We now have an explicit equation for the heteroclinic connection from $R = (1-B^{-1},0)$ to $S = (0,-\ln B)$:

$$x = x(q_x) = 1 - \frac{e^{-q_x}}{B}.$$

After integration by parts, we have

$$\Phi(0,0) = -\int_0^{-\ln B} x(q)dq = \frac{1-B}{B} + \ln B. \tag{15.3.14}$$

In the regime $0 < \delta \ll 1$ with $\delta = B - 1$, we have $\Phi(0,0) \approx \delta^2/2$ and the mean extinction time $\tau_e = 1/\lambda^{(0)}$ is

$$\tau_e \sim e^{K\delta^2/2}. \tag{15.3.15}$$

For comparison, note that when persisters are also taken into account and phenotypic switching is slow, the extinction time is increased according to [569] (see below)

$$\tau_e \sim e^{K\delta^2(1/2+\alpha/\beta)}. \tag{15.3.16}$$

That is, spending more time in the persister state delays extinction.

Now suppose that an environmental catastrophe occurs, in which either the reproduction rate or the carrying capacity temporarily reduces [34]. The latter can be modeled by taking $Bx(1-x) \to Bx(f(t)-x)$ with a time-dependent factor such that $f(\pm\infty) = 1$. Suppose for the sake of illustration that

$$f(t) = \begin{cases} 1 & \text{if } t < 0 \text{ or } t > T \\ 0 & \text{if } 0 < t < T \end{cases}. \tag{15.3.17}$$

That is, the carrying capacity drops instantaneously to zero at $t = 0$ and subsequently jumps back to the original value at $t = T$. The effective Hamiltonian system now switches between the unperturbed Hamiltonian (15.3.13) and the perturbed Hamiltonian

$$\widehat{H}(x,q_x) = -Bx^2(e^{q_x}-1) + x(e^{-q_x}-1). \tag{15.3.18}$$

Each of the Hamiltonians is an integral of the motion during the associated time interval. Hence, the maximum likelihood fluctuational path to extinction can be determined by matching three separate trajectories corresponding to the pre-catastrophe, catastrophe, and post-catastrophe stages. The trajectories for $t \in (-\infty,0)$ and $t \in (T,\infty)$ are zero energy and thus belong to the original heteroclinic connection from $R$ to $S$. On the other hand, the catastrophe path for $t \in [0,T]$ has some nonzero energy $E_c$, which is unknown a priori. In particular, the points $s_1,s_2$ where the nonzero energy path intersects the heteroclinic connection depend on $E_c$. The full solution is obtained by imposing continuity of $x(t)$ and $q_x(t)$ at the intersection points and imposing the condition that the duration of the catastrophe is $T$. One thus finds that the modified quasi-potential or action is

$$\widehat{\Phi}(0,0) = \Phi(0,0) - \Delta A - E_c(T)T, \qquad (15.3.19)$$

where $\Delta A$ is the shaded region in Fig. 15.11. The reduction in the action leads to an exponential decrease in the mean extinction time. However, this is mitigated by the presence of a persister subpopulation, as shown in Ref [569].

**Two-population model in the slow switching regime.** Following [569], let us return to the full two-population model, and suppose that the system is close to the bifurcation point $B = 1$ where all fixed points merge. Again set $\delta = B - 1 \ll 1$. One finds that $x, \beta y/\alpha, |q_x|, |q_y| \sim \delta$ so that all exponentials in the Hamiltonian (15.3.9) can be Taylor expanded. Finally, suppose that switching between the two phenotypes is rare such that $\alpha, \beta \ll \delta \ll 1$. Under these various assumptions, the Hamiltonian can be approximated as

$$H(x,y,q_x,q_y) = xq_x(q_x - x + \delta) - (\alpha x - \beta y)(q_x - q_y) + O(\delta^4). \qquad (15.3.20)$$

Hamilton's equations become

$$\dot{x} = x(2q_x - x + \delta) - (\alpha x - \beta y), \qquad (15.3.21a)$$

$$\dot{y} = \alpha x - \beta y, \qquad (15.3.21b)$$

$$\dot{q}_x = -q_x(q_x - 2x + \delta) + \alpha(q_x - q_y), \qquad (15.3.21c)$$

$$\dot{q}_y = -\beta(q_x - q_y). \qquad (15.3.21d)$$

The fixed points are $O = (0,0,0,0)$, $R = (\delta, \alpha\delta/\beta, 0, 0)$ and $S = (0,0,-\delta,-\delta)$.

Hamilton's equations can be analyzed using a separation of time scales. Introducing the rescalings $x = \delta X$, $y = \delta Y$, $q_x = \delta Q_X$, $q_y = \delta Q_Y$ and $t = T/\delta$, we have

$$\frac{dX}{dT} = X(2Q_X - X + 1) - \varepsilon(\Gamma X - Y), \qquad (15.3.22a)$$

$$\frac{dY}{dT} = \varepsilon(\Gamma X - Y), \qquad (15.3.22b)$$

$$\frac{dQ_X}{dT} = -Q_X(Q_X - 2X + 1) + \varepsilon\Gamma(Q_X - Q_Y), \qquad (15.3.22c)$$

$$\frac{dQ_Y}{dT} = -\varepsilon(Q_X - Q_Y), \qquad (15.3.22d)$$

where $\varepsilon = \beta/\delta$ and $\Gamma = \alpha/\beta$. It is clear that $(X, Q_X)$ can be treated as fast variables that evolve on a unit time scale, whereas $(Y, Q_Y)$ act as slow variables evolving on the slow time scale $T \sim 1/\varepsilon$. Introduce the slow time variable $\tau = \varepsilon T$ and consider perturbation expansions of the form

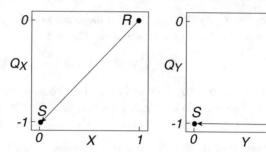

Fig. 15.12: Leading-order projections of optimal fluctuational path in the $(X, Q_X)$-plane and the $(Y, Q_Y)$-plane.

$$X = X_0(T) + \varepsilon X_1(T, \tau) + \ldots, \quad Q_X = Q_{X0}(T) + \varepsilon Q_{X1}(T, \tau) + \ldots \quad (15.3.23a)$$
$$Y = Y_0(\tau) + \varepsilon Y_1(T, \tau) + \ldots, \quad Q_Y = Q_{Y0}(\tau) + \varepsilon Q_{Y1}(T, \tau) + \ldots \quad (15.3.23b)$$

Substituting into the Hamilton equation (15.3.22) and collecting terms in powers of $\varepsilon$ leads to a hierarchy of PDEs (see Ex. 15.3). The $O(1)$ equations are

$$\frac{dX_0}{dT} = X_0(2Q_{X0} - X_0 + 1), \quad \frac{dQ_{X0}}{dT} = -Q_{X0}(Q_{X0} - 2X_0 + 1),$$

and we recover a one-population model. The zeroth-order Hamiltonian is

$$H_0 = X_0 Q_{X0}(Q_{X0} - X_0 + 1),$$

so that the fluctuational path projected on to the $(X - Q_X)$-plane is the straight line $Q_{X0} = X_0 - 1$. The solutions along this path take the form (see Ex. 15.3)

$$X_0(T) = \frac{1}{1 + e^T}, \quad Q_{X0}(T) = \frac{-1}{1 + e^{-T}}. \quad (15.3.24)$$

The slow persister variables appear at $O(\varepsilon)$ according to

$$\frac{dY_0}{d\tau} + Y_0(\tau) = \Gamma X_0(T), \quad \frac{dQ_{X0}}{d\tau} - Q_{Y0}(\tau) = -Q_{X0}(T).$$

For finite $\tau > 0$ ($\tau < 0$) we can take $T \to \infty$ ($T \to -\infty$) in the fast variables so that we can replace the fast variables by Heaviside functions: $X_0(T) \to H(-\tau)$ and $Q_{X0}(T) \to -H(\tau)$. We thus obtain the solutions (see Ex. 15.3)

$$Y_0(\tau) = \begin{cases} \Gamma & \text{for } \tau \leq 0 \\ \Gamma e^{-\tau} & \text{for } \tau \geq 0 \end{cases}, \quad Q_{Y0}(\tau) = \begin{cases} -e^{\tau} & \text{for } \tau \leq 0 \\ -1 & \text{for } \tau \geq 0 \end{cases}. \quad (15.3.25)$$

Thus, the projection of the fluctuational path in the $(Y, Q_Y)$-plane consists of a vertical line joining the points $(0, \Gamma)$ to $(\Gamma, -1)$ and a horizontal line joining $(\Gamma, -1)$ to $(0, -1)$, see Fig. 15.12. Evaluating the area integrals in equation (15.3.11) yields the result $\Phi(0,0) = 1/2 + \Gamma$, and hence we obtain the mean extinction time (15.3.16).

## 15.4 Bacterial quorum sensing

Quorum sensing (QS) is a form of collective chemical sensing and response that is correlated to population density. Many species of bacteria use QS to coordinate various types of behavior including the production and release of various effector molecules that aid in bioluminescence, biofilm formation, virulence, and antibiotic resistance, for example. [225, 261, 327, 552, 628, 710, 851, 926, 936]. In an analogous fashion, some social insects use QS to determine where to nest [211]. QS also has several useful applications outside the biological realm, for example, in computing and robotics. Roughly speaking, QS can function as a decision-making process in any decentralized system, provided that individual components have (i) some mechanism for determining the number or density of other components they interact with and (ii) a stereotypical response once some threshold has been reached.

In the case of bacteria, QS involves the production and extracellular secretion of certain signaling molecules called autoinducers. Each cell also has receptors that can specifically detect the signaling molecule (inducer) via ligand-receptor binding, which then activates transcription of certain genes, including those for inducer synthesis. However, since there is a low likelihood of an individual bacterium detecting its own secreted inducer, the cell must encounter signaling molecules secreted by other cells in its environment in order for gene transcription to be activated. When only a few other bacteria of the same kind are in the vicinity (low bacterial population density), diffusion reduces the concentration of the inducer in the surrounding medium to almost zero, resulting in small amounts of inducer being available to bind to receptors. On the other hand, as the population grows, the concentration of the inducer passes a threshold, causing more inducer to be synthesized. This generates a positive feedback loop that fully activates the receptor, and induces the upregulation of other specific genes. Hence, all of the cells initiate transcription at approximately the same time, resulting in some form of coordinated behavior. The basic process at the single-cell level is shown in Fig. 15.13.

**Quorum and diffusion sensing.** One of the potential problems with the above interpretation of QS is that the "decision" of a bacterium to alter behavior when a critical

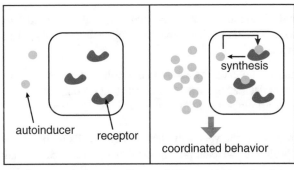

autoinducer  receptor

low population density    high population density

synthesis

coordinated behavior

Fig. 15.13: A schematic illustration of quorum sensing at the single-cell level.

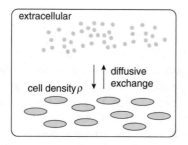 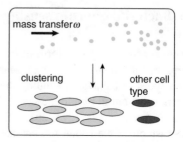

Fig. 15.14: A schematic illustration of various factors that can affect autoinducer sensing besides cell density: mass transfer, another cell type releasing the same autoinducer, and changes in cell density due to clustering rather than changes in cell number.

population density has been reached is not based on perfect information. This is a consequence of the fact that there are several other factors that could affect the autoinducer concentration, which acts as a proxy for cell density. These include the diffusion and advection of autoinducer away from a local population (mass transfer), the spatial distribution of the population (e.g., clustering), degradation, and the production of the same autoinducer by a third party [407, 940], see Fig. 15.14. This observation led to the proposal that QS is in fact diffusion sensing (DS) [755]. That is, the function of secreted autoinducers is to determine whether or not secreted effector molecules (implementing the response) would rapidly diffuse and/or advect away from the cell. This would allow an individual bacterium to identify situations in which the probability that effector molecules disappear is low so that effector secretion is efficient. Interestingly, QS and DS also invoke different evolutionary hypotheses. A basic feature of QS is that bacteria sense their population density in order to engage in social behavior, which suggests that sensing evolved due to group benefits. On the other hand, DS proposes that sensing is a basic activity of single cells in order to detect mass-transfer constraints on effective secretion. Thus, DS assumes that sensing evolved because of a direct fitness benefit for individuals. From an evolutionary perspective, QS presents a problem common to all theories of cooperation, namely, the existence of cheaters. That is, mutants can arise that benefit from public goods (autoinducers and effectors) without paying the cost of producing them.

However, the fundamental distinction between QS and DS has been called into question [732, 940], following a mathematical model of QS evolution [141]. The latter makes four key assumptions:

1. The production of autoinducer molecules is costly to the cells that produce them.
2. The production of effector molecules is costly to the cells that produce them.
3. The production of effectors provides a benefit to the local social group of interacting cells.
4. The benefit of effector production can vary with the number of cells in the social group, which is defined according to the local population density over which social interactions take place.

This model can interpolate between social and non-social QS by varying the size of the social group, and can take into account the complexities of autoinducer sensing by varying the form of the benefit function in assumption 4. Within this modeling framework, QS and DS are distinguished by the size of the social group. In the following, we will ignore some of the controversies regarding the interpretation of QS, and focus on the role of population density.

**Mathematical models.** Most models of bacterial QS are based on deterministic ordinary differential equations (ODEs), in which both the individual cells and the extracellular medium are treated as well-mixed compartments (fast diffusion limit) [186, 230, 246, 330, 362, 467, 621, 923]. (For a discussion of spatial models that take into account bulk diffusion of the autoinducer in the extracellular domain; see Refs. [246, 367, 514, 647, 648] and Sect. 15.4.5. Spatial models of QS also arise within the context of biofilms [24, 193, 319, 924]; see Sect. 15.6.) Suppose that there are $N$ cells labeled $i = 1, \ldots, N$. Let $U(t)$ denote the concentration of signaling molecules in the extracellular space at time $t$ and let $u_i(t)$ be the corresponding intracellular concentration within the $i$-th cell. Suppose that there are $K$ other chemical species within each cell, which together with the signaling molecule comprise a regulatory network. Let $\mathbf{v}_i = (v_{i,1}, \ldots, v_{i,K})$ with $v_{i,k}$ the concentration of species $k$ within the $i$-th cell. A deterministic model of QS can then be written in the general form [230]

$$\frac{du_i}{dt} = F(u_i, \mathbf{v}_i) - \kappa(u_i - U), \tag{15.4.1a}$$

$$\frac{dv_{i,k}}{dt} = G_k(u_i, \mathbf{v}_i), \tag{15.4.1b}$$

$$\frac{dU}{dt} = \frac{\alpha}{N} \sum_{j=1}^{N} \kappa(u_j - U) - \gamma U. \tag{15.4.1c}$$

Here $F(u, \mathbf{v})$ and $G_k(u, \mathbf{v})$ are the reaction rates of the regulatory network based on mass-action kinetics, the term $\kappa(u_j - U)$ represents the diffusive exchange of signaling molecules across the membrane of the $i$-th cell, and $\gamma$ is the rate of degradation of extracellular signaling molecules. Finally, $\alpha = V_{\text{cyt}}/V_{\text{ext}}$ is a cell density parameter equal to the ratio of the total cytosolic and extracellular volume. Note that $V_{\text{cyt}} = v_{\text{cyt}}N$, where $v_{\text{cyt}}$ is the single-cell volume. We can also write

$$\alpha = \frac{\rho}{1-\rho}, \quad \kappa = \frac{\delta}{\rho}, \tag{15.4.2}$$

where $\rho$ is the volume fraction of cells and $\delta$ is the effective particle conductance, which is independent of $\rho$.

From a dynamical systems perspective, two basic forms of collective behavior are typically exhibited by equation (15.4.1): either the population acts as a biochemical switch [246, 450, 467, 923] or as a synchronized biochemical oscillator [186, 230, 621]. In the former case, two distinct mechanisms for a biochemical switch have been identified. The first mechanism involves the occurrence of bistability in a gene regulatory network, as exemplified by the mathematical model of QS in the bacterium *Pseudomonas aeruginosa* developed by Dockery and Keener [246]. *P. aeruginosa* is a human pathogen that monitors its cell density in order to control the release of various virulence factors [225, 261]. That is, if a small number of bacteria released toxins then this could easily be neutralized by an efficient host response, whereas the effectiveness of the response would be considerably diminished if toxins were only released after the bacterial colony has reached a critical size via QS. Multiple steady-steady states have also been found in a related ODE model of QS in the bioluminescent bacteria *V. fisheri* [467]. In this system, QS limits the production of bioluminescent luciferase to situations where cell populations are large; this saves energy since the signal from a small number of cells would be invisible and thus useless. Recent experimental studies of QS in the bacterial species *V. harveyi* and *V. cholerae* [552, 851, 936] provide evidence for an alternative switching mechanism, which can provide robust switch-like behavior without bistability. In these QS systems, two or more parallel signaling pathways control a gene regulatory network via a cascade of phosphorylation-dephosphorylation cycles (PdPCs); see Sect. 10.1.1. Within the context of QS, the binding of an autodinducer to its cognate receptor switches the receptor from acting like a kinase to one acting like a phosphatase. Thus, the PdPCs are driven by the level of autoinducer, which itself depends on the cell density. One source of the switch-like behavior is thus ultrasensitivity of the PdPCs [66, 339, 351, 740, 743].

In the following, we derive conditions for the global convergence of the general QS system (15.4.1), and then consider models of switching in *P. aeruginosa* and *V. harveyi* under the assumption that they operate in the regime of global convergence

### 15.4.1 Global convergence of a quorum sensing network

The global convergence properties of QS networks, where coupling between nodes in the network is mediated by a common environmental variable, has been analyzed within the context of nonlinear dynamical systems in Ref. [777]. These authors consider a more general class of model than given by equation (15.4.1), including non-diffusive coupling and non-identical cells. In order to apply their analysis to the specific system (15.4.1), it is necessary to review some basic results of nonlinear contraction theory [570]. Consider the $n$-dimensional dynamical system

$$\frac{d\mathbf{x}}{dt} = \mathbf{f}(\mathbf{x},t), \quad \mathbf{x} \in \mathbb{R}^n, \tag{15.4.3}$$

with $\mathbf{f}: \mathbb{R}^n \to \mathbb{R}^n$ a smooth nonlinear vector field. Introduce the vector norm $|\mathbf{x}|$ for $\mathbf{x} \in \mathbb{R}^n$ and let $\|A\|$ be the induced matrix norm for an arbitrary square matrix $A$, that is,

$$\|A\| = \sup\{|A\mathbf{x}| : \mathbf{x} \in \mathbb{R}^n \text{ with } |\mathbf{x}| = 1\}.$$

Some common examples are as follows:

$$|\mathbf{x}|_1 = \sum_{j=1}^{n} |x_j|, \quad \|A\|_1 = \max_{1 \leq j \leq n} \sum_{i=1}^{n} |a_{ij}|,$$

$$|\mathbf{x}|_2 = \left( \sum_{j=1}^{n} |x_j|^2 \right)^{1/2}, \quad \|A\|_2 = \sqrt{\lambda_{\max}(A^\top A)},$$

$$|\mathbf{x}|_\infty = \max_{1 \leq j \leq n} |x_j|, \quad \|A\|_\infty = \max_{1 \leq i \leq n} \sum_{j=1}^{n} |a_{ij}|,$$

where $A^\top$ is the transpose of $A$ and $\lambda_{\max}(A^\top A)$ is the largest eigenvalue of the positive semi-definite matrix $A^\top A$. Define the associated matrix measure $\mu$ as

$$\mu(A) = \lim_{h \to 0^+} \frac{1}{h}(\|I + hA\| - 1),$$

where $I$ is the identity matrix. For the three above norms on $\mathbb{R}^n$, the associated matrix measures are

$$\mu_1(A) = \max_{1 \leq j \leq n} \{a_{jj} + \sum_{i \neq j} |a_{ij}|\},$$

$$\mu_2(A) = \max_{1 \leq i \leq n} \{\lambda_i([A + A^\top]/2)\},$$

$$\mu_\infty(A) = \max_{1 \leq i \leq n} \{a_{ii} + \sum_{j \neq i} |a_{ij}|\}.$$

Given these definitions, the basic contraction theorem is as follows [777]:

**Theorem 15.1.** *The n-dimensional dynamical system (15.4.3) is said to be contracting if any two trajectories, starting from different initial conditions, converge exponentially to each other. A sufficient condition for a system to be contracting is the existence of some matrix measure $\mu$ for which there exists a constant $\lambda > 0$ such that*

$$\mu(J(\mathbf{x},t)) \leq -\lambda, \quad J_{ij} = \frac{\partial f_i}{\partial x_j} \tag{15.4.4}$$

*for all $\mathbf{x},t$. The scalar $\lambda$ defines the rate of contraction.*

In order to apply the above result to QS, we consider the related concept of a partial contraction [777]. First, rewrite equation (15.4.1) in the form

$$\frac{d\mathbf{x}_i}{dt} = \mathbf{f}(\mathbf{x}_i) - \kappa((\mathbf{x}_i)_1 - U)\mathbf{e}_1, \quad i = 1,\ldots,N, \quad \frac{dU}{dt} = \frac{\alpha\kappa}{N} \sum_{j=1}^{N} ((\mathbf{x}_j)_1 - U) - \gamma U,$$

$$\tag{15.4.5a}$$

with $\mathbf{x}_i = (u_i, \mathbf{v}_i) \in \mathbb{R}^{1+K}$, $(\mathbf{x}_i)_1 = u_i$, $\mathbf{f} = (F, G_1,\ldots,G_K)$ and $\mathbf{e}_1 = (1,0,\ldots,0)$. From the contraction theorem and the notion of partial contraction, one can show that the global convergence condition

$$|\mathbf{x}_i(t) - \mathbf{x}_j(t)| \to 0 \text{ as } t \to \infty$$

holds provided that $\mathbf{f}(\mathbf{x}) - \kappa(\mathbf{x})_1 \mathbf{e}_1$ is contracting. The proof follows from considering the reduced non-autonomous virtual system

$$\dot{\mathbf{y}} = \mathbf{f}(\mathbf{y}) - \kappa(\mathbf{y})_1 \mathbf{e}_1 + \kappa U(t) \mathbf{e}_1,$$

where the solution $U(t)$ of the full system treated as an external input. Setting $\mathbf{y}(t) = \mathbf{x}_i(t)$ in the virtual system recovers the dynamics of the $i$-th cell. Hence, $\mathbf{x}_i(t)$ for $i = 1, \ldots, N$ are particular solutions of the virtual system so that if the virtual system is contracting in $\mathbf{y}$ for arbitrary $U(t)$, then all of its solutions converge exponentially toward each other, including the solutions $\mathbf{x}_i(t)$. In this asymptotic limit, we effectively have a single cell diffusively coupled to the extracellular medium, that is $u_i(t) \to u(t)$ and $\mathbf{v}_i(t) \to \mathbf{v}(t)$ with

$$\frac{du}{dt} = F(u, \mathbf{v}) - \kappa(u - U), \tag{15.4.6a}$$

$$\frac{dv_k}{dt} = G_k(u, \mathbf{v}), \tag{15.4.6b}$$

$$\frac{dU}{dt} = \alpha \kappa(u - U) - \gamma U. \tag{15.4.6c}$$

### 15.4.2 Bistability in a model of Pseudomonas aeruginosa *quorum sensing*

In *P. aeruginosa*, there are two QS systems working in series, known as the *las* and *rhl* system, respectively. Following Ref. [246], we consider the upstream *las* system, which is composed of *lasI*, the autoinducer synthase gene responsible for synthesis of the autoinducer 3-oxo-C12-HSL via the enzymatic action of the protein LasI, and the *lasR* gene that codes for transcriptional activator protein LasR, see Fig. 15.15. Positive feedback occurs due to the fact that LasR and 3-oxo-C12-HSL can form a dimer, which promotes both *lasR* and *lasI* activity. (We ignore an additional negative feedback loop that is thought to play a relatively minor role in QS.) Exploiting the fact that the lifetime of each type of mRNA is much shorter than its corresponding protein, we can eliminate the mRNA dynamics, and write down a system of ODES

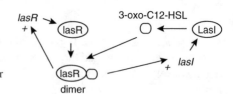

Fig. 15.15: Simplified regulatory network for the *las* system in *P. aerginosa*.

for the concentrations of LasR, the dimer LasR/3-oxo-C12-HSL and the autoinducer 3-oxo-C12-HSL, which we denote by $R, P$ and $A$, respectively. The resulting system of ODES for mass action kinetics thus take the form [246]

$$\frac{dP}{dt} = k_{RA}RA - k_P P, \tag{15.4.7a}$$

$$\frac{dR}{dt} = -k_{RA}RA + k_P P - k_R R + V_R \frac{P}{K_R + P} + R_0, \tag{15.4.7b}$$

$$\frac{dA}{dt} = -k_{RA}RA + k_P P + V_A \frac{P}{K_A + P} + A_0 - k_A A. \tag{15.4.7c}$$

Here, $k_A, k_R$ are the rates of degradation of $A, R$, $k_{RA}, k_P$ are the rates of production and degradation of the dimer $P$, and $A_0, R_0$ are baseline rates of production of $A, R$. Finally, the positive feedback arising from the role of the dimer $P$ as an activator protein that enhances the production of LasR and arising from the activation of R and A (with the latter mediated by lasI) is taken to have Michaelis-Menten form (see Sect. 1.4). Dockery and Keener [246] carry out a further reduction by noting that the formation and degradation of dimer is much faster than transcription and translation so that we can assume $P$ is in quasi-steady state: $P = (k_{RA}/k_P)RA = kRA$. Then, we have

$$\frac{dR}{dt} = -k_R R + V_R \frac{kRA}{K_R + kRA} + R_0, \tag{15.4.8a}$$

$$\frac{dA}{dt} = V_A \frac{kRA}{K_A + kRA} + A_0 - k_A A. \tag{15.4.8b}$$

Comparison of equation (15.4.8) with the general kinetic system in equations (15.4.6a,b), shows that we have one auxiliary species and we can make the following identifications (after dropping the $k = 1$ index): $u = A, v = R$ with

$$F = V_A \frac{uv}{K_A + kuv} + A_0 - k_A u,$$

$$G = -k_R v + V_R \frac{kuv}{K_R + kuv} + R_0.$$

Now suppose equation (15.4.8b) is diffusively coupled to an extracellular concentration along the lines of (15.4.6a-15.4.6c), and as a further simplification take $U$ to be in quasi-equilibrium. The latter conditions allows us to carry out a phase-plane analysis without changing the essential behavior of the system. The only modification is that the last term on the right-hand side of (15.4.8b) is transformed according to the scheme $k_A \to k_A + d(\rho)$, with

$$d(\rho) \equiv \frac{\delta}{\rho} \frac{\gamma}{\gamma + \delta/(1 - \rho)},$$

where we have used equation (15.4.2). QS thus arises due to the dependence of the effective degradation rate $d(\rho)$ on $\rho$. More specifically, when $\rho$ is small the

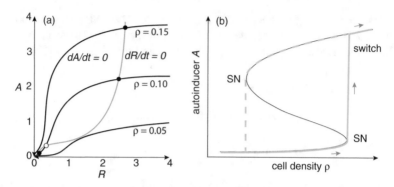

Fig. 15.16: Bistability in planar model of *las* system in *P. aerginosa.*. (a) Nullcline $\dot{R} = 0$ is shown by the gray curve and the $\rho$-dependent nullclines $\dot{A} = 0$ are shown by the black curves for three different values of $\rho$. It can be seen that for intermediate $\rho$ values there are two stable fixed points separated by an unstable fixed point. (b) The appearance and subsequent disappearance of a stable/unstable pair of fixed points via saddle-node bifurcations can switch the system from a low activity state to a high activity state. Decreasing the density again leads to hysteresis. Parameter values are $V_R = 2.0$, $V_A = 2.0$, $K_R = 1.0$, $K_A = 1.0$, $R_0 = 0.05$, $A_0 = 0.05$, $\delta = 0.2$, $\gamma = 0.1$, $k_R = 0.7$, and $k_A = 0.02$.

rate $d(\rho)$ is large, whereas $d(\rho)$ is small when $\rho \to 1$. In Fig. 15.16(a), we plot nullclines of the system at various values of $\rho$, illustrating that bistability occurs at intermediate values of $\rho$. In this parameter regime, the system acts like a bistable switch, where one stable state has low levels of autoinducer and the other high levels of autoinducer [246]. Switching occurs due to a saddle-node bifurcation as illustrated in Fig. 15.16(b). When noise is included, the resulting steady-state distribution exhibits peaks around the stable fixed points of the deterministic system, and the transition between a unimodal and a bimodal distribution is used as one indictator of a stochastic bifurcation [343, 835].

### 15.4.3 *Ultrasensitivity in a model of* V. harveyi *quorum sensing*

The bioluminescent bacterium *V. harveyi* has three parallel QS systems, each consisting of a distinct autoinducer (HAI-1, AI-2, CAI-1), cognate receptor (LuxN, LuxPQ, CqsS), and associated enzyme that helps produce the autoinducer, see Fig. 15.17. (The human pathogen *V. cholerae* has a similar QS network, except that there could be up to four parallel pathways [480].) Each autoinducer is freely exchanged between the intracellular and extracellular domains. Extracellular diffusion at low (high) cell densities leads to small (large) intracellular autoinducer concentrations, resulting in a low (high) probability that the autoinducer can bind to its cognate receptor. The receptors target downstream DNA-binding regulatory proteins LuxU and LuxO. At low cell densities, the receptors act as kinases and the resulting phosphorylation of LuxU and LuxO activates transcription of the genes encoding five

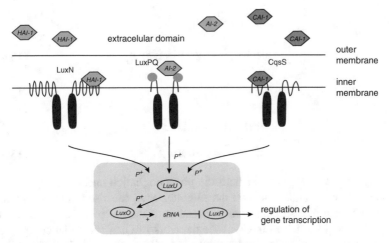

Fig. 15.17: Summary diagram of the *V. harveyi* QS circuit. Three phosphorylation cascades work in parallel to control the ratio of LuxO to LuxO-P based on local cell-population density. Five sRNA, qrr1-5, then regulate expression of QS target genes including the master transcriptional regulator LuxR, which upregulates downstream factors.

regulatory small non-coding RNAs (sRNAs) termed Qrr1-Qrr5, which destabilize the transcriptional activator protein LuxR. This prevents the activation of target genes responsible for the production of various proteins, including bioluminescent luciferase. Hence, at low cell densities the bacteria do not bioluminesce. On the other hand, at high cell densities the receptors switch from kinases to phosphatases, significantly reducing the levels of LuxU-P and LuxO-P. The sRNAs are thus no longer expressed, allowing the synthesis of LuxR and the expression of biolumi-nescence, for example. Both the phosphorylation-dephosphorylation cascades and the sRNA regulatory network provide a basis for a sharp, sigmoidal response of the concentration of LuxR to smooth changes in cell density.

**Single-cell model.** Following Ref. [118], we analyze the occurrence of ultrasen-sitivity at the single- cell level by focusing on a single phosphorylation pathway and adapting the Goldbeter-Koshland of phosphorylation-dephosphorylation cycles (PdPCs) [351]; see Sect. 10.1.1. (For a corresponding model of switching due to the action of sRNAs see Hunter et al. [450].) In particular, we consider the phosphorylation-dephosphorylation of LuxU by the enzymatic action of a partic-ular QS receptor, which is denoted by $R$ when acting as a kinase and by $\widehat{R}$ when it is it is bound by an autoinducer ($A$) and acts like a phosphatase. Denoting the protein LuxU by W, we define the following reaction schemes:

$$W + R \underset{d_1}{\overset{a_1}{\rightleftharpoons}} WR \overset{k_1}{\rightarrow} W^* + R, \tag{15.4.9a}$$

$$W^* + \widehat{R} \underset{d_2}{\overset{a_2}{\rightleftharpoons}} W^*\widehat{R} \overset{k_2}{\rightarrow} W + \widehat{R}, \tag{15.4.9b}$$

$$R + A \underset{k_-}{\overset{k_+}{\rightleftharpoons}} \widehat{R}, \tag{15.4.9c}$$

$$WR + A \overset{k_+}{\rightarrow} W^* + \widehat{R}, \tag{15.4.9d}$$

$$W^*\widehat{R_i} \overset{k_-}{\rightarrow} W + R + A. \tag{15.4.9e}$$

For simplicity, we assume that both the phosphorylation and the dephosphorylation steps are irreversible, and that the autoinducer $A_i$ can also bind to the complex $WR_i$, resulting in the simultaneous conversion of the receptor from a kinase to a phosphatase and LuxU to LuxU-P. (See the work of Qian and collaborators for an analysis of more detailed, reversible models of PdPCs [339, 740, 743].) Introducing the concentrations $u = [A]$, $w = [W]$, $w^* = [W^*]$, $r = [R] + [WR]$, $\widehat{r} = [\widehat{R}] + [W^*\widehat{R}]$, $v = [WR]$ and $v^* = [W^*\widehat{R}]$, the corresponding kinetic equations for a single cell with a fixed intracellular concentration of autoinducer, are as follows. First, the PdPC kinetics evolve according to equation (10.1.1), with $[E_1] = r - v$, $[E_2] = \widehat{r} - v^*$, $w_1 = v$, $w_2^* = v^*$:

$$\frac{dw}{dt} = -a_1 w(r - v) + d_1 v + k_2 v^*, \tag{15.4.10a}$$

$$\frac{dv}{dt} = a_1 w(r - v) - (d_1 + k_1)v, \tag{15.4.10b}$$

$$\frac{dw^*}{dt} = -a_2 w^*(\widehat{r} - v^*) + d_2 v^* + k_1 v, \tag{15.4.10c}$$

$$\frac{dv^*}{dt} = a_2 w^*(\widehat{r} - v^*) - (d_2 + k_2)v^*. \tag{15.4.10d}$$

However, in contrast to the standard Goldbeter-Koshland model, the concentrations of kinase and phosphatase also evolve

$$\frac{dr}{dt} = k_-\widehat{r} - k_+ur, \quad \frac{d\widehat{r}}{dt} = -k_-\widehat{r} + k_+ur. \tag{15.4.10e}$$

These are supplemented by the conservation equations

$$W_T = w + w^* + v + v^*, \quad R_T = r + \widehat{r}, \tag{15.4.11}$$

where $R_T$ is the total concentration of receptors and $W_T$ is the total concentration of LuxU.

For simplicity, we will assume that the conversion of the receptors from kinase to phosphatase activity is independent of the PdPC. That is, we ignore any positive feedback pathways, in which the regulation of gene expression by the phosphorylation/dephsophorylation of LuxU alters the production of the autoinducer [676].

This allows us to treat the receptor-ligand dynamics given by equation (15.4.10e), or subsequent extensions, independently of the PdPC dynamics given by equations (15.4.10a)-(15.4.10d). In particular, for fixed concentration $u$, we can take the concentration of kinases and phosphatases to be at equilibrium:

$$r_{eq} = \frac{k_-}{k_+ u + k_-} R_T \equiv R(u), \quad \widehat{r}_{eq} = R_T - R(u). \tag{15.4.12}$$

The system of equations (15.4.10a)-(15.4.10e) then reduces to the classical Goldbeter-Koshland model of PdPCs [351], and we can apply their analysis based on a generalization of Michaelis-Menten kinetics. The first step is to assume that the concentration of $W$ and $W^*$ is much larger than that of the receptor, that is, $W_T \gg R_T$ or equivalently $W_T = w + w^*$. This implies that the time scale for the dynamics of the complexes $WR$ and $W^*\widehat{R}$ is much faster than that for the dynamics of $W$ and $W^*$. Performing a separation of time scales, we can treat the concentrations $w$ and $w^*$ as constants when analyzing equations (15.4.10b) and (15.4.10b) , while we can take the steady-state values of the concentrations $v, v^*$ when solving equations. (15.4.10a) and (15.4.10d). Hence, setting $dv/dt = 0$ and $r = R(u)$ in (15.4.10b), we can solve for $v$ in terms of $w$. Similarly, setting $dv^*/dt = 0$ and $\widehat{r} = R_T - R(u)$ in (15.4.10d) we can solve for $v^*$ in terms of $w^*$. We thus obtain the reduced kinetic scheme

$$W \underset{f_2(w^*)}{\overset{f_1(w)}{\rightleftharpoons}} W^*, \tag{15.4.13}$$

with

$$f_1(w) = \frac{k_1 R(u) w}{K_1 + w}, \quad f_2(w^*) = \frac{k_2 [R_T - R(u)] w^*}{K_2 + w^*}, \tag{15.4.14}$$

and

$$K_1 = \frac{d_1 + k_1}{a_1}, K_2 = \frac{d_2 + k_2}{a_2}.$$

Imposing the conservation condition $W_T = w + w^*$ thus yields the single independent kinetic equation (analogous to equation (10.1.5))

$$\frac{dw^*}{dt} = f_1(W_T - w^*) - f_2(w^*). \tag{15.4.15}$$

The steady-state concentration $w^*_{eq}$ of LuxU-P is thus obtained by solving $f_1(W_T - w^*) = f_2(w^*)$, which yields a quadratic equation for $w^*$. Taking the positive root, and expressing it as a function of the fixed autoinducer concentration $u$, we find that

$$\frac{[LuxU-P]}{[LuxU-P] + [LuxU]} = \phi(V(u)), \tag{15.4.16}$$

with

$$V(u) = \frac{k_1 R(u)}{k_2 [R_T - R(u)]} = \frac{k_-}{k_+ u} \frac{k_1}{k_2}, \tag{15.4.17}$$

and $\phi$ given by

$$\phi = \frac{-B + \sqrt{B^2 + 4AC}}{2AW_T}, \tag{15.4.18}$$

for $V \neq 1$, where

$$A = V(u) - 1, \quad B = \frac{1}{W_T}(K_1 + K_2 V(u)) - (V(u) - 1), \quad C = \frac{K_2}{W_T}V(u).$$

A plot of $\phi$, as a function of the autoinducer concentration $u$ is shown in Fig. 15.18(a) for $K_1 = K_2 = K$. At low values of $K$, there is a sharp change from high to low levels of modified protein over a very small change in $u$ (ultrasensitivity); this corresponds to a regime in which the two enzymes are saturated. On the other hand, for large values of $K$, the curve is relatively shallow, and one obtains a response similar to first-order kinetics.

**Population model.** Now consider a population of $N$ identical cells that are coupled via a common extracellular domain due to the transfer of the autoinducer $A$ across the cell membrane. Equation (15.4.10e) for a single cell is then replaced by a system of equations of the form

$$\frac{du_i}{dt} = \Gamma + k_-(R_T - r_i) - k_+ u_i r_i - \kappa(u_i - U), \tag{15.4.19a}$$

$$\frac{dr_i}{dt} = k_-(R_T - r_i) - k_+ u_i r_i \quad i = 1, \ldots, N, \tag{15.4.19b}$$

$$\frac{dU}{dt} = \alpha \kappa(u_{av} - U) - \gamma U, \quad u_{av} = \frac{1}{N}\sum_{j=1}^{N} u_j, \tag{15.4.19c}$$

where $u_{av}$ is the population-averaged intracellular concentration of $A$, $U$ is the extracellular concentration of $A$, and $\Gamma$ is the rate of production of $A$ due to the action of enzymes.

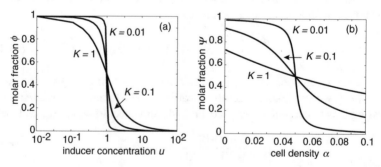

Fig. 15.18: Molar fraction of modified protein $W^*$ at steady state as a function of (a) the autoinducer concentration $u$ and (b) cell density $\alpha$ for different values of $K$, with $K = K_1 = K_2$, $k_- k_1 = k_+ k_2$, $\Gamma = 1$, $\kappa = 10$, and $\alpha_c = 0.05$.

In order to apply the contraction theorem to equation (15.4.19), we fix $i$, set $n = 2$, and consider the virtual system

$$\frac{d\mathbf{y}}{dt} = \mathbf{f}(\mathbf{y}, t) + \Gamma \begin{pmatrix} 1 \\ 0 \end{pmatrix}, \tag{15.4.20}$$

where

$$f_1(\mathbf{y}, t) = k_-(R_T - y_2) - k_+ y_1 y_2 - \kappa(y_1 - U(t)), \tag{15.4.21a}$$

$$f_2(\mathbf{y}, t) = k_-(R_T - y_2) - k_+ y_1 y_2. \tag{15.4.21b}$$

Note that here the extracellular concentration $U(t)$ is treated as a common time-dependent input. The Jacobian of the virtual system is

$$J(\mathbf{y}, t) = \begin{pmatrix} -k_+ y_2 - \kappa & -k_- - k_+ y_1 \\ -k_+ y_2 & -k_- - k_+ y_1 \end{pmatrix}. \tag{15.4.22}$$

Suppose that we take the $\ell_2$ norm for $\mathbf{y}$. In order to determine the corresponding matrix measure $\mu_2(J)$, we need to calculate the eigenvalues of the symmetric matrix

$$\frac{J + J^\top}{2} = \begin{pmatrix} -k_+ y_2 - \kappa_i & -[k_- + k_+(y_1 + y_2)]/2 \\ -[k_- + k_+(y_1 + y_2)]/2 & -k_- - k_+ y_1 \end{pmatrix}.$$

These are given by

$$\lambda_\pm = \frac{1}{2}\left[ -b(\mathbf{y}) \pm \sqrt{b(\mathbf{y})^2 - c(\mathbf{y})} \right],$$

with

$$b(\mathbf{y}) = \kappa + k_- + k_+(y_1 + y_2),$$

$$c(\mathbf{y}) = 4(\kappa + k_+ y_2)(k_- + k_+ y_1) - [k_- + k_+(y_1 + y_2)]^2.$$

Since $b(\mathbf{y}) > 0$, the matrix measure is

$$\mu_2(J) = \max\{\lambda_\pm\} = \lambda_+.$$

Hence, the virtual system is contracting provided that $c(\mathbf{y}) > 0$. Proceeding along identical lines to Ref. [118], it can be proven that $c(\mathbf{y})$ is positive definite provided that the following conditions hold:

$$kR_T < k_- < 4\kappa, \tag{15.4.23a}$$

and

$$k_- + k_+ \Theta < 2\kappa, \qquad \Theta \equiv \frac{(\Gamma + k_- R_T)(\alpha\kappa + \gamma)}{\kappa\gamma}. \tag{15.4.23b}$$

Suppose that conditions (15.4.23) hold so that the system (15.4.19) is globally convergent. It follows that $u_i(t) \to u(t)$ and $r_i(t) \to r(t)$ in the limit $t \to \infty$, with $(u, r)$ evolving according to the effective single-cell model

$$\frac{du}{dt} = \Gamma + k_-[R_T - r] - k_+ur - \kappa(u - U), \qquad (15.4.24a)$$

$$\frac{dr}{dt} = k_-[R_T - r] - k_+ur, \qquad (15.4.24b)$$

$$\frac{dU}{dt} = \alpha\kappa(u - U) - \gamma U. \qquad (15.4.24c)$$

Equations (15.4.24a)-(15.4.24c) have a unique stable fixed point $u_{eq}$ with

$$u_{eq} = \Gamma\frac{(\alpha\kappa + \gamma)}{\kappa\gamma} \equiv \psi(\alpha). \qquad (15.4.25)$$

Putting $u = u_{eq}$ in equation (15.4.16) finally shows that

$$\frac{[\text{LuxU-P}]}{[\text{LuxU-P}]+[\text{LuxU}]} = \phi\left(\frac{k_-}{k_+\psi(\alpha)}\frac{k_1}{k_2}\right) \equiv \Psi(\alpha). \qquad (15.4.26)$$

In order that the system exhibit switch-like behavior as a function of cell density $\alpha$, we require a critical value $\alpha_c$, $0 < \alpha_c < \infty$ such that (for $\Gamma = 1$)

$$\chi \equiv \frac{k_-}{k_+}\frac{k_1}{k_2} = \frac{\alpha_c}{\gamma} + \frac{1}{\kappa},$$

that is, $\alpha_c = \gamma(\chi - \kappa^{-1})$. Assuming that $\chi = 1$ and taking $\alpha_c = 0.05$ [230], we require $\kappa > 1$ and $\gamma = \gamma_c = 0.05(1 - \kappa^{-1})$. For low cell densities ($\alpha \ll \alpha_c$), we have $\Psi(\alpha) \approx 1$, which follows from the functional form of $\phi$, see Fig. 15.18(a). Hence the fraction of phosphorylated LuxU-P is high, which ultimately means that the expression of the gene regulator protein LuxR is suppressed. On the other hand, for large cell densities ($\alpha \gg \alpha_c$) we find that $\Psi(\alpha) \approx 0$. Now the fraction of phosphorylated LuxU-P is small, allowing the expression of LuxR and downstream gene regulatory networks. The $\alpha$-dependence is illustrated in Fig. 15.18(b) with $\alpha$ and $u$ linearly related.

Finally, recall from Sect. 10.1.3 that one of the characteristic features of ultrasensitive biochemical signaling networks is that they tend to amplify noise [66, 339, 555, 818]. On the other hand, collective behavior at the population level, as exhibited by QS networks, can mitigate the effects of noise [777, 853]. The effects of cell density on the amplification of intrinsic fluctuations in the V. harveyi QS model can be investigated using linear response theory [118], along analogous lines to Sect. 10.1.3.

Remark 15.1. In the above analysis, we ignored one of the important characteristics of QS in bacteria such as V. harveyi and V. cholerae, namely, the existence of several parallel autoinducer pathways [926] that are integrated into a single phosphorylation-dephosphorylation cycle, see Fig. 15.17. In particular, V. harveyi uses three distinct autoinducers, AI-1, AI-2,

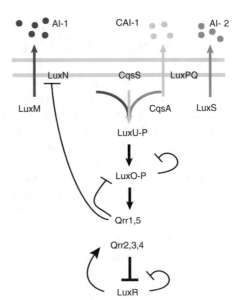

Fig. 15.19: Schematic diagram of feedback loops that are currently known in *V. harveyi* QS. The cognate receptors of the autoinducers AI-1, CAI-1, and AI-2 are LuxN, CqsS, and LuxPQ, respectively. The arrows feeding into LuxU represent the kinase action of the receptors. The corresponding regulators of the three autoinducers are LuxM, CqsA, and LuxS.

and CAI-1, each of which appears to encode distinct environmental information [867]. (The human pathogen *V. cholerae* has up to four parallel pathways [480].) That is, the autoinducers have varying levels of bacterial species specificity: AI-1 is specific to *V. harveyi* and thus acts as an intraspecies signal, CA1-1 is produced primarily by vibrios, suggesting that it is an intragenera signal, whereas A1-2 is prevalent across many different bacteria thus acting as a nonspecific interspecies signal. Although many types of bacteria use several parallel autoinducer pathways, it is still not clear what environmental information each of these pathways carries and how this information is integrated to determine the QS response. In the case of *V. harveyi*, it has been observed experimentally that, at the population level, cells tightly synchronize their response to a wide range of environmental conditions (different relative fractions of the three types of autoinducer), and that the signals from at least two of the autoinducers (A1-1 and AI-2) are combined strictly additively within the shared PdPC pathway [571]. Assuming that a growing population of cells varies in its level of diversity, it has been hypothesized that the use of multiple autoinducers allows the population to synchronize gene expression over a sequence of distinct developmental stages [571]. Mathematical modeling suggests that QS may use feedback loops to "decode" the integrated signals by actively changing the sensitivity in different pathways, which allows bacteria to have a finer discrimination of their social and physical environment [294, 295, 450]. Experimentally, it has been shown that there are at least five feedback loops in *V. harveyi* QS circuits [867, 887], see Fig. 15.19. These can be partitioned into a *qrr*-LuxO-LuxU component, a *qrr*-LuxR component, and a third component involving *qrr* 1-4 repressing the production of receptor LuxN.

### 15.4.4 PDE-ODE model of quorum sensing in a population of cells coupled by bulk diffusion

So far, we have considered ODE models of quorum sensing, consisting of a population of cellular compartments diffusively coupled to a well-mixed extracellular compartment. We now turn to a class of PDE-ODE models (see also Sect. 11.3.2), in which a population of spatially segregated dynamically active cells are coupled through passive bulk diffusion. Taking each cell radius to be $O(\varepsilon)$ with $\varepsilon \ll 1$ relative to the size of the extracellular domain, one can use the method of matched asymptotic expansions to construct steady-state solutions and to formulate a spectral problem that characterizes the linear stability properties of the steady-state solutions. Similar techniques were used to study narrow escape problems in Chap. 6. For concreteness, we follow Gou and Ward [368] by considering a two-dimensional model, which relates to quorum sensing behavior on thin substrates. For analogous studies of quorum sensing in three-dimensional domains see [647, 648].

Consider $N$ cells labeled $j = 1, \ldots, N$ with spatial extent $\Omega_j$ inside a 2D domain $\Omega$, see Fig. 15.20. Let $U(\mathbf{x}, t)$ denote the concentration of autoinducer in the extracellular domain $\Omega \setminus \bigcup_{j=1}^{N} \Omega_j$, which evolves according to the PDE model

$$\frac{\partial U}{\partial t} = D\nabla^2 U - \gamma U, \quad \mathbf{x} \in \Omega \setminus \bigcup_{j=1}^{N} \Omega_j; \quad \frac{\partial U}{\partial n} = 0, \quad \mathbf{x} \in \partial\Omega, \qquad (15.4.27a)$$

$$D\frac{\partial U}{\partial n_j} = \beta(U - u_j), \quad \mathbf{x} \in \partial\Omega_j, \quad j = 1, \ldots, N. \qquad (15.4.27b)$$

Here $\partial_n$ and $\partial_{n_j}$ denote the outward normal of $\partial\Omega$ and $\partial\Omega_j$, respectively, $D$ is the extracellular diffusivity and $\gamma$ is the rate of degradation in the bulk. For simplicity, the cellular compartments $\Omega_j$ are taken to be disks of radius $\sigma$ centered at $\mathbf{x}_j \in \Omega$ for $j = 1, \ldots, N$. As in the case of the general network model (15.4.1), let $u_j$ be the intracellular autoinducer concentration of the $j$-th cell, and suppose that there are $K$ other chemical species within each cell, which together with the signaling molecule comprise a regulatory network. Let $\mathbf{z}_j = (u_j, v_{j,1} \ldots, v_{j,K}) \in \mathbb{R}^{1+K}$. Following equation (15.4.5), we have a system of ODEs for the cell dynamics of the form

Fig. 15.20: Schematic diagram showing a population of cells labeled $j = 1, \ldots, N$ that can exchange an autodinducer $A$ with the extracellular domain $\Omega$. The extracellular concentration of $A$ is denoted by $U(\mathbf{x}, t)$ and the intracellular concentration of $A$ in the $j$-th cell is denoted by $u_j(t)$.

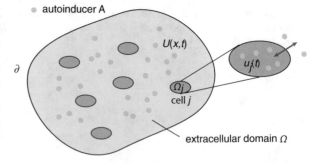

$$\frac{d\mathbf{z}_j}{dt} = \mathbf{f}(\mathbf{z}_j) - \frac{\kappa \mathbf{e}_1}{2\pi\sigma} \int_{\partial\Omega_j} (u_j - U) ds \tag{15.4.27c}$$

for $j = 1,\ldots,N$, with $(\mathbf{z}_j)_1 = u_j$, $\mathbf{f} = (F, G_1, \ldots, G_K)$, and $\mathbf{e}_1 = (1,0,\ldots,0)$. (For simplicity, we assume that the mass-action kinetics are the same for all cells, that is, the form of $\mathbf{f}$ is independent of $j$.) Note that the integral on the right-hand side is the number of autoinducer molecules leaving the cell per second. Also, note that integrating equation (15.4.27b) over the domain $\Omega \setminus \cup_j \partial\Omega_j$ and imposing the boundary conditions gives

$$\frac{dU_{\text{tot}}}{dt} = \beta \sum_{j=1}^{N} \int_{\partial\Omega_j} (u_j - U) ds - \gamma U_{\text{tot}}. \tag{15.4.28}$$

Dividing through by the volume $|\Omega|$ and comparing with equation (15.4.1c) suggests identifying $2\pi\sigma\beta$ with $\widehat{\kappa} = \alpha\kappa|\Omega|/N$.

We now impose the condition that the cells are small relative to the size of the domain by setting $\varepsilon = \sigma/L \ll 1$, where $L$ is a typical radius of $\Omega$. We also fix the units of length and time by setting $L = 1$ and $\gamma = 1$. We thus obtain the system of equations

$$\frac{\partial U}{\partial t} = D\nabla^2 U - U, \quad \mathbf{x} \in \Omega \setminus \bigcup_{j=1}^{N} \Omega_j; \quad \frac{\partial U}{\partial n} = 0, \quad \mathbf{x} \in \partial\Omega, \tag{15.4.29a}$$

$$2\pi\varepsilon D \frac{\partial U}{\partial n_j} = \widehat{\kappa}(U - u_j), \quad \mathbf{x} \in \partial\Omega_j, \quad j = 1,\ldots,N, \tag{15.4.29b}$$

$$\frac{d\mathbf{z}_j}{dt} = \mathbf{f}(\mathbf{z}_j) - \frac{\mathbf{e}_1 \kappa}{2\pi\varepsilon} \int_{\partial\Omega_j} (u_j - U) ds, \quad j = 1,\ldots,N. \tag{15.4.29c}$$

Finally, we assume that the cells are spatially well separated in the sense that $|\mathbf{x}_i - \mathbf{x}_j| = O(1)$ for $i \neq j$ and $\text{dist}(\mathbf{x}_j, \partial\Omega) = O(1)$ for $j = 1,\ldots,N$ as $\varepsilon \to 0$. Following Gou and Ward [368], we now use matched asymptotic expansions to construct the steady-state solution of equation (15.4.29), in order to deal with the sharp spatial gradient of $U$ in an $O(\varepsilon)$ neighborhood of each cell, see also Chap. 6.

*Asymptotic analysis of steady-state solution.* In the inner region around the $j$-th cell, we introduce the stretched coordinates $\mathbf{y} = \varepsilon^{-1}(\mathbf{x} - \mathbf{x}_j)$ and set $U_j(\mathbf{y}) = U(\mathbf{x}_j + \varepsilon\mathbf{y})$. Equations (15.4.29a,b) become

$$D\nabla_{\mathbf{y}}^2 U_j = 0, \quad |\mathbf{y}| > 1; \quad 2\pi D \frac{\partial U_j}{\partial n_j} = \widehat{\kappa}(U_j - u_j), \quad |\mathbf{y}| = 1. \tag{15.4.30}$$

Assuming a radially symmetric solution $U_j = U_j(\rho)$, where $\rho = |\mathbf{y}|$, we have

$$\frac{\partial^2 U_j}{\partial\rho^2} + \frac{1}{\rho}\frac{\partial U_j}{\partial\rho} = 0, \quad 1 < \rho < \infty; \quad 2\pi D \frac{\partial U_j}{\partial\rho} = \widehat{\kappa}(U_j - u_j), \quad \rho = 1. \tag{15.4.31}$$

The inner solution thus takes the form

$$U_j = S_j \log\rho + \chi_j, \quad \chi_j = \frac{1}{\kappa}(2\pi DS_j + \widehat{\kappa}u_j) \tag{15.4.32}$$

for $j = 1, \ldots, N$. The constants $S_j$ will be determined by matching the inner solutions to the outer solution. Given the steady-state solution $U_j$, the intracellular concentrations $\mathbf{z}_j$ are obtained, at least in principle, by solving the nonlinear algebraic system

$$\mathbf{f}(\mathbf{z}_j) + \frac{2\pi D\kappa S_j}{\widehat{\kappa}} \mathbf{e}_1 = 0. \tag{15.4.33}$$

As far as the outer solution is concerned, each cell shrinks to a point $\mathbf{x}_j \in \Omega$ as $\varepsilon \to 0$. Each point $\mathbf{x}_j$ effectively acts as a point source that generates a logarithmic singularity resulting from the asymptotic matching of the outer solution to the far-field behavior of the inner solution. Thus, the outer solution satisfies

$$\nabla^2 U(\mathbf{x}) - \frac{1}{D}U(\mathbf{x}) = 0, \quad \mathbf{x} \in \Omega \backslash \{\mathbf{x}_1, \ldots, \mathbf{x}_N\}; \quad \partial_n U = 0, \quad \mathbf{x} \in \partial\Omega, \tag{15.4.34a}$$

and

$$U(\mathbf{x}) \sim \frac{S_j}{v} + S_j \log |\mathbf{x} - \mathbf{x}_j| + \chi_j, \quad \text{as } \mathbf{x} \to \mathbf{x}_j, \quad j = 1, \ldots, N, \tag{15.4.34b}$$

where $v = -1/\log(\varepsilon)$. The outer problem can be solved in terms of the Neumann Green's function $G$ satisfying

$$\nabla^2 G(\mathbf{x}, \mathbf{x}_j) - \frac{1}{D}G(\mathbf{x}, \mathbf{x}_j) = -\delta(\mathbf{x} - \mathbf{x}_j), \quad \mathbf{x} \in \Omega, \tag{15.4.35a}$$

$$\partial_n G(\mathbf{x}, \mathbf{x}_j) = 0, \quad \mathbf{x} \in \partial\Omega. \tag{15.4.35b}$$

As $\mathbf{x} \to \mathbf{x}_j$, the Green's function has the local behavior

$$G(\mathbf{x}, \mathbf{x}_j) \sim -\frac{1}{2\pi} \log |\mathbf{x} - \mathbf{x}_j| + R_j(\mathbf{x}_j) + o(1) \quad \text{as } \mathbf{x} \to \mathbf{x}_j,$$

where $R_j(\mathbf{x}_j)$ is the regular part of $G(\mathbf{x}, \mathbf{x}_j)$ at $\mathbf{x} = \mathbf{x}_j$. It follows that the outer solution can be expressed as

$$U(\mathbf{x}) = -2\pi \sum_{j=1}^{N} S_j G(\mathbf{x}, \mathbf{x}_j). \tag{15.4.36}$$

Finally, expanding $U$ as $\mathbf{x} \to \mathbf{x}_j$, and equating the resulting expression with equation (15.4.34b), yields the following algebraic system for the unknowns $S_j$:

$$\left(1 + \frac{2\pi Dv}{\widehat{\kappa}} + 2\pi v R_j(\mathbf{x}_j)\right) S_j + 2\pi v \sum_{k \neq j} G(\mathbf{x}_j, \mathbf{x}_k) S_k = -v u_j. \tag{15.4.37}$$

*Large diffusion limit (existence).* It is not possible in general to use equations (15.4.37) and (15.4.33) to derive explicit conditions for the existence of a steady-state solution. (The corresponding eigenvalue problem that determines linear stability is also intractable.) However, progress can be made in the fast diffusion limit [368]. More specifically, suppose that $D = O(v^{-1}) \gg 1$ and set $D = D_0/v$. The equation for the Green's function becomes

$$\nabla^2 G(\mathbf{x}, \mathbf{x}_j) - \frac{v}{D_0}G(\mathbf{x}, \mathbf{x}_j) = -\delta(\mathbf{x} - \mathbf{x}_j), \quad \mathbf{x} \in \Omega; \quad \partial_n G(\mathbf{x}, \mathbf{x}') = 0, \quad \mathbf{x} \in \partial\Omega.$$

Since this has no solution when $v = 0$, we introduce the perturbation series expansion

$$G = v^{-1} G_{-1} + G_0 + v G_1 + v^2 G_2 + \ldots.$$

Substituting into the Green's function equation and equating powers of $v$ generates a hierarchy of equations for $G_l$, $l = -1, 0, 1, \ldots$. In the limit $v \to 0$, we obtain the following two term expansion:

$$G(\mathbf{x}, \mathbf{x}_j) = \frac{D_0}{v|\Omega|} + G_0(\mathbf{x}, \mathbf{x}_j), \quad R_j = \frac{D_0}{v|\Omega|} + R_{0,j}, \qquad (15.4.38)$$

where $G_0(\mathbf{x}, \mathbf{x}_j)$, with regular part $R_{0,j}$, is the Neumann's Green's function satisfying

$$\nabla^2 G_0(\mathbf{x}, \mathbf{x}_j) = \frac{1}{|\Omega|} - \delta(\mathbf{x} - \mathbf{x}_j), \quad \mathbf{x} \in \Omega, \qquad (15.4.39a)$$

$$\partial_n G(\mathbf{x}, \mathbf{x}_j) = 0, \quad \mathbf{x} \in \partial\Omega, \quad \int_\Omega G_0 d\mathbf{x} = 0, \qquad (15.4.39b)$$

and

$$G_0(\mathbf{x}, \mathbf{x}_j) \sim -\frac{1}{2\pi} \log |\mathbf{x} - \mathbf{x}_j| + R_{0,j}, \quad \mathbf{x} \to \mathbf{x}_j.$$

Substituting the expansion into equations (15.4.37) and (15.4.33) with $D = D_0/v$ gives

$$\left(1 + \frac{2\pi D_0}{\widehat{\kappa}} + 2\pi v R_{0,j}\right) S_j + \frac{2\pi D_0}{|\Omega|} \sum_{l=1}^{N} S_l + 2\pi v \sum_{k \neq j} G_0(\mathbf{x}_j, \mathbf{x}_k) S_k = -v u_j, \quad (15.4.40a)$$

$$\mathbf{f}(\mathbf{z}_j) + \frac{2\pi D_0 \kappa S_j}{\widehat{\kappa} v} \mathbf{e}_1 = 0. \qquad (15.4.40b)$$

The leading order solution to equation (15.4.40a) when $v \ll 1$ is thus

$$S_j = v S_{j,0} + O(v^2), \quad \mathbf{z}_j = \mathbf{z}_{j,0},$$

where

$$\left(1 + \frac{2\pi D_0}{\widehat{\kappa}}\right) S_{j,0} + \frac{2\pi D_0}{|\Omega|} \sum_{l=1}^{N} S_{l,0} = -u_{j,0}, \quad \mathbf{f}(\mathbf{z}_{j,0}) + \frac{2\pi D_0 \kappa S_{j,0}}{\widehat{\kappa}} \mathbf{e}_1 = 0. \quad (15.4.41)$$

The important point to note is that to leading order $S_j$ and $\mathbf{z}_j$ are independent of the spatial configuration of the cells within $\Omega$. In particular, there exists a homogeneous collective steady state $S_j = \overline{S}$, $\mathbf{z}_j = \overline{\mathbf{z}}$ with

$$\left(1 + \frac{2\pi D_0}{\kappa\alpha} + \frac{2\pi N D_0}{|\Omega|}\right) \overline{S} = -\overline{u}, \quad \mathbf{f}(\overline{\mathbf{z}}) + \frac{2\pi D_0 \kappa \overline{S}}{\widehat{\kappa}} \mathbf{e}_1 = 0. \qquad (15.4.42)$$

We thus obtain the system of equations

$$F(\overline{u}, \overline{\mathbf{v}}) - \Gamma\overline{u} = 0, \quad G_k(\overline{u}, \overline{\mathbf{v}}) = 0 \qquad (15.4.43)$$

for $k = 1, \ldots, K$, where

$$\Gamma = \frac{2\pi D_0 \kappa/\widehat{\kappa}}{1 + 2\pi D_0/\widehat{\kappa} + 2\pi N D_0/|\Omega|\gamma}. \qquad (15.4.44)$$

Assuming that the second and third terms in the denominator are much larger than one, the dependence on $D_0$ vanishes and

$$\chi \approx \frac{\gamma\kappa}{\gamma + \alpha\kappa}, \qquad (15.4.45)$$

where $\alpha = N v_{\text{cyt}}/V_{\text{ext}}$. We thus recover the steady-state equations of the ODE network model (15.4.6).

*Linear stability analysis.* Assuming that there exists a steady-state solution to equations (15.4.37) and (15.4.33), the next step is to investigate its linear stability. Let us denote a steady-state solution by $U^*(\mathbf{x})$ and $\mathbf{z}_j^*$ for $j = 1, \ldots, N$ and consider the perturbations

$$U(\mathbf{x},t) = U^*(\mathbf{x}) + e^{\lambda t}\eta(\mathbf{x}), \quad \mathbf{z}_j(t) = \mathbf{z}_j^* + e^{\lambda t}\zeta_j, \quad \zeta_j = (\phi_j, \eta_{j,1}, \ldots \eta_{j,K}).$$

Substituting this perturbation into equation (15.4.29) leads to the following eigenvalue problem [368]:

$$\lambda\eta = D\nabla^2\eta - \eta, \quad \mathbf{x} \in \Omega \setminus \bigcup_{j=1}^{N}\Omega_j; \quad \frac{\partial\eta}{\partial n} = 0, \quad \mathbf{x} \in \partial\Omega, \tag{15.4.46a}$$

$$2\pi\varepsilon D\frac{\partial\eta}{\partial n_j} = \widehat{\kappa}(\eta - \phi_j), \quad \mathbf{x} \in \partial\Omega_j, \quad j = 1, \ldots, N, \tag{15.4.46b}$$

$$\lambda\zeta_j = \mathbf{J}^{(j)}\zeta_j - \frac{\kappa\mathbf{e}_1}{2\pi\varepsilon}\int_{\partial\Omega_j}(\phi_j - \eta)\,ds, \quad j = 1, \ldots, N, \tag{15.4.46c}$$

where $\mathbf{J}^{(j)}$ is the Jacobian of $\mathbf{F}$ evaluated at $\mathbf{z}_j^*$. These equations can also be studied using matched asymptotic expansions. The analysis of the inner solution proceeds as before, and we find that

$$\eta_j = c_j\log\rho + B_j, \quad B_j = \chi_j = \frac{1}{\widehat{\kappa}}(2\pi Dc_j + \widehat{\kappa}\phi_j). \tag{15.4.47}$$

Substituting into equation (15.4.46c) gives

$$\left(\mathbf{J}^{(j)} - \lambda\mathbf{I}\right)\zeta_j + \frac{2\pi D\kappa c_j}{\widehat{\kappa}}\mathbf{e}_1 = 0, \quad j = 1, \ldots, N. \tag{15.4.48}$$

Similarly, the outer solution becomes

$$\eta(\mathbf{x}) = -2\pi\sum_{j=1}^{N}G_\lambda(\mathbf{x}, \mathbf{x}_j), \tag{15.4.49}$$

where we now have an eigenvalue-dependent Green's function

$$\nabla^2 G_\lambda - \sqrt{\frac{1+\lambda}{D}}G_\lambda = -\delta(\mathbf{x} - \mathbf{x}_j), \quad \mathbf{x} \in \Omega; \quad \partial_n G(\mathbf{x}, \mathbf{x}_j) = 0, \quad \mathbf{x} \in \partial\Omega. \tag{15.4.50}$$

Matching the singularity condition with the far-field behavior of the inner solution leads to a system of equations for $c_j$

$$\left(1 + \frac{2\pi Dv}{\widehat{\kappa}} + 2\pi v R_{\lambda,j}\right)c_j + 2\pi v\sum_{k\neq j}G_\lambda(\mathbf{x}_j, \mathbf{x}_k)c_k = -v\phi_j, \tag{15.4.51}$$

where $R_{\lambda,j}$ is the regular part of $G_\lambda(\mathbf{x}, \mathbf{x}_j)$ at $\mathbf{x} = \mathbf{x}_j$.

Assuming that $\lambda$ is not an eigenvalue of the Jacobian $\mathbf{J}^{(j)}$, we can solve equation (15.4.48) for the vector $\zeta_j$ in terms of $c_j$:

$$\zeta_j = \frac{2\pi D\kappa c_j}{\widehat{\kappa}}(\lambda\mathbf{I} - \mathbf{J}^{(j)})^{-1}\mathbf{e}_1.$$

Projecting onto the first component of $\zeta_j$ shows that

$$\phi_j = \frac{2\pi D\kappa c_j}{\widehat{\kappa}}\mathbf{e}_1^\top(\lambda\mathbf{I} - \mathbf{J}^{(j)})^{-1}\mathbf{e}_1.$$

Finally, substituting for $\phi_j$ into equation (15.4.51), we obtain the homogeneous linear system $\mathbf{M}(\lambda)\mathbf{c} = 0$, with

$$M_{ij} = \left(1 + \frac{2\pi D\nu}{\widehat{\kappa}} + 2\pi\nu R_{\lambda,j} + \frac{2\pi\nu D\kappa}{\widehat{\kappa}}\mathcal{K}_j\right)\delta_{i,j} + 2\pi\nu G_\lambda(\mathbf{x}_j,\mathbf{x}_k)(1 - \delta_{j,k}), \quad (15.4.52)$$

and

$$\mathcal{K}_j = \mathbf{e}_1^\top(\lambda\mathbf{I} - \mathbf{J}^{(j)})^{-1}\mathbf{e}_1.$$

In the limit $\varepsilon \to 0$, $\lambda$ is a discrete eigenvalue of the linearized problem (15.4.46) if and only if $\lambda$ is a root of the transcendental equation $\det\mathbf{M}(\lambda) = 0$, and the stationary solution $(U^*, \mathbf{z}_1^*, \ldots, \mathbf{z}_N^*)$ is stable provided that $\mathrm{Re}(\lambda) < 0$.

*Large diffusion limit (stability).* The eigenvalue problem $\mathbf{M}(\lambda)\mathbf{c} = 0$, with $\mathbf{M}$ given by equation (15.4.52) can also be analyzed in the large diffusion limit. Again we set $D = D_0/\nu \gg 1$ and expand the Green's function $G_\lambda$

$$G_\lambda(\mathbf{x}, \mathbf{x}_j) = \frac{D_0}{\nu(1+\lambda)|\Omega|} + G_0(\mathbf{x}, \mathbf{x}_j). \quad (15.4.53)$$

Similarly,

$$M_{ij}(\lambda) = a(\lambda)\delta_{i,j} + \frac{b(\lambda)}{N} + O(\nu), \quad (15.4.54)$$

with

$$a(\lambda) = 1 + \frac{2\pi D}{\widehat{\kappa}} + \frac{2\pi D\kappa}{\widehat{\kappa}}\mathcal{K}, \quad b(\lambda) = \frac{2\pi D_0}{(1+\lambda)|\Omega|}. \quad (15.4.55)$$

Here

$$\mathcal{K} = \mathbf{e}_1^\top(\lambda\mathbf{I} - \mathbf{J})^{-1}\mathbf{e}_1,$$

where $\mathbf{J}$ is the Jacobian evaluated at the homogeneous solution $\bar{\mathbf{z}}$. The eigenvalue problem thus reduces to

$$a(\lambda)c_j + \frac{b(\lambda)}{N}\sum_{l=1}^N c_l = 0. \quad (15.4.56)$$

This naturally leads to the distinction between a synchronous eigenmode $c_j = c$ for all $j = 1, \ldots, N$ and $N-1$ asynchronous eigenmodes $\mathbf{q}_r$, $r = 2, \ldots, N$, with $\sum_{j=1}^N(\mathbf{q}_r)_j = 0$. It follows that $\lambda$ is a discrete eigenvalue of a synchronous mode if $a(\lambda) + b(\lambda) = 0$, whereas an eigenvalue of an asynchronous mode satisfies $a(\lambda) = 0$.

Gou and Ward [368] used the reduced spectral problem to investigate whether or not temporal oscillations can be triggered by the cell-bulk coupling. In the case of one-component cells, no oscillations can occur (see Ex. 15.4). On the other hand, for certain two-component cellular models, they find that when the number of cells exceeds some critical number, passive bulk diffusion can initiate synchronous collective oscillations through a Hopf bifurcation that would otherwise not occur in the absence of diffusive coupling. A number of quorum sensing systems exhibit such oscillations. For example, elevated levels of the autoinducer signal cAMP can induce colonies of *Dictyostelium discoideum* to exhibit synchronous oscillations that precede the chemotactic migration of amoeba towards some organizing center [376, 687]. Moreover, starved yeast cells can exhibit synchronous oscillations in glycolytic intermediates, which are triggered through the bulk diffusing chemical Acetaldehyde [230]. Interesting open questions concern how bulk diffusion effects some of the other phenomena highlighted in this section, including bistability and ultrasensitivity, and how the local stability results of the PDE-ODE model relate to the global stability analysis of the network models.

## 15.5 Dynamical quorum sensing and the synchronization of genetic oscillators

Synthetic versions of bacterial QS have recently been developed in order to construct coupled systems of genetic oscillators, with the goal of enhancing the oscillatory response via synchronization. Such a mechanism was originally proposed in a number of theoretical studies [329, 608, 980]. An experimental realization of such a system was subsequently constructed by Danino et al. [221], which we now describe in more detail. The basic circuit diagram is shown in Fig. 15.21.

Each cell in the population produces a signaling molecule N-acyl-homoserine lactone (AHL) that can exit the cell, diffuse in the extracellular medium, and subsequently enter other cells. Two intracellular signaling proteins are used to create and degrade AHL, which are given by LuxI and AiiA, respectively. The genes encoding these two proteins are placed downstream of a promoter that is upregulated by AHL via a constitutively produced protein LuxR. Hence, there exists a dual feedback loop involving both positive and negative feedback paths, which means that an isolated cell can support oscillations (see Chap. 5). The diffusive coupling mediated by AHL then produces well-synchronized oscillations within a population of cells, once the cell density is high enough. Let $i = 1, \ldots, N$ label the $N$ cells of a population. Denote the protein concentrations of AiiA and LuxI in the $i$-th cell by $a_i$ and $b_i$, respectively. The mass-action kinetic equations for these proteins take the form

$$\frac{da_i}{dt} = \frac{\kappa_{0A} + \kappa_A u_i(t-\tau)^2}{K^2 + u_i(t-\tau)^2} - \frac{V_A a_i(t)}{K_m + a_i(t) + b_i(t)}, \tag{15.5.1a}$$

$$\frac{db_i}{dt} = \frac{\kappa_{0I} + \kappa_I u_i(t-\tau)^2}{K^2 + u_i(t-\tau)^2} - \frac{V_I b_i(t)}{K_m + a_i(t) + b_i(t)}, \tag{15.5.1b}$$

where $u_i(t)$ is the intracellular concentration of AHL and $\tau$ is a time delay. Here $\kappa_{0s}$ and $\kappa_s$, $s \in \{A, I\}$, are the basal and upregulated rates of protein synthesis. The

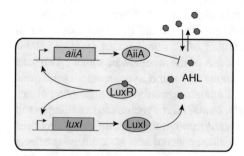

Fig. 15.21: Circuit diagram of a synthetic genetic oscillator that produces an autoinducer AHL, which can be exchanged with the extracellular domain and thus mediate diffusion-based cell-to-cell communication. LuxI and AiiA provide positive and negative feedback by producing and degrading AHL, respectively. AHL upregulates the promoters of LuxI and AiiA via constitutively produced LuxR.

proteins are enzymatically degraded according to Michaelis-Menten kinetics. The dynamical equation for $u_i$ couples to the extracellular AHL concentration $U$ as follows:

$$\frac{du_i}{dt} = \frac{\kappa b_i(t)}{K_u + b_i(t)} - \frac{\gamma a_i(t) u_i(t)}{K_H + u_i(t)} + d_1(U(t) - u_i(t)), \tag{15.5.2a}$$

$$\frac{dU}{dt} = -d_2 \sum_{j=1}^{N}(U(t) - u_j(t)) - \mu U(t). \tag{15.5.2b}$$

The first two terms in the dynamical equation for $u_i$ represent the effects of LuxI and AiiA on the production and degradation of AHL, respectively, whereas the third term is the diffusive exchange of AHL with the extracellular medium at a rate $d_1$. A similar term appears in the equation for $U$, with a scaled rate $d_2$, together with a term representing the loss of AHL at a rate $\mu$ due to fluid flow out of the system, for example. For simplicity, the extracellular concentration is taken to be spatially uniform.

### 15.5.1 Limit cycle oscillators coupled through an external medium

One of the distinctive features of the above system is that, the oscillators are coupled via an external medium, rather than directly, which is the more common form of coupling found in other cellular systems such as neural networks [284]. This has motivated a number of theoretical studies of simplified oscillator models coupled via quorum sensing, under the simplifying assumption that the extracellular concentration is spatially uniform [758, 805, 806, 814, 905]. (Including the effects of diffusion in the extracellular medium can be developed along the lines outlined in Refs. [366, 368] and Sect. 15.4.4.)

Consider a population of $N$ genetic oscillators that are diffusively coupled to an external medium. Following Ref. [805], suppose that each oscillator is close to a Hopf bifurcation point (see Box 5A), so that its state at time $t$ can be represented by a complex amplitude $z_j(t)$, $j = 1, \ldots, N$. It follows that the state of the environment can be represented by a complex variable $Z(t)$, which diffusively couples to each $z_j$ with a coupling strength $D$. Let $\omega_j$ be the natural frequency of the $j$-th oscillator. The frequencies $\omega_j$ are randomly drawn from a distribution $h(\omega)$, which is taken to be an even function about a mean frequency $\omega_0$. Finally, when the autoinducer molecules leave a cell and enter the medium, their concentration is diluted by a factor $\alpha = V_{int}/V_{ext}$, which is the ratio of the cell volume to the total extracellular volume. Introducing the dimensionless cell density $\rho = \alpha N$ and moving to a rotating frame with frequency $\omega_0$, the dynamics of the system can be represented by the equations

$$\frac{dz_j}{dt} = (\lambda_0 + i\omega_j - |z_j|^2)z_j - D(z_j - Z), \tag{15.5.3a}$$

$$\frac{dZ}{dt} = \frac{\rho D}{N}\sum_{j=1}^{N}(z_j - Z) - (\gamma + i\omega_0)Z. \tag{15.5.3b}$$

In the rotating frame, the frequencies $\omega_j$ are drawn from an even distribution $g(\omega) = h(\omega - \omega_0)$ with zero mean.

Note that the density factor $\rho$ in equation (15.5.3b) is the origin of the density-dependent collective response of bacterial QS systems. Here, we are interested in determining conditions under which the population of oscillators synchronizes. Intuitively speaking, increasing the effective coupling $\rho D$ should favor synchronization, whereas having a wider distribution of frequencies $\omega_j$ should favor desynchronization. These opposing effects are well known within the theory of oscillator synchronization [533, 728]. It is difficult to analyze the existence and stability of synchronous oscillations directly. However, insights can be gained by considering the stability of the so-called amplitude death state $z(t) = Z(t) = 0$ [805].

**Linear stability analysis of the amplitude death state.** Setting $z_j(t) = e^{\mu t}v_j$ and $Z(t) = e^{\mu t}V$ in the linearized equations gives

$$\mu v_j = (\lambda_0 + i\omega_j - D)v_j + DV,$$

$$\mu V = \frac{\rho D}{N}\sum_{j=1}^{N}v_j - (\rho D + \gamma + i\omega_0)Z.$$

In vector notation with $\mathbf{v} = (v_1,\dots,v_N,V)^\top$, we have the eigenvalue equation $\mu \mathbf{v} = \mathbf{Mv}$ with

$$\mathbf{M} = \begin{pmatrix} \lambda_0 - D + i\omega_1 & 0 & \cdot & \cdot & 0 & D \\ 0 & \lambda_0 - D + i\omega_2 & \cdot & \cdot & 0 & D \\ \cdot & & \cdot & \cdot & & \cdot \\ \cdot & & \cdot & \cdot & & \cdot \\ 0 & 0 & \cdot & \cdot\ \lambda_0 - D + i\omega_N & D \\ \frac{\rho D}{N} & \frac{\rho D}{N} & \cdot & \cdot & \frac{\rho D}{N} & -(\rho D + \gamma + i\omega_0) \end{pmatrix}.$$

Linear stability of the amplitude state requires $\mathrm{Re}[\mu_k] < 0$ for all the eigenvalues $\mu_k$, $k = 1,\dots,N$.

Since $\sum_k \mu_k = \mathrm{Tr}[\mathbf{M}]$, a necessary condition for stability is $\mathrm{Re}[\mathrm{Tr}[\mathbf{M}]] < 0$, that is,

$$\lambda_0 - D < \frac{\rho D + \gamma}{N} \to 0 \tag{15.5.4}$$

for large $N$, where we have made the approximation

$$\sum_{j=1}^{N}\omega_j \approx \int_{-\infty}^{\infty}g(\omega)d\omega = 0.$$

The eigenvalues are determined from the characteristic equation $\det(\mu \mathbf{I} - \mathbf{M}) = 0$, which yields

$$[\mu + \rho D + \gamma + i\omega_0] \prod_{j=1}^{N}[\mu - (\lambda_0 - D + i\omega_j)] = \frac{\rho D^2}{N} \sum_{s=1}^{N} \prod_{j=1, j \neq s}^{N} [\mu - (\lambda_0 - D + i\omega_j)].$$

This can be rearranged to give

$$\mu + \rho D + \gamma + i\omega_0 = \frac{\rho D^2}{N} \sum_{s=1}^{N} \frac{1}{\mu - (\lambda_0 - D + i\omega_j)}. \tag{15.5.5}$$

Now taking the thermodynamic limit $N \to \infty$ for fixed density $\rho$, we have

$$\mu + \rho D + \gamma + i\omega_0 = \rho D^2 \int_{-\infty}^{\infty} \frac{g(\omega)}{\mu - (\lambda_0 - D + i\omega)} d\omega. \tag{15.5.6}$$

In order to determine the stability boundary of the amplitude death state, set $\mu = ib$ for real $b$ and equate real and imaginary parts. This yields the pair of coupled integral equations [805]

$$\frac{\rho D + \gamma}{\rho D^2} = \int_{-\infty}^{\infty} g(\omega) \frac{D - \lambda_0}{(D - \lambda_0)^2 + (b - \omega)^2} d\omega, \tag{15.5.7a}$$

$$\frac{b + \omega_0}{\rho D^2} = -\int_{-\infty}^{\infty} g(\omega) \frac{b - \omega}{(D - \lambda_0)^2 + (b - \omega)^2} d\omega, \tag{15.5.7b}$$

with $\lambda_0 < D$. Note that in the limit $D \to \lambda_0$ we can use the definition of principal part in complex contour integration (see Box 10A). That is, setting $D - \lambda_0 = \varepsilon$, we have for $\varepsilon \to 0^+$

$$\frac{\rho D + \gamma}{\rho D^2} + i\frac{b + \omega_0}{\rho D^2} = i\int_{-\infty}^{\infty} g(\omega) \frac{\omega - b - i\varepsilon}{\varepsilon^2 + (\omega - b)^2} d\omega$$

$$= i\int_{-\infty}^{\infty} g(\omega) \frac{1}{\omega - (b - i\varepsilon)} d\omega$$

$$= i\left[\mathscr{P}\int_{-\infty}^{\infty} \frac{g(\omega)}{\omega - b} d\omega - i g(b)\right].$$

We have used the standard identity

$$\frac{1}{x + i\varepsilon} = \mathscr{P}(1/x) - i\pi\delta(x),$$

where $\mathscr{P}$ denotes principal part. Hence, in the limit $D \to \lambda_0$, the phase boundary is determined by the equations

$$\frac{\rho D + \gamma}{\rho D^2} = \pi g(b), \qquad \frac{b + \omega_0}{\rho D^2} = \mathscr{P}\int_{-\infty}^{\infty} \frac{g(\omega)}{\omega - b} d\omega. \tag{15.5.8}$$

In the case of a homogeneous population, $g(\omega) = \delta(\omega)$, equation (15.5.7) reduces to the conditions

$$\frac{\rho D + \gamma}{\rho D^2} = \frac{D - \lambda_0}{(D - \lambda_0)^2 + b^2}, \qquad \frac{b + \omega_0}{\rho D^2} = -\frac{b}{(D - \lambda_0)^2 + b^2}. \qquad (15.5.9)$$

Eliminating $b$ from these equations then yields a phase boundary in the $(\rho, D)$-plane, as illustrated in Fig. 15.22 for various degradation rates $\gamma$ and $\omega_0 = \lambda_0 = 1$. One finds that crossing a phase boundary from below by increasing $\rho$ or $D$ results in a transition from amplitude death to synchronized oscillations with collective frequency $\omega_0 + b$ (in the lab frame). It can be seen that one way to eliminate oscillations is to increase the degradation rate $\gamma$. This can be understood by noting that stability of the amplitude death state requires $\lambda_0 - D < 0$. In this regime, an uncoupled cell ($D = 0$) cannot oscillate since we require $\lambda_0 > 0$, so the existence of collective oscillations requires diffusing coupling with the external variable. However, if the degradation rate of $Z$ is too large, then it cannot be maintained at sufficient levels to support oscillations. Another limiting case is the high density limit $\rho \to \infty$, whereby the left-hand side of equation (15.5.7b) vanishes. Since $g(\omega) = g(-\omega)$, this implies that $b = 0$. Substitution into equation (15.5.7a) yields the single stability condition

$$\frac{1}{D} = \int_{-\infty}^{\infty} g(\omega) \frac{D - \lambda_0}{(D - \lambda_0)^2 + \omega^2} d\omega.$$

In this case, the phase boundary is independent of $\gamma$ and the mechanism for destroying oscillations is due to the heterogeneity in the oscillator frequencies.

Finally, note that when $D < \lambda_0$ the amplitude death state is no longer stable. However, there is now another type of solution besides synchrony, namely, an incoherent state, in which individual oscillators rotate in an unsynchronized fashion. Analytical insights into the existence and stability of the incoherent state can be obtained in the regime $\lambda_0 \gg D$ for which the oscillators are weakly coupled. In that case, equations (15.5.3) reduce to a version of the Kuramoto model [806].

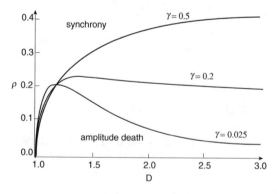

Fig. 15.22: Sketch of phase boundaries for homogeneous oscillators. Phase boundary between the amplitude death state (below a boundary curve) and the synchronized state (above a boundary curve) in the $(\rho, D)$-plane for various values of the degradation rate $\gamma$. Other parameters are $\omega_0 = \lambda_0 = 1$. Note that a stable amplitude state can only exist when $D > 1$. [Redrawn from Schwab et al. [805].]

## 15.5.2 *Kuramoto model with coupling via an external medium*

Suppose that equation (15.5.3a) is rewritten in polar coordinates with $z_j = r_j e^{i\theta_j}$ and $Z = Re^{i\Phi}$:

$$\frac{dr_j}{dt} = (\lambda_0 - D - r_j^2)r_j + DR\cos(\Phi - \theta_j), \qquad (15.5.10a)$$

$$\frac{d\theta_j}{dt} = \omega_j + \frac{DR}{r_j}\sin(\Phi - \theta_j). \qquad (15.5.10b)$$

The main simplification in the large $\lambda_0$ regime is to note that $r_j \approx \sqrt{\lambda_0}$ in steady state for all $j = 1,\dots,N$. Thus, we can effectively eliminate the dynamics of the amplitudes $r_j$. After performing the rescalings $r_j \to r_j/\sqrt{\lambda_0}$ and $R \to R/\sqrt{\lambda_0}$, equations (15.5.3) reduce to [806]

$$\frac{d\theta_j}{dt} = \omega_j + DR\sin(\Phi - \theta_j), \qquad (15.5.11a)$$

$$\frac{dZ}{dt} = \frac{\rho D}{N}\sum_{j=1}^{N}(e^{i\theta_j} - Z) - (\gamma + i\omega_0)Z. \qquad (15.5.11b)$$

This is almost identical to the Kuramoto model of globally coupled phase oscillators [533], except that $R$ is the real amplitude of the medium rather than the amplitude of the Kuramoto order parameter. The latter is defined according to

$$\bar{z}(t) := \frac{1}{N}\sum_{j=1}^{N}e^{i\theta_j(t)} = r(t)e^{i\phi(t)}. \qquad (15.5.12)$$

This has a geometric interpretation as the centroid of the phases with $\phi(t)$ equal to the average phase and $r(t)$ a measure of the degree of phase-coherence, see Fig. 15.23. A completely incoherent state corresponds to the case $\bar{z} = 0$, whereas a completely synchronized state satisfies $\bar{z} = 1$. Given the definition of $\bar{z}$, equation (15.5.11b) simplifies as

$$\frac{dZ}{dt} = \rho D(\bar{z} - Z) - (\gamma + i\omega_0)Z. \qquad (15.5.13)$$

Following the analysis of the classical Kuramoto model [533], see Ex. 15.5, the transition from incoherent to partially coherent states can be investigated in the thermodynamic limit $N \to \infty$ by considering a continuum model. That is, let $f(\theta, \omega, t)d\theta$ denote the fraction of oscillators with frequency $\omega$ that have a phase in the interval $[\theta, \theta + d\theta]$ at time $t$. Since the total number of oscillators is fixed, we have the continuity or Liouville equation

$$\frac{\partial f}{\partial t} = -\frac{\partial}{\partial \theta}(\dot{\theta}f)). \qquad (15.5.14)$$

Substituting for $\dot{\theta}$ using equation (15.5.11a) then yields

$$\frac{\partial f}{\partial t} = -\frac{\partial}{\partial \theta}\left[\omega f + \frac{D}{2i}(Ze^{-i\theta} - Z^* e^{i\theta})f\right].$$

(15.5.15)

This is supplemented by the normalization condition

$$\int_0^{2\pi} f(\theta, \omega, t)d\theta = g(\omega),$$

(15.5.16)

where $g(\omega)$ is the even, zero mean distribution of frequencies. It follows that $f$ can be expanded as a Fourier series in $\theta$

$$f(\theta, \omega, t) = \frac{g(\omega)}{2\pi}\left(1 + \sum_{n=1}^{\infty}\left[f_n(\omega, t)e^{in\theta} + \text{c. c.}\right]\right).$$

(15.5.17)

As it stands, solving the initial value problem for $f$ involves solving an infinite hierarchy of equations for the coefficients $f_n$. However, a major simplification can be achieved by assuming that the system evolves on a low-dimensional manifold of states first introduced by Ott and Antonsen [700]. (Numerical simulations of the full system indicate that the dynamics is captured by such a reduction [806].) The Ott-Antonsen ansatz is to take $f_n = \alpha^n$ so that

$$f(\theta, \omega, t) = \frac{g(\omega)}{2\pi}\left(1 + \sum_{n=1}^{\infty}\left[\alpha^n(\omega, t)e^{in\theta} + \text{c. c.}\right]\right).$$

(15.5.18)

Substituting this ansatz into the continuity equation (15.5.15) gives

$$\frac{\partial \alpha}{\partial t} + i\omega\alpha + \frac{D}{2}(Z\alpha^2 - Z^*) = 0.$$

(15.5.19)

Moreover, since the order parameter can be written as

$$\bar{z}(t) = \int_{-\infty}^{\infty}\int_0^{2\pi} f(\theta, \omega, t)e^{i\theta}d\theta d\omega,$$

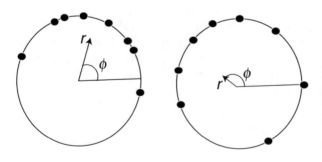

Fig. 15.23: Complex order parameter $z(t) = r(t)e^{i\phi(t)}$ for two distributions of phases on the circle. The state shown on the left-hand side is more coherent.

the Ott-Antonsen ansatz implies that

$$\bar{z}(t) = \int_{-\infty}^{\infty} g(\omega)\alpha^*(\omega,t)d\omega. \tag{15.5.20}$$

**Stability of the incoherent state.** In the fully incoherent state, the phase of the oscillators is uniformly distributed so that $f = g(\omega)/2\pi$. This corresponds to setting $\alpha = 0$ in the Fourier series expansion (15.5.18). It immediately follows that we have the solution $Z = \bar{z} = 0$. (Such a solution is distinct from the amplitude death state as the individual cells oscillate, but are incoherent so that the order parameter $\bar{z} = 0$.) In order to calculate the linear stability of the incoherent state, take $\alpha, Z, \bar{z} = O(\varepsilon)$ with $0 < \varepsilon \ll 1$. Linearize equation (15.5.19) with $\alpha = A_0 e^{(\lambda+i\Omega)t}$, $Z = Z_0 e^{(\lambda-i\Omega)t}$ and $\bar{z} = B_0 e^{(\lambda-i\Omega)t}$, where $\lambda, \Omega$ are real numbers. This results in the equation

$$A_0 = \frac{DZ_0^*}{2[\lambda + i(\Omega + \omega)]}.$$

Similarly, equations (15.5.13) and (15.5.20) yield

$$(\lambda - i\Omega)Z_0 = \rho DB_0 - (\rho D + \gamma + i\omega_0)Z_0,$$

with

$$B_0 = \int_{-\infty}^{\infty} g(\omega)A_0^*(\omega)d\omega = \int_{-\infty}^{\infty} g(\omega)\frac{DZ_0}{2[\lambda - i(\Omega + \omega)]}d\omega.$$

We thus obtain the following complex-valued equation for $\lambda$ and $\Omega$:

$$\lambda + i(\omega_0 - \Omega) + \rho D + \gamma = \frac{\rho D^2}{2}\int_{-\infty}^{\infty}\frac{g(\omega)}{[\lambda - i(\Omega + \omega)]}d\omega. \tag{15.5.21}$$

Finally, in order to determine the stability boundary, set $\lambda = \varepsilon$ with $\varepsilon \to 0^+$ and equate real and imaginary parts

$$2\frac{\rho D + \gamma}{\rho D^2} = \pi g(\Omega), \quad 2\frac{\omega_0 - \Omega}{\rho D^2} = \mathscr{P}\int_{-\infty}^{\infty}\frac{g(\omega)}{\omega + \Omega}d\omega. \tag{15.5.22}$$

This pair of implicit equations determines the stability phase boundary in $(\rho, D)$-plane for the incoherent state. In the case of a Lorentzian distribution of frequencies

$$g(\omega) = \frac{1}{\pi}\frac{\Delta}{\omega^2 + \Delta^2}, \tag{15.5.23}$$

the principal part can be evaluated explicitly using contour integration, and one finds the following equation for the phase boundary [806]

$$\frac{\Delta\omega_0^2}{\Delta + \rho D + \gamma}\left(1 - \frac{\Delta}{\Delta + \rho D + \gamma}\right) + \Delta(\rho D + \gamma) - \frac{\rho D^2}{2} = 0. \tag{15.5.24}$$

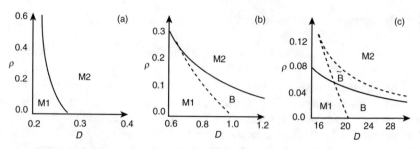

Fig. 15.24: Sketch of phase diagrams for incoherent and partially coherent oscillatory solutions of the Kuramoto model with coupling to an external medium. Parameters are $\Delta = 0.1$ and $\gamma = 0$ and (a) $\omega_0 = 0.0577$, $(b)\omega_0 = 0.4$, and (c) $\omega_0 = 10$. Labeled regions are incoherent oscillations (M1), coherent oscillations (M2), bistability between incoherent and coherent oscillations (B), and bistability between large and small amplitude coherent oscillations. [Redrawn from Schwab et al. [806].]

**Existence of partially coherent states.** Let us now search for solutions of the form

$$\alpha = A(\omega)e^{i\Omega t}, \quad Z = Z_0 e^{-i\Omega t}, \quad \bar{z} = \bar{z}_0 e^{-i\Omega t}, \tag{15.5.25}$$

where $\Omega$ is the collective frequency of a partially coherent state. Substitution into equation (15.5.19) gives

$$iA(\omega + \Omega) + \frac{D}{2}(Z_0 A^2 - Z_0^*) = 0.$$

This quadratic equation in $A$ has the roots

$$A_\pm(\omega) = \frac{-i(\omega + \Omega) \pm \sqrt{-(\omega + \Omega)^2 + D^2 R^2}}{DZ_0}, \tag{15.5.26}$$

with $R = |Z_0|$. A necessary condition that the solution stays on the Ott-Antonsen manifold is that $|A(\omega)| \leq 1$ so that the Fourier series in (15.5.18) does not diverge. This is guaranteed by choosing the positive root $A = A_+(\omega)$ when $-DR < \omega + \Omega$ and $A = A_-(\omega)$ when $\omega + \Omega < -DR$. The next step is to substitute the assumed partially incoherent state solutions (15.5.25) into equation (15.5.13) to obtain

$$[i(\omega_0 - \Omega) + \rho D + \gamma]R^2 = DZ_0 \rho \int_{-\infty}^{\infty} g(\omega)A(\omega + \Omega)d\omega. \tag{15.5.27}$$

Combining this equation with the solution for $A(\omega)$ then yields a dispersion relation between $\Omega$ and $R$ for partially coherent states.

In the case of the Lorentzian distribution (15.5.23), it is possible to calculate explicit dynamical equations for the partially coherent states. First, substituting (15.5.23) into equation (15.5.20) and evaluating the resulting contour integral establishes that

$$\bar{z}(t) = \alpha^*(-i\Delta, t).$$

Hence, evaluating equation (15.5.19) at $\omega = -i\Delta$ yields the following equation for the order parameter $\bar{z}$:

$$\frac{\partial \bar{z}}{\partial t} + \Delta \bar{z} + \frac{D}{2}(Z^* \bar{z}^2 - Z) = 0. \tag{15.5.28}$$

In order to determine the existence of a partially coherent state, substitute the trial solution $\bar{z}(t) = re^{-i\Omega t}$ into equation (15.5.13) to find that

$$Z(t) = \frac{\rho D r}{i(\omega_0 - \Omega) + \rho D + \gamma} e^{-i\Omega t}.$$

Inserting the form of $\bar{z}$ and $Z$ back into equation (15.5.28) then determines an equation relating $r$ and $\Omega$

$$(\Delta - i\Omega)r + \frac{r\rho D^2}{2}\left[\frac{r^2}{\rho D + \gamma - i(\omega_0 - \Omega)} - \frac{1}{\rho D + \gamma + (\omega_0 - \Omega)}\right] = 0. \tag{15.5.29}$$

One solution of equation (15.5.29) is the incoherent state $r = 0$, which can either be stable or unstable, as determined by the stability phase boundary condition (15.5.24). Dividing equation (15.5.29) by $r$ and equating real and imaginary parts yields a pair of equations for the pair $(r, \Omega)$. Eliminating $\Omega$ leads to a cubic equation for $q = r^2$. This implies that there can exist from zero to three partially coherent states, depending on the parameter range. However, the stability of such solutions has to be determined numerically. Example phase diagrams are shown in Fig. 15.24.

## 15.6 Biofilms

Microorganisms such as bacteria and other prokaryotes constitute most of the biomass on earth, and are found virtually everywhere, including the atmosphere, oceans, soils, rocks, ice, and volcanoes. It is thought that most of the microbial biomass is located in close-knit communities known as biofilms, which consist of a large number of cells living within a self-secreted polymer-rich matrix. Bacterial biofilms are of crucial importance in a wide variety of environmental, industrial, and medical systems. For example, soil bacteria play a primary role in the nitrogen cycle and the breakdown of complex organic contaminants. On the other hand, bacterial growth is a major source of infections in the medical and pharmaceutical industries, and a primary cause of human diseases such as tooth decay, cystic fibrosis, and urinary infections.

Bacteria typically form a biofilm by attaching to some surface and then growing to form a layer of thickness of order 100 $\mu m$. Growth is supported by the supply of nutrients from fluid flowing across the free surface of the biofilm. This leads to a nutrient concentration gradient within the biofilm, so that as the biofilm thickens, growth tends to be localized at the surface of the biofilm. Moreover, bacteria at the base of the biofilm become starved of nutrient, which may result in cell death and loss of adhesion to the surface. Thus, parts of the biofilm may break off. In other cases, biofilm growth may restrict fluid flow resulting in self-limiting growth due to the resulting restriction in the supply of nutrient. As bacteria proliferate, they produce a polymer-rich matrix known as extracellular polymeric substance (EPS), which consists primarily of long-chain polysaccharides together with various proteins, lipids and nucleic acids. In a mature bacterial biofilm, the cellular content is typically only 2-5 % by volume, while the extracellular component is characterized by a viscous swollen gel (see Chap. 12) consisting of 1-5% EPS in water.

One mathematical approach to investigating the growth of biofilms is to use continuum models; see the reviews [514, 602, 919]. The starting point for such models is to consider nutrient diffusing into the growing biofilm and being absorbed. Early models focused on one spatial dimension [765, 920], but were superseded by three-dimensional models in order to account for the heterogeneous growth of biofilms that includes mushroom-like structures and fingering [247, 264, 513]. A more biophysically detailed class of model was introduced by Cogan and Keener [200, 201], who explicitly treated the biofilm as a biological gel consisting of EPS and water. They used a two-fluid continuum model and the classical theory of polymers to investigate the existence and stability of a one-dimensional steady state with constant interface growth velocity. As highlighted by Winstanley et al. [315, 944], an implicit assumption of the original two-fluid model is that there is an approximate balance between viscous stress and osmotic pressure; the latter authors analyze the model in a different asymptotic regime. Further extensions of the two-fluid model have been developed by Zhang et al. [976, 977]. There have also been a number of continuum models that include the effects of QS [24, 193, 272, 319, 924]. Finally, note that an alternative approach to modeling biofilm growth is in terms of discrete biomass models, involving either cellular automata or agent-based models; see for example [7, 176, 726, 727]. In this section, we will focus on continuum models of biofilm growth.

### 15.6.1  Nutrient-based model of 1D growth

Following Dockery and Klapper [247], consider a single-species biofilm occupying the domain $0 < z < h(t)$, where $h(t)$ is the thickness of the biofilm at time $t$ and there is a flat surface (wall) at $z = 0$, see Fig. 15.25. The model is effectively 1D,

Fig. 15.25: Schematic diagram of a biofilm occupying a domain of thickness $h$ and supplied by nutrient diffusing from the bulk fluid through a boundary layer of thickness $L$.

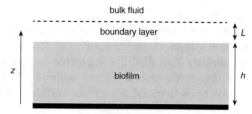

since the thickness is taken to be independent of the other spatial coordinates $x$ and $y$. The region $z > h(t)$ corresponds to the bulk fluid region, which supplies nutrient such as oxygen to the biofilm. Let $c(z,t)$ denote the concentration of a substrate in the domain $z > 0$. The bulk fluid is itself partitioned into a well-mixed region with constant substrate concentration $c_0$ and a diffusive boundary layer of thickness $L$ adjoining the biofilm. Within the biofilm, the nutrient substrate is consumed by bacteria at a concentration-dependent rate $r(c)$. Denoting the diffusivity of substrate molecules by $D_0$, we have [247]

$$\frac{\partial c}{\partial t} = D_0 \frac{\partial^2 c}{\partial z^2} - H(h-z)r(c), \quad 0 < z < h+L, \tag{15.6.1}$$

where $H$ is the Heaviside function. The positive consumption rate satisfies $r(0) = 0$ and is typically taken to be the Hill function $[r(c) = kc/(c_h + c)]$. For simplicity, the diffusivity is assumed to be the same in the biofilm and boundary layer. However, it is straightforward to extend the model to include a discontinuity in the diffusivity at the interface by introducing appropriate matching conditions [247]; see below. Since the diffusive time scale $h^2/D_0$ is usually much smaller than other relevant time scales such as the rate of growth of the colony, we can focus on the quasi-steady-state approximation

$$\frac{\partial^2 c}{\partial z^2} = H(h-z)r(c), \ 0 < z < h+L; \quad \frac{\partial c}{\partial z}(0,t) = 0, \quad c(h+L,t) = c_0. \tag{15.6.2}$$

Absorption of nutrient by bacterial cells drives biofilm growth at a rate $g$, which is taken to be of the form $g = g(c) = (r(c) - b)\Gamma$, where $\Gamma$ is a yield coefficient and $b$ is a base subsistence level. This growth generates a pressure $p(z,t)$ that deforms the biofilm at a velocity $u(z,t)$. Ignoring inertial terms, the pressure stress is balanced by friction according to Darcy's law [247]

$$\frac{1}{\eta}u(x,t) = -\frac{\partial p}{\partial z}, \quad 0 < z < h, \tag{15.6.3}$$

where $\eta$ is a friction coefficient. Since the biofilm mainly consists of water it can be treated as an incompressible fluid (see Box 10B), which means that

$$\frac{\partial u}{\partial z} = g, \quad 0 < z < h. \tag{15.6.4}$$

Combining equations (15.6.3) and (15.6.4) gives

$$\frac{\partial^2 p}{\partial z^2} = -\frac{g}{\eta}, \quad 0 < z < h, \quad \left.\frac{\partial p}{\partial z}\right|_{z=0} = 0, \quad p(h,t) = 0. \tag{15.6.5}$$

The boundary condition at the wall reflects the fact that $u(0,t) = 0$. Finally, note that the biofilm-bulk fluid interface moves as

$$\frac{dh}{dt} = u(h,t) = -\eta\frac{\partial p}{\partial z}(h,t) = \int_0^h g(c(z,t))dz. \tag{15.6.6}$$

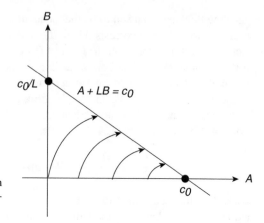

Fig. 15.26: Phase-plane   representation
of solutions to the boundary value problem given by equation (15.6.7).

That is, the thickness of the biofilm changes in order to allow for the new growth or
shrinkage.

   Under the quasi-steady-state approximation, suppose that we fix time $t$ and hence
$h$, and determine $c$ as a function of $h$. (We drop the explicit dependence on $t$.) For
$h < z < L$ we have $d^2c/dz^2 = 0$, so that

$$c(z) = c(h) + (z - h)c'(h).$$

It follows that within the biofilm the concentration satisfies the boundary value problem

$$\frac{d^2c}{dz^2} = \frac{r(c)}{D_0}, \quad c'(0) = 0, \quad c(h) + Lc'(h) = c_0. \tag{15.6.7}$$

This can be reformulated in terms of an effective planar dynamical system with $z$
playing the role of a "time" coordinate. In particular, setting $A = c, B = c'$, we have

$$\frac{dA}{dz} = B, \quad \frac{dB}{dz} = \frac{r(A)}{D_0}, \tag{15.6.8}$$

together with the initial conditions $A(0) = A_0, 0 < A_0 < c_0$, and $B(0) = 0$. It remains
to determine $A_0$ and thus the trajectory in the positive quadrant of the phase plane.
This is achieved by imposing the "final condition" that after a time $h$ the trajectory
reaches the line $A(h) + LB(h) = c_0$, see Fig. 15.26. Under mild smoothness conditions on the consumption rate function $r(c)$, one finds that an orbit starting at the
origin takes an infinite time to reach the line $A + LB = c_0$, whereas an orbit starting
at $A = c_0$ takes zero time. These two cases correspond to a layer of infinite thickness
and a degenerate layer, respectively. Using a continuity argument, it then follows
that there exists a solution for a given depth $h$, although it need not be unique unless
$r$ is monotone [247].

As a simple example, suppose that $r(c) = \alpha c$, where $\alpha$ has units of inverse time. The solution of the boundary value problem (15.6.7) is of the form

$$c(z) = C\left[e^{z/\xi} + e^{-z/\xi}\right], \quad \xi = \sqrt{\frac{D_0}{\alpha}},$$

where we have imposed the boundary condition $c'(0) = 0$. The boundary condition at $z = h$ then determines the integration constant $C$:

$$C = c_0 \left[(1 + L/\xi)e^{h/\xi} + (1 - L/\xi)e^{-h/\xi}\right]^{-1}.$$

We can thus write the solution as

$$c(z) = \frac{e^{-(h-z)/\xi} + e^{-(h+z)/\xi}}{(1 + L/\xi) + (1 - L/\xi)e^{-2h/\xi}} c_0, \quad 0 < z < h. \tag{15.6.9}$$

It follows that the substrate is concentrated in a layer of thickness $\xi$ at the top of the biofilm. If we also take the growth rate to be a linear function, $g(c) = \beta c$, and assume that $\xi \ll h$, then the interface equation becomes

$$\frac{dh}{dt} \approx \frac{\beta c_0}{1 + L/\xi} \int_0^h e^{-(h-z)/\xi} dz \approx \frac{\beta \xi c_0}{1 + L/\xi}. \tag{15.6.10}$$

That is, to leading order, $h$ increases linearly in time.

*Stability analysis of one-dimensional interfacial growth.* The one-dimensional front solution obtained above can be incorporated into a three-dimensional model in which the interface surface is now $h = h(x,y,t)$. The three-dimensional versions of equations (15.6.3) and (15.6.4) are

$$\mathbf{u} = -\eta \nabla p, \quad \nabla \cdot \mathbf{u} = \beta c, \tag{15.6.11}$$

which can be combined to yield the following equation for the pressure:

$$\nabla^2 p = -\frac{\beta c}{\eta}. \tag{15.6.12}$$

For simplicity, we take the wall to be at $z = -\infty$ and consider a linear growth function $g(c) = \beta c$. The equations for the substrate become (after allowing for a discontinuity in the diffusivity at the interface)

$$\frac{\partial c}{\partial t} = D_0 \nabla^2 c - \alpha c, \quad -\infty < z < h, \tag{15.6.13a}$$

$$\frac{\partial c}{\partial t} = D_1 \nabla^2 c, \quad h < z < h + L, \tag{15.6.13b}$$

together with the boundary conditions

$$\mathbf{n} \cdot \nabla c\big|_{z=-\infty} = 0, \quad c\big|_{z=h} = 1, \tag{15.6.14}$$

and the interface matching conditions

$$c\big|_{z=h^-} = c\big|_{z=h^+}, \quad D_0 \nabla c\big|_{z=h} = D_1 \nabla c\big|_{z=h}. \tag{15.6.15}$$

We have set the bulk fluid concentration to unity and, for simplicity, assumed a linear consumption rate $r(c) = \alpha c$. It is now possible to investigate the emergence of heterogeneities at the interface by analyzing the linear stability of a one-dimensional solution $c_0(z)$ [247]. The latter is obtained from equation (15.6.9) by setting $h = h_0 \gg \xi$ and allowing for a discontinuity in the diffusivities

$$c_0(z) = \Gamma e^{-(h_0-z)/\xi}, \quad -\infty < z < h_0, \tag{15.6.16a}$$

$$c_0(z) = 1 - \frac{\Gamma}{K\xi}(h_0 + L - z), \quad h_0 < z < h_0(t) + L, \tag{15.6.16b}$$

where $K = D_1/D_0$ and $\Gamma = 1/(1 + L/K\xi)$. Again, we are looking at a quasi-steady-state solution at fixed $t$.

Now suppose that

$$c(x,y,z) = c_0(z) + \varepsilon c_1(z)e^{i\mathbf{k}\cdot\mathbf{x}}, \ p(x,y,z) = p_0(z) + \varepsilon p_1(z)e^{i\mathbf{k}\cdot\mathbf{x}},$$
$$h(x,y) = h_0 + \varepsilon h_1 e^{i\mathbf{k}\cdot\mathbf{x}},$$

where $c_0, p_0, h_0$ denote the 1D front solution with $p_0$ and $h_0$ determined from equations (15.6.5) and (15.6.10), respectively, $\mathbf{k} = (k_x, k_y)$ and $\mathbf{x} = (x, y)$. Substituting these expansions into the quasi-steady-state version of (15.6.13) and only keeping $O(\varepsilon)$ terms gives (see Ex. 15.6)

$$0 = \frac{\partial^2 c_1}{\partial z^2} - k^2 c_1 - \frac{\alpha}{D_0}c_1, \quad -\infty < z < h_0, \tag{15.6.17a}$$

$$0 = \frac{\partial^2 c_1}{\partial z^2} - k^2 c_1, \quad h_0 < z < h_0 + L, \tag{15.6.17b}$$

together with the boundary conditions

$$c_1(h_0 + L) = \frac{\partial c_1}{\partial z}(-\infty) = 0, \tag{15.6.17c}$$

and the interface conditions

$$h_1 \frac{\partial c_0}{\partial z}(h_0^+) + c_1(h_0^+) = h_1 \frac{\partial c_0}{\partial z}(h_0^-) + c_1(h_0^-), \tag{15.6.17d}$$

$$K\frac{\partial c_1}{\partial z}(h_0^+) = h_1 \frac{\partial^2 c_0}{\partial z^2}(h_0^-) + \frac{\partial c_1}{\partial z}(h_0^-). \tag{15.6.17e}$$

We have used $\partial^2 c_0/\partial z^2 = 0$ for $z = h_0^+$. Using the zeroth-order solution (15.6.16a), the interfacial conditions reduce to the form

$$\frac{h_1 \Gamma}{K\xi}(1 - K) + c_1(h_0^+) = c_1(h_0^-), \quad K\frac{\partial c_1}{\partial z}(h_0^+) = \frac{h_1 \Gamma}{\xi^2} + \frac{\partial c_1}{\partial z}(h_0^-).$$

The solutions of equation (15.6.17) has the form

$$c_1(z) = Ae^{-\kappa(h_0-z)}, \quad -\infty < z < h_0, \tag{15.6.18a}$$

$$c_1(z) = B\sinh(k[h_0 + L - z]), \quad h_0 < z < h_0 + L, \tag{15.6.18b}$$

where $\kappa^2 = k^2 + \xi^{-2}$. Imposing the interfacial conditions then determines the coefficients $A$ and $B$; see Ex. 15.6

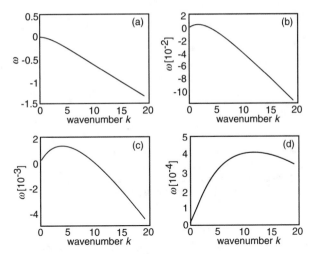

Fig. 15.27: Example dispersion curves for $\omega(k)$ given by equation (15.6.23) for decreasing values of $\xi$ and $K = 8$, $L = 2$, $\beta = \alpha = 1$ (non-dimensionalized). $2\log_{10}\xi = -(2+n) - \log_{10}2.5$ for (a) $n = 0$, (b) $n = 1$, (c) $n = 2$, and (d) $n = 3$. [Redrawn from Dockery and Klapper [247].]

$$A = \frac{\Gamma h_1}{\xi}\mathscr{A}(k), \quad \mathscr{A}(k) = -\frac{(k\xi)^{-1}\tanh(kL) + K - 1}{k^{-1}\kappa\tanh(kL) + K}, \tag{15.6.19a}$$

$$B = \frac{\Gamma h_1}{\xi}\mathscr{B}(k), \quad \mathscr{B}(k) = \frac{1}{\cosh(kL)}\frac{-(\kappa\xi)^{-1} - 1 + K^{-1}}{Kk\kappa^{-1} + \tanh(kL)}. \tag{15.6.19b}$$

It follows from equation (15.6.12) that the equation for the pressure term $p_1$ takes the form

$$\frac{\partial^2 p_1}{\partial z^2} - k^2 p_1 = -\frac{\beta}{\eta}c_1, \tag{15.6.20}$$

with the boundary conditions

$$\frac{\partial p_1}{\partial z}(-\infty, t) = 0, \quad p(h) \approx p_1(h_0) + h_1\frac{\partial p_0}{\partial z}(h_0) = 0.$$

From equation (15.6.17), we see that

$$p_1(z) = -\frac{\xi^2\beta}{\eta}c_1(z) + Fe^{k(z-h_0)}, \tag{15.6.21}$$

with $F$ determined from the boundary condition at $z = h_0$

$$F = \frac{\xi^2\beta}{\eta}c_1(h_0) + \frac{\beta h_1\xi\Gamma}{\eta}, \tag{15.6.22}$$

after substituting the explicit solution for $p_0$.

Finally, from equation (15.6.6), we have the kinematic equation

$$\frac{dh_0}{dt} + \varepsilon\frac{dh_1}{dt} = \beta\int_{-\infty}^{h_0+\varepsilon h_1}[c_0(z,t) + \varepsilon c_1(z,t)]dz,$$

where the time-dependence of $c$ is due to the slow evolution of the thickness $h(t)$. Imposing the kinematic equation for $h_0(t)$, we obtain the $O(\varepsilon)$ equation

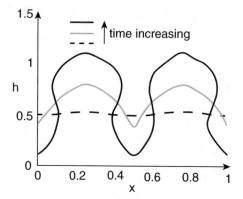

Fig. 15.28: Illustrative sketch of the evolution of the biofilm interface due to a fingering instability. The time scale for the development of these protrusion is of order $10^6$s.

$$\frac{dh_1}{dt} = \beta h_1 c_0(h_0,t) + \int_{-\infty}^{h_0} c_1(z,t)dz = \beta h_1 c_0(h_0,t) - \eta \frac{\partial p_1}{\partial z}(h_0,t)$$

$$= \beta h_1 \Gamma + \xi^2 \beta(\kappa - k)A(t) - k\beta h_1 \xi \Gamma = \frac{\Gamma \beta h_1(t)}{\xi}\left(\frac{\mathscr{A}(k)}{\kappa + k} + \xi - k\xi^2\right),$$

since $\kappa^2 - k^2 = \xi^{-2}$. That is

$$\frac{dh_1}{dt} = \omega(k)h_1, \quad \omega_1(k) = \frac{\Gamma \beta}{\xi}\left(\frac{\mathscr{A}(k)}{\kappa + k} + \xi - k\xi^2\right). \qquad (15.6.23)$$

Note that when $\xi \ll L$, we have

$$\omega(k) \to \frac{\Gamma \beta}{L/\xi + K} \approx 0 \text{ as } k \to 0.$$

Some example dispersion curves are shown in Fig. 15.27, which show a band of unstable modes whose dominant wavenumber $k_c$ increases as $\xi^{-1}$ increases. Numerical simulations of a two-dimensional version of the model show that instabilities of the flat interface result in the growth of bubble regions indicative of a fingering instability [246]. This is illustrated in Fig. 15.28. The time scale of growth is of the order $10^6 - 10^7$s [514].

## 15.6.2 Polymer-solvent gel model of biofilm growth

We now turn to the polymer-solvent model of Cogan and Keener [200, 201]; see also Refs. [315, 944, 976, 977]. Again consider a biofilm in a domain $0 < z < h$ with a wall at $z = 0$ and nutrient-rich water above the interface, see Fig. 15.25. However, rather than simply keeping track of the nutrient concentration within the biofilm, the model also treats the biofilm as a swollen gel consisting of a matrix of EPS in water. Let $\phi$ be the volume fraction of polymer, and $1 - \phi$ the volume fraction of water. (For simplicity the contribution to the biomass from cells is taken to be negligible.) The mass conservation equations for polymer and solvent take the form

$$\frac{\partial \phi}{\partial t} + \nabla \cdot (\phi \mathbf{v}) = g(\phi, c), \tag{15.6.24a}$$

$$-\frac{\partial \phi}{\partial t} + \nabla \cdot [(1 - \phi)\mathbf{w}] = 0, \tag{15.6.24b}$$

where $\mathbf{v}$ and $\mathbf{w}$ denote the velocities of the polymer and solvent, and $g$ is a growth term that depends on $\phi$ and the concentration $c$ of some nutrient. Since the nutrient is dissolved in the solvent, the nutrient flux has contributions from both solvent motion and diffusion. It is assumed that both fluxes are scaled by the volume fraction of solvent so that

$$\mathbf{J}_c = (1 - \phi)[\mathbf{w}c - D\nabla c].$$

It follows that the mass conservation equation for the nutrient substrate is

$$\frac{\partial (1 - \phi)c}{\partial t} + \nabla \cdot \{(1 - \phi)[c\mathbf{w} - D\nabla c]\} = -r(\phi, c), \tag{15.6.25}$$

where $r(\phi, c)$ is the rate of nutrient consumption. Analogous to the previous nutrient model, the growth and consumption rates are taken to be of the form

$$g = \frac{G\phi c}{K + c}, \quad r = \frac{R\phi c}{K + c}. \tag{15.6.26}$$

Note that the volume fraction of EPS is taken to be a proxy for the bacterial concentration, under the assumptions that bacteria are distributed uniformly throughout the matrix and the production of EPS is proportional to the bacterial growth rate.

In order to complete the description of the model, it is necessary to specify the various forces acting on the polymer-solvent gel, which are taken to be balanced if inertial terms are ignored. Both the polymer and solvent are subject to viscous stresses, hydrostatic pressure $p$, and a frictional drag term due to the relative motion of the polymer and solvent. However, there is an additional force acting on the polymer due to osmotic pressure. Recall from the Flory-Huggins theory in Sect. 12.4 and Sect. 13.1 that the osmotic pressure is given by (see equation (12.4.10))

$$\Pi(\phi) = \frac{k_B T}{v_c}(-\ln(1 - \phi) - \phi - \chi \phi^2), \tag{15.6.27}$$

where $v_c$ is the volume of each monomer, $\phi$ is the volume fraction of EPS, and $\chi$ is an interaction energy. We are assuming that the number of monomers $M$ is large, which holds for long-chained polysaccharides. The force or momentum balance equations thus take the form

$$0 = \mu \nabla \cdot [\phi(\nabla \mathbf{v} + \nabla \mathbf{v}^\top)] - f\phi(1 - \phi)(\mathbf{v} - \mathbf{w}) - \nabla \Pi - \phi \nabla p, \tag{15.6.28a}$$

$$0 = \mu_w \nabla \cdot [(1 - \phi)(\nabla \mathbf{w} + \nabla \mathbf{w}^\top)] + f\phi(1 - \phi)(\mathbf{v} - \mathbf{w}) - (1 - \phi)\nabla p. \tag{15.6.28b}$$

Here, $\mu$ is the long-time viscosity of the EPS, $\mu_w$ is the viscosity of water, and $f$ is a frictional drag coefficient. The term $\nabla \mathbf{v} + \nabla \mathbf{v}^\top$ is the stress tensor (see Box 10B and 12B) with components

$$\sigma_{ij} = \frac{\partial v_i}{\partial x_j} + \frac{\partial v_j}{\partial x_i},$$

and similarly for $\nabla \mathbf{w} + \nabla \mathbf{w}^\top$.

Equations (15.6.24)–(15.6.28) together with appropriate boundary conditions specify the polymer-solvent gel model of a biofilm [200]. Cogan and Keener analyzed the model in an asymptotic regime where the dominant force balance is between viscous stress and osmotic pressure. In particular, they obtained a one-dimensional steady-state solution with constant interface growth, and examined the instability of this solution to spatially varying perturbations along the lines of the nutrient-based model of Ref. [247]. A more recent analysis of Winstanley *et al.* [944] considered analogous solutions in an alternative regime where viscous forces are negligible.

### 15.6.3 Biofilms and quorum sensing

The QS state of bacteria also influences biofilm formation [507, 645]. For example, *Pseudomonas aeruginosa* forms biofilms at high cell density in response to autoinducer accumulation and detection, whereas biofilm growth in *V. cholerae* occurs at low cell density; autoinducer accumulation and detection now represses biofilm formation. It follows that the mass transfer of extracellular autoinducer due to fluid flow has opposite effects in *Pseudomonas aeruginosa* and *V. cholerae*. In particular, a higher cell density is required for biofilm formation in the former case in order to compensate for the loss of extracellular autoinducer. On the other hand, autoinducer removal by fluid flow relieves repression in the latter case, promoting biofilm formation at lower cell densities. It also follows that cells close to the biofilm interface experience a different flow regime from internally residing cells. This means that cells in different regions of the biofilm can implement different forms of QS-driven gene expression. In other words, the flow environment determines spatial fate decisions, resulting in biofilm heterogeneity.

Another source of heterogeneity within biofilms is stochasticity [645]. That is, a subpopulation of cells can be in the QS on state, whereas the remaining population is in the QS off state. The growth of the biofilm can then alter the relative abundance of the two populations via an increased production of QS. We previously mentioned that phenotypic heterogeneity occurs in *V. Harveyi* biofilms [19]. Another example has been found in the early stages of QS-driven biofilm development in *Pseudomonas putida*, where only a subpopulation of cells produce autoinducer [163]. However, the stochastic production of the QS signal now acts primarily as a self-regulatory mechanism rather than inducing neighboring cells to produce autoinducer. More specifically, QS induces the expression of a biosurfactant known as pitisolvin, which adheres to the surface of producer cells and causes them to disperse from the community. This has the effect of delaying population-wide QS in the immature biofilm (negative feedback). More generally, one expects stochastic effects to contribute significantly to QS heterogeneity when bacterial populations are in a low density state, where only a few cells are producing and/or responding to autoinducers. Following Sect. 15.3, one possible mechanism for heterogeneity is noise-induced switching of bistable gene networks. An alternative mechanism has been proposed in a recent modeling study [48], based on the balance between the fitness advantage gained by nonproducers who avoid the costly production of autoinducers, and the persistence of producers due to QS.

There have been several mathematical studies of QS in biofilms using continuum reaction-diffusion models; see for example Refs. [24, 193, 272, 319, 924]. These typically include subpopulations of active and inactive bacterial cells and a single QS autoinducer that is produced at different rates by the two subpopulations. For the sake of illustration, consider a macroscopic model for QS and population growth in a well-mixed domain due to Ward *et al.* [923]. Motivated by bistability in the model of *P. aeruginosa*, the bacteria are partitioned into two subpopulations consisting of upregulated and downregulated cells, respectively, with corresponding cell densities $C_u$ and $C_d$. The downregulated cells are upregulated at a rate $\alpha U$, where $U$ is the extracellular concentration of autoinducer. For simplicity, the intracellular and extracellular autoinducer concentrations are taken to be proportional to each other, and the detailed chemical kinetics of QS is ignored. (The model thus considers a longer time scale than kinetic models.) The autoinducer is produced by both upregulated and downregulated cells at the rates $\kappa_u$ and $\kappa_d$, respectively, with $\kappa_d \ll \kappa_u$. The final major component of the model is the specification of cell division. This is assumed to occur at a rate $r$, with upregulated cells producing on average $\gamma$ upregulated cells and $2 - \gamma$ downregulated cells, whereas the daughter cells of downregulated cells are themselves downregulated. The parameter $\gamma$ is taken to be approximately unity. The resulting system of ODEs takes the form

$$\frac{dC_d}{dt} = r(C_d + (2 - \gamma)C_u)F(C_d + C_u) - \alpha U C_d + \beta C_u, \tag{15.6.29a}$$

$$\frac{dC_u}{dt} = r(\gamma - 1)C_u F(C_d + C_u) + \alpha U C_d - \beta C_u, \tag{15.6.29b}$$

$$\frac{dU}{dt} = \kappa_u C_u + \kappa_d C_d - \alpha U C_d - \lambda U. \tag{15.6.29c}$$

Here $\beta$ is the spontaneous rate of downregulation, $F$ is a dimensionless bacterial growth function and the QS concentration decreases either due to forming a complex during the upregulation of cells at the rate $\alpha A C_d$ or due to some form of degradation at a rate $\lambda$. Note that adding equations (15.6.29a,b) gives

$$\frac{dC}{dt} = rCF(C), \quad C = C_u + C_d. \tag{15.6.30}$$

Hence, it is simpler to take $C_u$, $C$, and $U$ as the dependent variables. Two possible choices for $F$ are $F = 1$ (exponential growth) and the logistic function $F(C) = 1 - C/K$, where $K$ is a carrying capacity. In the latter case, $f(C) = CF(C)$ has an unstable fixed point at $C = 0$ and a stable fixed point at $C = K$. Using a multiple time scale analysis for $\varepsilon = \kappa_d/\kappa_u \ll 1$, Ward *et al.* showed that starting from a small colony of downregulated cells under exponential growth, the level of upregulation remains very low until a time of order $\ln(1/\varepsilon)$, beyond which there is a rapid increase in upregulated cells. Moreover, the parameter $\Theta = r^{-1}(\kappa_u + r(\gamma - 1) - \beta)$ determines whether or not all cells ultimately become upregulated. Total upregulation occurs for $\Theta > 0$, whereas partial upregulation is found for $\Theta < 0$.

Ward *et al.* [924] extended their ODE model to include advection and diffusion terms, in order to consider the growth of a densely packed biofilm spreading out on a surface as a thin film. That is, they took

$$\frac{dC_{d,u}}{dt} \rightarrow \frac{\partial C_{d,u}}{\partial t} + \nabla \cdot (\mathbf{v} C_{d,u}), \quad \frac{dA}{dt} \rightarrow \frac{\partial A}{\partial t} + \nabla \cdot (\mathbf{v} A) - D\nabla^2 A,$$

where $\mathbf{v} = (v_x, v_y, 0)$ and $\nabla = (\partial/\partial x, \partial/\partial y, 0)$ with $x, y$ the coordinates along the surface that the biofilm is spreading. Again they showed that there is a phase of biofilm maturation during which the cells remain in the downregulated state followed by a rapid switch to the upregulated state. The latter occurred throughout the biofilm except at the leading edges. They also analyzed the spread of the biofilm in terms of a traveling wave. An analogous model was developed by Chopp *et al.* [193], except that they considered the growth of a biofilm normal to a surface and included the effects of a diffusing nutrient. The latter authors showed that QS is upregulated once the biofilm grows to a critical depth. They also found that for QS to be upregulated near the base of the biofilm, cells in oxygen-deficient regions of the biofilm must still be synthesizing the autoinducer. Subsequent models of QS in biofilms have included additional features such as the effects of anti-QS drugs, growth of the EPS, QS regulated EPS production, and QS induced biofilm detachment [23, 24, 272, 319].

**Mechanical effects in dense cell populations.** Another emergent feature of bacterial colonies such as biofilms is the spatial organization at high densities due to the "contact biomechanics" arising from cellular growth and division. At low densities, one expects the main form of cell-to-cell communication to be mediated by chemical signaling. However, as bacteria aggregate and form dense populations, direct physical interactions play an increasingly strong role [722]. One of the first studies of this phenomenon was the spatial organization of nonmotile *E. coli* growing in a controlled microfluidic environment [910]; a related study was carried out in Ref. [190]. It was observed that growth and expansion of the colony of cells resulted in a dynamical transition from an isotropic disordered phase to a phase characterized by orientational alignment of the rod-like cells, often referred to as a nematic phase in physics. Indeed, continuum models of liquid crystals [251] can be adapted to model phase ordering in bacterial colonies [910]. There are now a growing number of modeling and imaging studies of bacterial spatial organization at high densities, involving both single-species [96, 255, 488, 966] and multi-species communities [960, 961]. One striking example of the latter is the observation of "flower–like" patterns in a colony grown from a mixture of motile and nonmotiles bacterial species [961]. The nonmotile bacteria accumulate at the boundary of the colony and trigger an instability resulting in the intricate patterns. One way to understand the mechanism underlying such an instability is to consider a geometrical model of the colony boundary motion, based on an approach originally developed for solidification patterns [138, 139]; see Ex. 15.7.

## 15.7 Motility-induced phase separation

In this final section, we consider an example of collective behavior in populations of motile bacteria [172, 856], which exhibits a form of multicellular liquid-liquid phase separation analogous to the cellular-level phase separation considered in Chap. 13. Each cell is modeled in terms of a symmetric one-dimensional run-and-tumble particle (RTP), see Sect. 10.4, with a velocity and tumbling rate that depend on particle position and cell density; the latter could arise from some form of quorum sensing. The separation into regions of high and low cell density in the absence of a chemotactic signal depends upon a positive feedback mechanism between two features: RTPs tend to accumulate in regions where they move more slowly, and they tend to move more slowly in regions of higher accumulation. From a more general perspective, populations of motile bacteria are an example of so-called "active matter" [172, 751, 768]. Active matter describes systems whose constituent elements consume energy in order to move or to exert mechanical forces, and are thus out-of-equilibrium. Examples within biology include flocks or herds of animals, collections of cells, and components of the cellular cytoskeleton. In many cases, the individual particles have an intrinsic orientation (e.g., fish, birds, rod-like cells, and macromolecules) and can exhibit long-range orientational interactions mediated by sensing or by hydrodynamic coupling to the surrounding medium [907]. Populations of RTPs tend to be simpler, since interactions mediated by quorum sensing signals tend to be isotropic. (Bacteria cannot directly detect vector quantities such as the direction of a chemical concentration gradient. Instead, they operate by integrating temporal information as they move.)

**Population model.** Let us return to the symmetric RTP model of bacterial chemotaxis given by equation (10.4.24), with a space-dependent speed $v(x)$ and tumbling rate $k(x)$. This can be reduced to an effective FP equation along the lines of Refs. [172, 801], see equation (10.4.25), that is

$$\frac{\partial p}{\partial t} = -\frac{\partial}{\partial x}\left[-D(x)\frac{\partial p(x)}{\partial x} + V(x)p(x)\right], \quad D(x) = \frac{v(x)^2}{2k(x)}, \quad V(x) = -\frac{v(x)v'(x)}{2k(x)}.$$
$$(15.7.1)$$

The conservation equation for $p$ can be rewritten as an Ito FP equation for the position density $p(x,t)$, with the corresponding Ito SDE

$$dX = A(X)dt + \sqrt{2D(X)}dW(t), \quad A(X) = V(X) + \frac{\partial D}{\partial X},$$
$$(15.7.2)$$

where $W(t)$ is a Wiener process.

Now consider a population of $N$ identical bacteria with positions $X_i(t)$, $i = 1, \ldots, N$, and introduce the particle density

$$\rho(x,t) = \sum_{i=1}^{N} \rho_i(x,t) = \sum_{i=1}^{N} \delta(x - X_i(t)).  \qquad (15.7.3)$$

(A smooth version of the density may then be constructed by coarse-graining.) Following Tailleur and Cates [856], suppose that the speed $v$ and the switching rate $k$ are both functions of cell position and functionals of the population density, with the latter dependence representing cell interactions. It follows that equation (15.7.2) is replaced by the set of coupled SDES

$$dX_i = A(X_i, [\rho])dt + \sqrt{2D(X_i, [\rho])}dW_i(t),  \qquad (15.7.4)$$

with

$$\langle dW_i(t) \rangle = 0, \quad \langle dW_i(t)dW_j(t') \rangle = \delta_{i,j}\delta(t - t')dt\,dt'.$$

It is important to note that in general $[\rho]$ depends on all of the particle positions $X_j$, $j = 1, \ldots, N$. That is, given some integral kernel $\psi$, we have

$$[\rho] = \int \rho(x,t)\psi(x)dx = \sum_{i=1}^{N} \psi(X_i(t)).$$

Hence, it follows from the chain-rule that in the definition of $A(X_i, [\rho])$

$$\frac{\partial D(X_i, [\rho])}{\partial X_i} = D'(X_i, [\rho]) + \frac{\partial \rho}{\partial x} \frac{\delta D(X_i, [\rho])}{\delta \rho(x,t)}\bigg|_{x=X_i}.  \qquad (15.7.5)$$

The next step is to construct an SPDE for the density $\rho$. Using Ito's lemma (see Sect. 2.2) and equation (15.7.2), for any differentiable function $f(X_i)$

$$df(X_i) = \left[ A(X_i, [\rho])\frac{\partial f}{\partial X_i} + D(X_i, [\rho])\frac{\partial^2 f}{\partial X_i^2} \right]dt + \sqrt{2D(X_i, [\rho])}\frac{\partial f}{\partial X_i}dW_i(t)$$

$$= \int \rho_i(x,t)[(Af' + Df'')dt + \sqrt{2D}f'dW_i]dx.$$

Integrating the right-hand side by parts, and using

$$df(X_i) = \int d\rho_i(x,t)f(x)dx,$$

we obtain the following SPDE for $\rho_i$:

$$d\rho_i = \left[ -\frac{\partial}{\partial x}(A\rho_i) + \frac{\partial^2}{\partial x^2}(D\rho_i) \right]dt - \frac{\partial}{\partial x}(\sqrt{2D}\rho_i)dW_i(t).$$

Summing both sides with respect to $i$ yields

$$d\rho = \left[-\frac{\partial}{\partial x}(A\rho) + \frac{\partial^2}{\partial x^2}(D\rho)\right] dt - \frac{\partial}{\partial x}\sqrt{2D}d\Phi(x,t),$$

where $d\Phi(x,t) = \sum_{i=1}^{N} \rho_i(x,t)dW_i(t)$. We now note that $\langle d\Phi(x,t)\rangle = 0$ and

$$\langle d\Phi(x,t)d\Phi(x',t')\rangle = \left\langle \sum_{i=1}^{N} \rho_i(x,t)dW_i(t) \sum_{i=1}^{N} \rho_j(x',t')dW_j(t') \right\rangle$$

$$= \sum_{i,j=1}^{N} \delta(x-X_i(t))\delta(x'-X_j(t'))\langle dW_i(t)dW_j(t')\rangle$$

$$= \sum_{i=1}^{N} \delta(x-X_i(t))\delta(x'-X_j(t'))\delta(t-t')dt\,dt'$$

$$= \delta(x-x')\delta(t-t')\sum_{i=1}^{N} \delta(x-X_i(t))dt\,dt' = \delta(x-x')\delta(t-t')\rho(x,t)dt\,dt'.$$

Hence, setting $d\Phi(x,t) = \sqrt{\rho}d\Lambda(x,t)$ we obtain the SPDE

$$d\rho = \left[-\frac{\partial}{\partial x}(A\rho) + \frac{\partial^2}{\partial x^2}(D\rho)\right] dt - \frac{\partial}{\partial x}\sqrt{2D\rho}\,d\Lambda(x,t), \qquad (15.7.6)$$

with

$$\langle d\Lambda(x,t)d\Lambda(x',t')\rangle = \delta(x-x')\delta(t-t')dt\,dt'.$$

Finally, substituting for $A$ using equations (15.7.2) and (15.7.5), and dropping a functional derivative term (which is equivalent to dropping self-density interactions [856]), we can rewrite the SPDE as

$$d\rho = -\frac{\partial}{\partial x}\left[\rho V - D\frac{\partial \rho}{\partial x}\right] dt - \frac{\partial}{\partial x}\sqrt{2D\rho}\,d\Lambda(x,t). \qquad (15.7.7)$$

**Phase separation.** In order to use the thermodynamic and kinetic theories of phase separation developed in Sect. 13.1, we will assume that $V$ and $D$ have a particular functional dependence on the density $\rho$ such that there exists a free energy functional $F_{ext}[\rho]$ for which [856]

$$\frac{V(x,[\rho])}{D(x,[\rho])} = -\partial_x \ln v(x,[\rho]) \equiv -\frac{1}{k_B T}\frac{\partial}{\partial x}\frac{\delta F_{ext}[\rho]}{\delta \rho(x,t)}. \qquad (15.7.8)$$

Dropping the noise term on the right-hand side of equation (15.7.7), we then see that the resulting deterministic equation can be written in the form (cf. equation (13.1.19))

$$\frac{\partial \rho}{\partial t} = -\frac{\partial}{\partial x}J(x,t), \quad J(x,t) = -\xi\rho(x,t)\frac{\partial}{\partial x}\mu(x,t), \qquad (15.7.9)$$

where $\xi = D/k_B T$ is a positive transport coefficient and the corresponding chemical potential is given by the functional derivative

$$\mu(x) = \frac{\delta F[\rho]}{\delta \rho(x)}, \tag{15.7.10}$$

with total free energy

$$F[\rho] = F_{ext}[\rho] + k_B T \int \rho(\ln\rho - 1)dx. \tag{15.7.11}$$

If the noise term were included in equation (15.7.7), then from the fluctuation-dissipation theorem, the stationary distribution of densities $\rho$ is given by the Boltzmann Gibbs formula

$$\mathcal{P}[\rho] \sim e^{-F[\rho]/k_B T}. \tag{15.7.12}$$

Suppose, for example, that the speed of a run depends locally on the cell density, $v = v(\rho(x))$. (One possible mechanism is that the bacteria measure the density in a local neighborhood via quorum sensing; see Sect. 15.4.) In this case, a free energy $F_{ext}$ satisfying equation (15.7.8) exists and

$$F[\rho] = \int f(\rho(x))dx, \quad f(\rho) = k_B T \left[ \rho(\ln\rho - 1) + \int_0^\rho \ln v(s)ds \right]. \tag{15.7.13}$$

Given the free energy density $f(\rho)$, we can now thermodynamically investigate the occurrence of phase separation along identical lines to Sect. 13.1. In particular, we require a range of uniform densities for which $f''(\phi) < 0$. This yields the condition

$$v'(\rho) < -v(\rho)/\rho, \tag{15.7.14}$$

which can also be understood using the following heuristic argument [856]. Generically, active particles accumulate in regions where they move more slowly. Therefore, suppose that there exists a uniform density $\rho_0$ satisfying $\rho_0 = c/v(\rho_0)$ for some constant $c$. Consider a small perturbation $\delta\rho(x)$ of the density, such that

$$v(\rho_0 + \delta\rho(x)) \approx v(\rho_0) + v'(\rho_0)\delta\rho(x).$$

This change in the speed induces a corresponding change in the density according to

$$\rho_0 + \delta\rho' \approx \frac{c}{v(\rho_0) + v'(\rho_0)\delta\rho} \approx \rho_0 \left( 1 - \frac{v'(\rho_0)}{v(\rho_0)}\delta\rho \right).$$

Hence, a linear instability is expected whenever $\delta\rho' > \delta\rho$, which will hold when the inequality (15.7.14) is satisfied. Numerical simulations confirm that spinodal decomposition can occur [856].

## 15.8 Exercises

---

**Problem 15.1. Age-structured model of asymmetric cell division.** Fill in the steps of the analysis of the asymmetric cell division model given by the equations (see Sect. 15.1)

$$\frac{\partial n}{\partial t} + \frac{\partial n}{\partial \tau} = -\alpha(\tau)n(t,\tau), \quad \frac{\partial n_q}{\partial t} + \frac{\partial n_q}{\partial \tau_q} = -\alpha_q(\tau_q)n_q(t,\tau_q),$$

and

$$n(t,0) = v\rho(\tau)N(t) + \rho_q(t)N_q(t), \quad n_q(t,0) = v_q\rho(t)N(t).$$

(a) Derive the following equations for the evolution of the total number of normal ($N(t)$) and quiescent cells ($N_q(t)$),

$$\frac{dN(t)}{dt} = (v-1)\rho(t)N(t) + \rho_q(t)N_q(t),$$

$$\frac{dN_q(t)}{dt} = v_q\rho(t)N(t) - \rho_q(t)N_q(t),$$

and obtain the corresponding equations for $G(t,\tau) = n(t,\tau)/N(t)$ and $G_q(t,\tau) = n_q(t,\tau)/N_q(t)$.

(b) Suppose that there exists a steady-state solution $G(t,\tau) = G^*(\tau), \rho(t) = \rho^*$ and $G_q(t,\tau_q) = G_q^*(\tau_q), \rho_q(t) = \rho_q^*$. Also assume that both cell populations grow exponentially at the same rate $k$ so that $\phi = N_q(t)/N(t)$ is a constant. Derive the equations

$$k = (v-1)\rho^* + \rho_q^*\phi = \rho^*(v-1+\phi\gamma),$$

$$k = v_q\frac{\rho^*}{\phi} - \rho_q^* = \rho^*\left(\frac{v_q - \phi\gamma}{\phi}\right).$$

(c) Repeating the steady-state analysis of the symmetric case, obtain the steady-state equation (15.1.23). Finally, solving these equations and imposing the normalization conditions $\int_0^\infty G^*(\tau)d\tau = 1, \int_0^\infty G_q^*(\tau)d\tau = 1$ derive equations

$$\left\langle e^{-k\tau}\right\rangle_\rho = 1 - \frac{k}{\rho^*(v+\phi\gamma)} = \frac{1}{v+\phi\gamma},$$

$$\left\langle e^{-k\tau_q}\right\rangle_{\rho_q} = 1 - \frac{k\phi}{\rho^* v_q} = \frac{\gamma\phi}{v_q}.$$

**Problem 15.2. Model of stochastic switching with reset in catastrophic environments [909].** Consider two microbial subpopulations $A$ and $B$ corresponding to distinct phenotypes that grow in an environment that occasionally undergoes a catastrophic event. Between catastrophes, the bacteria grow exponentially at rates $\gamma_A$ and $\gamma_B$ with $\gamma_A > \gamma_B$. Thus under normal conditions, subpopulation $A$ is fitter.

Let $n_A$ and $n_B$ be the number of microbes in the two subpopulations, which evolve between catastrophes according to the pair of equations

$$\frac{dn_A}{dt} = \gamma_A n_A + k_B n_B - k_A n_A,$$

$$\frac{dn_B}{dt} = \gamma_B n_B - k_B n_B + k_A n_A,$$

with $k_A, k_B$ the rates of phenotypic switching. Whenever a catastrophe occurs, the population size $n_A$ decreases instantaneously to some new value $n'_A < n_A$, with a reset probability $v(n'_A|n_A)$. Finally, suppose that the rate of catastrophes depends on the population size according to some environmental response function $\beta(n_A, n_B)$. (This could mimic, for example, the host immune response.)

(a) Following along similar lines to Sect. 15.3.1, show that the fitness

$$f(t) = \frac{n_A(t)}{N(t)}, \quad N(t) = n_A(t) + n_B(t)$$

evolves according to

$$\frac{df}{dt} = -\Delta\gamma(f - f_+)(f - f_-) \equiv V(f),$$

where $f_\pm$ are the roots of the quadratic equation

$$f^2 - \left(1 - \frac{k_A + k_B}{\Delta\gamma}\right)f - \frac{k_B}{\Delta\gamma} = 0,$$

and $\Delta\gamma = \gamma_A - \gamma_B > 0$. Hence, show that in the absence of catastrophes, the system converges towards the fitness value $f_+$.

(b) Suppose that

$$v(n'_A|n_A) = \frac{1}{n_A}\mathscr{F}(n'_A/n_A),$$

with $\int_0^x \mathscr{F}(x)dx = 1$. That is, when a catastrophe occurs, the fit population $A$ is reduced by a random factor generated from the distribution $\mathscr{F}$. In terms of the fitness $f$, we have $f \to f'$ with $f' = n'_A/(n'_A + n_B)$. The corresponding distribution of jumps in fitness, $\mu(f'|f)$, then satisfies $\mu(f'|f)df' = v(n'_A|n_A)dn'_A$. Hence derive the result

$$\mu(f|f') = \mathscr{F}\left(\frac{f'(1-f)}{f(1-f')}\right)\frac{1-f}{(1-f')^2 f}, \quad f > f'.$$

For the particular choice, $\mathscr{F}(x) = (\alpha + 1)x^\alpha$ with $\alpha > -1$, show that

$$\mu(f'|f) = \frac{d}{df'}\frac{m(f')}{m(f)}, \quad m(f) = \left(\frac{f}{1-f}\right)^{1+\alpha}.$$

(c) Suppose that the rate of catastrophes only depends on the fitness $f$, $\beta = \beta(f)$. Using Fig. 15.29, explain why the steady-state density of fitness, $p(f)$, satisfies the flux balance condition

$$V(f)p(f) = \int_f^{f^+} \int_0^f \beta(f')p(f')\mu(f''|f')df''\,df'.$$

Substituting the explicit expression for $v$ from part (b) show that

$$V(f)p(f) = \int_f^{f^+} \beta(f')p(f')\frac{m(f)}{m(f')}df'.$$

Dividing through by $m(f)$ and differentiating both sides with respect to $f$ obtain the solution

$$p(f) = \frac{C}{V(f)}\left(\frac{f}{1-f}\right)^{1+\alpha}\exp\left(-\frac{\beta(f)}{V(f)}\right).$$

The constant $C$ can then be determined from the normalization of $p(f)$.

(d) Suppose that $\beta(f)$ is given by the parameterized sigmoid function

$$\beta_\lambda(f) = \frac{\xi}{2}\left(1 + \frac{f - f^*}{\sqrt{\lambda^2 + (f - f^*)^2}}\right), \quad 0 \le f \le 1.$$

Sketch $\beta_\lambda(f)$ for $\xi = 1$, $f^* = 0.5$ and various values of $\lambda$. Using this function in the formula for $p(f)$ in part (c), plot the average population fitness

$$\langle f \rangle = \int_0^1 fp(f)df$$

as a a function of the switching rate $k_A$ for $\lambda = 1, 0.1, 0.01$. Take $k_B = 0.1$, $\Delta\gamma = 1$ and $\alpha = -0.99$. Hence, show that for sufficiently small $\lambda$, there is a local maximum at a nonzero switching rate $k_A$, implying that switching into the slow-growing state represents an optimal strategy.

**Problem 15.3. WKB analysis of population extinction.** Consider the quasi-stationary master equation

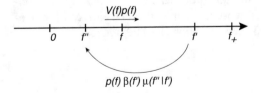

Fig. 15.29: Illustration of the flux balance condition.

$$0 = B(n-1)\left(1 - \frac{n-1}{K}\right)\Pi_{n-1,m} + (n+1)\Pi_{n+1,m} - Bn\left(1 - \frac{n}{K}\right)\Pi_{n,m} - n\Pi_{n,m}$$
$$+ \alpha(n+1)\Pi_{n+1,m-1} + \beta(m+1)P_{n-1,m+1} - \alpha n\Pi_{n,m} - \beta(1 - \delta_{n,K})m\Pi_{n,m}.$$

(a) Substitute the WKB solution

$$\Pi_{n,m} = e^{-K\Phi(x,y)},$$

into the master equation with $x, y$ treated as continuous variables. Taylor expanding with respect to $\varepsilon = 1/K$, multiplying both sides by $e^{K\Phi(x,y)}$, and collecting the $O(1)$ terms, derive the Hamilton-Jacobi equation

$$H(x, y, \partial_x \Phi, \partial_y \Phi) = 0,$$

with effective Hamiltonian given by equation (15.3.9).

(b) Consider Hamilton's equation (15.3.22) in the slow switching regime:

$$\frac{dX}{dT} = X(2Q_X - X + 1) - \varepsilon(\Gamma X - Y),$$
$$\frac{dY}{dT} = \varepsilon(\Gamma X - Y),$$
$$\frac{dQ_X}{dT} = -Q_X(Q_X - 2X + 1) + \varepsilon\Gamma(Q_X - Q_Y),$$
$$\frac{dQ_Y}{dT} = -\varepsilon(Q_X - Q_Y).$$

Using the method of multiple scales and the perturbation expansions (15.3.23), derive the $O(1)$ and $O(\varepsilon)$ equations for the fast and slow variables, respectively. Hence, obtain the leading order solutions

$$X_0(T) = \frac{1}{1 + e^T}, \quad Q_{X0}(T) = \frac{-1}{1 + e^{-T}}.$$

and

$$Y_0(\tau) = \begin{cases} \Gamma & \text{for } \tau \leq 0 \\ \Gamma e^{-\tau} & \text{for } \tau \geq 0 \end{cases}, \quad Q_{Y0}(\tau) = \begin{cases} -e^\tau & \text{for } \tau \leq 0 \\ -1 & \text{for } \tau \geq 0 \end{cases}.$$

**Problem 15.4. PDE-ODE model of quorum sensing.** Consider a non-dimensionalized PDE-ODE model of quorum sensing, in which each cell is a one-component chemical compartment. That is, equation (15.4.29) reduces to

$$\tau \frac{\partial U}{\partial t} = D\nabla^2 U - U, \quad \mathbf{x} \in \Omega; \quad \frac{\partial U}{\partial n} = 0, \quad \mathbf{x} \in \partial\Omega,$$

$$\varepsilon D \frac{\partial U}{\partial n_j} = d_1 U - d_2 u_j, \quad \mathbf{x} \in \partial\Omega_j, \quad j = 1,\ldots,N,$$

$$\frac{du_j}{dt} = F(u_j) - \frac{1}{\varepsilon\tau} \int_{\partial\Omega_j} (d_2 u_j - d_1 U)\,ds, \quad j = 1,\ldots,N.$$

(a) Consider the linear stability of the steady-state state $u_j = \bar{u}$ for all $j$ and $U = U^*(\mathbf{x})$. Following the stability analysis of Sect. 15.4.5, show that in the large diffusion limit, $D = D_0/v \gg 1$, the eigenvalue problem takes the form

$$a(\lambda)c_j + \frac{b(\lambda)}{N} \sum_{l=1}^{N} c_l = 0$$

to leading order in $v$ with

$$a(\lambda) = 1 + \frac{D_0}{d_1} + \frac{2\pi d_2 D_0}{d_1 \tau} \frac{1}{\lambda - F'(\bar{u})}, \quad b(\lambda) = \frac{2\pi D_0}{(1+\tau\lambda)|\Omega|}.$$

(b) Show that the eigenvalue problem for the synchronous eigenmode, $c_j =$ for all $j$, reduces to a quadratic

$$\lambda^2 - \lambda p_1 + p_2, \quad p_1 = F'(\bar{u}) - \frac{A}{\tau} - \frac{B}{\tau}, \quad p_2 = \frac{1}{\tau}\left(\frac{A}{\tau} - BF'(\bar{u})\right),$$

where

$$A = \frac{2\pi d_2 D_0}{d_1 + D_0} > 0, \quad B = 1 + \frac{2\pi d_1 D_0}{|\Omega|(d_1 + D_0)} > 1.$$

Establish that if $p_1 = 0$ then $p_2 < 0$ and thus deduce that a Hopf bifurcation associated with a synchronous eigenmode cannot occur.

(c) Show that the steady-state solution is stable to synchronous oscillations if and only if $p_1 < 0$ and $p_2 > 0$, that is,

$$\tau F'(\bar{u}) < \min(A+B, A/B) = A/B,$$

since $B > 1$.

(d) From the eigenvalue equation $a(\lambda) = 0$, show that no Hopf bifurcation can occur involving an asynchronous eigenmode, and that the steady state is stable with respect to asynchronous modes when the condition of part (c) holds. Finally, establish that $A/B$ is a monotone increasing function of $D_0$, so that the effect of faster diffusion is to extend the stability region of the steady state.

**Problem 15.5. The classical Kuramoto model.** The classical Kuramoto model considers the synchronization properties of a system of weakly-coupled, near identical limit-cycle oscillators with sinusoidal phase interaction function

$$\frac{d\theta_i}{dt} = \omega_i + \frac{K}{N}\sum_{j=1}^{N}\sin(\theta_j - \theta_i)$$

for $i = 1,\ldots N$, where $K \geq 0$ is the coupling strength and $\omega_i$ is the natural frequency of the $i$-th oscillator. The frequencies $\omega_i$ are assumed to be distributed according to a zero mean probability density $g(\omega)$ with (i) $g(-\omega) = g(\omega)$ and (ii) $g(0) \geq g(\omega)$ for all $w \in [0, \infty)$.

(a) Introducing the complex order parameter

$$z(t) = r(t)e^{i\psi(t)} \equiv \frac{1}{N}\sum_{j=1}^{N}e^{i\theta_j(t)},$$

show that the phase equation can be rewritten as

$$\frac{d\theta_i}{dt} = \omega_i + Kr\sin(\psi - \theta_i),$$

with $r, \psi$ determined self-consistently.

(b) Consider a steady-state solution $r(t) = R$ and $\psi(t) = 0$. Show that are two types of solution depending on whether $|\omega_i| \leq KR$ (phase-locked solution) or $|\omega_i| > KR$ (drift solution).

(c) Let $\rho(\theta, t, \omega)$ denote the fraction of oscillators that lie between $\theta$ and $\theta + d\theta$ at time $t$. The continuity or Liouville equation

$$\frac{\partial \rho}{\partial t} = -\frac{\partial}{\partial \theta}(\rho\dot{\theta}).$$

Show that for the drifting population the steady-state density is

$$\rho_{\text{drift}}(\theta, \omega) = \frac{C}{|\omega - KR\sin\theta|},$$

with $C$ determined by the normalization condition $\int_{-\pi}^{\pi}\rho(\theta, \omega)d\theta = 1$ for each $\omega$. On the other hand, for the locked subpopulation, $\rho_{\text{lock}}(\theta, \omega) = \delta(\theta - \theta(\omega))$ where $KR\sin\theta(\omega) = \omega$.

(d) Given the steady-state solution, the self-consistency condition for $R$ can be expressed as

$$R = \langle e^{i\theta}\rangle_{\text{lock}} + \langle e^{i\theta}\rangle_{\text{drift}}.$$

Using properties of $g$ and $\rho$ show that $\langle e^{i\theta}\rangle_{\text{drift}} = 0$, and hence

$$R = KR\int_{-\pi/2}^{\pi/2}\cos^2(\theta)g(KR\sin\theta)d\theta.$$

(e) The zero solution $R = 0$ with $\rho(\theta, \omega) = 1/2\pi$ exists for all $K$. Show that a second branch of partially synchronized solutions satisfying

$$1 = K \int_{-\pi/2}^{\pi/2} \cos^2(\theta) g(KR\sin\theta) d\theta,$$

bifurcates from $R = 0$ at the critical coupling

$$K_c = \frac{2}{\pi g(0)}.$$

Assuming that $g''(0) < 0$, Taylor expand the above equation to second order in $R$ to show that for small, positive $\Delta K = K - K_c$

$$R \approx \frac{4}{K_c^2} \sqrt{\frac{K - K_c}{-\pi g''(0)}}.$$

**Problem 15.6. Nutrient-based model of a biofilm.** Consider the stability analysis of a flat, growing interface for the nutrient-based model given by equations (15.6.11)–(15.6.15).

(a) Linearizing the three-dimensional model about the one-dimensional quasi-steady-state solution $c_0(z)$, derive the corresponding system of equations for the first-order perturbation $c_1(z)$ given by equation (15.6.17).

(b) Given the zeroth-order front solution (15.6.16a) obtain the solutions

$$c_1(z) = Ae^{-\kappa(h_0-z)}, \quad 0 < z < h_0$$
$$c_1(z) = B\sinh(k[h_0 + L - z]), \quad h_0 < z < h_0 + L,$$

where

$$\kappa^2 = k^2 + \xi^{-2}.$$

(c) Imposing the interfacial conditions (15.6.17d,e), determine the coefficients $A$ and $B$

$$A = \frac{\Gamma h_1}{\xi} \mathscr{A}(k), \quad \mathscr{A}(k) = -\frac{(k\xi)^{-1}\tanh(kL) + K - 1}{k^{-1}\kappa\tanh(kL) + K},$$

$$B = \frac{\Gamma h_1}{\xi} \mathscr{B}(k), \quad \mathscr{B}(k) = \frac{1}{\cosh(kL)} \frac{-(\kappa\xi)^{-1} - 1 + K^{-1}}{Kk\kappa^{-1} + \tanh(kL)}.$$

**Problem 15.7. Geometric formulation of interface dynamics.** Consider a system such as a bacterial colony spreading across a two-dimensional substrate with the boundary of the system described by a closed curve $\mathbf{x}(t,\sigma) \in \mathbb{R}^2$, where $0 \leq \sigma \leq 1$ and $\mathbf{x}(t,0) = \mathbf{x}(t,1)$. In order to determine the evolution of the interface, it is necessary to know the normal velocity

$$\hat{\mathbf{n}} \cdot \frac{\partial \mathbf{x}}{\partial t} = \frac{F}{\gamma} \equiv U,$$

where $\gamma$ is a friction coefficient and $F$ is taken to be a local function of the interfacial position. The kinematic method for analyzing interfacial dynamics is to introduce a time-dependent parameterization of the interface so that the total time derivative of $\mathbf{x}$ is orthogonal to the interface [138, 139]. That is, $\sigma = \sigma(t)$ such that

$$\frac{d\mathbf{x}}{dt} = \hat{\mathbf{n}}U, \qquad \frac{d\mathbf{x}}{dt} \cdot \frac{\partial \mathbf{x}}{\partial \sigma} = 0.$$

Here $\boldsymbol{\tau} = \partial \mathbf{x}/\partial \sigma$ is the tangent vector to the interface at $\mathbf{x}$. The next step is to express the dynamics in terms of parameter-independent geometrical quantities such as arc length and curvature. Following Box 12A, the arc length is given by

$$s(\sigma) = \int_0^{\sigma} \sqrt{g(\sigma')} d\dot{\sigma}', \qquad g = \frac{\partial \mathbf{x}}{\partial \sigma} \cdot \frac{\partial \mathbf{x}}{\partial \sigma}.$$

It follows that the unit tangent vector is

$$\hat{\boldsymbol{\tau}} = \frac{\partial \mathbf{x}}{\partial s} = \frac{1}{\sqrt{g}} \boldsymbol{\tau}.$$

The curvature is then defined according to

$$\kappa = -\hat{\mathbf{n}} \cdot \frac{\partial^2 \mathbf{x}}{\partial s^2}.$$

(Note that choice of sign differs from the one used in Box 12A.)

(a) Write the kinematic equation as

$$\frac{\partial \mathbf{x}}{\partial t} + \frac{ds}{dt} \frac{\partial \mathbf{x}}{\partial s} = \hat{\mathbf{n}}U.$$

Taking the inner product with respect to $\hat{\boldsymbol{\tau}}$ and rearranging show that

$$\frac{ds}{dt} = -\frac{\partial \mathbf{x}}{\partial t} \cdot \frac{\partial \mathbf{x}}{\partial s}.$$

Hence, obtain the result

$$\frac{\partial}{\partial s} \dot{s} = \kappa U.$$

and thus derive the operator identity

$$\frac{\partial}{\partial s} \frac{d}{dt} = \frac{d}{dt} \frac{\partial}{\partial s} + \kappa U \frac{\partial}{\partial s}.$$

(b) Suppose that the curve is represented using the complex coordinate $z = x + iy$. Given two vectors $\mathbf{x}_j = (x_j, y_j)$, $j = 1, 2$, the inner product can be expressed as

$$\mathbf{x}_1 \cdot \mathbf{x}_2 = x_1 x_2 + y_1 y_2 = \text{Re}[(x_1 + iy_1)(x_2 - iy_2)] = \text{Re}[z_1 \bar{z}_2].$$

Show that in this representation, the tangent vector, unit normal and curvature are given by

$$\tau = \sqrt{g}e^{i\theta}, \quad n = -ie^{i\theta}, \quad \kappa = \frac{\partial \theta}{\partial s}.$$

Here $\theta$ is the direction of the curve at the point $\mathbf{x}$, that is, $\tan \theta = dy/dx$.

(c) The interface equation can now be written as $\dot{z} = nU$. Differentiating both sides with respect to $\sigma$ and reversing the order of differentiation then yields

$$\frac{d\tau}{dt} = \frac{\partial}{\partial \sigma}(nU).$$

Substituting the complex representations of $\tau$ and $n$ into the above equation, and equating real and imaginary parts, derive the pair of equations

$$\frac{dg}{dt} = 2g\kappa U, \quad \frac{d\theta}{dt} = -\frac{\partial U}{\partial s}.$$

From a geometrical perspective, these equations imply that a curve segment of given tangent $\tau$ and metric $g$ rotates by $-\partial U/\partial s$ and dilates by $\sqrt{g}\kappa U$. Differentiating the equation for $d\theta/dt$ with respect to $s$ and reversing the order of differentiation using part (a), derive the following evolution equation for the curvature:

$$\frac{d\kappa}{dt} = -\left[\kappa^2 + \frac{\partial^2}{\partial s^2}\right]U.$$

(d) Following the analysis of pattern formation in a multi-species bacterial colony [961], suppose that the force $F$ depends on the local curvature according to $F = F_0 - \mu\kappa$. Furthermore, assume that the friction coefficient is a linear function of the nonmotile $E.$ $coli$ population, $\gamma = 1 + \alpha c$ with $c = c_0/\sqrt{g}$. The kinematic equations become

$$\frac{d\kappa}{dt} = -\left[\kappa^2 + \frac{\partial^2}{\partial s^2}\right]\frac{F_0 - \mu\kappa}{1 + \alpha c_0/\sqrt{g}},$$

$$\frac{dg}{dt} = 2g\kappa\frac{F_0 - \mu\kappa}{1 + \alpha c_0/\sqrt{g}}.$$

Performing a linear stability analysis about a flat interface ($\kappa = 0, g = 1$) by setting

$$\kappa = (\Delta\kappa)e^{iks+\lambda t}, \quad g = 1 + (\Delta g)e^{iks+\lambda t},$$

obtain a pair of dispersion relations $\lambda = \lambda_\pm(k)$, Hence, establish that for positive $\mu, \alpha$ one of the two eigenvalues is always positive, indicating an instability.

# Chapter 16
# Stochastic reaction–diffusion processes

In Chap. 11, we considered various analytically tractable models of intracellular pattern formation and waves based on deterministic RD equations. A complementary approach is to develop more realistic multi-scale computational models, which include details of the structure of individual macromolecules, the biochemical network of signaling pathways, the aqueous environment of the cytoplasm, the mechanical properties of the cytoskeleton, and the geometry of the cell [74, 278, 317, 846]. A major challenge in the computational modeling of RD systems is how to efficiently couple stochastic chemical reactions involving low copy numbers with diffusion in complex environments. One approach is to consider a spatial extension of the Gillespie algorithm for well-mixed chemical reactions [347, 348] (Chap. 5) using a mesoscopic compartment-based method, although there are subtle issues with regards to choosing the appropriate compartment size [438, 459, 460, 890]. Alternatively, one can combine a coarse-grained deterministic RD model in the bulk of the domain with individual particle-based Brownian dynamics in certain restricted regions [18, 279, 280, 318]; in this case, considerable care must be taken in the choice of boundary conditions at the interface between the two domains.

In this chapter, we focus on some analytical approaches to studying stochastic RD processes. We begin in Sect. 16.1 by investigating the effects of intrinsic noise on spontaneous pattern formation using an RD master equation. The latter is obtained by discretizing space and treating spatially discrete diffusion as a hopping reaction. Carrying out a linear noise approximation of the master equation leads to an effective Langevin equation, whose power spectrum provides a means of extending the definition of a Turing instability to stochastic systems, namely, in terms of the existence of a peak in the power spectrum at a nonzero spatial frequency. [79, 154, 155, 580, 604, 803, 952]. One thus finds that noise can significantly extend the range over which spontaneous patterns occur. One possible limitation of fluctuation-driven patterns in the weak noise regime (where the system-size expansion holds) is that the amplitude of the patterns could be too small to be detectable. Following Ref. [81], we show how the interplay between intrinsic noise

P. Bressloff, *Stochastic Processes in Cell Biology*, Interdisciplinary Applied Mathematics 41, https://doi.org/10.1007/978-3-030-72519-8_16

and transient growth of perturbations can amplify the patterns. The source of transient growth is the presence of a non-normal matrix in the linear evolution operator.

We then turn to another major topic in the analysis of RD master equations, namely, statistical field theory [37, 164, 487, 859, 861, 988]. This is a powerful method for investigating the effects of fluctuations and spatial correlations on the long-time asymptotics of RD processes. It is based on the construction of a Doi–Peliti path integral representation of the RD master equation along the lines outlined in Sect. 8.2. Taking the continuum limit of the resulting path integral action yields a statistical field theoretic description of the RD process. Using the canonical example of pair annihilation with diffusion, we show how well-known techniques in statistical field theory, such as moment generating functionals, diagrammatic perturbation expansions, and renormalization theory, can capture the asymptotic decay of the system. Finally, in Sect. 16.3 we review a formal perturbation method for analyzing traveling front solutions in stochastic RD equations [29, 235, 708, 784, 798].

## 16.1 Stochastic pattern formation and the RD master equation

Consider the general two-species RD equation

$$\frac{\partial u_j}{\partial t} = D_j \nabla^2 u_j + \sum_{r=1}^{M} S_{jr} f_r(u_1, u_2) \tag{16.1.1}$$

for $j = 1, 2$, where $S_{jr}$ and $f_r$ are, respectively, the stoichiometric coefficients and affinities for the $M$ reactions, and $D_j$ are the diffusivities. In the absence of diffusion (a well-mixed compartment), the corresponding stochastic version of the model is described by a chemical master equation as detailed in Chap. 5. In order to incorporate diffusion into a corresponding RD master equation, we follow McKane *et. al* [604] by partitioning the cell into small domains with centers at discrete lattice points $\ell$. For simplicity, it is assumed that the domain centers are distributed on a regular $d$-dimensional hypercubic lattice $\mathcal{L}$ ($d = 1, 2, 3$) with lattice spacing $h$ and coordination number $z$. The spatially discrete version of the RD system is

$$\frac{du_{\ell,j}}{dt} = F_{\ell,j}(\mathbf{u}_1, \mathbf{u}_2) := \sum_{r=1}^{p} S_{jr} f_r(u_{\ell,1}, u_{\ell,2}) + \alpha_j \Delta u_{\ell,j} \tag{16.1.2}$$

for $j = 1, 2$. Here $\alpha_j = D_j/h^2$ and $\Delta$ is the discrete Laplacian defined as

$$\Delta u_\ell = \sum_{\ell' \in \partial \ell} [u_{\ell'} - u_\ell], \tag{16.1.3}$$

where $\ell' \in \partial \ell$ indicates that we are summing over nearest neighbors of $\ell$ on the lattice. The RD equation (16.1.1) is recovered in the continuum limit $h \to 0$ and $\alpha_j \to \infty$ with $\alpha_j h^2$ fixed and $u_{\ell,j}(t) \to u_j(\mathbf{x}, t)$. Now let $n_{\ell,j}$ denote the number of X molecules ($j = 1$) and Y molecules ($j = 2$) in the domain with center at $\ell$ and introduce the local densities $u_{\ell,j} = n_{\ell,j}/\Omega$ with $\Omega = h^d$. A major advantage of discretizing space is that diffusion can now be represented by a set of hopping reactions and treated on an equal footing with the chemical reactions in (16.1.1). The full set of reactions at site $\ell$ are

$$(n_{\ell,1}, n_{\ell,2}) \xrightarrow{T_{\ell,r}} (n_{\ell,1} + S_{1,r}, n_{\ell,2} + S_{2,r}), \quad r = 1, \ldots, M, \tag{16.1.4a}$$

$$(n_{\ell,j}, n_{\ell',j}) \xrightarrow{T^j_{\ell,\ell'}} (n_{\ell,j} - 1, n_{\ell',j} + 1), \quad \ell' \in \partial\ell, \tag{16.1.4b}$$

$$(n_{\ell,j}, n_{\ell',j}) \xrightarrow{T^j_{\ell',\ell}} (n_{\ell,j} + 1, n_{\ell',j} - 1), \quad \ell' \in \partial\ell, \tag{16.1.4c}$$

with

$$T_{\ell,r} = f_r(n_{\ell,1}/\Omega, n_{\ell,2}/\Omega) \text{ for } r = 1, \ldots, M, \quad T^j_{\ell,\ell'} = \frac{\alpha_j}{\Omega} n_{\ell,j}. \tag{16.1.5}$$

If $N$ denotes the total number of lattice points, then there are $2N$ "chemical species" labeled by $(\ell, j)$. Moreover, we have $zN$ hopping reactions and $MN$ local biochemical reactions between the X and Y molecules, so that the total number of reactions is $R = (z + M)N$. The corresponding RD master equation has the general form

$$\frac{dP(\mathbf{n}_1, \mathbf{n}_2, t)}{dt} = \Omega \sum_{q=1}^{R} \left( \prod_{i=1,2} \prod_{\ell} \mathbb{E}^{-\widehat{S}_{i,\ell;q}} - 1 \right) \widehat{f}_q(\mathbf{n}_1/\Omega, \mathbf{n}_2/\Omega) P(\mathbf{n}_1, \mathbf{n}_2, t),$$

$$\tag{16.1.6}$$

where $\mathbf{n}_j = \{n_{\ell,j}, \ell \in \mathcal{L}\}$, and

$$\widehat{f}_q = f_r(u_{\ell,1}, u_{\ell,2}), \quad \widehat{S}_{\ell,i;q} = S_{ir} \text{ for } q = (r, \ell),$$

$$\widehat{f}_q = \alpha_j u_{\ell,j}, \quad \widehat{S}_{\ell,j;q} = -1, \widehat{S}_{\ell',j;q} = 1 \text{ for } q = (j, \ell, \ell'), \quad \ell' \in \partial\ell.$$

### 16.1.1 Linear noise approximation of RD master equation

The RD master equation (16.1.6) is clearly difficult to analyze. However, if $\Omega$ is sufficiently large then we can carry out a system-size expansion along the lines outlined in Sect. 5.2 to derive a corresponding FP equation given by

$$\frac{\partial p}{\partial t} = -\sum_{j,\ell} \frac{\partial [F_{\ell,j}(\mathbf{u}_1, \mathbf{u}_2) p(\mathbf{u}_1, \mathbf{u}_2, t)]}{\partial u_{\ell,j}} + \frac{1}{2\Omega} \sum_{\ell,\ell'} \sum_{j,j'} \frac{\partial^2 [D_{\ell,j;\ell',j'}(\mathbf{u}_1, \mathbf{u}_2) p(\mathbf{u}_1, \mathbf{u}_2, t)]}{\partial u_{\ell,j} \partial u_{\ell',j'}},$$

$$\tag{16.1.7}$$

where

$$D_{\ell,j;\ell',j'} = \delta_{\ell,\ell'} C^\ell_{jj'} + \alpha_j \Gamma^j_{\ell\ell'} \delta_{j,j'}, \tag{16.1.8}$$

with

$$C^\ell_{jj'} = \sum_{r=1}^{M} S_{jr} S_{j'r} f_r(u_{\ell,1}, u_{\ell,2}), \tag{16.1.9}$$

and

$$\Gamma^j_{\ell\ell} = \sum_{\ell' \in \partial\ell} (u_{\ell,j} + u_{\ell',j}), \quad \Gamma^j_{\ell\ell'} = -(u_{\ell,j} + u_{\ell',j}) \text{ for all } \ell' \in \partial\ell. \tag{16.1.10}$$

Further simplification can be achieved if we linearize the corresponding Langevin equation about a homogeneous steady state $(u_1^*, u_2^*)$ by setting

$$\frac{n_{\ell,j}}{\Omega} = u_{\ell,j} = u_j^* + \frac{1}{\sqrt{\Omega}} v_{\ell,j}.$$

This yields

$$\frac{dv_{\ell,j}(t)}{dt} = \sum_{\ell',j'} A_{\ell,j;\ell'j'} v_{\ell',j'}(t) + \eta_{\ell,j}(t), \tag{16.1.11}$$

with white noise terms satisfying

$$\langle \eta_{\ell,j}(t) \rangle = 0, \quad \langle \eta_{\ell,j}(t) \eta_{\ell',j'}(t') \rangle = D_{\ell,j;\ell',j'} \delta(t - t').$$

Here

$$\sum_{\ell',j'} A_{\ell,j;\ell'j'} v_{\ell',j'} = \sum_{r=1}^{M} S_{jr} \frac{\partial f_r(u_1^*, u_2^*)}{\partial u_{j'}^*} v_{\ell,j'} + \alpha_j \Delta v_{\ell,j} \delta_{j,j'}, \tag{16.1.12}$$

and

$$D_{\ell,j;\ell',j'}^* = \delta_{\ell,\ell'} C_{jj'}^* + \alpha_j (\Gamma^*)^j_{\ell\ell'} \delta_{j,j'}, \tag{16.1.13}$$

with

$$C_{jj'}^* = \sum_{r=1}^{M} S_{jr} S_{j'r} f_r(u_1^*, u_2^*), \tag{16.1.14}$$

and

$$(\Gamma^*)^j_{\ell\ell} = 2zu_j^*, \quad (\Gamma^*)^j_{\ell\ell'} = -2u_j^* \text{ for all } \ell' \in \partial\ell. \tag{16.1.15}$$

Considerable insight into the behavior of the system can now be obtained by transforming to Fourier space [580, 604]. For simplicity, consider a 1D lattice with periodic boundary conditions, $v_{\ell+N} = v_\ell$ for $\ell = 1, \ldots, N$ and set the lattice spacing $h = 1$. Introduce the discrete Fourier transforms

$$V(k) = \sum_{\ell} e^{-ik\ell} v_\ell, \quad v_\ell = \frac{1}{N} \sum_k e^{ik\ell} V(k),$$

with $k = 2\pi m/N, m = 0, \ldots, N-1$. For a regular 1D lattice, the Fourier transform of the discrete Laplacian operator is

$$\sum_{\ell} e^{-ik\ell} \Delta v_\ell = \sum_{\ell} e^{-ik\ell} [v_{\ell+1} + v_{\ell-1} - 2v_\ell] = 2[\cos(k) - 1] V(k). \tag{16.1.16}$$

The discrete Fourier transform of the Langevin equation is

$$\frac{dV_j(k,t)}{dt} = \sum_{j'} A_{jj'}(k)V_{j'}(k,t) + \eta_j(k,t), \qquad (16.1.17)$$

with

$$A_{jj'}(k) = \sum_{r=1}^{M} S_{jr}\frac{\partial f_r(u_1^*,u_2^*)}{\partial u_{j'}^*} + 2\alpha_j[\cos(k) - 1]\delta_{j,j'}, \qquad (16.1.18)$$

$$\langle \eta_j(k,t) \rangle = 0, \quad \langle \eta_j(k,t)\eta_{j'}(k',t') \rangle = \hat{D}_{jj'}(k,k')\delta(t-t').$$

Moreover, using the identity $\sum_{\ell} e^{i(k-k')\ell} = N\delta_{k,k'}$, we have

$$\hat{D}_{jj'}(k,k') = \sum_{\ell,\ell'} e^{-ik\ell} e^{-ik'\ell'} D_{\ell,j;\ell',j'}^* = \sum_{\ell,\ell'} e^{-ik\ell} e^{-ik'\ell'} \left[ \delta_{\ell,\ell'} C_{jj'}^* + \alpha_j(\Gamma^*)_{\ell\ell'}^j \delta_{j,j'} \right]$$

$$= N\delta_{k,-k'}\left( C_{jj'}^* + 4\alpha_j u_j^*[1-\cos(k)]\delta_{j,j'} \right) := N\delta_{k,-k'}C_{jj'}(k).$$

The factor of $N$ can be eliminated by rescaling time $t$ and $V_j(k)$ appropriately. Now Fourier transforming the Langevin equation with respect to time gives

$$\sum_l \Phi_{jl}(k,\omega)V_l(k,\omega) = \eta_j(k,\omega),$$

with

$$\Phi_{jl}(k,\omega) = -i\omega\delta_{j,l} - A_{jl}(k),$$

and

$$\langle \eta_j(k,\omega) \rangle = 0, \quad \langle \eta_j(k,\omega)\eta_{j'}(k',\omega') \rangle = \delta_{k,-k'}C_{jj'}(k)\delta(\omega+\omega').$$

Hence,

$$\langle V_i(k,\omega)V_i(k',\omega') \rangle = \left\langle \left[ \sum_l \Phi_{il}^{-1}(k,\omega)\eta_l(k,\omega) \right]\left[ \sum_j \Phi_{ij}^{-1}(k',\omega')\eta_j(k',\omega') \right] \right\rangle$$

$$= \delta_{k,-k'}\delta(\omega+\omega')\sum_{l,j} \Phi_{il}^{-1}(k,\omega)C_{lj}(k)\Phi_{ij}^{-1}(-k,-\omega')$$

$$= \delta_{k,-k'}\delta(\omega+\omega')\sum_{l,j} \Phi_{il}^{-1}(k,\omega)C_{lj}(k)(\Phi^\dagger)_{ji}^{-1}(k,\omega).$$

Defining the power spectrum of the $k$th eigenmode by

$$\langle V_i(k,\omega)V_i(k',\omega') \rangle = S_i(k,\omega)\delta_{k,-k'}\delta(\omega+\omega'),$$

we deduce that

$$S_i(k,\omega) = \sum_{l,j} \Phi_{il}^{-1}(k,\omega)C_{lj}(k)(\Phi^\dagger)_{ji}^{-1}(k,\omega). \qquad (16.1.19)$$

*Example 16.1. Turing instability in the Brusselator model.* In order to illustrate the above theory, consider a spatially extended version of the Brusselator model introduced in Chap. 5, which is given by the RD system

$$\frac{\partial u_1}{\partial t} = D_1 \nabla^2 u + a - (b+1)u_1 + u_1^2 u_2, \tag{16.1.20a}$$

$$\frac{\partial u_2}{\partial t} = D_2 \nabla^2 u_2 + bu_1 - u_1^2 u_2. \tag{16.1.20b}$$

Linearizing about the spatially homogeneous steady state $u_1^* = a, u_2^* = b/a$ by setting $u_j(\mathbf{x},t) = u_j^* + \xi_j(\mathbf{x})e^{\lambda t}$ and expanding to first order in $\xi_j$, leads to the eigenvalue equation

$$\lambda \begin{pmatrix} \xi_1(\mathbf{x}) \\ \xi_2(\mathbf{x}) \end{pmatrix} = \begin{pmatrix} b-1+D_1\nabla^2 & a^2 \\ -b & -a^2+D_2\nabla^2 \end{pmatrix} \begin{pmatrix} \xi_1(\mathbf{x}) \\ \xi_2(\mathbf{x}) \end{pmatrix}. \tag{16.1.21}$$

This has eigensolutions of the form $\xi_j(\mathbf{x}) = \widehat{\xi}_j(k)e^{-i\mathbf{k}\cdot\mathbf{x}}$, such that $\nabla^2 \to -k^2$ and $\lambda$ is the solution to

$$(b-1-D_1k^2-\lambda)(a^2+D_2k^2+\lambda)-a^2b = 0.$$

That is, $\lambda = \lambda_\pm(k)$ with

$$\lambda_\pm(k) = \frac{1}{2}\left[\Gamma(k) \pm \sqrt{\Gamma(k)^2 + 4\Lambda(k)}\right], \tag{16.1.22}$$

where

$$\Gamma(k) = b-1-a^2-(D_1+D_2)k^2, \quad \Lambda(k) = (b-1-D_1k^2)(a^2+D_2k^2)-a^2b. \tag{16.1.23}$$

Recall from Sect. 11.2 that one of the conditions for a Turing instability is that the homogeneous fixed point should be stable with respect to homogeneous ($k = 0$) perturbations. Setting $k = 0$ in equation (16.1.22), we thus require $\mathrm{Re}[\lambda_\pm(0)] < 0$, where

$$\lambda_\pm(0) = \frac{1}{2}\left[b-1-a^2 \pm ia\right]. \tag{16.1.24}$$

This yields the necessary condition $b < 1 + a^2$. Now observe that as $k$ increases from zero $\Gamma(k)$ becomes more negative. Therefore, in order that the fixed point becomes unstable due to the growth of some non-zero frequency mode ($k \neq 0$), we require $\Lambda(k)$ to become positive. The critical wavenumber $k_c$ for a Turing instability is thus given by the condition $\Lambda(k_c) = 0$, that is,

$$[(b-1)D_v - a^2 D_u]k_c^2 = a^2.$$

Hence, a Turing instability will occur provided that

$$1 + \frac{D_u}{D_v}a^2 < b < 1 + a^2,$$

which immediately implies that $D_u < D_v$, that is, $Y$ molecules diffuse more quickly than $X$ molecules. If $b$ is taken to be a bifurcation parameter, then increasing $b$ from zero will lead to a Turing instability at the critical value $b_c = 1 + D_u a^2/D_v$, beyond which spatially periodic eigenmodes with spatial frequencies around $k_c$ will start to grow, see Fig. 16.1. As the amplitude of these eigenmodes increases beyond the linear regime, saturating nonlinearities of the full system will typically stabilize the resulting patterns, whose fundamental wavelength will be approximately given by $2\pi/k_c$.

The spatially discrete version of the RD system (16.1.20) is

$$\frac{du_{\ell,1}}{dt} = a - (b+1)u_{\ell,1} + u_{\ell,1}^2 u_{\ell,2} + \alpha_1 \Delta u_{\ell,1}, \tag{16.1.25a}$$

$$\frac{du_{\ell,2}}{dt} = bu_{\ell,1} - u_{\ell,1}^2 u_{\ell,2} + \alpha_2 \Delta u_{\ell,2}. \tag{16.1.25b}$$

The full set of reactions at site $\ell$ is

$$(n_{\ell,1},n_{\ell,2}) \xrightarrow{T_{\ell,1}} (n_{\ell,1}+1,n_{\ell,2}), \quad (n_{\ell,1},n_{\ell,2}) \xrightarrow{T_{\ell,2}} (n_{\ell,1}-1,n_{\ell,2}+1),$$

$$(n_{\ell,1},n_{\ell,2}) \xrightarrow{T_{\ell,3}} (n_{\ell,1}+1,n_{\ell,2}-1), \quad (n_{\ell,1},n_{\ell,2}) \xrightarrow{T_{\ell,4}} (n_{\ell,1}-1,n_{\ell,2}),$$

$$(n_{\ell,j},n_{\ell',j}) \xrightarrow{T^j_{\ell,\ell'}} (n_{\ell,j}-1,n_{\ell',j}+1), \quad \ell' \in \partial\ell,$$

$$(n_{\ell,j},n_{\ell',j}) \xrightarrow{T^j_{\ell',\ell}} (n_{\ell,j}+1,n_{\ell',j}-1), \quad \ell' \in \partial\ell,$$

such that

$$\sum_r S_{1r} f_r(u_{\ell,1},u_{\ell,2}) = a - (b+1)u_{\ell,1} + u_{\ell,1}^2 u_{\ell,2}, \tag{16.1.26}$$

$$\sum_r S_{2r} f_r(u_{\ell,1},u_{\ell,2}) = b u_{\ell,1} - u_{\ell,1}^2 u_{\ell,2}, \tag{16.1.27}$$

and $\sum_{r=1}^4 S_{jr} S_{j'r} f_r(u_{\ell,1},u_{\ell,2}) = C^\ell_{jj'}$ with

$$\mathbf{C}^\ell = \begin{pmatrix} a+(b+1)u_{\ell,1}+u_{\ell,1}^2 u_{\ell,2} & -b u_{\ell,1} - u_{\ell,1}^2 u_{\ell,2} \\ -b u_{\ell,1} - u_{\ell,1}^2 u_{\ell,2} & b u_{\ell,1} + u_{\ell,1}^2 u_{\ell,2} \end{pmatrix}. \tag{16.1.28}$$

Carrying out the linear noise approximation and transforming to Fourier space then leads to the Langevin equation (16.1.17) with drift matrix

$$\mathbf{A}(k) = \begin{pmatrix} b-1+2\alpha_1[\cos(k)-1] & a^2 \\ -b & -a^2+2\alpha_2[\cos(k)-1] \end{pmatrix}, \tag{16.1.29}$$

and diffusion matrix

$$\mathbf{C}(k) = \begin{pmatrix} 2(b+1)a+4\alpha_1 u_1^*[1-\cos(k)] & -2ba \\ -2ba & 2ba+4\alpha_2 u_2^*[1-\cos(k)] \end{pmatrix}, \tag{16.1.30}$$

with $u_1^* = a, u_2^* = b/a$. If one now plots the power spectrum $S_i(k,\omega)$ defined by equation (16.1.19) as a function of $k$, one finds that for a range of parameter values the power spectrum shows a peak at $k \neq 0, \omega = 0$, indicating the presence of a stochastic Turing pattern

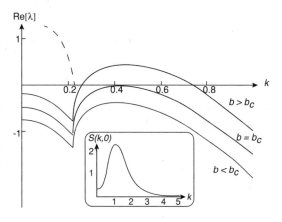

Fig. 16.1: Sketch of dispersion curves for the deterministic Brusselator model, showing $\mathrm{Re}[\lambda(k)]$ as a function of $k$ for the most unstable eigenvalue. Parameter values are $a = 1.5, D_1 = 2.8, D_2 = 22.4$. Three different values for $b$ are used: $b = 1.8 < b_c$, $b = b_c = 2.34$ and $b = 2.9 > b_c$. Dashed line shows imaginary part of $\lambda(k)$, which is approximately the same for all given $b$ values. [Insert: Sketch of the power spectrum $S(k,0)$ for species $X$ in the stochastic Brusselator model. Redrawn from [604]].

[79, 154, 155, 580, 604, 803, 952], see insert of Fig. 16.1. Moreover, one can compute the region of parameter space for which the spectrum has a peak at non-zero $k$, and demonstrate that these regions are significantly larger than the region over which a deterministic Turing pattern occurs. It also possible to extend the analysis to the case of Turing-Hopf bifurcations, where there is a peak in the power spectrum at $k \neq 0, \omega \neq 0$, which can result in either standing or traveling waves [80, 803].

A second example is considered in Ex. 16.1 and Ex. 16.2 in the case of an RD equation for a membrane-bound signaling molecule with positive feedback, which is used to study cell polarization in budding yeast. A third example is developed in Ex. 16.3, which is a multicellular version of the LEGI model of spatial gradient sensing introduced in Sect. 10.6. Finally, note that one important extension of the above spectral analysis would be to determine the effects of intrinsic noise on the selection and stability of patterns. One way to approach this would be to construct a stochastic version of the amplitude equations derived using weakly nonlinear analysis, as illustrated in Ex. 11.4. However, it is not clear to what extent a diffusion approximation of the RD master equation is valid close to a bifurcation point where weakly nonlinear analysis can be applied. A related issue is to what extent symmetries of the underlying deterministic RD system persist in the presence of intrinsic noise.

### 16.1.2 Non-normality, transient dynamics, and the amplification of stochastic patterns

Although noise-induced pattern formation can broaden the parameter region over which a Turing pattern exists, the amplitude of the patterns are $O(\Omega^{-1/2})$, where $\Omega$ is the system size. Hence, the amplitude of a pattern in the fluctuation-driven regime is expected to be much smaller than in the deterministic regime, and thus might not be observable. However, it turns out that giant amplification of fluctuation-driven patterns can occur in cases where there is an interplay between intrinsic noise and transient growth of perturbations about a spatially uniform state [81]. In the following, we present the theory of transient dynamics and then apply it to pattern forming instabilities.

**Transient growth and non-normal matrices.** Consider the linear system in $\mathbb{R}^n$

$$\frac{dx}{dt} = \mathbf{A}\mathbf{x}, \quad \mathbf{x}(0) = \mathbf{x}_0,$$

which has the formal solution $\mathbf{x}(t) = e^{\mathbf{A}t}\mathbf{x}_0$. If all the eigenvalues of $\mathbf{A}$ have negative real part, then $\mathbf{x}(t) \to 0$ as $t \to \infty$, and so the equilibrium solution $\mathbf{x}^* = 0$ is asymptotically stable. Suppose that the eigenvalues $\lambda_1(\mathbf{A}), \ldots, \lambda_n(\mathbf{A})$ are ordered such that

$$0 > \operatorname{Re}(\lambda_1(\mathbf{A})) > \operatorname{Re}(\lambda_2(\mathbf{A})) \geq \ldots \geq \operatorname{Re}(\lambda_n(\mathbf{A})).$$

For the sake of illustration, we assume that there exists a unique eigenvalue $\lambda_1$ with the largest real part, which we will refer to as the principal eigenvalue. It follows that for almost all initial conditions,

$$\lim_{t \to \infty} e^{-\lambda_1(\mathbf{A})t} \mathbf{x}(t) = \mathbf{v}_1,$$

where $\mathbf{v}_1$ is the corresponding principal eigenvector. Recall from Box 5A that one application of a linear ODE is to determine the local stability of a fixed point of a nonlinear system, with the matrix $\mathbf{A}$ corresponding to the Jacobian. It follows that in the case of a stable fixed point, the asymptotic behavior of a perturbation about this fixed point is dominated by the principal eigenvalue and eigenvector. However, there are a growing number of examples in fluid dynamics [884, 885], ecology [665, 858], and biological pattern formation [666] where short-term transient behavior may be dramatically different from the long-time behavior. That is, although a perturbation of a stable equilibrium eventually decays to zero, there can be rapid initial growth of the perturbation that persists for long times. From a mathematical perspective, such transient growth cannot be detected by looking at the spectrum of the matrix $\mathbf{A}$, even though it is a consequence of the linear dynamics. (Nonlinearities may amplify the effects, but do not cause them.) Therefore, a number of new quantities have been introduced that characterize the transient dynamics. In particular, one crucial feature of the matrix $\mathbf{A}$ that can support transient growth is non-normality[1].

Following Ref. [665], we define reactivity as the maximum initial amplification rate of the linear system in response to all initial perturbations $\mathbf{x}_0$:

$$\text{reactivity} = \max_{\|\mathbf{x}_0\| \neq 0} \left[ \frac{1}{\|\mathbf{x}\|} \frac{d\|\mathbf{x}\|}{dt} \bigg|_{t=0} \right]. \tag{16.1.31}$$

In order to evaluate the right-hand side, note that

$$\frac{d\|\mathbf{x}\|}{dt} = \frac{d\sqrt{\mathbf{x} \cdot \mathbf{x}}}{dt} = \frac{\mathbf{x} \cdot \dot{\mathbf{x}} + \dot{\mathbf{x}} \cdot \mathbf{x}}{2\|\mathbf{x}\|} = \frac{\mathbf{x} \cdot (\mathbf{A} + \mathbf{A}^\top)\mathbf{x}}{2\|\mathbf{x}\|}.$$

It follows that

$$\frac{1}{\|\mathbf{x}\|} \frac{d\|\mathbf{x}\|}{dt} = \frac{\mathbf{x} \cdot H(\mathbf{A})\mathbf{x}}{\|\mathbf{x}\|^2}, \quad H(\mathbf{A}) = \frac{\mathbf{A} + \mathbf{A}^\top}{2}, \tag{16.1.32}$$

with $H(\mathbf{A})$ called the symmetric or Hermitian part of the matrix $\mathbf{A}$. Hence, the reactivity is given by the maximum of the Rayleigh quotient of the matrix $H(\mathbf{A})$:

$$\text{reactivity} = \max_{\|\mathbf{x}_0\| \neq 0} \frac{\mathbf{x}_0 \cdot H(\mathbf{A})\mathbf{x}_0}{\|\mathbf{x}_0\|^2}. \tag{16.1.33}$$

---

[1] A normal real matrix $\mathbf{A}$ is one for which $\mathbf{A}\mathbf{A}^\top = \mathbf{A}^\top\mathbf{A}$. Examples include symmetric ($\mathbf{A}^\top = \mathbf{A}$) and skew-symmetric ($\mathbf{A}^\top = -\mathbf{A}$) matrices. A normal matrix has an orthogonal set of eigenvectors and is diagonalizable by a unitary matrix ($\mathbf{U}\mathbf{A}\mathbf{U}^\dagger = \mathbf{A}_{\text{diag}}$, $\mathbf{U}^\dagger = \mathbf{U}^{-1}$). However, the eigenvalues of a normal but non-symmetric matrix may be complex. Finally, the real part of the principal eigenvalue of a normal matrix $\mathbf{A}$ is equal to the principal eigenvalue of the Hermitian part $H(\mathbf{A}) = (\mathbf{A} + \mathbf{A}^\top)/2$.

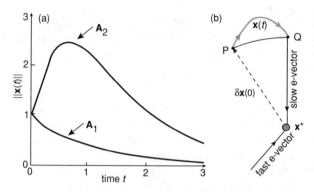

Fig. 16.2: Transient amplification and nonorthogonality. (a) Nonreactive dynamics (matrix $\mathbf{A}_1$) versus reactive dynamics (matrix $\mathbf{A}_2$) in response to the same initial condition $x_1(0) = \cos(2\pi/5)$, $x_2(0) = \sin(2\pi/5)$. In both cases, the solution approaches zero asymptotically. (b) A stable equilibrium $\mathbf{x}^*$ is subject to a perturbation $\delta\mathbf{x}(0)$. The deterministic trajectory $\mathbf{x}(t)$ is initially parallel to the fast eigenvector before eventually relaxing to follow the slow eigenvector. From P to Q the trajectory satisfies $\|\mathbf{x}(t)\| > \|\delta\mathbf{x}(0)\|$.

In particular, it can be shown that the Rayleigh coefficient is maximized by $\mathbf{x}_0 = \mathbf{u}_1$, the eigenvector corresponding to the largest eigenvalue $\lambda_1$ of $H(\mathbf{A})$, such that

$$\text{reactivity} = \lambda_1(H(\mathbf{A})). \qquad (16.1.34)$$

As $H(\mathbf{A})$ is a symmetric matrix, all of its eigenvalues are real and it has an orthogonal set of eigenvectors.

The important point to note is that if $\mathbf{A}$ is non-normal, then $H(\mathbf{A})$ and $\mathbf{A}$ will typically have different eigenvalues, which means that although the principal eigenvalue $\lambda_1(\mathbf{A})$ has negative real part, it is possible that $\lambda_1(H(\mathbf{A}) > 0$. In other words, even though an equilibrium is asymptotically stable it can also be reactive, in the sense that some arbitrarily small perturbations will initially grow in amplitude. As a simple example [665], consider a planar linear system for which $\mathbf{A}$ is either

$$\mathbf{A}_1 = \begin{pmatrix} -1 & 1 \\ 0 & -2 \end{pmatrix} \quad \text{or} \quad \mathbf{A}_2 = \begin{pmatrix} -1 & 10 \\ 0 & -2 \end{pmatrix}.$$

Although both matrices have the same eigenvalues $\lambda = -1, -2$, one finds that $\lambda_1(H(\mathbf{A}_1)) = -0.79$ (nonreactive), whereas $\lambda_1(H(\mathbf{A}_2)) = 3.52$ (reactive). The comparison of the dynamical responses to the same perturbation is illustrated in Fig. 16.2(a). A demonstration of why $\mathbf{A}$ should be non-normal rather than simply not symmetric is presented in Fig. 16.2(b).

**Non-normality and stochastic dynamics.** Following Ref. [81], we now turn to a linear SDE of the form (Chap. 2)

$$dX_i = \sum_{j=1}^{n} A_{ij}X_j dt + \sum_{j=1}^{n} B_{ij}dW_j(t), \tag{16.1.35}$$

where $W_j(t)$ are independent Wiener processes, and all eigenvalues of $\mathbf{A}$ have negative real part. The corresponding FP equation is

$$\frac{\partial p(\mathbf{x},t)}{\partial t} = -\sum_{i,j=1}^{K} A_{ij}\frac{\partial x_j p(\mathbf{x},t)}{\partial x_i} + \frac{1}{2}\sum_{i,j=1}^{K} D_{ij}\frac{\partial^2 p(\mathbf{x},t)}{\partial x_i \partial x_j}, \tag{16.1.36}$$

with $D_{ij} = \sum_{k=1}^{n} B_{ik}B_{jk}$. The stationary density is given by

$$p_s(\mathbf{x}) = \frac{1}{\sqrt{\det(2\pi\Sigma)}} \exp\left(-\frac{1}{2}\mathbf{x}^\top \Sigma^{-1}\mathbf{x}\right), \tag{16.1.37}$$

with the covariance matrix satisfying the Ricatti equation

$$\mathbf{A}\Sigma + \Sigma\mathbf{A}^T + \mathbf{B}\mathbf{B}^T = 0. \tag{16.1.38}$$

In order to explore the effects of non-normality on the stochastic system, we compute the mean-squared norm

$$\langle \|\mathbf{x}\|^2 \rangle = \int p_s(\mathbf{x})\|\mathbf{x}\|^2 d\mathbf{x}.$$

Substituting for $p_s(\mathbf{x})$ and using properties of multidimensional Gaussians leads to the compact result [81] $\langle \|\mathbf{x}\|^2 \rangle = \mathrm{Tr}\Sigma$. For the sake of illustration, suppose that the noise matrix is a multiple of the identity matrix, $\mathbf{B}\mathbf{B}^\top = \sigma^2\mathbf{I}$. Introducing the matrix

$$\mathbf{G} = -\frac{\sigma^2}{2}\Sigma^{-1}\mathbf{A}^{-1},$$

the Ricatti equation can be rewritten as

$$\frac{1}{2}(\mathbf{G}^{-1} + (\mathbf{G}^{-1})^\top) = \mathbf{I}.$$

Note that $\mathbf{G}\mathbf{A}$ is symmetric. The mean-squared norm can now be written as

$$\langle \|\mathbf{x}\|^2 \rangle = -\frac{\sigma^2}{2}\mathrm{Tr}(\mathbf{A}^{-1}\mathbf{G}^{-1}) = -\frac{\sigma^2}{2}\mathscr{H}(\mathbf{A})\mathrm{Tr}(\mathbf{A}^{-1}), \tag{16.1.39}$$

where

$$\mathscr{H}(\mathbf{A}) = \frac{\mathrm{Tr}(\mathbf{A}^{-1}\mathbf{G}^{-1})}{\mathrm{Tr}(\mathbf{A}^{-1})}. \tag{16.1.40}$$

In Ref. [81], the factor $\mathscr{H}(\mathbf{A})$ is defined to be the non-normality index for the stochastic system. This is motivated as follows. If $\mathbf{A}$ is symmetric, $\mathbf{A}^\top = \mathbf{A}$, then $\Sigma^{-1} = -2\mathbf{A}/\sigma^2$ and $\mathbf{G} = \mathbf{I}$. This means that $\mathscr{H}(\mathbf{A}) = 1$ and

$$\langle \|\mathbf{x}\|^2 \rangle = -\frac{\sigma^2}{2}\mathrm{Tr}(\mathbf{A}^{-1}) \le \frac{n\sigma^2}{2|\lambda_1|}, \tag{16.1.41}$$

where $\lambda_1$ is the principle eigenvalue of $\mathbf{A}$ (the negative eigenvalue with the smallest absolute value). The upper bound is consistent with the idea that the size of fluctuations in the large-time limit is dominated by the dynamics along the principal eigenvector. However, for non-normal matrices one can have $\mathscr{H}(\mathbf{A}) > 1$ such that the upper bound is exceeded. This implies that the spectral properties of $\mathbf{A}$ are not sufficient to determine the size of fluctuations, which may be greater than expected (noise amplification).

The non-normality index $\mathscr{H}(\mathbf{A})$ can be calculated explicitly when $n = 2$ (two dimensions). First, using the fact that for a $2 \times 2$ matrix, the trace of the inverse is equal to the trace over the determinant, we have

$$\langle \|\mathbf{x}\|^2 \rangle = -\frac{1}{2}\sigma^2 \frac{\mathrm{Tr}(\mathbf{GA})}{\det(\mathbf{G})\det(\mathbf{A})}.$$

From the definition of $\mathbf{G}$, it follows that $\mathrm{Tr}(\mathbf{GA}) = -\sigma^2 \mathrm{Tr}(\Sigma^{-1})/2$. Also, from the Ricatti equation, $\sigma^2 \mathrm{Tr}(\Sigma^{-1}) = -2\mathrm{Tr}(\mathbf{A})$ and thus $\mathrm{Tr}(\mathbf{GA}) = \mathrm{Tr}(\mathbf{A})$. Combining these various results establishes that for $n = 2$

$$\langle \|\mathbf{x}\|^2 \rangle = -\frac{1}{2}\sigma^2 \det(\mathbf{G}^{-1})\mathrm{Tr}(\mathbf{A}^{-1}), \tag{16.1.42}$$

and thus $\mathscr{H}(\mathbf{A}) = \det(\mathbf{G}^{-1})$. In the two-dimensional case, the Ricatti equation has the explicit solution

$$\Sigma = \frac{[\mathbf{A} - (\mathrm{Tr}\mathbf{A})\mathbf{I}]\mathbf{D}[(\mathrm{Tr}\mathbf{A})\mathbf{I} - \mathbf{A}]^\top - (\mathrm{Det}\mathbf{A})\mathbf{D}}{2(\mathrm{Tr}\mathbf{A})(\mathrm{Det}\mathbf{A})}.$$

One can thus calculate $\mathbf{G}$ explicitly in terms of the elements of $\mathbf{A}$ and thus show that

$$\mathscr{H}(\mathbf{A}) = 1 + \frac{(A_{12} - A_{21})^2}{(A_{11} + A_{22})^2}.$$

Suppose that the eigenvalues $\lambda_1, \lambda_2$ are real, so that that their normalized eigenvectors satisfy $\mathbf{v}_1 \cdot \mathbf{v}_2 = \cos\theta$ for some $\theta$. Using the identities

$$\lambda_1 + \lambda_2 = A_{11} + A_{22}, \quad \lambda_1\lambda_2 = A_{11}A_{22} - A_{12}A_{21},$$

and

$$\cot^2\theta = \frac{\cos^2\theta}{1 - \cos^2\theta} = \frac{(A_{11} - A_{22})^2}{(A_{11} - A_{22})^2 + 4A_{12}A_{21}},$$

one finally finds that [81]

$$\mathscr{H}(\mathbf{A}) = 1 + \cot^2\theta \left(\frac{\lambda_1 - \lambda_2}{\lambda_1 + \lambda_2}\right)^2. \tag{16.1.43}$$

This result illustrates the two key components of transient noise amplification: nonorthogonal eigenvectors ($\theta \neq \pi/2$) and a separation of time scales ($\lambda_1 \neq \lambda_2$). In the case of the deterministic system, suitable initial conditions are also required to achieve transient growth.

**Amplification of noise-induced patterns.** We now explore how the effects of non-normality in non-spatial systems carry over to pattern forming systems. Consider the $m$-species RD equation

$$\frac{\partial u_i}{\partial t} = f_i(\mathbf{u}) + D_i \nabla^2 u_i, \quad i = 1, \ldots, m. \tag{16.1.44}$$

Suppose that $\mathbf{u}^*$ is an asymptotically stable fixed point of the non-spatial model ($\mathbf{D} = 0$). That is, the eigenvalues of the Jacobian $A_{ij} = \partial f_i / \partial u_j|_{\mathbf{u}=\mathbf{u}^*}$ all have negative real part. Linearizing the full RD equation about the homogeneous solution $\mathbf{u}(x,t) = \mathbf{u}^*$ yields the linear equation

$$\frac{\partial v_i}{\partial t} = \sum_{j=1}^{m} A_{ij} v_j + D_i \nabla^2 v_i.$$

Introducing the Fourier transform $\tilde{\mathbf{v}}(\mathbf{k},t) = \int_{-\infty}^{\infty} e^{i\mathbf{k}\cdot\mathbf{x}} \mathbf{v}(\mathbf{x},t) d\mathbf{x}$, we then have

$$\frac{d\tilde{\mathbf{v}}(\mathbf{k},t)}{dt} = \mathbf{J}(k)\tilde{\mathbf{v}}(\mathbf{k},t), \quad \mathbf{J}(k) = \mathbf{A} - k^2 \mathbf{D}. \tag{16.1.45}$$

For each $k$, let $\lambda_1(k)$ be the principal eigenvalue of $\mathbf{J}(k)$. If $\mathrm{Re}(\lambda(k)) < 0$ for all $k$, then $\lim_{t\to\infty} \|\tilde{\mathbf{v}}(\mathbf{k},t)\| = 0$ and hence $\lim_{t\to\infty} \|\mathbf{v}(\mathbf{x},t)\| = 0$. In other words, $\mathbf{u}^*$ is stable with respect to all spatially varying perturbations. On the other hand, if $\mathrm{Re}(\lambda(k)) > 0$ for some $k$, then the fixed point is unstable with respect to perturbations with this wavenumber, which will grow and produce a spatial pattern.

It turns out that the largest eigenvalue of $H(\mathbf{A})$ must be positive for $\mathbf{J}(k)$ to have an eigenvalue with positive real part, that is, reactivity is a prerequisite of Turing instabilities in deterministic RD systems [666]. First, from the analysis of reactivity in non-spatial models, we have

$$\frac{d\|\tilde{\mathbf{v}}\|}{dt} \leq \lambda_1(H(\mathbf{J}))\|\tilde{\mathbf{v}}\|$$

for each $\mathbf{k}$. If $\lambda_1(H(\mathbf{J})) < 0$ or all $k$ then $\lim_{t\to\infty} \|\tilde{\mathbf{v}}(\mathbf{k},t)\| = 0$ for all $\mathbf{k}$ and a Turing instability cannot occur. Moreover,

$$H(\mathbf{J}(k)) = H(\mathbf{A}) - k^2 \mathbf{D}.$$

A classical result from linear algebra is that the largest eigenvalue of a sum of two symmetric matrices is less than or equal to the sum of the largest eigenvalue of each matrix (Weyl's theorem):

$$\lambda_1(H(\mathbf{A} - k^2\mathbf{D}) \leq \lambda_1(H(\mathbf{A})) - k^2 D_{\min}, \tag{16.1.46}$$

where $D_{\min}$ is the smallest diffusion coefficient. Hence, if $\lambda_1(H(\mathbf{A})) < 0$ then $\lambda_1(H(\mathbf{J}(k))) < 0$ for all $k$, and a Turing instability cannot occur.

Next we turn to a stochastic RD equation of the form

$$\frac{\partial u_i}{\partial t} = f_i(\mathbf{u}) + D_i \nabla^2 u_i + \sigma \eta_i(\mathbf{x},t), \quad i = 1,\ldots,m, \tag{16.1.47}$$

with

$$\langle \eta_i(\mathbf{x},t) \rangle = 0, \quad \langle \eta_i(\mathbf{x},t)\eta_j(\mathbf{x}',t') \rangle = \delta_{i,j}\delta(t-t')\delta(\mathbf{x}-\mathbf{x}').$$

(It is convenient here to write the SPDE in terms of Gaussian white noise rather than more rigorously in terms of Wiener processes.) Fourier transforming gives

$$\frac{d\tilde{\mathbf{v}}}{dt} = \mathbf{J}(k)\tilde{\mathbf{v}} + \sigma\tilde{\xi}(\mathbf{k},t). \tag{16.1.48}$$

For fixed $\mathbf{k}$, this equation is identical in form to equation (16.1.35), except that now the vectors are complex-valued. It turns out that the analysis of the effects of non-normality on complex-valued SDEs is almost identical to that of real-valued SDEs, except that one now has twice the number of degrees of freedom [81]. Following Sect. 16.1, suppose the parameters of the deterministic RD equation lie outside the domain for a Turing instability. In order that the system supports stochastic patterns, the principal branch $\lambda_1(k)$ of $\mathbf{J}(k)$ should have a maximum at some value $k = k_0$. That is,

$$\Delta \equiv \mathrm{Re}(\lambda_1(\mathbf{J}(k_0))) - \mathrm{Re}(\lambda_1(\mathbf{A})) > 0, \quad \mathrm{Re}(\lambda_1(\mathbf{J}(k_0))) < 0. \tag{16.1.49}$$

This condition also requires that $\mathbf{A}$ is a non-normal matrix. This follows from adding $k_0^2 D_{\min} - \mathrm{Re}(\lambda_1(\mathbf{A}))$ to both sides of equation (16.1.46):

$$\lambda_1(H(\mathbf{A})) - \mathrm{Re}(\lambda_1(\mathbf{A})) \geq \Delta + k_0^2 D_{\min}. \tag{16.1.50}$$

Note that the left-hand side is a measure of the degree of non-normality, only vanishing when $\mathbf{A}$ is normal. We conclude that, in the case of stochastic patterns, the non-normality of $\mathbf{A}$ may be sufficient to amplify the noise so that the analog of the inequality (16.1.41) with $\sigma \sim \Omega^{-1/2}$ is exceeded. An explicit example can be found in Ref. [81].

## 16.2 Statistical field theory and the renormalization group

In Chap. 8 we used path integrals to analyze escape problems in the weak noise limit, where least action paths corresponded to solutions of a Hamilton–Jacobi equation, consistent with a large deviation principle. In this section, we explore another important application of path integral methods, namely, studying the large-time asymptotics of reaction–diffusion processes using techniques borrowed from statistical field theory. This is a vast subject and we can only touch on some basic ideas here. For detailed surveys, see the books by Zinn-Justin [988], Atland and Simon [37], Kamenev [487], Täuber [861], as well as the review articles [164, 859]. Most biological applications of the statistical field theory of RD processes have been in population biology and ecology; see, for example, [183, 629, 860, 862]. Similar methods have been applied to neural network master equations [110, 146–149] and SPDEs describing collective cell motion [675] or bacterial population growth [341, 425].

### 16.2.1 Fluctuations and universality in RD systems

A canonical model for developing statistical field theory is a reaction–diffusion (RD) system involving the single-species pair annihilation process with reaction rate $\lambda$

$$A + A \xrightarrow{\lambda} \emptyset + \text{diffusion.}$$

The concentration of particles at time $t$ evolves according to the RD equation

$$\frac{\partial u}{\partial t} = D\nabla^2 u - \lambda u^2. \tag{16.2.1}$$

The second term on the right-hand side represents pair annihilation using mass-action kinetics. It is straightforward to show that equation (16.2.1) has the trivial solution $u(x,t) \to 0$ pointwise as $t \to \infty$. The interest in the above system is not, therefore, with regards the steady-state solution (in contrast to the pattern forming systems of Chap. 11), but the approach to steady state. Suppose that we ignore diffusion by taking a uniform initial density and solve the resulting ODE $\dot{u} = -\lambda u^2$ under the initial condition $u(0) = u_0$. This gives

$$u(t) = \frac{1}{1/u_0 + \lambda t}, \tag{16.2.2}$$

which exhibits the asymptotic behavior $u(t) \sim t^{-1}$ that is independent of the initial density (see also Ex. 16.4). On the other hand, Monte Carlo simulations of a spatially discretized version of the model, in which particles randomly hop between neighboring sites on a lattice, reveal more complex behavior that is dependent on the spatial dimension $d$ [757]:

$$u \sim t^{-1/2} \ (d = 1), \quad u \sim t^{-1} \ln t \ (d = 2), \quad u \sim t^{-1} \ (d > 2). \tag{16.2.3}$$

Hence the RD equation only captures the asymptotic dynamics in dimensions higher than two. This is a consequence of the fact that (16.2.1) is a macroscopic equation for the mean particle density $u(x,t)$, which ignores any spatial fluctuations and statistical correlations between particles.

One way to understand the breakdown of the macroscopic RD equation at $d = 2$ is to recall that random walks display certain universal properties. In particular, an unbiased random walk on a $d$-dimensional lattice is recurrent if $d \le 2$ and transient if $d > 2$ (see Chap. 2). In light of this, consider a decay process such as $A + A \to \emptyset$ where at large times the surviving particles are separated by large distances. This means that the probability of a pair of diffusing particles coming into close proximity to annihilate each other is strongly dependent on whether diffusion is recurrent or transient. For $d \le 2$ (recurrent diffusion) a pair of particles find each other with probability 1, even if they are represented by points in a continuum limit. Hence, the effective diffusion-limited reaction rate will be governed by universal features of diffusion. On the other hand, if $d > 2$ (transient diffusion) then the probability of point particles meeting vanishes. That is, for any reaction to occur, the particles must have a finite size or reaction radius, or be placed on a lattice. The effective reaction rate will then depend on the microscopic details of the short-distance spatial regularization, meaning that a degree of universality is lost. The occurrence of universal behavior at and below some dimension $d_c$, known as the upper critical dimension, is typically indicative of a breakdown of macroscopic mean-field equations due to the effects of statistical fluctuations. In the case of binary reactions such as pair annihilation, $d_c = 2$.

As a second example, consider the two species annihilation process

$$A + B \xrightarrow{\lambda} \emptyset + \text{ diffusion.}$$

Again the upper critical dimension is $d_c = 2$. For $d > 2$, the large-time asymptotic behavior can be captured by the RD equations

$$\frac{\partial u}{\partial t} = D \nabla^2 u - \lambda uv, \quad \frac{\partial v}{\partial t} = D \nabla^2 v - \lambda uv, \quad (16.2.4)$$

where $u(\mathbf{x},t)$ and $v(\mathbf{x},t)$ are the concentrations of $A$ and $B$, respectively. Substituting the pair of equations and setting $\chi = u - v$ gives

$$\frac{\partial \chi}{\partial t} = D \nabla^2 \chi,$$

which implies that the difference in the concentrations evolves according to pure diffusion. In particular, we have the conservation equation $\int \chi(\mathbf{r},t)d\mathbf{x} = \chi_0$. The slow diffusive relaxation of $\chi$ means that there is nontrivial behavior above the upper critical dimension. In order to establish this, consider uniform initial densities so that $u$ evolves according to the ODE

$$\frac{du}{dt} = -\lambda u(u - \chi_0).$$

For $\chi_0 > 0$ it is clear that $u \to \chi_0$ exponentially as $t \to \infty$, whereas $v(t) \to 0$. The converse holds if $\chi_0 < 0$. On the other hand, if $\chi_0 = 0$ (equal initial numbers of A and B molecules), then $u(t) \sim v(t) \sim t^{-1}$ as in single-species pair annihilation. However, the system is sensitive to spatial fluctuations in the initial difference density $\chi(\mathbf{x},t)$, which determine the long-time behavior of the process. In particular, the system can segregate into separate $A$ and $B$ rich domains such that $\int (u+v)dx = \int |u-v|dx/2$, which occurs at the relaxation rate $\sim t^{-d/4}$. Hence, for $d < 4$ this is considerably slower than the one predicted by the uniform rate equations and thus dominates for $t \to \infty$. Such behavior is not indicative of a critical upper dimension $d_c = 4$ rather than $d_c = 2$, since it is a property of the mean-field equations.

A powerful method for investigating the effects of fluctuations in reaction–diffusion processes with $d \leq d_c$ is statistical field theory. The basic idea is to spatially discretize the RD system and to treat the diffusive hopping between neighboring lattice sites as additional single-step reactions. The stochastic dynamics can then be formulated in terms of an RD master equation along the lines of Sect. 16.1. The advantage of such a formulation is that one can construct a Doi–Peliti path integral representation of the RD master equation that, on taking the continuum limit, yields a statistical field theoretic description of the RD process. This then allows well-established field theoretic methods to be used to investigate the role of fluctuations and to compute universal quantities [37, 164, 487, 859, 861, 988]. First, a systematic perturbation expansion of the associated moment generating functional may be performed using diagrammatic methods (Feynman diagrams) and causal Green's functions. However, this results in integral expressions that diverge at small (ultraviolet) or large (infrared) spatial and temporal scales. The former occurs above the upper critical dimension $d_c$ and relates to the transient nature of diffusion in higher dimensions. These divergences can at least be partially handled using renormalization theory. This combines dimensional analysis with scaling theory to extend perturbation results that hold at intermediate time and length scales to regimes where perturbation theory breaks down. In the following, we develop these ideas using the canonical example of pair annihilation.

## *16.2.2 Path integral formulation of the RD master equation*

Consider a $d$-dimensional square lattice in which each lattice site $\ell \in \mathbf{Z}^d$ is occupied by $n_\ell$ molecules of species A. Let $\Omega = h^d$ denote the hyper-volume of each fundamental domain with $h$ the lattice spacing. The spatially discretized version of equation (16.2.1), after setting $u_\ell = n_\ell/\Omega$ and treating $n_\ell$ as a continuous variable, is

$$\frac{dn_\ell}{dt} = \alpha \sum_{\ell' \in \partial \ell} [n_{\ell'} - n_\ell] + \lambda n_\ell(n_\ell - 1), \quad \alpha = \frac{D}{h^2}. \tag{16.2.5}$$

We have rescaled the reaction rate according to $\lambda \to \lambda/h^d$. If one views equation (16.2.5) as the rate equation of a Markov process with discrete variables $\mathbf{n}(t) = \sum_\ell n_\ell(t)\mathbf{e}_\ell$, then the associated master equation is

$$\frac{dP(\mathbf{n},t)}{dt} = \alpha \sum_{\langle \ell,\ell' \rangle} [(n_\ell+1)P(\mathbf{n}+\mathbf{e}_\ell-\mathbf{e}_{\ell'},t) - n_\ell P(\mathbf{n},t)]$$

$$+ \alpha \sum_{\langle \ell,\ell' \rangle} [(n_{\ell'}+1)P(\mathbf{n}-\mathbf{e}_\ell+\mathbf{e}_{\ell'},t) - n_{\ell'} P(\mathbf{n},t)]$$

$$+ \lambda \sum_{\ell} [(n_\ell+2)(n_\ell+1)P(\mathbf{n}+2\mathbf{e}_\ell,t) - n(n-1)P(\mathbf{n},t)], \quad (16.2.6)$$

where $\langle \ell,\ell' \rangle$ indicates that we are summing over nearest neighbors on the lattice without double counting. Note that $\mathbf{n}+\mathbf{e}_\ell-\mathbf{e}_{\ell'} = (\ldots,n_\ell+1,n_{\ell'}-1,\ldots)$, etc. The first and second lines represent single-step hopping transitions $\ell \to \ell'$ and $\ell' \to \ell$, respectively, while the third line represents pair annihilation $A+A \to \emptyset$ at each lattice site.

The next step is to convert the master equation to operator form by introducing Doi–Peliti annihilation and creation operators at each lattice site. Define the state vector by

$$|\psi(t)\rangle = \sum_{\mathbf{n}} P(\mathbf{n},t)|\mathbf{n}\rangle = \sum_{\mathbf{n}} P(\mathbf{n},t) \prod_{\ell} (a_\ell^\dagger)^{n_\ell} |0\rangle, \quad (16.2.7)$$

where $\sum_{\mathbf{n}} = \prod_{\ell \in \mathbf{Z}^d} \sum_{n_\ell \geq 0}$. Differentiating with respect to time and plugging in the master equation gives

$$\frac{d}{dt}|\psi(t)\rangle = \alpha \sum_{\mathbf{n}} \sum_{\langle \ell,\ell' \rangle} [(n_\ell+1)P(\mathbf{n}+\mathbf{e}_\ell-\mathbf{e}_{\ell'},t) - n_\ell P(\mathbf{n},t)] \prod_{\ell} (a_\ell^\dagger)^{n_\ell} |0\rangle$$

$$+ \alpha \sum_{\mathbf{n}} \sum_{\langle \ell,\ell' \rangle} [(n_{\ell'}+1)P(\mathbf{n}-\mathbf{e}_\ell+\mathbf{e}_{\ell'},t) - n_\ell P(\mathbf{n},t)] \prod_{\ell} (a_\ell^\dagger)^{n_\ell} |0\rangle$$

$$+ \lambda \sum_{\mathbf{n}} \sum_{\ell} [(n_\ell+2)(n_\ell+1)P(\mathbf{n}+2\mathbf{e}_\ell,t) - n_\ell(n_\ell-1)P(\mathbf{n},t)] \prod_{\ell} (a_\ell^\dagger)^{n_\ell} |0\rangle.$$

Focusing on the hopping transition $\ell \to \ell'$, we have

$$\sum_{n_\ell,n_{\ell'}} (n_\ell+1)P(\mathbf{n}+\mathbf{e}_\ell-\mathbf{e}_{\ell'},t)(a_\ell^\dagger)^{n_\ell}(a_{\ell'}^\dagger)^{n_{\ell'}}|0\rangle$$

$$= \sum_{n_\ell,n_{\ell'}} P(\mathbf{n}+\mathbf{e}_\ell-\mathbf{e}_{\ell'},t) a_{\ell'}^\dagger a_\ell (a_\ell^\dagger)^{n_\ell+1}(a_{\ell'}^\dagger)^{n_{\ell'}-1}|0\rangle$$

$$= \sum_{n_\ell,n_{\ell'}} P(\mathbf{n},t) a_{\ell'}^\dagger a_\ell (a_\ell^\dagger)^{n_\ell}(a_{\ell'}^\dagger)^{n_{\ell'}}|0\rangle.$$

The second step uses the identity $a|n+1\rangle = (n+1)|n\rangle$ and the third step relabels the integers. Similarly,

$$\sum_{n_\ell,n_{\ell'}} n_\ell P(\mathbf{n},t)(a_\ell^\dagger)^{n_\ell}(a_{\ell'}^\dagger)^{n_{\ell'}}|0\rangle = \sum_{n_\ell,n_{\ell'}} P(\mathbf{n},t) a_\ell^\dagger a_\ell (a_\ell^\dagger)^{n_\ell}(a_{\ell'}^\dagger)^{n_{\ell'}}|0\rangle.$$

The corresponding terms associated with the hopping transition $\ell' \to \ell$ are obtained by simply interchanging $\ell$ and $\ell'$. Hence the net contribution to the master equation from hopping reactions can be written as

$$-\alpha \sum_{\mathbf{n}} P(\mathbf{n},t) \sum_{\langle \ell,\ell' \rangle} (a_{\ell'}^{\dagger} - a_{\ell}^{\dagger})(a_{\ell} - a_{\ell'})|n\rangle.$$

Finally, the pair annihilation terms for the lattice site $\ell$ can be rewritten as

$$\sum_{n_\ell} [(n_\ell+2)(n_\ell+1)P(\mathbf{n}+2\mathbf{e}_\ell,t) - n_\ell(n_\ell-1)P(\mathbf{n},t)] \prod_{\ell} (a_\ell^{\dagger})^{n_\ell}|0\rangle$$

$$= \sum_{n_\ell} \left[ P(\mathbf{n}+2\mathbf{e}_\ell,t)a_\ell^2(a_\ell^{\dagger})^{n_\ell+2} - P(\mathbf{n},t)a_\ell^{\dagger}a_\ell(a_\ell^{\dagger}a_\ell-1)(a_\ell^{\dagger})^{n_\ell} \right]|0\rangle$$

$$= \sum_{n_\ell} P(\mathbf{n},t) \left[ a_\ell^2 - a_\ell^{\dagger}a_\ell(a_\ell^{\dagger}a_\ell-1) \right] (a_\ell^{\dagger})^{n_\ell}|0\rangle.$$

Combining these various results leads to the operator version of the master equation:

$$\frac{d}{dt}|\psi(t)\rangle = \widehat{H}|\psi(t)\rangle, \tag{16.2.8}$$

where

$$\widehat{H} = -\alpha \sum_{\langle \ell,\ell' \rangle} (a_{\ell'}^{\dagger} - a_{\ell}^{\dagger})(a_{\ell'} - a_{\ell}) + \lambda \sum_{\ell} \left[ a_\ell^2 - a_\ell^{\dagger}a_\ell(a_\ell^{\dagger}a_\ell-1) \right]. \tag{16.2.9}$$

Now recall from Sect. 8.2 that in order to construct the path integral representation of the master equation, it is necessary to normal-order the operator $\widehat{H}$ by moving all creation operators to the left of all annihilation operators. Hence, the normal-ordered version of the operator is

$$\widehat{H} = -\alpha \sum_{\langle \ell,\ell' \rangle} (a_{\ell'}^{\dagger} - a_{\ell}^{\dagger})(a_{\ell'} - a_{\ell}) + \lambda \sum_{\ell} [1 - (a_{\ell}^{\dagger})^2]a_{\ell}^2. \tag{16.2.10}$$

Formally speaking, the solution to the operator version of the master equation can be written as

$$|\psi(t)\rangle = e^{\widehat{H}t}|\psi(0)\rangle, \tag{16.2.11}$$

and the expectation value of some physical quantity such as the number $N(t)$ expressed as

$$\langle N_\ell(t)\rangle = \langle 0| \left\{ \prod_{\ell} \exp(a_\ell) \right\} a_\ell^{\dagger}a_\ell e^{\widehat{H}t}|\psi(0)\rangle, \tag{16.2.12}$$

where

$$\langle \emptyset| = \langle 0| \prod_{\ell} \exp(a_\ell) \tag{16.2.13}$$

is the projection state. In the following, it will be convenient to shift the term $e^{\sum_\ell a_\ell}$ to the right by using the identity

$$e^a f(a^{\dagger}) = f(a^{\dagger}+1)e^a,$$

for an arbitrary function $f$. In particular,

$$\langle N_\ell(t)\rangle = \langle 0|a_\ell^\dagger a_\ell e^{\widehat{H}'t} e^{\sum_\ell a_\ell}|\psi(0)\rangle, \tag{16.2.14}$$

with[2]

$$\widehat{H}' = -\alpha \sum_{\langle \ell,\ell'\rangle} (a_{\ell'}^\dagger - a_\ell^\dagger)(a_{\ell'} - a_\ell) - \lambda \sum_\ell [(a_\ell^\dagger)^2 + 2a_\ell^\dagger]a_\ell^2. \tag{16.2.15}$$

It is common to take the initial state to be a product of Poisson distributions,

$$P(\mathbf{n},0) = \prod_\ell \frac{r_0^{n_\ell} e^{-r_0}}{n_\ell!},$$

where $\langle n_\ell(0)\rangle = r_0$. Then, for each $\ell$,

$$\sum_{n_\ell} \frac{r_0^{n_\ell} e^{-r_0}}{n_\ell!}|n_\ell\rangle = \sum_{n_\ell} \frac{[r_0 a_\ell^\dagger]^{n_\ell} e^{-r_0}}{n_\ell!}|0\rangle = e^{-r_0} e^{r_0 a_\ell^\dagger}|0\rangle,$$

which is a coherent state. It follows that

$$e^{\sum_\ell a_\ell}|\psi(0)\rangle = e^{r_0 \sum_\ell a_\ell^\dagger}|0\rangle. \tag{16.2.16}$$

Repeating the steps in the derivation of the path integral for a birth–death process, see Sect. 8.2, we obtain the following path integral expression for the mean number of particles at lattice site $\ell_0$

$$\langle N_{\ell_0}(t)\rangle = \left(\int \prod_\ell \mathcal{D}\varphi_\ell \mathcal{D}\varphi_\ell^*\right) \varphi_{\ell_0}(t) e^{-S[\varphi,\varphi^*]}, \tag{16.2.17}$$

where $S$ is the action

$$S[\varphi,\varphi^*] = \int_0^t d\tau \left( \sum_\ell \varphi_\ell^* \partial_\tau \varphi_\ell + \alpha \sum_{\langle \ell,\ell'\rangle} (\varphi_{\ell'}^* - \varphi_\ell^*)(\varphi_{\ell'} - \varphi_\ell) \right.$$
$$\left. + \lambda \sum_\ell [(\varphi_\ell^*)^2 + 2\varphi_\ell^*]\varphi_\ell^2 \right) - r_0 \sum_\ell \varphi_\ell^*(0). \tag{16.2.18}$$

The final step in the construction of a statistical field description of pair annihilation with diffusion is to take the continuum limit of the above path integral and action by letting the lattice spacing $h \to 0$ such that $\sum_\ell h^d \to \int d\mathbf{x}$. This requires using the unscaled rate $\lambda$, setting $\alpha = D/h^2$ and redefining the fields as follows:

$$h^{-d}\varphi_\ell(t) \to \phi(\mathbf{x},t), \quad \varphi_\ell^*(t) \to \widetilde{\phi}(\mathbf{x},t). \tag{16.2.19}$$

We thus arrive at the field theoretic path integral

---

[2] The so-called Doi-shift $a^\dagger \to a^\dagger + 1$ was not needed in the weak noise limit of Sect. 8.2, since we rescaled the occupation numbers by the system size $N$, so that the shift $1/N$ becomes negligible.

$$\langle \phi(\mathbf{y},t) \rangle = \mathcal{N} \int \mathcal{D}[\phi(\mathbf{x},t)] \mathcal{D}[\widetilde{\phi}(\mathbf{x},t)] \, \phi(\mathbf{y},t) e^{-S[\phi,\widetilde{\phi}]}, \qquad (16.2.20)$$

with

$$S[\phi,\widetilde{\phi}] = \int_{\mathbb{R}^d} d\mathbf{x} \int_0^t d\tau \left( \widetilde{\phi}(\partial_\tau - D\nabla^2)\phi + \lambda [\widetilde{\phi}^2 + 2\widetilde{\phi}]\phi^2 - \rho_0 \widetilde{\phi} \delta(\tau) \right). \quad (16.2.21)$$

We have used

$$\frac{1}{h^2} \sum_{\ell' \in \partial \ell} (\varphi_{\ell'}^* - \varphi_\ell^*)(\varphi_{\ell'} - \varphi_\ell) \rightarrow \nabla \widetilde{\phi} \cdot \nabla \phi,$$

and performed an integration by parts. The factor $\mathcal{N}$ ensures the normalization $\langle \emptyset | \mathbf{n} \rangle = \sum_{\mathbf{n}} P(\mathbf{n},t) = 1$. Hence,

$$\mathcal{N} = \int \mathcal{D}[\phi(\mathbf{x},t)] \mathcal{D}[\widetilde{\phi}(\mathbf{x},t)] e^{-S[\phi,\widetilde{\phi}]}. \qquad (16.2.22)$$

(In Ex. 16.5 the corresponding action for two-species annihilation is derived.)
**Equivalent Langevin equation.** Formally speaking, the action (16.2.21) is identical to the one that would be obtained by considering an Ito SPDE of the form

$$\frac{\partial \phi}{\partial t} = D\nabla^2 \phi(\mathbf{x},t) - 2\lambda \phi(\mathbf{x},t)^2 + \xi(\mathbf{x},t), \qquad (16.2.23)$$

with

$$\langle \xi(\mathbf{x},t) \rangle = 0, \quad \langle \xi(\mathbf{x},t)\xi(\mathbf{x}',t') \rangle = -2\lambda \phi^2 \delta(t-t')\delta(\mathbf{x}-\mathbf{x}'). \qquad (16.2.24)$$

This follows from spatially discretizing the SPDE to obtain an SDE for the variables $\phi_\ell(t)$, and then constructing the corresponding path integral along the lines of Sect. 8.1 with $\widetilde{\phi}_\ell(t)$ identified as a "momentum" variable. The latter contributes a term to the action of the form $-\lambda \widetilde{\phi}_\ell^2 \phi^2$ due to the multiplicative noise. (The negative sign in the noise correlations means that they are complex-valued, which limits the phenomenological interpretation of the Langevin equation.) Taking expectations of the SPDE gives

$$\frac{\partial \langle \phi \rangle}{\partial t} = D\nabla^2 \langle \phi \rangle - 2\lambda \langle \phi \rangle^2. \qquad (16.2.25)$$

The mean-field approximation $\langle \phi^2 \rangle = \langle \phi \rangle^2$ then recovers the deterministic PDE (16.2.1) with $\langle \phi \rangle = u$. Also note that the SPDE has the formal solution

$$\phi(\mathbf{x},t) = \int d\mathbf{x}' \, dt' \, G_0(\mathbf{x}-\mathbf{x}',t-t')[-\lambda \phi(\mathbf{x}',t')^2 + \xi(\mathbf{x}',t') + \rho_0 \delta(t')] \quad (16.2.26)$$

under the initial condition $\phi(\mathbf{x},0) = \rho_0$, where $G_0$ is Green's function

$$\left[\frac{\partial}{\partial \tau} - D\nabla^2\right] G_0(\mathbf{x} - \mathbf{x}', \tau - \tau') = \delta(\mathbf{x} - \mathbf{x}')\delta(\tau - \tau'). \tag{16.2.27}$$

Equation (16.2.26) could then be used to develop a perturbation expansion of $\phi$ in powers of $\lambda$, from which corresponding perturbation expansions of $\langle \phi \rangle$ (and higher-order moments) could be obtained [164]. However, the simple exact mapping between the RD master equation and the Langevin equation is specific to binary reactions. Therefore, we will calculate $\langle \phi \rangle$ using the path integral (16.2.20).

### 16.2.3 Diagrammatic expansion of the path integral

It turns out to be more convenient to work with the generating functional

$$Z[J,\tilde{J}] = \int \mathcal{D}[\phi(\mathbf{x},t)]\, \mathcal{D}[\tilde{\phi}(\mathbf{x},t)]\, \mathrm{e}^{-S[\phi,\tilde{\phi}] - \int [J\tilde{\phi} + \tilde{J}\phi]\,d\tau\,d\mathbf{x}}. \tag{16.2.28}$$

We can then determine $\langle \phi(\mathbf{y},t) \rangle$ according to the functional derivative (Box 12C)

$$\langle \phi(\mathbf{y},t) \rangle = -\left.\frac{\delta \ln Z[\phi,\tilde{\phi}]}{\delta \tilde{J}(\mathbf{y},t)}\right|_{J=\tilde{J}=0}. \tag{16.2.29}$$

The diagrammatic method for evaluating $Z$ is based on the observation that the functional $Z_0 = Z_{\lambda=0}$ can be calculated exactly using functional Gaussian integrals. One can then determine $Z$ by performing a perturbation expansion around the free diffusion case. Setting $\lambda = 0$ in the action of equation (16.2.28) gives

$$Z_0[J,\tilde{J}] = \int \mathcal{D}[\phi(\mathbf{x},t)]\, \mathcal{D}[\tilde{\phi}(\mathbf{x},t)]\, \mathrm{e}^{-S_0[\phi,\tilde{\phi}] - \int [J\tilde{\phi} + \tilde{J}\phi]\,d\tau\,d\mathbf{x}}, \tag{16.2.30}$$

with

$$S_0[\phi,\tilde{\phi}] = \int_{\mathbb{R}^d} d\mathbf{x} \int_0^t d\tau\, \tilde{\phi}\, \mathbb{L}_0 \phi + \rho_0 \int_{\mathbb{R}^d} d\mathbf{x}\, \tilde{\phi}(\mathbf{x},0), \quad \mathbb{L}_0 = \frac{\partial}{\partial \tau} - D\nabla^2. \tag{16.2.31}$$

Identifying the inverse operator $\mathbb{L}_0^{-1}$ with Green's function $G_0(\mathbf{x},t|\mathbf{x}',t')$ of equation (16.2.27), we can formally evaluate the (complex) functional Gaussian integral to find

$$Z_0[J,\tilde{J}] = Z_0[0,0]\exp\left(\int d\mathbf{x}\,d\mathbf{x}' \int d\tau\,d\tau'\, \tilde{J}(\mathbf{x},\tau)G_0(\mathbf{x}-\mathbf{x}',\tau-\tau')J(\mathbf{x}',\tau')\right)$$

$$\times \exp\left(\rho_0 \int d\mathbf{x}\,d\mathbf{x}' \int d\tau\, \tilde{J}(\mathbf{x},\tau)G_0(\mathbf{x}-\mathbf{x}',\tau)\right). \tag{16.2.32}$$

Green's function $G_0$, also known as the free diffusion propagator, is most easily evaluated in Fourier space. That is, Fourier transforming equation (16.2.27) shows that

$$G_0(\mathbf{x},t) = \int \frac{d\mathbf{k}}{(2\pi)^d} \int \frac{d\omega}{2\pi} e^{-i\omega t - i\mathbf{k}\cdot\mathbf{x}} \widehat{G}_0(\mathbf{k},\omega), \quad \widetilde{G}_0(\mathbf{k},\omega) = \frac{1}{-i\omega + Dk^2}.$$
(16.2.33)

Inverting the temporal Fourier transform, we obtain the causal Green's function for diffusion,

$$\widehat{G}_0(\mathbf{k},t) = \Theta(t)e^{-Dk^2 t},$$
(16.2.34)

where $\Theta$ is the Heaviside step function with $\Theta(0) = 0$.

Returning to the full generating functional (16.2.28), we separate out the diffusive and non-diffusive parts:

$$Z[J,\tilde{J}] = \int \mathcal{D}[\phi(\mathbf{x},t)]\,\mathcal{D}[\tilde{\phi}(\mathbf{x},t)]\,e^{-S_0[\phi,\tilde{\phi}] - \int [J\tilde{\phi} + \tilde{J}\phi]d\tau dx}\,e^{-\lambda S_I[\phi,\tilde{\phi}]},$$
(16.2.35)

where

$$S_I[\phi,\tilde{\phi}] = \int dx d\tau [\tilde{\phi}^2 + 2\tilde{\phi}]\phi^2.$$
(16.2.36)

Hence, $Z$ can be rewritten as

$$Z[J,\tilde{J}] = \exp\left(-\lambda S_I[\delta/\delta\tilde{J},\delta/\delta J]\right) Z_0[J,\tilde{J}].$$
(16.2.37)

In order to interpret the expression on the right-hand side, imagine Taylor expanding the exponential using $e^{-\lambda S_I} = \sum_{n=0}^{\infty}(-\lambda S_I)^n/n!$. Each term in the expansion involves products of functional derivatives acting on the Gaussian functional $Z_0$, which leads to products of free diffusion Green's functions $G_0$ after setting $J = \tilde{J} = 0$. Such an expansion can be carried out systematically using Feynman diagrams, originally introduced within the context of quantum theory [307]. We will develop this diagrammatic expansion to $O(\lambda^2)$ for the path integral representation of $\langle\phi(\mathbf{y},t)\rangle$ [859]. The latter takes the form

$$\langle\phi(\mathbf{y},t)\rangle = -\frac{1}{Z[0,0]}\frac{\delta Z}{\delta\tilde{J}(\mathbf{y},t)}\bigg|_{J=\tilde{J}=0}$$

$$= -\frac{1}{Z[0,0]}\exp\left(-\lambda S_I[\delta/\delta\tilde{J},\delta/\delta J]\right)\frac{\delta Z_0}{\delta\tilde{J}(\mathbf{y},t)}\bigg|_{J=\tilde{J}=0}.$$
(16.2.38)

Expanding the exponential operator in (16.2.38) to $O(\lambda^2)$ gives

$$\langle\phi(\mathbf{y},t)\rangle = \rho_0\widehat{G}_0(0,t) - \frac{\lambda}{Z_0[0,0]}\int dx_1 d\tau_1 \left[\frac{\delta^4}{\delta J_1^2 \delta\tilde{J}_1^2} + 2\frac{\delta^3}{\delta J_1 \delta\tilde{J}_1^2}\right]\frac{\delta Z_0}{\delta\tilde{J}(\mathbf{y},t)}\bigg|_{J=\tilde{J}=0}$$

$$+ \frac{\lambda^2}{2Z_0[0,0]}\int dx_1 d\tau_1 \int dx_2 d\tau_2 \left[\frac{\delta^4}{\delta J_1^2 \delta\tilde{J}_1^2} + 2\frac{\delta^3}{\delta J_1 \delta\tilde{J}_1^2}\right]$$

$$\times \left[ \frac{\delta^4}{\delta J_2^2 \delta \tilde{J}_2^2} + 2 \frac{\delta^3}{\delta J_2 \delta \tilde{J}_2^2} \right] \frac{\delta Z_0}{\delta \tilde{J}(\mathbf{y},t)} \bigg|_{J=\tilde{J}=0} + O(\lambda^3), \qquad (16.2.39)$$

where

$$J_j = J(\mathbf{x}_j, \tau_j), \quad \tilde{J}_j = \tilde{J}(\mathbf{x}_j, \tau_j).$$

*Calculation of terms to $O(\lambda^2)$.* In order to evaluate the various functional derivatives of $Z_0$ at $J = \tilde{J} = 0$ we will make multiple use of the identities

$$\int d\mathbf{x} G_0(\mathbf{y} - \mathbf{x}, t) = \widehat{G}_0(0, t) = \Theta(t), \quad G_0(0, 0) = 0.$$

The second condition reflects the causal nature of the Green's function, namely, the presence of the Heaviside function $\Theta(t)$ in equation (16.2.34). The following results then hold:

$$\frac{\delta Z_0}{\delta \tilde{J}_1} = Z_0 \int d\mathbf{x} d\tau G_0(\mathbf{x}_1 - \mathbf{x}, \tau_1 - \tau) J(\mathbf{x}, \tau) + \rho_0 \widehat{G}_0(0, \tau_1) Z_0$$
$$\rightarrow \rho_0 \widehat{G}_0(0, \tau_1) Z_0(0, 0) \text{ as } J, \tilde{J} \rightarrow 0,$$

$$\frac{\delta Z_0}{\delta J_1} = Z_0 \int d\mathbf{x} d\tau \tilde{J}(\mathbf{x}, \tau) G_0(\mathbf{x} - \mathbf{x}_1, \tau - \tau_1) \rightarrow 0 \text{ as } J, \tilde{J} \rightarrow 0$$

$$\frac{\delta^2 Z_0}{\delta J_2 \delta \tilde{J}_1} = \frac{\delta Z_0}{\delta J_2} \left[ \int d\mathbf{x} d\tau G_0(\mathbf{x}_1 - \mathbf{x}, \tau_1 - \tau) J(\mathbf{x}, \tau) + \rho_0 \widehat{G}_0(0, \tau_1) \right] + Z_0 G_0(\mathbf{x}_1 - \mathbf{x}_2, \tau_1 - \tau_2)$$
$$\rightarrow G_0(\mathbf{x}_1 - \mathbf{x}_2, \tau_1 - \tau_2) Z_0(0, 0) \text{ as } J, \tilde{J} \rightarrow 0,$$

$$\frac{\delta^2 Z_0}{\delta \tilde{J}_1 \delta \tilde{J}_2} = \frac{\delta Z_0}{\delta \tilde{J}_2} \left[ \int d\mathbf{x} d\tau G_0(\mathbf{x}_1 - \mathbf{x}, \tau_1 - \tau) J(\mathbf{x}, \tau) + \rho_0 \widehat{G}_0(0, \tau_1) \right]$$
$$\rightarrow \rho_0^2 \widehat{G}_0(0, \tau_1) \widehat{G}_0(0, \tau_2) Z_0(0, 0) \text{ as } J, \tilde{J} \rightarrow 0$$

$$\frac{\delta^3 Z_0}{\delta J_2 \delta \tilde{J}_1^2} = \frac{\delta^2 Z_0}{\delta J_2 \delta \tilde{J}_1} \left[ \int d\mathbf{x} d\tau G_0(\mathbf{x}_1 - \mathbf{x}, \tau_1 - \tau) J(\mathbf{x}, \tau) + \rho_0 \widehat{G}_0(0, \tau_1) \right]$$
$$+ \frac{\delta Z_0}{\delta \tilde{J}_1} G_0(\mathbf{x}_1 - \mathbf{x}_2, \tau_1 - \tau_2)$$
$$\rightarrow 2\rho_0 \widehat{G}_0(0, \tau_1) G_0(\mathbf{x}_1 - \mathbf{x}_2, \tau_1 - \tau_2) Z_0(0, 0) \text{ as } J, \tilde{J} \rightarrow 0$$

and

$$\frac{\delta^3 Z_0}{\delta J_2^2 \delta \tilde{J}_1} = \frac{\delta^2 Z_0}{\delta J_2^2} \left[ \int d\mathbf{x} d\tau G_0(\mathbf{x}_1 - \mathbf{x}, \tau_1 - \tau) J(\mathbf{x}, \tau) + \rho_0 \widehat{G}_0(0, \tau_1) \right]$$
$$+ \frac{\delta Z_0}{\delta J_2} G_0(\mathbf{x}_1 - \mathbf{x}_2, \tau_1 - \tau_2) \rightarrow 0 \text{ as } J, \tilde{J} \rightarrow 0.$$

Looking at these and higher-order functional derivatives leads to the following set of rules:

1. Each quartic differential $\delta^4 / \delta J_i^2 \delta \tilde{J}_i^2$ (four vertex) produces a factor $-\lambda \int d\mathbf{x}_i d\tau_i$.
2. Each third-order differential $\delta^3 / \delta J_i \delta \tilde{J}_i^2$ (three vertex) produces a factor $-2\lambda \int d\mathbf{x}_i d\tau_i$.

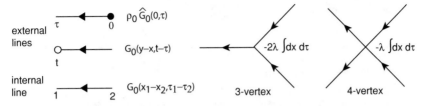

Fig. 16.3: Propagators and vertices used to construct Feynman diagrams.

3. Consider all possible pairings (contractions) $\delta^2/\delta J_i \delta \tilde{J}_j$ with $i \neq j$, each of which contributes a factor $G_0(\mathbf{x}_j - \mathbf{x}_i, \tau_j - \tau_i)$ that propagates forward in time from $(\mathbf{x}_i, \tau_i)$ to $(\mathbf{x}_j, \tau_j)$.

4. One of these propagators should connect to the final point $(\mathbf{y}, t)$.

5. Products $G_0(\mathbf{x}_j - \mathbf{x}_i, \tau_j - \tau_i) G_0(\mathbf{x}_i - \mathbf{x}_j, \tau_i - \tau_j) = 0$.

6. Any unmatched derivative $\delta/\delta \tilde{J}_k$ contributes a factor $\rho_0 \widehat{G}(0, \tau_k)$.

7. Any unmatched derivative $\delta/\delta J_k$ contributes a factor of zero.

These rules can be summarized diagrammatically using Feynman diagrams. These are constructed from the basic elements shown in Fig. 16.3, which are patched together using the above contraction rules. The diagrams contributing to $\langle \phi(\mathbf{y}, t) \rangle$ up to $O(\lambda^2)$ are shown in Fig. 16.4 with time flowing from right to left. From these diagrams we deduce that

$$\langle \phi(\mathbf{y}, t) \rangle = \rho_0 \widehat{G}_0(0, t) - 2\lambda \int d\mathbf{x}_1 d\tau_1 [\rho_0 \widehat{G}_0(0, \tau_1)]^2 G_0(\mathbf{y} - \mathbf{x}_1, t - \tau_1)$$
$$+ \frac{1}{2} \left[ -2\lambda \int d\mathbf{x}_1 d\tau_1 \right] \left[ -2\lambda \int d\mathbf{x}_2 d\tau_2 \right] [\rho_0 \widehat{G}_0(0, \tau_2)]^2 [\rho_0 \widehat{G}_0(0, \tau_1)][4G_0(\mathbf{x}_1 - \mathbf{x}_2, \tau_1 - \tau_2)]$$
$$+ \frac{1}{2} \left[ -2\lambda \int d\mathbf{x}_1 d\tau_1 \right] \left[ -\lambda \int d\mathbf{x}_2 d\tau_2 \right] [\rho_0 \widehat{G}_0(0, \tau_2)]^2$$
$$\times [4G_0(\mathbf{x}_1 - \mathbf{x}_2, \tau_1 - \tau_2)^2 G_0(\mathbf{y} - \mathbf{x}_1, t - \tau_1)] + O(\lambda^3). \tag{16.2.40}$$

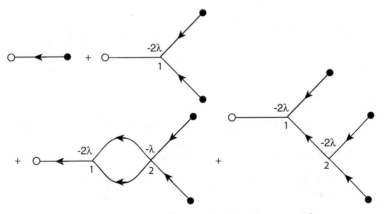

Fig. 16.4: Diagrams contributing to $\langle \phi(\mathbf{y}, t) \rangle$ up to $O(\lambda^2)$.

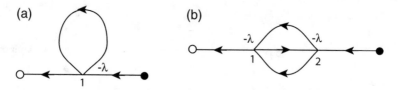

Fig. 16.5: Examples of diagrams not contributing to $\langle\phi(\mathbf{y},t)\rangle$.

Note the additional combinatorial factors arising from performing the contractions. Suppose that we only keep the so-called tree diagrams, that is, diagrams without any closed loops. Evaluating the first two lines of equation (16.2.40) and generalizing to higher-order tree diagrams leads to the geometric series

$$\langle\phi(\mathbf{y},t)\rangle_{\text{tree}} = \rho_0 - 2\lambda\rho_0^2 t + 4\lambda^2\rho_0^3 t^2 + \ldots = \frac{\rho_0}{1+2\lambda\rho_0 t}. \tag{16.2.41}$$

We thus recover the spatially uniform solution of mean field theory. The effects of fluctuations are thus determined by the loop diagrams. Before proceeding to calculate the one-loop contribution given by the third line of equation (16.2.40), we highlight some of the diagrams that do not contribute to $\langle\phi(\mathbf{y},t)\rangle$, which are shown in Fig. 16.5. The first diagram is zero because it involves a single time and $G_0(0,0) = 0$, whereas the second vanishes because it involves a line propagating backwards in time.

The simplest method for evaluating the expression on the third line of equation (16.2.40), which we denote by $I_{\text{loop}}$, is to use Fourier transforms. That is,

$$I_{\text{loop}} = 2\lambda^2\rho_0^2 \int_0^t d\tau_1 \int_0^{\tau_1} d\tau_2 \widehat{G}_0(0,t-\tau_1) I(\tau_1-\tau_2), \tag{16.2.42}$$

where

$$I(\tau) = 2\int \frac{d\mathbf{k}}{(2\pi)^d} \int d\mathbf{x}_1 d\mathbf{x}_2 e^{i\mathbf{k}\cdot\mathbf{x}_1} G_0(\mathbf{x}_1-\mathbf{x}_2,\tau)^2 \tag{16.2.43}$$

$$= \int \frac{d\mathbf{k}}{(2\pi)^d} \int d\mathbf{x}_1 d\mathbf{x}_2 e^{i\mathbf{k}\cdot\mathbf{x}_1} \left\{ \int \frac{d\mathbf{k}'}{(2\pi)^d} e^{-i\mathbf{k}'\cdot(\mathbf{x}_1-\mathbf{x}_2)} e^{-Dk'^2\tau} \int \frac{d\mathbf{k}''}{(2\pi)^d} e^{-i\mathbf{k}''\cdot(\mathbf{x}_1-\mathbf{x}_2)} e^{-Dk''^2\tau} \right\}.$$

($I(t)$ represents the so-called truncated one-loop diagram in which the incoming and outgoing propagators have been cut.) Reversing the order of integration and using

$$\int d\mathbf{k} e^{-i\mathbf{k}\cdot(\mathbf{x}-\mathbf{x}')} = (2\pi)^d \delta(\mathbf{x}-\mathbf{x}'),$$

we have

$$I(\tau) = \int \frac{d\mathbf{k}'}{(2\pi)^d} e^{-2Dk'^2\tau} = 2(8\pi D\tau)^{-d/2} \tag{16.2.44}$$

Substituting back into equation (16.2.42) shows that

$$I_{\text{loop}} = \frac{4\lambda^2\rho_0^2}{(8\pi D)^{d/2}} \int_0^t d\tau_1 \int_0^{\tau_1} \frac{d\tau_2}{(\tau_1-\tau_2)^{d/2}} = \frac{16\lambda^2\rho_0^2}{(8\pi D)^{d/2}} \frac{t^{2-d/2}}{(2-d)(4-d)}, \quad d \neq 2,4.$$

Comparison with the result $I_{\text{tree}} = -2\lambda\rho_0^2 t$ for the corresponding tree diagram (second graph in Fig. 16.4), shows that the loop diagram has an effective dimensionless coupling constant

$$\lambda_{\text{eff}} \sim \frac{\lambda}{D^{d/2}} t^{1-d/2}. \tag{16.2.45}$$

This indicates that for $d < d_c = 2$, the perturbation expansion is finite at small times but blows up at large times, whereas the opposite holds when $d > d_c$. In the critical two-dimensional case one finds that the effective coupling diverges as $(\lambda/D)\ln(Dt)$ for both $t \to 0$ and $t \to \infty$.

## 16.2.4 Renormalization

So far we have found that in order to go beyond mean-field theory, it is necessary to extend the diagrammatic expansion beyond tree contributions to include corrections associated with loop diagrams. However, this leads to divergent integrals whose asymptotic behavior is dependent on the spatial dimension. For $d \geq 2$ we have so-called ultraviolet (UV) divergences, since they arise in the short-time (high-frequency) regime and are a consequence of taking the integrals over the wavectors $\mathbf{k}$ to be unbounded. One way to remove such infinities is to introduce a short-distance cut-off, which is equivalent to taking $|\mathbf{k}| \leq \Lambda < \infty$. As we illustrate below, a more convenient mathematical way of isolating the UV singularities is to allow non-integer spatial dimensions by taking $d = d_c - \varepsilon$ with $d_c = 2$ (dimensional regularization). Having regularized the theory with respect to UV divergences, we still have to deal with the corresponding infrared (IR) divergences, which occur in the asymptotic limit $t \to \infty$. This is necessary in order to obtain the correct asymptotic limits in (16.2.3).

A powerful method for handling IR divergences is to use renormalization theory [37, 164, 859, 861, 988]. This combines dimensional analysis with scaling theory to extend perturbation results that hold at intermediate time and length scales to regimes where perturbation theory breaks down. The basic idea is to rewrite a given perturbation expansion of an expectation such as $\langle\phi\rangle$ in terms of appropriately renormalized parameters and fields that depend on some arbitrary length scale $\kappa^{-1}$. Requiring that the resulting expression for $\langle\phi\rangle$ is independent of $\kappa$ leads to a set of characteristic differential equations describing how renormalized quantities vary with $\kappa$, and thus determining their behavior in the IR limit $\kappa \to 0$. This is formally achieved by means of the Callan–Symanzik renormalization group (RG) flow equations. The existence of a stable fixed point in the IR limit is indicative of a form of scale invariance or universality: the theory on large length and time scales is independent of microscopic details. In practice, the renormalization procedure is only tractable at the lowest dimension that yields UV singularities, which is known as the upper critical dimension $d_c$. For the given system it also coincides with the highest dimension where IR infinities occur. In order to calculate infrared scaling behavior for $d < d_c$ it is necessary to carry out a dimensional expansion with respect to $\varepsilon = d_c - d$ (which could be $O(1)$).

The first step in any renormalization procedure is to use dimensional arguments to identify which quantities require renormalization and to determine the Feynman diagrams that are the source of UV and IR infinities. If the number of renormalized parameters is finite, then the field theory is said to be renormalizable. The example of pair annihilation with diffusion is particularly simple, since the reaction rate $\lambda$ is the only renormalized quantity. Moreover, the set of Feynman diagrams responsible for the singularities in $\langle \phi \rangle$ are the loop diagrams shown in 16.6. These (truncated) diagrams can be evaluated explicitly to obtain the series

$$\Gamma_3(\tau_1 - \tau_2) = 2\lambda \left[ \delta(\tau_1 - \tau_2) - 2\lambda^2 I(\tau_1 - \tau_2) \right.$$
$$\left. + 4\lambda^3 \int_{\tau_1}^{\tau_2} I(\tau_1 - \tau) I(\tau - \tau_2) d\tau - \dots \right], \tag{16.2.46}$$

with $I(\tau)$ given by equation (16.2.44). The obvious method for summing this series of convolutions is to use Laplace transforms. However, this is where we encounter the issue of UV and IR singularities. That is,

$$\tilde{I}(s) = \int_0^\infty e^{-s\tau} I(\tau) d\tau = \frac{1}{(8\pi D)^{d/d_c}} \Gamma(\varepsilon/d_c) s^{-\varepsilon/d_c}, \tag{16.2.47}$$

where $\Gamma(z)$ is the Gamma function

$$\Gamma(z) = \int_0^\infty x^{z-1} e^{-x} dx. \tag{16.2.48}$$

The latter has poles at $z = 0, -1, -2, \dots$ with $\Gamma(z) = 1/z + \gamma + \dots$ for $z \to 0$ and Euler's constant $\gamma$. Hence, the IR and UV divergences are neatly separated. The former occurs in the limit $s \to 0$ (large-time limit) when $\varepsilon > 0$ or $d < 2$, whereas the latter correspond to poles in the Gamma function at $\varepsilon = 0, -1, \dots$ or $d = 2, 3, \dots$. In particular, by allowing fractional dimensions it is possible to regularize or isolate the UV singularities. Laplace transforming the series expansion of $\Gamma_3$ now leads to a geometric series, which can be summed to give

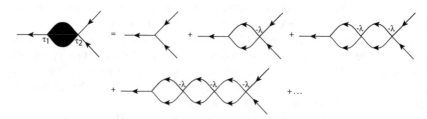

Fig. 16.6: Loop diagrams contributing to singularities in the effective 3-vertex coupling.

$$\widetilde{\Gamma}_3(s)/2 = \frac{\lambda}{1 + 2\lambda \widetilde{I}(s)}. \tag{16.2.49}$$

We can interpret $\widetilde{\Gamma}_3/2$ as an effective 3-vertex coupling parameter, $\lambda \rightarrow \widetilde{\Gamma}_3/2$, which includes the singularities associated with $\widetilde{I}(s)$. In order to properly perform a perturbation expansion away from the IR and UV limits, one should use a dimensionless parameter $g$. This can be achieved by introducing the reference wavenumber $\kappa$ and setting $g_0 = (\lambda/D)\kappa^{-2\varepsilon/d_c}$. We also use $\kappa$ to define an intermediate time scale $t_0 = 1/D\kappa^2$ and set $s = 1/t_0$. In this regime, perturbation theory holds and we define a renormalized dimensionless parameter

$$g_R = \frac{\lambda \kappa^{-2\varepsilon/d_c}/D}{1 + 2\lambda \widetilde{I}(D\kappa^2)} = \frac{g_0}{1 + g_0 B\Gamma(\varepsilon/d_c)}, \quad B = \frac{2}{(8\pi)^{d/d_c}}. \tag{16.2.50}$$

Now suppose that we rewrite the perturbation expansion of $\langle \phi \rangle$ in terms of the renormalized parameter $g_R$ and the reference wavevector $\kappa$. Then

$$\langle \phi \rangle = \overline{\phi}(t, \rho_0, \kappa, g_R).$$

Since $\kappa$ does not appear in the unrenormalized theory, any statistical quantity should be independent of $\kappa$. Therefore, we require

$$0 = \kappa \frac{d\overline{\phi}}{d\kappa} = \left[\kappa \frac{\partial}{\partial \kappa} + \beta_g(g_R) \frac{\partial}{\partial g_R}\right] \overline{\phi}, \tag{16.2.51}$$

where we have introduced the RG $\beta$ function

$$\beta_g(g_R) = \kappa \frac{dg_R}{d\kappa} = 2g_R \left[-\frac{\varepsilon}{d_c} + B\Gamma(1 + \varepsilon/d_c)g_R\right] = -\varepsilon g_R \left[1 - g_R B\Gamma(\varepsilon/2)\right]. \tag{16.2.52}$$

From dimensional analysis,

$$\overline{\phi}(t, n_0, \kappa, g_R) = \kappa^d F(t/t_0, \rho_0/\kappa^d, g_R),$$

which eliminates one of the independent variables and yields the Callan–Symanzik equation for the mean density:

$$\left[2Dt \frac{\partial}{D\partial t} - \rho_0 d \frac{\partial}{\partial \rho_0} + \beta_g(g_R) \frac{\partial}{\partial g_R} + d\right] \overline{\phi} = 0. \tag{16.2.53}$$

Equation (16.2.53) can be solved using the method of characteristics (Box 2C), which determines the flow of the renormalized parameter $g_R$ under changes in scale. Introducing the flow parameter $\ell$, we have the characteristic equations

$$\frac{dt}{d\ell} = 2t, \quad \frac{d\widetilde{\rho}_0}{d\ell} = -\widetilde{\rho}_0 d, \quad \frac{d\widetilde{g}_R}{d\ell} = -\beta_g(\widetilde{g}_R), \quad \frac{d\overline{\phi}}{d\ell} = -\overline{\phi} d.$$

The negative sign in the third equation reflects the fact that the IR limit is attained for $\ell \to \infty$, whereas the original definition of the RG $\beta$ function was in terms of $\kappa$ which scales as $\kappa/\ell$. The first equation relates $\ell$ to the time $t$ according to $t = t_0 e^{2(\ell-1)}$. We take $\ell = 1$ to be the initial renormalization point that lies outside the IR-singular regime, for which a perturbation expansion of $\overline{\phi}$ is valid. It follows that along a characteristic curve

$$\overline{\phi}(\widetilde{\rho}_0(t), \widetilde{g}_R(t)) = (t/t_0)^{-d/2} \overline{\phi}(t_0, \rho_0, g_R),$$

where

$$\widetilde{\rho}(t) = \rho_0 e^{-d[\ell-1]} = (t/t_0)^{-d/2} \rho_0,$$

and $\widetilde{g}_R(t)$ is the solution to the equation

$$2t \frac{d\widetilde{g}_R}{dt} = -\beta_g(\widetilde{g}_R). \tag{16.2.54}$$

Substituting for $\beta_g$ using equation (16.2.52) gives (for $\varepsilon \neq 0$)

$$\widetilde{g}_R(t) = g_R^* \left[ 1 + \frac{g_R^* - g_R}{g_R} \left(\frac{t_0}{t}\right)^{\varepsilon/d_c} \right]^{-1}, \quad g_R^* = \frac{1}{B\Gamma(\varepsilon/2)}. \tag{16.2.55}$$

Above the critical dimension ($\varepsilon < 0$), the renormalized coupling constant flows to zero algebraically as $t^{-|\varepsilon|/d_c}$, which means that the effective reaction rate $\lambda(t) \to$ constant, as assumed in mean-field theory. On the other hand, below the critical dimension, $\widetilde{g}_R(t) \to g_R^*$ as $t \to \infty$. That is, the system is said to flow to a nontrivial stable IR fixed point and exhibits universal behavior that is independent of $\rho_0$ and has the scaling $\overline{\phi} \sim t^{-d/2}$. Finally, at the critical dimension,

$$2t \frac{d\widetilde{g}_R}{dt} = -B\widetilde{g}_R^2, \tag{16.2.56}$$

which has the solution

$$\widetilde{g}_R(t) = \frac{4\pi}{\ln t}. \tag{16.2.57}$$

The corresponding asymptotic behavior of $\overline{\phi}$ is then $\sim t^{-1} \ln t$. Hence, the scaling behavior of (16.2.3) is recovered.

## 16.3 Traveling waves in stochastic RD equations

In Sect. 11.6 we presented various examples of RD equations supporting traveling wave solutions, and distinguished between bistable waves and unstable waves (pulled fronts). Here we briefly describe a formal perturbation method for analyzing the effects of weak extrinsic noise on the propagation of bistable waves, which has been developed by a number of authors [29, 235, 708, 784, 798]. Consider a scalar

SPDE of the form

$$dU(x,t) = \left[ \frac{\partial^2}{\partial x^2} U(x,t) + F(U(x,t)) \right] dt + \sqrt{\varepsilon} dW(x,t), \qquad (16.3.1)$$

where $W(x,t)$ is a space-dependent Wiener process with zero mean and

$$\langle dW(x,t)dW(x',t) \rangle = 2C([x-x']/\lambda)dt\,dt'. \qquad (16.3.2)$$

The parameter $\lambda$ is the spatial correlation length of the noise such that $C(x/\lambda) \to \delta(x)$ in the limit $\lambda \to 0$, and $\varepsilon$ determines the strength of the noise, which is assumed to be weak. Suppose that $F(U) = U(U-a)(1-U)$ for $0 < a, 1$ so that in the absence of noise, there exists a traveling front solution $U_0(\xi)$ with $\xi = x - c_0 t$ and

$$c_0 \frac{dU_0}{d\xi} + \frac{d^2 U_0^2}{d\xi^2} + F(U_0(\xi)) = 0. \qquad (16.3.3)$$

The basic idea underlying the formal perturbation analysis is to observe that the fluctuating term in (16.3.1) generates two distinct phenomena that occur on different time scales: a diffusive-like displacement of the front from its uniformly translating position at long time scales, and fluctuations in the front profile around its instantaneous position at short time scales [29, 235, 708, 784, 798], see Fig. 16.7. This

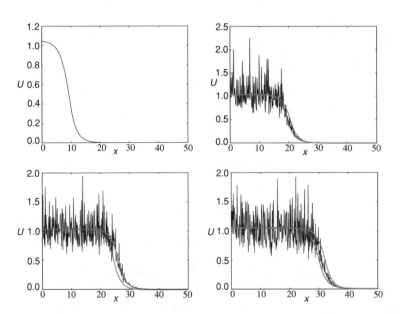

Fig. 16.7: Illustration of a stochastic traveling front for a bistable system. The wave profile is shown at successive times, with the initial profile given by equation (16.3.3). The deterministic part $U_0$ of the stochastic wave is shown by the blue curves and the corresponding solution in the absence of noise ($\varepsilon = 0$) is shown by the red curves.

motivates expressing the solution $U$ as a combination of a fixed wave profile $U_0$ that is displaced by an amount $\Delta(t)$ from its uniformly translating position $x - c_0 t$ and a time-dependent fluctuation $\Phi$ in the front shape about the instantaneous position of the front:

$$U(x,t) = U_0(\xi - \Delta(t)) + \varepsilon^{1/2}\Phi(\xi - \Delta(t),t). \tag{16.3.4}$$

It turns out that if $U_0$ is chosen to satisfy (16.3.3) then to leading order, the stochastic variable $\Delta(t)$ undergoes unbiased Brownian motion:

$$\langle \Delta(t) \rangle = 0, \quad \langle \Delta(t)^2 \rangle = 2\varepsilon Dt, \tag{16.3.5}$$

with a diffusion coefficient $D$ calculated below. Thus $\Delta(t)$ represents the effects of slow fluctuations, whereas $\Phi$ represents the effects of fast fluctuations.

*Calculation of diffusion coefficient.* We outline the steps in the derivation of the SDE for the phase $\Delta(t)$. The analysis uses various results from linear operator theory, see Boxes 2D and 11G. First, substitute the decomposition (16.3.4) into equation (16.3.1):

$$\begin{aligned}
&- c_0 U_0'(\xi - \Delta(t))dt - U_0'(\xi - \Delta(t))d\Delta(t) + \varepsilon^{1/2}d\Phi(\xi - \Delta(t),t) \\
&- \varepsilon^{1/2}\left[ c_0\Phi'(\xi - \Delta(t),t)dt + \Phi'(\xi - \Delta(t),t)d\Delta(t) \right] \\
&= \frac{d^2}{d\xi^2}\left[ U_0(\xi - \Delta(t)) + \varepsilon^{1/2}\Phi(\xi - \Delta(t),t) \right]dt \\
&+ F(U_0(\xi - \Delta(t)) + \varepsilon^{1/2}\Phi(\xi - \Delta(t),t))dt + \varepsilon^{1/2}dW(\xi + c_0 t,t).
\end{aligned} \tag{16.3.6}$$

Introduce the series expansion $\Phi = \Phi_0 + \sqrt{\varepsilon}\Phi_1 + O(\varepsilon)$. Self-consistency of this asymptotic expansion will then determine $\Delta(t)$. Substituting for $\Phi$ into equation (16.3.6), Taylor expanding the nonlinear function $F$, and imposing the homogeneous equation for $U_0$ leads to the following equation for $\Phi_0$:

$$\begin{aligned}
&- c_0 U_0'(\xi - \Delta(t))dt - U_0'(\xi - \Delta(t))d\Delta(t) + \varepsilon^{1/2}d\Phi_0(\xi - \Delta(t),t) \\
&- \varepsilon^{1/2}\left[ c_0\Phi_0'(\xi - \Delta(t),t)dt + \Phi_0'(\xi - \Delta(t),t)d\Delta(t) \right] \\
&= \frac{d^2}{d\xi^2}\left[ U_0(\xi - \Delta(t)) + \varepsilon^{1/2}\Phi_0(\xi - \Delta(t),t) \right]dt \\
&+ \varepsilon^{1/2}F'(U_0(\xi - \Delta(t)))\Phi_0(\xi - \Delta(t),t)dt + \varepsilon^{1/2}dW(\xi + c_0 t,t).
\end{aligned}$$

Shifting $\xi \to \xi + \Delta(t)$ and dividing through by $\varepsilon^{1/2}$ then gives

$$d\Phi_0(\xi,t) = \mathbb{L}\Phi_0(\xi,t)dt + \varepsilon^{-1/2}U_0'(\xi)d\Delta(t) + d\widetilde{W}(\xi,t), \tag{16.3.7}$$

where $\widetilde{W}(\xi,t) = W(\xi + \Delta(t) + c_0 t,t)$ and $\mathbb{L}$ is the non-self–adjoint linear operator

$$\mathbb{L}A(\xi) = \frac{d^2 A(\xi)}{d\xi^2} + c_0\frac{dA(\xi)}{d\xi} + F'(U_0(\xi))A(\xi) \tag{16.3.8}$$

for any function $A(\xi) \in L^2(\mathbb{R})$. It can be shown that the operator $\mathbb{L}$ has a 1D null space spanned by $U_0'(\xi)$. (The fact that $U_0'(\xi)$ belongs to the null space follows immediately from differentiating equation (16.3.3) with respect to $\xi$.) We then have the solvability condition for the existence of a bounded solution of equation (16.3.7), namely, that the inhomogeneous part is orthogonal to all elements of the null space of the adjoint operator $\mathbb{L}^\dagger$. The latter is defined with respect to the inner product

$$\int_{-\infty}^{\infty} B(\xi)\mathbb{L}A(\xi)d\xi = \int_{-\infty}^{\infty} \left[\mathbb{L}^\dagger B(\xi)\right]A(\xi)d\xi, \tag{16.3.9}$$

where $A(\xi)$ and $B(\xi)$ are arbitrary integrable functions. Hence,

$$\mathbb{L}^\dagger B(\xi) = \frac{d^2 B(\xi)}{d\xi^2} - c_0 \frac{dB(\xi)}{d\xi} + F'(U_0(\xi))B(\xi). \tag{16.3.10}$$

It can be proven that $\mathbb{L}^\dagger$ also has a one-dimensional null-space, that is, it is spanned by some function $\mathcal{V}(\xi)$. The solvability condition reflects the fact that the homogeneous system ($\varepsilon = 0$) is marginally stable with respect to uniform translations of a front. This means that the linear operator $\mathbb{L}$ has a simple zero eigenvalue whilst the remainder of the spectrum lies in the left-half complex plane. Hence, perturbations of $\Phi_0$ that lie in the null-space will not be damped and thus $\Phi_0$ will be unbounded in the large $t$ limit unless these perturbations vanish identically.

Taking the inner product of both sides of equation (16.3.7) with respect to $\mathcal{V}(\xi)$ leads to the solvability condition

$$\int_{-\infty}^{\infty} \mathcal{V}(\xi)\left[\varepsilon^{-1/2}U_0'(\xi)d\Delta(t) + d\widetilde{W}(\xi,t)\right]d\xi = 0. \tag{16.3.11}$$

It follows that, to leading order, $\Delta(t)$ satisfies the nonlinear SDE

$$d\Delta(t) = \varepsilon^{1/2}d\widehat{W}(t), \tag{16.3.12}$$

where

$$\widehat{W}(t) = -\frac{\displaystyle\int_{-\infty}^{\infty} \mathcal{V}(\xi)\widetilde{W}(\xi,t)d\xi}{\displaystyle\int_{-\infty}^{\infty} \mathcal{V}(\xi)U_0'(\xi)d\xi}. \tag{16.3.13}$$

Note that

$$\langle d\widehat{W}(t)\rangle = 0, \quad \langle d\widehat{W}(t)d\widehat{W}(t')\rangle = 2D\delta(t-t')dt'dt, \tag{16.3.14}$$

with $D$ the effective diffusivity

$$D = \frac{\displaystyle\int_{-\infty}^{\infty}\int_{-\infty}^{\infty} \mathcal{V}(\xi)\mathcal{V}(\xi')\langle d\widetilde{W}(\xi,t)d\widetilde{W}(\xi',t)\rangle d\xi d\xi'}{\left[\displaystyle\int_{-\infty}^{\infty} \mathcal{V}(\xi)U_0'(\xi)d\xi\right]^2}$$

$$= \frac{\displaystyle\int_{-\infty}^{\infty}\int_{-\infty}^{\infty} \mathcal{V}(\xi)C(\xi-\xi')\mathcal{V}(\xi')d\xi'd\xi}{\left[\displaystyle\int_{-\infty}^{\infty} \mathcal{V}(\xi)U_0'(\xi)d\xi\right]^2}. \tag{16.3.15}$$

Although the above analysis is based on a formal perturbation calculation, rather than rigorous analysis, it does capture the effects of weak external noise on front propagation in a variety of reaction–diffusion models [708, 784]. Note, however, that one class of front solution where the method breaks down is a pulled front, which propagates into an unstable rather than a metastable state and whose dynamics is dominated by the linear spreading of small perturbations within the leading edge of the front [898]; see Box 11H. The rigorous treatment of SPDEs is much more complicated than SDEs because one has to keep track of the regularity of solutions with respect to both time and space.

**Stimulus-locked fronts.** Now suppose that there is a weak external input $\sqrt{\varepsilon}I(x - ct)$ with $I$ given by a positive, bounded, monotonically decreasing function of amplitude $I_0 = I(-\infty) - I(\infty)$. In the absence of noise, the resulting inhomogeneous PDE can support a traveling front that locks to the stimulus, provided that the stimulus speed $c$ is sufficiently close to the natural speed $c_0$ of spontaneous fronts, that is,

$$c = c_0 + \sqrt{\varepsilon}c_1.$$

Equation (16.3.1) becomes

$$dU(x,t) = \left[ \frac{\partial^2}{\partial x^2} U(x,t) + F(U(x,t)) \right] dt + \sqrt{\varepsilon}I(x-ct) + \sqrt{\varepsilon}dW(x,t). \quad (16.3.16)$$

Repeating the steps in the derivation of equation (16.3.12), we obtain the nonlinear SDE (see Ex. 16.6)

$$d\Delta(t) + \varepsilon^{1/2}G(\Delta(t))dt = \varepsilon^{1/2}d\widehat{W}(t), \quad (16.3.17)$$

where

$$G(\Delta) = \frac{\int_{-\infty}^{\infty} \mathcal{V}(\xi)[I(\xi + \Delta) + c_1 U_0'(\xi)]d\xi}{\int_{-\infty}^{\infty} \mathcal{V}(\xi)U_0'(\xi)d\xi}. \quad (16.3.18)$$

Suppose that there exists a unique phase $\Delta = \xi_0$ for which $G(\xi_0) = 0$ and $G'(\xi_0) > 0$. This represents a stable stimulus-locked front in the absence of noise, with $\xi_0$ the relative shift of the stimulus-locked front and the input. Taylor expanding about this solution by setting $\varepsilon^{1/2}Y(t) = \Delta(t) - \xi_0$ with $Y(t) = O(1)$, we obtain the Ornstein–Uhlenbeck (OU) process

$$dY(t) + \varepsilon^{1/2}AY(t)dt = d\widehat{W}(t), \quad (16.3.19)$$

with

$$A = G'(\xi_0) = \frac{\int_{-\infty}^{\infty} \mathcal{V}(\xi)I'(\xi + \xi_0)d\xi}{\int_{-\infty}^{\infty} \mathcal{V}(\xi)U_0'(\xi)d\xi}.$$

Using standard properties of an OU process

$$\langle \Delta(t) \rangle = \xi_0 \left[ 1 - e^{-\sqrt{\varepsilon}At} \right] + \Delta(0)e^{-\sqrt{\varepsilon}At}, \quad (16.3.20a)$$

$$\langle \Delta(t)^2 \rangle - \langle \Delta(t) \rangle^2 = \frac{\sqrt{\varepsilon}D}{A} \left[ 1 - e^{-2\sqrt{\varepsilon}At} \right]. \quad (16.3.20b)$$

In particular, the variance approaches a constant $\sqrt{\varepsilon}D/A$ in the large $t$ limit and the mean converges to the fixed point $\xi_0$.

## 16.4 Exercises

---

**Problem 16.1. RD master equation.** Consider a 1D version of the RD model introduced in Ex. 11.7:

$$\frac{\partial u}{\partial t} = D\frac{\partial^2 u}{\partial x^2}u + K_+ n_c u - K_- u, \quad x \in [0,L].$$

Assume periodic boundary conditions. Discretize the model by taking $x = mh$, $m = 1, \ldots, N$, and setting $u(mh,t) = u_\ell(t)$. Identifying the system size $\Omega$ with the number of molecules in the cell, the discretized RD equation takes the form

$$\frac{du_\ell(t)}{dt} = A_\ell(\mathbf{u})u_\ell(t),$$

where

$$A_\ell(\mathbf{u}) = K_+ u_\ell(1 - \sum_\ell u_\ell) - K_- u_\ell + \alpha \Delta u_\ell,$$

with $\Delta u_\ell = u_{\ell+1} + u_{\ell-1} - 2u_\ell$.

(a) Identify the various chemical and hopping reactions for the lattice model, and thus determine the set of propensities and stoichiometric coefficients. Write down the corresponding RD master equation.

(b) Perform a system-size expansion of the master equation to derive the FP equation

$$\frac{\partial p}{\partial t} = -\sum_\ell \frac{\partial [A_\ell(\mathbf{u})p(\mathbf{u},t)]}{\partial u_\ell} + \frac{1}{2\Omega}\sum_{\ell,\ell'} \frac{\partial^2 [D_{\ell\ell'}(\mathbf{u})p(\mathbf{u},t)]}{\partial u_\ell \partial u_{\ell'}},$$

where

$$D_{\ell\ell'}(\mathbf{u}) = \left[ K_+ u_\ell(1 - \sum_\ell u_\ell) + K_- u_\ell + O(\alpha) \right]\delta_{\ell,\ell'}.$$

**Problem 16.2. Stochastic model of cell polarization.** Consider the RD Langevin equation corresponding to the FP equation of Ex. 16.1:

$$\frac{du_\ell(t)}{dt} = A_\ell(\mathbf{u})u_\ell(t) + \Omega^{-1/2}\eta_\ell(t),$$

where

$$A_\ell(\mathbf{u}) = K_+ u_\ell(1 - \sum_\ell u_\ell) - K_- u_\ell + \alpha \Delta u_\ell,$$

with $\Delta u_\ell = u_{\ell+1} + u_{\ell-1} - 2u_\ell$, and the Gaussian white noise terms satisfy

$$\langle \eta_\ell(t) \rangle = 0, \quad \langle \eta_\ell(t)\eta_{\ell'}(t') \rangle = D_{\ell\ell'}\delta(t-t'),$$

such that

$$D_{\ell\ell'}(\mathbf{u}) = \left[ K_+ u_\ell \left( 1 - \sum_\ell u_\ell \right) + K_- u_\ell \right] \delta_{\ell,\ell'}.$$

Assume periodic boundary conditions, $u_{\ell+Nh} = u_\ell$ for $\ell = 1,\ldots,N$ and lattice spacing $h$.

(a) Introduce the discrete Fourier transforms

$$U(k,t) = \sum_\ell e^{-ik\ell} u_\ell(t), \quad u_\ell(t) = \frac{1}{N} \sum_k e^{ik\ell} U(k,t),$$

with $k = 2\pi m/N, m = 0,\ldots,N-1$. Show that the discrete Fourier transform of the Langevin equation is

$$\frac{dU(k,t)}{dt} = A(k,t)U(k,t) + \Omega^{-1/2}\eta(k,t),$$

with

$$\langle \eta(k,t) \rangle = 0, \quad \langle \eta(k,t)\eta(k',t') \rangle = D(t)U(k+k',t)\delta(t-t'),$$

and

$$A(k,t) = [K_+(1-U(0,t)) - K_- + 2\alpha(\cos k - 1)]$$
$$D(t) = K_+(1-U(0,t)) + K_-.$$

(b) Show that for $K_+ > K_-$, the deterministic system ($\Omega \to \infty$) has a stable homogeneous fixed point $U(k,t) = u^* \delta_{k,0}$ with $u^* = 1 - K_-/K_+$. (In real space this is $u_\ell(t) = u^*/N$.)

(c) Eliminating the $k = 0$ mode by setting $U(0,t) = u^*$ leads to an effective FP equation for the modes $k \neq 0$ given by

$$\frac{\partial P}{\partial t} = -\sum_{k\neq 0} \frac{\partial[A^*(k)U(k)P(\mathbf{U},t)]}{\partial U(k)} + \frac{1}{2\Omega} \sum_{k,k'\neq 0} \frac{\partial^2[D^*(k,k')P(\mathbf{U},t)]}{\partial U(k)\partial U(k')},$$

where

$$A^*(k) = 2\alpha(\cos k - 1), D^*(k,k') = 2K_- \left[ u(k+k') - \frac{u(k)u(k')}{u^*} \right].$$

Taking first and second moments of the FP equation, derive the moment equations

$$\frac{d\langle U(k,t) \rangle}{dt} = 2\alpha \langle U(k,t) \rangle (\cos(k) - 1),$$

and

$$\frac{d\langle U(k,t)U(k',t)\rangle}{dt} = \frac{K_-}{\Omega}\langle U(k+k',t)\rangle$$

$$+ \left[2\alpha(\cos k + \cos k' - 2) - \frac{K_-}{\Omega u^*}\right]\langle U(k,t)U(k',t)\rangle.$$

(d) The first moment equation implies that $\langle \widehat{U}(k,t)\rangle \to 0$ as $t \to 0$ for all $k \neq 0$. Now use the second moment equation to show that all two-point correlations $\langle \widehat{U}(k,t)\widehat{U}(k',t)\rangle \to 0$ as $t \to 0$ unless $k' = -k$, in which case

$$\langle U(k,t)U(-k,t)\rangle = \langle |\widehat{U}(k,t)|^2\rangle \to \frac{(u^*)^2}{1 + 4\alpha\Omega(1-\cos k)u^*/K_-}.$$

**Remark 16.1.** One application of the final result is to a stochastic model of cell polarization [604]. First note that for long wavelengths relative to the lattice spacing $h$ (small $k$), we obtain the Lorentzian distribution

$$\langle |\widehat{U}(k)|^2\rangle \approx \frac{(u^*)^2}{1 + \Gamma^2 k^2}, \quad \Gamma = \sqrt{2\alpha\Omega u^*/K_-}.$$

Taking averages of the identity

$$\sum_{\ell'} u_{\ell'}u_{\ell+\ell'} = \frac{1}{N}\sum_k e^{ik\ell}|u(k)|^2.$$

This expresses the two-point spatial correlation function in terms of the Fourier transform of $|u(k)|^2$. It follows that

$$\sum_{\ell'}\langle u_{\ell'}u_{\ell+\ell'}\rangle \approx \frac{1}{N}\sum_k e^{ik\ell}\frac{(u^*)^2}{1+\Gamma^2 k^2} = \frac{1}{2\pi}\sum_k e^{ik\ell}\frac{(u^*)^2}{1+\Gamma^2 k^2}\Delta k.$$

For large $N$ (small $\Delta k$) we can approximate the sum over $k$ by an integral, which shows that

$$\sum_{\ell'}\langle u_{\ell'}u_{\ell+\ell'}\rangle \approx \frac{1}{2\Gamma}e^{-|\ell|/\Gamma}.$$

Hence, the averaged correlation function is a decaying exponential with spatial correlation length $\Gamma$, which establishes that there exist localized states provided that $\Gamma \ll 2\pi$ [604].

**Problem 16.3. Multicellular LEGI model.** Consider a 1D chain of $N$ cells aligned parallel to a spatial concentration gradient. Assume that the signal concentration varies linearly along the chain according to $c_\ell = c_N - a\gamma(N-\ell)$, where $a$ is the cell size, $\gamma$ is the concentration gradient and $c_\ell$ is the local concentration in the vicinity of the $\ell$th cell. Let $x_\ell$ and $y_{\ell n}$ denote the concentrations of the active excitatory and inhibitory enzymes in the $\ell$th cell. Both enzymes are assumed to deactivate at a rate $k$ and activate at a rate $\mu$ that is proportional to $c_\ell$. While the local excitatory enzyme is confined to each cell, the global inhibitory enzyme can hop between neighboring cells at a rate $\gamma$. In particular, the kinetic equation for the inhibitory enzyme becomes

$$\frac{dy_j}{dt} = -ky_j + \mu c_j + \gamma(y_{j-1} + y_{j+1} - 2y_j), \quad 1 < j < N,$$

where $y_j$ is the concentration in the $j$th cell. (We don't explicitly write down the boundary equations for $y_1, y_N$; boundary effects are negligible when $N$ is large.) Let $n_j$ denote the corresponding number of molecules so that $y_j = n_j/\Omega$, where $\Omega = a^3$ specifies the cell volume (non-dimensionalized). The full set of chemical reactions is

$$n_j \overset{\mu\Omega c_j}{\to} n_j + 1, \quad n_j \overset{kn_j}{\to} n_j - 1,$$

$$(n_j, n_{j'}) \overset{\gamma m_j}{\to} (n_j - 1, n_{j'} + 1), \quad j' = j \pm 1,$$

$$(n_j, n_{j'}) \overset{\gamma m_{j'}}{\to} (n_j + 1, n_{j'} - 1), \quad j' = j \pm 1.$$

Let $(a, j)$ label the reactions associated with activation ($a = 1$) and inactivation ($a = 2$), respectively. Similarly, let $(j, j')$ label the pair of hopping reactions out of the $j$th cell and $(j', j)$ the corresponding hopping reactions into the cell with $j' = j \pm 1$.
(a) Write down the stoichiometric coefficients $S_{q;l}$ and propensities $f_q(\mathbf{n})$ for the various reactions and thus construct the chemical master equation

$$\frac{dP(\mathbf{n}, t)}{dt} = \Omega \sum_q \left( \prod_l \mathbb{E}^{-\hat{S}_{q;l}} - 1 \right) f_q(\mathbf{n}/\Omega) P(\mathbf{n}, t),$$

where $\mathbf{n}_j = \{n_1, \ldots, N_n\}$, and the sum is over the set of reactions

$$q \in \{(1, j), (2, j), (j, j'), (j', j); 1 \le j \le N, j' = j \pm 1\}.$$

(b) Carry out a system-size expansion of the master equation and derive the corresponding FP equation. Hence, show that the stochastic dynamics can be approximated by the Langevin equation

$$\frac{dy_j}{dt} = -ky_j + \mu \bar{c}_j + \gamma(y_{j-1} + y_{j+1} - 2y_j) + \xi_j/\sqrt{\Omega}, \quad 1 < j < N,$$

with the white noise terms satisfying

$$\langle \xi_l(t) \rangle = 0, \quad \langle \xi_l(t) \xi_{l'}(t') \rangle = D_{ll'} \delta(t - t'), \quad D_{ll'} = \sum_q S_{q;l} S_{q;l'} f_q(\bar{\mathbf{y}}).$$

Here $\bar{\mathbf{y}}$ denotes the steady-state solution of the deterministic kinetic equations:

$$\bar{y}_\ell = \frac{\mu}{k} \sum_{\ell'} M_{\ell\ell'}^{-1} c_{\ell'},$$

where $\mathbf{M}$ is the tridiagonal matrix

$$M_{\ell\ell'} = (1 + 2\gamma/k)\delta_{\ell,\ell'} - (\gamma/k)(\delta_{\ell',\ell-1} + \delta_{\ell',\ell+1}).$$

(c) Substituting for the various stoichiometric coefficients and propensities using part (a), show that

$$D_{ll'} = \delta_{l,l'}(\mu\Omega c_l + k\bar{y}_l + \gamma[2\bar{y}_l + \bar{y}_{l+1} + \bar{y}_{l-1}]) - \gamma[\bar{y}_l + \bar{y}_{l-1}]\delta_{l',l-1} - \gamma[\bar{y}_l + \bar{y}_{l+1}]\delta_{l',l+1}.$$

**Problem 16.4. Annihilation process.** Consider the multiple particle annihilation process

$$kA \rightarrow \emptyset + \text{diffusion}$$

with integer $k \geq 2$. The corresponding RD equation for the concentration $u(\mathbf{x},t)$ is

$$\frac{\partial u}{\partial t} = D\nabla^2 u - \lambda u^k.$$

Show that for a uniform initial density $U(\mathbf{x},t) = \rho_0$ we have the solution

$$u(t) = \frac{\rho_0}{[1 + \rho_0^{k-1}k(k-1)\lambda t]^{1/(k-1)}},$$

and determine the large-time behavior. For $k = 2$ the upper critical dimension is $d_c = 2$, which corresponds to the dimension beyond which a pair of diffusing, well-separated point particles meet with zero probability. What is the critical dimension when $k = 3$?

**Problem 16.5. Two-species annihilation.** Consider the two-species pair annihilation reaction (no diffusion)

$$A + B \rightarrow \emptyset.$$

We will construct the Doi–Peliti operator for the master equation along the lines of Sect. 8.2.
(a) Write down the master equation for the probability distribution $P(n_A, n_B, t)$.
(b) Introduce the annihilation–creation operator pairs $(a, a^\dagger)$ and $(b, b^\dagger)$ with commutation relations

$$[a, a^\dagger] = 1 = [b, b^\dagger], \quad [a, b^\dagger] = 0 = [b, a^\dagger].$$

Consider the Fock space spanned by the vectors

$$|n_A, n_B\rangle = a^{\dagger n_A} b^{\dagger n_B}|0 >,$$

and define the state vector

$$|\psi(t)\rangle = \sum_{n_A \geq 0} \sum_{n_B \geq 0} P((n_A, n_B, t)|n_A, n_B\rangle.$$

Differentiating both sides with respect to $t$ and using the master equation of part (a), derive the normal-ordered operator equation

$$\frac{d}{dt}|\psi(t)\rangle = \widehat{H}|\psi(t)\rangle, \quad \widehat{H} = \lambda(ab - a^\dagger b^\dagger ab).$$

(c) Now include hopping reactions on a lattice for each species along identical lines to pair annihilations and thus show that the corresponding operator for the RD master equation is

$$\widehat{H} = -\frac{D}{h^2}\sum_{\langle \ell,\ell'\rangle}(a^\dagger_{\ell'} - a^\dagger_\ell)(a_{\ell'} - a_\ell) - \frac{D}{h^2}\sum_{\langle \ell,\ell'\rangle}(b^\dagger_{\ell'} - b^\dagger_\ell)(b_{\ell'} - b_\ell) + \lambda(ab - a^\dagger b^\dagger ab).$$

(d) Following identical steps to the construction of the continuum action (16.2.21) for pair annihilation, derive the path integral action

$$S[\phi,\widetilde{\phi},\psi,\widetilde{\psi}] = \int_{\mathbb{R}^d} dx \int_0^t d\tau \bigg(\widetilde{\phi}(\partial_\tau - D\nabla^2)\phi + (\widetilde{\psi}(\partial_\tau - D\nabla^2)\psi$$

$$+ \lambda[\widetilde{\phi} + \widetilde{\psi}]\phi\psi + \lambda\widetilde{\phi}\widetilde{\psi}\phi\psi - \rho_A\widetilde{\phi}\delta(\tau) - \rho_B\widetilde{\psi}\delta(\tau)\bigg).$$

**Problem 16.6. Stimulus-locked wave.** Consider the stochastic RD equation (16.3.16) for stimulus-locked waves. Decompose the solution as

$$U(x,t) = U_0(\xi - \widehat{\Delta}(t)) + \varepsilon^{1/2}\Phi(\xi - \widehat{\Delta}(t),t),$$

where $\widehat{\Delta}(t) = \Delta(t) + (c - c_0)t$. Here $c_0$ is the natural wave speed when $\varepsilon = 0$ and $c$ is the speed of the external traveling stimulus. Repeating the steps in the derivation of equation (16.3.12), derive the nonlinear SDE for the phase $\Delta(t)$:

$$d\Delta(t) + \varepsilon^{1/2}G(\Delta(t))dt = \varepsilon^{1/2}d\widehat{W}(t),$$

where

$$\langle d\widehat{W}(t)\rangle = 0, \quad \langle d\widehat{W}(t)d\widehat{W}(t')\rangle = 2D\delta(t - t')dt'dt,$$

$$G(\Delta) = \frac{\displaystyle\int_{-\infty}^{\infty}\mathcal{V}(\xi)[I(\xi + \Delta) + c_1 U_0'(\xi)]d\xi}{\displaystyle\int_{-\infty}^{\infty}\mathcal{V}(\xi)U_0'(\xi)d\xi},$$

and

$$D = \frac{\displaystyle\int_{-\infty}^{\infty}\int_{-\infty}^{\infty}\mathcal{V}(\xi)C(\xi - \xi')\mathcal{V}(\xi')d\xi'd\xi}{\left[\displaystyle\int_{-\infty}^{\infty}\mathcal{V}(\xi)U_0'(\xi)d\xi\right]^2}.$$

# References

1.    Acar, M., Mettetal, J.T., van Oudenaarden, A.: Stochastic switching as a survival strategy in fluctuating environments. Nat. Gen. **40**, 471–475 (2008)
2.    Agmon, N., Szabo, A.: Theory of reversible diffusion-influenced reactions. J. Chem. Phys. **92**, 5270–5284 (1990)
3.    Alberts, B., Johnson, A., Lewis, J., Raff, M., Walter, K.R.: Molecular Biology of the Cell, 5th edn. Garland, New York (2008)
4.    Alikakos, N.D., Fusco, G., Karali, G.: Ostwald ripening in two dimensions - the rigorous derivation of the equations from the Mullins-Sekerka dynamics. J. Diff. Eqs. **205**, 1–49 (2004)
5.    Alipour, E., Marko, J.F.: Self-organization of domain structures by DNA-loop-extruding enzymes. Nucleic Acids Res. **40**, 11202–11212 (2012)
6.    Allard, J., Mogilner, A.: Traveling waves in actin dynamics and cell motility. Curr. Opin. Cell Biol. **25**, 107–115 (2012)
7.    Alpkvist, E., Picioreanu, C., van Loosdrecht, M.C.M., Heyden, A.: Three-dimensional biofilm model with individual cells and continuum EPS matrix. Biotechnol. Bioeng. **94**, 961–979 (2006)
8.    Alt, W.: Biased random walk model for chemotaxis and related diffusion approximation. J. Math. Biol. **9**, 147–177 (1980)
9.    Altschuler, S.J., Angenent, S.B., Wang, Y., Wu, L.F.: On the spontaneous emergence of cell polarity. Nature **454**, 886–890 (2008)
10.   Amir, A.: Cell size regulation in bacteria. Phys. Rev. Lett. **112**, 208102 (2014)
11.   Amitai, A., Kupka, I., Holcman, D.: Computation of the mean first-encounter time between the ends of a polymer chain. Phys. Rev. Lett. **109**, 108302 (2012)
12.   Amitai, A., Holcman, D.: Polymer model with long-range interactions: Analysis and applications to the chromatin structure. Phys. Rev. E **88**, 052604 (2013)
13.   Amitai, A., Holcman, D.: Polymer physics of nuclear organization and function. Phys. Rep. **678**, 1–83 (2017)
14.   Amodeo, A.A., Skotheim, J.M.: Cell-size control. Cold Spring Harb. Perspect. Biol. **8**, a019083 (2016)
15.   Anderson, D.F.: A modified next reaction method for simulating chemical systems with time dependent propensities and delays. J. Chem. Phys. **127**, 214107 (2007)
16.   Anderson, J.K., Smith, T.G., Hoover, T.R.: Sense and sensibility: flagellum-mediated gene regulation. Trends in Microbiology **18**, 30–37 (2011)

© Springer Nature Switzerland AG 2021
P. Bressloff, *Stochastic Processes in Cell Biology*, Interdisciplinary Applied Mathematics 41, https://doi.org/10.1007/978-3-030-72519-8

17. Anderson, D. F., Kurtz, T. G.: Continuous time Markov chain models for chemical reaction networks In: Design and Analysis of Biomolecular Circuits pp. 3-42, (2011)

18. Andrews, S., Bray, D.: Stochastic simulation of chemical reactions with spatial resolution and single molecule detail. Phys. Biol. **1**, 137–151 (2004)

19. Anetzberger, C., Pirch, T., Jung, K.: Heterogeneity in quorum sensing-regulated bioluminescence in Vibrio harveyi. Molecular Microbiology **73**, 267–277 (2009)

20. Angelani, L.: Run-and-tumble particles, telegrapher's equation and absorption problems with partially reflecting boundaries. J. Phys. A **48**, 495003 (2015)

21. Angelani, L.: Confined run-and-tumble swimmers in one dimension. J. Phys. A **50**, 325601 (2017)

22. Angeli, D., De Leenheer, P., Sontag, E.D.: A petri net approach to the study of persistence in chemical reaction networks. Math. Biosci. **210**, 598–618 (2007)

23. Anguige, K., King, J.R., Ward, J.P.: Modeling antibiotic- and anti-quorum sensing treatment of a spatially structured Pseudomonas aeruginosa population. J. Math. Biol. **51**, 557–594 (2005)

24. Anguige, K., King, J.R., Ward, J.P.: A multi-phase mathematical model of quorum-sensing in a maturing Pseudomonas aeruginosa biofilm. Math. Biosci. **203**, 240–276 (2006)

25. Aquino, G., Endres, R.G.: Increased accuracy of ligand sensing by receptor diffusion on cell surface. Phys. Rev. E **82**, 041902 (2010)

26. Aquino, G., Wingreen, N.S., Endres, R.G.: Know the single-receptor sensing limit? Think again. J. Stat. Phys. **162**, 1353–1364 (2016)

27. Arimura, N., Kaibuchi, K.: Neuronal polarity: from extracellular signals to intracellular mechanisms. Nat. Rev. Neurosci **8**, 194–205 (2007)

28. Arkin, A., Ross, J., McAdams, H.H.: Stochastic kinetic analysis of developmental pathway bifurcation in phage infected escherichia coli cells. Genetics **149**, 1633–1648 (1998)

29. Armero, J., Casademunt, J., Ramirez-Piscina, L., Sancho, J.M.: Ballistic and diffusive corrections to front propagation in the presence of multiplicative noise. Phys. Rev. E **58**, 5494–5500 (1998)

30. Aronson, D.G., Weinberger, H.F.: Multidimensional nonlinear diffusion arising in population genetics. Adv. Math. **30**, 33–76 (1978)

31. Arshavsky, V.Y., Lamb, T.D., Pugh, E.N., Jr.: G proteins and phototransduction. Annu. Rev. Physiol. **64**, 153–187 (2002)

32. Ashe, H.L., Briscoe, J.: The interpretation of morphogen gradients. Development. **133**, 385–394 (2006)

33. Assaf, M., Meerson, B.: Spectral theory of metastability and extinction in birth-death systems Phys. Rev. Lett. **97**, 200602 (2006)

34. Assaf, M., Kamenev, A., Meerson, B.: Population extinction risk in the aftermath of a catastrophic event Phys. Rev. E **79**, 011127 (2009)

35. Assaf, M.: Meerson, B: Extinction of metastable stochastic populations Phys. Rev. E **81**, 021116 (2010)

36. Assaf, M., Roberts, M., Luthey-Schulten, Z., Goldenfeld, N.: Extrinsic noise driven phenotype switching in a self-regulating gene. Phys. Rev. Lett. **111**, 058102 (2013)

37. Altland, A., Simons, B. D. *Condensed Matter Field Theory 2nd ed* Cambridge University Press (2010)

38. Balaban, N.Q., Merrin, J., Chait, R., Kowalik, L., Leibler, S.: Bacterial persistence as a phenotypic switch. Science **205**, 1578–1579 (2004)

39. Balakrishnan, V., Chaturvedi, S.: Persistent diffusion on a line. Physica A **148**, 581–596 (1988)

40. Ball, J.M., Carr, J., Penrose, O.: The cluster equations: Basic properties and asymptotic behaviour of solutions. Comm. Math. Phys. **104**, 657–692 (1986)

41. Ballesteros-Yanez, I., Benavides-Piccione, R., Elston, G.N., Yuste, R., DeFelipe, J.: Density and morphology of dendritic spines in mouse neocortex. Neuroscience **138**, 403–409 (2006)

42. Banani, S.F., Lee, H.O., Hyman, A.A., Rosen, M.K.: Biomolecular condensates: organizers of cellular biochemistry. Nat. Rev. Mol. Cell Biol. **18**, 285–298 (2017)

43. Banjade, S, Rosen, M. K.: Phase transitions of multivalent proteins can promote clustering of membrane receptors. eLife **3** e04123 (2014)

44. Barkai, N., Leibler, S.: Robustness in simple biochemical networks. Nature **387**, 913–917 (1997)

45. Barkai, N., Shilo, B.Z.: Robust generation and decoding of morphogen gradients. Cold Spring Harb. Perspect. Biol. **1**, a001362 (2009)

46. Barnhart, E.L., Lee, K.C., Keren, K., Mogilner, A., Theriot, J.A.: An adhesion-dependent switch between mechanisms that determine motile cell shape. PLoS Biol. **9**, e1001059 (2011)

47. Barrio, B.A., Varea, C., Aragon, J.L., Maini, P.K.: A two-dimensional numerical study of spatial pattern formation in interacting Turing systems. Bull. Math. Biol. **61**, 483 (1999)

48. Bauer, M., Knebel, J., Lechner, M., Pickl, P., Frey, E.: Ecological feedback in quorum-sensing microbial populations can induce heterogeneous production of autoinducers. eLife **6** e25773 (2017)

49. Baumgart, T., Capraro, B.R., Zhu, C., Das, S.L.: Thermodynamics and mechanics of membrane curvature generation and sensing by proteins and lipids. Annu. Rev. Phys. Chem. **62**, 483–506 (2011)

50. Baylor, D.A., Lamb, T.D., Yau, K.-W.: Responses of retinal rods to single photons. J. Physiol. **288**, 613–634 (1979)

51. Beard, D.A., Qian, H.: Chemical Biophysics: Quantitative Analysis of Cellular Systems. Cambridge Univ. Press, Cambridge, UK (2008)

52. Becker, R., Döring, W.: Kinetische behandlung der keimbildung in "ubers"attigten d"ampfen. Ann. Phys. **24**, 719–752 (1935)

53. Belan, S.: Restart could optimize the probability of success in a Bernouilli trial. Phys. Rev. Lett. **120**, 080601 (2018)

54. Bell, G. I., Anderson, E. C.: Cell growth and division: I. A mathematical model with applications to cell volume distributions in mammalian suspension cultures. Biophys J. **7** 329-351 (1967)

55. Bell, G.I.: Models for specific adhesion of cells to cells. Science **200**, 618–627 (1978)

56. Bement, W.M., Leda, M., Moe, A.M., Kita, A.M., Larson, M.E., Golding, A.E., Pfeuti, C., Su, K.C., Miller, A.L., Goryachev, A.B., von Dassow, G.: Activator-inhibitor coupling between Rho signalling and actin assembly makes the cell cortex an excitable medium. Nat. Cell Biol. **17**, 1471–1483 (2015)

57. Benishou, O., Coppey, M., Moreau, M., Oshanin, G.: Kinetics of diffusion-limited catalytically activated reactions: An extension of the Wilemski-Fixman approach. J. Chem. Phys. **123**, 194506 (2005)

58. Ben-Naim, A.: Cooperativity and Regulation in Biochemical Processes. Kluwer Academic, New York (2010)

59. Berezhkovskii, A.M., Coppey, M., Shvartsman, S.Y.: Signaling gradients in cascades of two-state reaction-diffusion systems. Proc. Natl. Acad. Sci. **106**, 1087–1092 (2009)

60. Berezhkovskii, A.M., Sample, C., Shvartsman, S.Y.: How long does it take to establish a morphogen gradient? Biophys. J. **99**, L59–L61 (2010)

61. Berezhkovskii, A.M., Sample, C., Shvartsman, S.Y.: Formation of morphogen gradients: local accumulation time. Phys Rev E **83**, 051906 (2011)

62. Berezhkovskii, A.M., Szabo, A.: Effect of ligand diffusion on occupancy fluctuations of cell-surface receptors. J. Chem. Phys. **139**, 121910 (2013)

63. Berg, H.C., Brown, D.A.: Chemotaxis in Escherichia coli analyzed by three-dimensional tracking. Nature **239**, 500–504 (1972)

64. Berg, H.C., Purcell, E.M.: Physics of chemoreception. Biophys. J. **20**, 93–219 (1977)

65. Berg, H.C.: Random Walks in Biology. University Press, Princeton (1983)

66. Berg, H.C.: Motile behavior of bacteria. Phys. Today **53**, 24–28 (2000)

67. Berg, O.G.: A model for the statistical fluctuations of protein numbers in a microbial population. J. Theor. Biol. **4**, 587–603 (1978)

68.  Berg, O. G., Paulsson, J., Ehrenberg. M.: Fluctuations and quality of control in biological cells: zero-order ultrasensitivity reinvestigated. Biophys. J. **79** 1228-1236 (2000)

69.  Berger, S.L.: The complex language of chromatin regulation during transcription. Nature **447**, 407–412 (2007)

70.  Berry, J., Brangwynne, C., Haataja, M.: P: Physical principles of intracellular organization via active and passive phase transitions. Rep. Prog. Phys. **81**, 046601 (2018)

71.  Besser, A., Safran, S.A.: Force-induced adsorption and anisotropic growth of focal adhesions. Biophys. J. **90**, 3469–3484 (2006)

72.  Best, J.: Doubly stochastic processes: an approach for understanding central nervous system activity. Proc. 3rd Int. Conf. Appl. Math, Simul. Modeling pp. 155-158 (2009)

73.  Beta, C., Kruse, K.: Intracellular oscillations and waves. Annu. Rev. Condens. Matter Phys. **8**, 239–264 (2017)

74.  Bhalla, U. S.: Signaling in small subcellular volumes. I. Stochastic and diffusion effects on individual pathways. Biophys. J. **87** 733-744 (2004)

75.  Bhatt, J.S., Ford, I.J.: Kinetics of heterogeneous nucleation for low mean cluster populations. J. Chem. Phys. **118**, 3166–3176 (2003)

76.  Bialek, W., Setayeshgar, S.: Physical limits to biochemical signaling. Proc. Natl. Acad. Sc. USA **102**, 10040–10045 (2005)

77.  Bialek, W.: Setayeshgar, S: Cooperativity, sensitivity and noise in biochemical signaling. Phys. Rev. Lett. **100**, 258101 (2008)

78.  Bialek, W.: Biophysics. Princeton University Press, Princeton (2012)

79.  Biancalani, T., Fanelli, D., Di Patti, F.: Stochastic Turing patterns in the Brusselator model. Phys. Rev. E **81**, 046215 (2010)

80.  Biancalani, T., Galla, T., McKane, A.J.: Stochastic waves in a Brusselator model with non-local interaction. Phys. Rev. E **84**, 026201 (2011)

81.  Biancalani, T., Jafarpour, F., Goldenfeld, N.: Giant amplification of noise in fluctuation-induced pattern formation. Phys. Rev. Lett. **118**, 018101 (2017)

82.  Bicek, A.D., Tuzel, E., Kroll, D.M., Odde, D.J.: Analysis of microtubule curvature. Methods Cell Biol. **83**, 237–268 (2007)

83.  Bicout, D.J.: Green's functions and first passage time distributions for dynamic instability of microtubules. Phys. Rev. E **56**, 6656–6667 (1997)

84.  Binder, B., A. Goede, N. Berndt, Holzhutter, H.-G.: A conceptual mathematical model of the dynamic self-organization of distinct cellular organelles. PLoS One **4** e8295 (2009)

85.  Bischoff, M., Gradilla, A. C. , Seijo, I., Andres, G., Rodriguez-Navas, C., Gonzalez-Mendez. L., Guerrero, I.: Cytonemes are required for the establishment of a normal Hedgehog morphogen gradient in Drosophila epithelia. Nat Cell Biol **15** 1269-1281 (2013)

86.  Blackburn, E.H., Epel, E.S., Lin, J.: Human telomere biology: A contributory and interactive factor in aging, disease risks, and protection. Science **350**, 1193–1198 (2015)

87.  Boal, D.: Mechanics of the Cell, 2nd edn. Cambridge University Press, Cambridge (2010)

88.  Boeckh, J., Kaissling, K.E., Schneider, D.: Insect olfactory receptors. Cold Spring Harbor Symp. Quant. Biol **30**, 1263–1280 (1965)

89.  Boettiger, D.: Mechanical control of integrin-mediated adhesion and signaling. Curr. Opin. Cell Biol. **24**, 592–599 (2012)

90.  Bollenbach, T., Kruse, K., Pantazis, P., Gonzalez-Gaitan, M., Julicher, F.: Robust formation of morphogen gradients. Phys. Rev. Lett. **94**, 018103 (2005)

91.  Bollenbach, T., Kruse, K., Pantazis, P., Gonzalez-Gaitan, M., Julicher, F.: Morphogen transport in epithelia. Phys. Rev. E **75**, 011901 (2007)

92.  Bouzas, P.R., Valderrama, M.J., Aguilera, A.M.: On the characteristic functional of a doubly stochastic Poisson process: Application to a narrow-band process. Appl. Math. Modelling **30**, 1021–1032 (2006)

93.  Bouzigues, C., Morel, M., Triller, A., Dahan, M.: Asymmetric redistribution of GABA receptors during GABA gradient sensing by nerve growth cones analyzed by single quantum dot imaging. Proc. Natl. Acad. Sci. USA **104**, 11251–11256 (2007)

94. Bovier, A., Eckhoff, M., Gayrard, V., Klein, M.: Metastability in reversible diffusion processes. I. Sharp asymptotics for capacities and exit times, J. Eur. Math. Soc. (JEMS) **6** 399-424 (2004)

95. Bovier, A., Gayrard, V., Klein, M.: Metastability in reversible diffusion processes. II. Precise asymptotics for small eigenvalues, J. Eur. Math. Soc. (JEMS) **7** 69-99 (2005)

96. Boyer, D., Mather, W., Mondragon-Palomino, O., Orozco-Fuentes, S., Danino, T., Hasty, J., Tsimring, L.S.: Buckling instability in ordered bacterial colonies. Phys. Biol. **8**, 026008 (2011)

97. Brangwynne, C.P., Koenderink, G.H., Barry, E., Dogic, Z., MacKintosh, F.C., Weitz, D.A.: Bending dynamics of fluctuating biopolymers probed by automated high-resolution filament tracking. Biophys. J. **93**, 346–359 (2007)

98. Brangwynne, C.P., et al.: Germline P granules are liquid droplets that localize by controlled dissolution/condensation. Science **324**, 1729–1732 (2009)

99. Brangwynne, C.P., Mitchison, T.J., Hyman, A.A.: Active liquid-like behavior of nucleoli determines their size and shape in Xenopus laevis oocytes. Proc. Natl Acad. Sci. **108**, 4334–4339 (2011)

100. Brangwynne, C.P., Tompa, P., Pappu, R.V.: Polymer physics of intracellular phase transitions. Nature Phys. **11**, 899–904 (2015)

101. Braun, E.: The unforeseen challenge: from genotype-to-phenotype in cell populations. Rep. Prog. Phys. **78**, 036602 (2015)

102. Brauns, F., Halatek, J., Frey, E.: Phase-space geometry of reaction-diffusion dynamics. arXiv:1812.08684 (2020)

103. Bray, D., Levin, M.D., Morton-Firth, C.J.: Receptor clustering as a cellular mechanism to control sensitivity. Nature **393**, 85–88 (1998)

104. Bray, D.: Cell Movements, 2nd edn. Garland, New York (2001)

105. Brenner, N., Farkash, K., Braun, E.: Dynamics of protein distributions in cell populations. Phys. Biol. **3**, 172–182 (2006)

106. Brenner, N., Shokef, Y.: Nonequilibrium statistical mechanics of dividing cell populations. Phys. Rev. Lett. **99**, 138102 (2007)

107. Brenner, N., Newman, C.M., Osmanovic, D., Rabin, Y., Salman, H., Stein, D.L.: Universal protein distributions in a model of cell growth and division. Phys. Rev. E **92**, 042713 (2015)

108. Bressloff, P.C., Cowan, J.D., Golubitsky, M., Thomas, P.J., Wiener, M.: Geometric Visual Hallucinations, Euclidean Symmetry and the Functional Architecture of Striate Cortex. Phil. Trans. Roy. Soc. Lond. B **356**, 299–330 (2001)

109. Bressloff, P.C.: A stochastic model of intraflagellar transport. Phys. Rev. E **73**, 061916 (2006)

110. Bressloff, P.C.: Stochastic neural field theory and the system-size expansion. SIAM J. Appl. Math **70**, 1488 (2009)

111. Bressloff, P.C.: Two-pool model of cooperative vesicular transport. Phys. Rev. E **86**, 031911 (2012)

112. Bressloff, P.C.: Propagation of CaMKII translocation waves in heterogeneous spiny dendrites. J. Math. Biol. **66**, 1499–1525 (2013)

113. Bressloff, P.C., Newby, J.M.: Stochastic models of intracellular transport. Rev. Mod. Phys. **85**, 135–196 (2013)

114. Bressloff, P.C.: Waves in Neural Media: From Single Neurons to Neural Fields. Springer, New York (2014)

115. Bressloff, P.C., Newby, J.M.: Stochastic hybrid model of spontaneous dendritic NMDA spikes. Phys. Biol. **11**, 016006 (2014)

116. Bressloff, P.C., Newby, J.M.: Path-integrals and large deviations in stochastic hybrid systems. Phys. Rev. E. **89**, 042701 (2014)

117. Bressloff, P.C.: Aggregation-fragmentation model of vesicular transport in neurons. J. Phys. A **49**, 145601 (2016)

118. Bressloff, P. C.: Ultrasensitivity and noise amplification in a model of *V. harveyi* quorum sensing Phys Rev E **93** 062418 (2016)

119.   Bressloff, P.C., Lawley, S.D.: Dynamically active compartments coupled by a stochastically gated gap unction. J. Nonlinear Sci. **27**, 1487–1512 (2017)

120.   Bressloff, P.C., Xu, B.: Stochastic active-transport model of cell polarization. SIAM J. Appl. Appl. Math. **75**, 652–678 (2015)

121.   Bressloff, P.C., Karamched, B.R.: Doubly stochastic Poisson model of flagellar length control. SIAM J. Appl. Math. **78**, 719–741 (2018)

122.   Bressloff, P.C., Lawley, S.D.: Temporal disorder as a mechanism for spatially heterogeneous diffusion. Phys. Rev. E **95**, 060101(R) (2017)

123.   Bressloff, P.C., Lawley, S.D.: Hybrid colored noise process with space-dependent switching rates. Phys. Rev. E **96**, 012129 (2017)

124.   Bressloff, P.C., Kim, H.: Bidirectional transport model of morphogen gradient formation via cytonemes. Phys. Biol. **15**, 026010 (2018)

125.   Bressloff, P.C., Lawley, S.D., Murphy, P.: Diffusion in an age-structured randomly switching environment. J. Phys. A **51**, 315001 (2018)

126.   Bressloff, P.C.: Kim, H, Search-and-capture model of cytoneme-mediated morphogen gradient formation. Phys. Rev. E **99**, 052401 (2019)

127.   Bressloff, P.C., Lawley, S.D., Murphy, P.: Protein concentration gradients and switching diffusions. Phys. Rev. E **99**, 032409 (2019)

128.   Bressloff, P.C.: Active suppression of Ostwald ripening: Beyond mean field theory. Phys. Rev. E **101**, 042804 (2020)

129.   Bressloff, P.C.: Stochastic resetting and the mean-field dynamics of focal adhesions. Phys. Rev. E **102**, 022134 (2020)

130.   Bressloff, P.C.: Search processes with stochastic resetting and multiple targets. Phys. Rev. E **102**, 022115 (2020)

131.   Bressloff, P.C.: Queueing theory of search processes with stochastic resetting. Phys. Rev. E **102**, 032109 (2020)

132.   Bressloff, P.C.: Modeling active cellular transport as a directed search process with stochastic resetting and delays. J. Phys. A **53**, 355001 (2020)

133.   Bressloff, P. C.: Directional search-and-capture model of cytoneme-based morphogenesis. SIAM J. Appl. Math **81**, 919–938 (2021)

134.   Bronshtein, I., Israel, Y., Kepten, E., Mai, S., Shav-Tal, Y., Barkai, E., Garini, Y.: Transient anomalous diffusion of telomeres in the nucleus of mammalian cells. Phys. Rev. Lett. **103**, 018102 (2009)

135.   Bronshtein, I., Kepten, E., Kanter, I., Berezin, S., Linder, M., Redwood, A.B., Mai, S., Gonzalo, S., Foisner, R., Shav-Tal, Y., Garini, Y.: Loss of lamin A function increases chromatin dynamics in the nuclear interior. Nat. Commun. **6**, 8044 (2015)

136.   Brooks, H.A., Bressloff, P.C.: A mechanism for Turing pattern formation with active and passive transport. SIAM J. Appl. Dyn. Syst. **15**, 1823–1843 (2016)

137.   Brooks, H. A., Bressloff, P. C.: Turing mechanism for homeostatic control of synaptic density in C. elegans. Phys. Rev. E **96** 012413 (2017)

138.   Brower, R.C., Kessler, D.A., Koplik, J., Levine, H.: Geometrical approach to moving-interface dynamics. Phys. Rev. Lett. **51**, 1111–1114 (1983)

139.   Brower, R.C., Kessler, D.A., Koplik, J., Levine, H.: Geometrical models of interface evolution. Phys. Rev. A **29**, 1335–1342 (1984)

140.   Brown, G.C., Kholodenko, B.N.: Spatial gradients of cellular phospho-proteins. FEBS Lett. **457**, 452–454 (1999)

141.   Brown, S.P., Johnstone, R.A.: Cooperation in the dark: signaling and collective action in quorum-sensing bacteria. Proc. Roy Soc. Lond. B **268**, 961–965 (2001)

142.   Bruinsma, R., Grosberg, A.Y., Rabin, Y., Zidovska, A.: Chromatin hydrodynamics. Biophys. J. **106**, 1871–1881 (2014)

143.   Brumley, D.R., Carrara, F., Hein, A.M., Yawata, Y., Levin, S.A., Stocker, R.: Bacteria push the limits of chemotactic precision to navigate dynamic chemical gradients. Proc. Natl. Acad. Sci. USA **116**, 10792–10797 (2019)

144. Brust-Mascher, I., Civelekoglu-Scholey, G., Kwon, M., Mogilner, A., Scholey, J.M.: Model for anaphase B: Role of three mitotic motors in a switch from poleward flux to spindle elongation. Proc. Natl. Acad. Sci. USA **101**, 15938–15943 (2004)

145. Buettner, F., Natarajan, K.N., Casale, F.P., Proserpio, V., et al.: Computational analysis of cell-to-cell heterogeneity in single-cell. Nat. Biotechnol. **33**, 155–160 (2015)

146. Buice, M., Cowan, J.D.: Field-theoretic approach to fluctuation effects in neural networks. Phys. Rev. E **75**, 051919 (2007)

147. Buice, M.A., Chow, C.C.: Correlations, fluctuations, and stability of a finite-size network of coupled oscillators. Phys. Rev. E **76**, 031118 (2007)

148. Buice, M., Cowan, J.D., Chow, C.C.: Systematic fluctuation expansion for neural network activity equations. Neural Comput. **22**, 377 (2010)

149. Buice, M.A., Chow, C.C.: Beyond mean field theory: statistical field theory for neural networks. Journal of Statistical Mechanics **P03003** (2013)

150. Burbank, K.S., Mitchison, T.J., Fisher, D.S.: Slide-and-cluster models for spindle assembly. Curr. Biol. **17**, 1373–1383 (2007)

151. Burns, M.E., Baylor, D.A.: Activation, deactivation, and adaptation in vertebrate, photoreceptor cells. Annu. Rev. Neurosci. **24**, 779–805 (2001)

152. Burns, M.E., Arshavsky, V.Y.: Beyond counting photons: trials and trends in vertebrate visual transduction. Neuron **48**, 387–401 (2005)

153. Buszczak, M., Inaba, M., Yamashita, Y.M.: Signaling by cellular protrusions: keeping the conversation private. Trends Cell Biol **26**, 526–534 (2016)

154. Butler, T.C., Goldenfeld, N.: Robust ecological pattern formation induced by demographic noise. Phys. Rev. E. **80**, 030902(R) (2009)

155. Butler, T.C., Goldenfeld, N.: Fluctuation-driven Turing patterns. Phys. Rev. E **84**, 011112 (2011)

156. Cahn, J.W. , Hilliard, J.: Free energy of a nonuniform system. I. Interfacial free energy. J. Chem. Phys. **28** 258-267 (1958)

157. Camalet, S., Duke, T., Julicher, F., Prost, J.: Auditory sensitivity provided by self-tuned critical oscillations of hair cells. Proc. Natl. Acad. Sci. USA **97**, 3183–3188 (2000)

158. Cameron, L.A., Giardini, P.A., Soo, F.S., Theriot, J.A.: Secrets of actin-based motility revealed by a bacterial pathogen. Nat. Rev. Mol. Cell Biol. **1**, 110–119 (2000)

159. Camley, B.A.: Collective gradient sensing and chemotaxis: modeling and recent developments. J. Phys. Cond. Matt. **30**, 223001 (2018)

160. Campas, O., Leduc, C., Bassereau, P., Joanny, J.- F., Prost. J.: Collective oscillations of processive molecular motors. Biophys. Rev. Lett. **4**, 163–178 (2009)

161. Capas, O., Sens, P.: Chromosome oscillations in mitosis. Phys. Rev. Lett. **97**, 128102 (2006)

162. Carazo-Salas, R.E., et al.: Ran-GTP coordinates regulation of microtubule nucleation and dynamics during mitotic-spindle assembly. Nat. Cell Biol. **3**, 228–234 (2001)

163. Carcamo-Oyarce, G., Lumjiaktase, P., Kummerli, R., Eberl, L.: Quorum sensing triggers the stochastic escape of individual cells from Pseudomonas putida biofilms. Nat. Commun. **6**, 5945–5954 (2015)

164. Cardy, J.: Reaction-diffusion processes. In: Nazarenko, S., Zaboronski, O.V. (eds.) Nonequilibrium Statistical Mechanics and Turbulence, London Mathematical Society Lecture Note Series 108–131. Cambridge University Press, Cambridge (2008)

165. Carlsson, A.E.: Growth of branched actin networks against obstacles. Biophys. J. **81**, 1907–1923 (2001)

166. Carlsson, A.E.: Membrane bending by actin polymerization. Curr. Opin. Cell Biol. **50**, 1–7 (2018)

167. Carr, J., Penrose, O.: Asymptotic behavior of solutions to a simplified Lifshitz-Slyozov equation. Physica D **124**, 166–176 (1998)

168. Carvalho, G., Guilhen, C., Balestrino, D., Forestier, C., Mathias, J.-D.: Relating switching rates between normal and persister cells to substrate and antibiotic concentrations: a mathematical modeling approach supported by experiments. Microbial Biotech. **10**, 1616–1627 (2017)

169. Carvalho, G., Balestrino, D., Forestier, C., Mathias, J.-D.: How do environment-dependent switching rates between susceptible and persister cells affect the dynamics of biofilms faced with antibiotics? Biofilms and Microbiomes **6**, 1–8 (2018)

170. Cassimeris, L., Rieder, C., Salmon, E.: Microtubule assembly and kinetochore directional instability in vertebrate monopolar spindles: implications for the mechanism of chromosome congression. J. Cell Sci. **107**, 285–297 (1994)

171. Cassimeris, L.: The oncoprotein 18/stathmin family of microtubule destabilizers. Curr. Opin. Cell Biol. **14**, 18–24 (2002)

172. Cates, M.E., Tailleur, J.: Motility-induced phase separation. Annu. Rev. Condens. Matter Phys. **6**, 219–244 (2015)

173. Caudron, M., Bunt, G., Bastiaens, P., Karsenti, E.: Spatial coordination of spindle assembly by chromosome-mediated signaling gradients. Science **309**, 1373–1376 (2005)

174. Celton-Morizur, S., Racine, V., Sibarita, J.B., Paoletti, A.: Pom1 kinase links division plane position to cell polarity by regulating Mid1p cortical distribution. J. Cell Sci. **119**, 4710–4718 (2006)

175. Cerone, L., Novák, B., Neufeld, Z.: Mathematical model for growth regulation of fission yeast *schizosaccharomyces pombe*. **7** e49675 PLoS One (2012)

176. Chambless, J.D., Hunt, S.M., Stewart, P.S.: A three-dimensional computer model of four hypothetical mechanisms protecting biofilms from antimicrobials. Appl. Environ. Microbiol. **72**, 2005–2013 (2006)

177. Chan, C.E., Odde, D.J.: Traction dynamics of filopodia on compliant substrates. Science **322**, 1687–1691 (2008)

178. Chandler, D.: Introduction to Modern Statistical Mechanics. Oxford University Press, USA (1987)

179. Chang, F., Martin, S.G.: Shaping fission yeast with microtubules. Cold Spring Harbor Perspectives in Biology **1** (2009)

180. Chang, J.B., Ferrell, J., Jr.: E: Mitotic trigger waves and the spatial coordination of the Xenopus cell cycle. Nature. **500**, 603–607 (2013)

181. Charlebois, D.A., Balazsi, G.: Modeling cell population dynamics. Silico Biol. **13**, 21–39 (2018)

182. Chen, H.Y.: Internal states of active inclusions and the dynamics of an active membrane. Phys. Rev. Lett. **92**, 168101 (2004)

183. Chen, S., Täuber, U.C.: Non-equilibrium relaxation in a stochastic lattice Lotka?Volterra model Phys. Biol. **13**, 025005 (2016)

184. Chen, W., Huang, H., Hatori, R., Kornberg, T.B.: Essential basal cytonemes take up Hedgehog in the Drosophila wing imaginal disc. Development **144**, 3134–3144 (2017)

185. Chevance, F.F.V., Hughes, K.T.: coordinating assembly of a bacterial macromolecular machine. Nat. Rev. Microbiology. **6**, 455–465 (2008)

186. Chiang, W.Y., Li, Y.X., Lai, P.Y.: Simple models for quorum sensing: Nonlinear dynamical analysis. Phys. Rev. E **84**, 041921 (2011)

187. Chickarmane, V., Kholodenko, B.N., Sauro, H.M.: Oscillatory dynamics arising from competition and multisite phoshphorylation. J. Theor. Biol. **244**, 68–76 (2007)

188. Chien, A.-C., Hill, N.S., Levin, P.A.: Cell size control in bacteria. Curr. Biol. **22**, R340–R349 (2012)

189. Chiou, J.-G., Balasubramanian, M.K., Lew, D.J.: Cell polarity in yeast. Ann. Rev. Cell Develop. Biol. **33**, 77–101 (2017)

190. Cho, H., Jonsson, H., Campbell, K., Melke, P., Williams, J.W., Jedynak, B., Stevens, A.M., Groisman, A., Levchenko, A.: Self-organization in high-density bacterial colonies: efficient crowd control. PLOS Biol. **5**, e302 (2007)

191. Choe, Y., Magnasco, M.O., Hudspeth, A.J.: A model for amplification of hair-bundle motion by cyclical binding of Ca2+ to mechanoelectrical transduction channels. Proc. Natl. Acad. Sci. USA **95**, 15321–15326 (1998)

192. Choi, C.K., Vicente-Manzanares, M., Zareno, J., Whitmore, L.A., Mogilner, A., Horwitz, A.R.: Actin and $\alpha$-actinin orchestrate the assembly and maturation of nascent adhesions in a myosin II motor-independent manner. Nat Cell Biol **10**, 1039–1050 (2008)

193. Chopp, D.L., Kirisits, M.L., Moran, B., Parsek, M.R.: The dependence of quorum sensing on the depth of a growing biofilm. Bull. Math. Biol. **65**, 1053–1079 (2003)

194. Chou, T., Greenman, C.D.: A hierarchical kinetic theory of birth, death and fission in age-structured interacting populations. J. Stat. Phys. **164**, 49–76 (2016)

195. Chuang, J., Kantor, Y., Kardar, M.: Anomalous dynamics of translocation. Phys. Rev. E **65**, 011802 (2002)

196. Civelekoglu-Scholey, G., Sharp, D.J., Mogilner, A., Scholey, J.M.: Model of chromosome motility in Drosophila embryos: adaptation of a general mechanism for rapid mitosis. Biophys. J. **90**, 3966–3982 (2006)

197. Civelekoglu-Scholey, G., Scholey, J.M.: Mitotic force generators and chromosome segregation. Cell. Mol. Life Sci. **67**, 2231–2250 (2010)

198. Cizeau, P., Viovy, J.L.: Modeling extreme extension of DNA. Biopolymers. **42**, 383–385 (1997)

199. Clausznitzer, D., Linder, B., Julicher, F., Martin, P.: Two-state approach to stochastic hair bundle dynamics. Phys. Rev. E **77**, 041901 (2008)

200. Cogan, N.G., Keener, J.P.: The role of the biofilm matrix in structural development. Math. Med. Biol. **21**, 147–166 (2004)

201. Cogan, N.G., Keener, J.P.: Channel formation in gels. SIAM J. Appl. Math. **65**, 1839–1854 (2005)

202. Cohen, N., Boyle, J.J.: Swimming at low Reynolds number: a beginners guide to undulatory locomotion. Contemp. Phys. **51**, 103–123 (2009)

203. Cohen, M., Georgiou, M., Stevenson, N.L., Miodownik, M., Baum, B.: Dynamicfilopodia transmit intermittent Delta-Notch signaling to drive pattern refinement during lateral inhibition. Dev. Cell **19**, 78–89 (2010)

204. Collet, J.F., Poupaud, F.: Existence of solutions to coagulation-fragmentation systems with diffusion. Transport Theory Statist. Phys. **25**, 503–513 (1996)

205. Collet, J.F., Poupaud, F.: Asymptotic behavior of solutions to the diffusive fragmentation-coagulation system. Phys. D. **114**, 123–146 (1998)

206. Collins, S.R., Yang, H.W., Bonger, K.M., Guignet, E.G., Wandless, T.J., Meyer, T.: Using light to shape chemical gradients for parallel and automated analysis of chemotaxis. Mol. Syst. Biol. **11**, 804 (2015)

207. Cook, P.R., Marenduzzo, D.: Entropic organization of interphase chromosomes. J Cell Biol. **186**, 825–834 (2009)

208. Coombs, D., Straube, R., Ward, M.J.: Diffusion on a sphere with localized Traps: Mean first passage time, eigenvalue asymptotics, and Fekete points. SIAM J. Appl. Math **70**, 302–332 (2009)

209. Cooper, S.: Distinguishing between linear and exponential cell growth during the division cycle: Single-cell studies, cell-culture studies, and the object of cell cycle research. Theor. Biol. Med. Model. **3**, 10 (2006)

210. Coppey, M., Boettiger, A.N., Berezhkovskii, A.M., Shvartsman, S.Y.: Nuclear trapping shapes the terminal gradient in the Drosophila embryo. Curr. Biol. **18**, 915–919 (2008)

211. Couzin, I.D.: Collective cognition in animal groups. Trends. Cogn. Sci. **13**, 36–43 (2009)

212. Coveney, P.V., Wattis, J.A.D.: A Becker-Döring model of self-reproducing vesicles. J. Chem. Soc. Faraday Trans. **102**, 233–246 (1998)

213. Cox, D.R.: Some statistical methods connected with series of events. Journal of Royal Statistical Society B. **17**, 129–164 (1955)

214. Cox, D.R., Isham, V.: Point Processes. Chapman and Hall (1980)

215. Cox, J.S., Chapman, R.E., Walter, P.: The unfolded protein response coordinates the production of endoplasmic reticulum protein and endoplasmic reticulum membrane. Mol. Biol. Cell **8**, 1805–1814 (1997)

216. Craig, E.M., Stricker, J., Gardel, M., Mogilner, A.: Model for adhesion clutch explains biphasic relationship between actin flow and traction at the cell leading edge. Phys. Biol. **12**, 035002 (2015)

217. Crampin, E.J., Gaffney, E.A., Maini, P.K.: Reaction and diffusion on growing domains: scenarios for robust pattern formation. Bull. Math. Biol. **61**, 1093–1120 (1999)

218. Cremer, T., Cremer, M.: Chromosome territories. Cold Spring Harb. Perspect. Biol. **2**, a003889 (2010)

219. Cross, M., Hohenberg, P.: Pattern formation outside of equilibrium. Rev. Mod. Phys. **65**, 851–1112 (1993)

220. Cytrynbaum, E.N., Scholey, J.M., Mogilner, A.: A force balance model of early spindle pole separation on drosophila embryos. Biophys. J. **84**, 757–769 (2003)

221. Danino, T., Mondragon-Palomino, O., Tsimring, L., Hasty, J.: A synchronized quorum of genetic clocks. Nature **463**, 326–330 (2010)

222. Danuser, G., Allard, J., Mogilner, A.: Mathematical modeling of eukaryotic cell migration: insights beyond experiments. Annu. Rev. Cell Dev. Biol. **29**, 501–528 (2013)

223. Das, M., Drake, T., Wiley, D.J., Buchwald, P., Vavylonis, D., Verde, F.: Oscillatory dynamics of Cdc42 GTPase in the control of polarized growth. Science **337**, 239–243 (2012)

224. Davis, M.H.A.: Piecewise-deterministic Markov processes: A general class of non-diffusion stochastic models. Journal of the Royal Society, Series B (Methodological) **46** 353-388 (1984)

225. Davies, D.G., Parsek, M.R., Pearson, J.P., Iglewski, B.H., Costerton, J.W., Greenberg, E.P.: The involvement of cell-to-cell signals in the development of bacterial biofilm. Science **280**, 295–298 (1998)

226. De, P.S., De, R.: Stick-slip dynamics of migrating cells on viscoelastic substrates. Phys. Rev. E **100**, 012409 (2019)

227. Deane, J.A., Cole, D.G., Rosenbaum, J.L.: Localization of intraflagellar transport protein IFT52 identifies basal body transitional fibers as the docking site for IFT particles. Curr. Biol. **11**, 1586–1590 (2001)

228. Delorme, V., Machacek, M., DerMardirossian, C., Anderson, K.L., Wittmann, T., Hanein, D., Waterman-Storer, C., Danuser, G., Bokoch, G.M.: Cofilin activity downstream of Pak1 regulates cell protrusion efficiency by organizing lamellipodium and lamella actin networks. Dev. Cell. **13**, 646–662 (2007)

229. Dembo, M., Torney, D. C., Saxman, K., Hammer, D.: The reaction-limited kinetics of membrane-to-surface adhesion and detachment. Proc. R. Soc. London, Ser. B **234** 55-83 (1988)

230. De Monte, S., d'Ovido, F., Dano, S., Sorensen, P.G.: Dynamical quorum sensing: population density encoded in cellular dynamics. Proc. Nat. Acad. Sci. **104**, 18377–18381 (2007)

231. Deneke, V.E., Melbinger, A., Vergassola, M., Di Talia, S.: Waves of Cdk1 activity in S phase synchronize the cell cycle in Drosophila embryos. Dev. Cell. **38**, 399–412 (2016)

232. Deneke, V.E., Di Talia, S.: Chemical waves in cell and developmental biology. J. Cell Biol. **217**, 1193–1204 (2018)

233. Den Blaauwen, T., Lindqvist, A., Lowe, J., Nanninga, N.: Distribution of the Escherichia coli structural maintenance of chromosomes (SMC)-like protein MukB in the cell. Mol. Microbiol. **42**, 11799–1188 (2001)

234. Deng, Q., Ramskold, D., Reinius, B., Sandberg, R.: Single-cell RNA-seq reveals dynamic, random monoallelic gene expression in mammalian cells. Science **343**, 193–196 (2014)

235. de Pasquale, F., Gorecki, J., Poielawski., J.: On the stochastic correlations in a randomly perturbed chemical front. J. Phys. A **25** 433 (1992)

236. de Ruyter van Steveninck, R.R., Laughlin, S.B.: The rate of information transfer at graded-potential synapses. Nature **379**, 642–645 (1996)

237. de Ruyter van Steveninck, R.R., Laughlin, S.B.: Light adaptation and reliability in blowfly photoreceptors. Int. J. Neural Syst. **7**, 437–444 (1996)

238. Deshpande, V.S., Mrksich, M., McMeeking, R.M., Evans, A.G.: A bio-mechanical model for coupling cell contractility with focal adhesion formation. J. Mech. Phys. Solids **56**, 1484–1510 (2008)

239. Dhar, A., Kundu, A., Majumdar, S.N., Sabhapandit, S., Schehr, G.: Run-and-tumble particle in one-dimensional confining potentials: Steady-state, relaxation, and first-passage properties. Phys. Rev. E **99**, 032132 (2019)

240. Diamant, H., Andelman, D.: Kinetics of surfactant adsorption at fluid-fluid interfaces. J. Phys. Chem. **100**, 13732–13742 (1996)

241. Dickinson, R.B., Tranquillo, R.T.: Transport equations and indices for random and biased cell migration based on single cell properties. SIAM J. Appl. Math. **55**, 1419–1454 (1995)

242. Dickinson, R.B., Purich, D.L.: Clamped-filament elongation model for actin-based motors. Biophys. J. **82**, 605–617 (2002)

243. Dickinson, R.B.: Models for actin polymerization. J. Math. Biol. **58**, 81–103 (2009)

244. DiMilla, P.A., Barbee, K., Lauffenburger, D.A.: Mathematical model for the effects of adhesion and mechanics on cell migration speed. Biophys. J. **60**, 15–37 (1991)

245. Dmitrieff, S., Sens, P.: Cooperative protein transport in cellular organelles. Phys. Rev. E **83**, 041923 (2011)

246. Dockery, J.D., Keener, J.P.: A mathematical model for quorum-sensing in Pseudomonas aeruginosa. Bull. Math. Biol. **63**, 95–116 (2001)

247. Dockery, J., Klapper, I.: Finger formation in biofilm layers. SIAM J. Appl. Math. **62**, 853–869 (2001)

248. Doering, C. R. , Sargsyan, K. V., Sander, L. M., Vanden-Eijnden, E.: Asymptotics of rare events in birth-death processes bypassing the exact solutions. J. Phys.: Condens. Matter **19** 065145 (2007)

249. Dogterom, M., Leibler, S.: Physical aspects of the growth and regulation of microtubule structures. Phys. Rev. Lett. **70**, 1347–1350 (1993)

250. Doi, M., Edwards, S.F.: The Theory of Polymer Dynamics. Oxford Science Publications, Oxford (1986)

251. Doi, M.: Soft Matter Physics. Oxford University Press, Oxford (2013)

252. D'Orsogna, M.R., Lakatos, G., Chou, T.: Stochastic self-assembly of incommensurate clusters. J. Chem. Phys. **136**, 084110 (2012)

253. D'Orsogna, M.R., Lei, Q., Chou, T.: First assembly times and equilibration in stochastic coagulation-fragmentation. J. Chem. Phys. **143**, 014112 (2015)

254. Drake, T., Vavylonis, D.: Cytoskeletal dynamics in fission yeast: a review of models for polarization and division. HFSP journal **4**, 122–130 (2010)

255. Drescher, K., et al.: Architectural transitions in Vibrio cholerae biofilms at singlecell resolution. Proc Natl Acad Sci USA **113**, E2066–E2072 (2016)

256. Dubrovskii, V.G.: Nucleation Theory and the Growth of Nanostructures. Springer (2014)

257. Dubuis, J.O., Tkacik, G., Wieschaus, E.F., Gregor, T., Bialek, W.: Positional information, in bits. Proc. Natl. Acad. Sci. USA **110**, 16301–16308 (2013)

258. Dueck, H., Khaladkar, M., Kim, T.K., Spaethling, J.M., et al.: Deep sequencing reveals cell-type-specific patterns of single-cell transcriptome variation. Genome Biol **16**, 122 (2015)

259. Dueck, H., Eberwine, J., Kim, J.: Variation is function: Are single cell differences functionally important? Bioessays **38**, 172–180 (2015)

260. Duke, T.A.J., Bray, D.: Heightened sensitivity of a lattice of membrane of receptors. Proc. Natl. Acad. Sci. USA **96**, 10104–10108 (1999)

261. Dunlap, P.V.: Quorum regulation of luminescence in Vibrio fischeri. J. Mol. Microbiol. Biotechnol. **1**, 5–12 (1999)

262. Dykman, M.I., Schwartz, I.B., Landsman, A.S.: Phys. Rev. Lett. **101**, 078101 (2008)

263. Earnshaw, B.A., Bressloff, P.C.: Diffusion-activation model of CaMKII translocation waves in dendrites. J. Comput. Neurosci. **28**, 77–89 (2010)

264. Eberl, H.J., Parker, D.F., Van Loosdrecht, M.C.M.: A new deterministic spatio-temporal continuum model for biofilm development. J. Theor. Med. **3**, 161–175 (2001)

265. Eberwine, J., Kim, J.: Cellular deconstruction: finding meaning in individual cell variation. Trends Cell Biol **25**, 569–578 (2015)

266. Eguiluz, V.M., Ospeck, M., Choe, Y., Hudspeth, A.J., Magnasco, M.O.: Essential nonlinearities in hearing. Phys. Rev. Lett. **84**, 5232–5235 (2000)

267. Elbaum-Garfinkle, S., et al.: The disordered P granule protein LAF-1 drives phase separation into droplets with tunable viscosity and dynamics. Proc. Natl. Acad. Sci. USA **112**, 7189–7194 (2015)

268. Eldar, A., Rosin, D., Shilo, B.-Z., Barkai, N.: Self-enhanced ligand degradation underlies robustness of morphogen gradients. Dev. Cell **5**, 635–646 (2003)

269. Elgart, V., Kamenev, A.: Rare event statistics in reaction-diffusion systems. Phys. Rev. E **70**, 041106 (2004)

270. Elgart, V., Kamenev, A.: Classification of phase transitions in reaction-diffusion models Phys. Rev. E **70**, 041106 (2006)

271. Ellison, D., Muglerc, A., Brennana, M.D., Leeb, S.H., Huebnere, R.J., Shamire, E.R., Wooa, L.A., Kima, J., Amarf, P., Nemenman, I., Ewalde, A.J., Levchenko, A.: Cell-cell communication enhances the capacity of cell ensembles to sense shallow gradients during morphogenesis. Proc. Natl. Acad. Sci. USA **13**, E689-695 (2016)

272. Emerenini, B.O., Hense, B.A., Kuttler, C., Eberl, H.J.: A mathematical model of quorum sensing induced biofilm detachment. PLoS One **10**, e0132385 (2015)

273. Endres, R.G., Wingreen, N.S.: Accuracy of direct gradient sensing by single cells. Proc. Natl. Acad. Sci. USA **105**, 15749–15754 (2008)

274. Endres, R.G., Wingreen, N.S.: Maximum likelihood and the single receptor. Phys. Rev. Lett. **103**, 158101 (2009)

275. Endres, R.G., Wingreen, N.S.: Accuracy of direct gradient sensing by cell-surface receptors. Prog. Biophys. Mol. Biol. **100**, 33–39 (2009)

276. England, J.L., Cardy, J.: Morphogen gradient from a noisy source. Phys. Rev. Lett. **94**, 078101 (2005)

277. Erban, R., Othmer, H.: From individual to collective behavior in bacterial chemotaxis. SIAM J. Appl. Math. **65**, 361–391 (2005)

278. Erban, R., Othmer, H.: From signal transduction to spatial pattern formation in E. coli: a paradigm for multi-scale modeling in biology. Multiscale Model. Simul. **3** 362-394 (2005)

279. Erban, R., Chapman, S.J.: Reactive boundary conditions for stochastic simulations of reaction-diffusion processes. Phys. Biol. **4**, 16–28 (2007)

280. Erban, R., Chapman, S.J.: Stochastic modelling of reaction-diffusion processes: algorithms for bimolecular reactions. Phys. Biol. **6**, 046001 (2009)

281. Erdmann, T., Howard, M., ten Wolde, P.R.: Role of spatial averaging in the precision of gene expression patterns. Phys. Rev. Lett. **103**, 2–5 (2009)

282. Erhardt, M., Singer, H.M., Wee, D.H., Keener, J.P., Hughes, K.T.: An infrequent molecular ruler controls flagellar hook length in Salmonella enterica. EMBO J. **30**, 2948–2961 (2011)

283. Ermentrout, G.B., Cowan, J.: A mathematical theory of visual hallucination patterns. Bio. Cybern. **34**, 137–150 (1979)

284. Ermentrout, G.B., Terman, D.: Mathematical Foundations of Neuroscience. Springer, Berlin (2010)

285. Evans, L.C., Sougandis, P.E.: A PDE approach to geometric optics for certain semilinear parabolic equations. Ind. Univ. Math. J. **38**, 141–172 (1989)

286. Evans, E.: Probing the relation between force-lifetime-and chemistry in single molecular bonds. Annu. Rev. Biophys. Biomol. Struct. **30**, 105–128 (2001)

287. Evans, L.D., Poulter, S., Terentjev, E.M., Hughes, C., Fraser, G.M.: A chain mechanism for flagellum growth. Nature **504**, 287–29 (2013)

288. Evans, E., Ritchie, K.: Dynamic strength of molecular adhesion bonds. Biophys. J. **72**, 1541–1555 (1997)

289. Evans, M.R., Majumdar, S.N., Schehr, G.: Stochastic resetting and applications. J. Phys. A: Math. Theor. **53**, 193001 (2020)

290. Facchetti, G., Chang, F., Howard, M.: Controlling cell size through sizer mechanisms. Curr. Opin. Cell Biol. **5**, 86–92 (2017)

291. Fain, G.L., Matthews, H.R., Cornwall, M.C., Koutalos, Y.: Adaptation in vertebrate photoreceptors. Physiol. Rev. **81**, 117–151 (2001)

292. Fairchild, C.L., Barna, M.: Specialized filopodia: at the tip of morphogen transport and vertebrate tissue patterning. Curr. Opin. Genet. Dev. **27**, 67–73 (2014)

293. Falahati, H., Haji-Akbari, A.: Thermodynamically driven assemblies and liquid-liquid phase separations in biology. Soft Matter. **15**, 1135–1154 (2019)

294. Fan, G., Bressloff, P.C.: Population Model of quorum sensing with multiple parallel pathways. Bull. Math. Biol. **79**, 2599–2626 (2017)

295. Fan, G., Bressloff, P.C.: Modeling the role of feedback in the adaptive response of bacterial quorum sensing. Bull. Math. Biol. **81**, 1479–1505 (2019)

296. Fancher, S., Mugler, A.: Fundamental limits to collective concentration sensing in cell populations. Phys. Rev. Lett. **118**, 078101 (2017)

297. Fange, D., and Elf, J.: Noise-induced Min phenotypes in E. coli. PLoS Comput. Biol. **2** e80 (2006)

298. Fedotov, S., Tanand, A., Zubarev, S.: Persistent random walk of cells involving anomalous effects and random death. Phys. Rev. E **91**, 042124 (2015)

299. Fedotov, S., Korabel, N.: Emergence of Levy walks in systems of interacting individuals. Phys. Rev. E **95**, 030107(R) (2017)

300. Feng, J., Kurtz, T.G.: Large Deviations for Stochastic Processes. American Mathematical Society (2006)

301. Ferenz, N.P., Paul, R., Fagerstrom, C., Mogilner, A., Wadsworth, P.: Dynein antagonozes eg5 by crosslinking and sliding antiparallel microtubules. Curr. Biol. **19**, 1833–1838 (2009)

302. Feric, M., Brangwynne, C.P.: A nuclear F-actin scaffold stabilizes ribonucleoprotein droplets against gravity in large cells. Nat. Cell Biol. **15**, 1253–1259 (2013)

303. Ferreira, T., Wilson, S.R., Choi, Y.G., Risso, D., Dudoit, S., Speed, T.P., Ngai, J.: Silencing of odorant receptor genes by G protein signaling ensures the expression of one odorant receptor per olfactory sensory neuron. Neuron **81**, 847–859 (2014)

304. Ferrel, J.E., Ha, S.H.: Ultrasensitivity part I: Michaelian responses and zero-order ultrasensitivity. Trends Biochem. Sci. **39**, 496–502 (2014)

305. Ferrel, J.E., Ha, S.H.: Ultrasensitivity part II: multisiste phosphorylation, stoichiometric inhibitors, and positive feeedback. Trends Biochem. Sci. **39**, 556–569 (2014)

306. Ferrel, J.E., Ha, S.H.: Ultrasensitivity part III: cascades, bistable switches, and oscillators. Trends Biochem. Sci. **38**, 612–618 (2014)

307. Feynman, R.P.: Space-time approach to non-relativistic quantum mechanics Rev. Mod. Phys. **20**, 367–387 (1948)

308. Fife, P. C.: *Mathematical Analysis of Reacting and Diffusing Systems* Springer Berlin (1979)

309. Filippov, A.E., Klafter, J., Urbakh, M.: Friction through dynamical formation and rupture of molecular bonds. Phys. Rev. Lett. **92**, 135503 (2004)

310. Fischer-Friedrich, E., Meacci, G., Lutkenhaus, J., Chate, H., Kruse, K.: Intra- and intercellular fluctuations in Min-protein dynamics decrease with cell length. Proc. Natl. Acad. Sci. USA **107**, 6134–6139 (2010)

311. Fisher, R.A.: The wave of advance of advantageous genes. Ann. Eugenics **7**, 353–369 (1937)

312. Flory, P.J.: Principles of Polymer Chemistry. Cornell University Press, Ithaca, NY (1953)

313. Fodor, E., Marchetti, M.C.: The statistical physics of active matter: From self-catalytic colloids to living cells. Physica A **504**, 106–120 (2018)

314. Forero, M., Thomas, W.E., Bland, C., Nilsson, L.M., Sokurenko, E.V., Vogel, V.A.: A catch-bond based nanoadhesive sensitive to shear stress. Nano Lett. **4**, 1593–1597 (2004)

315. Fowler, A.C., Kyrke-Smith, T.M., Winstanley, H.F.: The development of biofilm architecture. Proc. R. Soc. A **472**, 20150798 (2016)

316. Fox, R.F., Lu, Y.N.: Emergent collective behavior in large numbers of globally coupled independent stochastic ion channels. Phys. Rev. E **49**, 3421–3431 (1994)

317. Franks, K.M., Bartol, T.M., Sejnowski, T.J.: An MCell model of calcium dynamics and frequency-dependence of calmodulin activation in dendritic spines. Neurocomputing. **38**, 9–16 (2001)

318. Franz, B., Flegg, M.B., Chapman, S.J., Erban, R.: Mutiscale reaction-diffusion algorithms: PDE-assisted Brownian dynamics. SIAM J. Appl. Math. **73**, 1224–1247 (2013)

319. Frederick, M.R., Kuttler, C., Hense, B.A., Eberl, H.J.: A mathematical model of quorum sensing regulated EPS production in biofilm communities. Theor. Biol. and Med. Model. **8**, 8 (2011)

320. Freidlin, M.I.: Limit theorems for large deviations and reaction-diffusion equations. Ann. Prob. **13**, 639–675 (1985)

321. Freidlin, M.I.: Geometric optics approach to reaction-diffusion equations. SIAM J. Appl. Math **46**, 222–232 (1986)

322. Friedman, N., Cai, L., Xies, X.S.: Linking stochastic dynamics to population distribution: an analytical framework of gene expression. Phys. Rev. Lett. **97**, 168302 (2006)

323. Fudenberg, G., Imakaev, M., Lu, C., Goloborodko, A., Abdennur, N., Mirny, L.A.: Formation of chromosomal domains by loop extrusion. Cell Rep. **15**, 2038–2049 (2016)

324. Fudenberg, G., Abdennur, N., Imakaev, M., Goloborodko, A., Mirny, L.A.: Emerging evidence of chromosome folding by loop extrusion. Cold Spring Harb. Symp. Quant. Biol. **82**, 45–55 (2017)

325. Fulinski, A.: Noise-stimulated active transport in biological cell membranes. Phys. Lett. A **193**, 267–273 (1994)

326. Fuller, D., Chen, W., Adler, M., Groisman, A., Levine, H., Rappel, W.-J., Loomis, W.F.: External and internal constraints on eukaryotic chemotaxis. Proc. Natl Acad. Sci. **107**, 9656–9659 (2010)

327. Fuqua, C., Winans, S.C., Greenberg, E.P.: Census and consensus in bacterial ecosystems: the LuxR-LuxI family of quorum-sensing transcriptional regulators. Annu. Rev. Microbiol. **50**, 727–751 (1996)

328. Ganai, N., Sengupta, S., Menon, G.I.: Chromosome positioning from activity-based segregation. Nucleic Acids Res. **42**, 4145–4159 (2014)

329. Garcia-Ojalvo, J., Elowitz, M.B., Strogatz, S.H.: Modeling a synthetic multicellular clock: Repressilators coupled by quorum sensing. Proc. Natl. Acad. Sci. U.S.A. **101**, 10955–10960 (2004)

330. Garde, C., Bjarnsholt, T., Givskov, M., Jakobsen, T.H., Hentze, M., Claussen, A., Sneppen, K., Ferkinghoff-Borg, J., Sams, T.: Quorum sensing regulation in Aeromonashydrophila. J. Mol. Biol. **396**, 849–857 (2010)

331. Gardel, M.L., Kasza, K.E., Brangwynne, C.P., Liu, J., Weitz, D.A.: Mechanical response of cytoskeletal networks. Methods Cell Biol. **89**, 487–519 (2008)

332. Gardel, M.L., Schneider, I.C., Aratyn-Schaus, Y., Waterman, C.M.: Mechanical integration of actin and adhesion dynamics in cell migration. Annu Rev. Cell Dev. Biol. **26**, 315–333 (2010)

333. Gardiner, C.W.: Handbook of Stochastic Methods, 4th edn. Springer, Berlin (2009)

334. Gardner, M.K., Odde, D.: Modeling of chromosome motility during mitosis. Curr. Opin. Cell Biol. **18**, 639–647 (2006)

335. Gardner, M.K., et al.: Tension-dependent regulation of microtubule dynamics at kinetochores can explain metaphase congress ion in yeast. Mol. Biol. Cell **16**, 3764–3775 (2005)

336. Gardner, M.K., Zanic, M., Gell, C., Bormuth, V., et al.: Depolymerizing kinesins Kip3 and MCAK shape cellular microtubule architecture by differential control of catastrophe. Cell **147**, 1092–103 (2011)

337. Gardner, M.K., Zanic, M., Howard, J.: Microtubule catastrophe and rescue. Curr. Opin. Cell Biol. **25**, 14–22 (2013)

338. Gartner, J., Freidlin, M.I.: On the propagation of concentration waves in periodic and random media. Soviet Math Dokl **20**, 1282–1286 (1979)

339. Ge, H., Qian, M.: Sensitivity amplification in the phosphorylation-dephosphorylation cycle: nonequilibrium steady states, chemical master equation and temporal cooperativity. J. Chem. Phys. **129**, 015104 (2008)

340. Ge, H., Qian, M., Qian, H.: Stochastic theory of nonequilibrium steady states. Part II: applications in chemical biophysics. Phys. Rep. **510** 87-118 (2012)

341. Gelimson, A., Golestanian, R.: Collective dynamics of dividing chemotactic cells. Phys. Rev. Lett. **114**, 028101 (2015)

342. Gerbal, F., Chaikin, P., Rabin, Y., Prost, J.: An elastic analysis of Listeria monocytogenes propulsion. Biophys. J. **79**, 2259–2275 (2000)

343. Ghosh, S., Pal, A.K., Bose, I.: Noise-induced regime shifts: A quantitative characterization Eur. Phys. J. E **36**, 123 (2013)

344. Ghosh, A., Gov, N.S.: Dynamics of active semiflexible polymers. Biophys. J. **107**, 1065–1073 (2014)

345. Giannone, G., Mege, R.M., Thoumine, O.: Multi-level molecular clutches in motile cell processes. Trends Cell Biol. **19**, 475–486 (2009)

346. Gierer, A., Meinhardt, H.: A theory of biological pattern formation. Kybernetik **12**, 30–39 (1972)

347. Gillespie, D.T.: Exact stochastic simulation of coupled chemical reactions. J. Phys. Chem. **81**, 2340–2361 (1977)

348. Gillespie, D.T.: Approximate accelerated stochastic simulation of chemically reacting systems. J. Chem. Phys. **115**, 1716–1733 (2001)

349. Giron, B., Meerson, B., Sasorov, P.V.: Weak selection and stability of localized distributions in Ostwald ripening. Phys. Rev. E **58**, 4213–4216 (1998)

350. Goehring, N.W., Trong, P.K., Bois, J.S., Chowdhury, D., Nicola, E.M., Hyman, A.A., Grill, S.W.: Polarization of PAR proteins by advective triggering of a pattern-forming system. Science **334**, 1137–1141 (2011)

351. Goldbeter, A., Koshland, D.E.: An amplified sensitivity arising from covalent modification in biological systems. Proc. Natl. Acad. Sci. USA **78**, 6840–6844 (1981)

352. Goldstein, S.: On diffusion by discontinuous movements and the telegraph equation. Quart. J. Mech. Appl. Math. **4**, 129–156 (1951)

353. Goldstein, B., Macara, I.G.: The PAR proteins: fundamental players in animal cell polarization. Dev. Cell. **13**, 609–622 (2007)

354. Goloborodko, A., Marko, J.F., Mirny, L.A.: Chromosome compaction by active loop extrusion. Biophys. J. **110**, 2162–2168 (2016)

355. Golubitsky, M., Stewart, I., Schaeffer, D.G.: Singularities and Groups in Bifurcation Theory II. Springer-Verlag, Berlin (1988)

356. Gomez, D., Marathe, R., Bierbaum, V., Klumpp, S.: Modeling stochastic gene expression in growing cells. J Theor Biol **348**, 1–11 (2014)

357. Gong, H., Guo, Y., Linstedt, A., Schwartz, R.: Discrete, continuous and stochastic models of protein sorting in the Golgi. Phys. Rev. E **81**, 011914 (2010)

358. Gopalakrishnan, M., Govindan, B.S.: A first-passage-time theory for search and capture of chromosomes by microtubules in mitosis. Bull. Math. Biol. **73**, 2483–2506 (2011)

359. Gordon, P., Sample, C., Berezhkovskii, A.M., Muratov, C.B., Shvartsman, S.Y.: Local kinetics of morphogen gradients. Proc Natl Acad Sci **108**, 6157–6162 (2011)

360. Goryachev, A.B., Toh, D.J., Wee, K.B., Lee, T., Zhang, H.B., Zhang, L.H.: Transition to quorum sensing in an Agrobacterium population: A stochastic model. PLoS Comput. Biol. **1**, e37 (2005)

361. Goryachev, A.B., Pokhilko, A.V.: Dynamics of Cdc42 network embodies a Turing-type mechanism of yeast cell polarity. FEBS Lett **582**, 1437–1443 (2008)

362. Goryachev, A.B.: Design principles of the bacterial quorum sensing gene networks. WIRE Syst. Biol. Med. **1**, 45–60 (2009)

363. Goryachev, A.B., Leda, M.: Many roads to symmetry breaking: molecular mechanisms and theoretical models of yeast cell polarity. Mol. Biol. Cell **28**, 370–380 (2017)

364. Goshima, G., Wollman, R., Stuurman, N., Scholey, J.M., Vale, R.D.: Length control of the metaphase spindle. Curr. Biol. **15**, 1979–1988 (2005)

365. Goshima, G., Scholey, J.M.: Control of Mitotic Spindle Length. Annu. Rev. Cell Dev. Biol. **26**, 21–57 (2010)

366. Gou, J., Li, Y.X., Nagata, W., Ward, M.J.: Synchronized oscillatory dynamics for a 1-D model of membrane kinetics coupled by linear bulk diffusion. SIAM J. Appl. Dyn. Sys. **14**, 2096–2137 (2015)

367. Gou, J., Ward, M.J.: Oscillatory dynamics for a coupled membrane-bulk diffusion model with Fitzhugh-Nagumo kinetics. SIAM J. Appl. Math. **76**, 776–804 (2016)

368. Gou, J., Ward, M.J.: Asymptotic analysis of a 2-D Model of dynamically active compartments coupled by bulk diffusion. J. Nonlinear Science. **26**, 979–1029 (2016)

369. Gov, N.S., Gopinathan, A.: Dynamics of membranes driven by actin polymerization. Biophys. J. **90**, 454–469 (2006)

370. Gov, N.S.: Guided by curvature: shaping cells by coupling curved membrane proteins and cytoskeletal forces. Roy. Soc. Phil Trans. B **373**, 20170115 (2018)

371. Govindan, B.S., Gopalakrishnan, M., Chowdhury, D.: Length control of microtubules by depolymerizing motor proteins. Europhys. Lett. **83**, 40006 (2008)

372. Gradilla, A.C., Guerrero, I.: Cytoneme-mediated cell-to-cell signaling during development. Cell Tissue Res. **352**, 59–66 (2013)

373. Grandell, J.: Doubly Stochastic Process, 1st edn. Springer-Verlang, New York (1976)

374. Greenman, C.D., Chou, T.: A kinetic theory for age-structured stochastic birth-death processes. Phys. Rev. E **93**, 012112 (2016)

375. Gregor, T., Tank, D.W., Wieschaus, E.F., Bialek, W.: Probing the limits to positional observation. Cell **130**, 153–164 (2007)

376. Gregor, T., Fujimoto, K., Masaki, N., Sawai, S.: The onset of collective behavior in social amoeba. Science. **328**, 1021–1025 (2010)

377. Grill, S.W., Kruse, K., Julicher, F.: Theory of mitotic spindle oscillations. Phys. Rev. Let. **94**, 108104 (2005)

378. Grimmett, G.R., Stirzaker, D.R.: Probability and Random Processes, 3rd edn. Oxford University Press, Oxford (2001)

379. Guo, B., Guilford, W.H.: Mechanics of actomyosin bonds in different nucleotide states are tuned to muscle contraction. Proc. Natl. Acad. Sci. U.S.A. **103**, 9844–9849 (2006)

380. Guo, W.H., Wang, Y.L.: Retrograde fluxes of focal adhesion proteins in response to cell migration and mechanical signals. Mol. Biol. Cell **18**, 4519–4527 (2007)

381. Gupta, M.L., Carvalho, P., Roof, D.M., Pellman, D.: Plus end-specific depolymerase activity of Kip3, a kinesin-8 protein, explains its role in positioning the yeast mitotic spindle. Nat. Cell Biol. **8**, 913–923 (2006)

382. Gupta, P.B., Fillmore, C.M., Jiang, G., Shapira, S.D., et al.: Stochastic state transitions give rise to phenotypic equilibrium in populations of cancer cells. Cell **146**, 633–644 (2011)

383. Gupta, A.: Stochastic model for cell polarity. Ann. Appl. Prob. **22**, 827–859 (2012)

384. Gupta, K.K., Li, C., Goodson, H.V.: Mechanism for the catastrophe-promoting activity of the microtubule destabilizer Op18/ stathmin. Proc. Natl. Acad. Sci. USA **110**, 20449–20454 (2013)

385. Haastert, P.J.V., Devreotes, P.N.: Chemotaxis: signaling the way forward. Nat. Rev. Mol. Cell Biol. **5**, 626–634 (2004)

386. Haastert, P.J.V., Postma, M.: Biased random walk by stochastic fluctuations of chemoattractant-receptor interactions at the lower limit of detection. Biophys. J. **93**, 1787–1796 (2007)

387. Hadjivasiliou, Z., Hunter, G.L., Baum, B.: A new mechanism for spatial pattern formation via lateral and protrusion-mediated lateral signalling. J. R. Soc. Interface **13**, 20160484 (2016)

388. Hadjivasiliou, Z., Moore, R.E., McIntosh, R., Galea, G.L., Clarke, J.D.W., Alexandre, P.: Basal protrusions mediate spatiotemporal patterns of spinal neuron differentiation. Dev. Cell **49**, 907–919 (2019)

389. Hagan, M.F., Baskaran, A.: Emergent self-organization in active materials. Curr. Opin. Cell Biol. **38**, 74–80 (2016)
390. Halatek, J., Frey, E.: Highly canalized MinD transfer and MinE sequestration explain the origin of robust minCDE-protein dynamics. Cell Reports **1**, 741–752 (2012)
391. Halatek, J., Brauns, F., Frey, E.: Self-organization principles of intracellular pattern formation. Phil. Trans. R. Soc. B **373**, 20170107 (2018)
392. Halatek, J., Frey, E.: Rethinking pattern formation in reaction-diffusion systems. Nat. Phys. **14**, 507–514 (2018)
393. Hall, A.: Rho GTPases and the actin cytoskeleton. Science **279**, 509–514 (1998)
394. Hamer, R.D., Nicholas, S.C., Tranchina, D., Liebman, P.A., Lamb, T.D.: Multiple steps of phosphorylation of activated rhodopsin can account for the reproducibility of vertebrate rod single-photon responses. J. Gen. Physiol. **122**, 419–444 (2003)
395. Hamer, R.D., Nicholas, S.C., Tranchina, D., Lamb, T.D., Jarvinen, J.L.: Toward a unified model of vertebrate rod phototransduction. Visual Neurosci. **22**, 417–436 (2005)
396. Hanggi, P., Grabert, H., Talkner, P., Thomas, H.: Bistable systems: master equation versus Fokker-Planck modeling. Phys. Rev. A **29**, 371–378 (1984)
397. Hanggi, P., Talkner, P., Borkovec, M.: Reaction rate theory: fifty years after Kramers. Rev. Mod. Phys. **62**, 251–341 (1990)
398. Hardie, R.C., Juusola, M.: Phototransduction in Drosophila. Curr. Opin. Neurobiol. **34**, 37–45 (2015)
399. Hawkins, R.J., Benichou, O., Piel, M., Voituriez, R.: Rebuilding cytoskeleton roads: active-transport-induced polarization of cell. Phys. Rev. E **80**, 040903(R) (2009)
400. Head, D.A., Levine, A.J., MacKintosh, F.C.: Deformation of cross-linked semiflexible polymer networks. Phys. Rev. Lett. **91**, 108102 (2003)
401. Head, D.A., Levine, A.J., MacKintosh, F.C.: Distinct regimes of elastic response and deformation modes of cross-linked cytoskeletal and semiflexible polymer networks. Phys. Rev. E **68**, 061907 (2003)
402. Heald, R., Khodjakov, A.: Thirty years of search and capture: The complex simplicity of mitotic spindle assembly. J. Cell Biol. **211**, 1103–1111 (2015)
403. Heberle, F. A., W. Feigenson, G. W.: Phase separation in lipid membranes. Cold Spring Harb. Perspect. Biol. **3** a004630 (2011)
404. Hecht, S., Shlaer, S., Pirenne, M.H.: Energy quanta and vision. J. Gen. Physiol. **25**, 819–840 (1942)
405. Hein, A.M., Brumley, D.R., Carrara, F., Stocker, R., Levin, S.A.: Physical limits on bacterial navigation in dynamic environments. J. Roy. Soc. Interface **13**, 20150844 (2016)
406. Heinrich, R., Rapoport, T.A.: Generation of nonidentical compartments in vesicular transport systems. J. Cell Biol. **168**, 271–280 (2005)
407. Hense, B.A., Kuttler, C., Muller, J., Rothballer, M. Hartmann, A., Kreft, J-U.: Does efficiency sensing unify diffusion and quorum sensing? Nat. Rev. Microbiol. **5** 230-239 (2007)
408. Hernandez-Verdun, D., et al.: The nucleolus: structure/function relationship in RNA metabolism. Wiley Interdiscip. Rev. RNA **1**, 415–431 (2010)
409. Heun, P., Laroche, T., Shimada, K., Furrer, P., Gasser, S.M.: Chromosome dynamics in the yeast interphase nucleus. Science **294**, 2181–2186 (2001)
410. Hill, T.L.: Theoretical problems related to the attachment of microtubules to kinetochores. Proc. Natl. Acad. Sci. U. S. A. **82**, 4404–4408 (1985)
411. Hillen, T.: A Turing model with correlated random walk. J. Math. Biol. **35**, 49–72 (1996)
412. Hillen, T., Othmer, H.: The diffusion limit of transport equations derived from velocity-jump processes. SIAM J. Appl. Math. **61**, 751–775 (2000)
413. Hillen, T., Painter, K.J.: A user's guide to PDE models for chemotaxis. J. Math. Biol. **58**, 183–217 (2009)
414. Hillen, T., Swan, A.: The diffusion limit of transport equations in biology In: *Mathematical Models and Methods for Living Systems*. L. Preziosi et al (eds.) 3-129 (2016)
415. Hinch, R., Chapman, S.J.: Exponentially slow transitions on a Markov chain: the frequency of calcium sparks. Eur. J. Appl. Math. **16**, 427–446 (2005)

416. Hingant, E., Yvinec, R.: Deterministic and stochastic Becker-Döring Equations: Past and recent mathematical developments. In: *Stochastic processes, multiscale modeling, and numerical methods for computational cellular biology.* D. Holcman (ed) Springer, Cham (2017)

417. Hinshelwood, C.N.: On the chemical kinetics of autosynthetic systems. J. Chem. Soc. **136**, 745–755 (1952)

418. Ho, P.-Y., Lin, J., Amir, A.: Modeling cell size regulation: from single-cell-level statistics to molecular mechanisms and population-level effects. Annu. Rev. Biophys. **47**, 251–71 (2018)

419. Hoerndli, F.J., Wang, R., Mellem, J.E., Kallarackal, A., Brockie, P.J., Thacker, C., Jensen, E., Madsen, D.M., Maricq, A.V.: Neuronal activity and CaMKII regulate kinesin-mediated transport of synaptic AMPARs. Neuron **86**, 457–474 (2015)

420. Hoerndli, F.J., Maxfield, D.A., Brockie, P.J., Mellem, J.E., Jensen, E., Wang, R., Madsen, D.M., Maricq, A.V.: Kinesin-1 regulates synaptic strength by mediating the delivery, removal, and redistribution of AMPA receptors. Neuron **80**, 1421–1437 (2013)

421. Hofmann, A., Makela, J., Sherratt, D. J., Heermann, D., Murray, S. M.: Self-organised segregation of bacterial chromosomal origins. eLife **8** e46564 (2019)

422. Holmes, W.R., Lin, B., Levchemko, A., Edelstein-Keshet, L.: Modelling cell polarization driven by synthetic spatially graded Rac activation. PLoS Comp. Biol. **8**, e1002366 (2012)

423. Holmes, W.R., Edelstein-Keshet, L.: A comparison of computational models for eukaryotic cell shape and motility. PLoS Comput. Biol. **8**, e1002793 (2012)

424. Holy, T.E., Leibler, S.: Dynamic instability of microtubules as an efficient way to search in space. Proc. Natl. Acad. Sci. U. S. A. **91**, 5682–5685 (1994)

425. Horowitz, J.R., Kardar, M.: Bacterial range expansions on a growing front: Roughness, fixation, and directed percolation. Phys. Rev. E **99**, 042134 (2019)

426. Hough, L.E., Schwabe, A., Glaser, M.A., McIntosh, J.R., Betterton, M.D.: Microtubule depolymerization by the kinesin-8 motor Kip3p: a mathematical model. Biophys. J. **96**, 3050–3064 (2009)

427. Howard, J.: *Mechanics of Motor Proteins and the Cytoskeleton.* Sinauer (2001)

428. Howard, M., Rutenberg, A. D., de Vet, S.: Dynamic compartmentalization of bacteria: accurate division in *E. coli*. Phys. Rev. Lett. **87** 278102 (2001)

429. Howard, M., Rutenberg, A.D.: Pattern formation inside bacteria: Fluctuations due to the low copy number of proteins. Phys. Rev. Lett. **90**, 128102 (2003)

430. Howard, M., Kruse, K.: Cellular organization by self-organization: Mechanisms and models for Min protein dynamics. J. Cell Biol. **168**, 533–536 (2005)

431. Howard, M.: How to build a robust intracellular concentration gradient. Trends Cell Biol. **22**, 311–317 (2012)

432. Howell, A.S., Savage, N.S., Johnson, S.A., Bose, I., Wagner, A.W., Zyla, T.R., Nijhout, H.F., Reed, M.C., Goryachev, A.B., Lew, D.J.: Singularity in polarization: rewiring yeast cells to make two buds. Cell **139**, 731–743 (2009)

433. Hoyle, R.: *Pattern Formation: An Introduction to Methods.* Cambridge Texts in Applied Mathematics. (2006)

434. Hoze, N., Holcman, D.: Kinetics of aggregation with a finite number of particles and application to viral capsid assembly. J. Math. Biol. **70**, 1685–1705 (2015)

435. Hu, B., Chen, W., Rappel, W.J., Levine, H.: Physical limits on cellular sensing of spatial gradients. Phys. Rev. Lett. **105**, 048104 (2010)

436. Hu, B., Fuller, D., Loomis, W.F., Levine, H., Rappel, W.J.: Phenomenological approach to eukaryotic chemotactic efficiency. Phys. Rev. E **81**, 031906 (2010)

437. Hu, B., Chen, W., Rappel, W.J., Levine, H.: How geometry and internal bias affect the accuracy of eukaryotic gradient sensing. Phys. Rev. E **83**, 021917 (2011)

438. Hu, J., Kang, H.-W., Othmer, H.G.: Stochastic analysis of reaction-diffusion processes. Bull. Math. Biol. **76**, 854–894 (2014)

439. Hu, B., Rappel, W.J., Levine, H.: How input noise limits biochemical sensing in ultrasensitive systems. Phys. Rev. E **90**, 032702 (2014)

440. Huang, K.C., Meir, Y., Wingreen, N.S.: Dynamic structures in Escherichia coli: Spontaneous formation of MinE rings and MinD polar zones. Proc. Natl. Acad. Sci. USA **100**, 12724–12728 (2003)

441. Huang, Q., Qian, H.: The dynamics of zeroth-order ultrasensitivity: a critical phenomenon in cell biology. Discr. Cont. Dyn. Sys. S **4**, 1457–1464 (2011)

442. Huber, F., Schnauss, J., Ronicke, S., Rauch, P., Muller, K., Futterer, C., Kas, J.E.: Emergent complexity of the cytoskeleton: from single filaments to tissue. Adv. Phys. **62**, 1–12 (2013)

443. Hudspeth, A.J.: Mechanical amplification of stimuli by hair cells. Curr. Opin. Neurobiol. **7**, 480–486 (1997)

444. Hudspeth, A.J.: Making an effort to listen: mechanical amplification in the ear. Neuron **59**, 530–545 (2008)

445. Hufton, P.G., Lin, Y.T., Galla, T., McKane, A.J.: Intrinsic noise in systems with switching environments. Phys. Rev. E **93**, 052119 (2016)

446. Hufton, P.G., Lin, Y.T., Galla, T.: Phenotypic switching of populations of cells in a stochastic environment. J. Stat. Mech. **023501** (2018)

447. Hughes, K.T.: Flagellum length control: How long is long enough? Current Biol. **27**, R413–R415 (2017)

448. Huh, D., Paulsson, J.: Non-genetic heterogeneity from stochastic partitioning at cell division. Nat Genet **43**, 95–100 (2011)

449. Huh, D., Paulsson, J.: Random partitioning of molecules at cell division. Proc. Natl. Acad. Sci. USA **108**, 15004–15009 (2011)

450. Hunter, G.A.M., Guevara Vasquez, F., Keener, J.P.: A mathematical model and quantitative comparison of the small RNA circuit in the Vibrio harveyi and Vibrio cholerae quorum sensing systems. Phys. Biol. **10**, 046007 (2013)

451. Hyman, A.A., Weber, C.A., Julicher, F.: Liquid-liquid phase separation in biology. Annu. Rev. Cell Dev. Biol. **30**, 39–58 (2014)

452. Iannelli, M., Milner, F.: The basic approach to age-structured population dynamics: models, methods and numerics. Lecture notes on mathematical modelling in the life sciences. Springer (2017)

453. Iarovaia, O.V., Minina, E.P., Sheval, E.V., Onichtchouk, D., Dokudovskaya, S., Razin, S.V., Vassetzky, Y.S.: Nucleolus: A central hub for nuclear functions. Trends Cell Biol. **29**, 647–659 (2019)

454. Iino, T.: Assembly of Salmonella flagellin in vitro and in vivo. J. Supramol. Struct. **2**, 372–384 (1974)

455. Inoue, S., Salmon, E.D.: Force generation by microtubule assembly/disassembly in mitosis and related movements. Mol. Cell Biol. **6**, 1619–1640 (1995)

456. Iron, D., Ward, M.J., Wei, J.: The stability of spike solutions to the one-dimensional Gierer-Meinhardt model. Physica D **150**, 25–62 (2001)

457. Iron, D., Ward, M.J.: The Dynamics of multi-spike solutions for the one-dimensional Gierer-Meinhardt model. SIAM J. Appl. Math. **62**, 1924–1951 (2002)

458. Iron, D., Wei, J., Winter, M.: Stability Analysis of Turing Patterns Generated by the Schnakenberg model. J. Math. Biol. **49**, 358–390 (2004)

459. Isaacson, S., Peskin, C.: Incorporating diffusion in complex geometries into stochastic chemical kinetics simulations. SIAM J. Sci. Comp. **28**, 47–74 (2006)

460. Isaacson, S.A.: The reaction-diffusion master equation as an asymptotic approximation of diffusion to a small target. SIAM J. Appl. Math. **7**, 77–111 (2009)

461. Ishihara, K., Nguyen, P. A., Wühr, M., Groen, A. C., Field, C. M., Mitchison, T. J.: Organization of early frog embryos by chemical waves emanating from centrosomes. Philos. Trans. R. Soc. Lond. B Biol. Sci. **369** 369 (2014)

462. Ishihara, S., Otsuji, M., Mochizuki, A.: Transient and steady state of mass-conserved reaction-diffusion systems. Phys. Rev. E **75**, 015203(R) (2007)

463. Israelachvili, J.N.: Intermolecular and Surface Forces, 2nd edn. Academic Press, New York (1991)

464. Iyer-Biswas, S., Crooks, G.E., Scherer, N.F., Dinner, A.R.: Universality in stochastic exponential growth. Phys. Rev. Lett. **113**, 028101 (2014)

465. Iyer-Biswas, S., Wright, C.S., Henry, J.T., Lo, K., Burov, S., Lin, Y., Crooks, G.E., Crosson, S., Dinner, A.R., Scherer, N.F.: Scaling laws governing stochastic growth and division of single cell bacteria. Proc. Natl. Acad. Sci. USA **111**, 155912–15917 (2014)

466. Jain, S., et al.: ATPase-modulated stress granules contain a diverse proteome and substructure. Cell **164**, 487–498 (2016)

467. James, S., Nilsson, P., James, G., Kjelleberg, S., Fagerstrom, T.: Luminescence control in the marine bacterium Vibrio fischeri: an analysis of the dynamics of lux regulation. J. Mol. Biol. **296**, 1127–1137 (2000)

468. Janetopoulos, C., Ma, L., Devreotes, P.N., Iglesias, P.A.: Chemoattractant-induced phosphatidylinositol 3,4,5- trisphosphate accumulation is spatially amplified and adapts, independent of the actin cytoskeleton. Proc. Natl. Acad. Sci. USA **101**, 8951–8956 (2004)

469. Janson, M.E., de Dood, M.E., Dogterom, M.: Dynamic instability of microtubules is regulated by force. J. Cell Biol. **161**, 1029–1034 (2003)

470. Jafarpour, F., Wright, C.S., Gudjonson, H., Ridebling, J., Dawson, E., Lo, K., Fiebig, A., Crosson, S., Dinner, A.R., Iyer-Biswas, S.: Bridging the timescales of single-cell and population dynamics. Phys. Rev. X **8**, 021007 (2018)

471. Jarsch, I.K., Daste, F., Gallop, J.L.: Membrane curvature in cell biology: an integration of molecular mechanisms. J. Cell. Biol. **214**, 375–387 (2016)

472. Jemseema, V., Gopalakrishnan, M.: Effects of aging in catastrophe on the steady state and dynamics of a microtubule population. Phys. Rev. E **91**, 052704 (2015)

473. Jilkine, A., Maree, A.F.M., Edelstein-Keshet, L.: Mathematical model for spatial segregation of the Rho-family GTPases based on inhibitory crosstalk. Bull. Math. Biol. **68**, 1169–1211 (2007)

474. Jilkine, A., Edelstein-Keshet, L.: A comparison of mathematical models for polarization of single eukaryotic cells in response to guided cues. PLoS Comput Biol **7**, e1001121 (2011)

475. Jilkine, A., Angenent, S.B., Wu, L.F., Altschuler, S.J.: A density-dependent switch drives stochastic clustering and polarization of signaling molecules. PLoS Comp. Biol. **7**, e1002271 (2011)

476. Joglekar, A.P., Hunt, A.J.: A Simple, Mechanistic Model for Directional Instability during Mitotic Chromosome Movements. Biophys. J. **83**, 42–58 (2002)

477. Johann, D., Erlenkamper, C., Kruse, K.: Length regulation of active biopolymers by molecular motors. Phys. Rev. Lett. **108**, 258103 (2012)

478. Johnson, J.M., Jin, M., Lew, D.J.: Symmetry breaking and the establishment of cell polarity in budding yeast. Curr. Opin. Gen. Dev. **21**, 740–746 (2011)

479. Jun, S., Si, F., Pugatch, R., Scott, M.: Fundamental principles in bacterial physiology-history, recent progress, and the future with focus on cell size control: a review. Rep. Prog. Phys. **81**, 056601 (2018)

480. Jung, S.A., Hawver, L.A., Ng, W.L.: Parallel quorum sensing signaling pathways in Vibrio cholerae Curr. Genet. **62**, 255 (2016)

481. Jurado, C., Haserick, J.R., Lee, J.: Slipping or gripping? Fluorescent speckle microscopy in fish keratocytes reveals two different mechanisms for generating a retrograde flow of actin. Mol. Biol. Cell **16**, 507–518 (2005)

482. Kabaso, D., Shlomovitz, R., Schloen, K., Stradal, T., Gov, N.S.: Theoretical model for cellular shapes driven by protrusive and adhesive forces. PLoS Comput Biol **7**, e1001127 (2011)

483. Kac, M.: On the distribution of certain Wiener functionals. Trans. Am. Math. Soc. **65**, 1–13 (1949)

484. Kac, M.: A stochastic model related to the telegrapher's equation. Rocky Mountain J. Math. **3**, 497–509 (1974)

485. Kaizu, K., de Ronde, W., Paijmans, J., Tkahashi, K., Tostevin, F., ten Wolde, P.R.: The Berg-Purcell limit revisited. Biophys. J. **106**, 976–985 (2014)

486. Kalab, P., Weis, K.: Heald, R: Visualization of a Ran-GTP gradient in interphase and mitotic Xenopus egg extracts. Science **295**, 2452–2456 (2002)

487. Kamenev, A.: *Field Theory of Non-Equilibrium Systems* Cambridge University Press, Cambridge (2014)
488. Karamched, B.R., Ott, W., Timofeyev, I., Alnahhas, R.N., Bennett, M.R., Josic, K.: Moran model of spatial alignment in microbial colonies. Physica D **395**, 1–6 (2019)
489. Kardar, M.: Statistical Physics of Particles. Cambridge University Press, Cambridge (2007)
490. Karmakar, R., Bose, I.: Graded and binary responses in stochastic gene expression. Phys. Biol. **1**, 197–204 (2004)
491. Katsura, I.: Determination of bacteriophage $\lambda$ tail length by a protein ruler. Nature **327**, 73–75 (1987)
492. Kavanagh, E.: Interface motion in the Ostwald ripening and chemotaxis systems. Master Thesis, University of British Columbia (2014)
493. Kay, S.: Fundamentals of Statistical Signal Processing: Estimation Theory. Prentice Hall, NJ (1993)
494. Kechagia, J.Z., Ivaska, J., Roca-Cusachs, P.: Integrins as biomechanical sensors of the microenvironment. Mol. Cell Biol. **20**, 457–473 (2019)
495. Keener, J.P.: How Salmonella typhimurium measures the length of flagellar filaments. Bull. Math. Biol. **68**, 1761–1778 (2006)
496. Keener, J.P., Sneyd, J.: Mathematical Physiology I: Cellular Physiology, 2nd edn. Springer, New York (2009)
497. Keener, J.P., Sneyd, J.: Mathematical Physiology II: Systems Physiology, 2nd edn. Springer, New York (2009)
498. Keener, J.P., Newby, J.M.: Perturbation analysis of spontaneous action potential initiation by stochastic ion channels. Phy. Rev. E **84**, 011918 (2011)
499. Keller, E., Segel, L.: Initiation of slime mold aggregation viewed as an instability. J. Theoret. Biol. **26**, 399–415 (1970)
500. Keller, E., Segel, L.: Model for chemotaxis. J. Theoret. Biol. **30**, 225–234 (1971)
501. Kepler, T.B., Elston, T.C.: Stochasticity in transcriptional regulation: origins, consequences, and mathematical representations. Biophys J. **81**, 3116–3136 (2001)
502. Keren, K., Pincus, Z., Allen, G.M., Barnhart, E.L., Marriott, G., Mogilner, A., Theriot, J.A.: Mechanism of shape determination in motile cells. Nature **453**, 475–481 (2008)
503. Keyfitz, N., Caswell, H.: *Appl. Math. Demogr.* 3rd edn. Springer, New York (2005)
504. Khasin, M., Dykman, M.I., Meerson, B.: Speeding up disease extinction with a limited amount of vaccine Phys. Rev. E **81**, 051925 (2010)
505. Kholodenko, B.N.: Spatially distributed cell signalling. FEBS Letters **583**, 4006–4012 (2009)
506. Kifer, Y.: Large deviations and adiabatic transitions for dynamical systems and Markov processes in fully coupled averaging. Memoirs of the AMS **201** issue 944 (2009)
507. Kim, M.K., Ingremeau, F., Zhao, A., Bassler, B.L., Stone, H.A.: Local and global consequences of flow on bacterial quorum sensing. Nat. Microbiol. **1**, 15005 (2016)
508. Kim, H., Bressloff, P. C.. Direct vs. synaptic contacts in a mathematical model of cytoneme-based morphogen gradient formation SIAM J. Appl. Math **78** 2323-2347 (2018)
509. Kim, H., Bressloff, P.C.: Impulsive signaling model of cytoneme-based morphogen gradient formation. Phys. Biol. **16**, 056005 (2019)
510. Kingman, J.F.C.: Poisson Processes. Clarendon Press, Oxford (1993)
511. Kirschner, M., Mitchison, T.: Beyond self-assembly: from microtubules to morphogenesis. Cell **45**, 329–342 (1986)
512. Klann, M., Koeppl, H., Reuss, M.: Spatial modeling of vesicle transport and the cytoskeleton: The challenge of hitting the right road. PLoS One **7**, e29645 (2012)
513. Klapper, I.: Effect of heterogeneous structure in mechanically unstressed biofilms on overall growth. Bull. Math. Biol. **66**, 809–824 (2004)
514. Klapper, I., Dockery, J.: Mathematical description of microbial biofilms. SIAM Rev. **52**, 221–265 (2010)
515. Klumpp, S., Zhang, Z., Hwa, T.: Growth rate-dependent global effects on gene expression in bacteria. Cell **139**, 1366 (2009)

516. Klumpp, S., Hwa, T.: Bacterial growth: global effects on gene expression, growth feedback and proteome partition. Curr. Opin. Biotechnol. **28**, 96–102 (2014)

517. Kolmogorff, A., Petrovsky, I., Piscounoff, N.: Étude de l'équation de la diffusion avec croissance de la quantité de matière et son application à un problème biologique. Moscow University Bull Math **1**, 1–25 (1937)

518. Kolokolnikov, T., Ward, M.J., Wei, J.: The existence and stability of spike equilibria in the one-dimensional Gray-Scott model: The low feed rate regime. Studies in Appl. Math. **115**, 21–71 (2005)

519. Kolokolnikov, T., Ward, M.J., Wei, J.: The existence and stability of spike equilibria in the one-dimensional Gray-Scott model: The pulse-splitting regime. Studies in Appl. Math. **115**, 258–293 (2005)

520. Kondo, S., Asai, R.: A reaction-diffusion wave on the skin of the marine angelfish Pomacanthus. Nature **376**, 765–768 (1995)

521. Konur, S., Rabinowitz, D., Fenstermaker, V.L., Yuste, R.: Systematic regulation of spine sizes and densities in pyramidal neurons. J. Neurobiol. **56**, 95–112 (2003)

522. Kornberg, T.B.: Cytonemes and the dispersion of morphogens. WIREs Dev Biol **3**, 445–463 (2014)

523. Kornberg, T.B., Roy, S.: Cytonemes as specialized signaling filopodia. Development **141**, 729–736 (2014)

524. Koster, G., VanDuijn, M., Hofs, B., Dogterom, M.: Membrane tube formation from giant vesicles by dynamic association of motor proteins. Proc. Natl. Acad. Sci. USA **100**, 15583–15588 (2003)

525. Kozlov, M.M., Campelo, F., Liska, N., Chernomordik, L.V., Marrink, S.J., McMahon, H.T.: Mechanisms shaping cell membranes. Curr. Opin. Cell. Biol. **29**, 53–60 (2014)

526. Kretschmer, S., Zieske, K., Schwille, P.: Large-scale modulation of reconstituted Min protein patterns and gradients by defined mutations in MinE's membrane targeting sequence. PLoS ONE **12**, e0179582 (2017)

527. Krumin, M., Shoham, S.: Generation of spike trains with controlled auto- and cross-correlation functions. Neural Comput. **21**, 1642–1664 (2009)

528. Kruse, K.: A dynamic model for determining the middle of Escherichia coli. Biophys. J. **82**, 618–627 (2002)

529. Kuan, H.-S., Betterton, M.D.: Biophysics of filament length regulation by molecular motors. Phys. Biol. **10**, 036004 (2013)

530. Kubo, R., Toda, M., Hahitsume, N.: Statistical Physics II: Nonequilibrium Statistical Mechanics. Springer (1991)

531. Kumar, N., Singh, A., Kulkarni, R.V.: Transcriptional bursting in gene expression: analytical results for general stochastic models. PLoS Comp. Biol. **11**, e1004292 (2015)

532. Kupiec, M.: Biology of telomeres: lessons from budding yeast. FEMS Microbiol. Rev. **38**, 144–171 (2014)

533. Kuramoto, Y.: Chemical Oscillations. Springer-Verlag, New-York, Waves and Turbulence (1984)

534. Kurtz, T.G.: Limit theorems and diffusion approximations for density dependent Markov chains. Math. Prog. Stud. **5**, 67–78 (1976)

535. Kurtz, T.G.: Representations of Markov processes as multiparameter changes. Ann. Prob. **8**, 682–715 (1980)

536. Kussell, E., Leibler, S.: Growth and information in fluctuating environments. Science **309**, 2075–2078 (2005)

537. Kussell, E., Kishony, R., Balaban, N.Q., Leibler, S.: Bacterial persistence: a model of survival in changing environments. Generics **169**, 1807–1814 (2005)

538. Ladoux, B., Nicolas, A.: Physically based principles of cell adhesion mechanosensitivity in tissues. Rep. Prog. Phys. **75**, 116601 (2012)

539. Lam, Y.W., Trinkle-Mulcahy, L., Lamond, A.I.: The nucleolus. J. Cell Sci. **118**, 1335–1337 (2005)

540. Landau, L.D., Lifshitz, E.M.: Theory of Elasticity, 3rd edn. Butterworth and Heinemann, Oxford, UL (1987)

541. Lander, A.D., Nie, W., Wan, F.Y.: Do morphogen gradients arise by diffusion? Dev Cell **2**, 785–796 (2002)

542. Lander, A.D., Lo, W.C., Nie, W., Wan, F.Y.: The measure of success: constraints, objectives, and tradeoffs in morphogen-mediated patterning. Cold Spring Harb Perspect Biol **1**, a002022 (2009)

543. Lander, A.D.: Pattern, growth and control. Cell **144**, 955–969 (2011)

544. Lando, D.: On Cox processes and credit risky securities. Review of Derivatives Research. **2**, 99–120 (1998)

545. Laurencot, Ph., Wrzosek, D.: The Becker-Döring model with diffusion. I. Basic properties of solutions, Colloq. Math. **75** 245-269 (1998)

546. Lawson, M.J., Drawert, B., Khammash, M., Petzold, L., Yi, T.-M.: Spatial stochastic dynamics enable robust cell polarization. PLoS Comp. Biol. **9**, e1003139 (2012)

547. Layton, A.T., Savage, N.S., Howell, A.S., Carroll, S.Y., Drubin, D.G., Lew, D.J.: Modeling vesicle traffic reveals unexpected consequences for Cdc42p-mediated polarity establishment. Curr. Biol. **21**, 184–194 (2011)

548. Leduc, C., et al.: Cooperative extraction of membrane nanotubes by molecular motors. Proc. Natl. Acad. Sci. USA **101**, 17096–17101 (2004)

549. Lee, C.F., Brangwynne, C.P., Gharakhani, J., Hyman, A.A., Julicher, F.: Spatial organization of the cell cytoplasm by position-dependent phase separation. Phys. Rev. Lett. **111**, 088101 (2013)

550. Lee, C.F., Wurtz, J.D.: Novel physics arising from phase transitions in biology. J. Phys. D **52**, 023001 (2019)

551. Lefebvre, P.A., Rosenbaum, J.L.: Regulation of the synthesis and assembly of ciliary and flagellar proteins during regeneration. Ann. Rev. Cell Biol. **2**, 517–546 (1986)

552. Lenz, D.D.H., Mok, K.C., Lilley, B.N., Kulkarni, R.V., Wingreen, N.S., Bassler, B.L.: The small RNA chaperone Hfq and multiple small RNAs control quorum sensing in Vibrio harveyi and Vibrio cholerae. Cell **118**, 69–82 (2004)

553. Levchenko, A., Iglesias, P.A.: Models of eukaryotic gradient sensing: Application to chemotaxis of amoebae and neutrophils. Biophys. J. **82**, 50–63 (2002)

554. Levin, S.A.: The problem of pattern and scale in ecology. Ecology **73**, 1943–1967 (1992)

555. Levine, J., Kueh, H.Y., Mirny, L.: Intrinsic fluctuations, robustness, and tunability in signaling cycles. Biophys. J. **92**, 4473–4481 (2007)

556. Lèvy, P.: Sur certaines processus stochastiques homogènes. Compos. Math. **7**, 283 (1939)

557. Lewis, P.A., Shedler, G.S.: Simulation of nonhomogeneous Poisson processes by thinning. Naval Research Logistics Quarterly **26**, 403–413 (1979)

558. Li, Y., Bhimalapuram, P., Dinner, A.R.: Model for how retrograde actin flow regulates adhesion traction stresses. J. Phys. Condens. Matter. **22**, 194113 (2010)

559. Li, P., et al.: Phase transitions in the assembly of multivalent signalling proteins. Nature **483**, 336–340 (2012)

560. Lieberman-Aiden, E., van Berkum, N.L., Williams, L., et al.: Comprehensive mapping of long-range interactions reveals folding principles of the human genome. Science. **326**, 289–293 (2009)

561. Lifshitz, I.M., Slyozov, V.V.: The kinetics of precipitation from supersaturated solid solutions. J. Phys. Chem. Solids. **19**, 35–50 (1961)

562. Likthman, A.E., Marques, C.M.: First-passage problem for the Rouse polymer chain: An exact solution. Europhys. Lett. **75**, 971–977 (2006)

563. Lin, C.H., Forscher, P.: Growth cone advance is inversely proportional to retrograde F-actin flow. Neuron **14**, 763–771 (1995)

564. Lin, Y.C., Koenderink, G.H., MacKintosh, F.C., Weitz, D.A.: Viscoelastic properties of microtubule networks. Macromolecules **40**, 7714–7720 (2007)

565. Lippincott-Schwartz, J., Roberts, T.H., Hirschberg, K.: Secretory protein trafficking and organelle dynamics in living cells. Ann. Rev. Cell Dev. Biol. **16**, 557–589 (2000)

566. Lippincott-Schwartz, J., Phair, R.D.: Lipids and cholesterol as regulators of traffic in the endomembrane system. Ann. Rev. Biophys. **39**, 559–578 (2010)

567. Liu, J., Deasi, A., Onunchic, J.N., Hwa, T.: An integrated mechanobiochemical feedback mechanism describes chromosome motility from prometaphase to anaphase in mitosis. Proc. Natl. Acad. Sci. USA **105**, 13752–13757 (2008)

568. Liu, A.P., Richmond, D.L., Maibaum, L., Pronk, S., Geissler, P.L., Fletcher, D.A.: Membrane-induced bundling of actin filaments. Nat. Phys. **4**, 789–793 (2008)

569. Lohmar, I., Meerson, B.: Switching between phenotypes and population extinction. Phys. Rev. E **84**, 051901 (2011)

570. Lohmiller, W., Slotine, J.J.E.: On contraction analysis of nonlinear systems. Automatica **34**, 683–696 (1998)

571. Long, T., Tu, K.C., Wang, Y., Mehta, P., Ong, N.P., Bassler, B.L., Wingreen, N.S.: Quantifying the integration of quorum-sensing signals with single-cell resolution. PLoS Biol. **7**, e1000068 (2009)

572. Loose, M., Fischer-Friedrich, E., Ries, J., Kruse, K., Schwille, P.: Spatial regulators for bacterial cell division self-organize into surface waves in vitro. Science **320**, 789–792 (2008)

573. Loose, M., Kruse, K., Schwille, P.: Protein self-organization: lessons from the Min system. Annu. Rev. Biophys. **40**, 315–336 (2011)

574. Loose, M., Fischer-Friedrich, E., Herold, C., Kruse, K., Schwille, P.: Min protein patterns emerge from rapid rebinding and membrane interaction of MinE. Nat. Struct. Mol. Biol. **18**, 577–583 (2011)

575. Losick, R., Desplan, C.: Stochasticity and cell fate. Science **320**, 65–68 (2008)

576. Lou, Y., Nie, Q., Wan, F.Y.M.: Nonlinear eigenvalue problems in the stability analysis of morphogen gradients. Stud. Appl. Math. **113**, 183–215 (2004)

577. Low, D.A., Weyand, N.J., Mahan, M.J.: Roles of DNA adenine methylation in regulating bacterial gene expression and virulence. Infect. Immun. **69**, 7197–7204 (2001)

578. Ludington, W.B., et al.: Avalanche-like behavior in ciliary import. Proc. Natl. Acad. Sci. U.S.A. **110**, 3925–3930 (2013)

579. Ludington, W.B., et al.: A systematic comparison of mathematical models for inherent measurement of ciliary length: how a cell can measure length and volume. Biophys. J. **108**, 1361–1379 (2015)

580. Lugo, C.A., McKane, A.J.: Quasi-cycles in a spatial predator-prey model. Phys. Rev. E **78**, 051911 (2008)

581. Luo, L.: Actin cytoskeleton regulation in neuronal morphogenesis and structural plasticity. Annu. Rev. Neurosci. **18**, 601–635 (2002)

582. MacKintosh, F.C., Kas, J., Janmey, P.A.: Elasticity of semiflexible biopolymer networks. Phys. Rev. Lett. **75**, 4425–4428 (1995)

583. Maly, I.V., Borisy, G.G.: Self-organization of a propulsive actin network as an evolutionary process. Proc. Natl. Acad. Sci. USA **98**, 11324–11329 (2001)

584. Manley, G.A.: Evidence for an active process and a cochlear amplifier in nonmammals. J. Neurophysiol. **86**, 541–549 (2001)

585. Marantan, A., Amir, A.: Stochastic modeling of cell growth with symmetric or asymmetric division. Phys. Rev. E **94**, 012405 (2016)

586. Marco, E., Wedlich-Soldner, R., Li, R., Altschuler, S.J., Wu, L.F.: Endocytosis optimizes the dynamic localization of membrane proteins that regulate cortical polarity. Cell **129**, 411–422 (2007)

587. Marcy, Y., Prost, J., Carlier, M.F., Sykes, C.: Forces generated during actin-based propulsion: a direct measurement by micromanipulation. Proc. Natl. Acad. Sci. USA **20**, 5992–5997 (2004)

588. Marder, E., Goaillard. J.-M.: Variability, compensation and homeostasis in neuron and network function. Nat. Rev. Neurosci. **7** 563-574 (2006)

589. Maree, A.F., Jilkine, A., Dawes, A., Grieneisen, V.A., Edelstein-Keshet, L.: Polarization and movement of keratocytes: A multiscale modelling approach. Bull Math Biol **68**, 1169–1211 (2006)

590. Markevich, N.I., Hoek, J.B., Kholodenko, B.N.: Signaling switches and bistability arising from multisite phosphorylation in protein kinase cascades. J. Cell Biol. **164**, 353–359 (2004)

591. Markin, V.S., Hudspeth, A.J.: Gating-spring models of mechanoelectrical transduction by hair cells of the internal ear. Annu. Rev. Biophys. Biomol. Struct. **24**, 59–83 (1995)

592. Marko, J.E., Siggia, E.D.: Bending and twisting elasticity of DNA. Macromolecules **28**, 8759–8770 (1995)

593. Marshall, W.F., Rosenbaum, J.L.: Intraflagellar transport balances continuous turnover of outer doublet microtubules: implications for flagellar length control. J. Cell Biol. **155**, 405–414 (2001)

594. Marshall, B.T., Long, M., Piper, J.W., Yago, T., McEver, R.P., Zhu, C.: Direct observation of catch bonds involving cell-adhesion molecules. Nature **423**, 190–193 (2003)

595. Marshall, W.F., Qin, H., Rodrigo Brenni, M., Rosenbaum, J.L.: Flagellar length control system: testing a simple model based on intraflagellar transport and turnover. Mol. Biol. Cell **16**, 270–278 (2005)

596. Martin, P., Mehta, A.D., Hudspeth, A.J.: Negative hair-bundle stiffness betrays a mechanism for mechanical amplification by the hair cell. Proc. Natl. Acad. Sci. USA **97**, 12026–12031 (2000)

597. Martin, P., Hudspeth, A.J., Julicher, F.: Comparison of a hair bundle's spontaneous oscillations with its response to mechanical stimulation reveals the underlying active process. Proc. Natl. Acad. Sci. USA **98**, 14380–14385 (2001)

598. Martin, P., Bozovic, D., Choe, Y., Hudspeth, A. J.: (2003). Spontaneous oscillation by hair bundles of the bullfrog's sacculus. J. Neurosci. **23** 4533-4454 (2003)

599. Martin, S.G., Berthelot-Grosjean, M.: Polar gradients of the DYRK-family kinase Pom1 couple cell length with the cell cycle. Nature. **459**, 852–856 (2009)

600. Martin, S.G., Arkowitz, R.A.: Cell polarization in budding yeast. FEMS Microbiol. Rev. **38**, 228–253 (2014)

601. Martins, B.M., Locke, J.C.: Microbial individuality: how single-cell heterogeneity enables population level strategies. Cell Regul **24**, 104–112 (2015)

602. Mazza, M.G.: The physics of biofilms - an introduction. J. Phys. D: Appl. Phys. **49**, 203001 (2016)

603. McIntosh, J.R., Molodtsov, M.I., Ataullakhanov, F.I.: Biophysics of mitosis. Q. Rev. Biophys. **45**, 147–207 (2012)

604. McKane, A.J., Biancalani, T., Rogers, T.: Stochastic pattern formation and spontaneous polarization: the linear noise approximation and beyond. Bull. Math. Biol. **76**, 895–921 (2014)

605. McKendrick, A.G.: Applications of mathematics to medical problems. Proc. Edinb. Math. Soc. **44**, 98 (1925)

606. Mclean, D.R., Graham, B.P.: Mathematical formulation and analysis of a continuum model for tubulin-driven neurite elongation. Proc. Roy. Soc. Lond. A **460**, 2437–2456 (2004)

607. McMahon, H.T., Gallop, J.L.: Membrane curvature and mechanisms of dynamic cell membrane remodelling. Nature **438**, 590–596 (2005)

608. McMillen, D., Kopell, N., Hasty, J., Collins, J.J.: Synchronizing genetic relaxation oscillators by intercell signaling. Proc. Natl. Acad. Sci. U.S.A. **99**, 679–684 (2002)

609. Meacci, G., Kruse, K.: Min-oscillations in Escherichia coli induced by interactions of membrane-bound proteins. Phys. Biol. **2**, 89–97 (2005)

610. Meinhardt, H., Gierer, A.: Pattern formation by local self activation and lateral inhibition. BioEssays **22**, 753–760 (2000)

611. Meinhardt, H.: Models of biological pattern formation: from elementary steps to the organization of embryonic axes. Curr. Top. Dev. Biol. **81**, 1–63 (2008)

612. Meinhardt, M., de Boer, P.A.J.: Pattern formation in Escherichia coli: A model for the pole-to-pole oscillations of min proteins and the localization of the division site. Proc. Natl. Acad. Sci. USA **98**, 14202–14207 (2001)

613. Melbinger, A., Reese, L., Frey, E.: Microtubule length regulation by molecular motors. Phys. Rev. Lett. **108**, 258104 (2012)

614. Mello, B.A., Tu, Y.: Quantitative modeling of sensitivity in bacterial chemotaxis: the role of coupling among different chemoreceptor species. Proc. Natl. Acad. Sci. USA **100**, 8223–8228 (2003)

615. Menchon, S.A., Gartner, A., Roman, P., Dotti, C.G.: Neuronal (bipolarity) as a self-organized process enhanced by growing membrane. PLoS one **6**, e24190 (2011)

616. Mendez, V., Fort, J., Rotstein, H.G., Fedotov, S.: Speed of reaction-diffusion fronts in spatially heterogeneous media. Phys. Rev. E **68**, 041105 (2003)

617. Mendez, V., Fedotov, S., Horsthemke, W.: Reaction-Transport Systems. Springer-Verlag, Berlin (2010)

618. Meyers, J., Craig, J., Odde, D.J.: Potential for control of signaling pathways via cell size and shape. Curr. Biol. **16**, 1685–1693 (2006)

619. Mileyko, Y., Joh, R.I., Weitz, J.S.: Small-scale copy number variation and large-scale changes in gene expression. Proc. Natl. Acad. Sci. USA **105**, 16659 (2008)

620. Mim, C., Unger, V.M.: Membrane curvature and its generation by BAR proteins. Trends. Biochem. Sci. **37**, 526–533 (2012)

621. Mina, P., di Bernardo, M., Savery, N.J., Tsaneva-Atanasova, K.: Modelling emergence of oscillations in communicating bacteria: a structured approach from one to many cells. J. Roy. Soc. Interface **10**, 20120612 (2013)

622. Mirny, L.A., Imakaev, M., Abdennur, N.: Two major mechanisms of chromosome organization. Curr. Opin. Cell Biol. **58**, 142–152 (2019)

623. Misteli, T.: Self-organization in cell architecture. J. Cell Biol. **155**, 181–186 (2001)

624. Mitchison, T.J., Kirschner, M.W.: Dynamic instability of microtubule growth. Nature **312**, 237–242 (1984)

625. Mitchison, J.M., Nurse, P.: Growth in cell length in the fission yeast Schizosaccharomyces pombe. J. Cell Sci. **75**, 357–376 (1985)

626. Mitchison, T.J., Kirschner, M.W.: Cytoskeletal dynamics and nerve growth. Neuron **1**, 761–772 (1988)

627. Mitchison, T.J.: Self-organization of polymer-motor systems in the cytoskeleton. Phil. Trans. R. Soc. Lond. B. Biol. Sci. **336**, 99–106 (1992)

628. Miyashiro, T., Ruby, E.G.: Shedding light on bioluminescence regulation in Vibrio fischeri. Mol. Microbiol. **84**, 795–806 (2012)

629. Mobilia, M., Georgiev, I.T., Täuber, U.C.: Phase transitions and spatio-temporal fluctuations in stochastic lattice Lotka?Volterra models. J. Stat. Phys. **128**, 447 (2007)

630. Modi, S., Vargas-Garcia, C.A., Ghusinga, K.R., Singh, A.: Analysis of noise mechanisms in cell-size control. Biophys J. **112**, 2408–2418 (2017)

631. Mogilner, A., Oster, G.: Cell motility driven by actin polymerization. Biophys. J. **71**, 3030–3045 (1996)

632. Mogilner, A., Edelstein-Keshet, L.: Regulation of actin dynamics in rapidly moving cells: a quantitative analysis. Biophys. J. **83**, 1237–1258 (2002)

633. Mogilner, A., Oster, G.: Force generation by actin polymerization II: the elastic ratchet and tethered filaments. Biophys. J. **84**, 1591–605 (2003)

634. Mogilner, A., Wollman, R., Civelekoglu-Sholey, G., Scholey, J.: Modeling mitosis. Trends Cell Biol. **16**, 89–96 (2006)

635. Mogilner, A.: Mathematics of cell motility: have we got its number? J. Math. Biol. **58**, 105–134 (2009)

636. Mogilner, A.: Craig, E: Towards a quantitative understanding of mitotic spindle assembly and mechanics. J. Cell Sci. **123**, 3435–3445 (2010)

637. Mora, T., Wingreen, N.S.: Limits of sensing temporal concentration changes by single cells. Phys. Rev. Lett. **104**, 248101 (2010)

638. Mori, Y., Jilkine, A., Edelstein-Keshet, L.: Wave-pinning and cell polarity from a bistable reaction-diffusion system. Biophys. J. **94**, 3684–3697 (2008)

639. Mori, Y., Jilkine, A., Edelstein-Keshet, L.: Asymptotic and bifurcation analysis of wave-pinning in a reaction-diffusion model for cell polarization. SIAM J. Appl. Math. **71**, 1401–1427 (2011)

640. Mori, F., Le Doussal, P., Majumdar, S.N., Schehr, G.: Universal survival probability for a d-dimensional run-and-tumble particle. POhys. Rev. Lett. **124**, 090603 (2020)

641. Morita, Y., Ogawa, T.: Stability and bifurcation of nonconstant solutions to a reaction-diffusion system with conservation of a mass. Nonlinearity **23**, 1387–1411 (2010)

642. Moseley, J.B., Mayeux, A., Paoletti, A., Nurse, P.: A spatial gradient coordinates cell size and mitotic entry in fission yeast. Nature. **459**, 857–860 (2009)

643. Motegi, F., Zonies, S., Hao, Y., Cuenca, A.A., Griffin, E., Seydoux, G.: Microtubules induce self-organization of polarized PAR domains in Caenorhabditis elegans zygotes. Nat. Cell Biol. **13**, 1361–1367 (2011)

644. Mugler, A., Levchenko, A., Nemenman, I.: Limits to the precision of gradient sensing with spatial communication and temporal integration. Proc. Natl. Acad. Sci. USA **13**, E689-695 (2016)

645. Mukherjee, S., Bassler, B. L.: Bacterial quorum sensing in complex and dynamically changing environments. Nat. Rev. Microbiol. **17** 371-382

646. Mulder, B.M.: Microtubules interacting with a boundary: Mean length and mean first-passage times. Phys. Rev. E **86**, 011902 (2012)

647. Muller, J., Kuttler, C., Hense, B.A., Rothballer, M., Hartmann, A.: Cell-cell communication by quorum sensing and dimension- reduction. J. Math. Biol. **53**, 672–702 (2006)

648. Muller, J., Uecker, H.: Approximating the dynamics of communicating cells in a diffusive medium by ODEs - homogenization with localization. J. Math. Biol. **67**, 1023–1065 (2013)

649. Muller, J., Hense, B.A., Fuchs, T.M., Utz, M., Potzsche, Ch.: Bet-hedging in stochastically switching environments. J. Th. Biol. **336**, 144–157 (2013)

650. Muller, P., Rogers, K.W., Yu, S.R., Brand, M., Schier, A.F.: Morphogen transport. Development **140**, 1621–1638 (2013)

651. Mullins, R.D., Heuser, J.A., Pollard, T.D.: The interaction of Arp2/3 complex with actin: nucleation, high affinity pointed end capping, and formation of branching networks of filaments. Proc. Natl. Acad. Sci. USA **95**, 6181–6186 (1998)

652. Munoz-Garcia, J., Neufeld, Z., Kholodenko, B.N.: Positional information generated by spatially distributed signaling cascades. PLoS Comp. Biol. **3**, e1000330 (2009)

653. Munoz-Garcia, J., Kholodenko, B.N.: Signaling and control from a systems perspective. Biochem. Soc. Trans. **38**, 1235–1241 (2010)

654. Munro, E., Nance, J., Priess, J. R.: Cortical flows powered by asymmetrical contraction transport PAR proteins to establish and maintain anterior-posterior polarity in the early C. elegans embryo. Dev. Cell **7** 413-424 (2004)

655. Murray, J. D.: *Mathematical Biology, Vols. I and II.* 3rd ed. Berlin: Springer (2008)

656. Murray, S.M., Sourjik, V.: Self-organization and positioning of bacterial protein clusters. Nature Phys. **13**, 1006–1013 (2017)

657. Subramanian, S., Murray, S. M.: Pattern selection in reaction diffusion systems. arXiv:2005.07940v2 (2020)

658. Muthukumar, M.: Polymer translocation through a hole. J. Chem. Phys. **111**, 10371–10374 (1999)

659. Muthukumar, M.: Polymer Translocation. CRC Press, Boca Raton, Florida (2011)

660. Nadrowski, B., Martin, P., Julicher, F.: Active hair-bundle motility harnesses noise to operate near an optimum of mechanosensitivity. Proc. Natl. Acad. Sci. USA **101**, 12195–12200 (2004)

661. Naoz, M., Manor, U., Sakaguchi, H., Kachar, B., Gov, N.S.: Protein localization by actin treadmilling and molecular motors regulates stereocilia shape and treadmilling rate. Biophys. J. **95**, 5706–5718 (2015)

662. Nedelec, F., Surrey, T., Maggs, A.C., Leibler, S.: Self-organization of microtubules and motors. Nature **389**, 305–308 (1997)

663. Nelson, P.: Biological Physics, 2nd edn. Freeman, W. H (2013)

664. Neu, J.C., Canizo, J.A., Bonilla, L.L.: Three eras of micellization. Phys. Rev. E **66**, 061406 (2002)

665. Neubert, M.G., Caswell, H.: Alternatives to resilience for measuring the responses of eco-logical systems to perturbations. Ecology **78**, 653–665 (1997)

666. Neubert, M.G., Caswell, H., Murray, J.: Transient dynamics and pattern formation: reactivity is necessary for Turing instabilities. Math. Biosci. **175**, 1–11 (2002)

667. Neukirchen, D., Bradke, F.: Neuronal polarization and the cytoskeleton. Sem. Cell Dev. Biol. **22**, 825–833 (2011)

668. Newby, J.M., Bressloff, P.C.: Quasi-steady state reduction of molecular-based models of directed intermittent search. Bull Math Biol **72**, 1840–1866 (2010)

669. Newby, J.M., Keener, J.P.: An asymptotic analysis of the spatially inhomogeneous velocity-jump process. SIAM Multiscale Model. Simul. **9**, 735–765 (2011)

670. Newby, J.M.: Isolating intrinsic noise sources in a stochastic genetic switch. Phys. Biol. **9**, 026002 (2012)

671. Newby, J.M., Bressloff, P.C., Keeener, J.P.: The effect of potassium channels on spontaneous action potential initiation by stochastic ion channels. Phys. Rev. Lett. **111**, 128101 (2013)

672. Newby, J.M., Chapman, S.J.: Metastable behavior in Markov processes with internal states: breakdown of model reduction techniques. J. Math Biol. **69**, 941–976 (2014)

673. Newby, J.M.: Spontaneous excitability in the Morris-Lecar model with ion channel noise. SIAM J. Appl. Dyn. Syst. **13**, 1756–1791 (2014)

674. Newby, J.M.: Bistable switching asymptotics for the self regulating gene. J. Phys. A **48**, 185001 (2015)

675. Newman, T. J., grima, R.: Many-body theory of chemotactic cell-cell interactions. Phys. Rev. E **70** 051916 (2004)

676. Ng, W.L., Bassler, B.L.: Bacterial quorum-sensing architectures. Annu. Rev. Genet. **43**, 197–222 (2009)

677. Nicolas, A., Geiger, B., Safran, S.A.: Cell mechanosensitivity controls the anisotropy of focal adhesions. Proc. Natl Acad. Sci. USA **101**, 12520–12525 (2004)

678. Nicolas, A., Safran, S.A.: Limitation of cell adhesion by the elasticity of the extracellular matrix. Biophys. J. **91**, 61–73 (2006)

679. Nicolas, A., Besser, A., Safran, S.A.: Dynamics of cellular focal adhesions on deformable substrates: consequences for cell force microscopy. Biophys. J. **95**, 527–539 (2008)

680. Niethammer, B.: Derivation of the LSW theory for Ostwald ripening by homogenization methods. Arch. Rat. Mech. Anal. **147**, 119–178 (1999)

681. Niethammer, B., Pego, R.L.: On the On the initial-value problem in the Lifshitz-Slyozov theory of Ostwald ripening. SIAM J. Math. Anal. **31**, 467–485 (2000)

682. Niethammer, B., Otto, F.: Ostwald Ripening: The screening length revisited. Calc. Var. and PDE **13**, 33–68 (2001)

683. Niethammer, B.: On the evolution of large clusters in the Becker-Döring model. J. Nonlin. Sci. **8**, 115–155 (2003)

684. Niethammer, P., Bastiaens, P., Karsenti, E.: Stathmin-tubulin interaction gradients in motile and mitotic cells. Science **303**, 1862–1866 (2004)

685. Niethammer, B., Otto, F., Velazquez J. J. L.: On the effect of correlations, fluctuations and collisions in Ostwald ripening. In A. Mielke (Ed.) *Analysis, Modeling and Simulation of Multiscale Problems.* pp. 501-530. Springer (2006)

686. Noble, J.V.: Geographic and temporal development of plagues. Nature **250**, 726–729 (1974)

687. Noorbakhsh, J., Schwab, D., Sgro, A., Gregor, T., Mehta, P.: Modeling oscillations and spiral waves in Dictyostelium populations. Phys. Rev. E. **91**, 062711 (2015)

688. Norman, T.M., Lord, N.D., Paulsson, J., Losick, R.: Stochastic switching of cell fate in microbes. Annu. Rev. Microbiol. **69**, 381–403 (2015)

689. Novak, B., Tyson, J.J.: Modeling the cell division cycle: M-phase trigger, oscillations, and size control. J. Theor. Biol. **165**, 101–134 (1993)

690. Novak, B., Tyson, J.J.: Design principles of biochemical oscillators. Nat Rev Mol Cell Biol. **9**, 981–991 (2008)

691. Bangasser, B.L., Rosenfeld, S.S., Odde, D.J.: Determinants of maximal force transmission in a motor-clutch model of cell traction in a compliant microenvironment. Biophys. J. **105**, 581 (2013)

692. Oksendal, B.: Stochastic Differential Equations: An Introduction with Applications, 5th edn. Springer (1998)

693. Olberding, J.E., Thouless, M.D., Arruda, E.M., Garikipati, K.: The non-equilibrium thermodynamics and kinetics of focal adhesion dynamics. PLoS One **5**, e12043 (2010)

694. Onsum, M., Rao, C.V.: A mathematical model for neutrophil gradient sensing and polarization. PLoS Comput Biol. **3**, e36 (2007)

695. Osmanovic, D., Rabin, Y.: Dynamics of active Rouse chains. Soft Matter **13**, 963–968 (2017)

696. Othmer, H., Dunbar, S., Alt, W.: Models of dispersal in biological systems. J. Math. Biol. **26**, 263–298 (1988)

697. Othmer, H.G., Hillen, T.: The diffusion limit of transport equations II: Chemotaxis equations. SIAM J. Appl. Math. **62**, 1222–1250 (2002)

698. Othmer, H.G., Painter, K., Umulis, D., Xue, C.: The intersection of theory and application in elucidating pattern formation in developmental biology. Math. Model. Nat. Phenom. **4**, 1–80 (2009)

699. Otsuji, M., Ishihara, S., Co, C., Kaibuchi, K., Mochizuki, A., Kuroda, S.: A mass conserved reaction-diffusion system captures properties of cell polarity. PLoS Comput Biol. **3**, e108 (2007)

700. Ott, E., Antonsen, T.M.: Low dimensional behavior of large systems of globally coupled oscillators. Chaos **18**, 037113 (2008)

701. Oxtoby, D.W.: Homogeneous nucleation: theory and experiment. J. Phys. Cond. Matter. **4**, 7627–7650 (1992)

702. Padte, N.N., Martin, S.G., Howard, M., Chang, F.: The cell-end factor pom1p inhibits mid1p in specification of the cell division plane in fission yeast. Curr. Biol. **16**, 2480–2487 (2006)

703. Painter, P.R., Marr, A.G.: Mathematics of microbial populations. Annu. Rev. Microbiol. **22**, 519–548 (1968)

704. Pal, A., Reuveni, S.: First passage under restart. Phys. Rev. Lett. **118**, 030603 (2017)

705. Pal, A., Kusmierz, L., Reuveni, S.: Home-range search provides advantage under high uncertainty. arXiv:1906.06987 (2020)

706. Paliwal, S., Ma, L., Krishnan, J., Levchenko, A., Iglesias, P. A.: Responding to directional cues: a tale of two cells. IEEE Cont. Syst. Mag. August 77-90 (2004)

707. Pan, K. Z., Saunders, T. E., Flor-Parra, I., Howard, M., Chang, F.: Cortical regulation of cell size by a sizer Cdr2p. eLife **3** e02040 (2014)

708. Panja, D.: Effects of fluctuations on propagating fronts. Phys. Rep. **393**, 87–174 (2004)

709. Panja, D., Barkema, G.T., Kolomeisky, A.B.: Through the eye of the needle: recent advances in understanding biopolymer translocation. J. Phys. Condens. Matter **25**, 413101 (2013)

710. Papenfort, K., Bassler, B.L.: Quorum sensing signal-response systems in Gram-negative bacteria. Nat. Rev. Microbiol. **14**, 577–588 (2016)

711. Parent, C.A., Devreotes, P.N.: A cell's sense of direction. Science **284**, 765–770 (1999)

712. Parmar, J.J., Woringer, M., Zimmer, C.: How the genome folds: the biophysics of four-dimensional chromatin organization. Annu. Rev. Biophys. **48**, 231–253 (2019)

713. Parmeggiani, A., Franosch, T., Frey, E.: Phase coexistence in driven one-dimensional transport. Phys. Rev. Lett. **90**, 086601 (2003)

714. Parsons, J.T., Horwitz, A.R., Schwartz, M.A.: Cell adhesion: integrating cytoskeletal dynamics and cellular tension. Nat. Rev. Mol. Cell Biol. **11**, 633–643 (2010)

715. Pastor, R.W., Zwanzig, R., Szabo, A.: Diffusion limited first contact of the ends of a polymer: Comparison of theory with simulation. J. Chem. Phys. **105**, 3878–3882 (1996)

716. Patra, P., Klumpp, S.: Population dynamics of bacterial persistence. PLoS One. **8**, e62814 (2013)

717. Pavin, N., Tolic, I.M.: Self-organization and forces in the mitotic spindle. Annu. Rev. Biophys. **45**, 279–298 (2016)

718. Penrose, O.: Metastable states for the Becker-Döring cluster equations. Comm. Math. Phys. **124**, 515–541 (1989)
719. Perelson, A.P., Coutsias, E.A.: A moving boundary model of acrosomal elongation. J. Math. Biol. **23**, 361–379 (1986)
720. Pereverzev, Y.V., Prezhdo, O.V., Forero, M., Sokurenko, E.V., Thomas, W.E.: The two-pathway model for the catch-slip transition in biological adhesion. Biophys. J. **89**, 1446–1454 (2005)
721. Perlman, I., Normann, R.A.: Light adaptation and sensitivity controlling mechanisms in vertebrate photoreceptors. Prog. Retinal Eye Research. **17**, 523–563 (1998)
722. Persat, A., Nadell, C.D., Kim, M.K., Ingremeau, F., Siryaporn, A., Drescher, K., Wingreen, N.S., Bassler, B.L., Gitai, Z., Stone, H.A.: The mechanical world of Bacteria. Cell **161**, 988–997 (2015)
723. Perthame, B.: Parabolic Equations in Biology. Springer, Switzerland (2015)
724. Peskin, C.P., Odell, G.M., Oster, G.F.: Cellular motions and thermal fluctuations: the brownian ratchet. Biophys. J. **65**, 316–324 (1993)
725. Phillips, R., Kondev, J., Theriot, J. Garcia, H.: *Physical biology of the cell 2nd. ed.* Garland Science (2012)
726. Picioreanu, C., van Loosdrecht, M.C.M., Heijnen, J.J.: Mathematical modeling of biofilm structure with a hybrid differential-discrete cellular automaton approach. Biotechnol. Bioeng. **58**, 101–116 (1998)
727. Picioreanu, C., Kreft, J.-U., van Loosdrecht, M.C.M.: Particle-based multidimensional multispecies biofilm model. Appl. Environ. Microbiol. **70**, 3024–3040 (2004)
728. Pikovsky, A., Rosenblum, M., Kurths, J.: Synchronization: A Universal Concept in Nonlinear Sciences. Cambridge nonlinear science series, Cambridge (2003)
729. Pillay, S., Ward, M.J., Peirce, A., Kolokolnikov, T.: An asymptotic analysis of the mean first passage time for narrow escape problems: Part I: Two-dimensional domains. SIAM Multiscale Model. Sim. **8**, 803–835 (2010)
730. Piras, V., Tomita, M., Selvarajoo, K.: Transcriptome-wide variability in single embryonic development cells. Sci Rep **4**, 7137 (2014)
731. Pollard, T.D., Borisy, G.G.: Cellular motility driven by assembly and disassembly of actin filaments. Cell **112**, 453–465 (2003)
732. Popat, R., Cornforth, D.M., McNally, L., Brown, S.P.: Collective sensing and collective responses in quorum-sensing bacteria. J. R. Soc. Interface **12**, 20140882 (2015)
733. Popkin, G.: The physics of life. Nature **529**, 16–18 (2016)
734. Porter, D., Stirling, D.S.: Integral equations: A Practical Treatment from Spectral Theory to Applications. Cambridge Texts in Applied Mathematics, Cambridge (1990)
735. Prezhdo, O.V., Pereverzev, Y.V.: Theoretical aspects of the biological catch bond. Acc. Chem Res. **42**, 693–703 (2009)
736. Prost, J., Manneville, J.-B., Bruinsma, R. 1998. Fluctuation-magnification of non-equilibrium membranes near a wall. Eur. Phys. J. B. **1** 465-480 (1998)
737. Ptashne, M.: Gene regulation by proteins acting nearby and at a distance. Nature **322**, 697–701 (1986)
738. Purcell, E.M.: Life at low Reynolds number. Am. J. Phys. **45**, 3–11 (1977)
739. Put, S., Sakaue, T., Vanderzande, C.: Active dynamics and spatially coherent motion in chromosomes subject to enzymatic force dipoles. Phy. Rev. E **99**, 032421 (2019)
740. Qian, H.: Thermodynamic and kinetic analysis of sensitivity amplification in biological signal transduction. Biophys. Chem. **105**, 585–593 (2003)
741. Qian, H.: Phosphorylation energy hypothesis: open chemical systems and their biological functions. Annu. Rev. Phys. Chem. **58**, 113–142 (2007)
742. Qian, H., Cooper, J.A.: Temporal cooperativity and sensitivity amplification in biological signal transduction. Biochemistry **47**, 2211–2220 (2008)
743. Qian, H.: Cooperativity in cellular biochemical processes. Annu. Rev. Biophys. **41**, 179–204 (2012)

744. Rading, M., Engel, T., Lipowsky, R., Valleriani, A.: Stationary size distributions of growing cells with binary and multiple cell division. J Stat Phys. **145**, 1–22 (2011)
745. Radmacher, M.: Studying the mechanics of cellular processes by atomic force microscopy. Methods Cell Biol. **83**, 347–372 (2007)
746. Rafelski, S.M., Theriot, J.A.: Crawling toward a unified model of cell motility: spatial and temporal regulation of actin dynamics. Annu. Rev. Biochem. **73**, 209–239 (2004)
747. Rafelski, S.M., Marshall, W.F.: Building the cell: design principles of cellular architecture. Mol. Cell Biol. **9**, 593–603 (2008)
748. Raman, R., Pinto, C.S., Sonawane, M.: Polarized organization of the cytoskeleton: regulation by cell polarity proteins. J. Mol. Biol. **430**, 3565–3584 (2018)
749. Ramaswamy, S., Toner, J.: Prost, J: Nonequilibrium fluctuations, travelling waves, and instabilities in active membranes. Phys. Rev. Lett. **84**, 3494–3497 (2000)
750. Ramaswamy, S., Rao, M.: The physics of active membranes. C. R. Acad. Sci. Paris **2**, 817–839 (2001)
751. Ramaswamy, S.: The mechanics and statistics of active matter. Annu. Rev. Condens. Matter Phys. **1**, 323–345 (2010)
752. Ramirez-Weber, F.A., Kornberg, T.B.: Cytonemes: cellular processes that project to the principal signaling center in Drosophila imaginal discs. Cell **97**, 599–607 (1999)
753. Rappel, W. J., Loomis, W. F.: Eukaryotic chemotaxis. Wiley Interdiscip. Rev.: Syst. Biol. Med. **1** 141-149 (2009)
754. Raskin, D.M., de Boer, P.A.: Rapid pole-to-pole oscillation of a protein required for directing division to the middle of Escherichia coli. Proc. Natl. Acad. Sci. USA **96**, 4971–4976 (1999)
755. Redfield, R.J.: Is quorum sensing a side effect of diffusion sensing? Trends Microbiol. **10**, 365–370 (2002)
756. Reese, L., Melbinger, A., Frey, E.: Crowding of molecular motors determines microtubule depolymerization Biophys. J. **101**, 2190–2200 (2011)
757. Reid, B.A., Tauber, U.C., Brunson, J.C.: Reaction-controlled diffusion: Monte Carlo simulations. Phys. Rev. E **64**, 046121 (2003)
758. Resmi, V., Ambika, G., Amritkar, R.E.: General mechanism for amplitude death in coupled systems. Phys. Rev. E **84**, 046212 (2011)
759. Ribeiro, A.S.: Dynamics and evolution of stochastic bistable gene networks with sensing in fluctuating environments. Phys. Rev. E **78**, 061902 (2008)
760. Rice, S.O.: Mathematical analysis of random noise. In: Wax, N. (ed.) Selected Papers on Noise and Stochastic Processes, pp. 133–294. Dover, New York (1954)
761. Rieder, C.L., Salmon, E.D.: Motile kinetochores and polar ejection forces dictate chromosome position on the vertebrate mitotic spindle. J. Cell Biol. **124**, 223–233 (1994)
762. Rieke, F., Baylor, D.: Single-photon detection by rod cells of the retina. Rev. Mod. Phys. **70**, 1027–1036 (1998)
763. Renault, T.T., Abraham, A.O., Bergmiller, T., Paradis, G., Rainville, S., Charpentier, E., Guet, C.C., Tu, Y., Namba, K., Keener, J.P., et al. Bacterial flagella grow through an injection-diffusion mechanism. eLife **6** e23136 (2017)
764. Rietkerk, M., van de Koppel, J.: Regular pattern formation in real ecosystems. Trends. Ecol. Evol. **23**, 169–175 (2007)
765. Rittmann, B.E., McCarty, P.L.: Model of steady-state-biofilm kinetics. Biotechnol. Bioeng. **22**, 2343–2357 (1980)
766. Robbins, J.R., Monack, D., McCallum, S.J., Vegas, A., Pham, E., et al.: The making of a gradient: IcsA (VirG) polarity in Shigella flexneri. Mol. Microbiol. **41**, 861–872 (2001)
767. Robert, L., Hoffmann, M., Krell, N., Aymerich, S., Robert, J., Doumic, M.: Division in Escherichia coli is triggered by a size-sensing rather than a timing mechanism. BMC Biol. **12**, 17 (2014)
768. Romanczuk, R., Bar, M., Ebeling, W., Lindner, B., Schimansky-Geier, L.: Active Brownian particles: From individual to collective stochastic dynamics. Eur. Phys. J. Special Topics **202**, 1–162 (2012)

769. Rongo, C., Kaplan, J.M.: CaMKII regulates the density of central glutamatergic synapses in vivo. Nature **402**, 195–199 (1999)

770. Rosa, A., Everaers, R.: Structure and dynamics of interphase chromosomes. PLoS Comp. Biol. **4**, e1000153 (2008)

771. Rose, J., Jin, S.X., Craig, A.M.: Heterosynaptic molecular dynamics: locally induced propagating synaptic accumulation of CaM Kinase II. Neuron **61**, 351–358 (2009)

772. Roshan, A., Jones, P.H., Greenman, C.D.: Exact, time-independent estimation of clone size distributions in normal and mutated cells. Roy. Soc. Interface **11**, 20140654 (2014)

773. Roy, S., Hsiung, F., Kornberg, T.B.: Specificity of Drosophila cytonemes for distinct signaling pathways. Science **332**, 354–358 (2011)

774. Roy, S., Kornberg, T.B.: Paracrine signaling mediated at cell-cell contacts. BioEssays **37**, 25–33 (2015)

775. Rubinstein, M., Colby, R.H.: Polymer Physics. Oxford University Press, Oxford (2003)

776. Rudner, D.Z., Losick, R.: Protein subcellular localization in bacteria. Cold. Spring. Harb. Perspect. Biol. **2**, a000307 (2010)

777. Russo, G., Slotine, J.J.E.: Global convergence of quorum sensing network. Phys. Rev. E **82**, 041919 (2010)

778. Rzadzinska, A.K., Schneider, M.E., Davies, C., Riordan, G.P., Kachar, B.: An actin molecular treadmill and myosins maintain stereocilia functional architecture and self-renewal. J. Cell Biol. **164**, 887–897 (2004)

779. Saakian, D.B.: Kinetics of biochemical sensing by single cells and populations of cells. Phys. Rev. E **96**, 042413 (2017)

780. Sabass, B., Schwarz, U. S.: Modeling cytoskeletal flow over the adhesion sites: competition between stochastic bond dynamics and intracellular relaxation J. Phys.: Condens. Matter **22** 194112 (2010)

781. Sabhapandit, S., Majumdar, S.N., Comtet, A.: Statistical properties of the paths of a particle diffusing in a one-dimensional random potential. Phys. Rev. E **73**, 051102 (2006)

782. Safran, S.: *Statistical Thermodynamics of Surfaces, Interfaces and Membranes*. Frontiers in Physics. Westview Press (2003)

783. Sagar, F.P., Scaal, M.: Signaling filopodia in vertebrate embryonic development. Cell Mol Life Sci **73**, 961–974 (2016)

784. Sagues, F., Sancho, J.M., Garcia-Ojalvo, J.: Spatiotemporal order out of noise. Rev. Mod. Phys. **79**, 829–882 (2007)

785. Saha, S., et al.: Polar positioning of phase-separated liquid compartments in cells regulated by an mRNA competition mechanism. Cell **166**, 1572–1584 (2016)

786. Saha, T., Galic, M.: Self-organization across scales: from molecules to organisms. Phil Trans. Roy. Soc. B **373**, 20170113 (2018)

787. Saha, S., Nagy, T.L., Weiner, O.D.: Joining forces: crosstalk between biochemical signalling and physical forces orchestrates cellular polarity and dynamics. Phil. Trans. R. Soc. B **373**, 20170145 (2018)

788. Saleh, B.E.A., Teich, M.C.: Multiplied-Poisson noise in pulse, particle, and photon detection. Proc. IEEE **70**, 229–245 (1982)

789. Samoilov, M., Plyasunov, S., Arkin, A.P.: Stochastic amplification and signaling in enzymatic futile cycles through noise-induced bistability with oscillations. Proc. Natl. AScad. Sci. USA **102**, 2310–2315 (2005)

790. Sanders, T.A., Llagostera, E., Barna, M.: Specialized filopodia direct long-range transport of SHH during vertebrate tissue patterning. Nature **497**, 628–632 (2013)

791. Sasai, M., Wolynes, P.G.: Stochastic gene expression as a many-body problem. Proc. Natl. Acad. Sci. **100**, 2374–2379 (2003)

792. Saunders, T.E., Pan, K.Z., Angel, A., Guan, Y., Shah, J.V., Howard, M., Chang, F.: Noise reduction in the intracellular pom1p gradient by a dynamic clustering mechanism. Dev. Cell **22**, 558–572 (2012)

793. Saunders, T.E.: Aggregation-fragmentation model of robust concentration gradient formation. Phys. Rev. E **91**, 022704 (2015)

794. Savage, N.S., Layton, A.T., Lew, D.J.: Mechanistic model of polarity in yeast. Mol. Biol. Cell **23**, 1998–2013 (2012)

795. Sazer, S., Schiessel, H.: The biology and polymer physics underlying large-scale chromosome organization. Traffic **19**, 87–104 (2018)

796. Schallamach, A.: A theory of dynamic rubber friction Wear **6**, 375–382 (1963)

797. Schaus, T.E., Borisy, G.G.: Performance of a population of independent filaments in lamellipodial protrusion. Biophys. J. **95** 1393-1411(2008)

798. Schimansky-Geier, L., Mikhailov, A.S., Ebeling., W.: Effects of fluctuations on plane front propagation in bistable nonequilibrium systems. Ann. Phys. **40** 277 (1983)

799. Schleif, R.: DNA looping. Annu. Rev. Biochem. **61**, 199–223 (1992)

800. Schmoller, K.M., Skotheim, J.M.: The biosynthetic basis of cell size control. Trends Cell Biol. **25**, 793–802 (2015)

801. Schnitzer, M.J.: Theory of continuum random walks and application to chemotaxis. Phys. Rev. E **48**, 2553–2568 (1993)

802. Scholey, J.M.: Intraflagellar transport. Annu. Rev. Cell Dev. Biol. **19**, 423–443 (2003)

803. Schumacher, L.J., Woolley, T.E., Baker, R.E.: Noise-induced temporal dynamics in Turing systems. Phys. Rev. E **87**, 042719 (2013)

804. Schuss, Z.: Theory and applications of stochastic processes: an analytical approach, *Applied mathematical sciences*. **170** Springer, New York (2010)

805. Schwab, D.J., Baetica, A., Mehta, P.: Dynamical quorum-sensing in oscillators coupled through an external medium. Physica D **241**, 1782–1788 (2012)

806. Schwab, D.J., Plunk, G.G., Mehta, P.: Kuramoto model with coupling through an external medium. Chaos **22**, 043139 (2012)

807. Schwabe, A., Bruggeman, F.J.: Contributions of cell growth and biochemical reactions to non-genetic variability of cells. Biophys J **107**, 301–313 (2014)

808. Schwarz, U.S., Safran, S.A.: Physics of adherent cells. Rev. Mod. Phys. **85**, 1327–1381 (2013)

809. Schweisguth, F., Corson, F.: Self-organization in pattern formation. Dev. Cell **49**, 659–677 (2019)

810. Schweitzer, F., Schimansky-Geier, L., Ebeling, W., Ulbricht, H.: A stochastic approach to nucleation in finite systems: Theory and computer simulations. Physica A **150**, 261–279 (1988)

811. Seger, J., Brockman, H. What is bet-hedging? In Oxford Surveys in *Evolutionary Biology* vol. 4 pp. 182-211. Oxford University Press, Cambridge, UK (1987)

812. Sens, P.: Rigidity sensing by stochastic sliding friction. Europhys. Lett. **104**, 38003 (2013)

813. Shapere, A., Wilczek, F.: Geometry of self-propulsion at low Reynolds number. J. Fluid Mech. **198**, 557–585 (1989)

814. Sharma, P.R., Kamal, N.K., Verma, U.K., Suresh, K., Thamilmaran, K., Shrimali, M.D.: Suppression and revival of oscillation in indirectly coupled limit cycle oscillators. Phys. Lett. A **380**, 3178–3184 (2016)

815. Shemesh, T., Geiger, B., Bershadsky, A.D., Kozlov, M.M.: Focal adhesions as mechanosensors: a physical mechanism. Proc. Natl. Acad. Sci. USA **102**, 12383–12388 (2005)

816. Shemesh, T., Bershadsky, A.D., Kozlov, M.M.: Physical model for self-organization of actin cytoskeleton and adhesion complexes at the cell front. Biophys. J. **102**, 1746–1756 (2012)

817. Shen, K., Meyer, T.: Dynamic control of CaMKII translocation and localization in hippocampal neurons by NMDA receptor stimulation. Science **284**, 162–166 (1999)

818. Shibata, T., Fujimoto, K.: Noisy amplification in ultrasensitive signal transduction Proc. Natl. Acad. Sci. USA **100**, 331–336 (2005)

819. Shinomoto, S., Yasuhiro, T.: Modeling spiking behavior of neurons with time-dependent Poisson processes. Phys. Rev. E **64**, 041910 (2001)

820. Shlomovitz, R., Gov, N.: Membrane waves driven by actin and myosin. Phys. Rev. Lett. **98**, 168103 (2007)

821. Shtylla, B., Keener, J.P.: A mechanomolecular model for the movement of chromosomes during mitosis driven by a minimal kinetochore bicyclic cascade. J. Theor. Biol. **263**, 455–70 (2010)

822. Shtylla, B., Keener, J.P.: A mathematical model for force generation at the kinetochore-microtubule interface. SIAM J. Appl. Math. **71**, 1821–1848 (2011)

823. Shvartsman, S.Y., Baker, R.E.: Mathematical models of morphogen gradients and their effects on gene expression. Rev. Dev Biol **1**, 715–730 (2012)

824. Simpson, M.J.: Exact solutions of linear reaction-diffusion processes on a uniformly growing domain: criteria for successful colonization. PLoS ONE **10**, e0117949 (2015)

825. Skibbens, R.V., Petrie-Skeen, V., Salmon, E.D.: Directional instability of kinetochore motility during chromosome congression and segregation in mitotic newt lung cells: a push pull mechanism. J. Cel lBiol. **122**, 859–875 (1993)

826. Slaughter, B.D., Smith, S.E., Li, R.: Symmetry breaking in the life cycle of budding yeast. Cold Spring Harb. Perspect. Biol. **1**, a003384 (2009)

827. Slaughter, B.D., Das, A., Schwartz, J.W., Rubinstein, B., Li, R.: Dual modes of Cdc42 recycling fine-tune polarized morphogenesis. Dev. Cell **17**, 823–835 (2009)

828. Slemrod, M.: Trend to equilibrium in the Becker-Döring cluster equations. Nonlinearity **2**, 429–443 (1989)

829. Smith, S.B., Cui, Y., Bustamante, C.: Overstretching B-DNA: the elastic response of individual double-stranded and single-stranded DNA molecules. Science **271**, 795–799 (1996)

830. Smith, S.B., Finzi, L., Bustamante, C.: Direct mechanical measurements of the elasticity of single DNA molecules by using magnetic beads. Science **258**, 1122–1126 (1992)

831. Smith, T., Fancher, S., Levchenko, A., Nemenman, I., Mugler, A.: Role of spatial averaging in multicellular gradient sensing. Phys. Biol. **13**, 035004 (2016)

832. Smits, W.K., Kuipers, O.P., Veening, J.W.: Phenotypic variation in bacteria: the role of feedback regulation. Nat. Rev. Microbiol. **4**, 259–271 (2006)

833. Sokolov, I.M.: Cyclization of a polymer: First-passage problem for a non-Markovian process. Phys. Rev. Lett. **90**, 080601 (2003)

834. Solovei, I., Kreysing, M., Lanctot, C., et al.: Nuclear architecture of rod photoreceptor cells adapts to vision in mammalian evolution. Cell. **137**, 356–368 (2009)

835. Song, C., Phenix, H., Abed, V., Scott, M., Ingalls, B.P., Kaern, M., Perkins, T.J.: Estimating the stochastic bifurcation structure of cellular networks. PLoS Comput. Biol. **6**, e1000699 (2010)

836. Sourjik, V., Berg, H.C.: Receptor sensitivity in bacterial chemotaxis. Proc. Natl. Acad. Sci. USA **99**, 123–127 (2002)

837. Sprague, B.L., Pearson, C.G., Maddox, P.S., Bloom, K.S., Salmon, E.D., Odde, D.J.: Mechanisms of microtubule-based kinetochore positioning in the yeast metaphase spindle. Biophys. J. **84**, 1–18 (2003)

838. Srinivasan, M., Walcott, S.: Binding site models for friction due to the formation and rupture of bonds: state-function formalism, force-velocity relations, response to slip velocity transients, and slip stability. Phys. Rev. E **80**, 046124 (2009)

839. Stadler, L., Weiss, M.: Non-equilibrium forces drive the anomalous diffusion of telomeres in the nucleus of mammalian cells. New J. Phys. **19**, 113048 (2017)

840. Stanganello, E., Hagemann, A.I., Mattes, B., Sinner, C., Meyen, D., Weber, S., Schug, A., Raz, E., Scholpp, S.: Filopodia-based Wnt transport during vertebrate tissue patterning. Nat. Commun. **6**, 5846 (2015)

841. Stanganello, E., Scholpp, S.: Role of cytonemes in Wnt transport. J. Cell Sci. **129**, 665–672 (2016)

842. Stauffer, E.A., Scarborough, J.D., Hirono, M., Miller, E.D., Shah, K., Mercer, J.A., Holt, J.R., Gillespie, P.G.: Fast adaptation in vestibular hair cells requires myosin-1c activity. Neuron **47**, 541–553 (2005)

843. Stephens, R.E.: Quantal tektin synthesis and ciliary length in sea-urchin embryos. J. Cell Sci. **92**, 403–413 (1989)

844. Stephens, D.J., Pepperkok, R.: Illuminating the secretory pathway: when do we need vesicles? J. Cell Sci. **114**, 1053–1059 (2001)

845. Steyger, P.S., Gillespie, P.G., Baird, R.A.: Myosin Ib is located at tip link anchors in vestibular hair bundles. J. Neurosci. **18**, 4603–4615 (1998)

846. Stolarska, M.A., Kim, Y., Othmer, H.G.: Multi-scale models of cell and tissue dynamics. Phil Trans. R. Soc. A **367**, 3525–3553 (2009)

847. Strick, T., Allemand, J.-F., Croquette, V., Bensimon, D.: Twisting and stretching single DNA molecules. Prog. Biophys. Mol. Biol. **74**, 115–140 (2000)

848. Stukalin, E.B., Aifuwa, I., Kim, J.S., Wirtz, D., Sun, S.X.: Age-dependent stochastic models for understanding population fluctuations in continuously cultured cells. Roy. Soc. Interface **10**, 20130325 (2013)

849. Sung, W., Park, P.J.: Polymer translocation through a pore in a membrane. Phys. Rev. Lett. **77**, 783–786 (1996)

850. Sutradhar, S., Paul, R.: Tug-of-war between opposing molecular motors explains chromosomal oscillation during mitosis. J. Theor. Biol. **334**, 56–69 (2014)

851. Swem, L. R., Swem, D. L., Wingreen, N. S., Bassler, B. L.: Deducing receptor signaling parameters from in vivo analysis: LuxN/AI-1 quorum sensing in Vibrio harveyi Cell **134** 461-473 (2008)

852. Szabo, A., Schulten, K., Schulten, Z.: First passage time approach to diffusion controlled reactions. J. Chem. Phys. **72**, 4350–4357 (1980)

853. Tabareau, N., Slotine, J.J., Pham, Q.C.: How synchronization protects from noise. PLoS Comput. Biol. **6**, e1000637 (2010)

854. Taheri-Araghi, S., Bradde, S., Sauls, J., Hill, N., Levin, P., Paulsson, J., Vergassola, M., Jun, S.: Cell-size control and homeostasis in bacteria Curr. Biol. **25**, 385–391 (2015)

855. Taillefumier, T., Wingreen, N.S.: Optimal census by quorum sensing. PLoS Comput. Biol. **11**, e1004238 (2015)

856. Tailleur, J., Cates, M.E.: Statistical mechanics of interacting run-and-tumble bacteria. Phys. Rev. Lett. **100**, 218103 (2008)

857. Tailleur, J., Cates, M.E.: Sedimentation, trapping, and rectification of dilute bacteria. Europhys. Lett. **86**, 60002 (2009)

858. Tang, S., Allesina, S.: Reactivity and stability of large ecosystems. Front. Ecol. Evol. **2**(21), 1–8 (2014)

859. Täuber, U.C., Howard, M., Vollmayr-Lee, B.P.: Applications of field-theoretic renormalization group methods to reaction-diffusion problems. J. Phys. A: Math. Gen. **38**, R79–R131 (2005)

860. Täuber, U.C.: Population oscillations in spatial stochastic Lotka?Volterra models: a field theoretic perturbational analysis. J. Phys. A: Math. Theor. **45**, 405002 (2012)

861. Täuber, U.C.: Critical Dynamics - A Field Theory Approach to Equilibrium and Non-Equilibrium Scaling Behavior. Cambridge University Press, Cambridge (2014)

862. Dobramysl, U., Mobilia, M., Plei,limg, M., Täuber, U. C.: Stochastic population dynamics in spatially extended predator-prey systems. J. Phys. A: Math. Theor. **51** 063001 (2018)

863. Teich, M.C., Saleh, B.E.A.: Inter-event-time statistics for shot-noise-driven self-exciting point processes in photon detection. J. Opt. Soc. Am. **71**, 771–776 (1981)

864. Teich, M.C., Heneghan, C., Lowen, S.B., Ozaki, T., Kaplan, E.: Fractal character of the neural spike train in the visual system of the cat. J. Opt. Soc. Am. A **14**, 529–546 (1997)

865. Teimouri, H., Kolomeisky, A.B.: Mechanisms of the formation of biological signaling profiles. J. Phys. A: Math. Theor. **49**, 483001 (2016)

866. Teimouri, H., Kolomeisky, A.B.: New model for understanding mechanisms of biological signaling: direct transport via cytonemes. J. Phys. Chem. Lett. **7**, 180–185 (2016)

867. Teng, S.W., Schaffer, J.N., Tu, K.C., Mehta, P., Lu, W., Ong, N.P., Bassler, B.L., Wingreen, N.S.: Active regulation of receptor ratios controls integration of quorum-sensing signals in Vibrio harveyi. Mol. Syst. Biol. **7**, 491 (2011)

868. ten Wolde, P.R., Becker, N.B., Ouldridge, T.E., Mugler, A.: Fundamental limits to cellular sensing. J. Stat. Phys. **162**, 1395–1424 (2016)

869. Thanbichler, M., Shapiro, L.: MipZ, a spatial regulator coordinating chromosome segregation with cell division in Caulobacter. Cell **126**, 147–162 (2006)

870. Thattai, M., van Oudenaarden, A.: Attenuation of noise in ultrasensitive signaling cascades.
Biophys. J. **82**, 2943–2950 (2001)

871. Thattai, M., van Oudenaarden, A.: Stochastic gene expression in fluctuating environments.
Genetics **167**, 523–530 (2004)

872. Theriot, J. A., T. J. Mitchison, L. G. Tilney, Portnoy, D. A.: The rate of actin-based motility
of intracellular Listeria monocytogenes equals the rate of actin polymerization. Nature. **357**
257-260 (1992)

873. Thomas, W.E., Trintchina, E., Forero, M., Vogel, V., Sokurenko, E.V.: Bacterial adhesion to
target cells enhanced by shear force. Cell **109**, 913–923 (2002)

874. Thomas, P., Popovic, N., Grima, R.: Phenotypic switching in gene regulatory networks Proc.
Natl. Acad. Sci. USA **111**, 6994–6999 (2014)

875. Thomas, P.: Analysis of cell size homeostasis at the single-cell and population level. Front.
Phys. **6**, 64 (2018)

876. Tilney, L.G., Portnoy, D.A.: Actin filaments and the growth, movement, and spread of the
intracellular bacterial parasite. Listeria monocytogenes. J. Cell Biol. **109**, 1597–1608 (1989)

877. Tindall, M.J., Porter, S.L., Maini, P.K., Gaglia, G., Armitage, J.P.: Overview of mathematical
approaches to model bacterial chemotaxis I: the single cell. Bull. Math. Biol. **70**, 1525–1569
(2008)

878. Tischer, C., ten Wolde, P.R., Dogterom, M.: Providing positional information with active
transport on dynamic microtubules Biophys. J. **99**, 726–35 (2010)

879. Tkacik, G., Callan, C.G., Jr., Bialek, W.: Information flow and optimization in transcriptional regulation. Proc. Natl. Acad. Sci. USA **105**, 12265–12270 (2008)

880. Toomre, D., Keller, P., White, J., Olivo, J.C., Simons, K.: Dual-color visualization of trans-
Golgi network to plasma membrane traffic along microtubules in living cells. J. Cell Sci.
**112**, 21–33 (1999)

881. Tostevin, F., Rein ten Wolde, P., Howard, M.: Fundamental limits to position determination
by concentration gradients. PLoS Comp. Biol. **3** e78 (2007)

882. Tostevin, F.: Precision of sensing cell length via concentration gradients. Biophys. J. **100**,
294–303 (2011)

883. Tostevin, F., ten Wolde, P.R.: Mutual information between input and output trajectories of
biochemical networks. Phys. Rev. Lett. **102**, 218101 (2011)

884. Trefethen, L.N., Trefethen, A.E., Reddy, A.E., Driscoll, T.A.: Hydrodynamic stability without eigenvalues. Science **261**, 578–584 (1993)

885. Trefethen, L.N., Embree, M.: Spectra and Pseudospectra: The Behavior of Nonnormal
Matrices and Operators. Princeton University Press, Princeton, NJ (2005)

886. Tsimiring, L.S.: Noise in biology. Rep. Prog. Phys. **77**, 026601 (2014)

887. Tu, K.C., Long, T., Svenningsen, S.L., Wingreen, N.S., Bassler, B.L.: Negative feedback
loops involving small regulatory RNAs precisely vontrol the Vibrio harveyi quorum-sensing
response. Molecular Cell **37**, 567–579 (2010)

888. Tu, Y.: Quantitative modeling of bacterial chemotaxis: signal amplification and accurate
adaptation. Ann. Rev. Biophys. **42**, 337–359 (2013)

889. Turing, A.M.: The chemical basis of morphogenesis. Philos. Trans. R. Soc. Lond. B **237**,
37–72 (1952)

890. Turner, T.E., Schnell, S., Burrage, K.: Stochastic approaches for modelling in vivo reactions.
Comp. Biol. Chem. **28**, 165–178 (2004)

891. Turner, L., Stern, A.S., Berg, H.C.: Growth of flagellar filaments of Escherichia coli is independent of filament length. J. Bacteriol. **194**, 2437–2442 (2012)

892. Tzou, J., Nec, Y., Ward, M.J.: The stability of localized spikes for the 1-D Brusselator
reaction-diffusion model. E. J. Appl. Math. **24**, 515–564 (2013)

893. Ueda, M., Sako, Y., Tanaka, T., Devreotes, P.N., Yanagida, T.: Single-molecule analysis of
chemotactic signaling in Dictyostelium cells. Science **294**, 864–867 (2001)

894. Ueda, M., Shibata, T.: Stochastic signal processing and transduction in chemotactic response of eukaryotic cells. Biophys. J. **93**, 11–20 (2007)
895. Vagne, Q., Sens, P.: Stochastic model of vesicular sorting in cellular organelles. Phys. Rev. Lett. **120**, 058102 (2018)
896. van den Engh, G., Sachs, R., Trask, B.J.: Estimating genomic distance from DNA sequence location in cell nuclei by a random walk model. Science. **257**, 1410–1412 (1992)
897. van der Lee, R., et al.: Classification of intrinsically disordered regions and proteins. Chem. Rev. **114**, 6589–6631 (2014)
898. van Saarloos, W.: Front propagation into unstable states. Phys. Rep. **386**, 29–222 (2003)
899. Varga, V., et al.: Yeast kinesin-8 depolymerizes microtubules in a length-dependent manner. Nature Cell Biol. **8**, 957–962 (2006)
900. Vasilopoulos, G., Painter, K.J.: Pattern formation in discrete cell tissues under long range filopodia-based direct cell to cell contact. Math. Biosci. **273**, 1–15 (2016)
901. Veening, J.W., Smits, W.K., Kuipers, O.P.: Bistability, epigenetics and bet-hedging in bacteria. Ann. Rev. Microbiol. **62**, 193–210 (2008)
902. Veksler, A., Gov, N.S.: Phase transitions of the coupled membrane-cytoskeleton modify cellular shape. Biophys. J. **93**, 3798–3810 (2007)
903. Vellela, M., Qian, H.: Stochastic dynamics and non-equilibrium thermodynamics of a bistable chemical system: the Schlögl model revisited. J. R. Soc. Interface **6**, 925–940 (2009)
904. Vergassola, M., Deneke, V.E., Di Talia, S.: Mitotic waves in the early embryogenesis of Drosophila: Bistability traded for speed. Proc. Natl. Acad. Sci. **115**, E2165–E2174 (2018)
905. Verma, U.K., Chaurasia, S.S., Sinha, S.: Explosive death in nonlinear oscillators coupled by quorum sensing. Phys. Rev. E **100**, 032203 (2019)
906. Vicente-Manzanares, M., Choi, C.K., Horwitz, A.R.: Integrins in cell migration-the actin connection. J. Cell Sci. **122**, 199–206 (2009)
907. Vicsek, T., Zafeiris, A.: Collective motion. Phys. Rep. **517**, 71–140 (2012)
908. Vilfan, A., Duke, T.: Two adaptation processes in auditory hair cells together can provide an active amplifier. Biophys. J. **85**, 191–203 (2003)
909. Visco, P., Allen, R.J., Majumdar, S.N., Evans, M.R.: Switching and growth for microbial populations in catastrophic responsive environments. Biophys. J. **98**, 1099–1108 (2010)
910. Volfson, D., Cookson, S., Hasty, J., Tsimring, L.S.: Biomechanical ordering of dense cell populations. Proc. Natl. Acad. Sci. USA **105**, 15346–15351 (2008)
911. Von Foerster, H.: Some remarks on changing populations, in *The Kinetics of Cellular Proliferation.* edited by F. Stohlman, Jr. Grune and Stratton, New York (1959)
912. Wagner, C.: Theorie der Alterung von Niederschlägen durch Umlösen. Z. Elektrochemie **65**, 581–594 (1961)
913. Walcott, S., Sun, S.X.: A mechanical model of actin stress fiber formation and substrate elasticity sensing in adherent cells. Proc. Natl Acad. Sci. USA **107**, 7757–7762 (2010)
914. Walcott, S., Kim, D.H., Wirtz, D., Sun, S.X.: Nucleation and decay initiation are the stiffness-sensitive phases of focal adhesion maturation. Biophys. J. **101**, 2919–2928 (2011)
915. Walgraef, D.: Spatio-Temporal Pattern Formation. Springer-Verlag, USA (1997)
916. Walter, W.: Nonlinear parabolic differential equations and inequalities. Disc. Cont. Dyn. Syst. **8**, 451–468 (2002)
917. Walther, G.R., Maree, A.F., Edelstein-Keshet, L., Grieneisen, V.A.: Deterministic versus stochastic cell polarisation through wave-pinning. Bull. Math. Biol. **74**, 2570–2599 (2012)
918. Wang, K., Rappel, W.-J., Kerr, R., Levine, H.: Quantifying noise levels of intracellular signals. Phys. Rev. E **75**, 061905 (2007)
919. Wang, Q., Zhang, T.: Review of mathematical models for biofilms. Solid State Comm. **150**, 1009–1022 (2010)
920. Wanner, O., Gujer, W.: A multispecies biofilm model. Biotechnol. Bioeng. **28**, 314–328 (1986)
921. Ward, M.J., Henshaw, W.D., Keller, J.B.: Summing logarithmic expansions for singularly perturbed eigenvalue problems. SIAM J. Appl. Math **53**, 799–828 (1993)

922. Ward, M.J., Wei, J.: Hopf bifurcation and oscillatory instabilities of spike solutions for the one-dimensional Gierer-Meinhardt model. J. Nonlin. Sci. **13**, 209–264 (2003)

923. Ward, J.P., King, J.R., Koerber, A.J., Williams, P., Croft, J.M., Sockett, R.E.: Mathematical modelling of quorum-sensing in bacteria. IMA J. Math. Appl. Med. **18**, 263–292 (2001)

924. Ward, J.P., King, J.R., Koerber, A.J., Williams, P., Croft, J.M., Sockett, R.E.: Early development and quorum sensing in bacterial biofilms. J. Math. Biol. **47**, 23–55 (2003)

925. Wartlick, O., Kicheva, A., Gonzalez-Gaitan, M.: Morphogen gradient formation. Cold Spring Harb. Perspect. Biol. **1**, a001255 (2009)

926. Waters, C.M., Basser, B.L.: Quorum sensing: Cell-to-cell communication in bacteria. Annu. Rev. Cell Dev. Biol. **21**, 319–346 (2005)

927. Wattis, J.A.D., Coveney, P.V.: Generalised nucleation theory with inhibition for chemically reacting systems. J. Chem. Phys. **106**, 9122–9140 (1997)

928. Wattis, J.A.D., King, J.R.: Asymptotic solutions of the Becker-Döring equations. J. Phys. A **31**, 7169–7189 (1998)

929. Wattis, J.A.D.: An introduction to the fundamental mathematical models of coagulation-fragmentation processes. Physica D **22**, 1–20 (2006)

930. Weber, S.C., Spakowitz, A.J., Theriot, J.A.: Bacterial chromosomal loci move subdiffusively through a viscoelastic cytoplasm. Phys Rev. Lett. **104**, 238102 (2010)

931. Weber, S.C., Spakowitz, A.J., Theriot, J.A.: Nonthermal ATP-dependent fluctuations contribute to the in vivo motion of chromosomal loci. Proc. Natl. Acad. Sci. **109**, 7338–7343 (2012)

932. Weber, C.A., Lee, C.F.: Jülicher, F: Droplet ripening in concentration gradients. New J. Phys. **19**, 053021 (2017)

933. Weber, C.A., Zwicker, D., Jülicher, F., Lee, C.F.: Physics of active emulsions. Rep. Prog. Phys. **82**, 064601 (2019)

934. Wedlich-Soldner, R., Wai, S.C., Schmidt, T., Li, R.: Robust cell polarity is a dynamic state established by coupling transport and GTPase signaling. J. Cell. Biol. **166**, 889–900 (2004)

935. Wedlich-Söldner, R., Betz, T.: Self-organization: the fundament of cell biology. Phil. Trans. R. Soc. B **373**, 20170103 (2018)

936. Wei, Y,. Ng, W.-L., Cong, J. Bassler, B. L.: Ligand and antagonist driven regulation of the Vibrio cholerae quorum-sensing receptor CqsS Mol. Microbiol. **83** 1095-1108 (2012)

937. Wienand, K., Frey, E., Mobilia, M.: Evolution of a fluctuating population in a randomly switching environment. Phys. Rev. Lett. **119**, 158301 (2017)

938. Weiss, G.H.: A perturbation analysis of the Wilemski-Fixman approximation for diffusion-controlled reactions. J. Chem. Phys. **80**, 2880–2887 (1984)

939. Welf, E.S., Johnson, H.E., Haugh, J.M.: Bidirectional coupling between integrin-mediated signaling and actomyosin mechanics explains matrix-dependent intermittency of leading-edge motility. Mol. Biol. Cell **24**, 3945–3955 (2013)

940. West, S.A., Winzer, K., Gardner, A., Diggle, S.P.: Quorum sensing and the confusion about diffusion. Trends Microbiol. **20**, 586–594 (2012)

941. Wilemski, G., Fixman, M.: Diffusion controlled intrachain reactions of polymers. I. Theory. J. Chem. Phys. **60**, 866–877 (1974)

942. Wilemski, G., Fixman, M.: Diffusion controlled intrachain reactions of polymers. II. Results for a pair of terminal reactive groups. J. Chem. Phys. **60** 878-890 (1974)

943. Willis, L., Huang, K.: Sizing up the bacterial cell cycle. Nat. Rev. Microbiol. **15**, 606–620 (2017)

944. Winstanley, H.F., Chapwanya, M., Fowler, A.C., McGuinness, M.J.: A polymer-solvent model of biofilm growth. Proc. R. Soc. A **467**, 1449–1467 (2011)

945. Wittmann, T., Waterman-Storer, C.M.: Cell motility: can Rho GTPases and microtubules point the way? J. Cell Sci. **114**, 3795–3803 (2001)

946. Wittmann, T., Bokoch, G.M., Waterman-Storer, C.M.: Regulation of leading edge microtubule and actin dynamics downstream of Rac1. J. Cell Biol. **161**, 845–851 (2003)

947. Wolf, D. M., Arkin, and A. P.: Fifteen minutes of fim: control of type 1 pili expression in *E. coli.* Omics **6** 91-114 (2002)

948. Wollman, R., Cytynbaum, E.N., Jones, J.T., Meyer, T., Scholey, J.M., Mogilner, A.: Efficient chromosome capture requires a bias in the 'search-and-capture' process during mitotic-spindle assembly. Current Biology **15**, 828–832 (2005)

949. Wolpert, L.: Positional information and the spatial pattern of cellular differentiation. J. Theor. Biol. **25**, 1–47 (1969)

950. Wolpert, L.: Principles of Development. Oxford Univ. Press, Oxford, UK (2006)

951. Woods, B., Lai, H., Wu, C.F., Zyla, T.R., Savage, N.S., Lew, D.J.: Parallel actin-independent recycling pathways polarize Cdc42 in budding yeast. Curr. Biol. **26**, 2114–2126 (2016)

952. Woolley, T.E., Baker, R.E., Gaffney, E.A., Maini, P.K.: Stochastic reaction and diffusion on growing domains: understanding the breakdown of robust pattern formation. Phys. Rev. E **84**, 046216 (2011)

953. Wren, K.N., Craft, J.M., Tritschler, D., Schauer, A., Patel, D.K., Smith, E.F., Porter, M.E., Kner, P., Lechtreck, K.F.: A differential cargo-loading model of ciliary length regulation by IFT. Curr. Biol. **23**, 2463–2471 (2013)

954. Wright, P.E., Dyson, H.J.: Intrinsically disordered proteins in cellular signalling and regulation. Nature Rev. Mol. Cell Biol. **16**, 18–29 (2015)

955. Wrzosek, D.: Existence of solutions for the discrete coagulation-fragmentation model with diffusion. Topol. Methods Nonlinear Anal. **9**, 279–296 (1997)

956. Wu, C. F., Savage, N. S., Lew, D. J.: Interaction between bud-site selection and polarity-establishment machineries in budding yeast. Philos. Trans. R. Soc. B Biol. Sci. **368** 20130006 (2013)

957. Wu, Y., Han, B., Li, Y., Munro, E., Odde, D.J., Griffin, E.E.: Rapid diffusion-state switching underlies stable cytoplasmic gradients in the Caenorhabditis elegans zygote. Proc. Natl. Acad. Sci. USA **115**, E8440 (2018)

958. Wurtz, J.D., Lee, C.F.: Chemical-reaction-controlled phase separated drops: formation, size selection and coarsening. Phys. Rev. Lett. **120**, 078102 (2018)

959. Xin, J.: Front propagation in heterogeneous media. SIAM Rev. **42**, 161–230 (2000)

960. Xiong, L., Cooper, R., Tsimiring, L.: Coexistence and pattern formations in bacterial mixtures with contact-dependent killing. Biophys. J. **114**, 1741–1750 (2018)

961. Xiong, L., Cao, Y., Cooper, R., Rappel, W.-J., Hasty, J., Tsimiring, L.: Flower-like patterns in multi-species bacterial colonies. eLife **9** e48885 (2020)

962. Xu, B., Bressloff, P.C.: Model of growth cone membrane polarization via microtubule length regulation. Biophys. J. **114**, 711–722 (2018)

963. Xu, B., Bressloff, P.C.: Modeling the dynamics of Cdc42 oscillation in fission yeast. Biophys. J. **76**, 1844–1870 (2016)

964. Xu, B., Jilkine, A.: A PDE-DDE model for cell polarization in fission yeast SIAM. J. Appl. Math **76**, 1844–1870 (2018)

965. Yang, H.W., Collins, S.R., Meyer, T.: Locally excitable Cdc42 signals steer cells during chemotaxis. Nat. Cell Biol. **18**, 191–201 (2016)

966. Yana, J., Sharoc, A. G., Stone, H. A., Wingreen, N. S., Bassler, B. L.: *Vibrio cholerae* biofilm growth program and architecture revealed by single-cell live imaging. E5337-E5343 (2016)

967. Yang, Y., Wu, M.: Rhythmicity and waves in the cortex of single cells. Phil. Trans. R. Soc. B **373**, 20170116 (2018)

968. Yau, K.-W., Hardie, R.C.: Phototransduction motifs and variations. Cell **139**, 246–264 (2009)

969. Ye, X.F.F., Stinis, P., Qian, H.: Dynamic looping of a free-draining polymer. SIAM J. Appl. Math. **78**, 104–123 (2018)

970. Yochelis, A., Ebrahim, S., Millis, B., Cui, R., Kachar, B., Naoz, M., Gov, N.S.: Self-organization of waves and pulse trains by molecular motors in cellular protrusions. Scientific Reports **5**, 13521 (2008)

971. Yvert, G.: "Particle genetics?": treating every cell as unique. Trends Genet. **30**, 49–56 (2014)

972. Yvinec, Y., D'Orsogna, M.R., Chou, T.: First passage times in homogeneous nucleation and self-assembly. J. Chem. Phys. **137**, 244107 (2012)

973. Zaikin, A.N., Zhabotinsky, A.M.: Concentration wave propagation in two-dimensional liquid-phase self-oscillating system. Nature **225**, 535–537 (1970)

974. Zeitz, M., Kierfeld, J.: Feedback mechanism for microtubule length regulation by stathmin gradients. Biophsy. J. **107**, 2860–2871 (2014)

975. Zelinski, B., Muller, N., Kierfeld, J.: Dynamics and length distribution of microtubules under force and confinement. Phys. Rev. E Stat. Nonlin. Soft Matter Phys. **86** 041918 (2012)

976. Zhang, T., Cogan, N. G., Wang, Q.: Phase field models for biofilms. I. Theory and one-dimensional simulations. SIAM J. Appl. Math. **69** 641-669 (2008)

977. Zhang, T., Cogan, N. G., Wang, Q.: Phase field models for biofilms. II. 2-D numerical simulations of biofilm-flow interaction. Commun. Comput. Phys. **4** 72-101 (2008)

978. Zhang, X.-J., Qian, H., Qian, M.: Stochastic theory of nonequilibrium steady states and its applications. Part I. Phys Rep. **510**, 1–89 (2012)

979. Zhang, C., Scholpp, S.: Cytonemes in development. Curr. Opin. Gen. Dev. **58**, 25–30 (2019)

980. Zhou, T., Zhang, J., Yuan, Z., Chen, L.: Synchronization of genetic oscillators. Chaos. **18**, 037126 (2008)

981. Zhou, D., Wang, Y., Wu, B.: A multi-phenotypic cancer model with cell plasticity. J. Theor. Biol. **357**, 35–45 (2014)

982. Zhou, H.-X., Nguemaha, V., Mazarakos, K., Qin, S.: Why do disordered and structured proteins behave differently in phase separation. Trends Biochem. Sci. **43**, 499–516 (2018)

983. Zhu, C., Yago, T., Lou, J.Z., Zarnitsyna, V.I., McEver, R.P.: Mechanisms for flow-enhanced cell adhesion. Ann. Biomed. Eng. **36**, 604–621 (2008)

984. Zidovska, A., Weitz, D.A., Mitchisona, T.J.: Micron-scale coherence in interphase chromatin dynamics. Proc. Natl. Acad. Sci. **110**, 15555–15560 (2013)

985. Ziebert, F., Aranson, I.S.: Effects of adhesion dynamics and substrate compliance on the shape and motility of crawling cells. PLoS One **8**, e64511 (2013)

986. Zilman, A., Ganusov, V.V., Perelson, A.S.: Stochastic models of lymphocyte proliferation and death. PLoS One **5**, e12775 (2010)

987. Zimmerberg, J., Kozlov, M.M.: How proteins produce cellular membrane curvature. Nat. Rev. Mol. Cell. Biol. **7**, 9–19 (2006)

988. Zinn-Justin, J.: *Quantum Field Theory and Critical Phenomena* Oxford Science Publications, Oxford (2002)

989. Zmurchok, C., Small, T., Ward, M., Edelstein-Keshet, L.: Application of quasi-steady state methods to nonlinear models of intracellular transport by molecular motors. Bull. Math. Biol. **79**, 1923–1978 (2017)

990. Zwicker, D., Decker, M., Jaensch, S., Hyman, A.A., Jülicher, F.: Centrosomes are autocatalytic drops of pericentriolar material organized by centrioles. Proc. Natl Acad. Sci. USA **111**, E2636-2645 (2014)

991. Zwicker, D., Hyman, A.A., Jülicher, F.: Suppression of Ostwald ripening in active emulsions Phys. Rev. E **92**, 012317 (2015)

# Index

© Springer Nature Switzerland AG 2021
P. Bressloff, *Stochastic Processes in Cell Biology*, Interdisciplinary Applied
Mathematics 41, https://doi.org/10.1007/978-3-030-72519-8

Printed in the United States
by Baker & Taylor Publisher Services